POTENTIAL FLOWS OF VISCOUS AND VISCOELASTIC FLUIDS

The goal of this book is to show how potential flows enter into the general theory of motions of viscous and viscoelastic fluids. Traditionally, the theory of potential flow is presented as a subject called "potential flow of an inviscid fluid"; when the fluid is incompressible, these fluids are, curiously, said to be "perfect" or "ideal." This type of presentation is widespread; it can be found in every book and in all university courses on fluid mechanics, but it is deeply flawed. It is never necessary and typically not useful to put the viscosity of fluids in potential (irrotational) flow to zero. The dimensionless description of potential flows of fluids with a nonzero viscosity depends on the Reynolds number, and the theory of potential flow of an inviscid fluid can be said to rise as the Reynolds number tends to infinity. The theory given here can be described as the theory of potential flows at finite and even small Reynolds numbers.

Daniel Joseph is a Professor of Aerospace Engineering and Mechanics at the University of Minnesota since 1963, where he has served as the Russel J. Penrose Professor and Regent's Professor. He is presently a Distinguished Adjunct Professor of Aerospace and Mechanical Engineering at the University of California, Irvine, and an Honorary Professor at Xi'an Jiaotong University in China. He has authored 10 patents, 400 journal articles, and six books. He is a Guggenheim Fellow; a member of the National Academy of Engineering, the National Academy of Sciences; and the American Academy of Arts and Sciences, G. I. Taylor Medalist, Society of Engineering Science; a Fellow of the American Physical Society; winner of the Timoshenko Medal of the ASME, the Schlumberger Foundation Award, the Bingham Medal of the Society of Rheology, and the Fluid Dynamics Prize of the APS. He is listed in Thompson Scientific-ISI's Highly Cited Researchers™.

Toshio Funada is a Professor of Digital Engineering at the Numazu College of Technology in Japan and has served as both Dean and Department Head. He received his doctorate at Osaka University under the guidance of Prof Kakatuni. He is an expert in the theory of stability, bifurcation, and dynamical systems. He has worked on the potential flows of viscous fluids with Professor Joseph for nearly ten years.

Jing Wang is a Post Doctoral Fellow at the University of Minnesota. He received the "Best Dissertation Award" in Physical Sciences and Engineering for 2006 at the University of Minnesota.

CAMBRIDGE AEROSPACE SERIES

Editors
Wei Shyy and Michael J. Rycroft

1. J. M. Rolfe and K. J. Staples (eds.): *Flight Simulation*
2. P. Berlin: *The Geostationary Applications Satellite*
3. M. J. T. Smith: *Aircraft Noise*
4. N. X. Vinh: *Flight Mechanics of High-Performance Aircraft*
5. W. A. Mair and D. L. Birdsall: *Aircraft Performance*
6. M. J. Abzug and E. E. Larrabee: *Airplane Stability and Control*
7. M. J. Sidi: *Spacecraft Dynamics and Control*
8. J. D. Anderson: *A History of Aerodynamics*
9. A. M. Cruise, J. A. Bowles, C. V. Goodall, and T. J. Patrick: *Principles of Space Instrument Design*
10. G. A. Khoury and J. D. Gillett (eds.): *Airship Technology*
11. J. Fielding: *Introduction to Aircraft Design*
12. J. G. Leishman: *Principles of Helicopter Aerodynamics*, 2nd Edition
13. J. Katz and A. Plotkin: *Low Speed Aerodynamics*, 2nd Edition
14. M. J. Abzug and E. E. Larrabee: *Airplane Stability and Control: A History of the Technologies that Made Aviation Possible*, 2nd Edition
15. D. H. Hodges and G. A. Pierce: *Introduction to Structural Dynamics and Aeroelasticity*
16. W. Fehse: *Automatic Rendezvous and Docking of Spacecraft*
17. R. D. Flack: *Fundamentals of Jet Propulsion with Applications*
18. E. A. Baskharone: *Principles of Turbomachinery in Air-Breathing Engines*
19. Doyle D. Knight: *Elements of Numerical Methods for High-Speed Flows*
20. C. Wagner, T. Huettl, and P. Sagaut: *Large-Eddy Simulation for Acoustics*
21. D. Joseph, T. Funada, and J. Wang: *Potential Flows of Viscous and Viscoelastic Fluids*

POTENTIAL FLOWS OF VISCOUS AND VISCOELASTIC FLUIDS

Daniel Joseph

University of Minnesota

Toshio Funada

Numazu College of Technology

Jing Wang

University of Minnesota

Shaftesbury Road, Cambridge CB2 8EA, United Kingdom

One Liberty Plaza, 20th Floor, New York, NY 10006, USA

477 Williamstown Road, Port Melbourne, VIC 3207, Australia

314–321, 3rd Floor, Plot 3, Splendor Forum, Jasola District Centre, New Delhi – 110025, India

103 Penang Road, #05–06/07, Visioncrest Commercial, Singapore 238467

Cambridge University Press is part of Cambridge University Press & Assessment, a department of the University of Cambridge.

We share the University's mission to contribute to society through the pursuit of education, learning and research at the highest international levels of excellence.

www.cambridge.org
Information on this title: www.cambridge.org/9780521873376

First published 2007

A catalogue record for this publication is available from the British Library

Library of Congress Cataloging-in-Publication data
Joseph, Daniel D.
Potential flows of viscous and viscoelastic fluids / Daniel D. Joseph, Toshio Funda, Jing Wang.
p. cm. – (Cambridge aerospace series ; 21)
Includes bibliographical references and index.
ISBN-13: 978-0-521-87337-6 (hardback)
ISBN-10: 0-521-87337-1 (hardback)
1. Viscous flow. 2. Viscoelasticity. I. Funada, Toshio, 1948–
II. Wang, Jing, 1979– III. Title. IV. Series.
QA929.J67 2007
532.0533–dc22 2006039193

ISBN 978-0-521-87337-6 Hardback

Contents

Preface

Potential flows of incompressible fluids with constant properties are irrotational solutions of the Navier–Stokes equations that satisfy Laplace's equation. How do these solutions enter into the general problem of viscous fluid mechanics? Under certain conditions, the Helmholtz decomposition says that solutions of the Navier–Stokes equations can be decomposed into a rotational part and an irrotational part satisfying Laplace's equation. The irrotational part is required for satisfying the boundary conditions; in general, the boundary conditions cannot be satisfied by the rotational velocity, and they cannot be satisfied by the irrotational velocity; the rotational and irrotational velocities are both required and they are tightly coupled at the boundary. For example, the no-slip condition for Stokes flow over a sphere cannot be satisfied by the rotational velocity; harmonic functions that satisfy Laplace's equation subject to a Robin boundary condition in which the irrotational normal and tangential velocities enter in equal proportions are required.

The literature that focuses on the computation of layers of vorticity in flows that are elsewhere irrotational describes boundary-layer solutions in the Helmholtz decomposed forms. These kinds of solutions require small viscosity and, in the case of gas–liquid flows, are said to give rise to weak viscous damping. It is true that viscous effects arising from these layers are weak, but the main effects of viscosity in so many of these flows are purely irrotational, and they are not weak.

The theory of purely irrotational flows of a viscous fluid is an approximate theory that works well especially in gas–liquid flows of liquids of high viscosity at low Reynolds numbers. The theory of purely irrotational flows of a viscous fluid can be seen as a very successful competitor to the theory of purely irrotational flows of an inviscid fluid. We have come to regard every solution of free-surface problems in an inviscid liquid as an opportunity for a new study. There are hundreds of such opportunities that are still available.

The theory of irrotational flows of viscous and viscoelastic liquids that is developed here is embedded in a variety of fluid mechanics problems ranging from cavitation, capillary breakup and rupture, Rayleigh–Taylor and Kelvin–Helmholtz instabilities, irrotational Faraday waves on a viscous fluid, flow-induced structure of particles in viscous and viscoelastic fluids, boundary-layer theory for flow over rigid solids, rising bubbles, and other topics. The theory of stability of free-surface problems developed here is a great improvement of what was available previously and could be used as supplemental text in courses on hydrodynamic stability.

We have tried to assemble here all the literature bearing on the irrotational flow of viscous liquids. For sure, it is not a large literature, but it is likely that despite an honest effort we missed some good works.

We are happy to acknowledge the contributions of persons who have helped us. Terrence Liao made very important contributions to our early work on this subject in the early 1990's. More recently, Juan Carlos Padrino joined our group and has made truly outstanding contributions to problems described here. In a sense, Juan Carlos could be considered to be an author of this book and we are lucky that he came along. We are indebted to G. I. Barenblatt and to K. R. Sreenivasan for their support and help in promoting viscous potential flow as a topic at the foundation of fluid mechanics. The National Science Foundation has supported our work from the beginning.

We worked day and night on this research; Funada in his day and our night and Joseph and Wang in their day and his night. The whole effort was a great pleasure.

List of Abbreviations

2D	two-dimensional
3D	three-demensional
BEM	boundary-element method
BU	Benjamin and Unsell
C/A	convective–absolute
c.c.	complex conjugate
DM	dissipation method
ES	exact solution
FHS	fully hydrodynamic system
FVF	fully viscous flow
IPF	inviscid potential flow
JBB	Joseph, Belanger, and Beavers
JBF	Joseph, Beavers, and Funada
KH	Kelvin–Helmholtz
KT	Kumar and Tuckerman
LHC	Longuet-Higgins and Cokelet
MVK	Miksis, Vanden-Broeck, and Keller
ODE	ordinary differential equation
PAA	polyacrylamide
PDE	partial differential equation
PISO	pressure implicit with splitting of operators
PNSCC	principal normal stress cavitation criterion
PO or PEO	polyox or polyethylene oxide
QUICK	quadratic upwind interpolation for convective kinematics (scheme)
RT	Rayleigh–Taylor
TVF	Taylor vortex flow
VCVPF	viscous correction of VPF
VPF	viscous potential flow

1

Introduction

The theory of potential flow is a topic in both the study of fluid mechanics and in mathematics. The mathematical theory treats properties of vector fields generated by gradients of a potential. The curl of a gradient vanishes. The local rotation of a vector field is proportional to its curl so that potential flows do not rotate as they deform. Potential flows are irrotational.

The mathematical theory of potentials goes back to the 18th century (see Kellogg, 1929). This elegant theory has given rise to jewels of mathematical analysis, such as the theory of a complex variable. It is a well-formed or "mature" theory, meaning that the best research results have already been obtained. We are not going to add to the mathematical theory; our contributions are to the fluid mechanics theory, focusing on effects of viscosity and viscoelasticity. Two centuries of research have focused exclusively on the motions of inviscid fluids. Among the 131,000,000 hits that come up under "potential flows" on Google search are mathematical studies of potential functions and studies of inviscid fluids. These studies can be extended to viscous fluids at small cost and great profit.

The fluid mechanics theory of potential flow goes back to Euler in 1761 (see Truesdell, 1954, §36). The concept of viscosity was not known in Euler's time. The fluids he studied were driven by pressures, not by viscous stresses. The effects of viscous stresses were introduced by Navier (1822) and Stokes (1845). Stokes (1851) considered potential flow of a viscous fluid in an approximate sense, but most later authors restrict their attention to "potential flow of an inviscid fluid." All the books on fluid mechanics and all courses in fluid mechanics have chapters on "potential flow of inviscid fluids" and none on the "potential flow of a viscous fluid."

An authoritative and readable exposition of irrotational flow theory and its applications can be found in chapter 6 of the book on fluid dynamics by Batchelor (1967). He speaks of the role of the theory of flow of an inviscid fluid:

> In this and the following chapter, various aspects of the flow of a fluid regarded as entirely inviscid (and incompressible) will be considered. The results presented are significant only inasmuch as they represent an approximation to the flow of a real fluid at large Reynolds number, and the limitations of each result must be regarded as important as the result itself.

In this book we consider irrotational flows of a viscous fluid. We are of the opinion that when one is considering irrotational solutions of the Navier–Stokes equations it is never necessary and typically not useful for one to put the viscosity to zero. This observation runs counter to the idea frequently expressed that potential flow is a topic that is useful only for inviscid fluids; many people think that the notion of a viscous potential flow is an oxymoron. Incorrect statements like "... irrotational flow implies inviscid flow but not the other way around" can be found in popular textbooks.

Irrotational flows of a viscous fluid scale with the Reynolds number as do rotational solutions of the Navier–Stokes equations generally. The solutions of the Navier–Stokes equations, rotational and irrotational, are thought to become independent of the Reynolds number at large Reynolds numbers. Unlike the theory of irrotational flows of inviscid fluids, the theory of irrotational flow of a viscous fluid can be considered as a description of flow at a finite Reynolds number.

Most of the classical theorems reviewed in chapter 6 of Batchelor's 1967 book do not require that the fluid be inviscid. These theorems are as true for viscous potential flow as they are for inviscid potential flow. Kelvin's minimum-energy theorem holds for the irrotational flow of a viscous fluid. The theory of the acceleration reaction leads to the concept of added mass; it follows from the analysis of unsteady irrotational flow. The theory applies to viscous and inviscid fluids alike.

It can be said that every theorem about potential flow of inviscid incompressible fluids applies equally to viscous fluids in regions of irrotational flow. Jeffreys (1928) derived an equation [his (20)] that replaces the circulation theorem of classical (inviscid) hydrodynamics. When the fluid is homogeneous, Jeffreys' equation may be written as

$$\frac{\mathrm{d}C}{\mathrm{d}t} = -\frac{\mu}{\rho}\oint \operatorname{curl}\boldsymbol{\omega}\cdot\mathrm{d}\mathbf{l}, \tag{1.0.1}$$

where

$$\boldsymbol{\omega} = \operatorname{curl}\mathbf{u}, \quad C(t) = \oint \mathbf{u}\cdot\mathrm{d}\mathbf{l},$$

is the circulation around a closed material curve drawn in the fluid. This equation shows that

> ... the initial value of $\mathrm{d}C/\mathrm{d}t$ around a contour in a fluid originally moving irrotationally is zero, whether or not there is a moving solid within the contour. This at once provides an explanation of the equality of the circulation about an aeroplane and that about the vortex left behind when it starts; for the circulation about a large contour that has never been cut by the moving solid or its wake remains zero, and therefore the circulations about contours obtained by subdividing it must also add up to zero. It also indicates why the motion is in general nearly irrotational except close to a solid of to fluid that has pass near one.

Saint-Vénant (1869) interpreted the result of Lagrange (1781) about the invariance of circulation $\mathrm{d}C/\mathrm{d}t = 0$ to mean that

> vorticity cannot be generated in the interior of a viscous incompressible fluid, subject to conservative extraneous force, but is necessarily diffused inward from the boundaries.

Circulation formula (1.0.1) is an important result in the theory of irrotational flows of a viscous fluid. A particle that is initially irrotational will remain irrotational in motions that do not enter into the vortical layers at the boundary.

1.1 Irrotational flow, Laplace's equation

A potential flow is a velocity field $\mathbf{u}(\phi)$ given by the gradient of a potential ϕ:

$$\mathbf{u}(\phi) = \nabla\phi. \tag{1.1.1}$$

Potential flows have a zero curl:

$$\operatorname{curl}\mathbf{u} = \operatorname{curl}\nabla\phi = 0. \tag{1.1.2}$$

Fields that are curl free, satisfying (1.1.2), are called irrotational.

Vector fields satisfying the equation

$$\operatorname{div}\mathbf{u} = 0 \tag{1.1.3}$$

are said to be solenoidal. Solenoidal flows that are irrotational are harmonic; the potential satisfies Laplace's equation:

$$\operatorname{div}\mathbf{u}(\phi) = \operatorname{div}\nabla\phi = \nabla^2\phi = 0. \tag{1.1.4}$$

The theory of irrotational flow needed in this book is given in many books; for example, Lamb (1932), Milne-Thomson (1968), Batchelor (1967), and Landau and Lifshitz (1987). No-slip cannot be enforced in irrotational flow. However, eliminating all the irrotational effects of viscosity by putting $\mu = 0$ to reconcile our desire to satisfy no-slip at the cost of real physics is like throwing out the baby with the bathwater. This said, we can rely on the book by Batchelor and others for the results we need in our study of irrotational flow of viscous fluids.

1.2 Continuity equation, incompressible fluids, isochoric flow

The equation governing the evolution of the density ρ,

$$\begin{aligned} &\frac{d\rho}{dt} + \rho\operatorname{div}\mathbf{u} = 0, \\ &\frac{d\rho}{dt} = \frac{\partial\rho}{\partial t} + (\mathbf{u}\cdot\nabla)\,\rho, \end{aligned} \tag{1.2.1}$$

is called the continuity equation. It guarantees that the mass of a fluid element is conserved.

If the fluid is incompressible, then ρ is constant and $\operatorname{div}\mathbf{u} = 0$. Flows of compressible fluids for which $\operatorname{div}\mathbf{u} = 0$ are called isochoric. Low-Mach-number flows are nearly isochoric. Incompressible and isochoric flows are solenoidal.

1.3 Euler's equations

Euler's equations of motion are given by

$$\begin{aligned} &\rho\frac{d\mathbf{u}}{dt} = -\nabla p + \rho\mathbf{g}, \\ &\frac{d\mathbf{u}}{dt} = \frac{\partial\mathbf{u}}{\partial t} + (\mathbf{u}\cdot\nabla)\,\mathbf{u}, \end{aligned} \tag{1.3.1}$$

where $\mathbf{g}$ is a body force per unit mass. If ρ is constant and $\mathbf{g} = \nabla G$ has a force potential, then

$$-\nabla p + \rho\mathbf{g} = -\nabla\hat{p}, \tag{1.3.2}$$

where $\hat{p} = p - \rho G$ can be called the pressure head. If $\mathbf{g}$ is gravity, then

$$G = \mathbf{g}\cdot\mathbf{x}.$$

To simplify the writing of equations we put the body force to zero except in cases for which it is important. If the fluid is compressible, then an additional relation, from thermodynamics, relating p to ρ is required. Such a relation can be found for isentropic flow or isothermal flow of a perfect gas. In such a system there are five unknowns, p, ρ, and $\mathbf{u}$, and five equations.

The effects of viscosity are absent in Euler's equations of motion. The effects of viscous stresses are absent in Euler's theory; the flows are driven by pressure. The Navier–Stokes equations reduce to Euler's equations when the fluid is inviscid.

1.4 Generation of vorticity in fluids governed by Euler's equations

Euler's equations (1.3.1) may be written as

$$\frac{\partial \mathbf{u}}{\partial t} - \mathbf{u} \times \boldsymbol{\omega} + \frac{1}{2}\nabla |\mathbf{u}|^2 = \mathbf{g} - \frac{\nabla p}{\rho}, \tag{1.4.1}$$

where we have used the vector identity

$$(\mathbf{u} \cdot \nabla)\mathbf{u} = \frac{1}{2}\nabla |\mathbf{u}|^2 - \mathbf{u} \times \boldsymbol{\omega} \tag{1.4.2}$$

and

$$\boldsymbol{\omega} = \boldsymbol{\omega}[\mathbf{u}] = \operatorname{curl} \mathbf{u}.$$

We can obtain the vorticity equation by forming the curl of (1.4.1):

$$\frac{\partial \boldsymbol{\omega}}{\partial t} + \mathbf{u} \cdot \nabla \boldsymbol{\omega} = \boldsymbol{\omega} \cdot \nabla \mathbf{u} - \operatorname{curl}\frac{\nabla p}{\rho} + \operatorname{curl} \mathbf{g}. \tag{1.4.3}$$

If $\operatorname{curl}(\nabla p/\rho) = \nabla \times [\frac{1}{\rho}p'(\rho)\nabla\rho] = 0$ and $\operatorname{curl}\mathbf{g} = 0$, then $\boldsymbol{\omega}[\mathbf{u}] = 0$; $\operatorname{curl}\mathbf{g} = 0$ if $\mathbf{g}$ is given by a potential $\mathbf{g} = \nabla G$. Flows for which $\operatorname{curl}(\nabla p/\rho) = 0$ are said to be barotropic. Barotropic flows governed by Euler's equations with conservative body forces $\mathbf{g} = \nabla G$ cannot generate vorticity. If the fluid is incompressible, then the flow is barotropic.

1.5 Perfect fluids, irrotational flow

Inviscid fluids that are also incompressible are called "perfect" or "ideal." Perfect fluids satisfy Euler's equations. Perfect fluids with conservative body forces give rise to Bernoulli's equation:

$$\nabla\left[\rho\left(\frac{\partial \phi}{\partial t} + \frac{1}{2}|\nabla\phi|^2\right) + p - G\right] = 0,$$

where

$$\rho\left(\frac{\partial \phi}{\partial t} + \frac{1}{2}|\nabla\phi|^2\right) + p - G = f(t). \tag{1.5.1}$$

We may absorb the function $f(t)$ into the potential $\hat{\phi} = \phi + \int^t f(t)\,\mathrm{d}t$ without changing the velocity:

$$\nabla\hat{\phi} = \nabla\left[\phi + \int^t f(t)\,\mathrm{d}t\right] = \nabla\phi = \mathbf{u}. \tag{1.5.2}$$

Bernoulli's equation relates p and ϕ before any problem is solved; actually $\text{curl}\,\mathbf{u} = 0$ is a constraint on solutions. The velocity $\mathbf{u}$ is determined by ϕ satisfying $\nabla^2\phi = 0$ and the boundary conditions.

1.6 Boundary conditions for irrotational flow

Irrotational flows have a potential, and if the flow is solenoidal the potential ϕ is harmonic, $\nabla^2\phi = 0$. This book reveals the essential role of harmonic functions in the flow of incompressible viscous and viscoelastic fluids. All the books on partial differential equations (PDEs) have sections devoted to the mathematical analysis of the Laplace's equation. Laplace's equation may be solved for prescribed data on the boundary of the flow region including infinity for flows on unbounded domains. It can be solved for Dirichlet data in which values of ϕ are prescribed on the boundary or for Neumann data in which the normal component of $\nabla\phi$ is prescribed on the boundary. Almost any combination of Dirichlet and Neumann data all over the boundary will lead to unique solutions of Laplace's equation.

It is important that unique solutions can be obtained when only one condition is prescribed for ϕ at each point on the boundary. The problem is overdetermined when two conditions are prescribed. We can solve the problem when Dirichlet conditions are prescribed or when Neumann conditions are prescribed but not when both are prescribed. In the case in which Neumann conditions are prescribed over the whole the boundary the solution is unique up to the addition of any constant. If the boundary condition is not specified at each point on the boundary, then the problem may be not overdetermined when two conditions are prescribed.

This point can be forcefully made within the framework of fluid dynamics. Consider, for example, an irrotational streaming flow over a body. The velocity $\mathbf{U}$ at infinity in the direction $\mathbf{x}$ is given by the potential $\phi = \mathbf{U} \cdot \mathbf{x}$. The tangential component of velocity on the boundary S of the body is given by

$$u_S = \mathbf{e}_S \cdot \mathbf{u} = \mathbf{e}_S \cdot \nabla\phi, \tag{1.6.1}$$

where $\mathbf{e}_S$ is a unit lying entirely in S. The normal component of velocity on S is given by

$$u_n = \mathbf{n} \cdot \mathbf{u} = \mathbf{n} \cdot \nabla\phi, \tag{1.6.2}$$

where $\mathbf{n}$ is the unit normal pointing from body to fluid.

Laplace's equation for streaming flow can be solved if

$$\phi \text{ is prescribed on } S, \tag{1.6.3}$$

or if

$$\mathbf{n} \cdot \nabla\phi \text{ is prescribed on } S. \tag{1.6.4}$$

It cannot be solved if ϕ and $\mathbf{n} \cdot \nabla\phi$ are simultaneously prescribed on S.

The prescription of the tangential velocity on S is a Dirichlet condition. If ϕ is prescribed on S the tangential derivatives can be computed. It is possible to solve Laplace's equation for a linear combination of ϕ and $\mathbf{n} \cdot \nabla\phi$; say $\mathbf{n} \cdot \nabla\phi + \alpha\phi$ is prescribed on S. This bind of condition is called a Robin boundary condition.

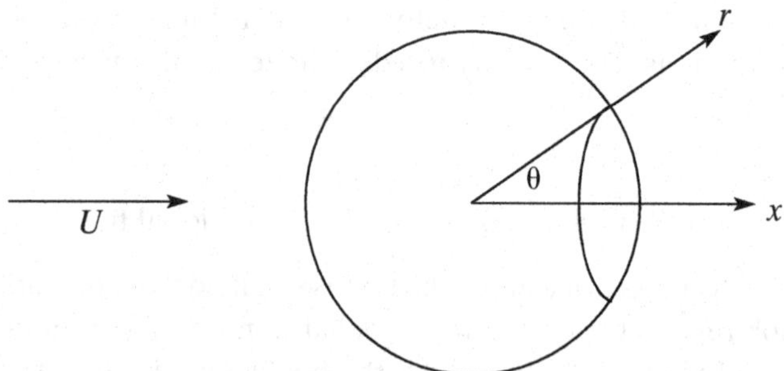

Figure 1.1. Axisymmetric flow over a sphere of radius a. The flow depends on the radius r and the polar angle θ.

The no-slip condition of viscous fluid mechanics requires that

$$u_S = 0, \quad u_n = 0 \tag{1.6.5}$$

simultaneously on S. These conditions cannot be satisfied by potential flow. In fact, these conditions cannot be satisfied by solutions of Euler's equations even when they are not irrotational.

The point of view that has been universally adopted by researchers and students of fluid mechanics for several centuries is that the normal component should be enforced so that the fluid does not penetrate the solid. The fluid must then slip at the boundary because there is no other choice. To reconcile this with no-slip, researchers put the viscosity to zero. The resolution of these difficulties lies in the fact that real flows have a nonzero rotational velocity at boundaries generated by the no-slip condition. The no-slip condition usually cannot be satisfied by the rotational velocity alone; the irrotational velocity is also needed (see chapter 4).

1.7 Streaming irrotational flow over a stationary sphere

A flow of speed U in the direction $x = r\cos\theta$ streams past a sphere of radius a (see figure 1.1). A solution ϕ of Laplace's equation $\nabla^2\phi = 0$ in spherical polar coordinates is

$$\phi = Ur\cos\theta + \frac{c}{r^2}\cos\theta. \tag{1.7.1}$$

The normal and tangential components of the velocity at $r = a$ are, respectively,

$$\frac{\partial\phi}{\partial r} = U\cos\theta - \frac{2c}{a^3}\cos\theta, \tag{1.7.2}$$

$$\frac{1}{a}\frac{\partial\phi}{\partial\theta} = -U\sin\theta - \frac{c}{a^3}\sin\theta. \tag{1.7.3}$$

If the normal velocity is prescribed to be zero on the sphere, then $c = Ua^3/2$ and

$$\phi_1 = U\left(r + \frac{a^3}{2r^2}\right)\cos\theta. \tag{1.7.4}$$

The corresponding tangential velocity is $-3U \sin\theta/2$. If the tangential velocity is prescribed to be zero on the sphere, then $c = -Ua^3$ and

$$\phi_2 = U\left(r - \frac{a^3}{r^2}\right)\cos\theta. \tag{1.7.5}$$

The corresponding normal velocity is $3U \cos\theta$.

The preceding analysis and uniqueness of solutions of Laplace's equation show that we may obtain the solution ϕ_1 with a zero normal velocity by prescribing the nonzero function $-\frac{3}{2}U \sin\theta$ for the tangential velocity. We may obtain the solution ϕ_2 with a zero tangential velocity by prescribing the nonzero function $3U \cos\theta$ for the normal velocity.

More complicated conditions for harmonic solutions on rigid bodies are encountered in exact solutions of the Navier–Stokes equations in which irrotational and rotational flows are tightly coupled at the boundary. The boundary conditions cannot be satisfied without irrotational flow and they cannot be satisfied by irrotational flow only [cf. equation (4.6.6)].

2

Historical notes

Potential flows of viscous fluids are an unconventional topic with a niche history assembled in a recent review article (Joseph, 2006a, 2006b).

2.1 Navier–Stokes equations

The history of Navier–Stokes equations begins with the 1822 memoir of Navier, who derived equations for homogeneous incompressible fluids from a molecular argument. Using similar arguments, Poisson (1829) derived the equations for a compressible fluid. The continuum derivation of the Navier–Stokes equation is due to Saint-Vénant (1843) and Stokes (1845). In his 1851 paper, Stokes wrote as follows:

> Let P_1, P_2, P_3 be the three normal, and T_1, T_2, T_3 the three tangential pressures in the direction of three rectangular planes parallel to the co-ordinate planes, and let D be the symbol of differentiation with respect to t when the particle and not the point of space remains the same. Then the general equations applicable to a heterogeneous fluid (the equations (10) of my former (1845) paper), are
>
> $$\rho\left(\frac{\mathrm{D}u}{\mathrm{D}t} - X\right) + \frac{\mathrm{d}P_1}{\mathrm{d}x} + \frac{\mathrm{d}T_3}{\mathrm{d}y} + \frac{\mathrm{d}T_2}{\mathrm{d}z} = 0, \tag{132}$$
>
> with the two other equations which may be written down from symmetry. The pressures P_1, T_1, etc. are given by the equations
>
> $$P_1 = p - 2\mu\left(\frac{\mathrm{d}u}{\mathrm{d}x} - \delta\right), \quad T_1 = -\mu\left(\frac{\mathrm{d}v}{\mathrm{d}z} + \frac{\mathrm{d}w}{\mathrm{d}y}\right), \tag{133}$$
>
> and four other similar equations. In these equations
>
> $$3\delta = \frac{\mathrm{d}u}{\mathrm{d}x} + \frac{\mathrm{d}v}{\mathrm{d}y} + \frac{\mathrm{d}w}{\mathrm{d}z}. \tag{134}$$

The equations written by Stokes in his 1845 paper are the same ones we use today:

$$\rho\left(\frac{\mathrm{d}\mathbf{u}}{\mathrm{d}t} - \mathbf{X}\right) = \operatorname{div}\mathbf{T}, \tag{2.1.1}$$

where $\mathbf{X}$ is presumably a body force, which was not specified by Stokes, and

$$\mathbf{T} = \left(-p - \frac{2}{3}\mu \operatorname{div}\mathbf{u}\right)\mathbf{1} + 2\mu\mathbf{D}[\mathbf{u}], \tag{2.1.2}$$

$$\frac{d\mathbf{u}}{dt} = \frac{\partial \mathbf{u}}{\partial t} + (\mathbf{u}\cdot\nabla)\,\mathbf{u}, \tag{2.1.3}$$

$$\mathbf{D}[\mathbf{u}] = \frac{1}{2}\left[\nabla\mathbf{u} + \nabla\mathbf{u}^T\right], \tag{2.1.4}$$

$$\frac{d\rho}{dt} + \rho \operatorname{div}\mathbf{u} = 0. \tag{2.1.5}$$

Stokes assumed that the bulk viscosity $-\frac{2}{3}\mu$ is selected so that the deviatoric part of $\mathbf{T}$ vanishes and trace $\mathbf{T} = -3p$.

Inviscid fluids are fluids with zero viscosity. Viscous effects on the motion of fluids were not understood before the notion of viscosity was introduced by Navier in 1822. Perfect fluids, following the usage of Stokes and other 19th-century English mathematicians, are inviscid fluids that are also incompressible. Statements like Truesdell's (1954),

> In 1781 Lagrange presented his celebrated velocity-potential theorem: if a velocity potential exists at one time in a motion of an inviscid incompressible fluid, subject to conservative extraneous force, it exists at all past and future times.

though perfectly correct, could not have been asserted by Lagrange, because the concept of an inviscid fluid was not available in 1781.

2.2 Stokes theory of potential flow of viscous fluid

The theory of potential flow of a viscous fluid was introduced by Stokes in 1851. All of his work on this topic is framed in terms of the effects of viscosity on the attenuation of small-amplitude waves on a liquid–gas surface. Everything he said about this problem is subsequently cited. The problem treated by Stokes was solved exactly by Lamb (1932), who used the linearized Navier–Stokes equations, without assuming potential flow.

Stokes' discussion is divided into three parts, discussed in §§51, 52, and 53 of his (1851) paper:

(1) The dissipation method in which the decay of the energy of the wave is computed from the viscous dissipation integral in which the dissipation is evaluated on potential flow (§51).

(2) The observation that potential flows satisfy the Navier–Stokes equations together with the notion that certain viscous stresses must be applied at the gas–liquid surface to maintain the wave in permanent form (§52).

(3) The observation that, if the viscous stresses required for maintaining the irrotational motion are relaxed, the work of those stresses is supplied at the expense of the energy of the irrotational flow (§53).

Lighthill (1978) discussed Stokes' ideas but he did not contribute more to the theory of irrotational motions of a viscous fluid. On page 234 he notes that

> Stoke's ingenious idea was to recognise that the average value of this rate of working
>
> $$2\mu\left[(\partial\phi/\partial x)\,\partial^2\phi/\partial x\partial z+(\partial\phi/\partial z)\,\partial^2\phi/\partial z^2\right]_{z=0}$$
>
> required to maintain the unattenuated irrotational motions of sinusoidal waves must exactly balance the rate at which the same waves when propagating freely would lose energy by internal dissipation!.

Lamb (1932) gave an exact solution of the problem considered by Stokes in which vorticity and boundary layers are not neglected. Wang and Joseph (2006a) did purely irrotational theories of Stokes' problem that are in good agreement with Lamb's exact solution.

In fact, Stokes' idea called "ingenious" by Lighthill has serious defects in implementation. The unattenuated irrotational waves move with a speed independent of viscosity as would be true for waves on an inviscid fluid. Lamb's 1932 application of Stokes' idea gives rise to a good approximation of the decay rates due to viscosity for long waves but does not correct the wave speeds for the effects of viscosity. Moreover, the cutoff between long and short waves, which is defined by a condition (say large viscosity) for which the wave speed vanishes and progressive waves become standing waves, cannot be obtained from the dissipation calculation proposed by Stokes and implemented by Lamb and all other authors. A correct irrotational approximation, called VCVPF, which gives an approximation to Lamb's exact solution for all wavenumbers, including the dependence of the wave speed on viscosity and the all important cutoff value, was implemented by Wang and Joseph (2006b). Padrino and Joseph (2007) have demonstrated that all these results may be obtained by a revised and rigorous application of the dissipation method to the problem of viscous decay of capillary-gravity waves. They show that the effects of viscosity on the speed of progressive waves are sensible for waves near the cutoff value which are not too long and they obtain the irrotational approximation of this cutoff value (see figure 14.14).

2.3 The dissipation method

In his 1851 paper, Stokes writes:

> 51. By means of the expression given in Art. 49, for the loss of *vis viva* due to internal friction, we may readily obtain a very approximate solution of the problem: To determine the rate at which the motion subsides, in consequence of internal friction, in the case of a series of oscillatory waves propagated along the surface of a liquid. Let the vertical plane of xy be parallel to the plane of motion, and let y be measured vertically downwards from the mean surface; and for simplicity's sake suppose the depth of the fluid very great compared with the length of a wave, and the motion so small that the square of the velocity may be neglected. In the case of motion which we are considering, $u\mathrm{d}x+v\mathrm{d}y$ is an exact differential $\mathrm{d}\phi$ when friction is neglected, and
>
> $$\phi = c\varepsilon^{-my}\sin(mx-nt)\,, \tag{140}$$
>
> where c, m, n are three constants, of which the last two are connected by a relation which it is not necessary to write down. We may continue to employ this equation as a near approximation

when friction is taken into account, provided we suppose c, instead of being constant, to be parameter which varies slowly with the time. Let V be the *vis viva* of a given portion of the fluid at the end of the time t, then

$$V = \rho c^2 m^2 \iiint e^{-2my} \mathrm{d}x\mathrm{d}y\mathrm{d}z. \tag{141}$$

But by means of the expression given in Art.49, we get for the loss of *vis viva* during the time $\mathrm{d}t$, observing that in the present case μ is constant, $w = 0$, $\delta = 0$, and $u\mathrm{d}x + v\mathrm{d}y = \mathrm{d}\phi$, where ϕ is independent of z,

$$4\mu\mathrm{d}t \iiint \left\{ \left(\frac{\mathrm{d}^2\phi}{\mathrm{d}x^2}\right)^2 + \left(\frac{\mathrm{d}^2\phi}{\mathrm{d}y^2}\right)^2 + 2\left(\frac{\mathrm{d}^2\phi}{\mathrm{d}x\mathrm{d}y}\right)^2 \right\} \mathrm{d}x\mathrm{d}y\mathrm{d}z$$

which becomes, on substituting for ϕ its value,

$$8\mu c^2 m^4 \mathrm{d}t \iiint e^{-2my} \mathrm{d}x\mathrm{d}y\mathrm{d}z.$$

But we get from (141) for the decrement of *vis viva* of the same mass arising from the variation of the parameter c

$$-2\rho m^2 c \frac{\mathrm{d}c}{\mathrm{d}t} \mathrm{d}t \iiint e^{-2my} \mathrm{d}x\mathrm{d}y\mathrm{d}z.$$

Equating the two expressions for the decrement of *vis viva*, putting for m its value $2\pi\lambda^{-1}$, where λ is the length of a wave, replacing μ by $\mu'\rho$, integrating, and supposing c_0 to be the initial value of c, we get[1]

$$c = c_0 e^{-\frac{16\pi^2\mu' t}{\lambda^2}}.$$

It will presently appear that the value of $\sqrt{\mu'}$ for water is about 0.0564, an inch and a second being the units of space and time. Suppose first that λ is two inches, and t ten seconds. Then $16\pi^2\mu' t\lambda^{-2} = 1.256$, and c: c_0::1:0.2848, so that the height of the waves, which varies as c, is only about a quarter of what it was. Accordingly, the ripples excited on a small pool by a puff of wind rapidly subside when the exciting cause ceases to act.

Now suppose that λ is to fathoms or 2880 inches, and that t is 86400 seconds or a whole day. In this case $16\pi^2\mu' t\lambda^{-2}$ is equal to only 0.005232, so that by the end of an entire day, in which time waves of this length would travel 574 English miles, the height would be diminished by little more than the one two hundredth part in consequence of friction. Accordingly, the long swells of the ocean are but little allayed by friction, and at last break on some shore situated at the distance of perhaps hundreds of miles from the region where they were first excited.

2.4 The distance a wave will travel before it decays by a certain amount

The observations made by Stokes about the distance a wave will travel before its amplitude decays by a given amount point the way to a useful frame for the analysis of the effects of viscosity on wave propagation. Many studies of nonlinear irrotational waves can be found in the literature, but the only study known to us of the effects of viscosity on the decay of these waves is due to Longuet-Higgins (1997), who used the dissipation method to determine the decay that is due to viscosity of irrotational steep capillary-gravity waves in deep water. He finds that the limiting rates of decay for small-amplitude solitary waves are

[1] In a footnote on page 624, Lamb notes that "Through an oversight in the original calculation the value $\lambda^2/16\pi^2\nu$ was too small by one half." The value 16 should be 8.

twice those for linear periodic waves computed by the dissipation method. The dissipation of very steep waves can be more than 10 times that of linear waves because of the sharply increased curvature in wave troughs. He assumes that the nonlinear wave maintains its steady form while decaying under the action of viscosity. The wave shape could change radically from its steady shape in very steep waves.

Stokes (1880) studied the motion of nonlinear irrotational gravity waves in two dimensions that are propagated with a constant velocity, and without change of form. This analysis led Stokes (1880) to the celebrated maximum wave, whose asymptotic form gives rise to a pointed crest of angle 120°. The effects of viscosity on such extreme waves has not been studied, but they may be studied by the dissipation method or the same potential flow theory used by Stokes (1851) for inviscid fluids with the caveat that the normal stress condition that p vanish on the free surface be replaced with the condition that

$$p + 2\mu \partial u_n / \partial n = 0$$

on the free surface with normal $\mathbf{n}$, where the velocity component $u_n = \partial\phi/\partial n$ is given by the potential.

3

Boundary conditions for viscous fluids

Boundary conditions for incompressible viscous fluids cannot usually be satisfied without contributions from potential flows [Joseph and Renardy, 1991, §1.2(d)].

(1) No-slip conditions are required at the boundary S of a rigid solid:

$$\mathbf{u} = \mathbf{U} \quad \text{for } \mathbf{x} \in S, \tag{3.0.1}$$

where $\mathbf{U}$ is the velocity of the solid.

(2) At the interface S between two fluids, the fluid velocities are continuous, the shear stress is continuous, and the stress is balanced by surface tension.

Now we express the condition just mentioned with equations. Let the position of the surface S as it moves through the surface be $F[\mathbf{x}(t), t] = 0$, where $\mathbf{x}(t) \in S$ for all t and $\dot{\mathbf{x}}(t) \in \mathbf{u}_S$ is the velocity of points of S. Because $F = 0$ is an identity in t, we have

$$\frac{\mathrm{d}F}{\mathrm{d}t} = \frac{\partial F}{\partial t} + \mathbf{u}_S \cdot \nabla F = 0; \tag{3.0.2}$$

in fact $\mathrm{d}^m F/\mathrm{d}t^m = 0$ for all m. Now, the normal $\mathbf{n}$ to S (figure 3.1) is $\mathbf{n} = \nabla F / |\nabla F|$; hence

$$\mathbf{u}_S \cdot \nabla F = (\mathbf{n} \cdot \mathbf{u}_S) |\nabla F| . \tag{3.0.3}$$

The tangential component of $\mathbf{u}_S$ is irrelevant, and

$$\mathbf{n} \cdot \mathbf{u}_S = \mathbf{n} \cdot \mathbf{u}, \tag{3.0.4}$$

where $\mathbf{u}$ is the velocity of a material point on $\mathbf{x} \in S$.

We now set a convention for jumps in the value of variables as we cross S at $\mathbf{x}$. Define

$$[[\bullet]] = (\bullet)_1 - (\bullet)_2 ,$$

where $(\bullet)_1$ means to evaluate $(\bullet)$ at $\mathbf{x}$ in the fluid 1 and likewise for 2.

Combining (3.0.2), (3.0.3), and (3.0.4), we get the kinematic equation for the evolution of F:

$$\frac{\mathrm{d}F}{\mathrm{d}t} = \frac{\partial F}{\partial t} + \mathbf{u} \cdot \nabla F = 0. \tag{3.0.5}$$

Because $[[\partial F/\partial t]] = 0$, we have

$$[[\mathbf{u} \cdot \mathbf{n}]] = [[\mathbf{u}]] \cdot \mathbf{n} = 0. \tag{3.0.6}$$

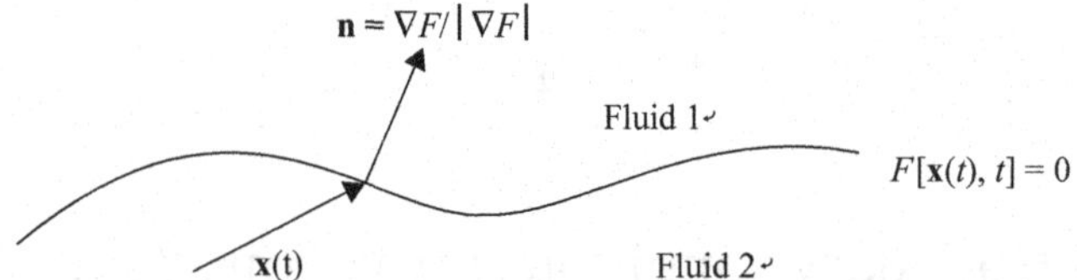

Figure 3.1. Interface S between two fluids.

The normal component of velocity is continuous. If the tangential component of velocity is also continuous across S, we have

$$\mathbf{n} \times [[\mathbf{u}]] = 0, \quad \mathbf{x} \in S. \tag{3.0.7}$$

Turning now to the stress,

$$\begin{aligned} \mathbf{T} &= -p\mathbf{1} + 2\mu\mathbf{D}[\mathbf{u}], \\ T_{ij} &= -p\delta_{ij} + \mu\left(\frac{\partial u_i}{\partial x_j} + \frac{\partial u_j}{\partial x_i}\right), \end{aligned} \tag{3.0.8}$$

we express the continuity of the shear stress as

$$\mathbf{e}_S \cdot [[2\mu\mathbf{D}[\mathbf{u}]]] \cdot \mathbf{n} = 0, \tag{3.0.9}$$

where $\mathbf{e}_S$ is a unit vector tangent to S; $\mathbf{e}_S \cdot \nabla F = 0$.

The condition that the jump in the normal component of the stress be balanced by a surface-tension force may be expressed as

$$\mathbf{n} \cdot [[\mathbf{T}]] \cdot \mathbf{n} = -[[p]] + 2\mathbf{n} \cdot [[\mu\mathbf{D}[\mathbf{u}]]] \cdot \mathbf{n} = 2H\gamma - \nabla_{\text{II}}\gamma, \tag{3.0.10}$$

where H is the mean curvature,

$$2H = \frac{1}{R_1} + \frac{1}{R_2}, \tag{3.0.11}$$

and R_1 and R_2 are principal radii of curvature. The surface gradient is

$$\nabla_{\text{II}} = \nabla - \mathbf{n}(\mathbf{n} \cdot \nabla). \tag{3.0.12}$$

The curvature $2H$ is determined by the surface divergence of the normal (see Joseph and Renardy, 1991):

$$2H = \nabla_{\text{II}} \cdot \mathbf{n}. \tag{3.0.13}$$

(3) Free surfaces. Free surfaces are two fluid interfaces for which one fluid is a dynamically inactive gas. The density of the gas is usually much less than the density of the liquid, and the viscosity of the gas is usually much less than the viscosity of the liquid. In some cases we can put the viscosity and density of the gas to zero; then we have a liquid–vacuum interface. This procedure works well for problems of capillary instability and Rayleigh–Taylor (RT) instability. However, for Kelvin–Helmholtz (KH) instabilities the kinematic viscosity $\nu = \mu/\rho$ is important, and ν for the liquid and gas are of comparable magnitude. The dynamic participation of the ambient is necessary for KH instability.

For free-surface problems with a dynamically inactive ambient, no condition on the tangential velocity of the interface is required; the shear stress vanishes,

$$2\mu \mathbf{e}_S \cdot \mathbf{D}[\mathbf{u}] \cdot \mathbf{n} = 0, \quad \mathbf{x} \in S, \tag{3.0.14}$$

and the normal stress balance is

$$-p + 2\mu \mathbf{n} \cdot \mathbf{D}[\mathbf{u}] \cdot \mathbf{n} = 2H\gamma - \nabla_{\mathrm{II}}\gamma. \tag{3.0.15}$$

Sometimes the sign of the mean curvature is ambiguous (we could have assigned **n** from 2 to 1). A quick way to determine the sign is to look at the terms,

$$-p = 2\left(\frac{1}{R_1} + \frac{1}{R_2}\right)\gamma,$$

making sure the pressure is higher inside the equivalent sphere.

Summarizing, the main equations on S are (3.0.5), (3.0.6), (3.0.7), (3.0.9), (3.0.10), (3.0.14), and (3.0.15). These equations are written for Cartesian, cylindrical, and polar spherical coordinates in Appendix A.

4

Helmholtz decomposition coupling rotational to irrotational flow

In this chapter we present the form of the Navier–Stokes equations implied by the Helmholtz decomposition in which the relation of the irrotational and rotational velocity fields is made explicit. The idea of self-equilibration of irrotational viscous stresses is introduced. The decomposition is constructed first by selection of the irrotational flow compatible with the flow boundaries and other prescribed conditions. The rotational component of velocity is then the difference between the solution of the Navier–Stokes equations and the selected irrotational flow. To satisfy the boundary conditions, the irrotational field is required, and it depends on the viscosity. Five unknown fields are determined by the decomposed form of the Navier–Stokes equations for an incompressible fluid: the three rotational components of velocity, the pressure, and the harmonic potential. These five fields may be readily identified in analytic solutions available in the literature. It is clear from these exact solutions that potential flow of a viscous fluid is required for satisfying prescribed conditions, such as the no-slip condition at the boundary of a solid or continuity conditions across a two-fluid boundary. The decomposed form of the Navier–Stokes equations may be suitable for boundary layers because the target irrotational flow that is expected to appear in the limit, say at large Reynolds numbers, is an explicit to-be-determined field. It can be said that equations governing the Helmholtz decomposition describe the modification of irrotational flow due to vorticity, but the analysis shows the two fields are coupled and cannot be completely determined independently.

4.1 Helmholtz decomposition

The Helmholtz decomposition theorem says that every smooth vector field $\mathbf{u}$, defined everywhere in space and vanishing at infinity, together with its first derivatives, can be decomposed into a rotational part $\mathbf{v}$ and an irrotational part $\nabla\phi$:

$$\mathbf{u} = \mathbf{v} + \nabla\phi, \tag{4.1.1}$$

where

$$\operatorname{div}\mathbf{u} = \operatorname{div}\mathbf{v} + \nabla^2\phi, \tag{4.1.2}$$

$$\operatorname{curl}\mathbf{u} = \operatorname{curl}\mathbf{v}. \tag{4.1.3}$$

This decomposition leads to the theory of the vector potential, which is not followed here. The decomposition is unique on unbounded domains, and explicit formulas for the scalar and vector potentials are well known. A framework for embedding the study of potential flows of viscous fluids, in which no special flow assumptions are made, is suggested by this decomposition (Joseph, 2006a, 2006c). If the fields are solenoidal, then

$$\operatorname{div} \mathbf{u} = \operatorname{div} \mathbf{v} = 0, \quad \nabla^2 \phi = 0. \tag{4.1.4}$$

Because ϕ is harmonic, we have from (4.1.1) and (4.1.4) that

$$\nabla^2 \mathbf{u} = \nabla^2 \mathbf{v}. \tag{4.1.5}$$

The irrotational part of $\mathbf{u}$ is on the null space of the Laplacian, but in special cases, such as plane shear flow, $\nabla^2 \mathbf{v} = 0$ but $\operatorname{curl} \mathbf{v} \neq 0$.

Unique decompositions are generated by solutions of Navier–Stokes equation (4.2.1) in the decomposed form of (4.2.2) for which the irrotational flows satisfy (4.1.1), (4.1.3), (4.1.4), and (4.1.5) and certain boundary conditions. The boundary conditions for the irrotational flows bear a heavy weight in all this. Simple examples of unique decomposition, taken from hydrodynamics, are presented later.

The decomposition of the velocity into rotational and irrotational parts holds at each and every point and varies from point to point in the flow domain. Various possibilities for the balance of these parts at a fixed point and the distribution of these balances from point to point can be considered:

(1) The flow is purely irrotational or purely rotational. These two possibilities do occur as approximations but are not typical.

(2) Typically the flow is mixed with rotational and irrotational components at each point.

4.2 Navier–Stokes equations for the decomposition

To study solutions of the Navier–Stokes equations, it is convenient to express the Navier–Stokes equations for an incompressible fluid,

$$\rho \frac{\mathrm{d}\mathbf{u}}{\mathrm{d}t} = -\nabla p + \mu \nabla^2 \mathbf{u}, \tag{4.2.1}$$

in terms of the rotational and irrotational fields implied by the Helmholtz decomposition,

$$\rho \frac{\partial \mathbf{v}}{\partial t} + \nabla \left(\rho \frac{\partial \phi}{\partial t} + \frac{\rho}{2} |\nabla \phi|^2 + p \right) + \rho \operatorname{div} (\mathbf{v} \otimes \nabla \phi + \nabla \phi \otimes \mathbf{v} + \mathbf{v} \otimes \mathbf{v}) = \mu \nabla^2 \mathbf{v} \tag{4.2.2}$$

or

$$\rho \frac{\partial v_i}{\partial t} + \frac{\partial}{\partial x_i} \left(\rho \frac{\partial \phi}{\partial t} + \frac{\rho}{2} |\nabla \phi|^2 + p \right) + \rho \frac{\partial}{\partial x_j} \left(v_j \frac{\partial \phi}{\partial x_i} + v_i \frac{\partial \phi}{\partial x_j} + v_i v_j \right) = \mu \nabla^2 v_i,$$

satisfying (4.1.4).

To solve this problem in a domain Ω, say, when the velocity $\mathbf{u} = \mathbf{V}$ is prescribed on $\partial\Omega$, we would need to compute a solenoidal field $\mathbf{v}$ satisfying (4.2.2) and a harmonic function ϕ satisfying $\nabla^2 \phi = 0$ such that

$$\mathbf{v} + \nabla \phi = \mathbf{V} \text{ on } \partial\Omega.$$

Because this system of five equations in five unknowns is just the decomposed form of the four equations in four unknowns that define the Navier–Stokes system for $\mathbf{u}$, it ought to be possible to study this problem with exactly the same mathematical tools that are used to study the Navier–Stokes equations.

In the Navier–Stokes theory for incompressible fluid, the solutions are decomposed into a space of gradients and its complement, which is a space of solenoidal vectors. The gradient space is not, in general, solenoidal because the pressure is not solenoidal. If it were solenoidal, then $\nabla^2 p = 0$, but $\nabla^2 p = -\text{div}\, \rho\mathbf{u}\cdot\nabla\mathbf{u}$ satisfies Poisson's equation.

It is not true that only the pressure is found on the gradient space. Indeed, equation (4.2.2) gives rise to Poisson's equation for the Bernoulli function, not just the pressure:

$$\nabla^2\left(\rho\frac{\partial\phi}{\partial t} + \frac{\rho}{2}|\nabla\phi|^2 + p\right) = -\rho\frac{\partial^2}{\partial x_i\partial x_j}\left(v_j\frac{\partial\phi}{\partial x_i} + v_i\frac{\partial\phi}{\partial x_j} + v_i v_j\right).$$

In fact, there may be hidden irrotational terms on the right-hand side of this equation.

The boundary condition for solutions of (4.2.1) is

$$\mathbf{u} - \mathbf{a} = 0 \text{ on } \partial\Omega,$$

where $\mathbf{a}$ is solenoidal field, $\mathbf{a} = \mathbf{V}$ on $\partial\Omega$. Hence,

$$\mathbf{v} - \mathbf{a} + \nabla\phi = 0. \tag{4.2.3}$$

The decomposition depends on the selection of the harmonic function ϕ; the traditional boundary condition,

$$\mathbf{n}\cdot\mathbf{a} = \mathbf{n}\cdot\nabla\phi \text{ on } \partial\Omega, \tag{4.2.4}$$

together with a Dirichlet condition at infinity when the region of flow is unbounded, and a prescription of the value of the circulation in doubly connected regions give rise to a unique ϕ. Then the rotational flow must satisfy

$$\mathbf{v}\cdot\mathbf{n} = 0, \quad \mathbf{e}_s\cdot(\mathbf{v} - \mathbf{a}) + \mathbf{e}_s\cdot\nabla\phi = 0 \tag{4.2.5}$$

on $\partial\Omega$. The $\mathbf{v}$ determined in this way is rotational and satisfies (4.1.3). However, $\mathbf{v}$ may contain other harmonic components.

Purely rotational velocities can be identified in the exact solutions exhibited in the examples in which the parts of the solution that are harmonic and the parts that are not are identified by inspection. Equation (4.6.6), in which the purely irrotational flow is identified by selection of a parameter α, is a good example. We have a certain freedom in selecting the harmonic functions used for the decomposition.

We can form a more general formulation of the boundary condition generating the potential in the Helmholtz decomposition by replacing Neumann condition (4.2.4) with a Robin condition,

$$\alpha\phi + \beta\mathbf{n}\cdot\nabla\phi = \mathbf{a}\cdot\mathbf{n}, \tag{4.2.6}$$

depending on two free parameters. The boundary conditions satisfied by the rotational velocity $\mathbf{v}$ are

$$\begin{aligned}\mathbf{v}\cdot\mathbf{n} + \mathbf{n}\cdot\nabla\phi &= \mathbf{a}\cdot\mathbf{n},\\ \mathbf{v}\cdot\mathbf{e}_s + \mathbf{e}_s\cdot\nabla\phi &= \mathbf{a}\cdot\mathbf{e}_s,\end{aligned} \tag{4.2.7}$$

where ϕ is a harmonic function satisfying (4.2.6). The free parameters α and β of equation (4.2.6) can be determined to obtain an optimal result, for which "optimal" is a concept that is in need of further elaboration. Ideally, we would like to determine α and β so that $\mathbf{v}$ is "purely rotational." The examples show that purely rotational flows exist in special cases. It remains to see if this concept makes sense in a general theory.

What is the value we add to solutions of Navier–Stokes equations (4.2.1) by solving them in the Helmholtz decomposed form[1] of (4.2.2)? Certainly it is not easier to solve for five rather than for four fields; if you cannot solve (4.2.1) then you certainly cannot solve (4.2.2). However, if the decomposed solution could be extracted from solutions of (4.2.1) or computed directly, then the form of the irrotational solution that is determined through coupling with the rotational solution and the changes in its distribution as the Reynolds number changes would be revealed. There is nothing approximate about this; it is the correct description of the role of irrotational solutions in the theory of the Navier–Stokes equations, and it looks different and is different than the topic "potential flow of an inviscid fluid" that we all learned in school.

The form (4.2.2) of the Navier–Stokes equations may be well suited to the study of boundary layers of vorticity with irrotational flow of viscous fluid outside. It may be conjectured that in such layers $\mathbf{v} \neq 0$, whereas $\mathbf{v}$ is relatively small in the irrotational viscous flow outside. Rotational and irrotational flows are coupled in the mixed inertial term on the left-hand side of (4.2.2). The irrotational flow does not vanish in the boundary layer, and the rotational flow, though small, probably will not be zero in the irrotational viscous flow outside. This feature is also in Prandtl's theory of boundary layers, but that theory is not rigorous, and the irrotational part is, so to say, inserted by hand and is not coupled to the rotational flow at the boundary. The coupling terms are of considerable interest, and they should play a strong role in the region of small vorticity at the edge of the boundary layer. The actions of irrotational flow in the exact boundary-layer solution of Hiemenz (1911) for a steady two-dimensional (2D) flow toward a "stagnation point" at a rigid body (Batchelor, 1967, 286–8), and of Hamel (1917) flow in diverging and converging channels in the Helmholtz decomposed form are discussed at the end of this chapter.

Effects on boundary layers on rigid solids arising from the viscosity of the fluid in the irrotational flow outside have been considered without the decomposition by Wang and Joseph (2006b, 2006c) and by Padrino and Joseph (2007).

4.3 Self-equilibration of the irrotational viscous stress

The stress in a Newtonian incompressible fluid is given by

$$\mathbf{T} = -p\mathbf{1} + \mu\left(\nabla\mathbf{u} + \mathbf{u}^T\right) = -p\mathbf{1} + \mu\left(\nabla\mathbf{v} + \nabla\mathbf{v}^T\right) + 2\mu\nabla\otimes\nabla\phi.$$

Most flows have an irrotational viscous stress. The term $\mu\nabla^2\mathbf{v}$ in (4.2.2) arises from the rotational part of the viscous stress.

The irrotational viscous stress $\boldsymbol{\tau}_I = 2\mu\nabla\otimes\nabla\phi$ does not give rise to a force-density term in (4.2.2). The divergence of $\boldsymbol{\tau}_I$ vanishes on each and every point in the domain V of flow.

[1] A cultured lady asked a famous conductor of Baroque music if J. S. Bach was still composing: "No madame, he is decomposing."

Even though an irrotational viscous stress exists, it does not produce a net force to drive motions. Moreover,

$$\int \operatorname{div} \boldsymbol{\tau}_I \mathrm{d}V = \oint \mathbf{n} \cdot \boldsymbol{\tau}_I \mathrm{d}S = 0. \tag{4.3.1}$$

The traction vectors $\mathbf{n} \cdot \boldsymbol{\tau}_I$ have no net resultant on each and every closed surface in the domain V of flow. We say that the irrotational viscous stresses, which do not drive motions, are self-equilibrated. Irrotational viscous stresses are not equilibrated at boundaries, and they may produce forces there.

4.4 Dissipation function for the decomposed motion

The dissipation function evaluated on the decomposed field (4.1.1) sorts out into rotational, mixed, and irrotational terms given by

$$\begin{aligned}\Phi &= \int_V 2\mu D_{ij} D_{ij} \mathrm{d}V \\ &= \int 2\mu D_{ij}[\mathbf{v}] D_{ij}[\mathbf{v}] \mathrm{d}V + 4\mu \int D_{ij}[\mathbf{v}] \frac{\partial^2 \phi}{\partial x_i \partial x_j} \mathrm{d}V + 2\mu \int \frac{\partial^2 \phi}{\partial x_i \partial x_j} \frac{\partial^2 \phi}{\partial x_i \partial x_j} \mathrm{d}V.\end{aligned} \tag{4.4.1}$$

Most flows have an irrotational viscous dissipation. In regions V' where $\mathbf{v}$ is small, we have approximately that

$$\mathbf{T} = -p\mathbf{1} + 2\mu \nabla \otimes \nabla \phi, \tag{4.4.2}$$

$$\Phi = 2\mu \int_{V'} \frac{\partial^2 \phi}{\partial x_i \partial x_j} \frac{\partial^2 \phi}{\partial x_i \partial x_j} \mathrm{d}V. \tag{4.4.3}$$

Equation (4.4.3) has been widely used to study viscous effects in irrotational flows since Stokes (1851).

4.5 Irrotational flow and boundary conditions

How is the irrotational flow determined? It frequently happens that the rotational motion cannot satisfy the boundary conditions; this well-known problem is associated with difficulties in forming boundary conditions for the vorticity. The potential ϕ is a harmonic function that can be selected so that the values of the sum of the rotational and irrotational fields can be chosen to balance prescribed conditions at the boundary. The allowed irrotational fields can be selected from harmonic functions that enter into the purely irrotational solution of the same problem on the same domain. A very important property of potential flow arises from the fact that irrotational viscous stresses do not give rise to irrotational viscous forces in equations of motion (4.2.2). The interior values of the rotational velocity are coupled to the irrotational motion through Bernoulli terms evaluated on the potential and inertial terms that couple the irrotational and rotational fields. The dependence of the irrotational field on viscosity can be generated by the boundary conditions.

4.6 Examples from hydrodynamics

4.6.1 Poiseuille flow

A simple example that serves as a paradigm for the relation of the irrotational and rotational components of velocity in all the solutions of Navier–Stokes equations is plane Poiseuille flow:

$$\begin{aligned}
\mathbf{u} &= \left[-\frac{P'}{2\mu}\left(b^2 - y^2\right),\ 0,\ 0\right], \\
\operatorname{curl}\mathbf{u} &= \left(0,\ 0,\ -\frac{P'y}{\mu}\right), \\
\mathbf{v} &= \left(\frac{P'}{2\mu}y^2,\ 0,\ 0\right), \\
\nabla\phi &= \left(-\frac{P'}{2\mu}b^2,\ 0,\ 0\right).
\end{aligned} \tag{4.6.1}$$

The rotational flow is a constrained field and cannot satisfy the no-slip boundary condition. To satisfy the no-slip condition we add the irrotational flow:

$$\frac{\partial\phi}{\partial x} = -\frac{P'}{2\mu}b^2.$$

The irrotational component is for uniform and unidirectional flow, chosen so that $\mathbf{u} = 0$ at the boundary.

4.6.2 Flow between rotating cylinders

$$\begin{aligned}
\mathbf{u} &= \mathbf{e}_\theta u\left(r\right), \\
\mathbf{u} &= \mathbf{e}_\theta \Omega_a a \text{ at } r = a, \\
\mathbf{u} &= \mathbf{e}_\theta \Omega_b b \text{ at } r = b, \\
u\left(r\right) &= Ar + \frac{B}{r}, \\
A &= -\frac{a^2\Omega_a - b^2\Omega_b}{b^2 - a^2}, \\
B &= \frac{\left(\Omega_a - \Omega_b\right)a^2b^2}{b^2 - a^2}, \\
\mathbf{v} &= \mathbf{e}_\theta Ar, \\
\frac{1}{r}\frac{\partial\phi}{\partial\theta} &= \frac{B}{r}.
\end{aligned} \tag{4.6.2}$$

The irrotational flow with $\mathbf{v} = 0$ is an exact solution of the Navier–Stokes equations with no-slip at boundaries.

4.6.3 Stokes flow around a sphere of radius a in a uniform stream U

This axisymmetric flow, described with spherical polar coordinates (r, θ, φ), is independent of φ. The velocity components can be obtained from a stream function $\psi(r, \theta)$ such that

$$\mathbf{u} = (u_r, u_\theta) = \frac{1}{r \sin\theta}\left(\frac{1}{r}\frac{\partial\psi}{\partial\theta}, -\frac{\partial\psi}{\partial r}\right). \tag{4.6.3}$$

The stream function can be divided into rotational and irrotational parts:

$$\psi = \psi_v + \psi_p, \quad E^2\psi_p = 0 \text{ with } E^2\psi = \frac{\partial^2\psi}{\partial r^2} + \frac{\sin\theta}{r^2}\frac{\partial}{\partial\theta}\left(\frac{1}{\sin\theta}\frac{\partial\psi}{\partial\theta}\right), \tag{4.6.4}$$

where ψ_p is the conjugate harmonic function. The potential ϕ corresponding to ψ_p is given by

$$\phi = \left(-\frac{A}{2r^2} + Cr\right) 2\cos\theta,$$

where the parameters A and C are selected to satisfy the boundary conditions on $\mathbf{u}$.

The Helmholtz decomposition of the solution of the problem of slow streaming motion over a stationary sphere is given by

$$\begin{aligned}
\mathbf{u} = (u_r, u_\theta) &= \left(v_r + \frac{\partial\phi}{\partial r}, v_\theta + \frac{1}{r}\frac{\partial\phi}{\partial\theta}\right),\\
u_r &= U\left[1 - \frac{3a}{2r} + \frac{1}{2}\left(\frac{a}{r}\right)^3\right]\cos\theta,\\
u_\theta &= -U\left[1 - \frac{3a}{4r} - \frac{1}{4}\left(\frac{a}{r}\right)^3\right]\sin\theta,\\
\operatorname{curl}\mathbf{u} &= U\left(0, 0, -\frac{3a}{2r^2}\sin\theta\right),\\
v_r &= -\frac{3}{2}\frac{a}{r}U\cos\theta,\\
v_\theta &= \frac{3}{4}\frac{a}{r}U\sin\theta,\\
\phi &= U\left(r - \frac{1}{4}\frac{a^3}{r^2}\right)\cos\theta,\\
p - p_\infty &= -\frac{3}{2}\frac{a}{r^2}\mu U\cos\theta.
\end{aligned} \tag{4.6.5}$$

The potential for flow over a sphere is

$$\phi = U\left(r - \frac{\alpha}{4}\frac{a^3}{r^2}\right)\cos\theta. \tag{4.6.6}$$

The normal component of velocity vanishes when $\alpha = -2$ and the tangential component vanishes when $\alpha = 4$. In the present case, to satisfy the no-slip condition, we take $\alpha = 1$. Both the rotational and irrotational components of velocity are required for satisfying the no-slip condition.

4.6.4 Streaming motion past an ellipsoid

The problem of the steady translation of an ellipsoid in a viscous liquid was solved by Lamb (1932, 604–5) in terms of the gravitational potential Ω of the solid and another harmonic function χ corresponding to the case in which $\nabla^2\chi = 0$, finite at infinity with $\chi = 1$ for the internal space, as stated by Lamb:

> If the fluid be streaming past the ellipsoid, regarded as fixed, with the general velocity U in the direction of x, we assume
>
> $$u = A\frac{\partial^2\Omega}{\partial x^2} + B\left(x\frac{\partial\chi}{\partial x} - \chi\right) + U, \quad v = A\frac{\partial^2\Omega}{\partial x\partial y} + Bx\frac{\partial\chi}{\partial y}, \quad w = A\frac{\partial^2\Omega}{\partial x\partial z} + Bx\frac{\partial\chi}{\partial z}.$$
>
> These satisfy the equation of continuity, in virtue of the relations
>
> $$\nabla^2\Omega = 0, \quad \nabla^2\chi = 0;$$
>
> and they evidently make $u = U$, $v = 0$, $w = 0$ at infinity. Again, they make
>
> $$\nabla^2 u = 2B\frac{\partial^2\chi}{\partial x^2}, \quad \nabla^2 v = 2B\frac{\partial^2\chi}{\partial x\partial y}, \quad \nabla^2 w = 2B\frac{\partial^2\chi}{\partial x\partial z}.$$

We note next that $\nabla^2\mathbf{v} = \nabla^2\mathbf{u}$; the irrotational part of $\mathbf{u}$ is on the null space of the Laplacian. It follows that the rotational velocity is associated with χ and is given by the terms proportional to B. The irrotational velocity is given by

$$\nabla\phi = \nabla\left(A\frac{\partial\Omega}{\partial x}\right).$$

The vorticity is given by

$$\boldsymbol{\omega} = -\mathbf{e}_y\left(2B\frac{\partial\chi}{\partial z}\right) + \mathbf{e}_z\left(2B\frac{\partial\chi}{\partial y}\right).$$

4.6.5 Hadamard–Rybyshinsky solution for streaming flow past a liquid sphere

As in the flow around a solid sphere, this problem is posed in spherical coordinates with a stream function and potential function related by (4.6.3), (4.6.4), and (4.6.5). The stream function is given by

$$\psi = f(r)\sin^2\theta, \tag{4.6.7}$$

$$f(r) = \frac{A}{r} + Br + Cr^2 + Dr^4. \tag{4.6.8}$$

The irrotational part of (4.6.7) is the part that satisfies $E^2\psi_p = 0$. From this it follows that

$$\psi_p = \left(\frac{A}{r} + Cr^2\right)\sin^2\theta, \tag{4.6.9}$$

$$\psi_v = \left(Br + Dr^4\right)\sin^2\theta. \tag{4.6.10}$$

The potential ϕ corresponding to ψ_p is given by (4.6.5).

The solution of this problem is determined by continuity conditions at $r = a$. The inner solution for $r < a$ is designated by $\bar{\mathbf{u}}, \bar{\mathbf{v}}, \bar{\phi}, \bar{\psi}, \bar{p}, \bar{\mu}, \bar{\rho}$.

The normal component of velocity,

$$v_r + \frac{\partial \phi}{\partial r} = \bar{v}_r + \frac{\partial \bar{\phi}}{\partial r}, \tag{4.6.11}$$

is continuous at $r = a$. The normal stress balance is approximated by a static balance in which the jump of pressure is balanced by surface tension so large that the drop is approximately spherical. The shear stress

$$\mu \mathbf{e}_\theta \cdot \left(\nabla \mathbf{v} + \nabla \mathbf{v}^T + 2\nabla \otimes \nabla \phi\right) \cdot \mathbf{e}_r = \bar{\mu} \mathbf{e}_\theta \cdot \left(\nabla \bar{\mathbf{v}} + \nabla \bar{\mathbf{v}}^T + 2\nabla \otimes \nabla \bar{\phi}\right) \cdot \mathbf{e}_r. \tag{4.6.12}$$

The coefficients A, B, C, and D are determined by the condition that $\mathbf{u} = U\mathbf{e}_x$ as $r \to \infty$ and continuity conditions (4.6.11) and (4.6.12). We find that, when $r \le a$,

$$\bar{f}(r) = \frac{\mu}{\mu + \bar{\mu}} \frac{1}{4} \frac{U}{a^2} \left(r^4 - a^2 r^2\right), \tag{4.6.13}$$

where the r^2 term is associated with irrotational flow and, when $r \ge a$,

$$f(r) = \frac{1}{2} U \left(r^2 - ar\right) + \frac{\bar{\mu}}{\bar{\mu} + \mu} \frac{1}{4} U \left(\frac{a^3}{r} - ar\right), \tag{4.6.14}$$

where the r^2 and a^2/r terms are irrotational.

4.6.6 Axisymmetric steady flow around a spherical gas bubble at finite Reynolds numbers

This problem is like the Hadamard–Rybyshinski problem, with the internal motion of the gas neglected, but inertia cannot be neglected. The coupling conditions reduce to

$$v_r + \frac{\partial \phi}{\partial r} = 0, \tag{4.6.15}$$

$$\mathbf{e}_\theta \cdot \left(\nabla \mathbf{v} + \nabla \mathbf{v}^T + 2\nabla \otimes \nabla \phi\right) \cdot \mathbf{e}_r = 0. \tag{4.6.16}$$

There is no flow across the interface at $r = a$, and the shear stress vanishes there.

The equations of motion are the r and θ components of (4.1.1) with time derivative zero. Since Levich (1949), it has been assumed that, at moderately large Reynolds numbers, the flow in the liquid is almost purely irrotational with a small vorticity layer where $\mathbf{v} \neq 0$ in the liquid near $r = a$. The details of the flow in the vorticity layer, the thickness of the layer, and the presence and variation of viscous pressure contribution all are unknown.

It may be assumed that the irrotational flow in the liquid outside the sphere can be expressed as a series of spherical harmonics. The problem then is to determine the participation coefficients of the different harmonics, the pressure distribution and the rotational velocity $\mathbf{v}$ satisfying continuity conditions (4.6.15) and (4.6.16). The determination of the participation coefficients may be less efficient than a purely numerical simulation of Laplace's equation outside a sphere subject to boundary conditions on the sphere that are coupled to the rotational flow. This important problem has not yet been solved.

4.6.7 Viscous decay of free-gravity waves

Flows that depend on only two space variables, such as plane flows or axisymmetric flows, admit a stream function. Such flows may be decomposed into a stream function and a potential function. Lamb (1932, §349) calculated an exact solution of the problem of the viscous decay of free-gravity waves as a free-surface problem of this type. He decomposes the solution into a stream function ψ and a potential function ϕ, and the solution is given by

$$u = \frac{\partial \phi}{\partial x} + \frac{\partial \psi}{\partial y}, \quad v = \frac{\partial \phi}{\partial y} - \frac{\partial \psi}{\partial x}, \quad \frac{p}{\rho} = -\frac{\partial \phi}{\partial t} - gy, \tag{4.6.17}$$

where

$$\nabla^2 \phi = 0, \quad \frac{\partial \psi}{\partial t} = \nu \nabla^2 \psi. \tag{4.6.18}$$

This decomposition is a Helmholtz decomposition; it can be said that Lamb solved this problem in the Helmholtz formulation.

Wang and Joseph (2006a) constructed a purely irrotational solution of this problem that is in very good agreement with the exact solution. The potential in the Lamb solution is not the same as the potential in the purely irrotational solution because they satisfy different boundary conditions. It is worth noting that viscous potential flow rather than inviscid potential flow is required for satisfying boundary conditions. The common idea that the viscosity should be put to zero to satisfy boundary conditions is deeply flawed. It is also worth noting that viscous component of the pressure does not arise in the boundary layer for vorticity in the exact solution; the pressure is given by (4.6.17)

4.6.8 Oseen flow

Steady streaming flow of velocity U (Milne-Thomson, 1968, 696–8, Lamb, 1932, 608–16) of an incompressible fluid over a solid sphere of radius a that is symmetric about the x axis satisfies

$$U \frac{\partial \mathbf{u}}{\partial x} = -\frac{1}{\rho} \nabla p + \nu \nabla^2 \mathbf{u}, \tag{4.6.19}$$

where $\operatorname{div} \mathbf{u} = 0$ and $U = 2k\nu$ (for convenience). The inertial terms in this approximation are linearized but not zero. The equations of motion in decomposed form are

$$\nabla \left(U \frac{\partial \phi}{\partial x} + \frac{p}{\rho} \right) + U \frac{\partial \mathbf{v}}{\partial x} = \nu \nabla^2 \mathbf{v}. \tag{4.6.20}$$

Because $\operatorname{div} \mathbf{v} = 0$ and ϕ is harmonic, $\nabla^2 p = 0$ and $U[(\partial \operatorname{curl} \mathbf{v})/\partial x] = \nu \nabla^2 \operatorname{curl} \mathbf{v}$. The rotational velocity is determined by a function χ:

$$\begin{aligned} \mathbf{v} &= \frac{1}{2k} \nabla \chi - \mathbf{e}_x \chi, \\ \chi &= \frac{a_0}{r} \exp\left[-kr\left(1 - \cos\theta\right)\right]. \end{aligned} \tag{4.6.21}$$

The potential ϕ is governed by the Laplace equation $\nabla^2 \phi = 0$, for which the solution is given by

$$\phi = Ux + \frac{b_0}{r} + b_1 \frac{\partial}{\partial x}\left(\frac{1}{r}\right) + \cdots + = Ur\cos\theta + \frac{b_0}{r} + b_1 \frac{-1}{r^2}\cos\theta + \cdots +,$$

where Ux is the uniform velocity term. With these, the velocity $u = (u_r,\ u_\theta)$ is expressed as

$$u_r = \frac{\partial \phi}{\partial r} + \frac{1}{2k}\frac{\partial \chi}{\partial r} - \cos\theta\chi, \quad u_\theta = \frac{1}{r}\frac{\partial \phi}{\partial \theta} + \frac{1}{2k}\frac{1}{r}\frac{\partial \chi}{\partial \theta} + \sin\theta\chi.$$

These should be zero at the sphere surface ($r = a$) to give

$$b_0 = -\frac{a_0}{2k} = -\frac{3Ua}{4k} = -\frac{3a\nu}{2}, \quad a_0 = \frac{3Ua}{2}, \quad b_1 = \frac{Ua^3}{4}.$$

The higher-order terms in the potential vanish.

To summarize, the decomposition of the velocity into rotational and irrotational components is

$$\begin{aligned}
\mathbf{u} = (u_r,\ u_\theta) &= \left(v_r + \frac{\partial \phi}{\partial r},\ v_\theta + \frac{1}{r}\frac{\partial \phi}{\partial \theta}\right), \\
\operatorname{curl}\mathbf{u} &= \operatorname{curl}\mathbf{v}, \\
v_r &= \frac{1}{2k}\frac{\partial \chi}{\partial r} - \chi\cos\theta, \\
v_\theta &= \frac{1}{2kr}\frac{\partial \chi}{\partial \theta} + \chi\sin\theta, \\
\phi &= Ux - \frac{3Ua}{4kr} - \frac{Ua^3}{4r^2}\cos\theta.
\end{aligned} \tag{4.6.22}$$

4.6.9 Flows near internal stagnation points in viscous incompressible fluids

The fluid velocity relative to uniform motion or rest vanishes at points of stagnation. These points occur frequently even in turbulent flow. When the flow is purely irrotational, the velocity potential ϕ can be expanded near the origin as a Taylor series:

$$\phi = \phi_0 + a_i x_i + \frac{1}{2}a_{ij}x_i x_j + O\left(x_i x_i^3\right), \tag{4.6.23}$$

where the tensor a_{ij} is symmetric. At the stagnation point $\nabla\phi = 0$, hence $a_i = 0$, and because $\nabla^2\phi = 0$ everywhere, we have $a_{ii} = 0$. The velocity is

$$\frac{\partial \phi}{\partial x_i} = a_{ij}x_j \tag{4.6.24}$$

at leading order. The velocity components along the principal axes (x, y, z) of the tensor a_{ij} are

$$u_1 = ax, \quad u_2 = by, \quad u_3 = -(a+b)z, \tag{4.6.25}$$

where a and b are unknown constants relating to the flow field. Irrotational stagnation points in the plane are saddle points, centers are stagnation points around which the fluid rotates. Saddle points and centers are embedded in vortex arrays.

Taylor vortex flow (TVF) is a 2D (x, y) array of counterrotating vortices (figure 4.1) whose vorticity decays in time are due to viscous diffusion ($\partial_t\omega = \nu\nabla^2\omega$). It is useful to introduce the stream function in the analysis of this type of fluid motion. It is well known

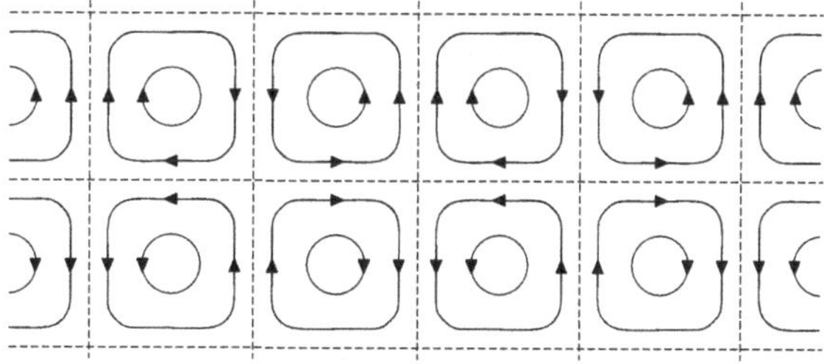

Figure 4.1. (After Taylor, 1923). Streamlines for a system of eddies dying down under the action of viscosity. All cell boundaries on which $\psi = 0$ are irrotational. Near each and every saddle point, say at $(x, y) = 0$, the velocity in and out of the saddle point is linear, say $(u, v) = \lambda(x, y)$. The vorticity in every quadrant of this generic saddle point is constant on hyperbolae (xy = constant).

that a 2D incompressible flow can be written in terms of a stream function $\psi(x, y, t)$ as

$$u_1 = \partial_y \psi, \quad u_2 = -\partial_x \psi. \tag{4.6.26}$$

When this representation is used, the vorticity equation becomes

$$(\partial/\partial t - \nu\nabla^2)(\nabla^2\psi) + \partial_y\psi\,\partial_x(\nabla^2\psi) - \partial_x\psi\,\partial_y(\nabla^2\psi) = 0. \tag{4.6.27}$$

Taylor (1923) showed that if $\psi(x, y, t)$ is in the form

$$\psi(x, y, t) = \exp(-\nu\alpha^2 t)\,\Psi(x, y) \tag{4.6.28}$$

and Ψ is a solution of the 2D equation for eigenmodes of the vibrating membrane

$$\nabla^2\Psi + \alpha^2\Psi = 0, \tag{4.6.29}$$

then,

$$u_1 = \exp(-\nu\alpha^2 t)\Psi_y, \quad u_2 = -\exp(-\nu\alpha^2 t)\Psi_x \tag{4.6.30}$$

is a solution of the time-dependent, nonlinear, incompressible Navier–Stokes equations. The corresponding pressure is

$$p = -\frac{\rho}{2}\exp(-2\nu\alpha^2 t)(|\nabla\Psi|^2 + \alpha^2\Psi^2), \tag{4.6.31}$$

and the vorticity is given by

$$\omega = \partial_y u_1 - \partial_x u_2 = \nabla^2\psi = -k^2\psi. \tag{4.6.32}$$

Consider the solution

$$\Psi = \frac{\omega_0}{k^2}\cos(k_x x)\cos(k_y y), \tag{4.6.33}$$

which satisfies (4.6.29) with $\alpha^2 = k^2 = k_x^2 + k_y^2$; k_x and k_y are the wavenumbers in the x and y directions and ω_0 is the initial maximum vorticity. This function has a maximum at the origin. From (4.6.30), by use of (4.6.33), the instantaneous velocity components are

$$\begin{aligned} u_1 &= -\omega_0\frac{k_y}{k^2}\exp(-\nu k^2 t)\cos(k_x x)\sin(k_y y), \\ u_2 &= -\omega_0\frac{k_x}{k^2}\exp(-\nu k^2 t)\sin(k_x x)\cos(k_y y), \end{aligned} \tag{4.6.34}$$

which gives the damped motion of a square array of counterrotating vortices (figure 4.1).

The vorticity vanishes at saddle points (X, Y). After expanding the solution in power series centered on a generic stagnation point we find that

$$u_1(X, Y, t) = \frac{\omega_0 k_y k_x}{k^2} X + \alpha_1 X^3 + \beta_1 X Y^2 + \cdots +,$$
$$u_2(X, Y, t) = -\frac{\omega_0 k_y k_x}{k^2} Y + \alpha_2 Y^3 + \beta_2 Y X^2 + \cdots +. \tag{4.6.35}$$

The vorticity vanishes at $(X, Y) = (0, 0)$ and appears first at second order:

$$\omega(X, Y, t) = -\omega_0 k_x k_y XY \exp\left(-\nu k^2 t\right) + \cdots +. \tag{4.6.36}$$

The Helmholtz decomposition of the local solution near $(X, Y) = (0, 0)$ is

$$u_1 = \frac{\partial \phi}{\partial X} + v_1, \quad u_2 = \frac{\partial \phi}{\partial Y} + v_2, \tag{4.6.37}$$

where

$$\left(\frac{\partial \phi}{\partial X}, \frac{\partial \phi}{\partial Y}\right) = \frac{\omega_0 k_y k_x}{k^2}(X, -Y), \tag{4.6.38}$$

and

$$(v_1, v_2) = \left(\alpha_1 X^3 + \beta_1 X Y^2 + \cdots +, \ \alpha_2 Y^3 + \beta_2 Y X^2 + \cdots +\right). \tag{4.6.39}$$

This solution is generally valid in eddy systems that are segregated into quadrants near the stagnation point as in figure 4.1.

The dissipation of vorticity at high Reynolds numbers is an important topic in the theory of turbulence. The dissipation integral is proportional to the viscosity but paradoxically this integral does not vanish as the Reynolds number tends to infinity. The role of potential flows in an eddy structure is also of interest. The TVF is not a turbulent flow but it does give rise to an interesting result on the role of stagnation points in the decay of the dissipation in the potential flow limit.

Consider an array of square cells of side l in a fixed square array of side $L = nl$. In the square array $k_x = k_y = \hat{k} = 2\pi/\lambda = \pi/l$. The dissipation of this array is computed as follows.

With the velocity field (4.6.34), the components of the strain-rate tensor are readily computed:

$$D_{11} = \frac{\partial u_1}{\partial x} = \frac{\omega_0 k_x k_y}{k^2} \exp(-\nu k^2 t) \sin(k_x x) \sin(k_y y),$$

$$D_{22} = \frac{\partial u_2}{\partial y} = -D_{11},$$

$$D_{12} = D_{21} = \frac{1}{2}\left(\frac{\partial u_1}{\partial_y} + \frac{\partial u_2}{\partial_x}\right) = \omega_0 \left(\frac{k_x^2 - k_y^2}{2k^2}\right) \exp(-\nu k^2 t) \cos(k_x x) \cos(k_y y).$$

The dissipation Φ is thus computed as

$$\begin{aligned}\Phi &= 2\mu \int_{-L}^{L}\int_{-L}^{L}\left[D_{11}^2 + D_{22}^2 + 2D_{12}^2\right]\mathrm{d}x\mathrm{d}y \\ &= 4\mu\frac{\omega_0^2}{k^4}\exp(-2\nu k^2 t)\int_{-L}^{L}\int_{-L}^{L}\left[k_x^2k_y^2\sin^2(k_y y)\sin^2(k_x x)\right. \\ &\quad \left.+\frac{(k_x^2-k_y^2)^2}{4}\cos^2(k_x x)\cos^2(k_y y)\right]\mathrm{d}x\mathrm{d}y \\ &= 4\mu\frac{\omega_0^2}{k^4}\exp(-2\nu k^2 t)L^2\left[k_x^2k_y^2 + \frac{(k_x^2-k_y^2)^2}{4}\right] = \omega_0^2\mu L^2\exp\left(-2\frac{\mu}{\rho}k^2 t\right),\end{aligned} \tag{4.6.40}$$

with $k^2 = k_x^2 + k_y^2 = 2\hat{k}^2$. Φ is maximum for a fixed t for a value of μ such that

$$\frac{\mathrm{d}}{\mathrm{d}\mu}\left[\mu\exp\left(-2\frac{\mu}{\rho}k^2 t\right)\right] = 0.$$

Hence

$$\mu_{\max} = \frac{\rho}{2k^2 t} = \frac{\rho\lambda^2}{16\pi^2 t}. \tag{4.6.41}$$

For this maximizing $\mu_{\max}$ we have

$$\Phi = \omega_0^2 L^2\frac{\rho}{2k^2 t}\exp(-1) = \omega_0^2 L^4\frac{\rho}{4n^2\pi^2 t}\exp(-1) = \frac{\omega_0^2 L^4}{16}\frac{\rho}{(N+1)^2\pi^2 t}\exp(-1), \tag{4.6.42}$$

where $n = 2(N+1)$ is the number of cells and N is the number of stagnation points in the array of side $L = nl$. The maximum dissipation of a Taylor vortex array of side L at a fixed time tends to zero as the number of stagnation points tends to infinity. The potential flows at the stagnation points dominate the dissipation as the number of cells increases and the size of the cells decreases in the limit (inviscid) in which $\mu_{\max}$ goes to zero as N goes to infinity.

4.6.10 Hiemenz boundary-layer solution for two-dimensional flow toward a "stagnation point" at a rigid boundary

Stagnation points on solid bodies are very important because the pressures at such points can be very high. However, stagnation-point flow cannot persist all the way to the boundary because of the no-slip condition. Hiemenz (1911) looked for a boundary-layer solution of the Navier–Stokes equations vanishing at $y = 0$ that tends to stagnation-point flow for large y, expressed as

$$\begin{aligned}\mathbf{u} &= (u, v), \\ \omega(\mathbf{u})\,\mathbf{e}_z &= \operatorname{curl}\mathbf{u}.\end{aligned} \tag{4.6.43}$$

The motion in the outer region is irrotational flow near a stagnation point at a plane boundary. The flow in the irrotational region is described by the stream function,

$$\psi = kxy,$$

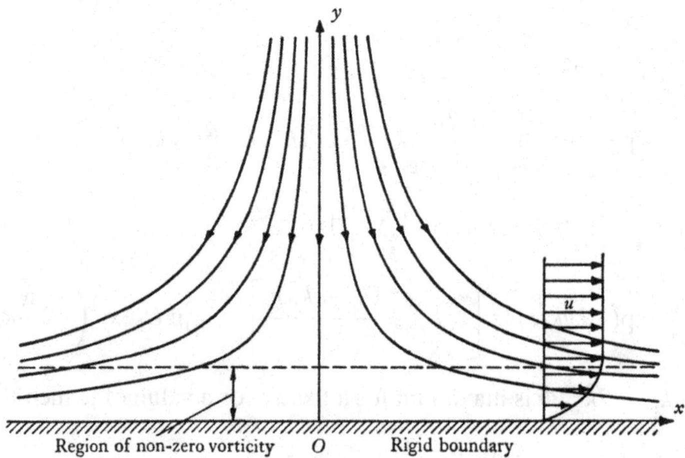

Figure 4.2. Steady 2D flow toward a stagnation point at a rigid boundary.

where x and y are rectilinear coordinates parallel and normal to the boundary (see figure 4.2), with the corresponding velocity distribution

$$u = kx, \quad v = -ky, \tag{4.6.44}$$

where k is a positive constant that, in the case of a stagnation point on a body fixed in a stream, must be proportional to the speed of the stream.

The next step is to determine the distribution of vorticity in the thin layer near the boundary from

$$u\frac{\partial \omega}{\partial x} + v\frac{\partial \omega}{\partial y} = \nu\left(\frac{\partial^2 \omega}{\partial x^2} + \frac{\partial^2 \omega}{\partial y^2}\right), \tag{4.6.45}$$

together with boundary conditions that $u = 0$ and $v = 0$ at $y = 0$ and that the flow tend to form (4.6.44) at the outer edge of the layer. Hiemenz found such a solution in the form

$$\psi = xf(y), \tag{4.6.46}$$

corresponding to

$$u = xf'(y), \quad v = -f(y),$$
$$\omega = \frac{\partial v}{\partial x} - \frac{\partial u}{\partial y} = -xf''(y),$$

where $f(y)$ is an unknown function and primes denote differentiation with respect to y, satisfying

$$-f'f'' + ff''' + \nu f^{iv} = 0,$$

and the boundary conditions

$$f = f' = 0 \text{ at } y = 0,$$
$$f \to ky \text{ as } y \to \infty.$$

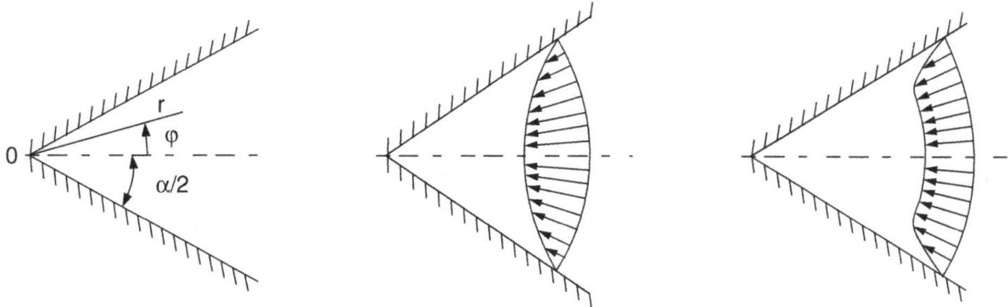

Figure 4.3. Hamel sink flow: (a) flow channel, (b) irrotational sink flow, (c) sink flow with rotational boundary layer.

Hiemenz (1911) showed that this system could be computed numerically and that it had a boundary-layer structure in the limit of small ν (figure 4.2).

Alternatively, we may decompose the solution relative to a stagnation point in the whole space. To satisfy the no-slip condition, the x component of stagnation-point flow on $y = 0$ must be put to zero by an equal and opposite rotational velocity. For the decomposed motion, we have

$$\omega(\mathbf{u}) = \omega(\mathbf{v}),$$

$$\mathbf{u} = (u,\ v) = \left(v_x + \frac{\partial\phi}{\partial x},\ v_y + \frac{\partial\phi}{\partial y}\right) = (v_x + kx,\ v_y - ky).$$

Instead of (4.6.45) we have the vorticity equation in the Helmholtz decomposed form:

$$(v_x + kx)\frac{\partial\omega(\mathbf{v})}{\partial x} + (v_y - ky)\frac{\partial\omega(\mathbf{v})}{\partial y} = \nu\nabla^2\omega(\mathbf{v}).$$

The solution of this problem is given by $\psi = xF(y)$, where $F(y) = f - ky$, $F(0) = 0$, and $F'(0) = -k$.

Analysis of a three-dimensional (3D) stagnation point in the axisymmetric case was given by Homann (1936).

4.6.11 Jeffrey–Hamel flow in diverging and converging channels

The problem is to determine the steady flow between two plane walls meeting at angle α shown in figure 4.3(a); see Batchelor (1967, 294–302) or Landau and Lifshitz (1987, 76–81). Batchelor's discussion of this problem is framed in terms of the solutions $\mathbf{u} = (v_r, v_z, v_\varphi) = (v[r, \varphi], 0, 0)$ of Navier–Stokes equations (4.2.1). The continuity equation, $(\partial rv/\partial r) = 0$, shows that

$$v = \tilde{v}(\varphi)/r. \tag{4.6.47}$$

The function $\tilde{v}(\varphi)$ is determined by an involved but straightforward nonlinear analysis leading to the line drawings shown in figures 4.3 and 4.4.

We next consider Helmholtz decomposition (4.1.1) of Hamel flow. Equation (4.6.47) shows that the only irrotational flow allowed in this decomposition is source or sink flow $\phi = C\log r$, where C is to be determined from the condition that

$$\tilde{v}(\varphi) + C = 0 \text{ at } \varphi = \pm\alpha.$$

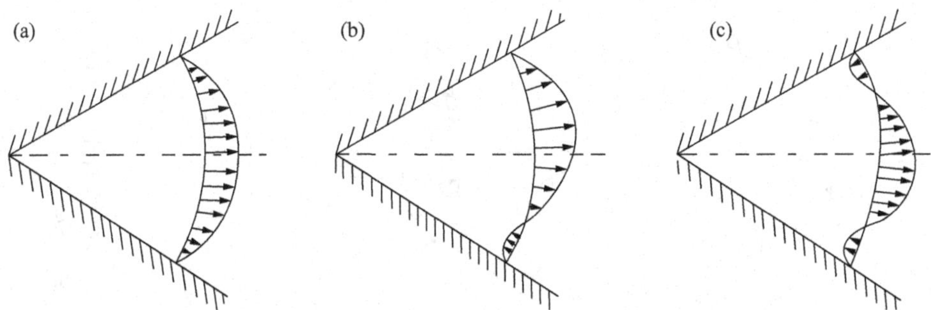

Figure 4.4. Hamel source flow: (a) irrotational source flow, (b) asymmetric rotational flow, (c) symmetric rotational flow.

The rotational field $\tilde{v}(\varphi)$ and constant C can be uniquely determined by an analysis like that given by Batchelor (1967).

4.6.12 An irrotational Stokes flow

An extreme example of an irrotational flow at finite Reynolds numbers is an irrotational flow at zero Reynolds number. Such an example has been given by Morrison (1970), who proves that, under very general conditions such as a uniform zeta potential, the slip-driven electro-osmotic flow about a charged particle is irrotational and the velocity field is proportional to the electric-field disturbance relative to the imposed field. The interaction between the charged particle and the irrotational flow was considered. Morrison showed that the viscous stresses that are due to the irrotational flow do not give rise to a drag or torque on the particle. This result applies to particles of any shape provided that the flow is irrotational and the forces and torques are computed from viscous stresses computed on the irrotational flow. Morrison's result can be said to be more paradoxical than D'Alembert's because the viscosity and viscous dissipation are not zero. The calculation of forces by use of the irrotational shear stress usually gives rise to erroneous results. A consequence of Morrison's uncorrected calculation of forces and torques is that the particle translates with the Smoluchowski velocity independently of its size, orientation, or presence of other particles and does not rotate. In this book we show that Morrison's result is valid for only Stokes flows, because in the presence of inertia the torque-free condition is ruined. So inertia here actually interrupts the irrotationality.

Yariv and Brenner (2003, 2004) showed that, if the zeta potential is nonuniform, the flow can be decomposed into a potential part and a rotational part, as in the Helmholtz decomposition discussed previously.

4.6.13 Lighthill's approach

A form of the Helmholtz decomposition that depends on the law of Biot–Savart was considered by Lighthill. He notes that Lamb (1932, chapter vii) has shown that, for *any* given solenoidal distribution of vorticity ω outside the body surface whose motion is prescribed, one and only one solenoidal velocity exists, tending to zero at infinity and with zero normal relative velocity at the solid surface. We have

$$\mathbf{v}_0 = \int \frac{\boldsymbol{\omega} \times \mathbf{r}}{4\pi r^3} \mathrm{d}\Omega, \tag{4.6.48}$$

where the integration is over the whole vorticity field and $\mathbf{r}$ is the position vector relative to the volume element $\mathrm{d}\Omega$. In general, $\mathbf{v}_0$ does not satisfy the boundary condition for $\mathbf{v}$. However, the difference $\mathbf{v} - \mathbf{v}_0$ must be irrotational (because $\operatorname{curl} \mathbf{v}_0 = \boldsymbol{\omega}$), and hence equal to grad ϕ for some potential ϕ. On the body surface, $\mathbf{u} = \mathbf{V}$

$$\frac{\partial \phi}{\partial n} = v_n - v_{0n} = \mathbf{V} \cdot \mathbf{n} - v_{0n}, \tag{4.6.49}$$

is prescribed, and also grad $\phi \to 0$ at infinity. Modulo an additive constant, just one ϕ satisfies these conditions. Lighthill notes that formulations (4.6.48) and (4.6.49) give rise to a tangential velocity at the surface that may not satisfy the no-slip condition. In fact, the Biot–Savart velocity will not in general satisfy the no-slip condition. The spurious slip velocity may be viewed as a vortex sheet on the surface of the body. To enforce the no-slip boundary condition, the vortex sheet is distributed diffusively into the flow, transferring the vortex sheet into an equivalent vortex layer by means of a vorticity flux. The no-slip condition therefore determines the vorticity flux, which is the strength of the spurious vortex sheet divided by the time increment. This construction of Lighthill is the basis for numerical constructions of Chorin (1973) to slightly viscous flow and Chorin (1978) to the boundary-layer equations.

Lighthill's Biot–Savart decomposition is the same as our PDE formulation (4.2.2)–(4.2.4) with ϕ determined by the normal component of the prescribed data and the rotational velocity $\mathbf{v}$ generated by solutions of (4.2.2) with the caveat that ϕ is determined by boundary conditions (4.2.5) that guarantee no-slip. We feel that our formulation has left open the question of "purely rotational" flows that possibly may be addressed by a better choice for generating the irrotational flow in the decomposition.

We have emphasized that irrotational functions are needed to satisfy the no-slip condition and that the boundary conditions for the rotational component of velocity need not be restricted to solenoidal velocities with zero relative normal component as is required by Lamb's theorem. The example of Stokes flow over a sphere is instructive because, for this example, everything is definite and cannot be manipulated. The rotational component is purely rotational; it has no irrotational part and it gives rise to the vorticity. It does not have a zero normal component as is required for the Biot–Savart approach. The irrotational components are required for satisfying the boundary condition. The irrotational flow does not have a zero normal component; rather, the irrotational flow satisfies a Robin condition in which the tangential and normal components that make up a linear combination have equal weight.

4.7 Conclusion

The Helmholtz decomposition gives rise to an exact theory of potential flow in the frame of the Navier–Stokes equations in which rotational and irrotational fields are tightly coupled and both fields depend on viscosity. This kind of theory leads to boundary layers of vorticity in asymptotic limits but the fields are always coupled. The exact theory is different from purely irrotational theories of the effect of viscosity that can lead to excellent but always approximate results.

Finally, we note that decomposition (4.1.1)–(4.1.5) does not depend on dynamics and may be applied to non-Newtonian fluids. Irrotational motions of non-Newtonian fluids are surely important, but the coupling of rotational and irrotational fields has not been studied or even identified as a subject worthy of study.

5

Harmonic functions that give rise to vorticity

The rotational component of the velocity in Stokes flow is generated by products of powers and harmonic functions. This is yet another way in which potential flow enters into the fluid mechanics of viscous fluids. The simplest case is for plane flow in two dimensions. Here we may satisfy the continuity equation,

$$\frac{\partial u}{\partial x} + \frac{\partial v}{\partial y} = 0, \tag{5.0.1}$$

with a stream function

$$(u,\ v) = \left(\frac{\partial \psi}{\partial y},\ -\frac{\partial \psi}{\partial x}\right). \tag{5.0.2}$$

The vorticity

$$\omega = \mathbf{e}_z \cdot \operatorname{curl} \mathbf{u} = \frac{\partial v}{\partial x} - \frac{\partial u}{\partial y} = -\nabla^2 \psi. \tag{5.0.3}$$

For irrotational flow, $\omega = 0$ and $\psi = \psi_I$ is harmonic:

$$\nabla^2 \psi_I = 0. \tag{5.0.4}$$

The Stokes equations for steady plane flow are

$$\begin{aligned} \mu \nabla^2 u &= \frac{\partial p}{\partial x}, \\ \mu \nabla^2 v &= \frac{\partial p}{\partial y}. \end{aligned} \tag{5.0.5}$$

Hence

$$\mu \nabla^2 \left(\frac{\partial v}{\partial x} - \frac{\partial u}{\partial y}\right) = \mu \nabla^2 \omega = 0. \tag{5.0.6}$$

The vorticity is a harmonic function.

Suppose that H is an arbitrary harmonic function and

$$\mathbf{r} = \mathbf{e}_x x + \mathbf{e}_y y. \tag{5.0.7}$$

Then

$$\nabla^2 (\mathbf{r} H) = 2 \nabla H. \tag{5.0.8}$$

Hence the function

$$\psi = \psi_I + xH_1 + yH_2, \tag{5.0.9}$$

where H_1 and H_2 are harmonic, satisfies

$$\omega = 2\frac{\partial H_1}{\partial x} + 2\frac{\partial H_2}{\partial y}, \tag{5.0.10}$$

where $\nabla^2\omega = 0$ and

$$\psi_I,\ H_1,\ H_2 \tag{5.0.11}$$

are selected to satisfy boundary conditions.

Lamb (1932) developed a general solution (§§335 and 336) of Stokes equation in terms of spherical harmonics. Lamb's arguments are very cumbersome. Weinberger developed a much simpler approach to this problem, which leads to Lamb's result (the text that follows is taken from a personal communication from Prof. Weinberger, 2006):

> The main tool is the identity
>
> $$(1 + r\partial/\partial r)\,\mathbf{v} = \text{grad}\,(\mathbf{r}\cdot\mathbf{v}) - \mathbf{r}\times\text{curl}\,\mathbf{v}, \tag{5.0.12}$$
>
> which holds for all vector fields, and is essentially Equation (5) of §335 [of Lamb (1932)].
>
> As in §335, we first look at the case of constant pressure. In this case
>
> $$\text{curl}\,(\text{curl}\,\mathbf{u}) = 0,$$
>
> and of course div (curl $\mathbf{u}$) $= 0$, so that curl $\mathbf{u}$ is the gradient of some harmonic function m. As Lamb points out, the function $l := \mathbf{r}\cdot\mathbf{u}$ is also harmonic. Thus, (5.0.12) with $\mathbf{v} = \mathbf{u}$ shows that
>
> $$(1 + r\partial/\partial r)\,\mathbf{u} = \text{grad}\,l - \mathbf{r}\times\text{grad}\,m. \tag{5.0.13}$$
>
> We note that if ϕ is a solution of the equation
>
> $$(r\partial/\partial r)\,\phi = l, \tag{5.0.14}$$
>
> then
>
> $$\text{grad}\,l = (1 + r\partial/\partial r)\,\text{grad}\,\phi. \tag{5.0.15}$$
>
> Similarly, we see that if
>
> $$(1 + r\partial/\partial r)\,\psi = m, \tag{5.0.16}$$
>
> then
>
> $$\mathbf{r}\times\text{grad}\,m = (1 + r\partial/\partial r)\,\mathbf{r}\times\text{grad}\,\psi. \tag{5.0.17}$$
>
> By substituting (5.0.15) and (5.0.17) into (5.0.13), we see that
>
> $$\mathbf{u} = \text{grad}\,\phi - \mathbf{r}\times\text{grad}\,\psi + \mathbf{w}\,(\Omega)\,/r, \tag{5.0.18}$$
>
> where $\mathbf{w}$ is an arbitrary vector-valued function of the angle variables in spherical coordinates. We now show that we can choose the solution ϕ (5.0.14) to be harmonic. By taking the Laplacian of both sides of (5.0.14), we see that
>
> $$(2 + r\partial/\partial r)\,(\nabla^2\phi) = 0.$$
>
> This shows that, if $\nabla^2\phi$ is zero at one point, it is zero on the radial line from the origin through this point. On the other hand, if we apply the operator $(1 + r\partial/\partial r)\,\psi$ to both sides of (5.0.14), we see that
>
> $$r^2\nabla^2\phi = (1 + r\partial/\partial r)\,l + \Delta_\Omega\psi,$$

where Δ_Ω is the surface Laplacian on the unit sphere. We can make $\nabla^2\phi = 0$ on a spherical surface $r = r_0$ by setting the right-hand side equal to zero and solving the resulting equation for $\phi(r_0, \Omega)$. This can easily be done if the intersection of the flow domain with the sphere $r = r_0$ has a nontrivial boundary. If it is the whole sphere, this problem can be solved if and only if the integral of $(1 + r\partial/\partial r)\, l$ over the whole sphere is zero. Because only the gradient of l appears in (5.0.14), we can add a constant to l to make this true. Thus we can assume without loss of generality that the solution ϕ of (5.0.14) is harmonic. In the same way, we arrive at the conclusion that ψ can be taken to be harmonic.

By taking the Laplacian and divergence of both sides of (5.0.18), we see that $\mathbf{w} = 0$. Thus we have shown that the velocity of any Stokes flow with constant pressure can be written in the form

$$\mathbf{u} = \operatorname{grad}\phi - \mathbf{r} \times \operatorname{grad}\psi, \tag{5.0.19}$$

where ϕ and ψ are harmonic. It is easily verified that for any harmonic ϕ and ψ this formula gives the velocity of a steady Stokes flow. This is the most general class of velocity fields that are self-equilibrating in the sense that they produce no body force.

Decomposition (5.0.19) of the velocity of a steady Stokes flow with constant pressure is unique. However, although this decomposition may be useful for domains that are nice in spherical coordinates, it is probably not very useful for dealing with domains with other symmetries or no symmetries.

To obtain a steady Stokes flow with variable pressure, it is necessary to obtain only one such solution for a prescribed pressure. Once this is done, we can obtain the general solution by adding solution (5.0.19) that corresponds to zero pressure. Following Lamb, we seek a solution in the form

$$\mathbf{u} = \operatorname{grad}\Lambda - H\mathbf{r},$$

where H is harmonic, but Λ is only biharmonic. The condition $\operatorname{div}\mathbf{u} = 0$ leads to the equation

$$\nabla^2\Lambda = (3 + r\partial/\partial r)\, H. \tag{5.0.20}$$

Then the condition $\nabla^2\mathbf{u} = \operatorname{grad} p$ leads to

$$\operatorname{grad}(3 + r\partial/\partial r)\, H - 2\operatorname{grad} H = \operatorname{grad} p,$$

or

$$(1 + r\partial/\partial r)\, H = p. \tag{5.0.21}$$

We again choose the harmonic solution of this equation. We still need to find a solution Λ of (5.0.20). We seek this solution in the form

$$\Lambda = r^2 Q,$$

where Q is harmonic. Then equation (5.0.20) becomes

$$(6 + 4r\partial/\partial r)\, Q = (3 + r\partial/\partial r)\, H. \tag{5.0.22}$$

As with the other equations, we can show that this one has a harmonic solution Q. We thus have the solution

$$\mathbf{u} = \operatorname{grad}\left(r^2 Q\right) - H\mathbf{r}, \tag{5.0.23}$$

with the harmonic functions Q and H previously defined in terms of the pressure p. It is easily verified that for any harmonic function H we can construct a harmonic function Q by solving (5.0.22), and that then (5.0.23) gives the velocity of a steady Stokes flow. By adding the right-hand side of (5.0.19), we obtain a representation

$$\mathbf{u} = \operatorname{grad}\phi - \mathbf{r} \times \operatorname{grad}\psi + \operatorname{grad}\left(r^2 Q\right) - H\mathbf{r} \tag{5.0.24}$$

of all steady Stokes flow in terms of three arbitrary harmonic functions.

The representations just derived show that velocity of Stokes flow may be decomposed into an irrotational part and a rotational part that are given by a composition of powers and harmonic functions. This is a Helmholtz decomposition with an additional feature that promotes the importance of harmonic functions in representing the rotational velocity.

The examples from hydrodynamics discussed in chapter 4 exhibit all the features specified in the representations in this chapter. The examples show that the irrotational and rotational components of velocity are required for satisfying the boundary conditions.

The rotational velocity of Poiseuille flow given by (4.6.1) is yH, where $H = -p'y/\mu$ is harmonic. The rotational velocity of (4.6.2) between rotating cylinders is proportional to r, which is r^2 times the irrotational velocity proportional to $1/r$. The rotational velocity for Stokes flow around a sphere given in (4.6.6) is proportional to $1/r$, which is r^2 times the irrotational velocity proportional to $1/r^3$. The rotational velocity for streaming flow past an ellipsoid is given by $x\nabla\chi$ when χ is harmonic. The flow over a liquid sphere associated with (4.6.13) and (4.6.14) is like the flow over a solid sphere; the rotational velocity is proportional to r^2 times the irrotational velocity.

The idea that the rotational velocity can be replaced with products of powers and harmonic functions applies to many flows that are not Stokes flow.

6

Radial motions of a spherical gas bubble in a viscous liquid

The problem of radial motion of a gas bubble in a liquid was first studied by Rayleigh (1917). The expanding bubble can be framed as a cavitation result induced by an underwater explosion. The contracting bubble can be viewed as the collapse of a cavitation bubble. Because the motion is purely radial, vorticity is not guaranteed and the effects of the viscosity of the liquid can be determined by potential flow. This problem was also considered by Batchelor (1967, 479) but, as in Rayleigh's work, with viscosity and surface tension neglected. Viscosity μ and surface-tension γ effects can be introduced into this problem without approximation because the motion is purely radial and irrotational. Although Plesset (1949) introduced a variable driving pressure and surface tension, the effects of surface tension were also introduced, and the effects of viscosity were first introduced by Poritsky (1951). The viscous terms were introduced by Poritsky (1951) and not by Plesset (1949). For this reason, we call equation (6.0.7) the Rayleigh–Poritsky equation rather than the Rayleigh–Plesset equation.

The Rayleigh–Poritsky equation arises from the normal stress balance. The velocity

$$u = R^2 \dot{R}/r^2 \tag{6.0.1}$$

is purely radial, R is the radius of the bubble, $\dot{R} = \mathrm{d}R/\mathrm{d}t$, and $u = \dot{R}$ when $r = R$. This is a potential flow,

$$\phi = -R^2 \dot{R}\frac{1}{r}. \tag{6.0.2}$$

The pressure in the liquid is given by Bernoulli's equation:

$$\frac{p - p_\infty}{\rho} = -\frac{\partial \phi}{\partial t} - \frac{1}{2}u^2 = \frac{2R\dot{R}^2 + R^2\ddot{R}}{r} - \frac{1}{2}\frac{R^4\dot{R}^2}{r^4}, \tag{6.0.3}$$

where p_∞ is the uniform pressure far from the bubble. At $r = R$, (6.0.3) reduces to

$$\rho\left(R\ddot{R} + \frac{3}{2}\dot{R}^2\right) = p - p_\infty. \tag{6.0.4}$$

The normal stress balance at $r = R$ is

$$\frac{2\gamma}{R} = T_{rr}(l) - T_{rr}(g), \tag{6.0.5}$$

where γ is the tension, l is the liquid, g is the gas, and $T_{rr}(g) = -p_b$, where p_b is bubble pressure and

$$T_{rr}(l) = -p + 2\mu\frac{\partial u}{\partial r}. \tag{6.0.6}$$

Combining now (6.0.4) and (6.0.6) with (6.0.1), we get

$$\frac{2\gamma}{R} = p_b - p_\infty - \rho\left(R\ddot{R} + \frac{3}{2}\dot{R}^2\right) - 4\mu\frac{\dot{R}}{R}. \tag{6.0.7}$$

Equation (6.0.7) is an evolution equation for the bubble radius $R(t)$. This equation is always called the Rayleigh–Plesset equation, but Plesset did not present or discuss this equation, which is given as (62) in the 1951 paper of Poritsky. It is well known that, when γ and μ are neglected, equation (6.0.7) can be formulated as an energy equation:

$$\frac{\mathrm{d}}{\mathrm{d}t}KE = (p_b - p_\infty)\,\dot{V},$$

where

$$KE = \frac{1}{2}\int_R^\infty \rho\left(\frac{\partial\phi}{\partial r}\right)^2 4\pi r^2 \mathrm{d}r,$$

$$\dot{V} = \frac{\mathrm{d}}{\mathrm{d}t}\left(\frac{4}{3}\pi R^3\right).$$

The equation

$$(p_b - p_\infty)\,\dot{V} = \frac{\mathrm{d}KE}{\mathrm{d}t} + D + 2\gamma\frac{\dot{V}}{R}, \tag{6.0.8}$$

where the dissipation

$$D = 2\mu\int_R^\infty \frac{\partial^2\phi}{\partial x_i \partial x_j}\frac{\partial^2\phi}{\partial x_i \partial x_j} 4\pi r^2 \mathrm{d}r = 16\pi\mu^2 R\dot{R}^2$$

follows from (6.0.7) after multiplication by $\dot{V}$.

The last term of (6.0.7) is from the viscous normal stress. This term converts to viscous dissipation in the energy balance.

Rayleigh's analysis and its extensions to viscous and surface-tension effects have applications to problems of cavitation, bubble collapse, and underwater explosions. For these applications the study of stability is important. Taylor (1950) showed that an interface is unstable when the light fluid is accelerated toward the dense fluid as is explained in §§9.1 and 9.2. The corresponding problem for a spherical interface was first discussed by Binnie (1953) who considered the collapse of vapor bubbles; his analysis is said to be flawed (see Plesset, 1954). The stability problem of this time-dependent problem is very complicated; normal modes cannot be used. An initial-value problem must be solved, and disturbances cannot be restricted to radial symmetry. Simplifying features are lost. Results by Birkhoff (1954) and Plesset (1954) are stated with many qualifiers and require special assumptions. The problem of stability of a spherical gas bubble in an inviscid liquid was considered by

Birkhoff (1954), who reduced the problem to the study of perturbations of the free surface expressed by Legendre polynomials:

$$r = b(t) + \sum_{n=1}^{\infty} b_n(t) P_n(\cos\varphi).$$

He derives coupled ordinary differential equations (ODEs) for $b(t)$ and $b_n(t)$ from the study of these equations. He finds stability during expansion and instability during collapse. Birkhoff's result is surprising. Collapsing bubbles for which the dense fluid is being accelerated toward the lighter vapor are unstable.

Birkhoff's result neglects the effects of viscosity. It would be of interest to see the effects of viscosity when irrotational theories are used.

7

Rise velocity of a spherical cap bubble

The theory of viscous potential flow (VPF) (Joseph, 2003a) was applied to the problem of finding the rise velocity U of a spherical cap bubble (Davies and Taylor, 1950; Batchelor, 1967). The rise velocity is given by

$$\frac{U}{\sqrt{gD}} = -\frac{8}{3}\frac{\nu(1+8s)}{\sqrt{gD^3}} + \frac{\sqrt{2}}{3}\left[1 - 2s - \frac{16s\sigma}{\rho g D^2} + \frac{32\nu^2}{gD^3}(1+8s)^2\right]^{1/2},$$

where $R = D/2$ is the radius of the cap, ρ and ν are the density and kinematic viscosity of the liquid, respectively, σ is surface tension, and $s = r''(0)/D$ is the deviation of the free surface,

$$r(\theta) = R + \frac{1}{2}r''(0)\theta^2 = R\left(1 + s\theta^2\right),$$

from perfect sphericity $r(\theta) = R$ near the stagnation point $\theta = 0$. The bubble nose is more pointed when $s < 0$ and blunted when $s > 0$. A more pointed bubble increases the rise velocity; the blunter bubble rises slower.

The result of Davies and Taylor (1950),

$$U = \frac{\sqrt{2}}{3}\sqrt{gD},$$

arises when all other effects vanish; if s alone is zero,

$$\frac{U}{\sqrt{gD}} = -\frac{8}{3}\frac{\nu}{\sqrt{gD^3}} + \frac{\sqrt{2}}{3}\left(1 + \frac{32\nu^2}{gD^3}\right)^{1/2}, \tag{7.0.1}$$

showing that viscosity slows the rise velocity. Equation (7.0.1) gives rise to a hyperbolic drag law,

$$C_D = 6 + 32/Re,$$

which agrees with data on the rise velocity of spherical cap bubbles given by Bhaga and Weber (1981).

7.1 Analysis

The spherical cap bubble (figure 7.1) arises in the motion of large gas bubbles, which take a lenticular shape. The analysis of the rise velocity of these bubbles that was given by

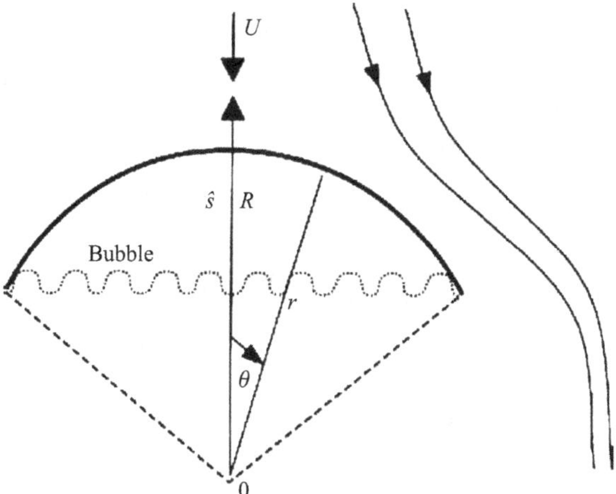

Figure 7.1. Spherical cap bubble. The rising bubble is viewed in a frame in which the bubble is stationary. The origin of z increasing is at the stagnation point $\hat{s}$. The surface of the cap is given by $z = -h(r, \theta) = -[R - r(\theta)\cos\theta]$. The cap is strictly spherical if $r(\theta) = R$ is constant.

Davies and Taylor (1950) is unusual because it is not computed from a balance of the drag and buoyant weight as it is for spherical gas bubbles (Levich, 1949; Moore, 1959, 1963; Taylor and Acrivos, 1964; Miksis, Vanden-Broeck, and Keller, 1982; Ryskin and Leal, 1984). Batchelor (1967) notes that "The remarkable feature of [the Davies–Taylor analysis] is that the speed of movement of the bubble is derived in terms of the bubble shape without any need for consideration of the mechanism of the retarding force which balances the effect of the buoyancy force on a bubble in steady motion."

The analysis of Davies and Taylor of the rise velocity of a spherical cap bubble is based on potential flow of an inviscid fluid. Here, we extend this analysis to viscous fluids with surface tension. The surface-tension effects enter only when the axisymmetric bubble is not spherical. The analysis of the irrotational effects of viscosity is clear but the analysis of surface tension is *ad hoc* because the deviation-from-sphericity parameter s is not calculated and its back influence to the potential is unknown. These defects are removed in the analysis of the rise velocity of an ellipsoidal bubble in chapter 8.

The velocity field on the gas bubble and the liquid is derived from a potential $\mathbf{u} = \nabla\phi$, $\nabla^2\phi = 0$. The velocity at $z = \infty$ is $-U$ (against z) and $\mathbf{g} = -\mathbf{e}_z g$. For steady flow,

$$\rho\mathbf{u}\cdot\nabla\mathbf{u} = -\nabla p - \rho\mathbf{e}_z g = -\nabla\Gamma, \tag{7.1.1}$$

where

$$\Gamma = p + \rho g z.$$

Equation (7.1.1) may be integrated because $\mathbf{u}\cdot\nabla\mathbf{u} = \nabla|\mathbf{u}|^2/2$, giving rise to a Bernoulli function in the liquid

$$\frac{\rho\,|\mathbf{u}|^2}{2} + \Gamma = \frac{\rho U^2}{2}, \tag{7.1.2}$$

and in the gas

$$\frac{\rho_G\,|\mathbf{u}|^2}{2} + \Gamma_G = C_G, \tag{7.1.3}$$

where C_G is an unknown constant.

We turn next to the normal stress balance,

$$-[[p]]+2\,[[\mu\mathbf{n}\cdot\mathbf{D}[\mathbf{u}]\cdot\mathbf{n}]]+\frac{2\sigma}{r(\theta)}=0, \tag{7.1.4}$$

where

$$[[\cdot]]=(\cdot)_G-(\cdot)_L$$

is evaluated on the free surface $r(\theta)=R(1+s\theta^2)$, σ is surface tension, μ is viscosity, and

$$\mathbf{n}\cdot\mathbf{D}[\mathbf{u}]\cdot\mathbf{n}=\frac{\partial u_n}{\partial n} \tag{7.1.5}$$

is the normal component of the rate of strain. Using (7.1.1), (7.1.4), and (7.1.5) we obtain

$$-[[\Gamma]]-[[\rho]]\,gh+2\left[\left[\mu\frac{\partial u_n}{\partial n}\right]\right]+\frac{2\sigma}{r}=0, \tag{7.1.6}$$

where $-h$ is the value of z on the free surface.

Following Davies and Taylor (1950), we assume that $\mathbf{u}$ may be approximated near the stagnation point on the bubble, which is nearly spherical, by the potential for the sphere; thus

$$\phi=-Ur\cos\theta\left(1+\frac{R^3}{2r^3}\right) \tag{7.1.7}$$

for the liquid. The form of ϕ in the gas will not be needed. From (7.1.7) we compute

$$u_r=\frac{\partial\phi}{\partial r}=-U\left(1-\frac{R^3}{r^3}\right)\cos\theta, \tag{7.1.8}$$

$$u_\theta=\frac{1}{r}\frac{\partial\phi}{\partial\theta}=U\sin\theta\left(1+\frac{R^3}{2r^3}\right), \tag{7.1.9}$$

$$\frac{\partial u_n}{\partial n}=\frac{\partial u_r}{\partial r}=-\frac{3UR^3}{r^4}\cos\theta. \tag{7.1.10}$$

Functions (7.1.8), (7.1.9), and (7.1.10) enter into the normal stress balance at $r=R(1+s\theta^2)$. This balance is to be satisfied near the stagnation point, for small θ, neglecting terms that go to zero faster than θ^2. At the free surface,

$$u_r=-U\left[1-\frac{1}{(1+s\theta^2)^3}\right]=-3Us\theta^2,\quad u_\theta=\frac{3}{2}U\theta,$$

$$\frac{\partial u_n}{\partial n}=-\frac{3U\left(1-\dfrac{\theta^2}{2}\right)}{R(1+s\theta^2)^4}=-\frac{3U}{R}\left[1-\left(4s+\frac{1}{2}\right)\theta^2\right], \tag{7.1.11}$$

$$u_r^2=0, \tag{7.1.12}$$

$$u_\theta^2=\frac{9}{4}U^2\theta^2, \tag{7.1.13}$$

$$h=R-r\cos\theta=R-R(1+s\theta^2)\left(1-\frac{\theta^2}{2}\right)=R\left(\frac{1}{2}-s\right)\theta^2. \tag{7.1.14}$$

The motion of the gas in the bubble is not known but it enters into (7.1.6) as the coefficient of ρ_G and μ_G, which are small relative to the corresponding liquid terms. Evaluating (7.1.2) and (7.1.3) on the free surface, with gas motion zero, we obtain

$$\Gamma = -\frac{9}{8}\rho U^2\theta^2 + \rho\frac{U^2}{2}, \tag{7.1.15}$$

$$\Gamma_G = C_G. \tag{7.1.16}$$

Using (7.1.11)–(7.1.16), we may rewrite (7.1.6) as

$$\begin{aligned} 0 &= -C_G + \Gamma + \rho g h - 2\mu\frac{\partial u_r}{\partial r} + \frac{2\sigma}{r} \\ &= -C_G + \frac{\rho U^2}{2} - \frac{9}{8}\rho U^2\theta^2 + \rho g R\left(\frac{1}{2} - s\right)\theta^2 \\ &\quad + \frac{6\mu U}{R}\left[1 - \left(4s + \frac{1}{2}\right)\theta^2\right] + \frac{2\sigma}{R}\left(1 - s\theta^2\right). \end{aligned} \tag{7.1.17}$$

The constant terms vanish:

$$C_G = \frac{\rho U^2}{2} + \frac{6\mu U}{R} + \frac{2\sigma}{R}. \tag{7.1.18}$$

The coefficient of θ^2 also vanishes:

$$\frac{9}{8}\rho U^2 + \frac{3\mu U}{R} + \frac{24\mu U s}{R} = \rho g\frac{R}{2} - s\left(\rho g R + \frac{2\sigma}{R}\right). \tag{7.1.19}$$

Surface tension, which balances the static pressure difference in a sphere or spherical cap, enters the formula for the velocity only when the axisymmetric cap is not spherical. When the spherical cap is perfectly spherical, as in the case treated by Davies and Taylor (1950), $s = 0$ and

$$U = -\frac{4}{3}\frac{\nu}{R} + \sqrt{\frac{4}{9}gR + \frac{16}{9}\frac{\nu^2}{R^2}}. \tag{7.1.20}$$

Equation (7.1.20) shows that the viscosity slows the rise velocity; when the viscosity is much larger than gravity,

$$U = \frac{1}{6}\frac{gR^2}{\nu}, \tag{7.1.21}$$

which is the velocity of a rising sphere computed by Moore (1959) from VPF balancing the drag with the buoyant weight.

The general solution of (7.1.19) with $D = 2R$ is

$$\frac{U}{\sqrt{gD}} = -\frac{8}{3}\frac{\nu\,(1+8s)}{\sqrt{gD^3}} + \frac{\sqrt{2}}{3}\left[1 - 2s - \frac{16s\sigma}{\rho g D^2} + \frac{32\nu^2}{gD^3}(1+8s)^2\right]^{1/2}. \tag{7.1.22}$$

It is convenient to write (7.1.22) in a dimensionless form,

$$Fr = -\frac{8\,(1+8s)}{3Re_G} + \frac{\sqrt{2}}{3}\left[1 - 2s - \frac{16s}{E\ddot{o}} + \frac{32}{Re_G^2}(1+8s)^2\right]^{1/2}, \tag{7.1.23}$$

where

$$Fr = \frac{U}{\sqrt{gD}}, \quad \text{Froude number;}$$

$$Re_G = \frac{\sqrt{gD^3}}{\nu}, \quad \text{gravity Reynolds number;}$$

$$E\ddot{o} = \frac{\rho g D^2}{\sigma}, \quad \text{Eötvös number.}$$

These three parameters are the only ones that enter into correlations for the rise velocity of long bubbles in round pipes (Wallis, 1969; Viana *et al.*, 2003).

It is of interest to consider different effects entering into the formula for the rise velocity U given by (7.1.22). The formula of Davies and Taylor,

$$U = \frac{\sqrt{2}}{3}\sqrt{gD}, \tag{7.1.24}$$

arises from (7.1.22) when $\nu = \sigma = s = 0$. Recall that $s = r''(0)/D$ represents the difference from an undeformed spherical cap. When $s = 0$ the cap is exactly spherical; when $s < 0$ the nose of the cap is more pointed than the spherical cap and when $s > 0$ the nose is blunter than a sphere. Davies–Taylor formula (7.1.24) arising from (7.1.22) in the asymptotic limit for large values of Re_G and $E\ddot{o}$,

$$U = \frac{\sqrt{2}}{3}(1 - 2s)\sqrt{gD}, \tag{7.1.25}$$

holds only when $s = 0$.

Unfortunately the analysis does not give the value of s; the shape of the nose is given when U is known, or, if s is known, then U is predicted. A more satisfying result would need to relax the assumption that velocity potential (7.1.7) does not change when the spherical cap is not exactly spherical. This kind of perturbation analysis requires global data and is well beyond what can at present be obtained by analysis. A formula [(8.2.39)] for an ellipsoidal bubble, replacing (7.1.23), was derived by Funada, Joseph, Maehara, and Yamashita (2004) and in §8.2.

7.2 Experiments

A review of experiments on the rise of spherical cap bubbles before 1973 together with an excellent collection of photographs can be found in the paper by Wegener and Parlange (1973). Reviews treating rising bubbles of all types were presented by Harper (1972) and Bhaga and Webber (1981).

A comparison of prediction (7.1.23) of the rise velocity of a spherical cap bubble arising from the application of VPF is unambiguous when the deviation s from sphericity vanishes. There are two cases in which $s = 0$; according to the analysis of Davies and Taylor (1950), the spherical cap that arises for large bubbles is one case; the other case includes the rise of small bubbles, or bubbles with large surface tension, which was considered by Levich (1949) and Moore (1959, 1963). In the first case, the sphericity arises from dynamics alone and the effects of surface tension on the spherical cap are negligible. In the case of small bubbles, or bubbles moving very slowly, surface tension can be important in keeping the bubble spherical, but the effect then of surface tension is absorbed totally by the pressure

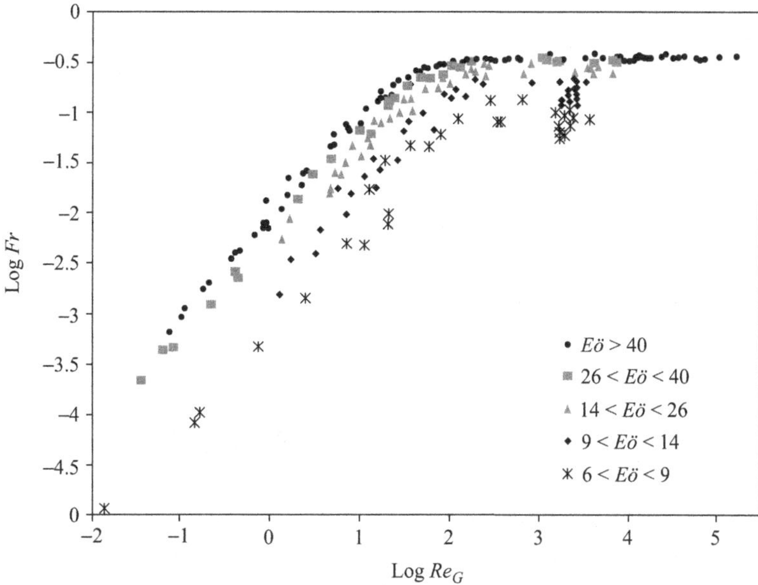

Figure 7.2. (After Viana *et al.*, 2003). Rise velocity Fr vs. Log Re_G for different Eötvös numbers for all published experiments on the rise velocity of Taylor bubbles in round pipes. The rise velocity is independent of $Eö$ when $Eö > 40$.

drop $[[p]] = 2\sigma/R$, as in (7.1.18), and does not enter the dynamics. The effects of surface tension on the rise velocity are associated with the deviation from sphericity; it is a shape effect because the net force and moment on a smooth bubble that are due to surface forces are zero (Hesla, Huang, and Joseph, 1993).

An efficient description of the effects affecting the deformation and rise velocity of gas bubbles in stagnant liquids can be carried out in terms of dimensionless parameters. Formula (7.1.23) expresses a functional relation among three parameters: the Froude number Fr, the gravity Reynolds number Re_G, and the Eötvös number $Eö$; these three parameters completely describe the rise velocity of Taylor bubbles, which are long gas bubbles capped by a spherical cap rising in tubes filled with stagnant liquid, which were discussed in the Davies–Taylor (1950) paper. Viana *et al.* (2003) correlated all the published data, 262 experiments, on the rise velocity, Fr, of Taylor bubbles in round pipes with a highly accurate rational fraction of power of the parameters Re_G and $Eö$; a graph of the data processed by them is shown in figure 7.2.

Other parameters are frequently used for the description of the rise velocity; these are

$$
\begin{aligned}
C_D &= \frac{4}{3}\frac{g d_e}{U^2}, && \text{drag coefficient;} \\
Re &= \frac{U d_e}{\nu}, && \text{Reynolds number;} \\
We &= \frac{\rho U^2 d_e}{\sigma}, && \text{Weber number;} \\
M &= \frac{g\mu^4}{\rho\sigma^3}, && \text{Morton number.}
\end{aligned}
\tag{7.2.1}
$$

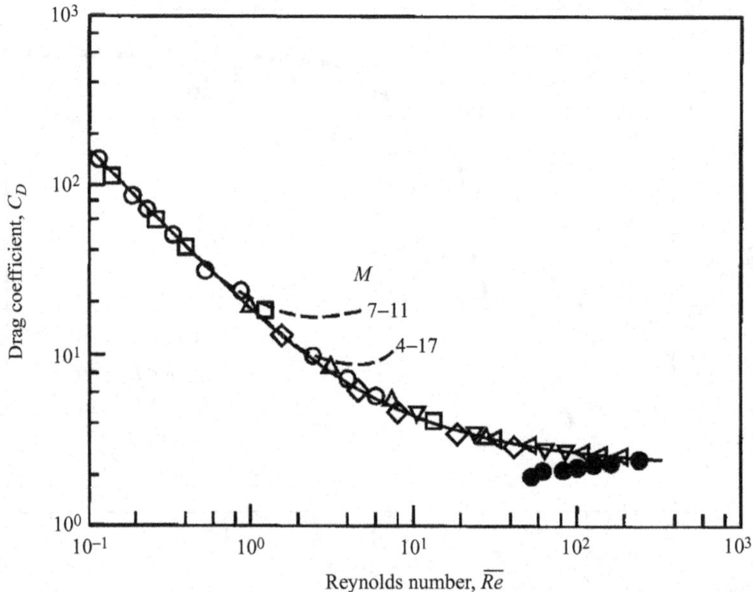

Figure 7.3. (After figure 7 in Bhaga and Weber, 1981). $\bar{C}_D$ $(= 4g\bar{d}/U^2)$ vs. $\overline{Re}$ $(= U\bar{d}/\nu)$. Taylor and Acrivos (1964) (- - -); $\bar{C}_D$ given by (7.2.2) (–); $M = 1.64 \times 10^{-3}$ (•).

The choice of effective diameter is important; two choices are made here: $d_e = D$, where d_e is the sphere diameter or the diameter of the spherical cap, $d_e = \bar{d}$, where $\bar{d}$ is the volume equivalent diameter defined by $V = [(\pi/6)\bar{d}^3]$, and Reynolds numbers by $Re = (UD/\nu)$, $\overline{Re} = (U\bar{d}/\nu)$, based on D and $\bar{d}$. Any pair of independent dimensionless parameters determines all the others in steady flow; as in (7.1.23), a dimensionless parameter involving U may be expressed in terms of two other independent parameters.

It is convenient to compare the theory developed here with experiments in which the surface tension, or parameters in which the surface tension is a factor, affects the rise velocity and those for which these parameters are not important; for example, according to figure 7.2, Fr is a function of Re_G alone when $E\ddot{o} > 40$.

In the case of a spherical cap bubble, limitations on the bubble size arise from several sources and restrict the values of the parameters that may be observed. Grace, Wairegi, and Brophy (1978) reported the maximum volume of air in bubbles that remain intact in five different liquids in a wide tank but did not identify which of the five were spherical caps. Batchelor (1987) looked at these data in terms of a stability theory that limits the maximum size bubbles. A numerical study by Bolton-Stone, Robinson, and Blake (1995) suggests that spherical cap bubbles arise only when the Eötvös number based on an equivalent spherical radius is less than about 32. For higher values of $E\ddot{o}$ an unstable toroidal bubble is formed before breakup.

Harper (1972) makes a distinction between high- and low-Morton-number (M) liquids. He characterizes the high-M liquids as those for which C_D decreases monotonically with $\overline{Re}$, as in figure 7.3, and the low-M liquids as those for which C_D vs. $\overline{Re}$ has a minimum. Bhaga and Weber (1981) identify this critical value as $M = 4 \times 10^{-3}$; a low-M liquid response is identified by the solid circle data points in figure 7.3.

The monotonic curve through the high-M open symbol points in figure 7.3 is described by an empirical formula,

$$\bar{C}_D = \left[(2.67)^{0.9} + \left(16/\overline{Re}\right)^{0.9}\right]^{1/0.9}, \tag{7.2.2}$$

in which $\bar{C}_D$ is determined by $\overline{Re}$ alone. The shapes of the bubbles change from spheres to spherical cap bubbles as $\overline{Re}$ increases because $\bar{C}_D$ and $\overline{Re}$ are defined here by use of the volume equivalent diameter $\bar{d}$, which decreases from the sphere diameter D at low $\overline{Re}$ to a value $\bar{d} < D$ for a spherical cap bubble, which can be estimated from the results of Davies and Taylor (1950) and those given here.

Consider the case in which the deviation from sphericity $s = 0$. In this case equation (7.1.20) holds. We may rewrite this equation as a drag relation,

$$C_D = 6 + 32/Re, \tag{7.2.3}$$

where $C_D = \frac{4}{3}\frac{gD}{U^3}$, $Re = \frac{UD}{\nu}$, and $D = 2R$ is the diameter of the spherical cap. The large-Re limit of (7.2.3), $C_D = 6$, is the value of drag coefficient that was established in the brilliant experiments by Davies and Taylor (1950). The asymptotic large-$\overline{Re}$ limit of (7.2.2) is $\bar{C}_D = 2.67$. These two limits should be the same; hence

$$\frac{\bar{C}_D}{C_D} = \frac{2.67}{6} = \frac{\bar{d}}{D}.$$

If this limit is a spherical cap bubble, the volume equivalent diameter is

$$\bar{d} = 0.445D \tag{7.2.4}$$

of the spherical cap diameter $d = 2R$. The ratio of the volume of the spherical cap bubble to the volume of a sphere of radius $R = D/2$ is

$$\bar{V} = \frac{\bar{d}^3 V}{D^3} = 0.0881. \tag{7.2.5}$$

The volume of the spherical cap bubble in this computation is a little less than 1/10 of the value of the volume of the sphere from which it is cut.

In comparing (7.2.2) and (7.2.3), we must convert $\overline{Re}$ to Re and, though this cannot be done generally, it can be done empirically for the limiting case of large Re for which $Re = 0.445Re$. In figure 7.4 we compare empirical relation (7.2.2) with a rescaled plot of (7.2.3):

$$\bar{C}_D = 0.445\left(6 + \frac{32}{Re}\right). \tag{7.2.6}$$

The agreement between (7.2.6) and (7.2.2) is spectacular but possibly misleading because the relation between $\bar{d}$ and D is not generally known.

A direct comparison of (7.2.3) with numerical results is tentative because the spherical cap bubble limit is beyond the capabilities of the numerical methods that have been applied to the problem. The calculations of Ryskin and Leal (1984) seem reliable and they work well for Weber and Reynolds numbers at which the spheres are greatly distorted. Their figure 1 is a plot of $\bar{C}_D$ vs. $\overline{We}$ with $\overline{Re}$ as a parameter that is in agreement with (7.2.2) for $\overline{Re} < 20$. The empirical formula could not be tested for the large values of We for which

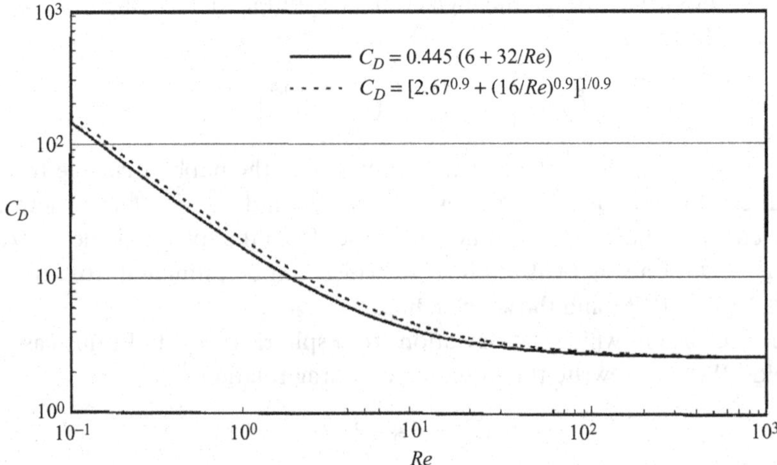

Figure 7.4. Comparison of empirical drag law (7.2.2) with theoretical drag law (7.2.6) scaled by the factor 0.445 required for matching the data in figure 7.3 with the experiments of Davies and Taylor (1950) at large Re.

the cap bubbles arise but the $\overline{Re}$ variation for smaller values of $\overline{We}$ is also consistent with (7.2.2).

The comparison of theory and experiment for the case in which the deviation s from sphericity is not zero is complicated. In this case the curvature of the nose of the bubble is different from the spherical radius R used in our calculations and those of Davies and Taylor. We have already noted that the analysis leading to s neglects some of the changes in the potential function that arise from the change in the shape of the domain. The computation or measurement of s will not be undertaken here but the comparison made in figures 1 and 2 of the paper by Miksis *et al.* (1982) for distorted bubbles in lenticular shape, but far from spherical caps, may point the way.

7.3 Conclusions

VPF is as a potential flow solution of the Navier–Stokes equations in which the vorticity vanishes and no-slip conditions at the interface are not enforced. This solution does not require that the viscosity be put to zero, and it is not a good idea to put it to zero. Using this theory, we extended the analysis of Davies and Taylor (1950) of the spherical cap bubble to include effects of viscosity, surface tension, and the deviation of the bubble nose from sphericity. The results of these analyses are then expressed by rise velocity formula (7.1.22) and drag formula (7.2.3). These formulas are in good agreement with the experiments of Bhaga and Weber (1981) at the high Morton numbers for which the cap bubbles arise, with the caveat that the conversion of the spherical radius to an effective volume equivalent radius be ambiguous. The possible effects of vorticity boundary layers on the rise velocity have not been analyzed here; it is not possible to use the same methods that work for spherical gas bubbles, and it is unlikely that the results of such an analysis would greatly improve the agreement between theory and experiments documented here.

8

Ellipsoidal model of the rise of a Taylor bubble in a round tube

The rise velocity of long gas bubbles (Taylor bubbles) in round tubes is modeled by an ovary ellipsoidal cap bubble rising in an irrotational flow of a viscous liquid. The analysis leads to an expression for the rise velocity that depends on the aspect ratio of the model ellipsoid and the Reynolds and Eötvös numbers. The aspect ratio of the best ellipsoid is selected to give the same rise velocity as the Taylor bubble at given values of the Eötvös and Reynolds numbers. The analysis leads to a prediction of the shape of the ovary ellipsoid that rises with same velocity as that of the Taylor bubble.

8.1 Introduction

The correlations given by Viana *et al.* (2003) convert all the published data on the normalized rise velocity $Fr = U/(gD)^{1/2}$ into analytic expressions for the Froude velocity versus buoyancy Reynolds number, $Re_G = [D^3 g(\rho_l - \rho_g)\rho_l]^{1/2}/\mu$ for fixed ranges of the Eötvös number, $E\ddot{o} = g\rho_l D^2/\sigma$, where D is the pipe diameter, ρ_l, ρ_g, and σ are densities and surface tension. Their plots give rise to power laws in $E\ddot{o}$; the compositions of these separate power laws emerge as bipower laws for two separate flow regions for large- and small-buoyancy Reynolds numbers. For large Re_G (>200) they find that

$$Fr = 0.34/(1 + 3805/E\ddot{o}^{3.06})^{0.58}. \tag{8.1.1}$$

For small Re_G (<10) they find

$$Fr = \frac{9.494 \times 10^{-3}}{(1 + 6197/E\ddot{o}^{2.561})^{0.5793}} Re_G^{1.026}. \tag{8.1.2}$$

The flat region for a high-buoyancy Reynolds number and the sloped region for a low-buoyancy Reynolds number is separated by a transition region ($10 < Re_G < 200$) that they describe by fitting the data to a logistic dose curve. Repeated applications of logistic dose curves lead to a composition of rational fractions of power laws. This leads to the following universal correlation:

$$Fr = \frac{0.34/(1 + 3805/E\ddot{o}^{3.06})^{0.58}}{\left\{1 + \left[\dfrac{Re_G}{31.08}\left(1 + \dfrac{778.76}{E\ddot{o}^{1.96}}\right)^{-0.49}\right]^{-1.45\left(1+\frac{7.22\times 10^{13}}{E\ddot{o}^{9.93}}\right)^{0.094}}\right\}^{0.71\left(1+\frac{7.22\times 10^{13}}{E\ddot{o}^{9.93}}\right)^{-0.094}}}. \tag{8.1.3}$$

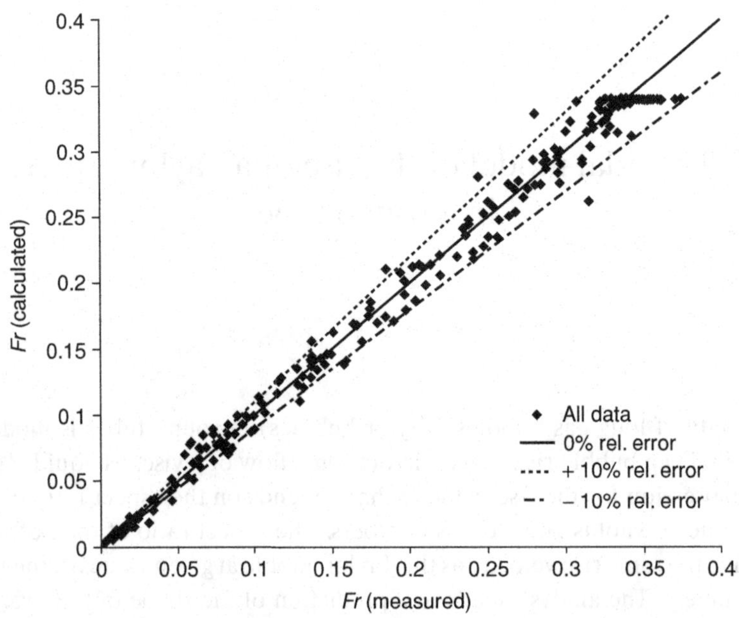

Figure 8.1. *Fr* predicted from (8.1.3) vs. experimental data ($Eö > 6$).

The performance of universal correlation (8.1.3) is evaluated in figure 8.1, where the values predicted by (8.1.3) are compared with those of the experiments. Almost all of the values fall within the 20% error line, and most of the data are within 10% of predicted values.

Formula (8.1.3) solves the problem of the rise velocity of Taylor bubbles in round pipes. This formula arises from processing data and not from flow fundamentals; one might say that the problem of the rise velocity has been solved without understanding.

8.1.1 Unexplained and paradoxical features

The teaching of fluid mechanics would lead one to believe that a bubble rising steadily in a liquid is in a balance of buoyant weight and drag. It is natural to think that the buoyant weight is proportional to the volume of gas, but the accurate formula [(8.1.3)] does not depend on the length of the bubble; this requires explanation.

Even the theoretical results are mysterious. The rise velocity U of the spherical cap bubble at high Reynolds number is accurately determined from a potential flow analysis of motion in an inviscid fluid by Davies and Taylor (1950) and in a viscous fluid by Joseph (2003a). Analysis of the rise velocity of Taylor bubbles in inviscid fluids based on shape of the bubble nose was given first by Dumitrescue (1943) and then by Davies and Taylor (1950).

In the analysis of Joseph (2003a), given in chapter 7, it was assumed that the bubble nose shape was given by

$$r(\theta) = R\left(1 + s\theta^2\right), \qquad (8.1.4)$$

where $s = r''(0)/2R$ is the deviation of the free surface from perfect sphericity.

The dependence of (8.1.4) on terms proportional to s is incomplete because the potential solution for a sphere and the curvature for a sphere were not perturbed. A complete formula [(8.2.39)] replacing (8.1.4) is derived in §8.2.

The Davies–Taylor (1950) result arises when all other effects vanish; if s alone is zero,

$$\frac{U}{\sqrt{gD}} = -\frac{8}{3}\frac{\nu}{\sqrt{gD^3}} + \frac{\sqrt{2}}{3}\left(1 + \frac{32\nu^2}{gD^3}\right)^{1/2}, \tag{8.1.5}$$

showing that viscosity slows the rise velocity. Equation (8.1.5) gives rise to a hyperbolic drag law,

$$C_D = 6 + 32/Re, \tag{8.1.6}$$

which agrees with data on the rise of spherical cap bubbles given by Bhaga and Weber (1981).

The velocity U given by (8.1.5) is not determined in any obvious way by a balance of buoyant weight and drag even when the viscosity is not zero. Batchelor (1967) notes that, in the inviscid case,

> ...the retarding force is evidently independent of the Reynolds number, and the rate of dissipation of mechanical energy is independent of viscosity, implying that the stesses due to turbulent transfer of momentum are controlling the flow pattern in the wake of the bubble.

Data from the experiments of Bhaga and Weber (1981) and Taylor bubble data of Viana *et al.* (2003) do not support the idea of turbulent transfer. The wake may be very turbulent as is true in water or apparently smooth and laminar as is true for bubbles rising in viscous oils, but this feature does not enter into any of the formulas for the rise velocity, empirical as in (8.1.3) or theoretical as in (8.1.5).

A related paradoxical property is that the Taylor bubble rise velocity does not depend on how the gas is introduced into the pipe. In the Davies–Taylor experiments the bubble column is open to the gas. In other experiments the gas is injected into a column whose bottom is closed.

It can be said, despite successes, that a good understanding of the fluid mechanics of the rise of cap bubbles and Taylor bubbles is not yet available.

8.1.2 Drainage

Many of the paradoxical features of the rise of Taylor bubbles can be explained by drainage, as in figure 8.2. The liquid at the wall drains under gravity with no pressure gradient. If the liquid is put into motion by a pressure gradient the gas bubble will deform and the film flow will not be governed by (8.1.7); the drain equation is

$$\frac{\mu}{r}\frac{\mathrm{d}}{\mathrm{d}r}\left(r\frac{\mathrm{d}u}{\mathrm{d}r}\right) = \rho_l g, \tag{8.1.7}$$

subject to no-slip at the wall and no shear at the bubble surface.

It can be argued that the cylindrical part of the long bubble is effectively not displacing liquid because the pressure does not vary along the cylinder. In this case the buoyant volume entering into the equation *buoyancy* = *drag* would be the vaguely defined hemisphere poking into the liquid at top. The source of drag is unclear; because shear stresses

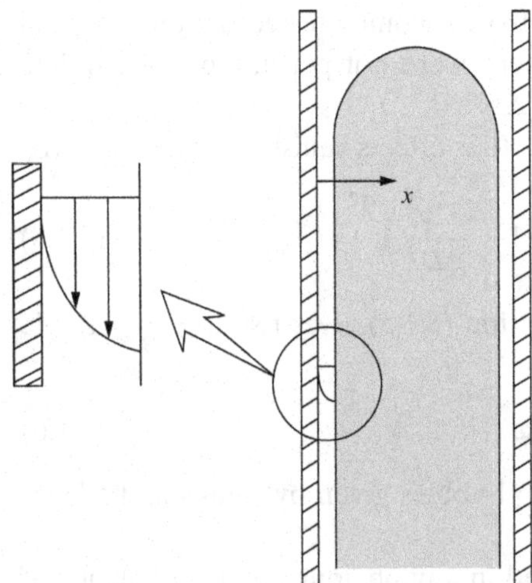

Figure 8.2. Drainage at the wall of a rising Taylor bubble. If U is added to this system the wall moves and the bubble is stationary.

do not enter, the drag ought to be determined by the vertical projections of normal stresses all around the bubble. This kind of analysis has not appeared in the literature. A different kind of analysis, depending on the shape of the bubble and sidewall drainage, has been applied. This kind of analysis leads ultimately to a formula for the rise velocity of the bubble nose. Apparently the shape of the bubble nose is an index of the underlying drag balance.

8.1.3 Brown's analysis of drainage

Brown (1965) put forward a model of the rise velocity of large gas bubbles in tubes. A similar model was given by Batchelor (1967). There are two elements for this model:

(1) The rise velocity is assumed to be given by $C\sqrt{gR_\delta}$, where $C\ (= 0.494)$ is an empirical constant, $R_\delta = R - \delta$ is the bubble radius, R is the tube radius, and δ is the unknown film thickness.

(2) It is assumed that the fluid drains in a falling film of constant thickness δ. The film thickness is determined by conserving mass: the liquid displaced by the rising bubble must balance the liquid draining at the wall.

After equating two different expressions for the rise velocity arising from (1) and (2), Brown finds that

$$U = 0.35V\sqrt{1 - \frac{2(\sqrt{1+\psi} - 1)}{\psi}}, \tag{8.1.8}$$

where

$$\psi = \left(14.5Re^2\right)^{1/3}, \quad Re = VD/\nu, \quad V = \sqrt{gD}. \tag{8.1.9}$$

The expression for the rise velocity, (8.1.8), does not account for effects of surface tension that are negligible when the bubble radius is large. The expression is in moderately good agreement with data, but not nearly as good as correlation formula (8.1.3).

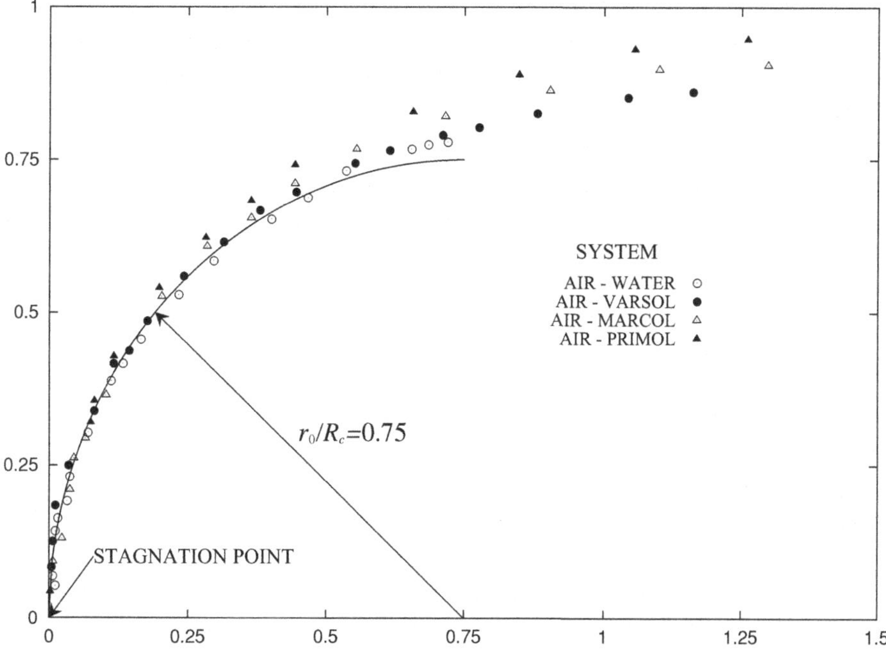

Figure 8.3. (After Brown, 1965). The profile of the cap of Taylor bubbles. The nose region is spherical with a radius r_0. For all the fluids, $r_0/R_c = 0.75$. The viscosities of water, Varsol, Marcol, and Primol apparently are 0.977, 0.942, 19.42, and 142.3 mPa s, respectively.

The rise velocity $C\sqrt{gR_\delta}$ is still determined by the bubble shape, but that shape is altered by drainage.

8.1.4 Viscous potential flow

We have already mentioned that correlation formula (8.1.3) accurately predicts the rise velocity and further improvement cannot be expected from modeling. Our understanding of the fluid mechanics underway is, however, far from complete. When surface tension is neglected, formula (8.1.5) extends the results of Dumitrescue (1943) and Davies and Taylor (1950) from inviscid fluids to viscous fluids by assuming that the cap of the bubble remains spherical, even at finite Reynolds number. The same extension to include the effects of viscosity in the formula for the rise velocity based on potential flow at the nose should be possible for Taylor bubbles if the nose remains spherical in viscous fluids. Brown says that figure 8.3 (note that he used R_c for $R - \delta$, which is R_δ in our nomenclature)

> . . . indicates that although the cavity shapes are different in the transition region, they are remarkably similar in the nose region. The second interesting fact . . . that the frontal radius of the cavity in normalized coordinates ($R_c = R - \delta$) which is the same for all liquids, is 0.75, the same value as was obtained in the analysis of bubbles in inviscid liquids.

The scatter in the data plotted in figure 8.3 and the data in the transition can be fit even better by the cap of an ovary ellipsoid that is nearly spherical (shown later in figure 8.9) arising from the analysis given in §8.2.

The effect of surface tension is to retard the rise velocity of Taylor bubbles in round tubes. Universal correlation (8.1.3) shows that the rise velocity decreases as the Eötvös number $E\ddot{o} = g\rho_l D^2/\sigma$ decreases for ever smaller values of D^2/σ. In fact, these kinds of bubbles, with large tensions in small tubes, do not rise; they stick in the pipe, preventing draining. If the radius of a stagnant bubble $R = 2\sigma/\Delta p$ with the same pressure difference Δp as in the Taylor bubble is larger than the tube radius, it will plug the pipe. White and Beardmore (1962) said that the bubble will not rise when $E\ddot{o} < 4$. This can be compared with the values 3.36 given by Hattori (1935), 3.37 given by Bretherton (1961), 5.8 given by Barr (1926), and 4 given by Gibson (1913).

A very convincing set of experiments showing the effect of drainage is reported for "Taylor bubbles in miniaturized circular and noncircular channels" by Bi and Zhao (2001). They showed that, for triangular and rectangular channels, elongated bubbles always rose upward even though the hydraulic diameter of the tube was as small as 0.866 mm, whereas in circular tubes the bubble motion stopped when $d \leq 2.9$ mm. They did not offer an explanation, but the reason is that surface tension cannot close the sharp corners where drainage can occur.

8.2 Ellipsoidal bubbles

Grace and Harrison (1967) studied the rise of an ellipsoidal bubble in an inviscid liquid. They sought to explain the influence of bubble shape on the rise velocity and concluded that elliptical cap and ovary ellipsoidal bubbles rise faster than the corresponding circular cap and spherical cap bubbles. They cited experimental data that they claim support their results. They say that "... bubbles take up elliptical shapes if they enclose a surface (e.g. a rod)." This statement is not correct because bubbles rising in the presence of a central rod are usually not axisymmetric, as can be seen in figure 8.4.

In this chapter we obtain expressions for the rise velocity of ovary and planetary ellipsoids. The ovary ellipsoid looks more like a long bubble than a planetary ellipsoid (figure 8.5). There is no way that the planetary ellipsoid can be fit to the data given by Viana *et al.* (2003), but the ovary ellipsoid can be made to fit the data of Viana *et al.* with one shape parameter for all cases. This is very surprising, as the Taylor bubble is not thought to be ellipsoidal and the dynamics of these bubbles is controlled by sidewall drainage that is entirely neglected in the following potential flow analysis of the rise of ellipsoidal bubbles in a viscous liquid.

Ovary and planetary ellipsoidal bubbles are shown in figure 8.5. We will be led by the analysis to cases in which the ovary ellipsoids are nearly spherical with $D = 2a$.

For axisymmetric flows of incompressible fluid around the ellipsoid of revolution, we can have the stream function and the velocity potential; then we have the solution that satisfies the kinematic condition at the surface of the bubble and the normal stress balance there that contains the viscous normal stress based on VPF.

8.2.1 Ovary ellipsoid

In an ellipsoidal frame (ξ, η, φ) on an ovary ellipsoid bubble moving with a uniform velocity U in a liquid, we have the stream function ψ and the velocity potential ϕ for axisymmetric flow:

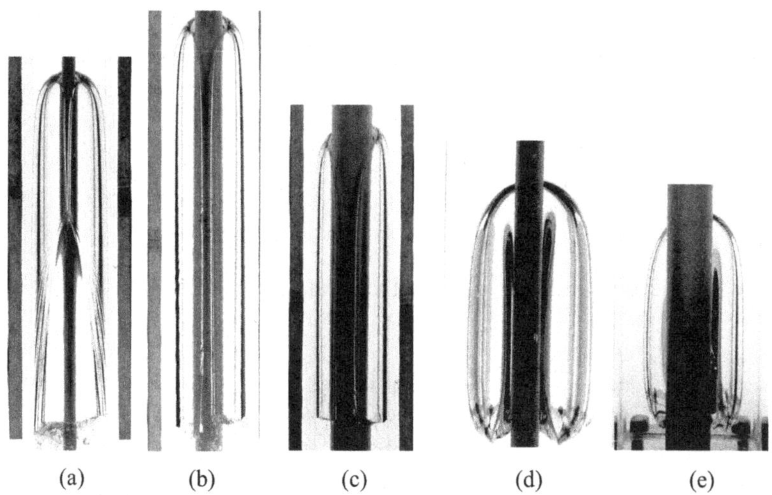

Figure 8.4. Photographs (unpublished, courtesy of F.Viana & R.Pardo) of Taylor bubbles rising in concentric annular space of 76.2 mm inside diameter pipe and different rod diameter (ID) filled with different viscous liquids: a) Water (1 mPa s, 997 kg/m^3), ID=12.7 mm; b) Water, ID=25.4 mm; c) Water, ID=38.1 mm; d) Silicone oil (1300 mPa s, 970 kg/m^3), ID=12.7 mm; e) Silicone oil (1300 mPa s, 970 kg/m^3), ID=25.4 mm. The gas bubbles do not wrap all the way around the inner cylinder; a channel is opened for liquid drainage.

$$\psi = \frac{1}{2} U c^2 \sin^2 \eta \left[\sinh^2 \xi - \frac{b^2}{a^2 K} \left(\cosh \xi + \sinh^2 \xi \ln \tanh \frac{\xi}{2} \right) \right], \tag{8.2.1}$$

$$\phi = U c \cos \eta \left[\cosh \xi - \frac{b^2}{a^2 K} \left(1 + \cosh \xi \ln \tanh \frac{\xi}{2} \right) \right], \tag{8.2.2}$$

with $c^2 = a^2 - b^2$, $e = c/a$, and K:

$$\begin{aligned} K &= e^2 \left(\frac{a}{c} + \frac{b^2}{c^2} \ln \frac{a+b-c}{a+b+c} \right) = \frac{1}{\cosh \xi_0} + \tanh^2 \xi_0 \ln \tanh \xi_0 / 2 \\ &= e + \left(1 - e^2\right) \tanh^{-1}(e). \end{aligned} \tag{8.2.3}$$

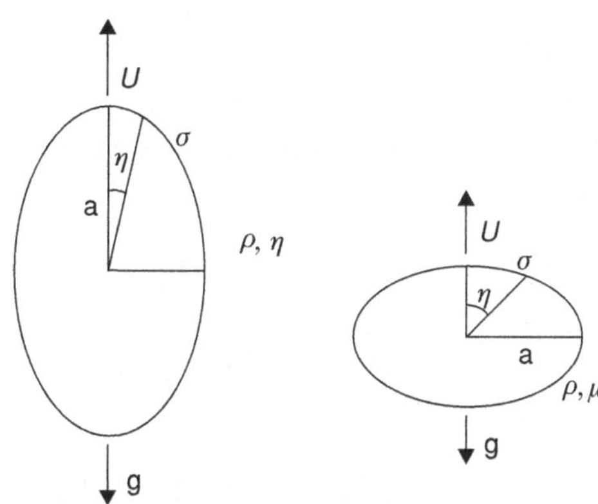

Figure 8.5. An ellipsoid bubble moving with a uniform velocity U in the z direction of Cartesian coordinates (x, y, z). An ovary ellipsoid is depicted in the left-hand side and a planetary ellipsoid is in the right-hand side, which are of the major semiaxis a, the minor semiaxis b, the aspect ratio $e = c/a$, $c^2 = a^2 - b^2$, and in a liquid (water) of density ρ, viscosity μ, with the surface tension σ at the surface given by $\xi = \xi_0$ and under the acceleration that is due to gravity g.

(In this case, we take $z = c\cosh\xi\cos\eta$, $\varpi = c\sinh\xi\sin\eta$, $x = \varpi\cos\varphi$, and $y = \varpi\sin\varphi$; $a = c\cosh\xi_0$ and $b = c\sinh\xi_0$.) Stream function (8.2.1) was derived from the article in §16.57 of Milne-Thomson's book (Milne-Thomson, 1968). The velocity $\mathbf{u} = (u_\xi, u_\eta)$ is expressed as

$$u_\xi = -\frac{1}{J}\frac{\partial\phi}{\partial\xi} = -\frac{1}{\varpi J}\frac{\partial\psi}{\partial\eta}$$
$$= -\frac{Uc}{J}\cos\eta\left[\sinh\xi - \frac{b^2}{a^2K}(\coth\xi + \sinh\xi\ln\tanh\xi/2)\right], \tag{8.2.4}$$

$$u_\eta = -\frac{1}{J}\frac{\partial\phi}{\partial\eta} = \frac{1}{J\varpi}\frac{\partial\psi}{\partial\xi} = \frac{Uc}{J}\sin\eta\left[\cosh\xi - \frac{b^2}{a^2K}(1 + \cosh\xi\ln\tanh\xi/2)\right]$$
$$\equiv \frac{Uc\sin\eta}{J}f_1(\xi), \tag{8.2.5}$$

$$\frac{1}{J}\frac{\partial u_\xi}{\partial\xi} = -\frac{1}{J^2}\frac{\partial^2\phi}{\partial\xi^2} + \frac{1}{J^3}\frac{\partial J}{\partial\xi}\frac{\partial\phi}{\partial\xi}$$
$$= -\frac{Uc}{J^2}\cos\eta\left[\cosh\xi - \frac{b^2}{a^2K}\left(1 - \frac{1}{\sinh^2\xi} + \cosh\xi\ln\tanh\xi/2\right)\right] - \frac{u_\xi}{J^2}\frac{\partial J}{\partial\xi}$$
$$= -\frac{Uc\cos\eta}{J^2}f_2(\xi) - \frac{u_\xi}{J^2}\frac{\partial J}{\partial\xi}, \tag{8.2.6}$$

$$\left(\frac{1}{J}\frac{\partial u_\xi}{\partial\xi}\right)_{\xi_0} = -Uc\cos\eta\left\{\frac{1}{J^2}\left[\cosh\xi - \frac{b^2}{a^2K}\left(1 - \frac{1}{\sinh^2\xi} + \cosh\xi\ln\tanh\xi/2\right)\right]\right\}_{\xi_0}$$
$$\equiv -\frac{Uc\cos\eta}{J_0^2}f_2(\xi_0), \tag{8.2.7}$$

with

$$J^2 = c^2\sinh^2\xi + c^2\sin^2\eta,\quad J_0 = (J)_{\xi_0},$$
$$f_1(\xi_0) = \cosh\xi_0 - \frac{b^2}{a^2K}(1 + \cosh\xi_0\ln\tanh\xi_0/2) = \frac{e^2}{e + (1 - e^2)\tanh^{-1}(e)} = f_1(e),$$
$$f_2(\xi_0) = \cosh\xi_0 - \frac{b^2}{a^2K}\left(1 - \frac{1}{\sinh^2\xi_0} + \cosh\xi_0\ln\tanh\xi_0/2\right) = f_2(e) = 2f_1(e). \tag{8.2.8}$$

From Joseph (2003a), the Bernoulli function is given by his (1.2) and (1.3):

$$\frac{\rho\,|\mathbf{u}|^2}{2} + \Gamma = \frac{\rho U^2}{2},\quad \frac{\rho_G\,|\mathbf{u}|^2}{2} + \Gamma_G = C_G. \tag{8.2.9}$$

Put $\rho_G\,|\mathbf{u}|^2 = 0$ in the gas. Then $\Gamma_G = C_G$ is constant. Boundary conditions at the surface of the ellipsoid (where $\xi = \xi_0$) are the kinematic condition and the normal stress balance:

$$u_\xi = 0,\quad -[[\Gamma]] - [[\rho]]\,gh + [[2\mu\mathbf{n}\cdot\mathbf{D}[\mathbf{u}]]]\cdot\mathbf{n} = -\sigma\nabla\cdot\mathbf{n}, \tag{8.2.10}$$

where $\Gamma = p + \rho g h$ as in Joseph (2003a), and the normal viscous stress $2\mu \mathbf{n} \cdot \mathbf{D}[\mathbf{u}] \cdot \mathbf{n}$ and the normal vector $\mathbf{n}$ satisfy the following relations:

$$[[\Gamma]] = \Gamma_G - \Gamma = C_G - \frac{\rho}{2}U^2 + \frac{\rho}{2}|\mathbf{u}|^2 = C_G - \frac{\rho}{2}U^2 + \frac{\rho}{2}(u_\eta)^2_{\xi_0}, \quad (8.2.11)$$

$$[[2\mu \mathbf{n} \cdot \mathbf{D}[\mathbf{u}]]] \cdot \mathbf{n} = -2\mu \left(\frac{1}{J}\frac{\partial u_\xi}{\partial \xi} + u_\eta \frac{c^2}{J^3} \cos\eta \sin\eta \right)_{\xi_0}, \quad (8.2.12)$$

$$\nabla \cdot \mathbf{n} = \frac{\coth \xi_0}{J_0^3}\left(J_0^2 + c^2 \sinh^2 \xi_0\right). \quad (8.2.13)$$

The normal stress balance is then expressed as

$$-C_G + \frac{\rho}{2}U^2 - \frac{\rho}{2}(u_\eta)^2_{\xi_0} + \rho g h - 2\mu \left(\frac{1}{J}\frac{\partial u_\xi}{\partial \xi} + u_\eta \frac{c^2}{J^3} \cos\eta \sin\eta \right)_{\xi_0} = -\sigma \nabla \cdot \mathbf{n}, \quad (8.2.14)$$

where the distance from the top of the ellipsoid bubble is

$$h = a(1 - \cos\eta) = c \cosh \xi_0 (1 - \cos\eta). \quad (8.2.15)$$

For small η, we have

$$\begin{aligned} J_0 &= c \sinh \xi_0 \left[1 + \frac{\eta^2}{2\sinh^2 \xi_0} + O(\eta^4) \right], \\ h &= c \cosh \xi_0 \left[\frac{1}{2}\eta^2 + O(\eta^4) \right], \\ (u_\eta)_{\xi_0} &= \frac{U f_1(\xi_0)}{\sinh \xi_0} \left[\eta + O(\eta^3) \right], \\ \left(\frac{1}{J}\frac{\partial u_\xi}{\partial \xi} \right)_{\xi_0} &= -\frac{U f_2(\xi_0)}{c \sinh^2 \xi_0} \left[1 - \left(\frac{1}{2} + \frac{1}{\sinh^2 \xi_0} \right) \eta^2 + O(\eta^4) \right], \\ \nabla \cdot \mathbf{n} &= \frac{2\cosh \xi_0}{c \sinh^2 \xi_0} \left[1 - \frac{\eta^2}{\sinh^2 \xi_0} + O(\eta^4) \right]. \end{aligned} \quad (8.2.16)$$

The substitution of (8.2.16) into normal stress balance (8.2.14) leads to a formula for the ovary bubble:

$$\begin{aligned} &-C_G + \frac{\rho}{2}U^2 - \frac{\rho}{2}(u_\eta)^2_{\xi_0} + \rho g h - 2\mu \left(\frac{1}{J}\frac{\partial u_\xi}{\partial \xi} + u_\eta \frac{c^2}{J^3} \cos\eta \sin\eta \right)_{\xi_0} \\ &= -\sigma \frac{2\cosh \xi_0}{c \sinh^2 \xi_0} \left(1 - \frac{\eta^2}{\sinh^2 \xi_0} \right). \end{aligned} \quad (8.2.17)$$

Thus we have C_G in $O(1)$,

$$C_G = \frac{\rho}{2}U^2 + \frac{2\mu U f_2(\xi_0)}{c \sinh^2 \xi_0} + \sigma \frac{2\cosh \xi_0}{c \sinh^2 \xi_0}, \quad (8.2.18)$$

and the following relation in $O(\eta^2)$,

$$-\frac{\rho}{2}\frac{U^2 f_1^2(\xi_0)}{\sinh^2 \xi_0} + \frac{\rho g c}{2} \cosh \xi_0 - \frac{2\mu U}{c \sinh^2 \xi_0} \left[f_2(\xi_0) \left(\frac{1}{2} + \frac{1}{\sinh^2 \xi_0} \right) + \frac{f_1(\xi_0)}{\sinh^2 \xi_0} \right] = \sigma \frac{2\cosh \xi_0}{c \sinh^4 \xi_0}. \quad (8.2.19)$$

8.2.2 Planetary ellipsoid

In an ellipsoidal frame (ξ, η, φ) on the planetary ellipsoid bubble moving with a uniform velocity U in a liquid, we have the stream function ψ and the velocity potential ϕ for axisymmetric flow:

$$\psi = \frac{1}{2}Uc^2 \sin^2\eta\left(\cosh^2\xi - \frac{\sinh\xi - \cosh^2\xi\cot^{-1}\sinh\xi}{e\sqrt{1-e^2} - \sin^{-1}e}\right), \tag{8.2.20}$$

$$\phi = Uc\cos\eta\left(\sinh\xi - \frac{1 - \sinh\xi\cot^{-1}\sinh\xi}{e\sqrt{1-e^2} - \sin^{-1}e}\right), \tag{8.2.21}$$

where $c^2 = a^2 - b^2$ and $e = c/a$. (In this case, we take $z = c\sinh\xi\cos\eta$, $\varpi = c\cosh\xi\sin\eta$, $x = \varpi\cos\varphi$, and $y = \varpi\sin\varphi$; $a = c\cosh\xi_0$ and $b = c\sinh\xi_0$.) The velocity $\mathbf{u} = (u_\xi, u_\eta)$ is given by

$$u_\xi = -\frac{1}{J}\frac{\partial\phi}{\partial\xi} = -\frac{Uc}{J}\cos\eta\left(\cosh\xi - \frac{\tanh\xi - \cosh\xi\cot^{-1}\sinh\xi}{e\sqrt{1-e^2} - \sin^{-1}e}\right), \tag{8.2.22}$$

$$u_\eta = -\frac{1}{J}\frac{\partial\phi}{\partial\eta} = \frac{Uc}{J}\sin\eta\left(\sinh\xi - \frac{1 - \sinh\xi\cot^{-1}\sinh\xi}{e\sqrt{1-e^2} - \sin^{-1}e}\right) \equiv \frac{Uc\sin\eta}{J}f_1(\xi), \tag{8.2.23}$$

$$\begin{aligned}
\frac{1}{J}\frac{\partial u_\xi}{\partial\xi} &= -\frac{1}{J^2}\frac{\partial^2\phi}{\partial\xi^2} + \frac{1}{J^3}\frac{\partial J}{\partial\xi}\frac{\partial\phi}{\partial\xi} \\
&= -\frac{Uc}{J^2}\cos\eta\left(\sinh\xi - \frac{1 + 1/\cosh^2\xi - \sinh\xi\cot^{-1}\sinh\xi}{e\sqrt{1-e^2} - \sin^{-1}e}\right) - \frac{u_\xi}{J^2}\frac{\partial J}{\partial\xi} \\
&= -\frac{Uc\cos\eta}{J^2}f_2(\xi) - \frac{u_\xi}{J^2}\frac{\partial J}{\partial\xi},
\end{aligned} \tag{8.2.24}$$

$$\begin{aligned}
\left(\frac{1}{J}\frac{\partial u_\xi}{\partial\xi}\right)_{\xi_0} &= -Uc\cos\eta\left[\frac{1}{J^2}\left(\sinh\xi - \frac{1 + 1/\cosh^2\xi - \sinh\xi\cot^{-1}\sinh\xi}{e\sqrt{1-e^2} - \sin^{-1}e}\right)\right]_{\xi_0} \\
&\equiv -\frac{Uc\cos\eta}{J_0^2}f_2(\xi_0),
\end{aligned} \tag{8.2.25}$$

with

$$J^2 = c^2\sinh^2\xi + c^2\cos^2\eta, \quad J_0 = (J)_{\xi_0},$$

$$f_1(\xi_0) = \sinh\xi_0 - \frac{1 - \sinh\xi_0\cot^{-1}\sinh\xi_0}{e\sqrt{1-e^2} - \sin^{-1}e} = \frac{-e^2}{e\sqrt{1-e^2} - \sin^{-1}(e)} = f_1(e), \tag{8.2.26}$$

$$f_2(\xi_0) = \sinh\xi_0 - \frac{1 + 1/\cosh^2\xi_0 - \sinh\xi_0\cot^{-1}\sinh\xi_0}{e\sqrt{1-e^2} - \sin^{-1}e} = f_2(e) = 2f_1(e).$$

Boundary conditions at the surface of the ellipsoid (where $\xi = \xi_0$) are the kinematic condition and the normal stress balance:

$$\begin{aligned}
&u_\xi = 0, \\
&-C_G + \frac{\rho}{2}U^2 - \frac{\rho}{2}u_\eta^2 + \rho gh - 2\mu\left(\frac{1}{J}\frac{\partial u_\xi}{\partial\xi} - u_\eta\frac{c^2}{J^3}\cos\eta\sin\eta\right) = -\sigma\nabla\cdot\mathbf{n},
\end{aligned} \tag{8.2.27}$$

where the distance from the top of the ellipsoid bubble and $\nabla \cdot \mathbf{n}$ are given, respectively, by

$$h = b\,(1 - \cos\eta) = c \sinh\xi_0\,(1 - \cos\eta)\,, \qquad \nabla \cdot \mathbf{n} = \frac{\tanh\xi_0}{J_0^3}\left(J_0^2 + c^2\cosh^2\xi_0\right). \tag{8.2.28}$$

For small η, we have

$$\begin{aligned}
J_0 &= c\cosh\xi_0\left[1 - \frac{\eta^2}{2\cosh^2\xi_0} + O(\eta^4)\right],\\
h &= c\sinh\xi_0\left[\frac{1}{2}\eta^2 + O(\eta^4)\right],\\
(u_\eta)_{\xi_0} &= \frac{Uf_1(\xi_0)}{\cosh\xi_0}\left[\eta + O(\eta^3)\right],\\
\left(\frac{1}{J}\frac{\partial u_\xi}{\partial \xi}\right)_{\xi_0} &= -\frac{Uf_2(\xi_0)}{c\cosh^2\xi_0}\left[1 - \left(\frac{1}{2} - \frac{1}{\cosh^2\xi_0}\right)\eta^2 + O(\eta^4)\right],\\
\nabla \cdot \mathbf{n} &= \frac{2\sinh\xi_0}{c\cosh^2\xi_0}\left[1 + \frac{\eta^2}{\cosh^2\xi_0} + O(\eta^4)\right].
\end{aligned} \tag{8.2.29}$$

Substitution of these into the normal stress balance leads to a formula for the planetary bubble:

$$\begin{aligned}
&-C_G + \frac{\rho}{2}U^2 - \frac{\rho}{2}(u_\eta)^2_{\xi_0} + \rho g h - 2\mu\left(\frac{1}{J}\frac{\partial u_\xi}{\partial \xi} - u_\eta\frac{c^2}{J^3}\cos\eta\,\sin\eta\right)_{\xi_0}\\
&= -\sigma\frac{2\sinh\xi_0}{c\cosh^2\xi_0}\left(1 + \frac{\eta^2}{\cosh^2\xi_0}\right).
\end{aligned} \tag{8.2.30}$$

Thus we have C_G in $O(1)$,

$$C_G = \frac{\rho}{2}U^2 + \frac{2\mu U f_2(\xi_0)}{c\cosh^2\xi_0} + \sigma\frac{2\sinh\xi_0}{c\cosh^2\xi_0}, \tag{8.2.31}$$

and the following relation in $O(\eta^2)$,

$$\begin{aligned}
&-\frac{\rho}{2}\frac{U^2 f_1^2(\xi_0)}{\cosh^2\xi_0} + \frac{\rho g c}{2}\sinh\xi_0 - \frac{2\mu U}{c\cosh^2\xi_0}\left[f_2(\xi_0)\left(\frac{1}{2} - \frac{1}{\cosh^2\xi_0}\right) - \frac{f_1(\xi_0)}{\cosh^2\xi_0}\right]\\
&= -\sigma\frac{2\sinh\xi_0}{c\cosh^4\xi_0}.
\end{aligned} \tag{8.2.32}$$

8.2.3 Dimensionless rise velocity

By taking the major axis $D = 2a$ as a representative length scale, $\sqrt{gD}$ as a: velocity scale, and $D/\sqrt{gD}$ as a time scale, we find the parameters involved in the expanded solution of dimensionless form:

Froude number, $Fr = \dfrac{U}{\sqrt{gD}}$; gravity Reynolds number, $Re_G = \dfrac{\sqrt{gD^3}}{\nu}$;

Eötvös number, $E\ddot{o} = \dfrac{\rho g D^2}{\sigma}$; aspect ratio, $e = \dfrac{c}{a} = \dfrac{1}{\cosh\xi_0}$.

In terms of these, the formula for the rise velocity of the ovary ellipsoid (which is now denoted by Fr) is given by (8.2.33) and that of the planetary ellipsoid is given by (8.2.34):

$$-Fr^2e^2 f_1^2(e) + \frac{1}{2}(1-e^2) - \frac{8Fr}{Re_G}e\left[f_2(e)\left(\frac{1}{2}+\frac{e^2}{1-e^2}\right) + \frac{e^2 f_1(e)}{1-e^2}\right] = \frac{8}{Eö}\frac{e^2}{1-e^2}, \tag{8.2.33}$$

$$-Fr^2e^2 f_1^2(e) + \frac{1}{2}\sqrt{1-e^2} - \frac{8Fr}{Re_G}e\left[f_2(e)\left(\frac{1}{2}-e^2\right) - e^2 f_1(e)\right] = -\frac{8}{Eö}e^2\sqrt{1-e^2}. \tag{8.2.34}$$

In these equations, the first term on the left-hand side denotes the kinetic energy that is due to the inertia (the pressure), the second is the gravity potential, and the third is the normal viscous stress; the right-hand side denotes the surface tension. The quadratic equations in Fr, (8.2.33) and (8.2.34), lead to the formula of the spherical bubble in the limit of $e \to 0$ ($\xi_0 \to \infty$ with a fixed).

The aspect ratio (or shape parameter) e is to be selected for a best fit to the experiment of Viana *et al.* There is no way that formula (8.2.34) for the planetary ellipsoid can be made to fit the data; for example, the dependence on an increase of $Eö$ is such as to reduce the rise velocity whereas an increase, compatible with (8.2.33), is observed. We now confine our attention to formula (8.2.33).

Formula (8.2.33) goes to the following equation in the limit $Re_G \to \infty$:

$$Fr_\infty = \frac{1}{ef_1(e)}\sqrt{\frac{1}{2}(1-e^2) - \frac{8}{Eö}\frac{e^2}{1-e^2}}. \tag{8.2.35}$$

For small Re_G, formula (8.2.33) may be approximated by a linear equation in Fr/Re_G to give the solution

$$Fr = \frac{(1-e^2)^2 - 16e^2/Eö}{f_2(e)(1+e^2) + 2e^2 f_1(e)}\frac{Re_G}{8e}, \tag{8.2.36}$$

whence $Fr \to 0$ as $Re_G \to 0$.

When $Fr = 0$, (8.2.33) is reduced to

$$Eö = 16\frac{e^2}{(1-e^2)^2}. \tag{8.2.37}$$

If we put $Eö = 4$ as noted in subsection 8.1.4, we have $e = 0.41$; thus $0.41 \le e < 1$ for $4 \le Eö$, which means that the bubble may be an ovary ellipsoid. It is noted here that (8.2.37) gives $Fr_\infty = 0$ in (8.2.35) and $Fr = 0$ in (8.2.36), which leads to the condition that $Eö \ge 16e^2/(1-e^2)^2$ for a positive or zero solution Fr to quadratic equation (8.2.33).

In the limit $e \to 0$ equation (8.2.33) describes the rise velocity of a perturbed spherical cap bubble. To obtain this perturbation formula we note that

$$\begin{gathered}\frac{c}{a} = \frac{1}{\cosh\xi_0} = e, \quad \frac{b}{c} = \sinh\xi_0 = \frac{\sqrt{1-e^2}}{e} \\ ef_1(e) = \frac{1}{2}ef_2(e) \sim \frac{3}{2}\left(1 - \frac{1}{5}e^2 - \frac{8}{175}e^4 + \cdots +\right).\end{gathered} \tag{8.2.38}$$

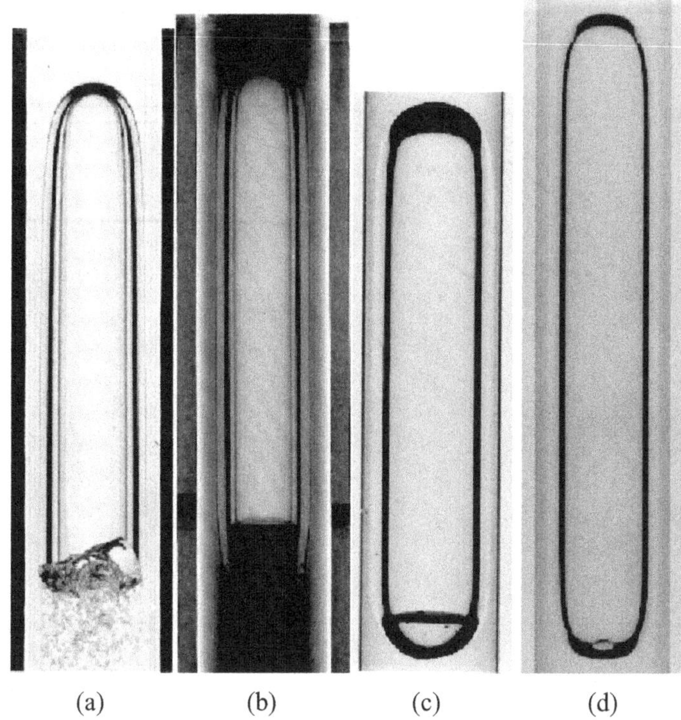

Figure 8.6. Photographs of Taylor bubbles rising through a 76.2-mm inside-diameter pipe filled with different viscosity liquids. This figure is taken from Viana *et al.* (2003).

After inserting these expressions into (8.2.33), retaining terms proportional to e^2, we find that

$$\frac{9}{4}Fr^2 + \frac{12Fr}{Re_G} - \frac{1}{2} = e^2\left\{\frac{8}{E\ddot{o}} + \frac{1}{2} - \frac{9}{10}Fr^2 + \frac{168}{5}\frac{Fr}{Re_G}\right\}. \tag{8.2.39}$$

The leading-order terms on the left-hand side were obtained by Joseph but the perturbation terms on the right-hand side are different. The curvature s in equation (8.1.4) is related to the aspect ratio by

$$s = -e^2/2\left(1 - e^2\right), \tag{8.2.40}$$

whence, to leading order, we find that $e^2 = -2s$.

8.3 Comparison of theory and experiment

Viana *et al.* (2003) made experiments of Taylor bubbles, as shown in figure 8.6. Their data may be expressed as functional relationships among three parameters, Fr, $E\ddot{o}$, and Re_G, as in equations (8.1.1)–(8.1.3).

In figure 8.7, we have plotted data Fr versus Re_G for 12 values of $E\ddot{o}$. It is important to note what was done with these data by Viana *et al.* (2003) and what we do with them here. Viana *et al.* (2003) identified a slope region for small $Re_G < 10$, a flat region for $Re_G > 200$ and a transition between. The flat region and the slope region give rise to power laws that

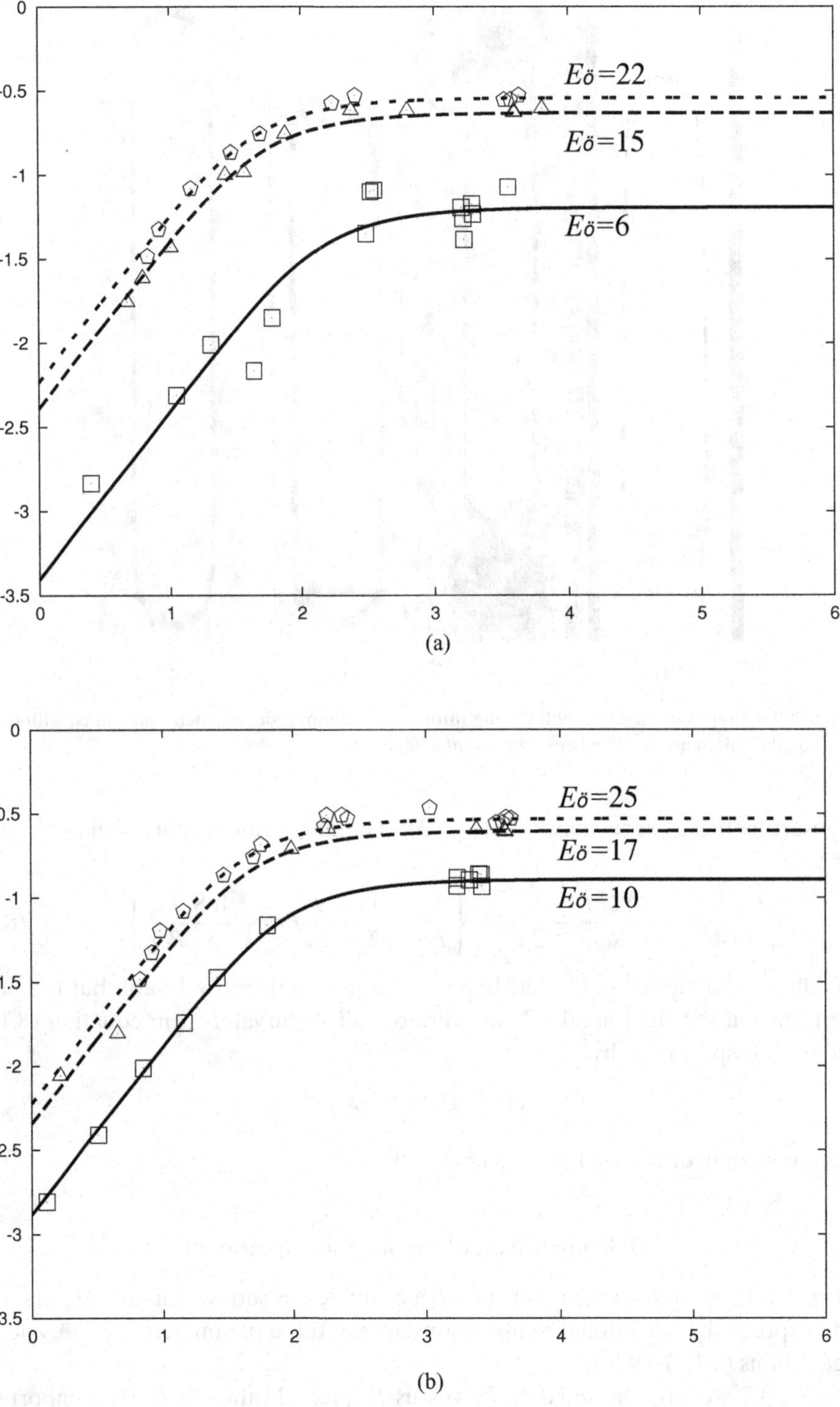

Figure 8.7. LogFr versus log Re_G for 12 values of $Eö$. The curves - - -, – – –, and —— are plots of (8.2.33) with $e(Eö)$ selected for best fit as described in table 8.1.

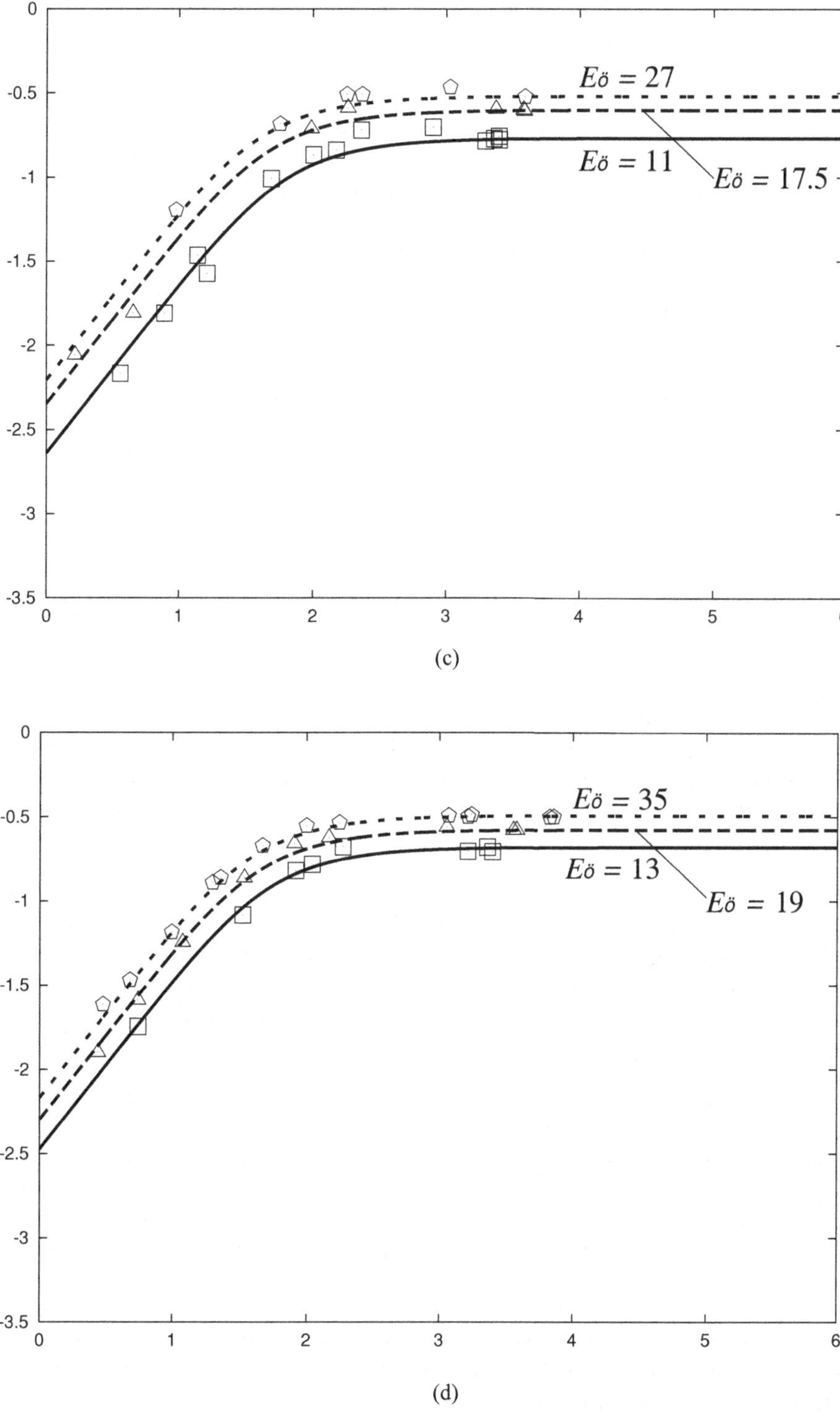

Figure 8.7 (*continued*)

Table 8.1. *Selection of e(Eö) for 12 cases of Eö*

The deviation is given by $\delta e = e - e_0$ with (8.3.1) and Fr_∞ is computed by (8.2.35)

Eö	*e*	δe	Fr_∞	$\log Fr_\infty$	Figure 8.7
6	0.471243	−0.042313	0.06391207	−1.194417	(a)
10	0.536543	−0.000228	0.128435	−0.891313	(b)
11	0.538667	−0.002551	0.170713	−0.767732	(c)
13	0.547626	−0.001474	0.209595	−0.678617	(d)
15	0.555870	−0.000073	0.233737	−0.631272	(a)
17	0.563567	0.001559	0.250282	−0.601570	(b)
17.5	0.567265	0.003855	0.250587	−0.601040	(c)
19	0.568630	0.003855	0.265555	−0.575845	(d)
22	0.571912	−0.002768	0.286857	−0.542334	(a)
25	0.580741	−0.000333	0.295449	−0.529518	(b)
27	0.583492	−0.001466	0.302880	−0.518728	(c)
35	0.593930	−0.004315	0.322398	−0.491607	(d)

were merged into the transition region by use of a logistic dose curve. This type of fitting was discussed briefly in §8.1 and extensively by Viana *et al.* (2003).

Here, in figure 8.7, we plot Fr versus Re_G data for 12 values of *Eö* but we fit these data with analytic expression (8.2.33) for an ovary ellipsoid rather than to power laws. The aspect ratio $e(Eö)$ is a fitting parameter and is listed in table 8.1, in which $e_0 = e(Eö)$ is the value of e in (8.2.33) selected as the value that most closely fit the data of Viana *et al.* (2003) for the 12 cases in figure 8.7. This fitting also gives the value Fr_∞ in (8.2.35), and it leads to the correlation

$$e_0 = Eö^{0.0866}/0.357, \tag{8.3.1}$$

where the deviation in given by $\delta e = e - e_0$. The success of this procedure is impressive.

The processing of data for table 8.1 is represented graphically in figure 8.8.

Table 8.1 shows that e_0 is a very weak function of $Eö \geq 15$ with data tending to a value $e_0 = 0.6$ corresponding to an ellipsoid with

$$b \cong 0.8a, \tag{8.3.2}$$

which, like the true Taylor bubble, has an almost spherical cap (see figure 8.9).

8.4 Comparison of theory and correlations

The processing of data on the rise velocity of Taylor bubbles in tubes of stagnant liquids by Viana *et al.* (2003) leads to correlation formula (8.1.3) with small errors described by figure 8.1. We may then propose that the rise velocity is accurately described by (8.1.3); however, the shape of the bubble nose is not predicted.

The theory of the rise of ovary ellipsoidal gas bubbles in viscous liquids leads to a rigorous prediction of the aspect ratio of the ovary ellipsoid that rises with exactly the same velocity Fr at given values of *Eö* and Re_G as the Taylor bubble. The value of e is

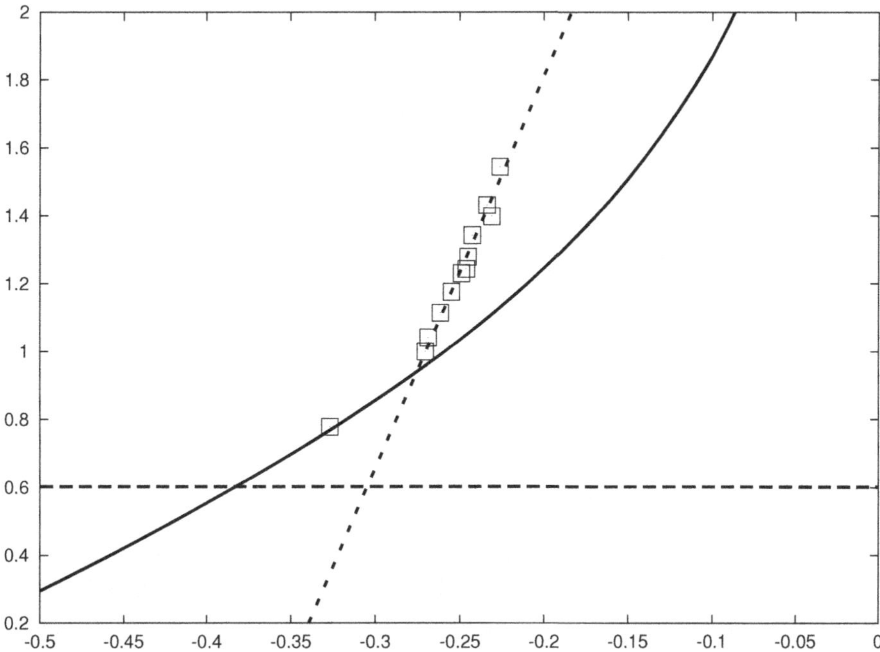

Figure 8.8. Log $E\ddot{o}$ versus $\log e$; □ denotes the data given in table 8.1. The solid curve denotes the border $Fr = 0$ given by (8.2.37), above which one positive solution of Fr may exist and below which there arise two negative solutions or complex solutions that are meaningless. The dashed line is for $E\ddot{o} = 4$. The dotted line denotes $\log e = 0.0865513 \log E\ddot{o} - 0.356762$ for which $e_0 = E\ddot{o}^{0.0866}/0.357$.

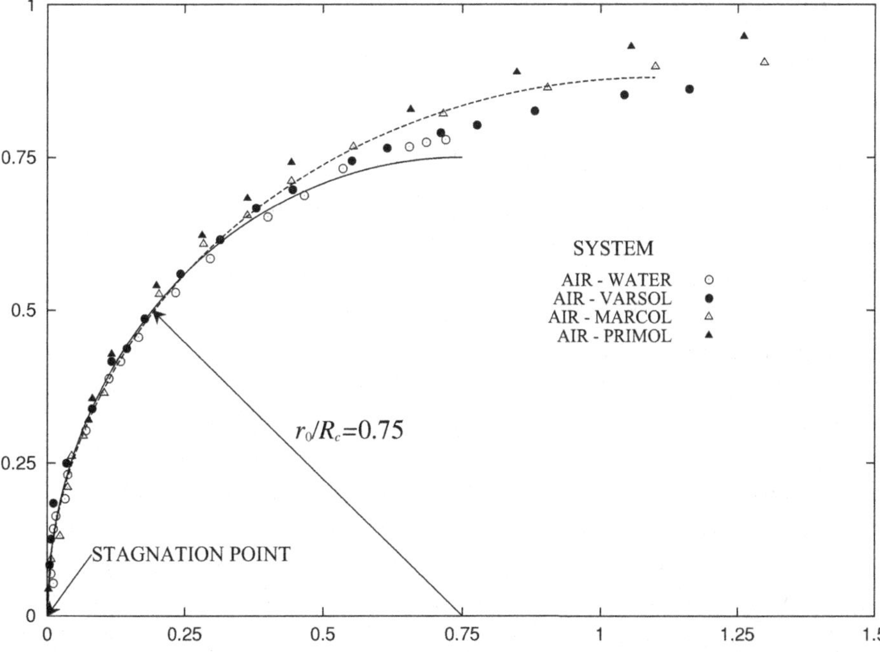

Figure 8.9. Comparison of Brown's (1965) measurements of the shape of a large Taylor bubble rising in a round tube with an ovary ellipsoid (denoted by the dashed curve) with $e = c/a = 0.6$, $b = a\sqrt{1-e^2} = 0.8a$, $a = 1.10$. Ovary ellipsoids with smaller e are more spherical.

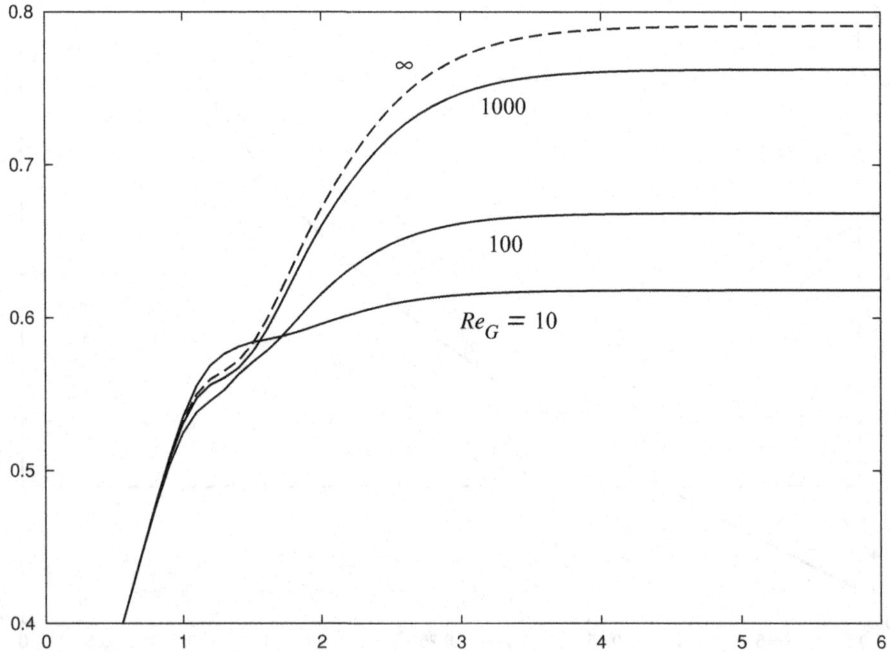

Figure 8.10. $e(Eö, Re_G)$ versus log $Eö$ for various values of Re_G. The curves —— are obtained by simultaneous equations (8.1.3) and (8.2.33). The solution of (8.4.1) and (8.4.2) is shown as - - - - -; all the solutions of (8.1.3) and (8.2.33) coincide with - - - - - when $Re_G > 10^4$.

determined by simultaneous solution of (8.2.33) and (8.1.3) for given values of Fr, Re_G, and $Eö$ (figure 8.10).

A slightly simpler solution can be written out for $Re_G > 200$ in which case (8.1.3) is replaced with the simpler formula

$$Fr = 0.34/\left(1 + 3805/Eö^{3.06}\right)^{0.58}, \tag{8.4.1}$$

which is to be solved simultaneously with (8.2.33) in the limit $Re_G \to \infty$:

$$\frac{-Fr^2e^6}{\left[e + (1 - e^2)\tanh^{-1}(e)\right]^2} + \frac{1}{2}\left(1 - e^2\right) = \frac{8}{Eö}\frac{e^2}{1 - e^2}, \tag{8.4.2}$$

for $e(Eö, Re_G)$ for given values of $Eö$ (figure 8.10).

8.5 Conclusion

A formula is derived that gives the rise velocity of an ellipsoidal gas bubble in a viscous liquid assuming that the motion of the liquid is irrotational. The rise velocity is expressed by a Froude number and it is determined by a Reynolds number, an Eötvös number, and the aspect ratio of the ellipsoid. The formula for the ovary ellipsoid was fit to the data of Viana *et al.* (2003), who correlated all the published data on the rise velocity of long gas bubbles in round tubes filled with viscous liquids. These data are accurately represented by our formula when the aspect ratio takes on certain values. The fitting generates a family of aspect ratios that depends strongly on the Eötvös number and less strongly on the Reynolds number; this shows that the change in the shape of the nose of the rising bubble is strongly influenced by surface tension. Our analysis completely neglects sidewall drainage induced by the rising bubble and cannot be a precise description of the dynamics. We have generated what might be called the ovary ellipsoid model of a Taylor bubble. This model is very simple and astonishingly accurate.

9

Rayleigh–Taylor instability of viscous fluids

Rayleigh–Taylor (hereafter RT) instabilities are very important and very pervasive. They are driven by acceleration when a liquid accelerates away from a gas or vacuum. The signature of this instability is the waves that corrugate the free surface at the instant of acceleration. Ultimately these waves will finger into the liquid, causing it to break up.

Rayleigh (1890) showed that a heavy fluid over a light fluid is unstable, as common experience dictates. He treated the stability of a heavy fluid over a light fluid without viscosity, and he found that a disturbance of the flat free surface grows exponentially as $\exp(nt)$, where

$$n = \left[\frac{kg\,(\rho_2 - \rho_1)}{\rho_1 + \rho_2}\right]^{1/2}, \tag{9.0.1}$$

where ρ_2 is the density of the heavy fluid, ρ_1 is the density of the light fluid, g is the acceleration that is due to gravity, and $k = 2\pi/\ell$ is the wavenumber, where ℓ is the wavelength. The instability described by (9.0.1) is catastrophic because the growth rate n tends to infinity, at any fixed time, no matter how small, as the wavelength tends to zero. The solutions are unstable to short waves even at the earliest times. Such disastrous instabilities are called "Hadamard unstable" and the initial-value problems associated with these instabilities are said to be "ill-posed" (Joseph and Saut, 1990). Ill-posed problems are disasters for numerical simulations: Because they are unstable to ever-shorter waves, the finer the mesh, the worse the result.

Nature will not allow such a singular instability; for example, neglected effects like viscosity and surface tension will enter the physics strongly at the shortest wavelength. These effects have been taken into account in the study of RT instability by Harrison (1908) and in the treatise of Chandrasekhar (1961). Surface tension eliminates the instability of the short waves; there is a finite wavelength depending strongly on viscosity as well as on surface tension for which the growth rate n is maximum. This is the wavelength that should occur in a real physical problem and would determine the wavelength on the corrugated fronts of breaking drops in a high-speed airflow.

Taylor (1950) extended Rayleigh's inviscid analysis to the case in which a constant acceleration of superposed fluids is taken into account. Assuming a constant value for the acceleration, Taylor (1950) showed that, when two superposed fluids of different densities are accelerated in a direction perpendicular to their interface, this surface is unstable if the acceleration is directed from the lighter to the heavier fluid. Taylor instability depends strongly on the value of the acceleration a; for example, if g in (9.0.1) is replaced with $a = 10^4 g$, the growth rate n is increased by a factor of 100.

9.1 Acceleration

The acceleration of the drop is a major factor in RT instability. It is instructive to see how the acceleration enters into the equations of motion. Suppose the laboratory frame is identified with $(\mathbf{X}, \hat{t})$ and the drop velocity is $\mathbf{v}(\mathbf{X}, \hat{t})$. Then we refer the equations of motion,

$$\rho\left(\frac{\partial \mathbf{v}}{\partial \hat{t}} + \mathbf{v} \cdot \frac{\partial \mathbf{v}}{\partial \mathbf{X}}\right) = \operatorname{div} \mathbf{T} + \rho \mathbf{g}, \quad \frac{\partial}{\partial \mathbf{X}} \cdot \mathbf{v} = 0, \tag{9.1.1}$$

to an accelerating frame in which the mass center of the drop is stationary, identified with $(\mathbf{x}, t)$ and $\mathbf{v}(\mathbf{X}, \hat{t}) = \mathbf{U}(\mathbf{x}, t) + \mathbf{V}(t)$, $\mathrm{d}\mathbf{X} = \mathrm{d}\mathbf{x} + \mathbf{V}(t)\mathrm{d}t$, $\mathrm{d}\hat{t} = \mathrm{d}t$. Then we find that

$$\rho\left(\frac{\partial \mathbf{U}}{\partial t} + \mathbf{U} \cdot \frac{\partial \mathbf{U}}{\partial \mathbf{x}}\right) + \rho \dot{\mathbf{V}} = \operatorname{div} \mathbf{T} + \rho \mathbf{g}, \quad \frac{\partial}{\partial \mathbf{x}} \cdot \mathbf{U} = 0, \tag{9.1.2}$$

where $\mathbf{T}$ is the stress tensor.

The term

$$\rho\left(\mathbf{g} - \dot{\mathbf{V}}\right) \tag{9.1.3}$$

enters into RT instability and $\dot{\mathbf{V}}$ dominates in the drop breakup problem because the initial velocity is very small and the initial acceleration is very large.

The RT instability occurs when $\dot{\mathbf{V}}$ is directed from the light to the heavy fluid as when the initially stationary drop is accelerated to the free-stream velocity in the high-speed airstream behind the shock in the shock tube or when a moving drop in a stagnant fluid is decelerated by the air to zero velocity. The analysis works well in air and liquid, where the density and viscosity of air can be neglected with only small error; there is RT instability in a vacuum because it is the drop acceleration term and not the material properties of air that induces the instability.

In the present application the acceleration of the drop is roughly proportional to the dynamic pressure that vanishes in a vacuum. Drop breakup in a vacuum could occur by acceleration that is due to gravity, say, in a rarified Jovian atmosphere.

9.2 Simple thought experiments

These experiments are embodied in the drawings and caption of figure 9.1. These experiments show the difference between the theory of Rayleigh (1890) who considered gravity g and Taylor (1950) who considered the effect of the acceleration $\dot{V}$. The accelerations destabilize the liquid–gas surfaces that accelerate away from the gas and stabilize the liquid–gas surfaces that accelerate toward the gas [figure 9.1(e)].

9.3 Analysis

The linear theory of stability superposes small disturbances on a basic flow solution of the equations of motion. Nonlinear terms in small-perturbation quantities are neglected.

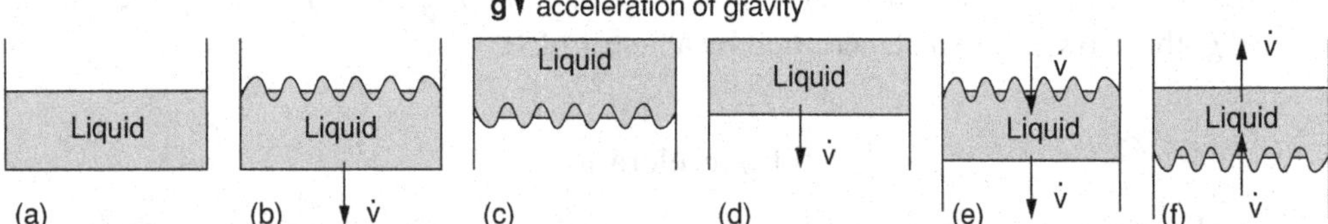

Figure 9.1. RT instability: (a) the liquid in the containers at rest is stable under gravity (Rayleigh, 1890) but if the container is turned upside down as in (c) the liquid falls out. The liquid at rest in container can be destabilized by downward acceleration of the liquid $a > g$ away from gas as in (b) and in the upside-down case the liquid can be prevented from falling out by accelerating downward with $\dot{V} > g$ as in (d). If we open up the container and accelerate the liquid downward with $\dot{V} > g$, the top surface that accelerates away from the gas is unstable but the bottom surface that accelerates into the gas is stabilized as in (e), with the opposite effect when the acceleration is reversed as in (f).

The properties of the basic flow are important. The basic flow for RT instability deserves consideration. Here we look for conditions under which corrugations appear on the flat free surface on a liquid moving into or away from a gas or vacuum. What is important here is the acceleration and the fact that shear stresses do not develop in the basic flow. This is, of course, even clearer if you take the gas away and leave a vacuum. In this chapter and in chapter 21 on RT instability of viscoelastic drops, we are obliged to consider RT instability on the front face of a flattened drops in which KH instability driven by shearing motion may play a minor role.

Lewis (1950) has shown experimentally that a layer of water driven by compressed air is unstable and that small disturbances, introduced at the air–water interface, grow so that it appears as though fingers of air move through the water layer. Lewis (1950) finds agreement with his experiments when the viscosity of the liquid is small. The peaks of water in his experiments are sharp, but the match between Taylor's theory and the Lewis experiment is not perfect because the regularizing effects of surface tension were neglected.

The breakup of water drops in high-speed air behind a shock wave in a shock tube was studied in excellent experiments by Engel (1958). Because the acceleration is perpendicular to the air–liquid interface and directed from gas to liquid, the accelerating liquid drop is unstable and is prey to the characteristic short-wave corrugation associated with this instability.

Joseph, Belanger, and Beavers (1999) (hereafter JBB) studied the breakup of liquid drops of high and low viscosity and viscoelastic drop in high air behind shock waves moving at Mach numbers $M_s = 2$ and $M_s = 3$. They measured acceleration and wavelengths on the corrugation drops.

The corrugations at the front of an unstable drop are driven toward the drop equator by shear flow of gas coming from the high-pressure stagnation point. This shear flow may also be subject to an instability of the KH type. Because the tangential velocity is zero at the stagnation point and small near the stagnation point, the KH instability may not interact too strongly with the RT instability.

Some details of the corrugations on the front of breaking drops can be predicted by mathematical analysis of the foregoing instabilities, but the effects of viscosity and surface tension must be taken into account.

9.3.1 Linear theory of Chandrasekhar

A thorough explanation and analysis of RT instability can be found in chapter X of the treatise by Chandrasekhar. Some effects of viscosity in special problems in RT instability have been considered by Plesset and Whipple (1974). In Chandrasekhar's analysis the two fluids are separated by an interface $z = h(x, y, t)$ that perturbs the plane $z = 0$. The Navier–Stokes equations for the perturbation velocity $\mathbf{u}$ and perturbation pressure p are solved above and below the plane $z = 0$. Gravity points down against the direction of z increasing. It is assumed that the heavy fluid is above the light fluid or that the light fluid is accelerated into the heavy fluid above (the drop is the heavy fluid and the gas flowing into and around it is the light fluid). At the flat interface at $z = 0$, the linearized kinematic equation is

$$w = \frac{\partial h}{\partial t}, \tag{9.3.1}$$

where

$$[|w|] = w_2 - w_1 \tag{9.3.2}$$

is the jump in the normal (z component) of velocity (w_2 is in $z > 0$, w_1 in $z < 0$). For the viscous case the jump in the tangential velocity is also zero:

$$[|w|] = [|u|] = [|v|] = 0, \tag{9.3.3}$$

where u is in the x direction and v is in the y direction. Because (9.3.3) holds for all x and y for $z = 0$, we may deduce that

$$\left[\left|\frac{\partial u}{\partial x} + \frac{\partial v}{\partial y}\right|\right] = -\left[\left|\frac{\partial w}{\partial z}\right|\right] = 0. \tag{9.3.4}$$

The continuity of the shear stress is expressed by

$$\left[\left|\mu\left(\frac{\partial u}{\partial z} + \frac{\partial w}{\partial x}\right)\right|\right] = 0, \quad \left[\left|\mu\left(\frac{\partial v}{\partial z} + \frac{\partial w}{\partial y}\right)\right|\right] = 0, \tag{9.3.5}$$

where μ is the viscosity.

The balance of normal stress is given by

$$-[|p|] + 2\left[\left|\mu\frac{\partial w}{\partial z}\right|\right] = -\gamma\left(\frac{\partial^2}{\partial x^2} + \frac{\partial^2}{\partial y^2}\right)h, \tag{9.3.6}$$

where γ is the surface tension. We may remove the hydrostatic pressure from the z equation of motion by introducing the dynamic pressure

$$\pi = p + \rho g h;$$

hence,

$$[|\pi|] = gh\,[|\rho|] + 2\left[\left|\mu\frac{\partial w}{\partial z}\right|\right] + \gamma\left(\frac{\partial^2}{\partial x^2} + \frac{\partial^2}{\partial y^2}\right)h. \tag{9.3.7}$$

After introducing normal modes proportional to

$$f(z)e^{nt+ik_x x+ik_y y}, \quad k^2 = k_x^2 + k_y^2, \tag{9.3.8}$$

where $f(z)$ is an amplitude function for π, u, v, and h that may be eliminated to find equation (40) on page 433 of Chandrasekhar (1961), we find

$$\begin{aligned}&\left[\left|\left\{\left[\rho-\frac{\mu}{n}\left(\frac{\mathrm{d}^2}{\mathrm{d}z^2}-k^2\right)\right]\frac{\mathrm{d}w}{\mathrm{d}z}-\frac{1}{n}\left[\left(\frac{\mathrm{d}^2}{\mathrm{d}z^2}+k^2\right)w\right]\frac{\mathrm{d}\mu}{\mathrm{d}z}\right\}\right|\right]\\ &=-\frac{k^2}{n^2}\left(g\,[|\rho|]-k^2\gamma\right)w-\frac{2k^2}{n}\left[\left|\mu\frac{\mathrm{d}w}{\mathrm{d}z}\right|\right].\end{aligned} \tag{9.3.9}$$

For the present problem for which the unperturbed states above and below $z=0$ are uniform $\mathrm{d}\mu/\mathrm{d}z=0$ and using (9.3.4), we find

$$\left[\left|\mu\frac{\mathrm{d}w}{\mathrm{d}z}\right|\right]=[|\mu|]\,\frac{\mathrm{d}w}{\mathrm{d}z}. \tag{9.3.10}$$

The entire problem is resolved by

$$\begin{aligned}w_1&=A_1e^{kz}+B_1e^{q_1z}\ (z<0),\\ w_2&=A_2e^{-kz}+B_2e^{-q_2z}\ (z>0),\end{aligned} \tag{9.3.11}$$

where

$$\begin{aligned}&q_1=\sqrt{k^2+n/\nu_1},\quad q_2=\sqrt{k^2+n/\nu_2},\\ &\nu_1=\mu_1/\rho_1,\quad \nu_2=\mu_2/\rho_2,\end{aligned} \tag{9.3.12}$$

provided that Chandrasekhar's equation (113) is satisfied:

$$\begin{aligned}&-\left\{\frac{gk}{n^2}\left[(\alpha_1-\alpha_2)+\frac{k^2\gamma}{g\,(\rho_1+\rho_2)}\right]+1\right\}(\alpha_2q_1+\alpha_1q_2-k)-4k\alpha_1\alpha_2+\frac{4k^2}{n}(\alpha_1\nu_1-\alpha_2\nu_2)\\ &\times\left\{(\alpha_2q_1-\alpha_1q_2)+k(\alpha_1-\alpha_2)\right\}+\frac{4k^3}{n^2}(\alpha_1\nu_1-\alpha_2\nu_2)^2(q_1-k)(q_2-k)=0,\end{aligned} \tag{9.3.13}$$

where

$$\alpha_1=\frac{\rho_1}{\rho_1+\rho_2},\quad \alpha_2=\frac{\rho_2}{\rho_1+\rho_2},\quad \alpha_1+\alpha_2=1.$$

When the light fluid is gas, with only a very small error, we may treat this as a vacuum. Hence we put

$$\begin{aligned}&\rho_1=\alpha_1=0,\quad \alpha_2=1,\\ &q_1\to k,\ \ \alpha_2q_1+\alpha_1q_2-k\to q_1-k,\\ &(\alpha_2q_1-\alpha_1q_2)+k(\alpha_1-\alpha_2)\to q_1-k,\end{aligned} \tag{9.3.14}$$

and, after factoring q_1-k in (9.3.13), we get

$$1-\frac{gk}{n^2}=-\frac{k^3\gamma}{n^2\rho_2}-\frac{4k^2}{n}\frac{\mu_2}{\rho_2}+\frac{4k^3}{n^2}\frac{\mu_2^2}{\rho_2^2}(q_2-k). \tag{9.3.15}$$

9.3.2 Viscous potential flow

RT instability may also be analyzed as a VPF (JBB, 1999). In this approach the perturbation velocity is given by a potential

$$\mathbf{u}=\nabla\phi, \tag{9.3.16}$$

and the Navier–Stokes equations are reduced to an identity, provided that the pressure is given by Bernoulli's equation,

$$p+\rho\frac{\partial\phi}{\partial t}+\rho g z=-\frac{\rho}{2}\,|\nabla\phi|^2. \tag{9.3.17}$$

In this linearization used to study instability the term on the right-hand side of (9.3.17) is put to zero. The interface condition requires continuity of tangential velocities, (9.3.3) and (9.3.4), and the continuity of shear stress (9.3.5) cannot be enforced in VPF. The pressure in the normal stress condition (9.3.6) is eliminated by use of (9.3.17); this gives rise to

$$\left[\left|\rho\frac{\partial\phi}{\partial t}\right|\right]+gh[|\rho|]+2\left[\left|\mu\frac{\partial w}{\partial z}\right|\right]+\gamma\left(\frac{\partial^2}{\partial x^2}+\frac{\partial^2}{\partial y^2}\right)h=0. \tag{9.3.18}$$

After introducing normal modes, writing (9.3.1) as $nh=w$, we get

$$n[|\rho\phi|]+\frac{w}{n}[|\rho|]g+2\left[\left|\mu\frac{\partial w}{\partial z}\right|\right]-\frac{\gamma k^2}{n}w=0. \tag{9.3.19}$$

For VPF, $\nabla^2\phi=0$ everywhere, (9.3.11) holds with $B_1=B_2=0$, and

$$\frac{\mathrm{d}w}{\mathrm{d}z}=\frac{\mathrm{d}^2\phi}{\mathrm{d}z^2}=k^2\phi. \tag{9.3.20}$$

After writing $[|\rho\phi|]=(1/k^2)[|\rho\,\mathrm{d}w/\mathrm{d}z|]$ in (9.3.19) we may verify that (9.3.9) and (9.3.19) are identical when (9.3.9) is evaluated for VPF. Further evaluation of (9.3.19) with normal modes gives rise to the dispersion relation

$$1=\frac{k}{n^2}\frac{\rho_2-\rho_1}{\rho_2+\rho_1}g-\frac{k^3\gamma}{n^2(\rho_2+\rho_1)}-\frac{2k^2}{n}\frac{\mu_2+\mu_1}{\rho_2+\rho_1}. \tag{9.3.21}$$

Equation (9.3.21) may be solved for the growth rate:

$$n=-k^2\frac{\mu_2+\mu_1}{\rho_2+\rho_1}\pm\left[k\frac{\rho_2-\rho_1}{\rho_2+\rho_1}g-\frac{k^3\gamma}{\rho_2+\rho_1}+k^4\left(\frac{\mu_2+\mu_1}{\rho_2+\rho_1}\right)^2\right]^{1/2}. \tag{9.3.22}$$

If the quantity under the root is negative, then the real part is negative and the interface is stable. For instability it is sufficient that

$$\sqrt{k\frac{\rho_2-\rho_1}{\rho_2+\rho_1}g+k^4\left(\frac{\mu_2+\mu_1}{\rho_2+\rho_1}\right)^2-\frac{k^3\gamma}{\rho_2+\rho_1}}>k^2\frac{\mu_2+\mu_1}{\rho_2+\rho_1}. \tag{9.3.23}$$

The border of instability $n=0$ for $k=k_c$ is given by

$$k_c=\sqrt{\frac{(\rho_2-\rho_1)g}{\gamma}}, \tag{9.3.24}$$

independent of viscosity.

Evaluation of the growth-rate formula with the gas neglected gives

$$n=-\frac{k^2\mu_2}{\rho_2}\pm\sqrt{kg-\frac{k^3\gamma}{\rho_2}+\frac{k^4\mu_2^2}{\rho_2^2}}. \tag{9.3.25}$$

It is of interest to compare growth-rate formula (9.3.25) for VPF with growth-rate formula (9.3.15) for Chandrasekhar's exact solution. These two formulas give exactly the same growth rate when

$$q_2 - k = k\sqrt{1 + \frac{n\rho_2}{\mu_2 k^2}} - k \cong \frac{n\rho_2}{2\mu_2 k}, \tag{9.3.26}$$

$$\frac{n\rho_2}{2\mu_2 k} \ll 1. \tag{9.3.27}$$

9.4 Comparison of theory and experiments

To evaluate the formulas for RT instability for the experiments of JBB we must

$$\text{replace } g \text{ with } a = \dot{V}. \tag{9.4.1}$$

Equation (9.3.24) then implies that

$$k_c = \frac{2\pi}{\lambda_c} = \sqrt{\frac{\rho_2 a}{\gamma}} \tag{9.4.2}$$

and waves with length

$$\lambda < \lambda_c = 2\pi\sqrt{\frac{\gamma}{\rho_2 a}} \tag{9.4.3}$$

are stable. Taking liberties with (9.4.3), we identify $\lambda = D$, where D is the diameter of drops with acceleration a that cannot be fragmented by RT instability. Applying this to our experiments (table 9.1) we get D between

$$23 \text{ and } 65\ \mu\text{m for } M_S = 3, \tag{9.4.4}$$

$$46 \text{ and } 135\ \mu\text{m for } M_S = 2. \tag{9.4.5}$$

A typical growth-rate curve for the parameters in (9.4.4) and in (9.4.5) is shown in figure 9.2.

9.5 Comparison of the stability theory with the experiments on drop breakup

JBB studied the breakup of liquid drops of high and low viscosity behind a shock wave in a shock tube moving with shock Mach number $M_s = 2$ and $M_s = 3$. The data for the experiments are listed in table 2 of JBB together with values of the dimensionless parameters. The Ohnesorge number Oh, the Weber number We, and the Reynolds number Re,

$$Oh = \frac{\mu_d}{(\rho_d D\gamma)^{1/2}}, \quad We = \frac{\rho V^2 D}{\gamma}, \quad Re = \frac{V D\rho}{\mu}, \tag{9.5.1}$$

are defined in terms of the initial drop diameter D, drop viscosity μ_d, surface tension γ, drop density ρ_d, and the free-stream values of the velocity V, viscosity μ, and density ρ.

The breakup of drops in a high-speed airstream behind a shock wave in a shock tube was photographed with a rotating-drum camera, giving one photograph every 5 μsec. From these photographs, JBB created movies of the fragmentation history of viscous drops of widely varying viscosity, and viscoelastic drops, at very high Weber and Reynolds numbers. Drops of the order of 1 mm are reduced to droplet clouds and possibly to vapor

Table 9.1. *Values of the wavenumber, wavelength, and growth rate of the most dangerous wave for the experimental conditions given in tables 1 and 2 of JBB*

Liquid	Viscosity ($kg\,m^{-1}\,sec^{-1}$)	Exact			Viscous potential		
		n (sec^{-1})	k (cm^{-1})	λ (mm)	n (sec^{-1})	k (cm^{-1})	λ (mm)
$M_s = 3$							
SO 10000	10	17790	9.5	6.61	19342	9.7	6.48
SO 6000	6	24673	14.45	4.35	26827	14.7	4.27
SO 5000	5	24787	15.86	3.96	26951	16.15	3.89
SO 4000	4	32550	20.3	3.10	35312	20.7	3.04
SO 3000	3	31507	23.08	2.72	34257	23.5	2.67
SO 1000	1	48769	49.68	1.26	53088	50.65	1.24
SO 100	0.1	132198	253.2	0.25	143699	259	0.24
Glycerine	1.49	38760	41.3	1.52	42141	42.1	1.49
2% PO	35	10492	3.95	15.91	11406	3.95	15.91
2.6% PSBA	1.13	34460	37.8	1.66	37467	38.5	1.63
2% PAA	0.96	28927	39.4	1.59	31451	40.15	1.56
Water	0.001	149632	531.95	0.12	151758	540.8	0.12
$M_s = 2$							
SO 4000	4	9868	11.2	5.61	10729	11.4	5.51
SO 3000	3	14388	15.6	4.03	15644	15.9	3.95
SO 1000	1	20018	31.8	1.98	21765	32.4	1.94
SO 100	0.1	52726	158	0.40	57304	161.9	0.39
Glycerine	1.49	25046	33.2	1.89	27231	33.8	1.86
2% PO	35	3723	2.35	26.74	4048	2.4	26.18
2.6% PSBA	1.13	16070	25.75	2.44	17472	26.25	2.39
Water	0.001	50971	260.9	0.24	51507	264.2	0.24

SO, silicone oil; PO, polyox; PSBA, poly(styrene-butyl acrylate); PAA, polyacrylamide.

in times less than 500 μsec. The movies may be viewed at http://www.aem.umn.edu/research/Aerodynamic_Breakup. They reveal sequences of breakup events that were previously unavailable for study. Bag and bag-and-stamen breakup can be seen at very high Weber numbers, in the regime of breakup previously called "catastrophic." The movies

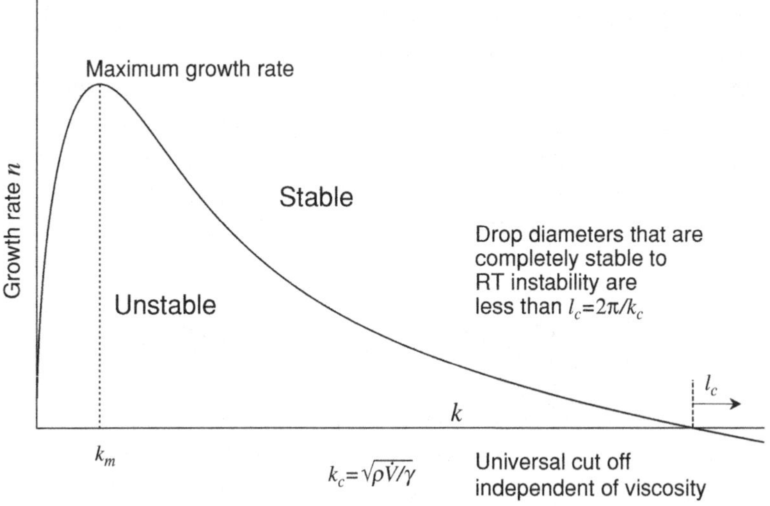

Figure 9.2. Growth-rate curve.

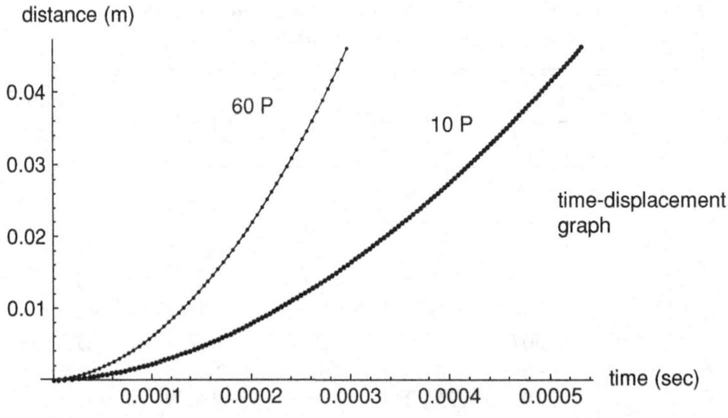

Liquid	Viscosity (Poise)	Initial accel. $M_S = 3.0$	(10^5m/sec^2) $M_S = 2.0$	Initial drop diameter (mm)
Silicone Oil	1	15.8	4.0	2.6
	10	11.12	2.92	2.6
	30	9.98	3.08	2.6
	40	12.06	2.02	2.5
	40	11.08	-	2.1
	50	8.99	-	2.5
	60	9.78	-	2.6
	100	7.73	-	2.6
2% Polyox	350	6.48	1.37	2.9
2% Polyacrylamide	9.6	4.92	-	3.2
2.6% PSBA / TBP	11.3	7.28	2.32	2.2
Glycerine	14.9	8.43	4.38	2.4
Water	0.01	6.47	1.52	2.5

Figure 9.3. Time-displacement graph giving the acceleration of drops from the experiment of JBB (1999). PSBA, poly(styrene-butyl acrylate); TBP, tri-*n*-butyl phosphate.

generate precise displacement-time graphs from which accurate values of acceleration (of orders of 10^4–10^5 times the acceleration of gravity) are computed. These large accelerations from gas to liquid put the flattened drops at high risk to RT instabilities. The most unstable RT wave fits nearly perfectly with waves measured on enhanced images of drops from the movies, but the effects of viscosity cannot be neglected.

The breakup of drops in a high-speed airstream exhibits universal features. First the shock wave passes over the drop. The shock wave does not have an important effect on the drop breakup. In the instant after the shock passes, the speed of the airstream steps up a high value. The drop flattens because the stagnation pressure, fore and aft, is very large before the boundary layer separates. At first, the flattened drop is stationary but it accelerates with huge accelerations larger than $10^4\, g$, from gas to liquid. Corrugations that are due to RT instability are seen on the front, but not on the back face. These corrugations are driven away from the high-pressure stagnation point on the front face to the equator, where they are stripped away from the drop. The waves on the front face can be said to be driven by a KH instability. Because the tangential velocity is zero at the stagnation point, the KH may not interact too strongly with the RT instability.

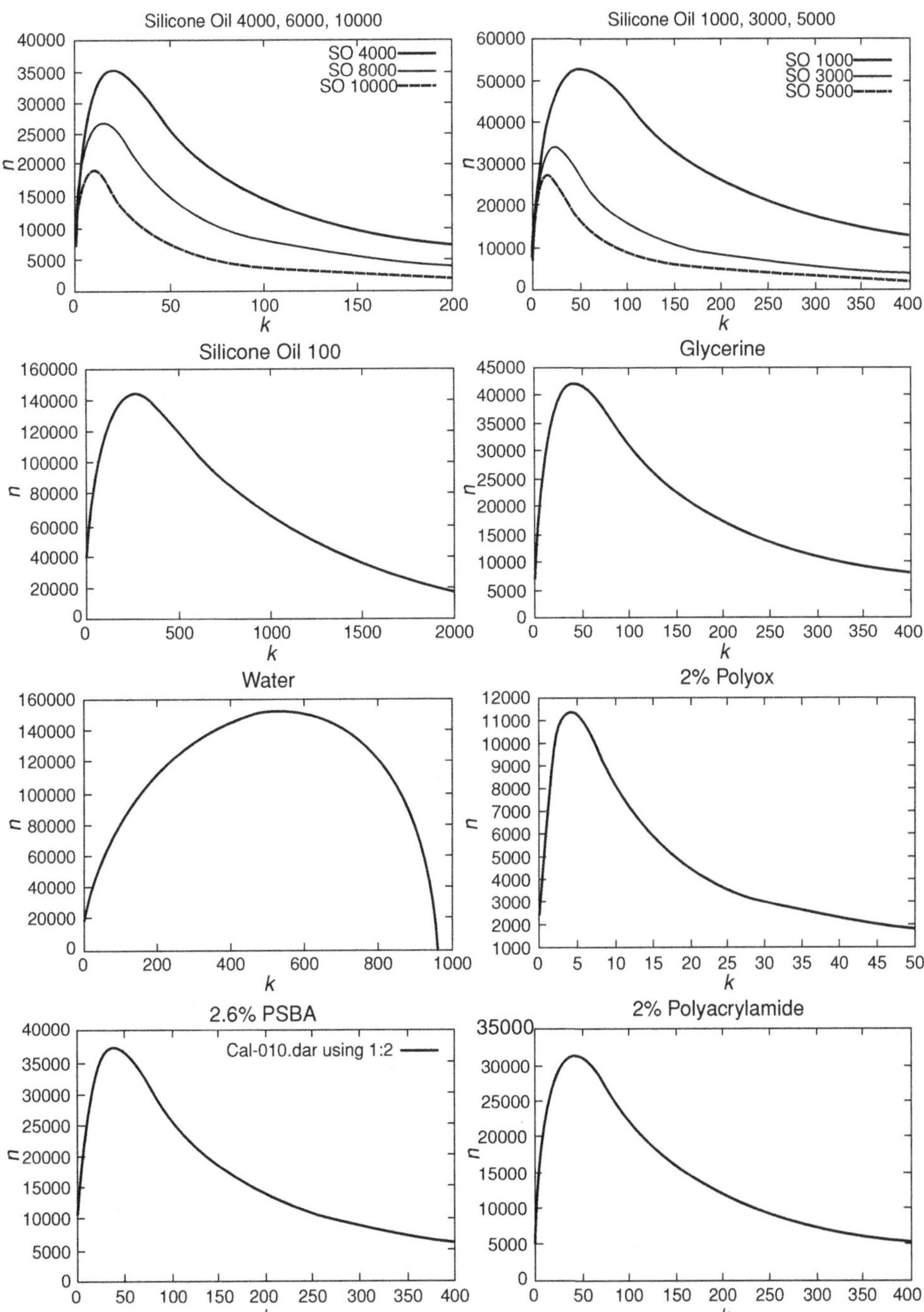

Figure 9.4. n (sec^{-1}) vs. k (cm^{-1}) for VPF; shock Mach number = 2.

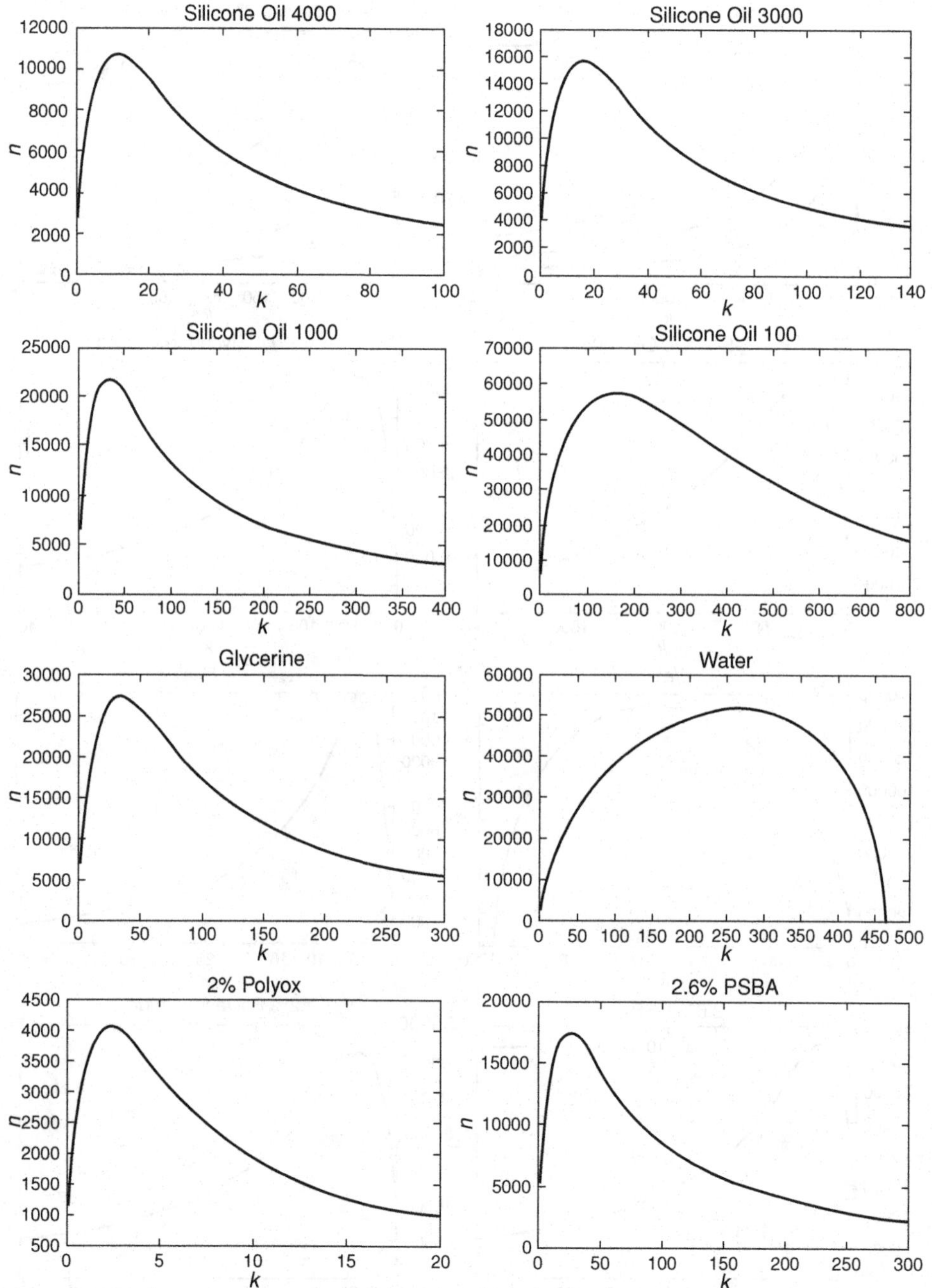

Figure 9.5. n (sec^{-1}) vs. k (cm^{-1}) for VPF; shock Mach number = 3.

A major achievement of the experimental study of JBB is the accurate measurement of acceleration by use of time-displacement data of the type shown in figure 9.3. More such time-displacement graphs can be found in JBB and in the paper on breakup of viscoelastic drops by Joseph, Beavers, and Funada (JBF) (2002).

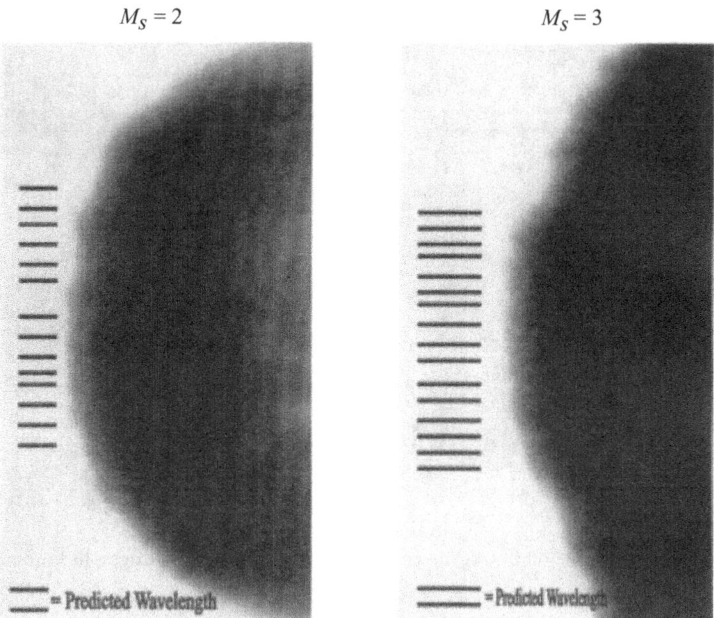

Figure 9.6. RT waves in water. The tick marks on the photographs locate wave troughs.

Growth-rate curves for the conditions specified in figure 9.3 are shown in figures 9.4 and 9.5. Values of the wavenumber, wavelength, and growth rate of the fastest-growing wave from figures 9.4 and 9.5 are given in table 9.1.

The parameters of the maximum growth rate computed by VPF are in excellent agreement with the exact solution.

9.6 Comparison of the measured wavelength of corrugations on the drop surface with the prediction of the stability theory

We have shown that the theoretical prediction about stability depends very strongly on viscosity and that this dependence is very well predicted by a purely irrotational theory. It remains to show that these predictions agree with experiments. JBB compared photos from their movies with the stability theory predicted in §9.3. No special choice was made in selecting the frames for the comparisons other than choosing ones from the early part of the motion that were well focused. The wavelike structure can be identified much more easily on the computer screen than in the images reproduced in figures 1–13 of JBB. In figure 9.6 the predicted and observed wavelengths on a water drop at $M_s = 2$ and 3 are compared. The tick marks on the photos identify wave troughs so that the predicted distance between the tick marks is the length of an unstable wave. The same kind of comparison is given in figure 9.7 for silicone oil and in figures 9.8 and 9.12 for glycerine. In figure 9.9 (which is figure 14 in JBF) we show that the analysis of RT waves can be enhanced by image processing.

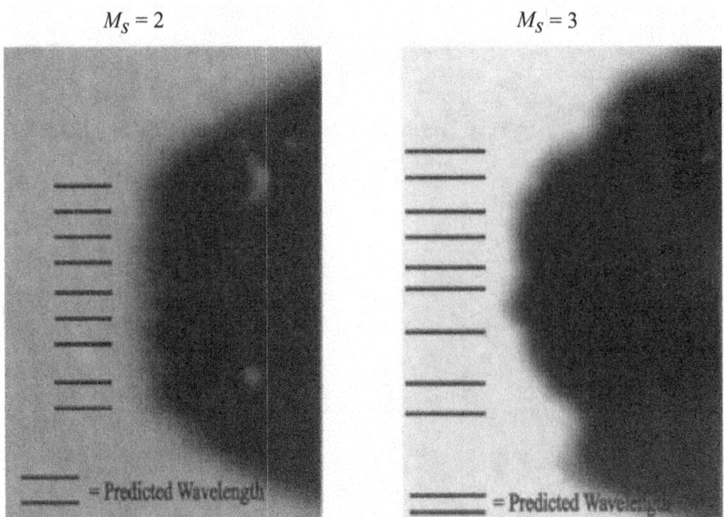

Figure 9.7. RT waves in silicone oil (0.1 kg/m sec). The tick marks on the photographs locate wave troughs.

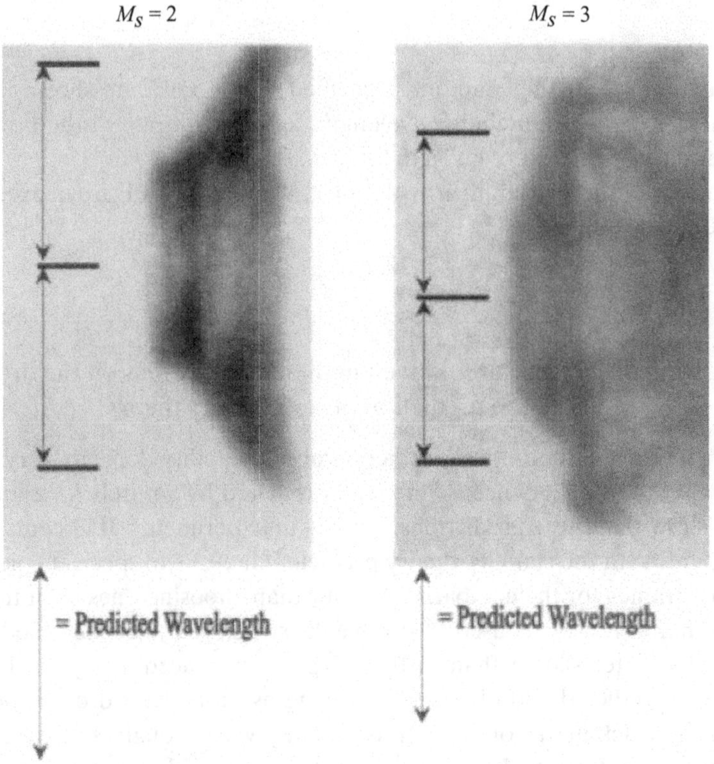

Figure 9.8. RT waves in glycerine. The tick marks on the photographs locate wave troughs.

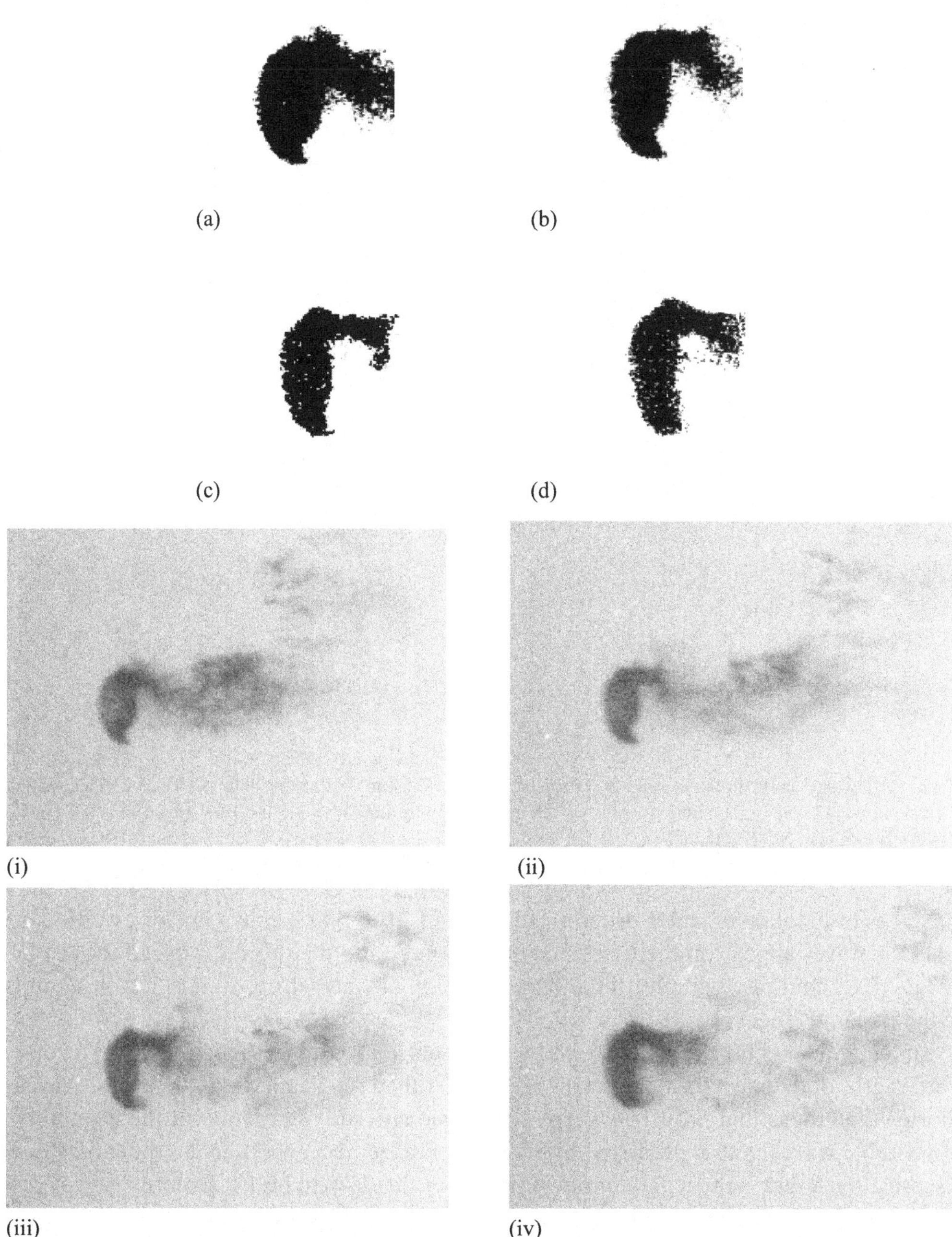

Figure 9.9. (JBF, figure 14). (a) RT waves in 2% aqueous polyox in the flow behind a Mach 2.9 shock wave. Time (in μsec) after passage of shock: (i) 30, (ii) 35, (iii) 40, (iv) 45. (b) Movie frames corresponding to the contrast-enhanced images of (a).

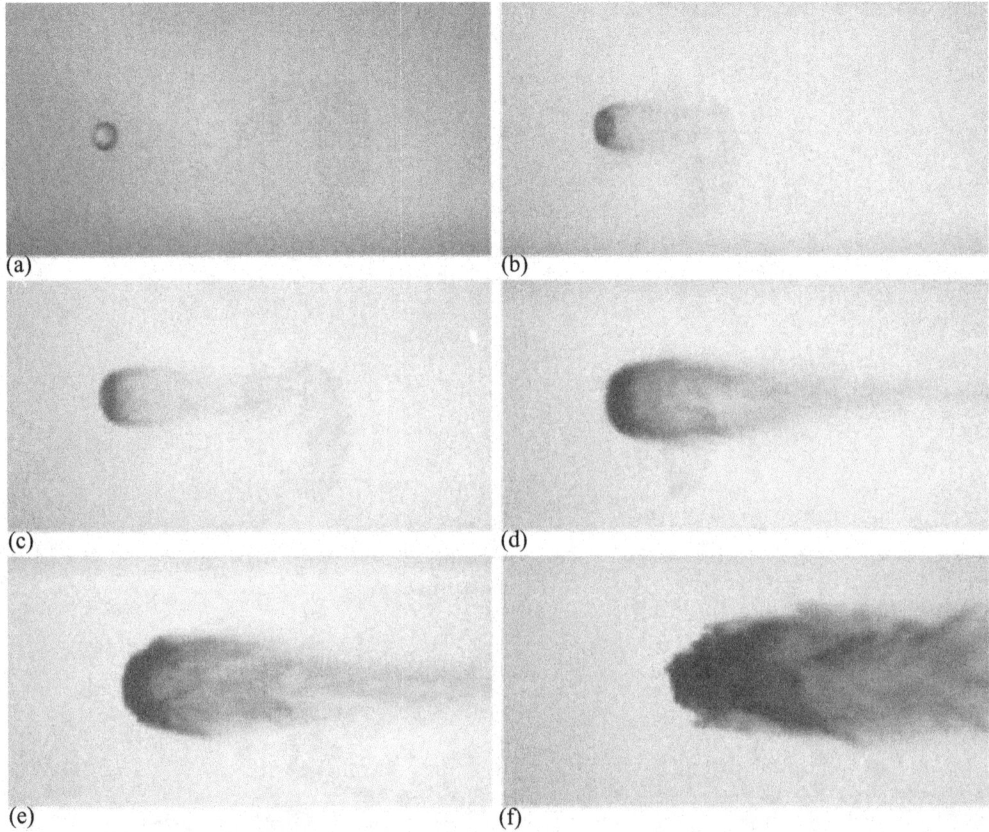

Figure 9.10. Stages in the breakup of a water drop (diameter = 2.6 mm) in the flow behind a Mach 2 shock wave. Air velocity = 432 m/sec; dynamic pressure = 158.0 kPa; Weber number = 11,700. Time (μsec): (a) 0, (b) 45, (c) 70, (d) 135, (e) 170, (f) 290.

Stages of breakup of water drops at $M_s = 2$ and $M_s = 3$ are shown in figures 9.10 and 9.11. RT waves appear on the front face immediately, and the drop is reduced to mist in 200 μs. Excellent photographs of the fragmentation of water drops to mist can be found in the paper of Engel (1958).

The agreements between theory and experiment may extend to high-viscosity drops. The lengths of the most unstable waves are greater than the diameter of the drops. Waves shorter than those that grow fastest grow more slowly. The retardation of the instability of unstable waves that is predicted by theory is realized by experiments; the larger the viscosity, the larger is the length of time it takes for the drop to break. Eventually they do break and the breakup looks like a bag breakup, as in figure 9.10.

9.7 Fragmentation of Newtonian and viscoelastic drops

The breakup of drops in a high-speed airflow is a fundamental problem of a two-phase fluid and it has given rise to a large literature. Most of the early literature focuses on drops of Newtonian liquids in subsonic airstreams, and is excellently reviewed by Pilch and Erdman (1987) and Hsiang and Faeth (1992). Shock tube studies of the breakup

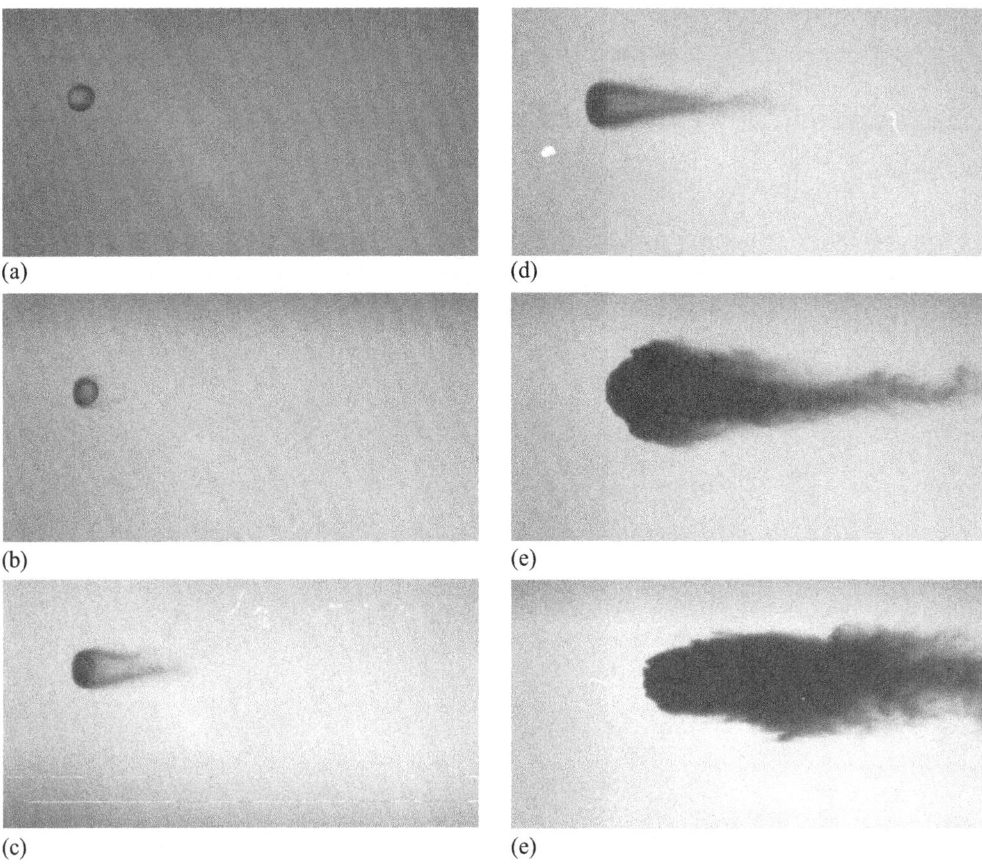

Figure 9.11. Stages in the breakup of a water drop (diameter = 2.5 mm) in the flow behind a Mach 3 shock wave. Air velocity = 764 m/sec; dynamic pressure = 606.4 kPa; Weber number = 43,330. Time (μsec): (a) 0, (b) 15, (c) 30, (d) 40, (e) 95, (f) 135.

of Newtonian drops, usually water, under high-subsonic and supersonic conditions were carried out by Hanson and Domich (1956), Engel (1958), Hanson, Domich, and Adams (1963), Ranger and Nicholls (1969), Reinecke and McKay (1969), Reinecke and Waldman (1970, 1975), Waldman, Reinecke, and Glenn (1972), Simpkins and Bales (1972), Wierzba and Takayama (1988), Yoshida and Takayama (1990), Hirahara and Kawahashi (1992), and others. The excellent study of Engel (1958) showed that water drops of millimeter diameter would be reduced to mist by the flow behind an incident shock moving at Mach numbers in the range 1.3–1.7. Many of the other aforementioned studies allude to the presence of large amounts of mist. Joseph, Huang, and Candler (1996) argued that mist could arise from condensed vapor under flash vaporization because of (1) low pressures at the leeside produced by rarefaction and drop acceleration, (2) high tensions produced by extensional motions in the liquid stripped from the drop, (3) the frictional heating by rapid rates of deformation, and (4) the heating of sheets and filaments torn from the drop by hot air. Although mist and vapor formation are not the focus of this study, it is relevant that the RT instability pumps fingers of hot gas, behind the shock, into the drop increasing both the frictional heating and the area of liquid surface exposed to hot gas. The

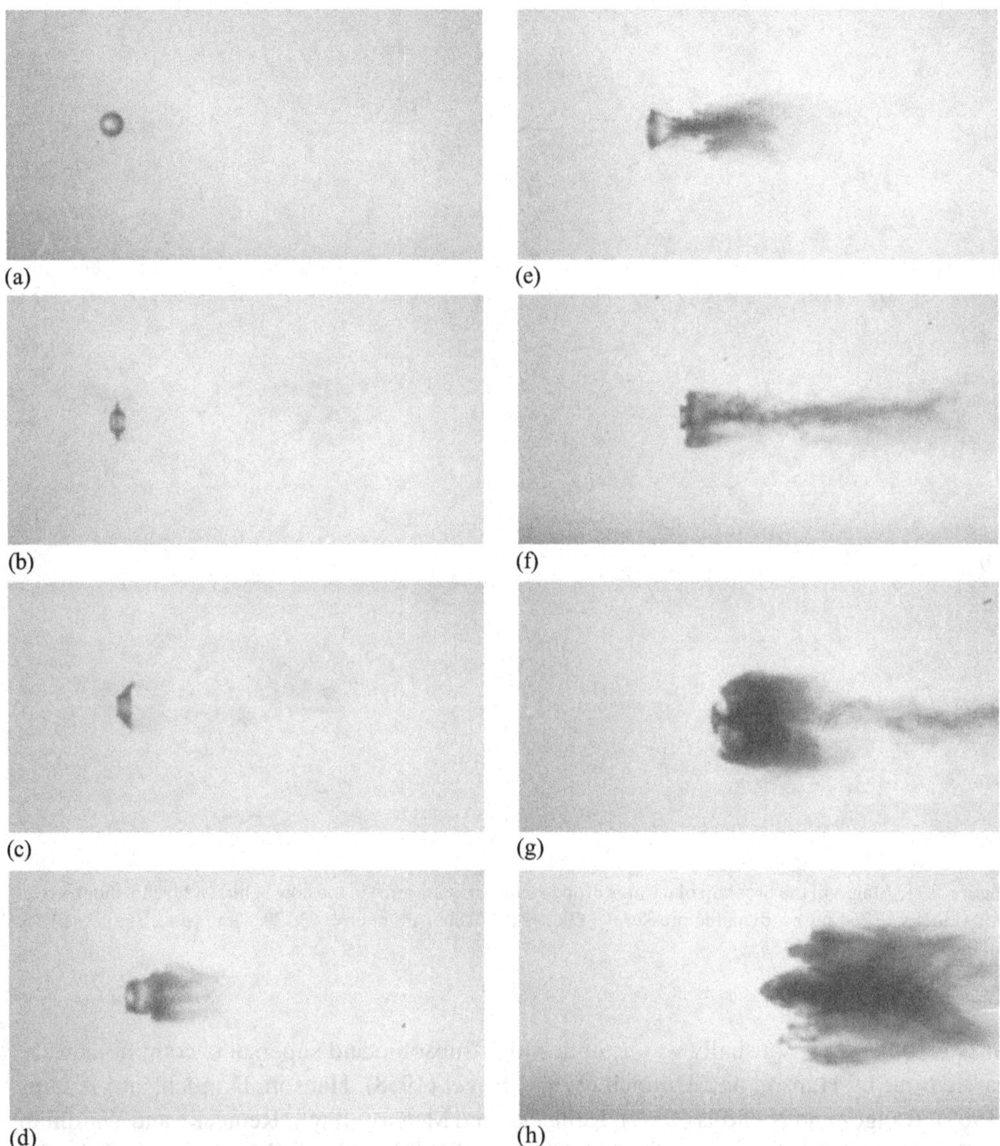

Figure 9.12. Stages in the breakup of a drop of glycerine (diameter = 2.4 mm) in the flow behind a Mach 3 shock wave. Air velocity = 758 m/sec; dynamic pressure = 554.0 kPa; Weber number = 42,220. Time (μsec): (a) 0, (b) 35, (c) 50, (d) 70, (e) 90, (f) 125, (g) 150, (h) 185.

recent and fairly extensive literature on atomization is well represented in the papers by Hsiang and Faeth (1992), Hwang, Liu, and Rietz (1996), and Faeth (1996). These results, and earlier drop breakup studies such as those by Krzeczkowski (1980), Wierzba (1990), Kitscha and Kocamustafaogullari (1989), and Stone (1994), are restricted to relatively low Weber and Reynolds numbers. The highest Weber and Reynolds data for drop breakup were reported by Hsiang and Faeth (1992), who worked under conditions for which the Weber numbers ranged from 0.5 to 1000 with Reynolds numbers from 300 to 1600.

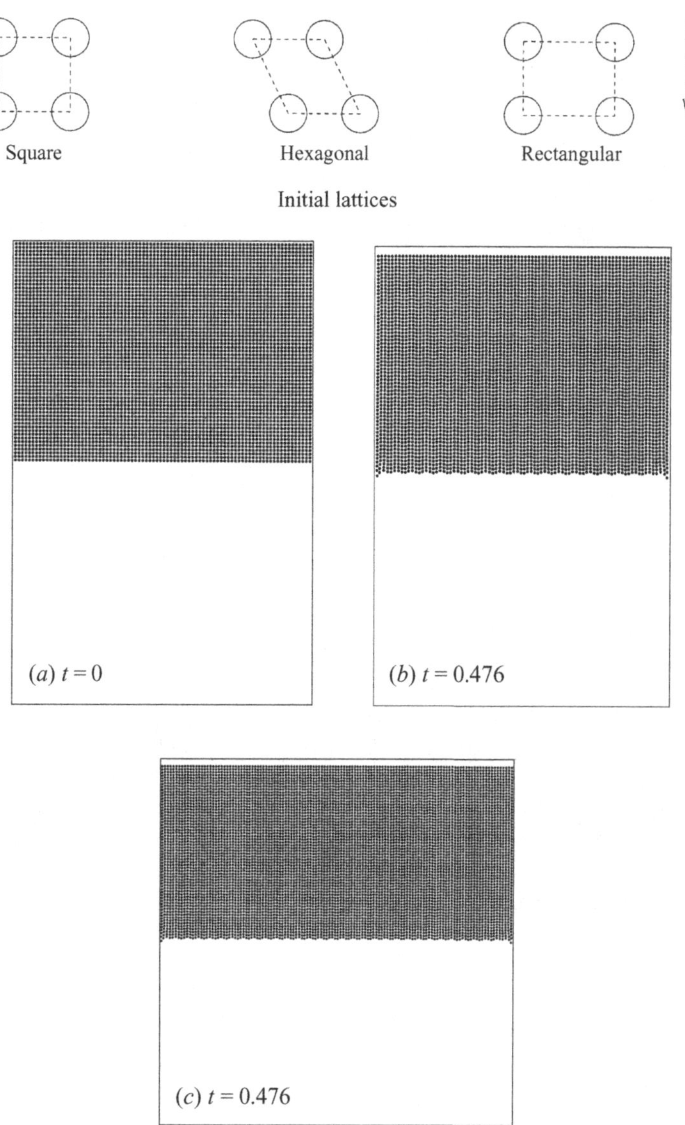

Figure 9.13. Pictures of the sedimentation of 5040 [(a), (b), width = 8 cm] and 7560 [(c), width = 12 cm] disks of diameter 14/192 cm in two dimensions. Different initial lattices were studied, here the initial lattices are in squares.

High-Weber-number drop breakup data were obtained by Engel (1958), Ranger and Nichols (1969), and Reinecke and Waldman (1970) for air and water only, so viscous effects could not be studied. In contrast, the data presented here cover a wide range of viscosities (from 0.001 to 35 $\text{kg}\,\text{m}^{-1}\,\text{sec}^{-1}$) and a wide range of high Weber numbers (from 11,700 to 169,000), Ohnesorge numbers (from 0.002 to 82.3), and Reynolds numbers (from 40,000 to 127,600), based on the free-stream conditions. Another feature that distinguishes the experiments of JBB from previous ones is that they recorded all of the data as real-time movies that may be seen under "video animations" at the Web address http://www.aem.umn.edu/research/Aerodynamic_Breakup/. The movies capture events that were previously unknown, such as bag breakup and bag-and-stamen breakup

Figure 9.14. Snapshots of the sedimentation of (a), (b) 4800 and (c), (d) 6400 disks in two dimensions (width = 8 cm). The diameter of disks in (a), (b), and (c) is 10/192 cm and the diameter of disks in (d) is 16/192 cm. The initial lattice is rectangular.

of high-viscosity drops at very high Weber numbers, whereas short-wave RT corrugations appear on water drops under similar free-stream conditions.

Only a few studies of the breakup of viscoelastic drops have been published: Lane (1951), Wilcox, June, Braun, and Kelly (1961), Matta and Tytus (1982), and Matta, Tytus, and Harris (1983). Matta and co-workers did studies at Mach numbers near one and less. They showed that threads and ligaments of liquid arise immediately after breakup, rather than the droplets that are seen in Newtonian liquids. The breakup of viscoelastic drops is shown in figure 9.9 and in many photographs shown by JBF (2002).

9.8 Modeling Rayleigh–Taylor instability of a sedimenting suspension of several thousand circular particles in a direct numerical simulation

Pan, Joseph, and Glowinski (2001) studied the sedimentation of several thousand circular particles in two dimensions using the method of distributed Lagrange multipliers for solid–liquid flow. The simulation gives rise to fingering, which resembles RT instabilities. The waves have a well-defined wavelength and growth rate that can be modeled as a conventional RT instability of heavy fluid above light. The heavy fluid is modeled as a composite solid–liquid fluid with an effective composite density and viscosity. Surface tension cannot enter this problem, and the characteristic short-wave instability is regularized by the viscosity of the solid–liquid dispersion. The dynamics of the RT instability are studied by use of VPF, generalizing the work of JBB to a rectangular domain bounded by solid walls; an exact solution is obtained. The model shows the same trends as the simulation but it allows the effective fluid to slip at the sidewalls. The model with slip at the sidewalls is very easily constructed, but an analytic solution with no-slip at the sidewall is not known and apparently cannot be found by the method of separation of variables. It is probable that the numerical simulation would be better modeled by a continuum model with no-slip at sidewalls.

Sample pictures of the simulation are shown in figures 9.13 and 9.14. The formation of RT instability follows the acceleration guidelines in the line drawing of figure 9.1(e); the top surface is flat and the bottom one is corrugated.

10

The force on a cylinder near a wall in viscous potential flows

We study the force on a 2D cylinder near a wall in two potential flows: the flow that is due to the circulation $2\pi\kappa$ about the cylinder and the uniform streaming flow with velocity V past the cylinder. The pressure is computed with Bernoulli's equation, and the viscous normal stress is calculated with VPF; the shear stress is ignored. The forces on the cylinder are computed by integration of the normal stress over the surface of the cylinder. In both of the two cases, the force perpendicular to the wall (lift) is due to only the pressure and the force parallel to the wall (drag) is due to only the viscous normal stress. Our results show that the drag on a cylinder near a wall is larger than on a cylinder in an unbounded domain. In the flow induced by circulation or in the streaming flow, the lift force is always pushing the cylinder toward the wall. However, when the two flows are combined, the lift force can be pushing the cylinder away from the wall or toward the wall.

10.1 The flow that is due to the circulation about the cylinder

Figure 10.1 shows a cylinder with radius a near the wall $x = 0$. Let b be the distance of the axis of the cylinder from the wall and $c = OP$, where OP is the tangent of the cylinder. The complex potential for the flow that is due to the anticlockwise circulation $2\pi\kappa$ about the cylinder is (Milne-Thomson, 1968, §6.53)

$$\omega = -2\kappa \cot^{-1} \frac{iz}{c}. \tag{10.1.1}$$

To facilitate the calculation, we introduce a parameter η, so that

$$a = c \operatorname{csch} \eta, \quad b = c \coth \eta. \tag{10.1.2}$$

We can also express c and η in terms of a and b:

$$c = \sqrt{b^2 - a^2}, \quad \eta = \cosh^{-1}(b/a). \tag{10.1.3}$$

The complex velocity is

$$u - iv = \frac{d\omega}{dz} = \frac{2ic\kappa}{c^2 - z^2}. \tag{10.1.4}$$

We separate the real and imaginary parts of the velocity to obtain

$$u = \frac{-4c\kappa xy}{(c^2 - x^2 + y^2)^2 + 4x^2y^2}, \quad v = \frac{-2c\kappa(c^2 - x^2 + y^2)}{(c^2 - x^2 + y^2)^2 + 4x^2y^2}. \tag{10.1.5}$$

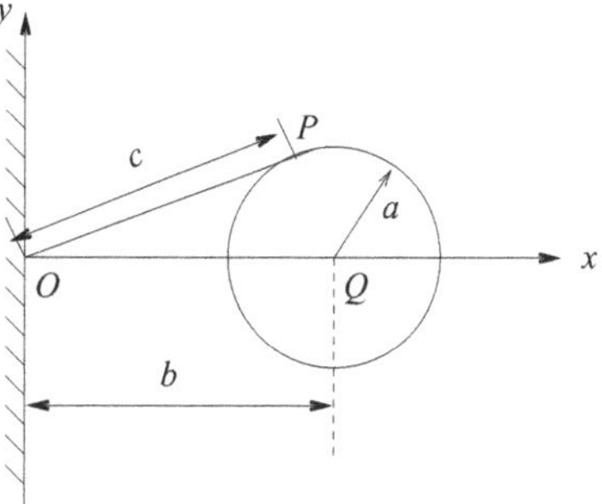

Figure 10.1. A cylinder with radius a near the wall $x = 0$.

The pressure can be computed with Bernoulli's equation:

$$p + \rho |u|^2/2 = \text{const} \Rightarrow -p = \rho |u|^2/2 - \text{const}. \tag{10.1.6}$$

The rate-of-strain tensor can be evaluated as

$$2\mathbf{D} = \begin{bmatrix} 2\partial u/\partial x & \partial u/\partial y + \partial v/\partial x \\ \partial u/\partial y + \partial v/\partial x & 2\partial v/\partial y \end{bmatrix}. \tag{10.1.7}$$

The surface of the cylinder can be expressed as

$$x = c \coth \eta + c \,\text{csch}\, \eta \cos \theta, \quad y = c \,\text{csch}\, \eta \sin \theta, \tag{10.1.8}$$

where $0 \leqslant \theta \leqslant 2\pi$. The normal on the surface of the cylinder is

$$\mathbf{n} = \cos \theta \mathbf{e}_x + \sin \theta \mathbf{e}_y. \tag{10.1.9}$$

The viscous normal stress at the surface is

$$\begin{aligned} \tau_{nn} &= \mathbf{n} \cdot 2\mu \mathbf{D} \cdot \mathbf{n} = 2\mu \left[\frac{\partial u}{\partial x} \cos^2 \theta + \left(\frac{\partial u}{\partial y} + \frac{\partial v}{\partial x} \right) \sin \theta \cos \theta + \frac{\partial v}{\partial y} \sin^2 \theta \right] \\ &= \frac{-2\mu\kappa \sin \theta \sinh^3 \eta}{c^2 (\cos \theta + \cosh \eta)^2}, \end{aligned} \tag{10.1.10}$$

and the pressure at the surface is

$$-p = \frac{\rho}{2} \frac{\kappa^2 \sinh^4}{c^2 (\cos \theta + \cosh \eta)^2} - \text{const}. \tag{10.1.11}$$

The forces on the cylinder by the normal stress are

$$F_x = \int_0^{2\pi} (-p + \tau_{nn}) \cos \theta \, c \,\text{csch}\, \eta \, d\theta, \quad F_y = \int_0^{2\pi} (-p + \tau_{nn}) \sin \theta \, c \,\text{csch}\, \eta \, d\theta. \tag{10.1.12}$$

We calculate the contributions to the forces by the pressure and viscous stress separately. The superscripts p and v stand for pressure and viscous stress, respectively. The forces by the pressure are

$$F_x^p = \int_0^{2\pi} \frac{\rho \kappa^2 \sinh^3 \eta}{2c} \frac{\cos \theta \, d\theta}{(\cos \theta + \cosh \eta)^2} = -\frac{\pi \rho \kappa^2}{c}, \tag{10.1.13}$$

$$F_y^p = \int_0^{2\pi} \frac{\rho \kappa^2 \sinh^3 \eta}{2c} \frac{\sin \theta \, d\theta}{(\cos \theta + \cosh \eta)^2} = 0, \tag{10.1.14}$$

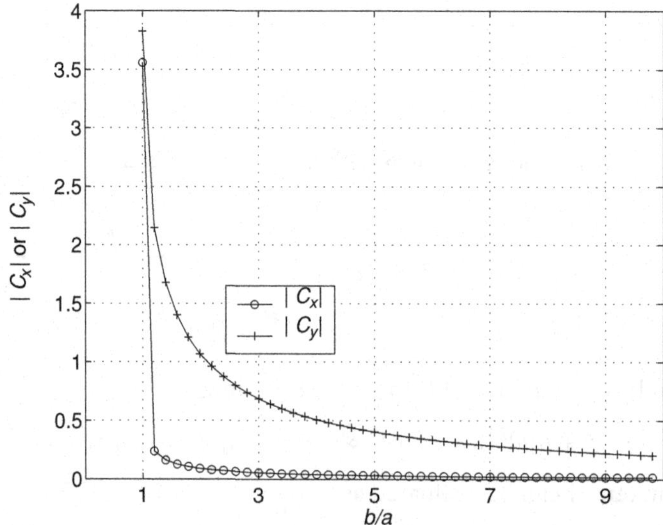

Figure 10.2. The absolute values of the coefficients for the forces $|C_x|$ and $|C_y|$ as functions of the ratio b/a. The Reynolds number is taken as 1 when C_y is computed.

which agree with the values computed with the theorem of Blasius (Milne-Thomson, 1968, §6.53). The forces by the viscous stress are

$$F_x^v = \int_0^{2\pi} -\frac{2\mu\kappa \sinh^2\eta}{c} \frac{\sin\theta\cos\theta\, d\theta}{(\cos\theta + \cosh\eta)^2} = 0, \tag{10.1.15}$$

$$F_y^v = \int_0^{2\pi} -\frac{2\mu\kappa \sinh^2\eta}{c} \frac{\sin\theta\sin\theta\, d\theta}{(\cos\theta + \cosh\eta)^2} = -\frac{2\pi\mu\kappa}{c}\left(1 - e^{-2\eta}\right). \tag{10.1.16}$$

Using equation (10.1.3), we can express the forces in terms of a and b:

$$F_x = F_x^p + F_x^v = -\frac{\pi\rho\kappa^2}{\sqrt{b^2 - a^2}}, \tag{10.1.17}$$

$$F_y = F_y^p + F_y^v = -\frac{2\pi\mu\kappa}{\sqrt{b^2 - a^2}}\left[1 - e^{-2\cosh^{-1}(b/a)}\right]. \tag{10.1.18}$$

We define a characteristic velocity $U = 2\pi\kappa/a$ and make the forces dimensionless:

$$C_x = \frac{F_x}{\frac{1}{2}\rho U^2 a} = -\frac{1}{2\pi\sqrt{(b/a)^2 - 1}}, \tag{10.1.19}$$

$$C_y = \frac{F_y}{\frac{1}{2}\rho U^2 a} = -\frac{4}{Re}\frac{1}{\sqrt{(b/a)^2 - 1}}\left[1 - e^{-2\cosh^{-1}(b/a)}\right], \tag{10.1.20}$$

where $Re = \rho U 2a/\mu$ is the Reynolds number. Here C_x is a function of the ratio b/a only; C_y is a function of b/a and Re. We plot $|C_x|$ and $|C_y|$ as functions of b/a in figure 10.2. The Reynolds number is taken as 1 when C_y is computed. When b/a approaches 1, meaning that the gap between the wall and the cylinder is very small, both of the two forces approach infinity. When b/a approaches infinity, both of the two forces approach zero, because the interaction between the cylinder and the wall is negligible.

10.2 The streaming flow past the cylinder near a wall

The domain is bounded by the plane $x = 0$ and the cylinder $(x-b)^2 + y^2 = a^2$ (figure 10.1). The uniform flow with the velocity V at infinity is streaming in the negtive direction of the y axis. The complex potential for the flow is (Milne-Thomson, 1968, 185)

$$\omega = iVz - 2iVa^2 z\left[\frac{1}{z^2-b^2} + \sum_{n=1}^{\infty}\frac{a^{2n}}{\prod_{r=0}^{n-1}(b+x_r)^2(z^2-x_n^2)}\right], \tag{10.2.1}$$

where $x_0 = b$, $x_n = b - a^2/(b+x_{n-1})$. The potential is represented by an infinite series that converges rapidly for moderate or large values of the ratio b/a. To carry out the calculation, we consider the terms up to $n = 3$ in (10.2.1):

$$\omega = iVz - 2iVa^2 z\left[\frac{1}{z^2-b^2} + \frac{a^2}{\beta_1(z^2-x_1^2)} + \frac{a^4}{\beta_2(z^2-x_2^2)} + \frac{a^6}{\beta_3(z^2-x_3^2)}\right], \tag{10.2.2}$$

where

$$x_1 = b - \frac{a^2}{2b}, \quad x_2 = b - \frac{a^2}{2b - a^2/(2b)}, \quad x_3 = b - \frac{a^2}{2b - \frac{a^2}{2b - a^2/(2b)}}; \tag{10.2.3}$$

$$\beta_1 = (2b)^2, \quad \beta_2 = (2b)^2\left(2b - \frac{a^2}{2b}\right)^2, \quad \beta_3 = (2b)^2\left(2b - \frac{a^2}{2b}\right)^2\left[2b - \frac{a^2}{2b - a^2/(2b)}\right]^2. \tag{10.2.4}$$

The complex velocity is $d\omega/dz$, and we separate the real and imaginary parts of it to obtain the velocities in the x and y directions:

$$\begin{aligned}
u = &-\frac{4a^2Vxy}{(x^2-y^2-b^2)^2+4x^2y^2} - \frac{4a^4Vxy}{\beta_1[(x^2-y^2-x_1^2)^2+4x^2y^2]}\\
&-\frac{4a^6Vxy}{\beta_2[(x^2-y^2-x_2^2)^2+4x^2y^2]} - \frac{4a^8Vxy}{\beta_3[(x^2-y^2-x_3^2)^2+4x^2y^2]}\\
&+\frac{8a^2Vxy[(x^2+y^2)^2-b^4]}{(b^4-2b^2x^2+x^4+2b^2y^2-6x^2y^2+y^4)^2+16x^2y^2(x^2-y^2-b^2)^2}\\
&+\frac{a^2}{\beta_1}\frac{8a^2Vxy[(x^2+y^2)^2-x_1^4]}{[(x_1^4-2x_1^2x^2+x^4+2x_1^2y^2-6x^2y^2+y^4)^2+16x^2y^2(x^2-y^2-x_1^2)^2]}\\
&+\frac{a^4}{\beta_2}\frac{8a^2Vxy[(x^2+y^2)^2-x_2^4]}{[(x_2^4-2x_2^2x^2+x^4+2x_2^2y^2-6x^2y^2+y^4)^2+16x^2y^2(x^2-y^2-x_2^2)^2]}\\
&+\frac{a^6}{\beta_3}\frac{8a^2Vxy[(x^2+y^2)^2-x_3^4]}{[(x_3^4-2x_3^2x^2+x^4+2x_3^2y^2-6x^2y^2+y^4)^2+16x^2y^2(x^2-y^2-x_3^2)^2]};
\end{aligned} \tag{10.2.5}$$

$$
\begin{aligned}
v = & -V + \frac{2a^2V(x^2-y^2-b^2)}{(x^2-y^2-b^2)^2+4x^2y^2} + \frac{2a^4V(x^2-y^2-x_1^2)}{\beta_1[(x^2-y^2-x_1^2)^2+4x^2y^2]} \\
& + \frac{2a^6V(x^2-y^2-x_2^2)}{\beta_2[(x^2-y^2-x_2^2)^2+4x^2y^2]} + \frac{2a^8V(x^2-y^2-x_3^2)}{\beta_3[(x^2-y^2-x_3^2)^2+4x^2y^2]} \\
& - \frac{4a^2V[b^4(x^2-y^2)-2b^2(x^2+y^2)^2+(x^2-y^2)(x^2+y^2)^2]}{(b^4-2b^2x^2+x^4+2b^2y^2-6x^2y^2+y^4)^2+16x^2y^2(x^2-y^2-b^2)^2} \\
& - \frac{a^2}{\beta_1}\frac{4a^2V[x_1^4(x^2-y^2)-2x_1^2(x^2+y^2)^2+(x^2-y^2)(x^2+y^2)^2]}{[(x_1^4-2x_1^2x^2+x^4+2x_1^2y^2-6x^2y^2+y^4)^2+16x^2y^2(x^2-y^2-x_1^2)^2]} \\
& - \frac{a^4}{\beta_2}\frac{4a^2V[x_2^4(x^2-y^2)-2x_2^2(x^2+y^2)^2+(x^2-y^2)(x^2+y^2)^2]}{[(x_2^4-2x_2^2x^2+x^4+2x_2^2y^2-6x^2y^2+y^4)^2+16x^2y^2(x^2-y^2-x_2^2)^2]} \\
& - \frac{a^6}{\beta_3}\frac{4a^2V[x_3^4(x^2-y^2)-2x_3^2(x^2+y^2)^2+(x^2-y^2)(x^2+y^2)^2]}{[(x_3^4-2x_3^2x^2+x^4+2x_3^2y^2-6x^2y^2+y^4)^2+16x^2y^2(x^2-y^2-x_3^2)^2]}.
\end{aligned}
\tag{10.2.6}
$$

We can then calculate the pressure by using (10.1.6) and the rate-of-strain tensor by using (10.1.7). At the surface of the cylinder, we have

$$x = b + a\cos\theta, \quad y = a\sin\theta, \quad \mathbf{n} = \cos\theta\mathbf{e}_x + \sin\theta\mathbf{e}_y. \tag{10.2.7}$$

The pressure and viscous normal stress at the surface are then obtained. The forces by the normal stress on the cylinder are

$$F_x = \int_0^{2\pi}(-p+\tau_{nn})\cos\theta\, a\, d\theta, \quad F_y = \int_0^{2\pi}(-p+\tau_{nn})\sin\theta\, a\, d\theta. \tag{10.2.8}$$

Again we compute the forces by the pressure and the viscous normal stress separately and use $\rho V^2 a/2$ to make the forces dimensionless. The coefficients for the forces by the pressure are

$$C_x^p = \frac{\int_0^{2\pi}(-p)\cos\theta\, a\, d\theta}{\rho V^2 a/2} = \frac{\int_0^{2\pi}(\rho|u|^2/2)\cos\theta\, a\, d\theta}{\rho V^2 a/2} = \frac{\int_0^{2\pi}(u^2+v^2)\cos\theta\, d\theta}{V^2}, \tag{10.2.9}$$

$$C_y^p = \frac{\int_0^{2\pi}(u^2+v^2)\sin\theta\, d\theta}{V^2}. \tag{10.2.10}$$

The coefficients for the forces by the normal viscous stress are

$$
\begin{aligned}
C_x^v &= \frac{\int_0^{2\pi}\tau_{nn}\cos\theta\, a\, d\theta}{\rho V^2 a/2} \\
&= \frac{8}{Re}\int_0^{2\pi}\left[\frac{\partial u}{\partial x}\cos^2\theta + \left(\frac{\partial u}{\partial y}+\frac{\partial v}{\partial x}\right)\sin\theta\cos\theta + \frac{\partial v}{\partial y}\sin^2\theta\right]\cos\theta\, a\, d\theta/V,
\end{aligned}
\tag{10.2.11}
$$

$$C_y^v = \frac{8}{Re}\int_0^{2\pi}\left[\frac{\partial u}{\partial x}\cos^2\theta + \left(\frac{\partial u}{\partial y}+\frac{\partial v}{\partial x}\right)\sin\theta\cos\theta + \frac{\partial v}{\partial y}\sin^2\theta\right]\sin\theta\, a\, d\theta/V, \tag{10.2.12}$$

where $Re = \rho V2a/\mu$. Because the expressions for the pressure and the viscous normal stress are very complicated, analytical expressions for C_x^p, C_y^p, C_x^v, and C_y^v are not obtained.

Table 10.1. *The coefficients for the forces on the cylinder by the pressure and the normal viscous stress obtained from numerical integration*

b/a	C_x^p	C_y^p	C_x^v	C_y^v
1.05	−10.20	0	0	$-67.94/Re$
1.1	−6.656	0	0	$-65.34/Re$
1.2	−3.712	0	0	$-61.82/Re$
1.3	−2.438	0	0	$-59.54/Re$
1.5	−1.307	0	0	$-56.74/Re$
2.0	−0.4595	0	0	$-53.64/Re$
10	-3.16×10^{-3}	0	0	$-50.4/Re$
30	-1.16×10^{-4}	0	0	$-50.28/Re$

Instead, numerical integration is used to obtain the values of these coefficients. We list the results of numerical integrations in table 10.1.

Table 10.1 shows that C_y^p and C_x^v are both zero; in other words, the lift force in the x direction is due to only the pressure, and the drag in the y direction is due to only the viscous normal stress. The lift force is in the negative x direction, pushing the cylinder toward the wall. This is because the velocity of the fluid in the gap between the cylinder and the wall is larger than that of exterior fluid, inducing a pressure gradient that gives rise to the lift force. The lift force has the largest magnitude when the gap between the cylinder and the wall is small ($b/a \sim 1$); it approaches zero when the cylinder is far away from the wall ($b/a \gg 1$). The drag is in the negative y direction, and its magnitude decreases as b/a increases. When the cylinder is far away from the wall, the magnitude of the drag approaches an asymptotic value $16\pi/Re \simeq 50.265/Re$. This asymptotic value is equal to the coefficient for the drag experienced by a cylinder in an unbounded uniform streaming flow. Therefore our results indicate that the presence of the wall increases the drag on the cylinder; when the cylinder is closer to the wall, the magnitude of the drag is larger.

10.3 The streaming flow past a cylinder with circulation near a wall

The combination of the two potentials (10.1.1) and (10.2.1) represents the uniform flow with the velocity V at infinity past a cylinder with circulation $2\pi\kappa$ near a wall. We calculate the force on the cylinder by the normal stress. The terms up to $n = 3$ in the infinite series are kept, and the potential is

$$\omega = -2\kappa \cot^{-1}\frac{iz}{c} + iVz - 2iVa^2 z\left[\frac{1}{z^2 - b^2} + \frac{a^2}{\beta_1(z^2 - x_1^2)} + \frac{a^4}{\beta_2(z^2 - x_2^2)} + \frac{a^6}{\beta_3(z^2 - x_3^2)}\right], \tag{10.3.1}$$

where $x_1, x_2, x_3, \beta_1, \beta_2$, and β_3 are defined in (10.2.3) and (10.2.4).

The velocity of the combined potential is the sum of the velocity induced by the circulation and that of the streaming flow. It follows that the viscous normal stress is also the sum of the two from the circulation and the streaming flow. The calculation of the

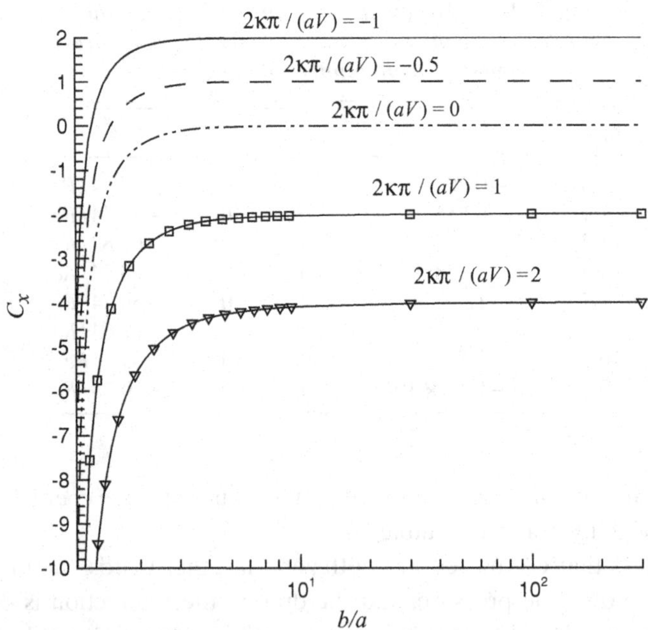

Figure 10.3. The lift coefficient C_x as a function of the ratio b/a. The five curves correspond to five values of the parameter $2\pi\kappa/(aV)$: −1, −0.5, 0, 1, and 2.

pressure is slightly complicated because the nonlinear term $|u|^2$ is involved. We skip the details of the calculation as they are similar to those in the previous sections.

There are two characteristic velocities here, V from the streaming flow and $U = 2\pi\kappa/a$ from the circulation. We introduce the parameter $2\pi\kappa/(aV)$; when $2\pi\kappa/(aV) = 1$, we have $V = U$. The coefficients for the forces are defined as

$$C_x = \frac{\int_0^{2\pi}(-p + \rho|u|^2/2)\cos\theta\, a\, d\theta}{\rho V^2 a/2}, \quad C_y = \frac{\int_0^{2\pi}(-p + \rho|u|^2/2)\sin\theta\, a\, d\theta}{\rho V^2 a/2}. \tag{10.3.2}$$

Our calculation shows that the lift force (C_x) is due to only the pressure, and the drag (C_y) is due to only the viscous normal stress, which is the same as in the previous two sections.

In the flow induced by circulation or in the streaming flow, the lift force is always pushing the cylinder toward the wall. However, when the two flows are combined, the lift force can be pushing the cylinder away from the wall or toward the wall. We plot C_x and C_y as functions of the ratio b/a in figures 10.3 and 10.4, respectively. Five values of the parameter $2\pi\kappa/(aV)$ are studied: −1, −0.5, 0, 1, and 2. The circulation is anticlockwise when $2\pi\kappa/(aV)$ is positive and clockwise when $2\pi\kappa/(aV)$ is negative.

The lift force is generated by the discrepancy between the fluid velocity in the gap and the velocity in the exterior. This discrepancy is determined by the effect of the wall and the circulation. The wall effect gives rise to a larger velocity in the gap than in the exterior. The smaller the gap, the larger the wall effect. When $\kappa = 0$, there is no circulation and the wall effect leads to a negative lift force. The anticlockwise circulation increases the velocity discrepancy generated by the wall effect. As a result, the magnitude of the lift force for $2\pi\kappa/(aV) = 1$ or 2 is larger than for $\kappa = 0$. On the contrary, the clockwise circulation offsets the velocity discrepancy by the wall effect. When the gap is small (b/a is close to 1),

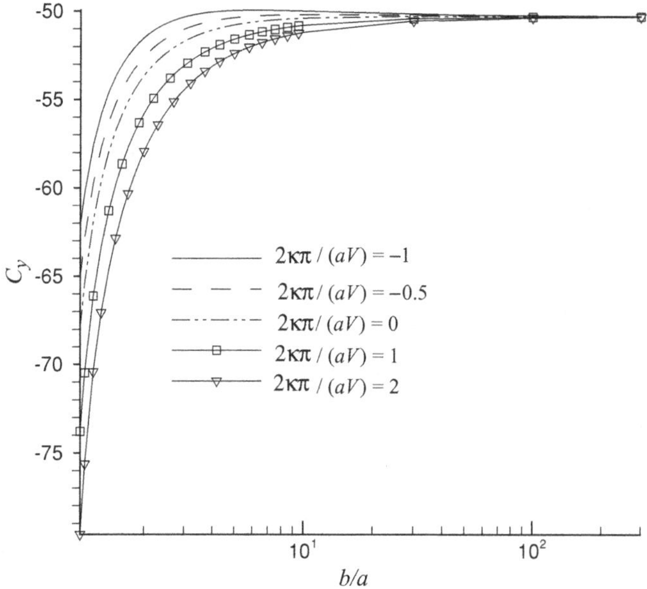

Figure 10.4. The drag coefficient C_y as a function of the ratio b/a. The five curves correspond to five values of the parameter $2\pi\kappa/(aV)$: -1, -0.5, 0, 1, and 2. The Reynolds number is taken as 1 here.

the fluid velocity in the gap is still larger than in the exterior but the difference is smaller because of the clockwise circulation. Therefore the lift force for $2\pi\kappa/(aV) = -0.5$ and -1 is still negative but with a smaller magnitude than for $\kappa = 0$. As b/a becomes larger, the wall effect is weaker and could be completely balanced by the clockwise circulation, leading to a zero lift force; the corresponding b/a may be called the critical gap. The critical gaps for $2\pi\kappa/(aV) = -0.5$ and -1 are approximately $b/a = 1.5$ and 1.23, respectively. When $b/a \gg 1$, the wall effect is negligible and the lift force is determined by the circulation alone. The asymptotic value of the lift force can be calculated with the well-known lift formula (Milne-Thomson, 1968, §7.12):

$$F_x = -2\pi\kappa\rho V, \tag{10.3.3}$$

which is for a cylinder with circulation in an unbounded uniform streaming flow. The lift coefficient is

$$C_x = \frac{F_x}{\rho V^2 a/2} = \frac{-2\pi\kappa\rho V}{\rho V^2 a/2} = -2\frac{2\pi\kappa}{aV}. \tag{10.3.4}$$

Figure 10.3 shows that the asymptotic values are achieved in our calculation.

The sum of drags from the circulation and from the streaming flow gives the total drag. When $\kappa = 0$, the drag is due to the streaming flow alone and is negative. The anticlockwise circulation increases the magnitude of the drag whereas the clockwise circulation decreases it. When $b/a \gg 1$, the drag from the circulation is negligible and the drag coefficient approaches the asymptotic value $16\pi/Re \simeq 50.265/Re$ (see figure 10.4).

We study the effects of the strength of the circulation on the lift and drag at a fixed value of b/a. The lift and drag coefficients are plotted as functions of $2\pi\kappa/(aV)$ in figures 10.5 and 10.6, respectively. We note that C_x is zero at two different values of $2\pi\kappa/(aV)$ in

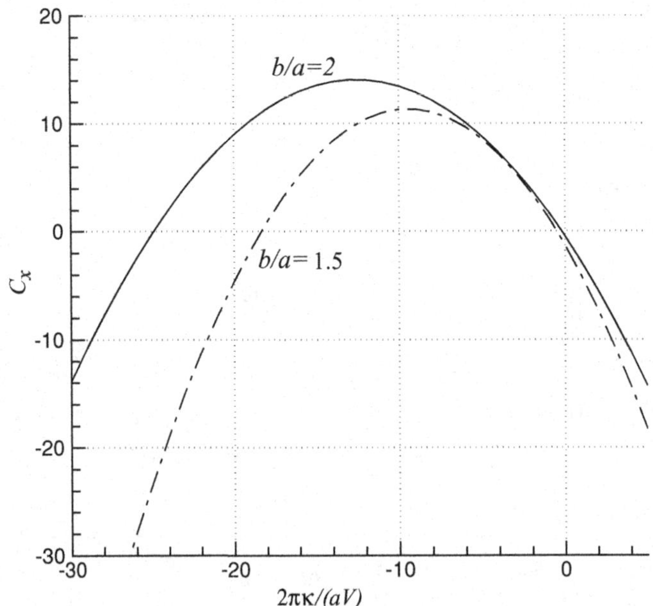

Figure 10.5. The lift coefficient C_x as a function of $2\pi\kappa/(aV)$. The two curves are for $b/a = 1.5$ and 2.

figure 10.5. When $2\pi\kappa/(aV)$ is negative with a large magnitude, the circulation dominates and the streaming flow can be neglected. The flow in the gap is faster than in the exterior because of the wall effect, leading to a negative lift force. When $2\pi\kappa/(aV)$ is slightly smaller than zero, the streaming flow dominates. Again, the velocity discrepancy gives rise

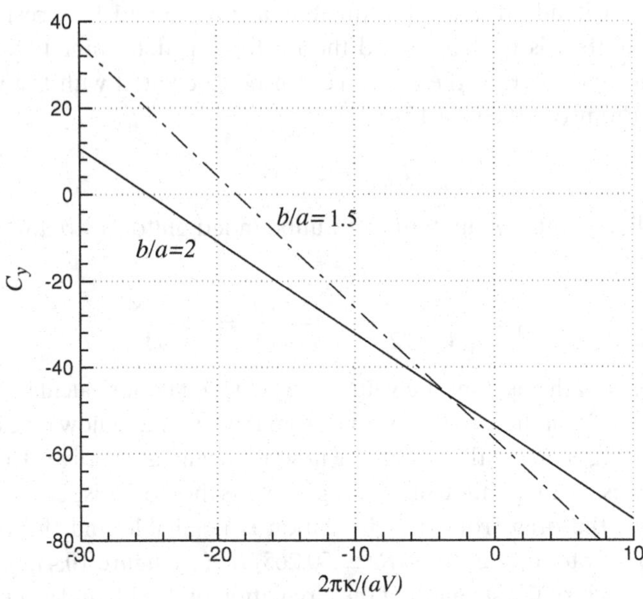

Figure 10.6. The drag coefficient C_y as a function of $2\pi\kappa/(aV)$. The two curves are for $b/a = 1.5$ and 2. The Reynolds number is taken as 1 here.

to a negative lift force. When $2\pi\kappa/(aV)$ is negative with an intermediate magnitude, the clockwise circulation overcomes the velocity discrepancy by the wall effect and the lift force is positive. The three regimes previously discussed are separated by two points at which the lift force is zero. When $2\pi\kappa/(aV)$ is positive, the lift force is always negative.

The drag has contributions from the circulation and the streaming flow. The part from the streaming flow is independent of $2\pi\kappa/(aV)$; the part from the circulation is given by (10.1.18) and the dimensionless form is

$$\frac{-2\pi\mu\kappa\left[1-e^{-2\cosh^{-1}(b/a)}\right]/\sqrt{b^2-a^2}}{\rho V^2 a/2} = -\frac{4}{Re}\frac{2\pi\kappa}{aV}\frac{\left[1-e^{-2\cosh^{-1}(b/a)}\right]}{\sqrt{(b/a)^2-1}}, \qquad (10.3.5)$$

where $Re = \rho V 2a/\mu$. It is clear from (10.3.5) that, when b/a and Re are fixed, the drag coefficient has a linear relationship with $2\pi\kappa/(aV)$, which can be seen in figure 10.6. The drag by the streaming flow is negative; its magnitude can be increased by an anticlockwise circulation $[2\pi\kappa/(aV) > 0]$ or decreased by a clockwise circulation $[2\pi\kappa/(aV) < 0]$. When $2\pi\kappa/(aV)$ is negative with a large magnitude, the sign of the drag is changed to positive.

11

Kelvin–Helmholtz instability

The instability of uniform flow of incompressible fluids in two horizontal parallel infinite streams of different velocities and densities is known as the Kelvin–Helmholtz (KH) instability. It is usual to study this instability by linearization of the nonlinear equations around the basic uniform flow followed by analysis of normal modes proportional to

$$\exp(\sigma t + ikx). \tag{11.0.1}$$

It is also usual to assume that the fluids are inviscid; if they were viscous the discontinuity of velocity could not persist. Moreover, if the basic streams are assumed uniform, then there are no rigorous ways to evaluate the shear and interfacial stresses. In practice, the streams being modeled are turbulent, and analytic studies of stability must make approximations. The usual procedure adopted by many authors, whose work has been reviewed by Mata *et al.* (2002), is to use empirical correlations for the evaluation of shear and interfacial stresses. The agreement with experiments achieved by this empirical approach is not compelling.

The study of KH instability by use of VPF does not allow for no-slip conditions or interfacial stresses but, unlike in the inviscid theory, the normal viscous stresses are well represented. There is an important viscous resistance to the up-and-down motion of the waves independent of the action of shear stresses.

KH instability is produced by the action of pressures on the perturbed interface; the instability is often discussed as being due to shear, but shear stresses are not the major actor here and in any event could not be inserted into an analysis based on potential flow even when viscosity is not neglected.

The formation of the instability may be described in terms of the action of pressure in the Bernoulli equation, which is small where the speed is great (at the top in figure 11.1) and large where it is small (in troughs).

An example of the KH instability that is due to wind shear in the atmosphere is shown in figure 11.2.

11.1 KH instability on an unbounded domain

The instability corresponding to figure 11.1 was considered by Drazin and Reid (1981); the fluids are inviscid and the flows irrotational. The flow domain is not bounded so that perturbations decay far away from the interface. Their analysis combines RT and KH instabilities, which are very different kinds of instability. In particular, RT instability

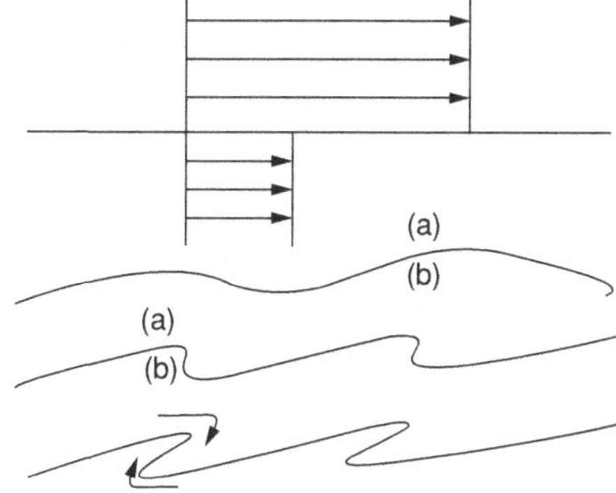

Figure 11.1. The speed of the fluid above is greater. The pressure is greater at crests (a) than at troughs (b). The upper part of the interface is carried by the upper fluid, causing the interface to overturn.

can occur in a vacuum but the pressures from the ambient in KH instability cannot be neglected.

Funada and Joseph (2001) did the analysis in which the fluid is viscous and the flow is irrotational and bounded in a channel. They confined their attention to the case in which air is above and RT instability for heavy fluid above does not occur. The results of Drazin and Reid can be obtained from the formulas of Funada and Joseph, which are developed in the sections that follow.

We can obtain the dispersion relation for stability analysis when the flow is irrotational and the fluids are viscous by following the analysis of Drazin and Reid (1981). The pressure balance is replaced with a viscous normal stress balance:

$$p_a' - 2\mu_a \frac{\partial w_a'}{\partial z} = p_l' - 2\mu_l \frac{\partial w_l'}{\partial z}, \tag{11.1.1}$$

and, taking the perturbation in the form of functions of z (the normal direction) times $\exp(\sigma t + ikx)$, we find the dispersion relation:

$$\begin{aligned}(\rho_a + \rho_l)\sigma^2 + 2\sigma\left[ik(\rho_a U_a + \rho_l U_l) + k^2(\mu_a + \mu_l)\right] - k^2\left(\rho_a U_a^2 + \rho_l U_l^2\right)\\ + 2ik^3(\mu_a U_a + \mu_l U_l) + (\rho_l - \rho_a)gk + \gamma k^3 = 0,\end{aligned} \tag{11.1.2}$$

Figure 11.2. KH instability rendered visible by clouds over Mount Duval in Australia.

where ρ, μ, U are respectively density, viscosity, velocity; γ is surface tension; and g is the gravitational constant. Gravity acts downward from side a toward side l. The real part of σ is the growth rate given as a function of wavenumber k.

11.2 Maximum growth rate, Hadamard instability, neutral curves

Equation (11.0.1) may be written as

$$\exp(\sigma_R t)\exp[i(\sigma_I t + kx)], \tag{11.2.1}$$

where

$$\sigma_R = \sigma_R(k, \text{ parameters}). \tag{11.2.2}$$

When the parameters are such that

$$\sigma_R(k) > 0, \tag{11.2.3}$$

the flow is KH unstable. When the parameters are such that

$$\sigma_R(k) = 0, \tag{11.2.4}$$

the flow is neutrally stable. Sometimes it is said that the flow is marginally stable.

11.2.1 Maximum growth rate

For unstable flows,

$$\sigma_{Rm} = \max_{k\in\mathbb{R}} \sigma_R(k) = \sigma_R(k_m), \tag{11.2.5}$$

where σ_{Rm} is the maximum growth rate for a certain set of parameters and k_m is the wavenumber of maximum growth. The wavelength $\lambda_m = 2\pi/k_m$ for the fastest-growing wave is usually observed in experiments. It is selected by nonlinear mechanisms that are not well understood.

11.2.2 Hadamard instability

Surface tension is very important; it stabilizes short waves. When $\gamma = 0$, the flow can be unstable as $k \to \infty$ or $\lambda \to 0$. If the viscosity is also zero, then the KH instability is Hadamard unstable with $e^{\alpha k t}$, $\alpha > 0$. This bad instability is worse than the RT instability, which grows with $\sqrt{k}$ instead of k. Hadamard instabilities are ubiquitous in fluid mechanics; they are very important and should be a central topic in the study of hydrodynamic instabilities. However, these instabilities are not discussed in any of the standard books on hydrodynamic stability. Flows that are Hadamard unstable cannot be realistically studied by numerical methods; the finer the mesh, the worse the result.

11.2.3 The regularization of Hadamard instability

In the mathematical literature, problems that are Hadamard unstable are shown to be ill-posed. A regularized problem is not catastrophically unstable to short waves; σ_{Rm}

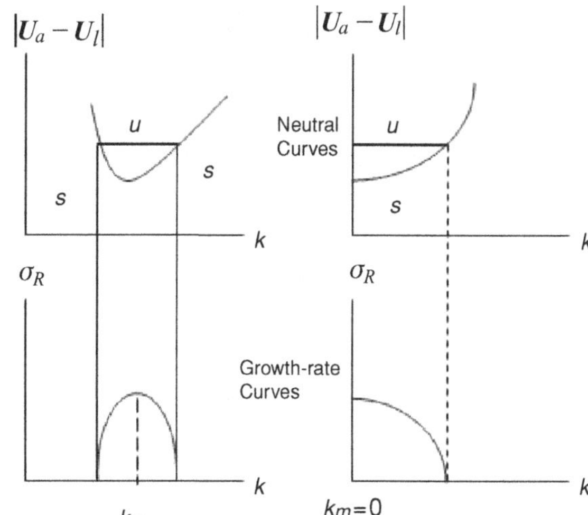

Figure 11.3. Neutral curves and growth-rate curves when surface tension $\gamma \neq 0$. (a) Flow is stable for short and long waves; (b) flow is stable for short waves but not long waves.

can be positive but it does not go to infinity as $k \to \infty$. Viscosity can be a regularizer. Really good regularizers, like surface tension, can stabilize short waves. The choice of regularizers should respect physics. It is not possible to present valid results for problems that are Hadamard unstable without understanding the physics that regularizes the awful instabilities. Unlike surface tension, viscosity will not cause the small waves to decay; they still grow but their growth is limited and the growth rate $Re\,[\sigma(k)]$ does not go to infinity with k as in Hadamard instability. The positive growth rate is given by

$$Re\,[\sigma_+] = \frac{\rho_a \mu_l^2 + \rho_l \mu_a^2}{2\,(\mu_l + \mu_a)^3}\,(U_a - U_l)^2\,, \quad k \to \infty.$$

11.2.4 Neutral curves

Neutral curves $|U_a - U_l|$ versus k for $\sigma_R = 0$ and growth-rate curves for an unstable value of $|U_a - U_l|$ are plotted in figures 11.3 and 11.4.

11.3 KH instability in a channel

Funada and Joseph (2001) studied the stability of stratified gas–liquid flow in a horizontal rectangular channel by using VPF. The analysis leads to an explicit dispersion relation in which the effects of surface tension and viscosity on the normal stress are not neglected but the effect of shear stresses is neglected. Formulas for the growth rates, wave speeds, and neutral stability curve are given in general and applied to experiments in air–water flows. The effects of surface tension are always important and determine the stability limits for the cases in which the volume fraction of gas is not too small. The stability criterion for VPF is expressed by a critical value of the relative velocity. The maximum critical value is when the viscosity ratio is equal to the density ratio; surprisingly the neutral curve for this viscous fluid is the same as the neutral curve for inviscid fluids. The maximum critical value of the velocity of all viscous fluids is given by inviscid fluids. For air at 20 °C and liquids with density $\rho = 1$ g/cm^3, the liquid viscosity for the critical conditions is 15 cP;

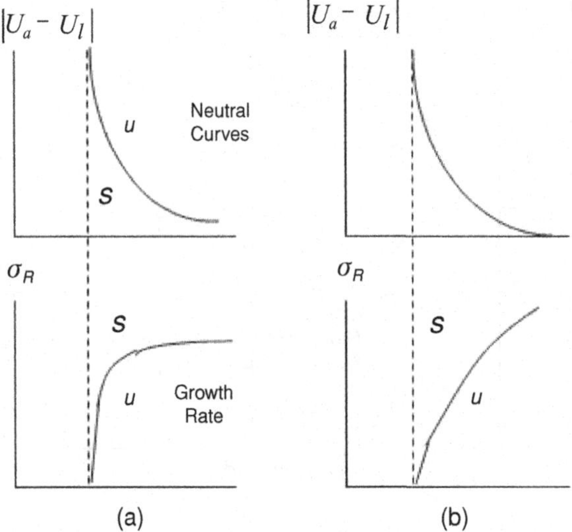

Figure 11.4. Neutral curves and growth-rate curves when $\gamma = 0$. (a) $\mu > 0$, the flow is unstable as $k \to 0$ but not Hadamard unstable; (b) $\mu = 0$, the flow is Hadamard unstable $\sigma_R \to \infty$ with k.

the critical velocity for liquids with viscosities larger than 15 cP is only slightly smaller but the critical velocity for liquids with viscosities smaller than 15 cP, like water, can be much lower. The viscosity of the liquid has a strong effect on the growth rate. The VPF theory fits the experimental data for air and water well when the gas fraction is greater than about 70%. It predicts the transition from smooth stratified to wavy stratified flow in a 0.508-m-diameter flow loop with air and 0.480 Pa s lube oil.

11.3.1 Formulation of the problem

A channel of rectangular cross section with height H and width W and of length L is set horizontally, in which a gas layer is over a liquid layer (see figure 11.5): the two-layer Newtonian incompressible fluids are immiscible. The undisturbed interface is taken at $z = 0$ on the z axis of Cartesian coordinates (x, y, z). We denote velocity by (u, w), pressure p, density ρ, viscosity μ, and acceleration that is due to gravity $(0, -g)$; the y component is ignored herein.

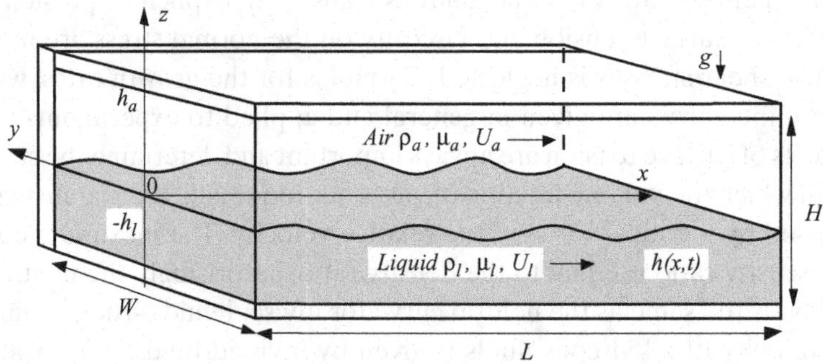

Figure 11.5. KH instability that is due to a discontinuity of velocity of air above liquid in a rectangular channel. The no-slip condition is not enforced in VPF so that the 2D solution satisfies the sidewall boundary conditions.

In the undisturbed state, the gas (air) with a uniform flow $(U_a, 0)$ is in $0 < z < h_a$, and the liquid with a uniform flow $(U_l, 0)$ is in $-h_l < z < 0$ $(H = h_l + h_a)$; the pressure has an equilibrium distribution that is due to the gravity. We consider KH instability of small disturbances to the undisturbed state.

The discontinuous prescription of data in the study of KH instability is a VPF solution of the Navier–Stokes in which no-slip conditions at the walls and no-slip and continuity of shear stress across the gas–liquid interface are neglected. Usually the analysis of KH instability is done with potential flow for an inviscid fluid, but this procedure leaves out certain effects of viscosity that can be included with complete rigor. An exact study of, say, air over water requires the inclusion of all of the effects of viscosity, and even the prescription of a basic flow is much more complicated. Problems of superposed viscous fluids have been considered, for example, in the monograph on two-fluid mechanics by Joseph and Renardy (1991) and more recently in the paper and references therein of Charru and Hinch (2000).

11.3.2 Viscous potential flow analysis

If the fluids are allowed to slip at the walls, then the 2D solution will satisfy the 3D equations, and we may reduce the analysis to flow between parallel plates. We found by computing that 3D disturbances are more stable than 2D ones. We now consider 2D disturbances, for which the velocity potential $\phi \equiv \phi(x, z, t)$ gives $(u, w) = \nabla\phi$.

The potential is subject to the equation of continuity:

$$\frac{\partial u}{\partial x} + \frac{\partial w}{\partial z} = 0 \to \frac{\partial^2 \phi}{\partial x^2} + \frac{\partial^2 \phi}{\partial z^2} = 0; \tag{11.3.1}$$

thus the potentials for the respective fluids are given by

$$\frac{\partial^2 \phi_a}{\partial x^2} + \frac{\partial^2 \phi_a}{\partial z^2} = 0 \quad \text{in } 0 < z < h_a, \tag{11.3.2}$$

$$\frac{\partial^2 \phi_l}{\partial x^2} + \frac{\partial^2 \phi_l}{\partial z^2} = 0 \quad \text{in } -h_l < z < 0. \tag{11.3.3}$$

Boundary conditions at the interface [at $z = h$, where $h \equiv h(x, t)$ is the interface elevation] are given by

$$\frac{\partial h}{\partial t} + U_a \frac{\partial h}{\partial x} = w_a, \quad \frac{\partial h}{\partial t} + U_l \frac{\partial h}{\partial x} = w_l, \tag{11.3.4}$$

and the conditions on the walls are given by

$$w_a = 0 \quad \text{at } z = h_a, \tag{11.3.5}$$

$$w_l = 0 \quad \text{at } z = -h_l. \tag{11.3.6}$$

The potential ϕ_a that satisfies (11.3.2) and (11.3.5) for the air and the potential ϕ_l that satisfies (11.3.3) and (11.3.6) for the liquid are given, respectively, by

$$\phi_a = A_a \cosh\left[k(z - h_a)\right] \exp\left(\sigma t + ikx\right) + \text{c.c.}, \tag{11.3.7a}$$

$$\phi_l = A_l \cosh\left[k(z + h_l)\right] \exp\left(\sigma t + ikx\right) + \text{c.c.}, \tag{11.3.7b}$$

and the interface elevation may be given by

$$h = A_0 \exp(\sigma t + ikx) + \text{c.c.}, \tag{11.3.7c}$$

where A_a, A_l, and A_0 denote the complex amplitude, and c.c. stands for the complex conjugate of the preceding expression; σ is the complex growth rate and $k > 0$ denotes the wavenumber; $i = \sqrt{-1}$. From kinematic conditions (11.3.4), we have the following equations for the complex amplitudes:

$$(\sigma + ikU_a) A_0 = -kA_a \sinh(kh_a), \tag{11.3.8a}$$

$$(\sigma + ikU_l) A_0 = kA_l \sinh(kh_l). \tag{11.3.8b}$$

The other boundary condition is the normal stress balance (with the normal viscous stress) at the interface:

$$-p_a + 2\mu_a \frac{\partial w_a}{\partial z} + \rho_a g h - \left(-p_l + 2\mu_l \frac{\partial w_l}{\partial z} + \rho_l g h\right) = -\gamma \frac{\partial^2 h}{\partial x^2}, \tag{11.3.9}$$

in which γ denotes the surface tension. Noting that the pressure is solely subject to the Laplace equation derived from the equation of motion for small disturbances, we may eliminate the pressure terms in (11.3.9) by using the equations of motion in which the viscous terms vanish identically when $\mathbf{u} = \nabla\phi$; $\mu\nabla^2\mathbf{u} = \mu\nabla\left(\nabla^2\phi\right) \equiv 0$. Thus p_a may be written, from the equation of motion in which the viscous term vanishes, as

$$\rho_a \left(\frac{\partial u_a}{\partial t} + U_a \frac{\partial u_a}{\partial x}\right) = -\frac{\partial p_a}{\partial x}, \tag{11.3.10a}$$

and, with the aid of the equation of continuity, we have the expression of p_a:

$$\rho_a \left(\frac{\partial^2 w_a}{\partial t \partial z} + U_a \frac{\partial^2 w_a}{\partial x \partial z}\right) = \frac{\partial^2 p_a}{\partial x^2}; \tag{11.3.10b}$$

the pressure p_l may be written as

$$\rho_l \left(\frac{\partial^2 w_l}{\partial t \partial z} + U_l \frac{\partial^2 w_l}{\partial x \partial z}\right) = \frac{\partial^2 p_l}{\partial x^2}. \tag{11.3.11}$$

Thus the normal stress balance is now written as

$$\begin{aligned} -\rho_a \left(\frac{\partial^2 w_a}{\partial t \partial z} + U_a \frac{\partial^2 w_a}{\partial x \partial z}\right) + 2\mu_a \frac{\partial^3 w_a}{\partial x^2 \partial z} + \rho_l \left(\frac{\partial^2 w_l}{\partial t \partial z} + U_l \frac{\partial^2 w_l}{\partial x \partial z}\right) \\ - 2\mu_l \frac{\partial^3 w_l}{\partial x^2 \partial z} - (\rho_l - \rho_a) g \frac{\partial^2 h}{\partial x^2} = -\gamma \frac{\partial^4 h}{\partial x^4}; \end{aligned} \tag{11.3.12}$$

hence we have the equation of σ, using (11.3.7a)–(11.3.8b):

$$\begin{aligned} &\left[\rho_a (\sigma + ikU_a)^2 + 2\mu_a k^2 (\sigma + ikU_a)\right] \coth(kh_a) \\ &+ \left[\rho_l (\sigma + ikU_l)^2 + 2\mu_l k^2 (\sigma + ikU_l)\right] \coth(kh_l) + (\rho_l - \rho_a) gk + \gamma k^3 = 0. \end{aligned} \tag{11.3.13}$$

11.3.2.1 *Dispersion relation*

From (11.3.13) the dispersion relation is given as

$$A\sigma^2 + 2B\sigma + C = 0, \tag{11.3.14}$$

where the coefficients A, B, and C are defined as

$$A = \rho_l \coth(kh_l) + \rho_a \coth(kh_a), \tag{11.3.15a}$$

$$\begin{aligned} B &= ik[\rho_l U_l \coth(kh_l) + \rho_a U_a \coth(kh_a)] + k^2[\mu_l \coth(kh_l) + \mu_a \coth(kh_a)] \\ &= B_R + iB_I, \end{aligned} \tag{11.3.15b}$$

$$\begin{aligned} C &= (\rho_l - \rho_a)gk - k^2\left[\rho_l U_l^2 \coth(kh_l) + \rho_a U_a^2 \coth(kh_a)\right] + \gamma k^3 \\ &\quad + 2ik^3[\mu_l U_l \coth(kh_l) + \mu_a U_a \coth(kh_a)] = C_R + iC_I. \end{aligned} \tag{11.3.15c}$$

The solution σ may be expressed as

$$\sigma = -\frac{B}{A} \pm \sqrt{\frac{B^2}{A^2} - \frac{C}{A}} \rightarrow \sigma_R + i\sigma_I = -\frac{B_R + iB_I}{A} \pm \frac{\sqrt{D}}{A}, \tag{11.3.16}$$

where D is given by

$$D = D_R + iD_I = (B_R + iB_I)^2 - A(C_R + iC_I), \tag{11.3.17a}$$

$$\begin{aligned} D_R &= \rho_l\rho_a(U_a - U_l)^2 k^2 \coth(kh_l)\coth(kh_a) + k^4[\mu_l \coth(kh_l) + \mu_a \coth(kh_a)]^2 \\ &\quad - [\rho_l \coth(kh_l) + \rho_a \coth(kh_a)]\left[(\rho_l - \rho_a)gk + \gamma k^3\right], \end{aligned} \tag{11.3.17b}$$

$$D_I = 2k^3(\rho_a\mu_l - \rho_l\mu_a)(U_a - U_l)\coth(kh_l)\coth(kh_a). \tag{11.3.17c}$$

When $\rho_a\mu_l = \rho_l\mu_a$ for which $D_I = 0$, and if $D_R \geq 0$, we have

$$\sigma_R = \frac{-B_R \pm \sqrt{D_R}}{A}, \tag{11.3.18a}$$

$$\sigma_I = -\frac{B_I}{A}. \tag{11.3.18b}$$

This is a typical case in which the real and imaginary parts of σ can be expressed most clearly.

If the top and bottom are far away, $h_l \to \infty$, $h_a \to \infty$, then (11.3.14) gives rise to

$$\begin{aligned} \sigma = &-\frac{ik(\rho_a U_a + \rho_l U_l) + k^2(\mu_a + \mu_l)}{(\rho_a + \rho_l)} \\ &\pm \left[\frac{\rho_a\rho_l k^2(U_a - U_l)^2}{(\rho_a + \rho_l)^2} - \frac{(\rho_l - \rho_a)gk + \gamma k^3}{(\rho_a + \rho_l)} + \frac{k^4(\mu_a + \mu_l)^2}{(\rho_a + \rho_l)^2}\right. \\ &\left. + 2ik^3\frac{(\rho_a\mu_l - \rho_l\mu_a)(U_a - U_l)}{(\rho_a + \rho_l)^2}\right]^{1/2}, \end{aligned}$$

which is reduced, for the particular case that $\rho_a\mu_l = \rho_l\mu_a$, to

$$\sigma_R = -\frac{k^2(\mu_a + \mu_l)}{(\rho_a + \rho_l)} \pm \left[\frac{\rho_a\rho_l k^2(U_a - U_l)^2}{(\rho_a + \rho_l)^2} - \frac{(\rho_l - \rho_a)gk + \gamma k^3}{(\rho_a + \rho_l)} + \frac{k^4(\mu_a + \mu_l)^2}{(\rho_a + \rho_l)^2}\right]^{1/2}, \tag{11.3.19a}$$

$$\sigma_I = -\frac{k(\rho_a U_a + \rho_l U_l)}{(\rho_a + \rho_l)}. \tag{11.3.19b}$$

Here, it is easy to find that $\sigma_R = 0$ gives a relation independent of viscosity. In other words, the relation holds even for inviscid fluids; this is helpful for the problem to be considered herein.

11.3.2.2 *Growth rates and wave speeds*

In terms of $\sigma = \sigma_R + i\sigma_I$, (11.3.14) is also written, for the real and imaginary parts, as

$$A\left(\sigma_R^2 - \sigma_I^2\right) + 2\left(B_R\sigma_R - B_I\sigma_I\right) + C_R = 0, \quad \sigma_I = -\frac{2B_I\sigma_R + C_I}{2\left(A\sigma_R + B_R\right)}. \tag{11.3.20}$$

Eliminating σ_I from the preceding equation, we have a quartic equation for σ_R:

$$a_4\sigma_R^4 + a_3\sigma_R^3 + a_2\sigma_R^2 + a_1\sigma_R + a_0 = 0, \tag{11.3.21}$$

where the coefficients are given as

$$a_4 = A^3, \quad a_3 = 4A^2B_R, \quad a_2 = 5AB_R^2 + AB_I^2 + A^2C_R, \tag{11.3.22a}$$

$$a_1 = 2B_R^3 + 2B_RB_I^2 + 2AB_RC_R, \quad a_0 = -\frac{1}{4}AC_I^2 + B_RB_IC_I + B_R^2C_R. \tag{11.3.22b}$$

Quartic equation (11.3.21) can be solved analytically. Neutral states for which $\sigma_R = 0$ are described in terms of the solution to the equation $a_0 = 0$. The peak value (*the maximum growth rate*) σ_m and the corresponding wavenumber k_m are obtained by the solution of (11.3.21). It is usually true, but unproven, that $\lambda_m = 2\pi/k_m$ will be the length of unstable waves observed in experiments.

The complex values of σ are frequently expressed in terms of a complex frequency ω with

$$\sigma_R + i\sigma_I = \sigma = -i\omega = -i\omega_R + \omega_I. \tag{11.3.23}$$

Hence

$$\sigma_R = \omega_I; \quad \sigma_I = -\omega_R. \tag{11.3.24}$$

The wave speed for the mode with wavenumber k is

$$\tilde{C}_R = \omega_R/k = -\sigma_I/k. \tag{11.3.25}$$

The set of wavenumbers for which flows are stable is also of interest. The wavelengths corresponding to these wavenumbers will not appear in the spectrum. *Cutoff wavenumbers* k_C separate the unstable and stable parts of the spectrum.

11.3.2.3 *Neutral curves*

Neutral curves define values of the parameters for which $\sigma_R(k) = 0$. We may obtain these curves by putting $a_0 = 0$:

$$-\frac{\rho_l\mu_a^2\coth(kh_l)\coth^2(kh_a) + \rho_a\mu_l^2\coth^2(kh_l)\coth(kh_a)}{\left[\mu_l\coth(kh_l) + \mu_a\coth(kh_a)\right]^2}kV^2 + (\rho_l - \rho_a)g + \gamma k^2 = 0, \tag{11.3.26}$$

where the relative velocity V is defined by $V \equiv U_a - U_l$. This equation may be solved for V^2, where

$$V^2(k) = \frac{[\mu_l \coth(kh_l) + \mu_a \coth(kh_a)]^2}{\rho_l \mu_a^2 \coth(kh_l)\coth^2(kh_a) + \rho_a \mu_l^2 \coth^2(kh_l)\coth(kh_a)} \frac{1}{k}\left[(\rho_l - \rho_a) g + \gamma k^2\right]. \tag{11.3.27}$$

The lowest point on the neutral curve $V^2(k)$ is

$$V_c^2 = \min_{k \geq 0} V^2(k) \equiv V^2(k_c), \tag{11.3.28}$$

where $\lambda_c = 2\pi/k_c$ is the wavelength that makes V^2 minimum. The flow is unstable when

$$V^2 = (-V)^2 > V_c^2. \tag{11.3.29}$$

This criterion is symmetric with respect to V and $-V$, depending on only the absolute value of the difference. This feature stems from Galilean invariance; the flow seen by an observer moving with the gas is the same as the one seen by an observer moving with the liquid.

By recalling the results obtained by computing, we find it interesting to note here that the 3D disturbances in the sense of the VPF lead to the relative velocity $V_{3\mathrm{D}}$, which can be expressed in terms of (11.3.27) as

$$V_{3\mathrm{D}}^2 \equiv \frac{(\mathbf{k} \cdot \mathbf{U}_a - \mathbf{k} \cdot \mathbf{U}_l)^2}{k_x^2} = \frac{k^2}{k_x^2} V^2(k), \tag{11.3.30}$$

only if we regard the 3D wavenumber vector $\mathbf{k} = (k_x, k_y)$ as

$$k = \sqrt{k_x^2 + k_y^2}, \quad k_y = 0, \pm\frac{\pi}{W}, \pm\frac{2\pi}{W}, \ldots. \tag{11.3.31}$$

It is evident in (11.3.30) that $V_{3\mathrm{D}}^2$ is larger than $V^2(k)$ if $k_y \neq 0$; the most dangerous 3D disturbance is 2D with $k_y = 0$.

11.3.3 KH instability of inviscid fluid

For inviscid fluids ($\mu_a = \mu_l = 0$), we have $B_R = 0$ and $C_I = 0$; thus $a_3 = a_1 = a_0 = 0$ and (11.3.21) reduces to

$$a_4 \sigma_R^4 + a_2 \sigma_R^2 = 0; \tag{11.3.32}$$

thus we have

$$a_4 \sigma_R^2 + a_2 = 0, \tag{11.3.33}$$

$$\sigma_I = -\frac{B_I}{A} = -\frac{k[\rho_l U_l \coth(kh_l) + \rho_a U_a \coth(kh_a)]}{\rho_l \coth(kh_l) + \rho_a \coth(kh_a)}. \tag{11.3.34}$$

It should be noted here that the neutral curve was given by the equation $a_0 = 0$ in the viscous potential analysis [(11.3.26) and (11.3.27)], whereas the neutral curve in this KH instability is given by the equation $a_2 = 0$. It is also noted that σ_I is the same as (11.3.18b), although σ_R may be different, in general, from (11.3.18a). However, the equation $\sigma_R = 0$ in (11.3.33) is the same $\sigma_R = 0$ in (11.3.18a) for the case of $\rho_a \mu_l = \rho_l \mu_a$.

From (11.3.33) with $a_2 < 0$, the growth rate σ_R is expressed as

$$\sigma_R = \pm\frac{\sqrt{\rho_l\rho_a k^2 \coth(kh_l)\coth(kh_a)\,V^2 - [\rho_l\coth(kh_l)+\rho_a\coth(kh_a)][(\rho_l-\rho_a)\,gk+\gamma k^3]}}{\rho_l\coth(kh_l)+\rho_a\coth(kh_a)}. \tag{11.3.35}$$

At the neutral state $\sigma_R = 0$ for which $a_2 = 0$, we have

$$\frac{\rho_l\rho_a k\coth(kh_l)\coth(kh_a)}{\rho_l\coth(kh_l)+\rho_a\coth(kh_a)}V^2 - \left[(\rho_l-\rho_a)g+\gamma k^2\right] = 0. \tag{11.3.36}$$

Instability arises if

$$V^2 > \left[\frac{\tanh(kh_l)}{\rho_l}+\frac{\tanh(kh_a)}{\rho_a}\right]\frac{1}{k}\left[(\rho_l-\rho_a)g+\gamma k^2\right] \equiv V_i^2(k). \tag{11.3.37}$$

In the stable case for which $a_2 > 0$, the wave velocity $\tilde{C}_R$ is given by

$$-k\tilde{C}_R = \sigma_I = -\frac{B_I}{A}\pm\sqrt{\frac{B_I^2}{A^2}+\frac{C_R}{A}}. \tag{11.3.38}$$

11.3.4 Dimensionless form of the dispersion equation

The dimensionless variables are designated with a caret and are

$$\hat{k} = kH, \qquad \hat{U}_a = \frac{U_a}{Q},$$
$$\hat{h}_a = \frac{h_a}{H} \equiv \alpha, \qquad \hat{U}_l = \frac{U_l}{Q},$$
$$\hat{h}_l = \frac{h_l}{H} = 1-\hat{h}_a, \qquad \hat{V} = \hat{U}_a - \hat{U}_l$$
$$\hat{\rho} = \frac{\rho_a}{\rho_l}, \qquad \hat{\sigma} = \frac{\sigma H}{Q},$$
$$\hat{\mu} = \frac{\mu_a}{\mu_l}, \qquad \theta = \frac{\mu_l}{\rho_l H Q},$$
$$\hat{\gamma} = \frac{\gamma}{\rho_l g H^2},$$

where

$$Q = \left[\frac{(1-\hat{\rho})gH}{\hat{\rho}}\right]^{1/2}.$$

The dimensionless form of (11.3.14) is given by

$$\begin{aligned}
&\left[\coth(\hat{k}\hat{h}_l) + \hat{\rho}\coth(\hat{k}\hat{h}_a)\right]\hat{\sigma}^2\\
&+2\hat{\sigma}\left\{i\hat{k}\left[\hat{U}_l\coth(\hat{k}\hat{h}_l)+\hat{\rho}\,\hat{U}_a\coth(\hat{k}\hat{h}_a)\right]+\theta\hat{k}^2\left[\coth(\hat{k}\hat{h}_l)+\hat{\mu}\coth(\hat{k}\hat{h}_a)\right]\right\}\\
&-\hat{k}^2\left[\hat{U}_l^2\coth(\hat{k}\hat{h}_l)+\hat{\rho}\,\hat{U}_a^2\coth(\hat{k}\hat{h}_a)\right]+2i\hat{k}^3\theta\left[\hat{U}_l\coth(\hat{k}\hat{h}_l)+\hat{\mu}\hat{U}_a\coth(\hat{k}\hat{h}_a)\right]\\
&+\hat{k}\left[1+\frac{\hat{\gamma}\hat{k}^2}{(1-\hat{\rho})}\right] = 0.
\end{aligned} \tag{11.3.39}$$

Expression (11.3.27) for the neutral curve $\hat{\sigma}_R(\hat{k}) = 0$ is written in dimensionless variables as

$$\hat{V}^2 = \frac{\left[\tanh(\hat{k}\hat{h}_a) + \hat{\mu}\tanh(\hat{k}\hat{h}_l)\right]^2}{\tanh(\hat{k}\hat{h}_a) + (\hat{\mu}^2/\hat{\rho})\tanh(\hat{k}\hat{h}_l)} \frac{1}{\hat{k}} \left[1 + \frac{\hat{\gamma}\hat{k}^2}{(1-\hat{\rho})}\right]. \tag{11.3.40}$$

Note that the growth-rate parameter $\theta = \mu_l/(\rho_l HQ)$, which depends linearly on the kinematic viscosity $\nu_l = \mu_l/\rho_l$ of the liquid, does not appear in (11.3.40). Note also that the value of $(1-\hat{\rho})$ is close to unity, as $\hat{\rho} = 0.0012$ for air–water.

We can obtain the neutral curves for an inviscid fluid (11.3.36) by putting $\hat{\mu} = \hat{\rho}$ or $\mu_l/\rho_l = \mu_a/\rho_a$. This gives from (11.3.40) the following expression;

$$\hat{V}^2 = \left[\tanh(\hat{k}\hat{h}_a) + \hat{\rho}\tanh(\hat{k}\hat{h}_l)\right] \frac{1}{\hat{k}} \left[1 + \frac{\hat{\gamma}\hat{k}^2}{(1-\hat{\rho})}\right], \tag{11.3.41}$$

which is the dimensionless form of (11.3.37). Although this reduction is immediate, it is surprising.

Evaluating (11.3.40) for $\hat{\mu} = 0$, we get

$$\hat{V}^2 = \tanh(\hat{k}\hat{h}_a) \frac{1}{\hat{k}} \left[1 + \frac{\hat{\gamma}\hat{k}^2}{(1-\hat{\rho})}\right]. \tag{11.3.42}$$

Evaluating (11.3.40) for $\hat{\mu} = \infty$, we get

$$\hat{V}^2 = \hat{\rho}\tanh(\hat{k}h_l) \frac{1}{\hat{k}} \left[1 + \frac{\hat{\gamma}\hat{k}^2}{(1-\hat{\rho})}\right]. \tag{11.3.43}$$

It is easy to verify that $\hat{V}^2$ is maximum at $\hat{\mu} = \hat{\rho}$ for inviscid fluids. Viscosity in VPF is destabilizing; however, large viscosities are less destabilizing than small viscosities.

Because $\hat{\rho} = 0.0012$, which is very small, the variation in the stability is large when $\hat{\mu}$ varies between $\hat{\rho}$ and ∞, and is very small when $\hat{\mu}$ varies between $\hat{\rho}$ and zero. The value $\hat{\mu} = 0.018 > \hat{\rho} = 0.0012$ and is in the interval in which $\hat{V}^2$ is rapidly varying (see figure 11.8 in the next subsection).

In the literature on gas–liquid flows, a long-wave approximation is often made to obtain stability limits. For long waves $\hat{k} \to 0$ and $\tanh(\hat{k}\hat{h}) \to \hat{k}\hat{h}$, and (11.3.40) reduces to

$$\hat{V}^2 = \frac{(\hat{h}_a + \hat{\mu}\hat{h}_l)^2}{\hat{h}_a + (\hat{\mu}^2/\hat{\rho})\hat{h}_l} \left[1 + \frac{\hat{\gamma}\hat{k}^2}{(1-\hat{\rho})}\right]. \tag{11.3.44}$$

The effect of surface tension disappears in this limit, but the effects of viscosity are important. To obtain the long-wave limit in the inviscid case, put $\hat{\mu} = \hat{\rho}$.

The regularization of short waves by surface tension is an important physical effect. For short waves, $\hat{k} \to \infty$, $\tanh(\hat{k}\hat{h}) \to 1$, and

$$\hat{V}^2 = \frac{(\hat{\mu}+1)^2}{1 + \hat{\mu}^2/\hat{\rho}} \frac{1}{\hat{k}} \left[1 + \frac{\hat{\gamma}\hat{k}^2}{(1-\hat{\rho})}\right]. \tag{11.3.45}$$

To obtain the short-wave limit in the inviscid case, put $\hat{\mu} = \hat{\rho}$.

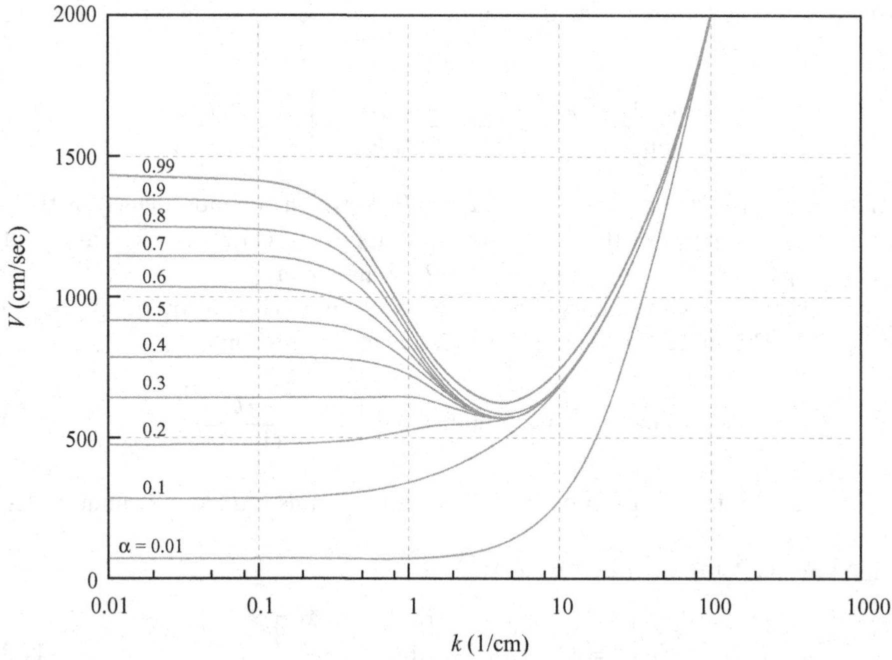

Figure 11.6. Neutral curves for air and water ($\hat{\mu} = 0.018$; see table 11.1 and figure 11.8); $\alpha = \hat{h}_a$ is the gas fraction. As in the usual manner, the disturbances will grow above the neutral, but decay below it. For α larger than about 0.2, there arises the critical velocity V_c below which all the disturbances will decay.

The effects of surface tension may be computed from (11.3.44) and (11.3.45). The stability limit for long waves, $\hat{k} \to 0$, is independent of $\hat{\gamma}$. For short waves, (11.3.44) has a minimum at $\hat{k} = \sqrt{(1 - \hat{\rho})/\hat{\gamma}}$ with a value there given by

$$\hat{V}^2 = \frac{2(\hat{\mu} + 1)^2}{1 + \hat{\mu}^2/\hat{\rho}} \sqrt{\frac{\hat{\gamma}}{(1 - \hat{\rho})}}. \tag{11.3.46}$$

Equation (11.3.46) shows that short waves are stabilized by increasing $\hat{\gamma}$. For small $\hat{\gamma}$, long waves are unstable.

11.3.5 The effect of liquid viscosity and surface tension on growth rates and neutral curves

For air and water at 20 °C,

$$\rho_a = 0.0012\,\text{g/cm}^3, \quad \rho_l = 1\,\text{g/cm}^3, \quad \hat{\rho} = \rho_a/\rho_l = 0.0012, \tag{11.3.47}$$

$$\mu_a = 0.00018\,\text{P}, \quad \mu_l = 0.01\,\text{P}, \quad \hat{\mu} = \mu_a/\mu_l = 0.018. \tag{11.3.48}$$

The surface tension of air and pure water is $\gamma = 72.8$ dyn/cm. Usually the water is slightly contaminated, and $\gamma = 60$ dyn/cm is more probable for the water–air tension in experiments. For all kinds of organic liquids a number like $\gamma = 30$ dyn/cm is a good approximation.

Neutral curves for $\hat{\mu} = 0.018$ (air–water) and $\hat{\mu} = \hat{\rho} = 0.0012$ (inviscid flow) and $\hat{\mu} = 3.6 \times 10^{-6}$ ($\mu_l = 50$ P) with $\gamma = 60$ dyn/cm are selected here; the former two are

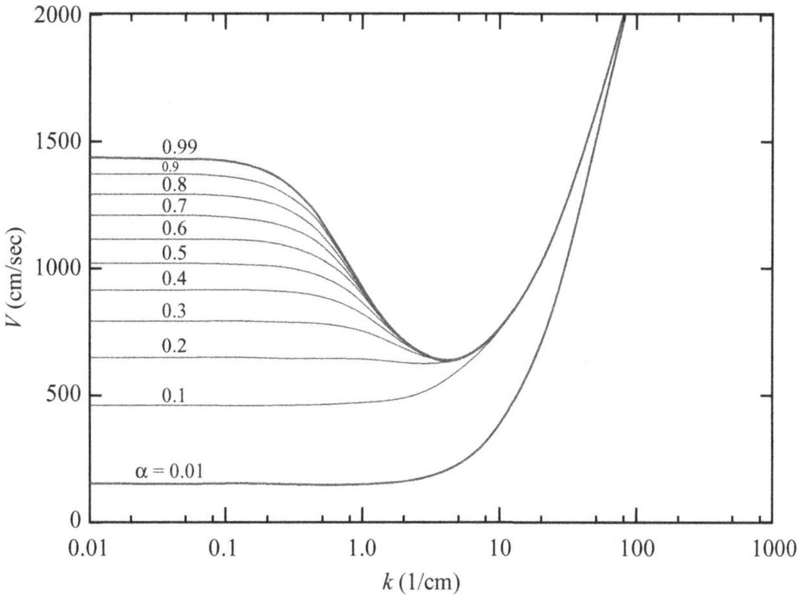

Figure 11.7. Neutral curves for inviscid fluids ($\hat{\mu} = \hat{\rho} = 0.0012$) for different gas fractions $\alpha = \hat{h}_a$. This neutral curve arose for the special case $\hat{\mu} = \hat{\rho} = 0.0012 = \mu_a/\mu_l$ with $\mu_a = 0.00018$ P; hence $\mu_l = 0.15$ P. Surprisingly, it is identical to the case $\mu_a = \mu_l = 0$ (see table 11.2 and figure 11.8). The neutral curves for viscous fluids with $\mu_l > 15$ cP are essentially the same as these (cf. tables 11.2 and 11.3).

shown in figures 11.6 and 11.7. The liquid viscosities $\mu_l = \rho_l\mu_a/\rho_a$ corresponding to these three cases are $\mu_l = 0.01$ P, 0.15 P, and 50 P. The neutral curves for $\hat{\mu} \geq \hat{\rho}$ are nearly identical. The neutral curves for $\hat{\mu} = 0.018$ (air–water) are to be compared with experiments. We have identified the minimum values of (11.3.40) over $\hat{k} \geq 0$ in the air–water case, and in tables 11.1, 11.2 and 11.3 the critical velocity $V_c = V(k_c)$, the critical wavenumber k_c (and wavelength $\lambda_c = 2\pi/k_c$), and associated wave speeds $\tilde{C}_{Rc} = \tilde{C}_R(k_c)$ are listed. In the tables, V_s and $\tilde{C}_{Rs}$ denote the values taken at $k = 10^{-3}$ cm^{-1}, which may be representative of values in the limit of long waves, $k \to 0$. The variation of the critical velocity with the viscosity ratio $\hat{\mu} = \mu_a/\mu_l$ for a representative gas fraction $\alpha = 0.5$ is shown in figure 11.8. The vertical line $\hat{\mu} = \hat{\rho}$ identifies the stability limit for inviscid fluids. Points to the left of this line have high liquid viscosities $\mu_l > 0.15$ P, and for points to the right, $\mu_l < 0.15$ P.

In all cases the critical velocity is influenced by surface tension; the critical velocity is given by long waves only when α is small (small air gaps). For larger values of α (say $\alpha > 0.2$), the most dangerous neutrally unstable wave is identified by a sharp minimum determined by surface tension, which is identified in table 11.1 [cf. equation (11.3.46)].

The growth rates depend strongly on the liquid viscosity, unlike the neutral curves. The most dangerous linear wave is the one whose growth rate σ_R is maximum at $k = k_m$,

$$\sigma_{Rm} = \sigma_R(k_m) = \max_{k \geq 0} \sigma_R(k), \tag{11.3.49}$$

with an associated wavelength $\lambda_m = 2\pi/k_m$ and wave speed $\tilde{C}_{Rm} = \tilde{C}_R(k_m)$. Typical growth-rate curves are shown in figure 11.9. Maximum growth-rate parameters for

Table 11.1. *Typical values of the neutral curves in figure 11.6 for air–water with* $\rho_a = 0.0012$ *g/cm*^{3}, $\mu_a = 0.00018$ *P,* $\rho_l = 1.0$ *g/cm*^{3}, $\mu_l = 0.01$ *P,* $g = 980.0$ *cm/sec*^{2}, $\gamma = 60.0$ *dyn/cm,* $H = 2.54$ *cm*

This table was based on the results of computation that the neutral curves with $\alpha = 0.1$ and 0.2 in figure 11.6 increase monotonically from the values V_s cm/sec at $k = 10^{-3}$ cm^{-1}; the curve with $\alpha = 0.3$ in figure 11.6 increases from the value V_s cm/sec at $k = 10^{-3}$ cm^{-1}, takes a maximum $V = 651.3$ cm/sec at $k = 0.692$ cm^{-1}, and then takes a minimum $V_c = 572.5$ cm/sec (the critical) at $k_c = 3.893$ cm^{-1}; for the other values of α, the corresponding curves give the critical V_c at k_c

$\hat{h}_a$	V_s (cm/sec)	$\tilde{C}_{Rs}$ (cm/sec)	k_c (cm^{-1})	λ_c (cm)	V_c (cm/sec)	$\tilde{C}_{Rc}$ (cm/sec)
0.01	76.04	198.6	0.649	9.676	72.92	155.9
0.1	285.6	43.22				
0.2	478.5	20.82				
0.3	643.4	12.50	0.692	9.076	651.3	9.432
			3.893	1.614	572.5	5.510
0.4	788.8	8.150	4.020	1.563	573.9	5.484
0.5	919.4	5.481	4.052	1.551	574.1	5.481
0.6	1039	3.676	4.052	1.551	574.1	5.479
0.7	1149	2.373	4.052	1.551	574.3	5.459
0.8	1252	1.389	4.117	1.526	575.7	5.319
0.9	1348	0.619	4.354	1.443	585.3	4.415
0.99	1430	0.056	4.150	1.514	628.0	0.585

$\hat{\mu} = 0.018$ (figure 11.9), $\hat{\mu} = \hat{\rho} = 0.0012$, $\mu_l = 15$ cP, and $\hat{\mu} = 3.6 \times 10^{-6}$ ($\mu_l = 50$ P) are listed for $V = 1500$ and 900 cm/sec in table 11.4.

11.3.6 Comparison of theory and experiments in rectangular ducts

Kordyban and Ranov (1970) and Wallis and Dobson (1973) are the only authors to report the results of experiments in rectangular ducts. Many other experiments have been

Table 11.2. *Typical values of the neutral curves in figure 11.7 for air–water (as inviscid fluids) with* $\rho_a = 0.0012$ *g/cm*^{3}, $\mu_a = 0.0$ *P,* $\rho_l = 1.0$ *g/cm*^{3}, $\mu_l = 0.0$ *P,* $g = 980.0$ *cm/sec*^{2}, $\gamma = 60.0$ *dyn/cm,* $H = 2.54$ *cm*

$\hat{h}_a$	V_s (cm/sec)	$\tilde{C}_{Rs}$ (cm/sec)	k_c (cm^{-1})	λ_c (cm)	V_c (cm/sec)	$\tilde{C}_{Rc}$ (cm/sec)
0.01	152.2	16.17	0.629	9.990	150.6	9.725
0.1	457.6	4.890				
0.2	645.3	3.082	2.990	2.101	619.8	0.818
0.3	789.5	2.204	3.924	1.601	634.4	0.764
0.4	911.2	1.637	4.020	1.563	635.7	0.762
0.5	1018	1.221	4.052	1.551	635.9	0.762
0.6	1115	0.892	4.052	1.551	635.9	0.762
0.7	1205	0.619	4.052	1.551	635.9	0.759
0.8	1288	0.386	4.052	1.551	635.9	0.738
0.9	1366	0.182	4.052	1.551	635.8	0.590
0.99	1432	0.017	4.052	1.551	635.6	0.078

Table 11.3. *Typical values of the neutral curves for air–high-viscosity liquid with* $\rho_a = 0.0012\ g/cm^3$, $\mu_a = 0.00018\ P$, $\rho_l = 1.0\ g/cm^3$, $\mu_l = 50.0\ P$, $g = 980.0\ cm/sec^2$, $\gamma = 60.0\ dyn/cm$, $H = 2.54\ cm$; *thus* $\hat{\mu} = 3.6 \times 10^{-6}$. *This corresponds to a high-viscosity case in figure 11.8*

The curves with $\hat{h}_a = 0.5$–0.8 take almost the same minimum value at $k = k_c$, though the values at $k = 10^{-3}\ cm^{-1}$ change as $V_s = 1018$–1287 cm/sec and $\tilde{C}_{Rs} = 0.0011-0.0003$ cm/sec; see table 11.4 for the maximum growth rate

$\hat{h}_a$	V_s (cm/sec)	$\tilde{C}_{Rs}$ (cm/sec)	k_c (cm^{-1})	λ_c (cm)	V_c (cm/sec)	$\tilde{C}_{Rc}$ (cm/sec)
0.01	144.0	0.1104				
0.1	455.2	0.0100				
0.2	643.7	0.0045	2.990	2.101	619.4	0.0012
0.3	788.4	0.0026	3.924	1.601	634.1	0.0011
0.4	910.4	0.0017	4.020	1.563	635.4	0.0011
0.5	1018	0.0011	4.052	1.551	635.5	0.0011
0.6	1115	0.0007	4.052	1.551	635.5	0.0011
0.7	1204	0.0005	4.052	1.551	635.5	0.0011
0.8	1287	0.0003	4.052	1.551	635.5	0.0011
0.9	1366	0.0001	4.052	1.551	635.5	0.0009
0.99	1432	1.1×10^{-5}	4.052	1.551	635.5	0.0001

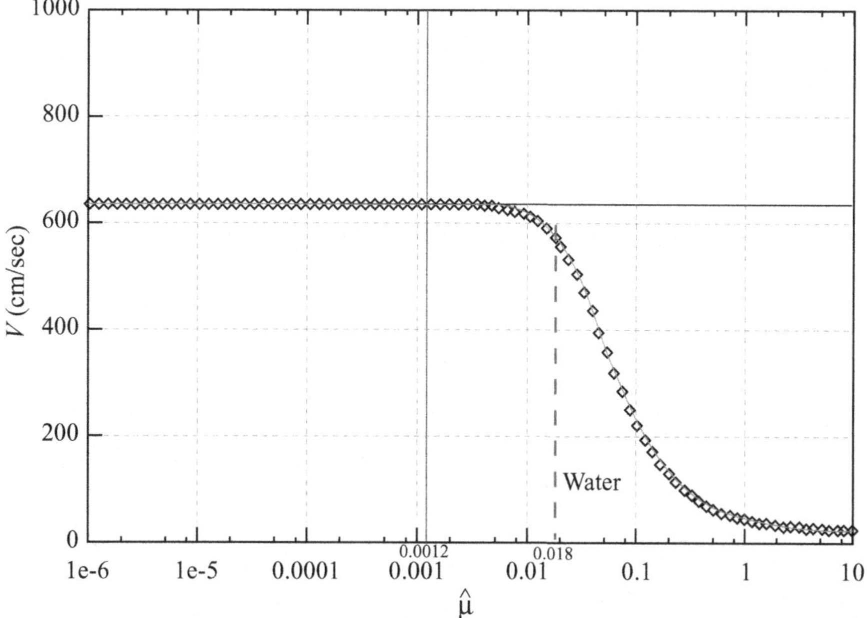

Figure 11.8. Critical velocity V vs. $\hat{\mu}$ for $\alpha = 0.5$. The critical velocity is the minimum value on the neutral curve. The vertical line is $\hat{\mu} = \hat{\rho} = 0.0012$ and the horizontal line at $V = 635.9$ cm/sec is the critical value for inviscid fluids. The vertical dashed line at $\hat{\mu} = 0.018$ is for air and water. Typical values for a high-viscosity liquid are given in table 11.3.

Table 11.4. *Wavenumber, wavelength, and wave speed for the maximum growth rate (11.3.49)*

$\hat{\mu}$	V(cm/sec)	$\hat{h}_a$	k_m (cm^{-1})	λ_m (cm)	σ_{Rm} (sec^{-1})	$\tilde{C}_{Rm}$ (cm/sec)
0.018	1500	0.01	29.90	0.2101	1448	3.044
		0.1–0.9	29.66	0.2118	872.5	2.049
		0.99	32.13	0.1955	706.2	1.454
	900	0.01	15.40	0.408	615.3	3.046
		0.1	10.00	0.628	167.7	1.183
		0.2	10.24	0.613	164.2	1.175
		0.3–0.8	10.24	0.613	164.2	1.174
		0.9	10.33	0.609	163.3	1.164
		0.99	11.36	0.553	84.66	0.367
0.0012	1500	0.01	26.95	0.233	1340	3.022
		0.1–0.9	27.17	0.231	768	1.798
		0.99	30.14	0.209	584.7	1.159
	900	0.01	14.45	0.435	585.2	3.064
		0.1	9.456	0.665	155.1	1.097
		0.2–0.7	9.685	0.649	151.0	1.079
		0.8	9.763	0.644	151.0	1.079
		0.9	9.841	0.638	149.9	1.064
		0.99	10.66	0.589	69.59	0.285
3.6×10^{-6}	1500	0.01	1.821	3.450	295.1	24.55
		0.1	0.916	6.861	60.04	4.495
		0.2	0.845	7.432	34.43	2.049
		0.3	3.087	2.035	21.96	0.086
		0.4–0.6	4.4–4.5	1.42–1.40	21.89	0.045–0.04
		0.7	4.679	1.343	21.85	0.040
		0.8	5.360	1.172	21.61	0.032
		0.9	7.743	0.812	20.21	0.017
		0.99	20.54	0.306	6.801	0.003
	900	0.01	1.323	4.750	145.9	19.64
		0.1	0.676	9.297	24.80	3.017
		0.2	0.581	10.82	10.48	1.199
		0.3	0.984	6.385	4.294	0.135
		0.4–0.6	4.02–4.08	1.56–1.51	4.86	0.010
		0.7	4.150	1.514	4.840	0.009
		0.8	4.460	1.409	4.735	0.008
		0.9	5.534	1.135	4.100	0.005
		0.99	7.994	0.786	0.741	0.001

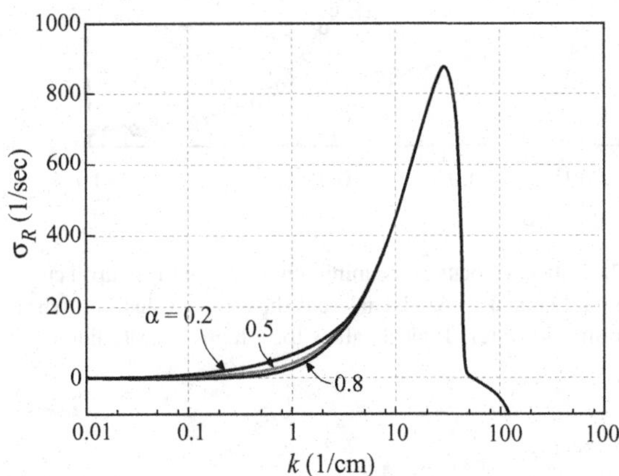

Figure 11.9. The real part of growth rate σ_R sec^{-1} vs. k cm^{-1} for $\hat{\mu} = 0.018$ (water, $\mu_l = 1$ cP), $V = 1500$ cm/sec. The graphs are, from top to bottom, $\alpha = 0.2,\ 0.5,\ 0.8$. The curves of σ_R sec^{-1} along the line of $V = 1500$ cm/sec in figure 11.6 are drawn here for respective values of α. Instability may arise for all the disturbances of wavenumbers below the cutoff wavenumber k_C. The maximum growth rate σ_{Rm} and the corresponding wavenumber $k_m = 2\pi/\lambda_m$ for $V = 1500$ and 900 cm/sec are listed with wave velocity $\tilde{C}_{Rm}$ in table 11.4.

carried out in round pipes; the results of these experiments are not perfectly matched to the analysis done here or elsewhere and will be discussed later.

All experimenters were motivated to understand the transition from stratified flow with a flat smooth interface to slug flow. They note that in many cases the first transition, which is studied here, is from smooth stratified flow to small-amplitude sinusoidal waves, called capillary waves by Wallis and Dobson (1973). The data given by these authors are framed as a transition to slug flow, though the criteria given are for the loss of stability of smooth stratified flow. The theoretical predictions are for the loss of stability, which may or may not be to slug flow.

Note also that all the linear theories that neglect viscosity overpredict the observed stability limit. Wallis and Dobson (1973) note that "... as a result of the present experiments it is our view that the various small wave theories are all inappropriate for describing 'slugging.' Slugging is the result of the rapid development of a large wave which rides over the underlying liquid and can eventually fill the channel to form a slug. ... " They also note that "it was found possible to produce slugs at air fluxes less than those predicted" by their empirical formula, $j^* < 0.5\alpha^{3/2}$. All this suggests that we may be looking at something akin to subcritical bifurcation with multiple solutions and hysteresis.

Turning next to linearized theory, we note that Wallis and Dobson (1973) do an inviscid analysis, stating that

> ... if waves are "long" ($kh_L \ll 1, kh_G \ll 1$) and surface tension can be neglected, the predicted instability condition is
>
> $$(v_G - v_L)^2 > (\rho_L - \rho_G)\, g \left(\frac{h_G}{\rho_G} + \frac{h_L}{\rho_L} \right). \tag{11.3.50}$$
>
> If $\rho_G \ll \rho_L$ and $v_L \ll v_G$ they may be simplified further to give
>
> $$\rho_G v_G^2 > g(\rho_L - \rho_G) h_G \tag{11.3.51}$$
>
> which is the same as
>
> $$j_G^* > \alpha^{3/2}. \tag{11.3.52}$$
>
> Here $\alpha = h_G/H$, and
>
> $$j_G^* = \frac{v_G \alpha \sqrt{\rho_G}}{\sqrt{gH(\rho_L - \rho_G)}} > \alpha^{3/2}.$$

Their criterion (11.3.50) is identical to our (11.3.44) for the long-wave inviscid case $\hat{\mu} = \hat{\rho}$ and $\hat{k} \to 0$. They compare their criterion (11.3.52) with transition observations that they call "slugging" and note that empirically the stability limit is well described by

$$j_G^* > 0.5\alpha^{3/2},$$

rather than by (11.3.52).

In figure 11.10 we plot j^* vs. α giving $j_G^* = \alpha^{3/2}$ and $0.5\alpha^{3/2}$; we give the results from our VPF theory for the inviscid case in table 11.2 and the air–water case in table 11.1, and we show the experimental results presented by Wallis and Dobson (1973) and Kordyban and Ranov (1970). Our theory fits the data better than that of Wallis and Dobson (1973), $j^* = \alpha^{3/2}$; it still overpredicts the data for small α but fits the large-α data quite well; we have good agreement when the water layer is small.

The predicted wavelength and wave speed in table 11.1 can be compared with experiments in principle, but in practice this comparison is obscured by the focus on the

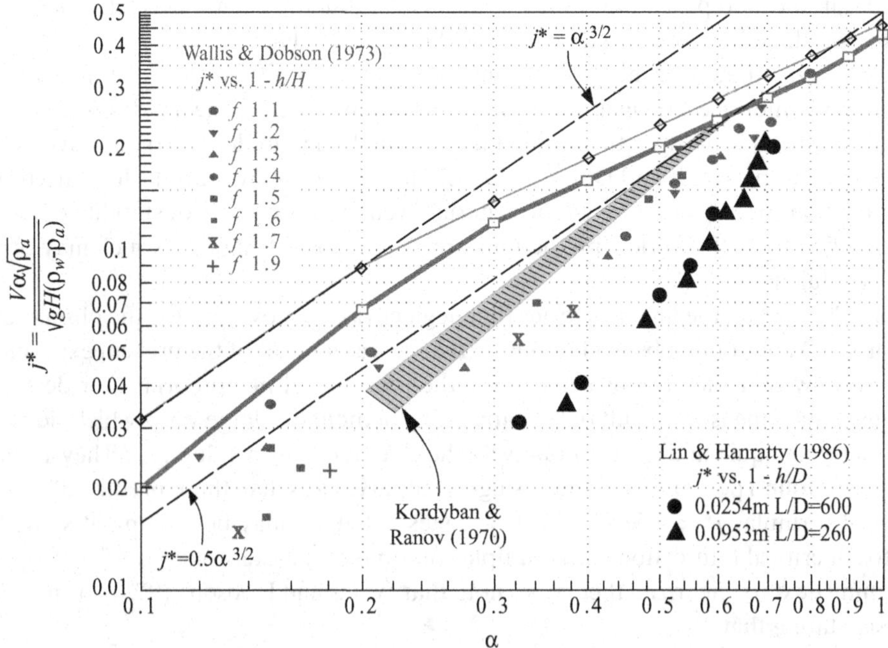

Figure 11.10. j^* vs. α is for marginal stability of air and water in a frame in which the water velocity is zero. The heavy line through □ = air–water, our result with $\gamma = 60$ dyn/cm from table 11.1; ◇ = inviscid fluid from table 11.2. $j^* = \alpha^{3/2}$ is the long-wave criterion for an inviscid fluid put forward by Wallis and Dobson (1973). $j^* = 0.5\alpha^{3/2}$ was proposed by them as best fit to experiments $f1.1$–$f1.9$ described in their paper. The shaded region is from experiments by Kordyban and Ranov (1970). Also shown are experimental data in rectangular conduits j^* vs. $1 - h/H = \alpha$ and in round pipes j^* vs. $1 - h/D = \alpha$ (Lin and Hanratty, 1986, figure 4).

formation of slugs. For example, Wallis and Dobson (1973) remark that "at a certain minimum air velocity, ripples appeared at the air entry end, and slowly spread down the channel. These waves were about 2-in. (0.05 m) long and were made up of long wave crests, with three or four capillary waves riding on the troughs. The long waves traveled faster than the capillary waves." The speeds of these long waves were reported by Wallis and Dobson (1973) to be less than 0.3 m/sec in all cases. Theoretical results from table 11.1 show that the wavelength λ_c increases with the water depth (as in the experiment) and the wave speed varies from 0.1 m/sec to 0.04 m/sec. The predicted spacing of the waves on average is about 1.5 cm/sec. The predicted wavelength and wave speed from VPF are apparently in good agreement with the waves Wallis and Dobson (1973) call capillary waves.

Observations similar to those of Wallis and Dobson (1973) just cited were made by Andritsos, Williams, and Hanratty (1989), who note that for high-viscosity liquid (80 cP) a region of regular 2D waves barely exists: "The first disturbances observed with increasing gas velocity are small-amplitude, small-wavelength, rather regular 2D waves. With a slight increase in gas velocity, these give way to a few large-amplitude waves with steep fronts and smooth troughs, and with spacing that can vary from a few centimeters to a meter."

11.3.7 Critical viscosity and density ratios

The most interesting aspect of our potential flow analysis is the surprising importance of the viscosity ratio $\hat{\mu} = \mu_a/\mu_l$ and density ratio $\hat{\rho} = \rho_a/\rho_l$; when $\hat{\mu} = \hat{\rho}$, equation (11.3.40) for marginal stability is identical to the equation for the neutral stability of an inviscid fluid even though $\hat{\mu} = \hat{\rho}$ in no way implies that the fluids are inviscid. Moreover, the critical velocity is a maximum at $\hat{\mu} = \hat{\rho}$; hence the critical velocity is smaller for *all* viscous fluids such that $\hat{\mu} \neq \hat{\rho}$ and is smaller than the critical velocity for inviscid fluids. All this may be understood by inspection of figure 11.8, which shows that $\hat{\mu} = \hat{\rho}$ is a distinguished value that can be said to divide high-viscosity liquids with $\hat{\mu} < \hat{\rho}$ from low-viscosity liquids. The stability limits of high-viscosity liquids can hardly be distinguished from each other whereas the critical velocities decrease sharply for low-viscosity fluids. This result may be framed in terms of the kinematic viscosity $\nu = \mu/\rho$ with high viscosities $\nu_l > \nu_a$. The condition $\nu_a = \nu_l$ can be written as

$$\mu_l = \mu_a \frac{\rho_l}{\rho_a}. \tag{11.3.53}$$

For air and water,

$$\mu_l = 0.15\ \mathrm{P}. \tag{11.3.54}$$

Hence $\mu_l > 0.15\,\mathrm{P}$ is a high-viscosity liquid and $\mu_l < 0.15\,\mathrm{P}$ is a low-viscosity liquid, provided that $\rho_l \approx 1\ \mathrm{g/cm^3}$.

Other authors have noted unusual relations between viscous and inviscid fluids. Barnea and Taitel (1993) note that "the neutral stability lines obtained from the viscous Kelvin–Helmholtz analysis and the inviscid analysis are quite different for the case of low liquid viscosities, whereas they are quite similar for high viscosity, contrary to what one would expect." Their analysis starts from a two-fluid model, and it leads to different dispersion relations; they do not obtain the critical condition $\hat{\mu} = \hat{\rho}$. Earlier, Andritsos *et al.* (1989) noted a "surprising result that the inviscid theory becomes more accurate as the liquid viscosity increases."

Andritsos and Hanratty (1987) have presented flow regime maps for flows in 2.52-cm and 9.53-cm pipe for fluids of different viscosity ranging from 1 to 80 cP. These figures present flow boundaries; the boundaries of interest to us are those that separate "smooth" flow from disturbed flow. Liquid holdups (essentially α) are not specified in these experiments. We extracted the smooth flow boundaries from figures in Andritsos and Hanratty (1987) and collected them in our figure 11.11. It appears from this figure that the boundaries of smooth flow for all the liquids with $\mu_l > 15$ cP are close together, but the boundary for water with $\mu_l = 1$ cP is much lower. The velocities shown in these figures are superficial velocities; the average velocities that could be compared with critical velocities in tables 11.1, 11.2, and 11.3 are larger than the superficial velocities and are significantly larger than those in the tables.

Even earlier, Francis (1951) observed that, even though the inviscid prediction of KH instability overestimates the onset for air over water, this prediction is in good agreement with experiments in rectangular ducts when air is above water.

11.3.8 Further comparisons with previous results

In practice, interest in the pipelining of gas–liquid flow is in round pipes. All experiments other than those of Kordyban and Ranov (1970) and Wallis and Dobson (1973), reviewed in subsection 11.3.6, have been carried out in round pipes. To our knowledge there is no other theoretical study in which the stability of stratified flow in a round pipe is studied without

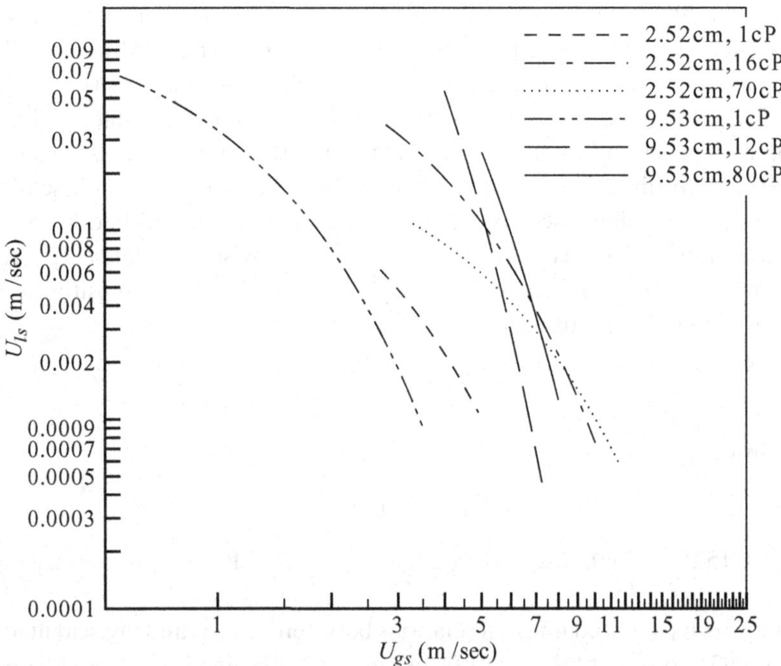

Figure 11.11. (After Andritsos and Hanratty, 1987.) The borders between smooth stratified flow and disturbed flow are observed in experiment. The water–air data are well below the cluster of high-viscosity data that are bunched together.

approximations. Theoretical studies of stability of stratified flow have been presented by Wallis (1969); Barnea (1991); Crowley, Wallis, and Barry (1992); Kordyban and Ranov (1970); Wallis and Dobson (1973); Taitel and Dukler (1976); Mishima and Ishii (1980); Lin and Hanratty (1986); Andritsos and Hanratty (1987); Andritsos *et al.* (1989); and Barnea and Taitel (1993). Viscosity is neglected by Kordyban and Ranov (1970), Wallis and Dobson (1973), Taitel and Dukler (1976), and Mishima and Ishii (1980). Surface tension is neglected by Wallis (1969), Kordyban and Ranov (1970), Wallis and Dobson (1973), Taitel and Dukler (1976), Mishima and Ishii (1980), and Lin and Hanratty (1986). Wallis (1969), Lin and Hanratty (1986), Wu *et al.* (1987), Barnea (1991), Crowley *et al.* (1992), and Barnea and Taitel (1993) use one or an other of the forms of two-fluids equations. In these equations averaged variables are introduced, the actual geometry is represented only so far as its area, and round, elliptical, or rectangular pipes with equal areas are equivalent. The effects of viscosity in these averaged models are introduced through empirical wall and interfacial fraction correlations. All these authors neglect the normal component of viscous stress (extensional stresses are neglected). The approach of Andritsos and Hanratty (1987) and Andritsos *et al.* (1989) is different; all the main physical effects are represented in analysis of the plane flow, which is later applied to flow in round pipes. The disturbance equations for the liquid are solved exactly except that the shear of basic liquid flow is neglected by use of a plug flow assumption. The effects of the gas on the liquid are represented through empirical correlations and further approximations are required for round pipes.

The viscous analysis of Andritsos and Hanratty (1987) for stability of stratified flow indicates that the critical velocity increases with increasing viscosity, unlike the present analysis, which predicts no such increase when $\nu_l > \nu_a$. The discrepancy may be due to the approximations made by Andritsos and Hanratty (1987).

Experiments on the stability of stratified flow have been reported by Kordyban and Ranov (1970), Wallis and Dobson (1973), Taitel and Dukler (1976), Lin and Hanratty (1986), Crowley *et al.* (1992), and Andritsos and Hanratty (1987). The experiments of Lin and Hanratty (1986) and Andritsos and Hanratty (1987) do not have data giving the height of the liquid and gas layers. Kordyban and Ranov (1970) and Wallis and Dobson (1973) did experiments in rectangular ducts, the geometry analyzed in this paper; the other experiments were done in round pipes. Authors Lin and Hanratty (1986), Crowley *et al.* (1992), and Andritsos and Hanratty (1987) are the only experimenters to report results for fluids with different viscosities.

It is difficult to compare the results of experiments in round pipes and rectangular channels. The common practice for round pipes is to express results in terms of h/D, where D is the pipe diameter and h is the height above the bottom of the pipe; h/H is the liquid fraction in rectangular pipes and $\alpha = 1 - h/H$ is the gas fraction, but h/D is not the liquid fraction in round pipes and $1 - h/D$ is not the gas fraction in round pipes. Lin and Hanratty (1986) presented experimental results for thin liquid films in round pipes giving (their figure 4) h/D vs. j^*; we converted their results to j^* vs. $1 - h/D$ and compared them in figure 11.10 with the results for rectangular pipes. The experimental results for round pipes are much lower than those for rectangular pipes. All this points to the necessity of taking care when comparing results between round and rectangular pipes and interpreting results of analysis for one experiment to another.

In general, we do not expect VPF analysis to work well in two-liquid problems; we get good results only when one of the fluids is a gas so that retarding effects of the second liquid can be neglected (see figure 11.12). However, the case of Holmboe waves studied by Pouliquen, Chomaz, and Huerre (1994) may have a bearing on the two-liquid case. These waves appear only near our critical condition of equal kinematic viscosity. They account for viscosity by replacing the vortex sheet with layers of constant vorticity across which no-slip conditions and the continuity of shear stress are enforced for the basic flow but the disturbance is inviscid. They did not entertain the notion that an inviscid analysis is just what would emerge from the condition of equal kinematic viscosity for VPF.

11.3.9 Nonlinear effects

None of the theories agree with experiments. Attempts to represent the effects of viscosity are only partial, as in our theory of VPF, or they require empirical data on wall and interfacial friction, which are not known exactly and may be adjusted to fit the data. Some choices for empirical inputs underpredict and others overpredict the experimental data.

It is widely acknowledged that nonlinear effects at play in the transition from stratified to slug flow are not well understood. The well-known criterion of Taitel and Dukler (1976), based on a heuristic adjustment of the linear inviscid long-wave theory for nonlinear effects, is possibly the most accurate predictor of experiments. Their criterion replaces $j^* = \alpha^{3/2}$ with $j^* = \alpha^{5/2}$. We can obtain the same heuristic adjustment for nonlinear effects on VPF by multiplying the critical value of velocity in table 11.1 by α. Plots of $j^* = \alpha^{3/2}$, $j^* = \alpha^{5/2}$ and the heuristic adjustment of VPF, together with the

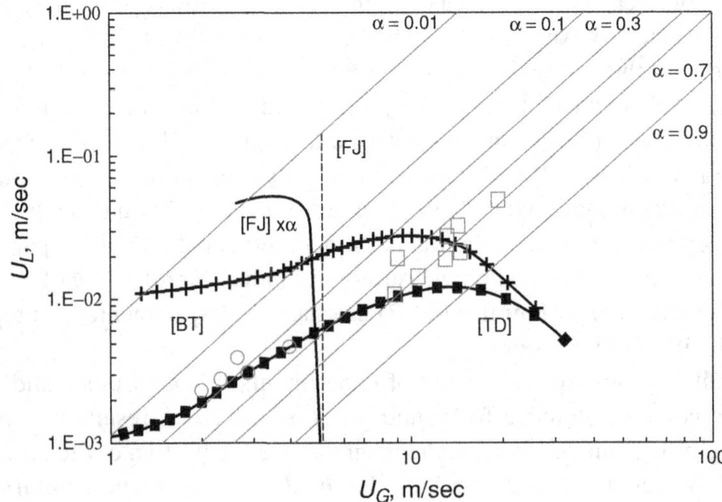

Figure 11.12. Local liquid velocity U_L versus local gas velocity U_G for PDVSA-Intevep data from 0.508-m inner-diameter flow loop with air and 0.404 Pa s lube oil. The identified flow patterns are SS (open circles), SW (open squares). Stratified-to-nonstratified flow transition theories after different authors are compared; TD, stars; BT, +; FJ, broken line; Funada and Joseph multiplied by α (2001), FJ × α, heavy curve. Constant void fraction α lines are indicated. Note that the curves FJ and FJ × α sharply drop around U_B 5 m/s, separating SS data from SW data.

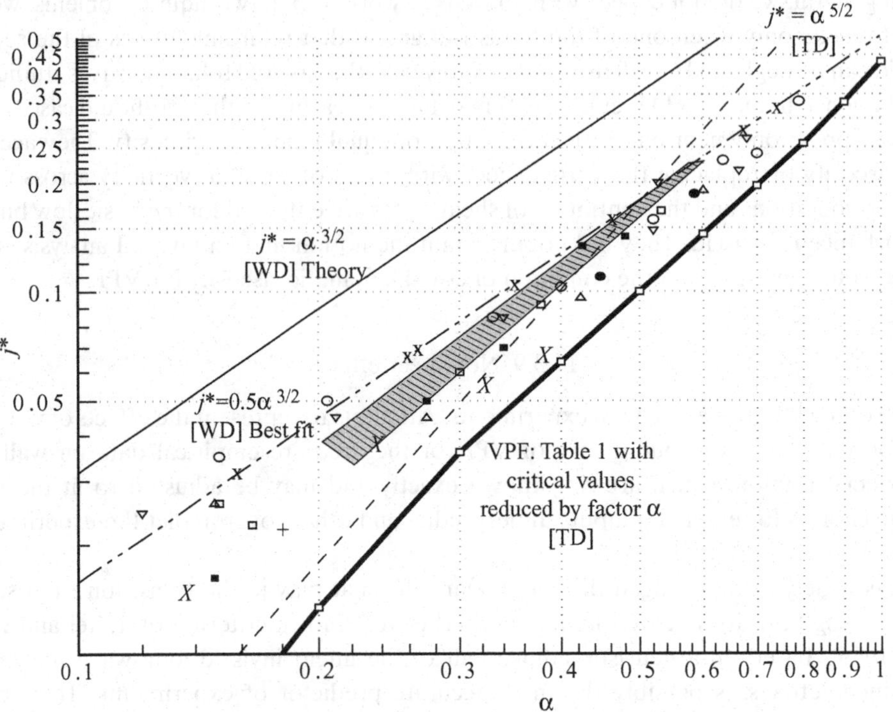

Figure 11.13. Nonlinear effects. The Taitel–Dukler (1976) correction (multiply by α).

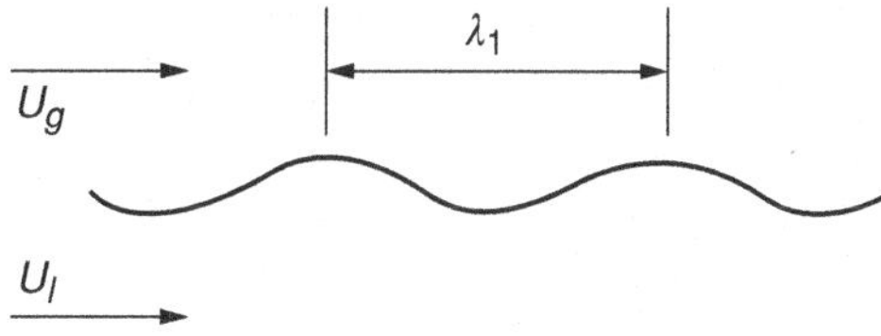

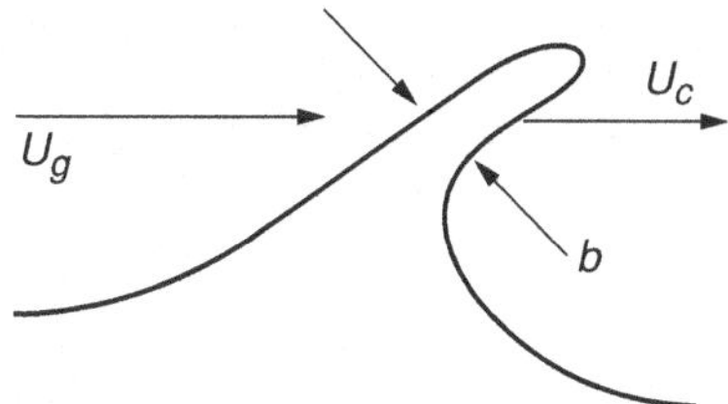

Figure 11.14. (Varga *et al.*, 2003). Schematic of the development of the liquid jet breakup process.

experimental values of Wallis and Dobson (1973) and Kordyban and Ranov (1970) are shown in figure 11.13. The good agreement in evidence there lacks a convincing foundation.

To summarize, we studied KH stability of superposed uniform streams in rectangular ducts by using VPF. VPFs satisfy the Navier–Stokes equations. Because the no-slip condition cannot be satisfied, the effects of shear stresses are neglected, but the effects of extensional stresses at the interface on the normal stresses are fully represented. The solution presented is complete and mathematically rigorous. The effects of shear stresses are neglected at the outset; after that, no empirical inputs are introduced. The main result of the analysis is the emergence of a critical value of velocity, discussed in §11.3.7. The main consequence of this result is that for air–liquid systems the critical values of velocity for liquids with viscosities greater than 15 cP are essentially independent of viscosity and the same as for an inviscid fluid, but for liquids with viscosities less than 15 cP the stability limits are much lower. The criterion for stability of stratified flow given by VPF is in good agreement with experiments when the liquid layer is thin, but it overpredicts the data when the liquid layer is thick. Although VPF neglects the effects of shear, the qualitative prediction of the peculiar effects of liquid viscosity has been obtained by other authors who used other methods of analysis in which shear is not neglected. A rather accurate predictor of experimental results is given by application of the Taitel and Dukler nonlinear correction factor to account for the effect of a finite-amplitude wave on the results of VPF.

11.3.10 Combinations of Rayleigh–Taylor and Kelvin–Helmholtz instabilities

Dispersion relations for the combined action of RT and KH instabilities of irrotational flows of inviscid fluids are presented in the first chapter of the book by Drazin and Reid (1981).

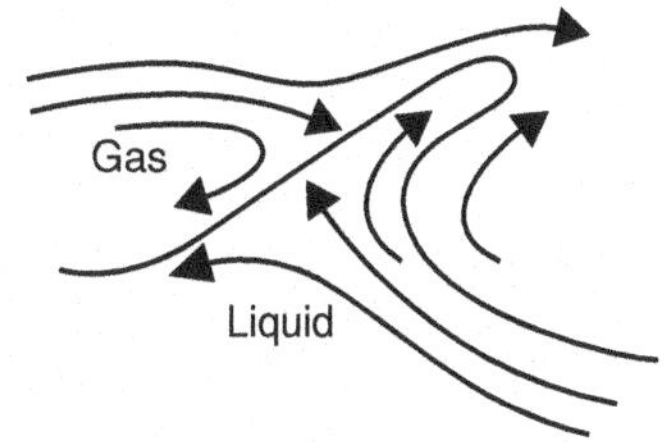

Figure 11.15. (Varga *et al.*, 2003). Sketch of the gas and liquid streamlines in the liquid tongue formation process.

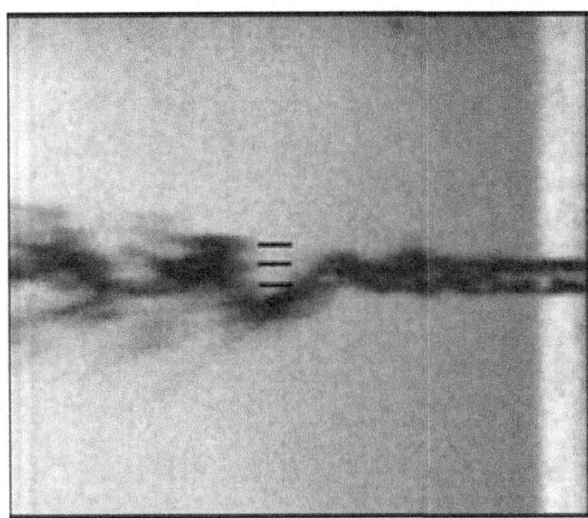

Figure 11.16. (Varga *et al.*, 2003). Instantaneous flow image with identified RT wavelengths, $D_l = 0.32$ mm, $We = 37$, $\lambda_{\text{measured}} = 200\ \mu\text{m}$ (cf. figures 9.6–9.8).

The analysis of the combined action of RT and KH instability can be readily extended to include important effects of viscosity in irrotational flows by techniques explained in chapters 9 and 11. Updating the potential flow analysis of the dispersion relation to include effects of viscosity by use of VPF is very easily done, and the results are only marginally more complicated than those given by the inviscid analysis but with a much richer physical content.

The extension of nonlinear analysis based on potential flow of inviscid fluids can also be extended to find the irrotational effects of viscosity. The combination of the two instabilities is ubiquitous in nonlinear problems. Joseph, Liao, and Saut (1992) considered the KH mechanism for side branching in the displacement of light with heavy fluid under gravity. They showed that the strongest Hadamard instabilities occur as KH instabilities associated with a velocity difference at the sides of falling fingers. The Hadamard instabilities are regularized by the addition of irrotational effects of viscosity and surface tension.

Another example of combined KH–RT instability goes in another way. Instead of generating a secondary KH instability on a primary RT instability generated by falling

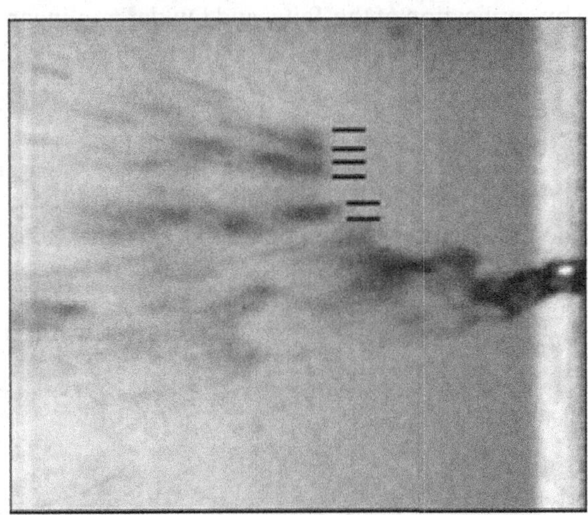

Figure 11.17. (Varga *et al.*, 2003). Instantaneous flow image with identified RT wavelengths, $D_l = 0.32$ mm, $We = 47$, $\lambda_{\text{measured}} = 185\ \mu\text{m}$ (cf. figures 9.6–9.8).

fingers, a secondary RT instability can develop on a primary KH instability in a small-diameter liquid jet exposed to a large-diameter high-speed gas jet, studied by Varga, Lasheras, and Hopfinger (2003). They proposed a breakup model based on the acceleration mechanism that breakup liquid drops in a high-speed airstream, studied in chapter 9. In their model, the RT instability is generated on the wave crests generated by a primary KH instability, as is shown in figures 11.14–11.17.

Varga *et al.* (2003) show that the mean droplet sizes scale well on the most unstable RT wavelength. The dependence of the droplet diameter on both the atomizing gas velocity and the liquid surface tension are successfully captured by the proposed breakup model.

12

Energy equation for irrotational theories of gas–liquid flow: viscous potential flow, viscous potential flow with pressure correction, and dissipation method

12.1 Viscous potential flow

We obtained the effects of viscosity on irrotational motions of spherical cap bubbles, Taylor bubbles in round tubes, and RT and KH instabilities described in previous chapters by evaluating the viscous normal stress on potential flow. In gas–liquid flows, the viscous normal stress does not vanish and it can be evaluated on the potential. It can be said that, in the case of gas–liquid flow, the appropriate formulation of the irrotational problem is the same as the conventional one for inviscid fluids with the caveat that the viscous normal stress is included in the normal stress balance. This formulation of VPF is not at all subtle; it is the natural and obvious way to express the equations of balance when the flow is irrotational and the fluid viscous.

In this chapter we use the acronym VPF, viscous potential flow, to stand for the irrotational theory in which the viscous normal stresses are evaluated on the potential.

In gas–liquid flows we may assume that the shear stress in the gas is negligible so that no condition need be enforced on the tangential velocity at the free surface, but the shear stress must be zero. The condition that the shear stress be zero at each point on the free surface is dropped in irrotational approximations. In general, you get an irrotational shear stress from the irrotational analysis. The discrepancy between this shear stress and the zero shear stress required for exact solutions will generate a vorticity layer. In many cases this layer is thin, and its influence on the bulk motion and on the irrotational effects of viscosity is minor. Exact solutions (ESs) cannot be the same as any irrotational approximation for cases in which the irrotational shear stress does not vanish.

12.2 Dissipation method according to Lamb

Another purely irrotational theory of flow of viscous fluids is associated with the dissipation method (DM). The DM extracts information from evaluating the equation governing the evolution of energy and dissipation on irrotational flow. VPF does not rely on arguments about energy and dissipation, and the results of the two irrotational theories, DM and VPF, are usually different.

The DM was used by Lamb (1932) to compute the decay of irrotational waves that is due to viscosity. This method can be traced back to Stokes (1851). A pressure correction was not introduced in these studies. Lamb presented some exact solutions from which the irrotational approximation can be evaluated and the nature of the boundary layer rigorously examined. Lamb (1932) studied the viscous decay of free oscillatory waves on deep water (§348) and small oscillations of a mass of liquid about the spherical form (§355) by using the DM. Lamb showed that in these cases the rate of dissipation can be calculated with sufficient accuracy by regarding the motion as irrotational.

Lamb's (1932) calculations are based on the assumption that the speed of a progressive wave is the same as in an inviscid fluid. Padrino (Padrino and Joseph 2007, §§14.2.4 and 14.3.1) noted that a more complete calculation of the DM yields a different result; the growth rates are the same as in Lamb's calculation, but the speed of the wave depends on the viscosity of the liquid.

12.3 Drag on a spherical gas bubble calculated from the viscous dissipation of an irrotational flow

The computation of the drag D on a sphere in potential flow by use of the DM seems to have been given first by Bateman in 1932 (see Dryden, Murnaghan, and Bateman 1956) and repeated by Ackeret (1952). They found that $D = 12\pi a\mu U$, where μ is the viscosity, a is the radius of the sphere, and U is its velocity. This drag is twice the Stokes drag and is in better agreement with the measured drag for Reynolds numbers in excess of about 8.

The drag on a spherical gas bubble of radius a rising in a viscous liquid at modestly high Reynolds numbers was calculated, with the DM by various investigators, beginning with Levich (1949), who obtained the value $12\pi a\mu U$ or equivalently the drag coefficient $48/Re$, where $Re = 2aU\rho/\mu$ is the Reynolds number, by calculating the dissipation of the irrotational flow around the bubble. Moore (1959) calculated the drag directly by integrating the pressure and viscous normal stress of the potential flow and neglecting the viscous shear stress (which physically should be zero), obtaining the value $8\pi a\mu U$.

12.4 The idea of a pressure correction

The discrepancy between two drag values led Batchelor, as reported by Moore (1963), to suggest the idea that discrepancy could be resolved by introducing a viscous correction to the irrotational pressure. We do not know whether we should praise or blame Batchelor for this idea as it comes to us secondhand. There is no doubt that the pressure in the liquid depends on the viscosity and it could take on its largest values in the boundary layer next to the free surface but the velocity there could also come to depend on the viscosity. Moore (1963) performed a boundary-layer analysis, and his pressure correction is readily obtained by setting $y = 0$ in his equation (2.37):

$$p_v = (4/Re)\,(1-\cos\theta)^2\,(2+\cos\theta)\,/\sin^2\theta, \tag{12.4.1}$$

which is singular at the separation point where $\theta = \pi$. The presence of separation is a problem for the application of boundary layers to the calculation of drag on solid bodies. To find the drag coefficient, Moore (1963) calculated the momentum defect and obtained the Levich value $48/Re$ plus contributions of the order of $Re^{-3/2}$ or lower.

Kang and Leal (1988a, 1988b) put the shear stress of the potential flow on the bubble surface to zero and calculated a pressure correction. They obtained the drag coefficient given by Levich's dissipation approximation by direct integration of the normal viscous stress and pressure over the bubble surface. They accomplished this by expanding the pressure correction as a spherical harmonic series and noting that only one term of this series contributes to the drag, no appeal to the boundary-layer approximation being necessary. A similar result was obtained by Joseph and Wang (2004), who used their theory subsequently described for computing a pressure correction.

The accepted idea, starting with Batchelor, is that the pressure correction is a real viscous pressure that varies from the pressure in the irrotational flow outside a narrow vorticity layer near the gas–liquid surface, to the required value at the interface. This boundary layer has not been computed. Maybe the idea is not quite correct. In the case of the decay of free irrotational gravity waves computed by Lamb (1932), a viscous pressure correction does not appear. The pressure correction computed by Wang and Joseph (2006a), as shown in §14.1, gives the same result as the DM and is in excellent agreement with the exact solution, but it is not a viscous correction in a boundary layer of the exact solution.

12.5 Energy equation for irrotational flow of a viscous fluid

Consider the equations of motion for an incompressible Newtonian fluid with gravity as a body force per unit mass:

$$\rho \frac{\mathrm{d}\mathbf{u}}{\mathrm{d}t} = -\nabla\Phi + \mu\nabla^2\mathbf{u}, \tag{12.5.1}$$

where

$$\Phi = p + \rho g z. \tag{12.5.2}$$

The stress is given by

$$\mathbf{T} = -p\mathbf{1} + \boldsymbol{\tau}, \tag{12.5.3}$$

where

$$\boldsymbol{\tau} = 2\mu\mathbf{D}[\mathbf{u}],$$
$$\nabla\cdot\boldsymbol{\tau} = \mu\nabla^2\mathbf{u}.$$

The mechanical-energy equation corresponding to (12.5.1) is given by

$$\frac{\mathrm{d}}{\mathrm{d}t}\int_\Omega \frac{\rho}{2}|\mathbf{u}|^2\,\mathrm{d}\Omega = \int_S \mathbf{n}\cdot\tilde{\mathbf{T}}\cdot\mathbf{u}\,\mathrm{d}S - 2\mu\int_\Omega \mathbf{D}:\mathbf{D}\,\mathrm{d}\Omega, \tag{12.5.4}$$

where S is the boundary of Ω, with outward normal $\mathbf{n}$. On solid boundaries, no-slip is imposed; say $\mathbf{u} = 0$ there, and on the free surface

$$z = \eta(x, y, t)$$

and the shear stress

$$\mathbf{n}\cdot\mathbf{T}\cdot\mathbf{e}_s = 0, \tag{12.5.5}$$

where $\mathbf{e}_s$ is any vector tangent to the free surface S_f. On S_f we have

$$\int_S \mathbf{n}\cdot\tilde{\mathbf{T}}\cdot\mathbf{u}\,\mathrm{d}S = \int_{S_f} \mathbf{n}\cdot\tilde{\mathbf{T}}\cdot\mathbf{u}\,\mathrm{d}S,$$

where

$$\tilde{\mathbf{T}} = \mathbf{T} - \rho g \eta \mathbf{1}, \tag{12.5.6}$$

$$\mathbf{n} \cdot \mathbf{T} \cdot \mathbf{n} = -\gamma \nabla_{\text{II}} \cdot \mathbf{n}. \tag{12.5.7}$$

Hence, on S_f,

$$\mathbf{n} \cdot \tilde{\mathbf{T}} \cdot \mathbf{n} = -\rho g \eta - \gamma \nabla_{\text{II}} \cdot \mathbf{n}, \tag{12.5.8}$$

and, because the shear stress vanishes on S_f,

$$\begin{aligned} \mathbf{n} \cdot \tilde{\mathbf{T}} \cdot \mathbf{u} &= \mathbf{n} \cdot \tilde{\mathbf{T}} \cdot (u_n \mathbf{n} + u_s \mathbf{e}_s) \\ &= -(\rho g \eta + \gamma \nabla_{\text{II}} \cdot \mathbf{n}) u_n. \end{aligned} \tag{12.5.9}$$

Hence, (12.5.4) may be written as

$$\frac{\text{d}}{\text{d}t} \int_{\Omega} \frac{\rho}{2} |\mathbf{u}|^2 \, \text{d}\Omega = -\int_{S_f} (\rho g \eta + \gamma \nabla_{\text{II}} \cdot \mathbf{n}) u_n \text{d}S_f - 2\mu \int_{\Omega} \mathbf{D} : \mathbf{D} \, \text{d}\Omega. \tag{12.5.10}$$

Equation (12.5.10) holds for viscous fluids satisfying Navier–Stokes equations (12.5.1) subject to vanishing shear stress condition (12.5.5).

We turn now to potential flow $\mathbf{u} = \nabla\phi$, $\nabla^2\phi = 0$. In this case, $\nabla^2\mathbf{u} = 0$, but the dissipation does not vanish. How can this be? In chapter 4 we showed that the irrotational viscous stress is self-equilibrated and does give rise to a force-density term $\nabla^2\nabla\phi = 0$; however, the power of self-equilibrated irrotational viscous stresses,

$$\int_{S_f} \mathbf{n} \cdot 2\mu \nabla \otimes \nabla\phi \cdot \mathbf{u} \text{d}S_f,$$

does not vanish, and it gives rise to an irrotational viscous dissipation:

$$\begin{aligned} 2\mu \int_{\Omega} \mathbf{D}[\mathbf{u}] : \mathbf{D}[\mathbf{u}] \text{d}\Omega &= 2\mu \int_{\Omega} \frac{\partial^2\phi}{\partial x_i \partial x_j} \frac{\partial^2\phi}{\partial x_i \partial x_j} \text{d}\Omega = 2\mu \int_S n_j \frac{\partial^2\phi}{\partial x_j \partial x_i} u_i \text{d}S \\ &= 2\mu \int_{S_f} \mathbf{n} \cdot \mathbf{D}[\nabla\phi] \cdot \mathbf{u} \text{d}S_f = 2\mu \int_{S_f} \mathbf{n} \cdot \mathbf{D} \cdot (u_n \mathbf{n} + u_s \mathbf{e}_s) \, \text{d}S_f \\ &= \int_{S_f} \left(2\mu \frac{\partial^2\phi}{\partial n^2} u_n + \tau_s u_s \right) \text{d}S_f, \end{aligned} \tag{12.5.11}$$

where τ_s is an irrotational shear stress,

$$\begin{aligned} \tau_s &= 2\mu \mathbf{n} \cdot \mathbf{D}[\nabla\phi] \cdot \mathbf{e}_s, \\ u_s &= \mathbf{u} \cdot \mathbf{e}_s. \end{aligned} \tag{12.5.12}$$

Turning next to the inertial terms, we have

$$(\mathbf{u}\cdot\nabla)\,\mathbf{u} = \nabla\frac{|\mathbf{u}|^2}{2},$$

$$\begin{aligned}\frac{d}{dt}\int_\Omega \frac{\rho}{2}|\mathbf{u}|^2 d\Omega &= \int_\Omega \rho\left(\mathbf{u}\cdot\frac{\partial\mathbf{u}}{\partial t}+\mathbf{u}\cdot\nabla\frac{|\mathbf{u}|^2}{2}\right)d\Omega \\ &= \int_\Omega \rho\left[\frac{\partial\phi}{\partial x_i}\frac{\partial}{\partial x_i}\frac{\partial\phi}{\partial t}+\nabla\cdot\left(\mathbf{u}\frac{|\mathbf{u}|^2}{2}\right)\right]d\Omega \\ &= \int_{S_f}\rho u_n\left(\frac{\partial\phi}{\partial t}+\frac{|\nabla\phi|^2}{2}\right)dS_f.\end{aligned} \tag{12.5.13}$$

Collecting results (12.5.9), (12.5.11), (12.5.12), and (12.5.13), we need to evaluate energy equation (12.5.4) for (12.5.1) when $\mathbf{u} = \nabla\phi$; we find that

$$\int_{S_f} u_n\left[\rho\left(\frac{\partial\phi}{\partial t}+\frac{|\nabla\phi|^2}{2}+g\eta\right)+2\mu\frac{\partial^2\phi}{\partial n^2}+\gamma\nabla_{\text{II}}\cdot\mathbf{n}\right]dS_f = -\int_{S_f}\tau_s u_s dS_f. \tag{12.5.14}$$

Equation (12.5.14) was derived by Joseph; it is the energy equation for the irrotational flow of a viscous fluid. As in the case of the radial motion of a spherical gas bubble, (6.0.8), the normal stress balance converts into the energy equation for viscous irrotational flow.

12.6 Viscous correction of viscous potential flow

The viscous pressure p_v for the viscous correction of VPF (VCVPF) can be defined by the equation

$$\int_{S_f}(-p_v)u_n dS_f = \int_{S_f}\tau_s u_s dS_f. \tag{12.6.1}$$

This equation can be said to arise from the condition that the shear stress should vanish, but it does not vanish in the irrotational approximation. Batchelor's idea, as reported by Moore (1963), is that the additional drag $4\pi a\mu U$ on a spherical gas bubble needed to obtain the Levich drag $12\pi a\mu U$ from the value $8\pi a\mu U$, computed with the viscous normal stress, arises from a real viscous pressure in a thin boundary layer at the surface of the bubble.

Our interpretation of Batchelor's idea is to replace the unwanted shear stress term in the energy equation with an additional contribution to the normal stress in the form of a viscous pressure satisfying (12.6.1). An identical interpretation of Batchelor's idea, with a different implementation, was proposed by Kang and Leal (1988b) and is discussed in §13.1.1.

Suppose now that there is such a pressure correction and Bernoulli equation

$$p_i+\rho\left(\frac{\partial\phi}{\partial t}+\frac{|\nabla\phi|^2}{2}+g\eta\right) = C$$

holds. Because

$$C\int_S \mathbf{u}\cdot\mathbf{n}dS = C\int_\Omega \nabla\cdot\mathbf{u}d\Omega = 0,$$

we obtain

$$\int_{S_f} u_n \left[-p_i - p_v + 2\mu \frac{\partial^2 \phi}{\partial n^2} + \gamma \nabla_{\text{II}} \cdot \mathbf{n} \right] \mathrm{d}S_f = 0. \tag{12.6.2}$$

The normal stress balance for VCVPF is

$$-p_i - p_v + 2\mu \frac{\partial^2 \phi}{\partial n^2} + \gamma \nabla_{\text{II}} \cdot \mathbf{n} = 0. \tag{12.6.3}$$

Equation (12.6.3) is a working equation in the VCVPF theory only if p_v can be calculated. In the problems discussed in this book, p_v and a velocity correction $\mathbf{u}_v$ are coupled in a linear equation,

$$\rho \frac{\partial \mathbf{u}_v}{\partial t} = -\nabla p_v + \mu \nabla^2 \mathbf{u}_v, \qquad \nabla \cdot \mathbf{u}_v = 0, \tag{12.6.4}$$

and because $\mathbf{u}_v$ is solenoidal, $\nabla^2 p_v = 0$, and p_v can be represented by a series of harmonics functions. In most of the problems in this book, orthogonality conditions show that all the terms in the series but one are zero and this one term is completely determined by (12.6.1). In other problems, such as the flow over bodies studied in chapter 13, (12.6.1) determines the coefficient of the term in the harmonic series that enters into the direct calculation of the drag; the other coefficients remain undetermined.

Having obtained p_v it is necessary to consider problem (12.6.4) for $\mathbf{u}_v$. This problem is overdetermined; (12.6.4) is a system of four equations for three unknowns. We must also satisfy the kinematic equation $\mathbf{u}_v = \partial \eta_v / \partial t$ and the normal stress boundary condition. Wang and Joseph (2006a) showed that a purely irrotational $\mathbf{u}_v$ and higher-order corrections satisfying (12.6.1) can be computed in the case in which the normal stress boundary condition is not corrected. Their solution procedure leads to a series of purely irrotational velocities in powers of ν, or reciprocal powers of the Reynolds number (see §14.12). This same kind of special irrotational solution in reciprocal powers of the Reynolds number was constructed by Funada, Saitoh, Wang and Joseph (2005) for capillary instability of a viscous fluid and by Wang *et al.* (2005b) for a capillary instability of a viscoelastic fluid.

There is no reason to think that the higher-order irrotational solutions generated by p_v have a hydrodynamical significance. In general, the solution of potential flow problems with pressure corrections can be expected to induce vorticity. This is the case for the nonlinear model of capillary collapse and rupture studied in §17.4.2. The flow from irrotational stagnation points of depletion to irrotational stagnation points of accumulation cannot occur without generating the vorticity given by (17.4.24) at leading order.

In the direct applications of the DM, a pressure correction is not introduced; the power of the irrotational shear stress on the right-hand side of (12.5.14) is computed directly. The energy equation for the system of VCVPF equations with p_v is the same as (12.5.14) when the power of p_v is replaced with the power of the shear stress by use of (12.6.1).

In the case of rising bubbles we may compute the same value for the drag in a direct computation in which the irrotational shear stress is neglected and p_v is included as we get from the DM by using (12.5.14) with $\tau_s \neq 0$ and $p_v = 0$.

The DM requires that we work with the power generated by the shear $u_s \tau_s$ and not with a direct computation by using p_v. The direct computation of drag with $p_v = 0$ and $\tau_s \neq 0$ given by potential flow leads to zero drag (13.1.8). The pressure correction allows us to do direct calculations without using the energy equations. The two methods lead to the same results by different routes. This feature of energy analysis is related to the fact that irrotational stresses are self-equilibrated (4.3.1) but do work.

The nature of the construction of VCVPF equations, which have the same energy equations as those used in the DM, guarantees that the same results will be obtained by these two methods.

The VCVPF theory leads to a revised normal stress balance that allows one to compute a new harmonic function, proportional to viscosity, which we have called a pressure correction. There is no guarantee that this harmonic function is related to a viscous component of the real pressure in a boundary layer or anywhere else; in fact, there is no explicit pressure function in the exact solution of Lamb (1932) for the viscous decay of gravity waves (see §14.1.3). However, in all cases, VCVPF leads to the same results as those of the disipation method DM. It is probably more accurate to call the pressure correction arising in VCVPF a viscous correction of the normal stress bazlance.

We have now arrived at the following classification of irrotational solutions:

IPF, inviscid potential flow;
VPF, viscous potential flow;
VCVPF, viscous correction of viscous potential flow;
DM, dissipation method.

Recognizing now that DM and VCVPF lead to the same results by different routes, even if p_v is not a true hydrodynamic pressure, we can introduce yet two more acronyms: VCVPF/DM and ES, exact solution.

VCVPF/DM and VPF are different irrotational solutions; they give rise to different results in nearly every instance except in the case of the limit of very high Reynolds numbers, at which they tend to the inviscid limit IPF. At finite-Reynolds-number VPF and VCVPF/DM can give significantly better results, closer to the exact solutions than IPF.

In the sequel, VCVPF means VCVPF/DM. We think that it is possible that either VPF or VCVPF gives (in different situations which we presently cannot specify *a priori* with precision) the best possible purely irrotational approximations to the exact solution. The problem of the best purely irrotational approximation of real flows is related to the problem of the optimal specification of the irrotational flow in the Helmholtz decomposition studied in chapter 4.

12.7 Direct derivation of the viscous correction of the normal stress balance for the viscous decay of capillary-gravity waves

J. C. Padrino has derived the viscous correction of the normal stress balance directly without making reference to connection formula (12.6.1). Consider the problem of "small" capillary-gravity waves about an infinite, plane free surface and assume irrotational motion of a Newtonian fluid in the domain $0 \leq x \leq \lambda$ and $-\infty < y \leq 0$, after linearization, with periodic boundary conditions at $x = 0$ and $x = \lambda$ (see also §14.3). The problem of decay of free-gravity waves is examined in §14.1. The integral involving the power of the shear stress at the interface $y = 0$ may be written as

$$\int_{S_f} \tau_s u_s \mathrm{d}S_f = \int_0^{\lambda} 2\mu \frac{\partial^2 \phi}{\partial x \partial y} \frac{\partial \phi}{\partial x} \mathrm{d}x. \tag{12.7.1}$$

For the velocity potential, in the interior of the domain of interest, we may write

$$\frac{\partial^2\phi}{\partial x\partial y}\frac{\partial\phi}{\partial x}=\frac{\partial}{\partial x}\left(\frac{\partial\phi}{\partial y}\right)\frac{\partial\phi}{\partial x}=\frac{\partial}{\partial x}\left(\frac{\partial\phi}{\partial y}\frac{\partial\phi}{\partial x}\right)-\frac{\partial\phi}{\partial y}\frac{\partial^2\phi}{\partial x^2}.$$

Then, at $y=0$,

$$\int_{S_f}\tau_s u_s \mathrm{d}S_f=2\mu\left(\frac{\partial\phi}{\partial y}\frac{\partial\phi}{\partial x}\right)\bigg|_0^\lambda-\int_0^\lambda 2\mu\frac{\partial^2\phi}{\partial x^2}\frac{\partial\phi}{\partial y}\mathrm{d}x.$$

The first term on the right-hand side vanishes because ϕ is periodic in x with period λ. Then (12.7.1) becomes

$$\int_{S_f}\tau_s u_s \mathrm{d}S_f=\int_0^\lambda\left(-2\mu\frac{\partial^2\phi}{\partial x^2}\right)\frac{\partial\phi}{\partial y}\mathrm{d}x. \tag{12.7.2}$$

Define

$$F\equiv 2\mu\frac{\partial^2\phi}{\partial x^2} \tag{12.7.3}$$

in $0\leq x\leq\lambda$ and $-\infty<y\leq 0$. Because ϕ satisfies Laplace's equation in the interior of this domain, so does F:

$$\nabla^2F=\nabla^2\left(2\mu\frac{\partial^2\phi}{\partial x^2}\right)=2\mu\frac{\partial^2}{\partial x^2}(\nabla^2\phi)=0.$$

Then, at $y=0$, we can write (12.7.2) as

$$\int_{S_f}\tau_s u_s \mathrm{d}S_f=\int_{S_f}(-F)u_n\mathrm{d}S_f, \tag{12.7.4}$$

for the linearized problem, with F defined in (12.7.3) such that

$$\nabla^2F=0 \tag{12.7.5}$$

in $0<x<\lambda$ and $-\infty<y<0$. Notice that the pressure correction introduced in §12.6 satisfies the same conditions that F satisfies, namely, (12.7.4) and (12.7.5). However, unlike the pressure correction, these conditions are not imposed on F but they are satisfied by F by construction as a result of the assumption of potential flow.

Substitution of (12.7.4) in the right-hand side of (12.5.14) and following the steps of §12.6 yields

$$-p_i-F+2\mu\frac{\partial^2\phi}{\partial n^2}+\gamma\nabla_{\mathrm{II}}\cdot\mathbf{n}=0 \tag{12.7.6}$$

at $y=0$ for the linearized problem. This is the same expression as (12.6.3) for the VCVPF, but obtained here without invoking the concept of pressure correction.

13

Rising bubbles

13.1 The dissipation approximation and viscous potential flow

Dissipation approximations have been used to calculate the drag on bubbles and drops; VCVPF can be used for the same purpose. The viscous pressure correction on the interface can be expressed by a harmonic series. The principal mode of this series is matched to the velocity potential, and its coefficient is explicitly determined. The other modes do not enter into the expression for the drag on bubbles and drops.

13.1.1 Pressure correction formulas

We consider separable solutions of $\nabla^2\phi = 0$. For simplicity, we consider axisymmetric or planar problems and use the orthogonal coordinate system (α, β); the gas–liquid interface is given by $\alpha = \text{const}$. The solution of the potential flow equations may be written as

$$\phi = h_k(\alpha)\, f_k(\beta), \tag{13.1.1}$$

where $f_k(\beta)$ is the kth mode of the surface harmonics. A pressure correction function that is periodic or finite at the gas–liquid interface may be expanded as a series of surface harmonics of integral orders:

$$-p_v = C_k f_k(\beta) + \sum_{j\neq k} C_j f_j(\beta), \tag{13.1.2}$$

where $f_j(\beta)$'s are surface harmonics and C_j's are constant coefficients.

Substitution of (13.1.2) into (12.6.1) leads to

$$C_k \int_A \mathbf{u}\cdot\mathbf{n} f_k(\beta)\,\mathrm{d}A + \sum_{j\neq k} C_j \int_A \mathbf{u}\cdot\mathbf{n} f_j(\beta)\,\mathrm{d}A = \mathcal{P}_s, \tag{13.1.3}$$

where $\mathcal{P}_s$ is defined as $\mathcal{P}_s = \int_A \mathbf{u}\cdot\mathbf{t}\tau_s\,\mathrm{d}A$. We assume that the normal velocity is orthogonal to $f_j(\beta)$, $j \neq k$:

$$\int_A \mathbf{u}\cdot\mathbf{n} f_j(\beta)\,\mathrm{d}A = 0 \quad \text{when} \quad j \neq k, \tag{13.1.4}$$

which is a verifiable condition and is confirmed in each example we consider here. Now equation (13.1.3) gives the coefficient C_k:

$$C_k = \frac{\mathcal{P}_s}{\int_A \mathbf{u}\cdot\mathbf{n} f_k(\beta)\,\mathrm{d}A}. \tag{13.1.5}$$

Using (13.1.5) for VCVPF we may write

$$-p = -p_i + \frac{\mathcal{P}_s f_k(\beta)}{\int_A \mathbf{u}\cdot\mathbf{n} f_k(\beta)\,\mathrm{d}A} + \sum_{j\neq k} C_j f_j(\beta). \tag{13.1.6}$$

The term $\mathcal{P}_s f_k(\beta)/\int_A \mathbf{u}\cdot\mathbf{n} f_k(\beta)\,\mathrm{d}A$ may be called the principal part of the viscous pressure correction; it is proportional to the power integral of the uncompensated irrotational shear stress. It is the only term in the pressure correction to enter into the power of traction integral and into the direct calculation of the drag on rising bubbles or drops. The principal part of the pressure correction is explicitly computable, as we shall see in the examples to follow. In general, the values of C_j, $j \neq k$ are not known, but for the special case of a rising spherical gas bubble, Kang and Leal (1988a) presented computable expressions for these coefficients.

Expression (13.1.6) completes the formulation of the pressure correction in VCVPF up to the principal part of the harmonic series.

We may compare the dissipation calculation and direct calculation of the drag by using VPF and VCVPF. Let D_1 be the drag calculated by the dissipation method DM,

$$D_1 = \mathcal{D}/U, \tag{13.1.7}$$

and D_2 be the drag by direct calculation,

$$D_2 = \int_A \mathbf{e}_x\cdot\mathbf{T}\cdot\mathbf{n}\,\mathrm{d}A = \int_A [\mathbf{e}_x\cdot\mathbf{n}(-p+\tau_n) + \mathbf{e}_x\cdot\mathbf{t}\tau_s]\,\mathrm{d}A, \tag{13.1.8}$$

where x is the direction of translation. The direct calculation using VPF leads to $D_2 = 0$ even though the dissipation is not zero, which is a known result (see, for example, Zierep, 1984; Joseph and Liao 1994a, 1994b). The dissipation approach involves a volume integral, and the direct calculation involves a surface integral. The solution to the Navier–Stokes equations in these nearly irrotational flows involves a leading-order term (the irrotational solution) and a viscous correction at the boundary. When the dissipation approach is used, the leading-order calculation involves only the irrotational result. However, the viscous correction has to be considered to obtain the leading-order result when the direct calculation is used. We shall focus on the direct calculation by using VCVPF in the following examples (§§13.2–13.4) and show that D_2 computed by using VCVPF here is equal to D_1 obtained by the DM in the literature.

13.2 Rising spherical gas bubble

Consider now a spherical gas bubble rising with a constant velocity $U\mathbf{e}_x$ in a viscous fluid, for which

$$\phi = -\frac{1}{2}Ua^3\frac{\cos\theta}{r^2}. \tag{13.2.1}$$

At the surface of the bubble, where $r = a$, we have

$$\begin{aligned} u_r &= U\cos\theta, \quad u_\theta = U\sin\theta/2; \\ \tau_{rr} &= -6\mu U\cos\theta/a, \quad \tau_{r\theta} = -3\mu U\sin\theta/a; \\ p_i &= p_\infty + \frac{\rho}{2}U^2\left(1 - \frac{9}{4}\sin^2\theta\right). \end{aligned} \tag{13.2.2}$$

The dissipation is given by $\mathcal{D} = 12\pi\mu a U^2$ and $\mathcal{P}_s = 4\pi\mu a U^2$.

The pressure correction may be expanded as a spherical surface harmonic series $\sum_{j=0}^{\infty} C_j P_j(\cos\theta)$. Substitution of u_r and p_v into (12.6.1) gives

$$-\int_{-1}^{1} UP_1(\cos\theta)\left[C_1 P_1(\cos\theta) + \sum_{j\neq 1} C_j P_j(\cos\theta)\right] 2\pi a^2\, \mathrm{d}(\cos\theta) = \mathcal{P}_s. \quad (13.2.3)$$

The coefficient C_1 is then determined, and the pressure correction is

$$-p_v = -3\mu U P_1(\cos\theta)/a + \sum_{j\neq 1} C_j P_j(\cos\theta), \quad (13.2.4)$$

which is the same as the pressure correction by Kang and Leal (1988a), who obtained it by means of a general relationship between the viscous pressure correction and the vorticity distribution for a spherical bubble in an arbitrary axisymmetric flow. Kang and Leal demonstrated that the drag from direct calculation using pressure correction (13.2.4) is $12\pi\mu a U$, the same as the drag by dissipation calculation.

13.3 Rising oblate ellipsoidal bubble

The equation of the ellipsoid (Moore, 1965) is

$$\frac{x^2+y^2}{b^2} + \frac{z^2}{a^2} = 1,$$

where $b \geq a$. Orthogonal ellipsoidal coordinates (α, β, ω) are related to (x, y, z) by

$$x = \kappa[(1+\alpha^2)(1-\beta^2)]^{\frac{1}{2}} \cos\omega,$$
$$y = \kappa[(1+\alpha^2)(1-\beta^2)]^{\frac{1}{2}} \sin\omega,$$
$$z = \kappa\alpha\beta.$$

The ellipsoid is given by $\alpha = \alpha_0$, provided that

$$\kappa\,(1+\alpha_0^2)^{\frac{1}{2}} = b, \quad \kappa\alpha_0 = a.$$

The potential for an oblate ellipsoid rising with a constant velocity $U\mathbf{e}_z$ is

$$\phi = -U\kappa q\beta(1-\alpha\cot^{-1}\alpha), \quad (13.3.1)$$

where $q(\alpha_0) = (\cot^{-1}\alpha_0 - \frac{\alpha_0}{1+\alpha_0^2})^{-1}$. The velocity components in the ellipsoidal coordinates are $(u_\alpha, u_\beta, 0)$, and at the surface of the ellipsoid, we have

$$u_\alpha = U\beta\sqrt{\frac{1+\alpha_0^2}{\alpha_0^2+\beta^2}}, \quad u_\beta = -Uq\sqrt{\frac{1-\beta^2}{\alpha_0^2+\beta^2}}(1-\alpha_0\cot^{-1}\alpha_0). \quad (13.3.2)$$

The normal stress $\tau_{\alpha\alpha}$ and shear stress $\tau_{\beta\alpha}$ are calculated with the potential flow, and their values at the surface of the ellipsoid are

$$\tau_{\alpha\alpha} = -2\mu\frac{U\beta q(1+2\alpha_0^2+\beta^2)}{(\alpha_0^2+\beta^2)^2\kappa(1+\alpha_0^2)}, \quad \tau_{\beta\alpha} = 2\mu\frac{Uq\alpha_0}{\kappa(\alpha_0^2+\beta^2)^2}\sqrt{\frac{1-\beta^2}{1+\alpha_0^2}}. \quad (13.3.3)$$

Then the power of the shear stress can be evaluated:

$$\mathcal{P}_s = -\int_A u_\beta\tau_{\beta\alpha}\,\mathrm{d}A = 4\mu\pi U^2\kappa q^2(1-\alpha_0\cot^{-1}\alpha_0)[\alpha_0 + (1-\alpha_0^2)\cot^{-1}\alpha_0]/\alpha_0^2. \quad (13.3.4)$$

Now we calculate the pressure correction p_v. Noting that ellipsoidal harmonics $P_j(\beta)$ (see Lamb, 1932) are appropriate in this case and potential (13.3.1) is proportional to $P_1(\beta) = \beta$, we write the pressure correction as

$$-p_v = C_1 P_1(\beta) + \sum_{j\neq 1} C_j P_j(\beta). \tag{13.3.5}$$

Inserting (13.3.5) into (12.6.1) and using $\mathrm{d}A = 2\pi\kappa^2(1+\alpha_0^2)^{\frac{1}{2}}(\alpha_0^2+\beta^2)^{\frac{1}{2}}\mathrm{d}\beta$, we obtain

$$-\int_A u_\alpha(-p_v)\,\mathrm{d}A = -2\pi\kappa^2 U(1+\alpha_0^2)\int_{-1}^{1} P_1(\beta)\left[C_1 P_1(\beta) + \sum_{j\neq 1} C_j P_j(\beta)\right]\mathrm{d}\beta = \mathcal{P}_s. \tag{13.3.6}$$

The terms P_j $(j \neq 1)$ do not contribute the integral; the coefficient C_1 is determined. Then the pressure correction is

$$-p_v = \frac{-3\mu U q^2}{\kappa(1+\alpha_0^2)\alpha_0^2}(1-\alpha_0\cot^{-1}\alpha_0)\left[\alpha_0 + (1-\alpha_0^2)\cot^{-1}\alpha_0\right] P_1(\beta) + \sum_{j\neq 1} C_j P_j(\beta). \tag{13.3.7}$$

At the limit $\alpha_0 \to \infty$, where the ellipsoid becomes a sphere, pressure correction (13.3.7) reduces to

$$\lim_{\alpha_0\to\infty} -p_v = -3\mu U\cos\theta/a + \sum_{j\neq 1} C_j P_j(\cos\theta), \tag{13.3.8}$$

with $\beta = \cos\theta$ at this limit being understood. This is in agreement with pressure correction (13.2.4) for the spherical gas bubble.

We calculate the drag by direct integration:

$$D_2 = \int_A \mathbf{e}_z\cdot\mathbf{e}_\alpha(-p_v + \tau_{\alpha\alpha})\,\mathrm{d}A = \frac{4\mu\pi U\kappa q^2}{1+\alpha_0^2}\left(\frac{1}{\alpha_0} + \frac{1-\alpha_0^2}{\alpha_0^2}\cot^{-1}\alpha_0\right), \tag{13.3.9}$$

which is in agreement with the dissipation calculation of Moore (1965).

13.4 A liquid drop rising in another liquid

The steady flow of a spherical liquid drop in another immiscible liquid (Harper and Moore, 1968) can be approximated by Hill's spherical vortex inside and potential flow outside. We use the superscript "o" for quantities outside the drop and "i" for quantities inside. The stream and potential functions of the outer flow are

$$\psi^o = \frac{1}{2}U\sin^2\theta\frac{a^3}{r}, \quad \phi = -\frac{1}{2}Ua^3\frac{\cos\theta}{r^2}, \tag{13.4.1}$$

respectively. The stream function for a Hill's vortex moving at a constant velocity relative to a fixed coordinate system is

$$\psi^i = \frac{3Ur^2}{4}\sin^2\theta\left(1-\frac{r^2}{a^2}\right) + \frac{1}{2}Ur^2\sin^2\theta = \frac{Ur^2}{4}\sin^2\theta\left(5-\frac{3r^2}{a^2}\right). \tag{13.4.2}$$

At the surface of the drop, where $r = a$, we have

$$u_r = u_r^o = u_r^i = U\cos\theta, \quad u_\theta = u_\theta^o = u_\theta^i = U\sin\theta/2, \tag{13.4.3}$$

$$\tau_{rr}^o = -6\mu^o U\cos\theta/a, \quad \tau_{r\theta}^o = -3\mu^o U\sin\theta/a, \tag{13.4.4}$$

$$\tau_{rr}^i = -6\mu^i U\cos\theta/a, \quad \tau_{r\theta}^i = 9\mu^i U\sin\theta/(2a). \tag{13.4.5}$$

Hill's vortex problem fits in the general framework discussed in this section in the sense that there is a shear stress discontinuity at the interface that needs to be resolved by the addition of a pressure correction to the irrotational pressure. However, it is somewhat different than gas–liquid interface problems because the shear stress inside the drop is not zero but is determined by Hill's vortex. We seek the expression for the pressure correction by comparing the VPF solution and the VCVPF solution. We proceed by calculating the total dissipation of the system, which is equal to the sum of the power of traction on the outer and inner liquids, $\mathcal{P}^o + \mathcal{P}^i$. There is only one way to calculate $\mathcal{P}^i$, but $\mathcal{P}^o$ may be evaluated on VPF or VCVPF. For VPF, τ_{rr}^o and $\tau_{r\theta}^o$ given by (13.4.4) are used to calculate $\mathcal{P}^o$ and

$$\mathcal{D} = \mathcal{P}^o + \mathcal{P}^i = -\int_A [u_r(-p_i + \tau_{rr}^o) + u_\theta\tau_{r\theta}^o]\,\mathrm{d}A + \mathcal{P}^i. \tag{13.4.6}$$

For VCVPF, a pressure correction is added to resolve the discontinuity between $\tau_{r\theta}^o$ and $\tau_{r\theta}^i$. Then the value of the shear stress at the interface is $\tau_{r\theta}^i$, not $\tau_{r\theta}^o$. The dissipation for VCVPF is

$$\mathcal{D} = \mathcal{P}^o + \mathcal{P}^i = -\int_A [u_r(-p_i - p_v + \tau_{rr}^o) + u_\theta\tau_{r\theta}^i]\,\mathrm{d}A + \mathcal{P}^i. \tag{13.4.7}$$

Because $\mathcal{D}, \mathcal{P}^i, p_i, \tau_{rr}^o$, and $\mathbf{u}$ are the same in both cases, we find that

$$\int_A u_r(-p_v)\,\mathrm{d}A = \int_A u_\theta(\tau_{r\theta}^o - \tau_{r\theta}^i)\,\mathrm{d}A. \tag{13.4.8}$$

Now we expand the pressure correction as a spherical surface harmonic series, and equation (13.4.8) becomes

$$\int_{-1}^{1} U\cos\theta\left[C_1\cos\theta + \sum_{j\neq 1} C_j P_j(\cos\theta)\right] 2\pi a^2\,\mathrm{d}(\cos\theta) = -4\pi a U^2(\mu^o + 3\mu^i/2). \tag{13.4.9}$$

The coefficient C_1 is then obtained, and the pressure correction is

$$-p_v = \frac{-3U}{a}\left(\mu^o + \frac{3\mu^i}{2}\right)\cos\theta + \sum_{j\neq 1} C_j P_j(\cos\theta). \tag{13.4.10}$$

If the inside liquid is gas, $\mu^i = 0$ and the first term of (13.4.10) becomes $-3\mu^o U\cos\theta/a$, which agrees with the first term of pressure correction (13.2.4) for a gas bubble. Pressure

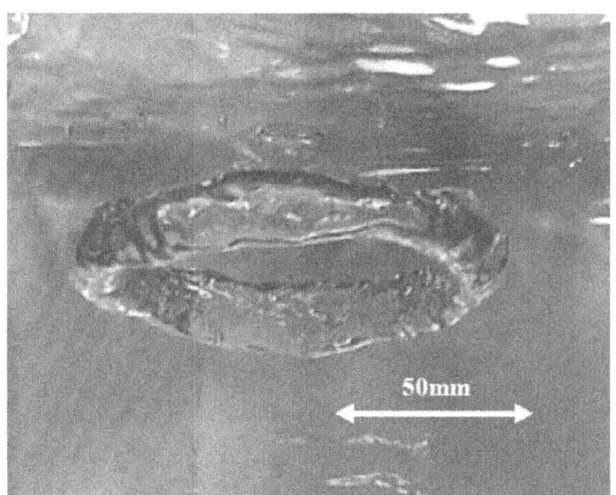

Figure 13.1. (Matsumoto *et al.* 1999). Photograph of a toroidal gas bubble in water.

correction (13.4.10) can also be tested by direct calculation of the drag D_2 on the drop:

$$D_2 = \int_A \mathbf{e}_x \cdot \mathbf{T} \cdot (-\mathbf{e}_r)\, \mathrm{d}A = -\int_A [\cos\theta(-p_v + \tau^o_{rr}) - \sin\theta\tau^i_{r\theta}]\, \mathrm{d}A = 12\pi a U \left(\mu^o + \frac{3\mu^i}{2}\right), \tag{13.4.11}$$

which is the same as the result from the dissipation approximation by Harper and Moore (1968).

13.5 Purely irrotational analysis of a toroidal bubble in a viscous fluid

Here we consider the problem of the rise of a toroidal gas bubble in a viscous liquid (figure 13.1) assuming that the motion is purely irrotational. Earlier, this same problem was studied by Pedley (1968), who considered the generation and diffusion of vorticity, and more recently by Lundgren and Mansour (1991), for the case in which the liquid is inviscid.

13.5.1 Prior work, experiments

Pedley rejected the purely irrotational analysis in which the viscous drag is computed from the viscous dissipation of the irrotational flow. The purely irrotational method works well for the study of drag on other bubbles considered in this chapter. Pedley argues that the toroidal bubble is an exception because of the way that vorticity is generated and diffused by the toroidal bubble. He calls attention to a difference between the equations of energy on which the irrotational theory is based and the equations of impulse in which he says the irrotational viscous drag does not appear. He says that "... It is also shown that if the flow is assumed to remain approximately irrotational in a viscous fluid, so that Lamb's formulae may again be used almost as they stand, the equations of impulse and energy yield conflicting results." The equations of impulse are not often used today, but similar conclusions follow from the Kutta–Joukowski approach used by Lundgren and Mansour (1991) [see the discussion following equation (13.5.49)].

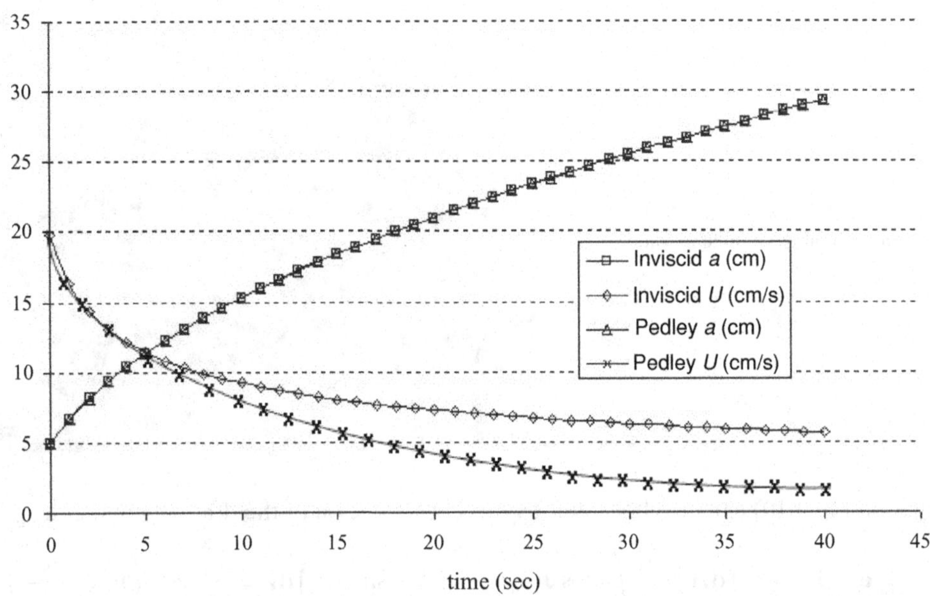

Figure 13.2. Evolution of the ring radius a and velocity U with time t, according to Pedley. The data for Pedley's solution for U are extracted from figure 3 in Pedley (1968). —□—, inviscid solution for a; —◇—, Pedley's viscous solution for a; —△—, inviscid solution for U; —×—, Pedley's viscous solution for U.

Pedley argues that the dissipation argument, which works well for other bubbles, fails for the toroidal bubble because the vorticity is contained as it diffuses and, following arguments presented by Moore (1963), there can be no drag until the vorticity diffuses outward from the bubble surface and forms a wake. The toroidal bubble may become unstable before this happens. Pedley gave an analysis of stability for an inviscid fluid. His analysis is extended to the case of irrotational flow of a viscous fluid.

Pedley showed that, in a viscous fluid, vorticity will continuously diffuse out from the bubble surface, with irrotational flow outside. Pedley's rotational solution does not differ greatly from the irrotational solution for an inviscid liquid (see figure 13.2)

Here we are going to do the irrotational analysis of the toroidal gas bubble in the case in which an irrotational viscous drag is added to force wrench in the impulse equation. In this case, the impulse equation and the energy equation governing the rise of the bubble are the same. The solution of this equation is computed; after a transient state the system evolves to a steady state in which the diameter, toroidal radius, and rise velocity are constant.

Experiments on vortex ring bubbles are sparse and inconclusive. All the experiments are for gas bubbles in water. These experiments suggest that gas bubbles rising in a large expanse of water do not reach a steady state before breaking up because of capillary instability.

Turner (1957) developed a theory for the motion of a buoyant ring in an inviscid liquid. The theory shows that the buoyant force acts to increase the impulse of a ring. The ring diameter increases as the ring rises. He also carried out experiments to verify the theory with small vortex rings formed in water, using methylated spirits and salt to produce the density differences.

Walters and Davidson (1963) observed that a rising toroidal bubble could be produced by release of a mass of gas in water. The form of the bubble is a vortex ring with a

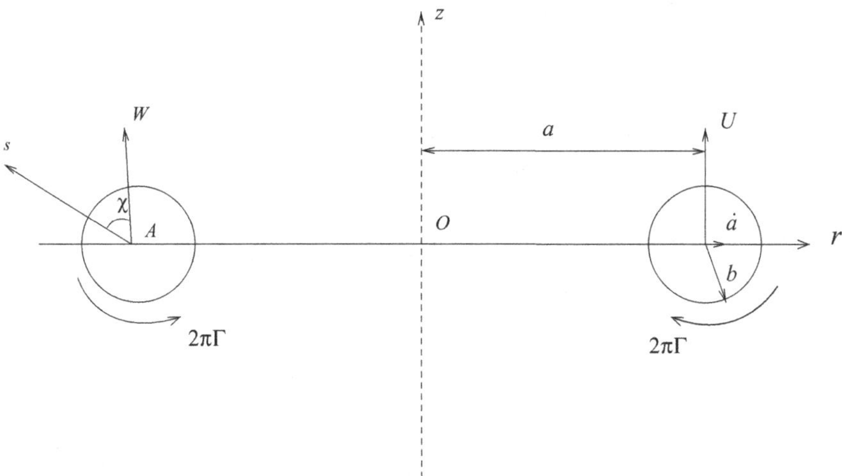

Figure 13.3. Meridional section of the toroidal bubble.

buoyant air core. They created vortex rings by rapidly opening and closing an air jet in the bottom of a water tank, obtaining ring bubbles from 6 to 110 cm^3. These were observed to spread as they rose, as Turner predicted, but there were no measurement of the rate of spreading.

Matsumoto, Kunugi, and Serizawa (1999) reported results for the numerical simulation of ring-type bubbles, and they also presented results for experiments on ring bubbles of air in water in a swimming pool of about 5 m in depth and in a small acrylic box. In a private communication, Serizawa noted that the ring bubbles in the acrylic box grew to 10 cm and were stable. However, the bubbles in the swimming pool grew to a few tens of centimeters in diameter, expanding outward and eventually breaking into air bubbles.

Lundgren and Mansour (1991) describe the ring bubbles as air vortex rings, with approximately constant core volumes perhaps as large as a half a liter, which expand to be rings of the order of 2 ft in diameter as they rise toward the free surface.

Experiments that give rise to reliable data with which various theories can be compared are not available. It would be particularly valuable if experiments could be carried out in highly viscous liquids, not only water. The effects of lateral boundaries on the rise velocity could also be considered.

13.5.2 The energy equation

We use the notation introduced by Pedley and summarized in figure 13.3. The radius of the ring is a, and the plane of this ring is horizontal. The center of this ring is instantaneously at point O, and axis Oz is the upward vertical. The cross section of the air core is assumed to be a circle with radius b. The circulation around the air core is $2\pi\Gamma$. (s, χ) are polar coordinates in a meridional plane, centered at the point A. The bubble is rising vertically with velocity U, and the radius a can be expanding at the same time. The volume of the bubble V is given by $2\pi^2 ab^2$.

Pedley analyzed the motion of a toroidal bubble by two methods, the impulse equation and the energy equation. The rate of change of the vertical impulse is

$$\mathrm{d}P/\mathrm{d}t = F, \tag{13.5.1}$$

where F is the resultant force on the bubble in the vertical direction. Pedley argued that, in this case, F is due to the buoyancy of the bubble and there is no viscous drag. The energy equation is

$$\frac{\mathrm{d}}{\mathrm{d}t}(T+\Omega) = -\mathcal{D}, \tag{13.5.2}$$

where T and Ω are the kinetic and potential energies, respectively, of the system and $\mathcal{D}$ is the dissipation. Pedley wrote the rate of change of potential energy as

$$\frac{\mathrm{d}\Omega}{\mathrm{d}t} = \frac{\mathrm{d}}{\mathrm{d}t}(\rho g V h) = -\rho g V U, \tag{13.5.3}$$

where ρ is the density of the ambient fluid and h is the depth of the bubble beneath a fixed reference level.

Pedley first considered the ambient fluid to be inviscid and the motion to be irrotational. Under these assumptions, the impulse and energy equations give rise to the same expressions for a and U as functions of t: a ultimately increases as $t^{1/2}$, and U decreases as $t^{-1/2}\ln t$. Pedley then considered viscous fluid and assumed that the flow remained approximately irrotational. He argued that the impulse equation would be the same as for the inviscid fluid; however, the energy equation would be changed because of the viscous dissipation. Thus the two methods yielded conflicting results, which led Pedley to conclude that the flow could not remain approximately irrotational. Pedley then investigated the diffusion of vorticity from the bubble surface and showed that the vorticity distribution became approximately Gaussian, with an effective radius b' that also increased as $t^{1/2}$. The solution for a as a function of t turned out to be the same as in the inviscid case, whereas U decreased as $t^{-1/2}$. Other results produced by Pedley's analysis include an evaluation of the effect of a hydrostatic variation in bubble volume and a prediction of the time when the bubble becomes unstable under the action of surface tension.

We consider the toroidal bubble under the assumption that the flow remains approximately irrotational and the ambient fluid is viscous. Our analysis is purely irrotational, and it is different than the one given by Pedley because we include the viscous drag on the bubble in impulse equation (13.5.1). This viscous drag can be computed with the DM, as was done by Levich (1949) in his calculation of the drag on a spherical gas bubble in the irrotational flow of a viscous fluid. When the drag is added, the inconsistency between the impulse equation and the energy equation disappears. Our equations predict that initially the bubble expands and rises in the fluid. The rise velocity U decreases as the ring radius a increases. Ultimately the viscous drag balances the buoyant force and a steady solution is achieved, for which a and U become constants.

In our calculation, we assume that the circulation Γ and the bubble volume V are constant and that the ratio $b/a \ll 1$. The same assumptions were adopted by Pedley (1968).

The flow around the toroidal bubble is approximated by the irrotational flow around a cylindrical bubble of radius b. The translational velocities of the bubble include the vertical velocity U and horizontal velocity $\dot{a}$ (figure 13.3). We denote the resultant velocity as

$$W = \sqrt{U^2+\dot{a}^2}. \tag{13.5.4}$$

The rise velocity U may be calculated from the condition that there is no normal velocity across the bubble surface. If the core cross section is taken to be circular and the quantity

ω/r uniform inside, where ω is the vorticity, Lamb (1932, §163) gives

$$U = \frac{\Gamma}{2a}\left[\ln\frac{8a}{b} - n\right], \tag{13.5.5}$$

where $n = 1/4$. Hicks (1884) gives the value $n = 1/2$ for his hollow vortex ring in which the core cross section is not circular and ω/r is not uniform. The irrotational flow around a cylindrical bubble of radius b is given by the velocity potential

$$\phi = -W\frac{b^2}{s}\cos\chi + \Gamma\chi, \tag{13.5.6}$$

where the coordinates s and χ are shown in figure 13.3. The velocities can be obtained from the potential

$$u_s = \frac{\partial\phi}{\partial s}, \quad u_\chi = \frac{1}{s}\frac{\partial\phi}{\partial\chi}. \tag{13.5.7}$$

The kinetic energy for a 2D section is

$$T' = \frac{1}{2}\rho\int_b^{C_0 a}\int_0^{2\pi}(u_s^2 + u_\chi^2)\, s\, ds d\chi \tag{13.5.8}$$

$$= \rho\pi\left[\frac{1}{2}W^2 b^2\left(1 - \frac{b^2}{C_0^2 a^2}\right) + \Gamma^2\ln\frac{C_0 a}{b}\right], \tag{13.5.9}$$

where the integration is from b to $C_0 a$ rather than b to ∞ because of the logarithmic singularity. Multiply by $2\pi a$, the circumference of the bubble, and obtain the total kinetic energy

$$T = \rho\pi^2 a W^2 b^2\left(1 - \frac{b^2}{C_0^2 a^2}\right) + 2\rho\pi^2 a\Gamma^2\ln\frac{C_0 a}{b}. \tag{13.5.10}$$

We may determine the value of C_0 by comparing (13.5.10) with the kinetic energy given by Pedley (1968). Pedley used Lamb's (1932) formula for the kinetic energy of an arbitrary axisymmetric distribution of azimuthal vorticity and obtained

$$T = 2\pi^2\rho a\Gamma^2\left[\ln\frac{8a}{b} - 2 + O\left(\frac{b^2}{a^2}\ln\frac{8a}{b}\right)\right] \approx 2\pi^2\rho a\Gamma^2\ln\left(\frac{8a}{b}e^{-2}\right). \tag{13.5.11}$$

The kinetic energy given by Pedley neglected the energy associated with the translation of the bubble. Comparing this with the energy associated with the circulation in (13.5.10), we obtain

$$C_0 = 8e^{-2} \approx 1.0827. \tag{13.5.12}$$

Because $b \ll a$, energy (13.5.10) is approximately

$$T = \rho\pi^2 a\left(W^2 b^2 + 2\Gamma^2\ln\frac{C_0 a}{b}\right). \tag{13.5.13}$$

The dissipation of potential flow (13.5.6) can be evaluated as

$$\mathcal{D}' = \int_0^{2\pi}\int_b^{\infty} 2\mu\mathbf{D}:\mathbf{D}\, s\, ds d\chi = 8\pi\mu W^2 + \frac{4\pi\mu\Gamma^2}{b^2}, \tag{13.5.14}$$

where $\mathbf{D}$ is the rate-of-strain tensor. The same dissipation was obtained by Ackeret (1952). After multiplying $2\pi a$, we obtain the total dissipation:

$$\mathcal{D} = 8\pi^2 \mu a \left(2W^2 + \Gamma^2/b^2\right). \tag{13.5.15}$$

The part of the dissipation associated with the circulation Γ in (13.5.15) is the same as that given by Pedley (1968). Pedley did not consider the dissipation associated with the translation W.

After inserting kinetic energy (13.5.13), potential energy (13.5.3), and dissipation (13.5.15) into energy equation (13.5.2), we obtain

$$\frac{1}{2}V\frac{dW^2}{dt} + \left(2\pi^2\Gamma^2\dot{a} - \frac{gV\Gamma}{2a}\right)\left(\ln\frac{8a}{b} - \frac{1}{2}\right) = -16\pi^2 \nu a \left(W^2 + \pi^2 a\Gamma^2/V\right), \tag{13.5.16}$$

where we have used $n = 1/2$ in the expression for U, (13.5.5). Because b can be eliminated from $V = 2\pi^2 ab^2$, a is the only unknown in equation (13.5.16). If the fluid is inviscid and the kinetic energy associated with translation is neglected, (13.5.16) reduces to

$$2\pi^2\Gamma^2\dot{a} - \frac{gV\Gamma}{2a} = 0 \quad \Rightarrow \quad a^2 = a_0^2 + \frac{gV}{2\pi^2\Gamma}t, \tag{13.5.17}$$

where a_0 is the value of a at $t = 0$. This equation is the same as the formula given by Pedley (1968) for both inviscid and viscous fluids.

The translational velocity W has the following form:

$$W^2 = \frac{\Gamma^2}{4a^2}\left(\ln\frac{8a}{b} - \frac{1}{2}\right)^2 + \dot{a}^2. \tag{13.5.18}$$

The derivative of W^2 with respect to time in (13.5.16) leads to a second-order ODE for a. We specify two initial conditions for a at $t = 0$. The first condition is

$$a = a_0 \quad \text{at} \quad t = 0. \tag{13.5.19}$$

The second condition is derived from (13.5.17):

$$\dot{a}(t = 0) = \frac{gV}{4\pi^2\Gamma a_0}. \tag{13.5.20}$$

This condition is justified if the energy and dissipation associated with the translation of the bubble are much smaller than those associated with the circulation and the viscous effects are small at $t = 0$. We will verify these conditions in §13.5.4 when we insert the parameters taken from the experiments of Walters and Davidson into our equation and compute the numerical results.

If the part of the energy and dissipation associated with W is relatively small at all times, (13.5.16) becomes

$$\left(2\pi^2\Gamma\dot{a} - \frac{gV}{2a}\right)\left(\ln\frac{8a}{b} - \frac{1}{2}\right) = -16\pi^4 \nu a^2\frac{\Gamma}{V}. \tag{13.5.21}$$

This is the same energy equation considered by Pedley (1968) under the assumption that the flow remains approximately irrotational in a viscous fluid. Equation (13.5.21) is a first-order ODE for a and condition (13.5.19) will suffice. In §13.5.4 we will show that the solutions of (13.5.16) and (13.5.21) are nearly the same when using the parameters taken from the experiments of Walters and Davidson.

13.5.3 The impulse equation

Pedley used Lamb's formula for the impulse of an arbitrary axisymmetric distribution of azimuthal vorticity and obtained

$$P = 2\pi^2 \rho a^2 \Gamma \left[1 + O\left(\frac{b^2}{a^2} \ln \frac{8a}{b} \right) \right]. \tag{13.5.22}$$

Pedley included the buoyant force in the impulse equation and argued that the drag does not enter into this problem even when the fluid is viscous. We propose to include the viscous drag on the bubble in the impulse equation; then the inconsistency between the impulse equation and the energy equation disappears.

The viscous drag on a bubble may be computed with the DM, which equates the power of the drag to the dissipation. Levich (1949) used the DM to calculate the drag on a spherical gas bubble rising in the irrotational flow of a viscous fluid. Ackeret (1952) used the DM to compute the drag on a rotating cylinder in a uniform stream. If the flow around the toroidal bubble is approximated by the irrotational flow around a cylindrical bubble with circulation, velocity potential (13.5.6) is the same as that used by Ackeret (1952). The irrotational dissipation is given by (13.5.15). In §13.5.4 we will show that the dissipation associated with the translation W is small and $\dot{a}$ is much smaller than U when the parameters taken from the experiments of Walters and Davidson are used. Therefore the drag in the vertical direction from the DM is approximately

$$D = \frac{\mathcal{D}}{U} = 8\pi^2 \mu a \frac{\Gamma^2}{b^2 U}. \tag{13.5.23}$$

With this viscous drag, impulse equation (13.5.1) becomes

$$2\pi^2 \rho \Gamma \frac{da^2}{dt} = \rho g V - 8\pi^2 \mu a \frac{\Gamma^2}{b^2 U}. \tag{13.5.24}$$

After using the expression for U, (13.5.5), we obtain

$$\left(2\pi^2 \Gamma \dot{a} - \frac{gV}{2a} \right) \left(\ln \frac{8a}{b} - \frac{1}{2} \right) = -16\pi^4 \nu a^2 \frac{\Gamma}{V}, \tag{13.5.25}$$

which is the same as energy equation (13.5.21). Thus the impulse equation with the viscous drag included is the same as the energy equation with the viscous dissipation.

13.5.4 Comparison of irrotational solutions for inviscid and viscous fluids

We solve equations (13.5.16) and (13.5.21) using the physical constants

$$g = 980\,\text{cm sec}^{-2}, \quad \nu = 0.011\,\text{cm}^2\,\text{sec}^{-1}, \tag{13.5.26}$$

and the parameters taken from the experiments of Walters and Davidson,

$$\Gamma = 50\,\text{cm}^2\,\text{sec}^{-1}, \quad V = 21\,\text{cm}^3, \quad a_0 = 2.5\,\text{cm}, \quad b_0 = 0.65\,\text{cm}. \tag{13.5.27}$$

These parameters, except a_0 and b_0, are the same as those used by Pedley (1968); Pedley estimated $a_0 = 5$ cm. Our estimate of a_0 is based on figure 7 of Walters and Davidson (1963), which shows photos of the toroidal bubble rising close to the free surface at the top of the tank. The caption in their figure 7 indicates that the ring radius is about 4 cm.

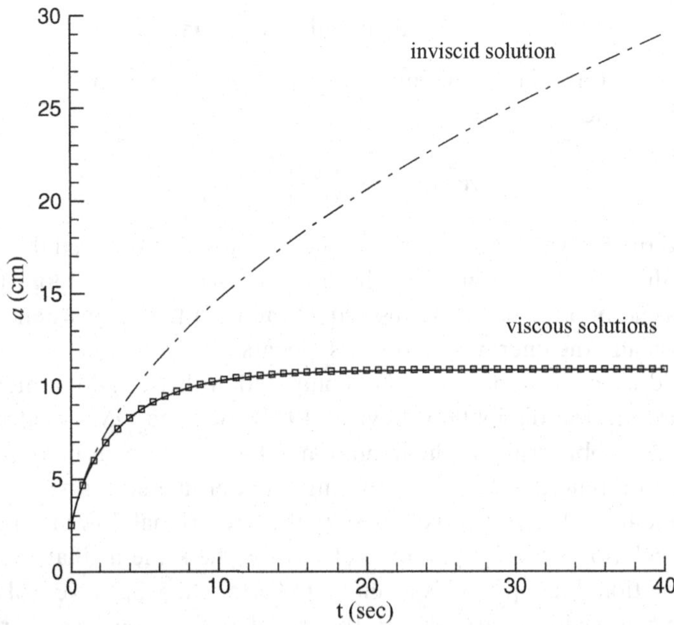

Figure 13.4. Evolution of the ring radius a with time. The dash–dotted curve represents the inviscid solution obtained by putting $\nu = 0$ in (13.5.16). The solid curve represents the viscous solution of (13.5.16) with initial conditions (13.5.19) and (13.5.20). The symbol □ represents the viscous solution of (13.5.21) with initial condition (13.5.19).

Because the ring expands as it rises, the initial ring radius should be less than 4 cm. We estimate a_0 to be 2.5 cm and $b_0/a_0 \approx 0.26$.

We solve equation (13.5.16) with initial conditions (13.5.19) and (13.5.20), and equation (13.5.21) with (13.5.19). To highlight the viscous effects, we also solve (13.5.16) with $\nu = 0$; the initial conditions for the inviscid equation are still (13.5.19) and (13.5.20). The solutions for a as a function of time are plotted in figure 13.4. The rise velocity U is then computed from (13.5.5) and plotted in figure 13.5. Integration of U gives rise to the height $h - h_0$ of rise, where h_0 is the initial position of the bubble. The plots for $h - h_0$ against a are given in figure 13.6.

The solution of the inviscid equation shows that a increases and U decreases with time. These solutions are similar to those obtained by Pedley (1968), who used irrotational flow of a inviscid fluid. The plot for $h - h_0$ against a from the inviscid solution demonstrates that the bubble rises and expands conically, which is consistent with the solutions of Turner (1957) and Pedley (1968). Figure 13.4 shows that the two viscous equations, (13.5.16) and (13.5.21), lead to almost the same solution. This demonstrates that the energy and dissipation associated with the translation W are relatively small and that the choice of initial condition (13.5.20) does not have substantial effects on the solution. Figure 13.7 shows the ratio between the ring expansion velocity $\dot{a}$ and the rise velocity U obtained from the solution of (13.5.16). This plot demonstrates that $\dot{a}$ is much smaller than U. Initially the viscous solution is similar to the inviscid solution; $\dot{a}$ increases and U decreases with time. As a/b^2 increases, viscous drag (13.5.23) increases and finally balances the buoyant force. Then a steady state is reached, in which the toroidal bubble keeps its shape

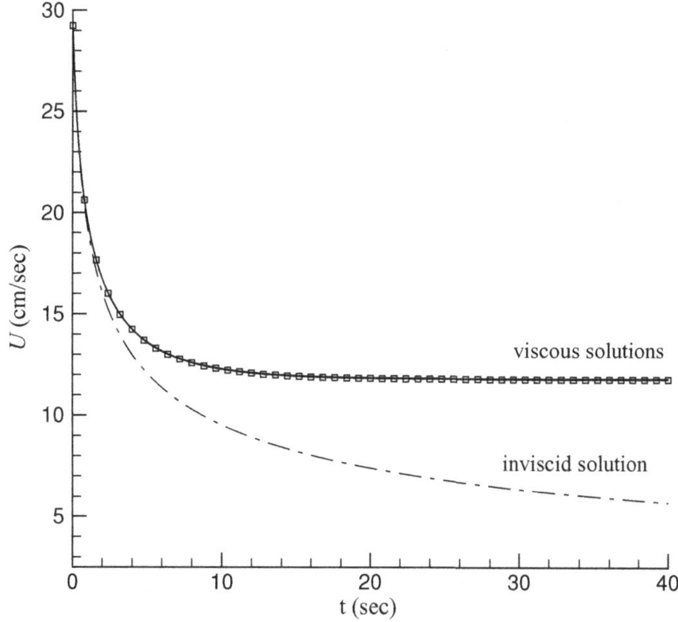

Figure 13.5. Evolution of the rise velocity U computed from (13.5.5) with time. The dash–dotted curve represents the inviscid solution obtained by putting $\nu = 0$ in (13.5.16). The solid curve represents the viscous solution of (13.5.16) with initial conditions (13.5.19) and (13.5.20). The symbol □ represents the viscous solution of (13.5.21) with initial condition (13.5.19).

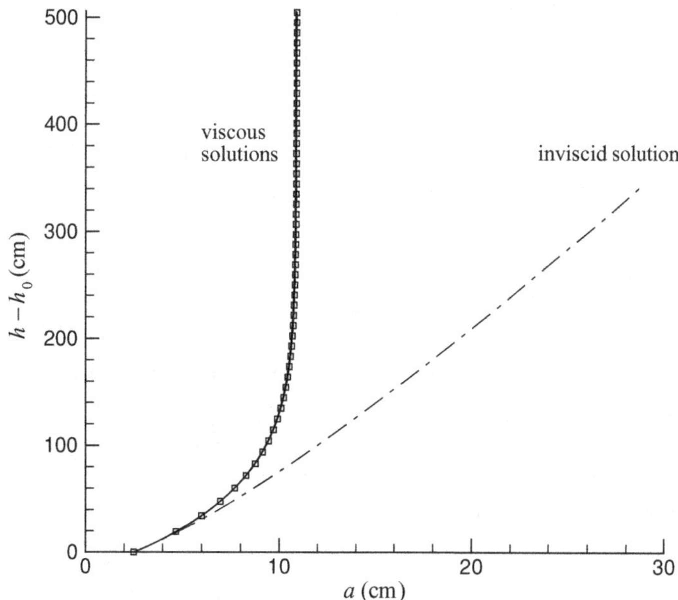

Figure 13.6. The height of rise $h - h_0$ against the ring radius a. The dash–dotted curve represents the inviscid solution obtained by putting $\nu = 0$ in (13.5.16). The solid curve represents the viscous solution of (13.5.16) with initial conditions (13.5.19) and (13.5.20). The symbol □ represents the viscous solution of (13.5.21) with initial condition (13.5.19).

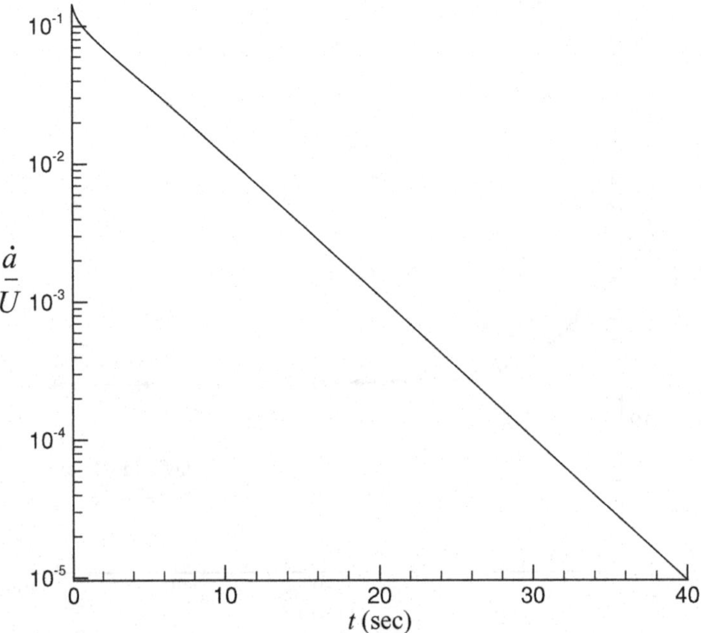

Figure 13.7. The ratio between ring expansion velocity $\dot{a}$ and rise velocity U obtained from the solution of (13.5.16).

and rises at a constant velocity. Our viscous solutions give $a = 10.9$ cm and $U = 11.8$ cm/s at the steady state. This steady-state solution is consistent with the classical description of the motion of a vortex ring. For example, Milne-Thomson (1968) says in §19.41,

> We may, however, observe that for points in the plane of the ring (considered as of infinitesimal cross-section) there is no radial velocity. This follows at once from the Biot and Savart principle, explained in 19.23. It therefore follows that the radius of the ring remains constant, and the ring moves forward with a velocity which must be constant since the motion must be steady relatively to the ring.

The comparison of our solutions with the experiments of Walters and Davidson (1963) is inconclusive. Walters and Davidson did not provide data for the ring radius or the rise velocity as a function of time. The tank used in their experiments is only 3 ft (91.44 cm) tall. Their figure 8 shows the computed circulation vs. time for only 0.6 sec. For such a short time, we cannot even tell the difference between the viscous and inviscid solutions in figures 13.4 and 13.5. It is not known whether the bubble will ultimately reach a steady state as predicted by our viscous solution or whether the ring radius will keep growing until instability occurs as in Pedley's solution.

13.5.5 Stability of the toroidal vortex

Ponstein (1959) and Pedley (1967) studied capillary stability of a vortex core with core curvature neglected. This was the problem studied by Rayleigh (1892) with viscosity neglected and in chapter 16 with viscosity included, in which a strong stabilizing of circulation is demonstrated. The swirl velocity $v_\theta = \Gamma_*/2\pi r$ for $r \gg a$ is irrotational. For axially symmetric disturbances of the form $\exp[i(kz - \omega t)]$, the dispersion relation is given by

$$\omega^2 = -\frac{ka\,K_1(ka)}{K_0(ka)}\left[\left(1 - k^2a^2\right)\frac{\gamma}{\rho a^3} - \frac{\Gamma_*^2}{4\pi^2 a^4}\right], \tag{13.5.28}$$

where K_1 and K_0 are Bessel functions.

The flow is stable when ω^2 is positive, and it is stable for all k when

$$\Gamma_*^2 \geq 4\pi^2 a\gamma/\rho. \tag{13.5.29}$$

Pedley obtained the same result and concluded that the vortex core would eventually become unstable because Γ_* decreases because of the action of viscosity and b decreases in his dynamic solution.

Pedley notes that his stability analysis ignores viscosity and "... is therefore likely to be valid only if disturbance time scales are much shorter than the viscous diffusion time $36^2/8\nu$." Here we construct a purely irrotational analysis of the effects of viscosity on the stability of the same basic state by using VPF. A viscous diffusion time does not enter into this analysis.[1]

13.5.5.1 *Basic flow*

We are considering the stability of a potential vortex with circulation $\Gamma_* = 2\pi\Gamma$. The velocity is purely azimuthal, $\mathbf{u} = \mathbf{e}_\theta V_\theta$:

$$V_\theta = \frac{\Gamma_*}{2\pi r} = \frac{\Gamma}{r}. \tag{13.5.30}$$

The pressure P is given by Bernoulli's equation:

$$\frac{P}{\rho} = \frac{P_\infty}{\rho} - \frac{1}{2}V_\theta^2 = \frac{P_\infty}{\rho} - \frac{1}{2}\frac{\Gamma_*^2}{4\pi^2 r^2}, \tag{13.5.31}$$

where P_∞ is a constant. The normal stress balance at $r = a$ is

$$P(a) - P_0 = -\frac{\gamma}{a}, \tag{13.5.32}$$

where P_0 is the uniform pressure in the vortex $r < a$.

13.5.5.2 *Small disturbances*

The velocity potential for a small disturbance of the basic flow satisfies Laplace's equation:

$$\frac{\partial^2\phi}{\partial r^2} + \frac{1}{r}\frac{\partial\phi}{\partial r} + \frac{1}{r^2}\frac{\partial^2\phi}{\partial\theta^2} + \frac{\partial^2\phi}{\partial z^2} = 0. \tag{13.5.33}$$

The free surface at $r = a$ is disturbed:

$$r = a + \delta(\theta, z, t). \tag{13.5.34}$$

The disturbed surface satisfies

$$\frac{\partial\phi}{\partial r} = \frac{\partial\delta}{\partial t} + \frac{V_\theta}{r}\frac{\partial\delta}{\partial\theta}. \tag{13.5.35}$$

[1] T. Funada, J. C. Padrino, and D. D. Joseph, 2006. Purely irrotational theories of capillary instability of viscous liquids. In preparation (see http://www.aem.umn.edu/people/faculty/joseph/ViscousPotentialFlow/).

The disturbed normal stress balance on $r = a$ is

$$\frac{\Gamma_*^2}{4\pi^2 a^2}\frac{\delta}{a} + \frac{p}{\rho} - 2\nu\frac{\partial^2\phi}{\partial r^2} - 4\nu\frac{\Gamma}{a^3}\frac{\partial\delta}{\partial\theta} = \frac{\gamma}{\rho}\left(\frac{\delta}{a^2} + \frac{\partial^2\delta}{\partial z^2} + \frac{1}{a^2}\frac{\partial^2\delta}{\partial\theta^2}\right). \tag{13.5.36}$$

We seek a solution of (13.5.33)–(13.5.36) in normal modes:

$$\phi = \phi_0(r)\exp(ikz - i\omega t + in\theta) + \text{c.c.}, \tag{13.5.37}$$

$$\delta = \delta_0 \exp(ikz - i\omega t + in\theta) + \text{c.c.} \tag{13.5.38}$$

We find that $\phi_0 = AI_n(kr) + BK_n(kr)$. Because $I_n(kr)$ is unbounded as $r \to \infty$,

$$\phi_0 = BK_n(kr). \tag{13.5.39}$$

Then

$$\frac{p}{\rho} = -\frac{\partial\phi}{\partial t} - \frac{\Gamma}{r^2}\frac{\partial\phi}{\partial\theta} = \left(i\omega - in\frac{\Gamma}{r^2}\right) BK_n(kr)\exp(ikz - i\omega t + in\theta) + \text{c.c.}, \tag{13.5.40}$$

and (13.5.35) implies that

$$\left(-i\omega + in\frac{\Gamma}{a^2}\right)\delta_0 = BkK_n'(ka). \tag{13.5.41}$$

Hence

$$\phi = \left(-i\omega + in\frac{\Gamma}{a^2}\right)\delta_0\frac{K_n(kr)}{kK_n'(ka)}\exp(ikz - i\omega t + in\theta) + \text{c.c.} \tag{13.5.42}$$

Normal stress condition (13.5.36) becomes

$$\begin{aligned}&\left(\omega - n\frac{\Gamma}{a^2}\right)^2\frac{K_n(ka)}{ka\,K_n'(ka)} + 2i\left(\omega - n\frac{\Gamma}{a^2}\right)\nu k^2\frac{K_n''(ka)}{ka\,K_n'(ka)} \\ &= \frac{\gamma}{\rho a^3}\left(1 - k^2a^2 - n^2\right) - \frac{\Gamma_*^2}{4\pi^2 a^4} + 4\nu\frac{in\Gamma}{a^4},\end{aligned} \tag{13.5.43}$$

which is a relation for the eigenvalue ω. By taking $n = 0$ in (13.5.43), we obtain the VPF dispersion relation for axisymmetric disturbances:

$$-\omega^2\frac{K_0(ka)}{ka\,K_1(ka)} - 2i\omega\nu k^2\left[\frac{K_0(ka)}{ka\,K_1(ka)} + \frac{1}{k^2a^2}\right] = \frac{\gamma}{\rho a^3}\left(1 - k^2a^2\right) - \frac{\Gamma^2}{a^4}, \tag{13.5.44}$$

where standard formulas for the modified Bessel functions have been used. Dispersion relation (13.5.44) gives the irrotational effects of viscosity on the stability of the basic flow. With $\nu = 0$, (13.5.44) reduces to the Ponstein–Pedley result, (13.5.28), for inviscid fluids. Introducing the following dimensionless parameters

$$k = \hat{k}/a, \qquad \omega = i\hat{\sigma}\sqrt{\frac{\gamma}{\rho a^3}}, \qquad \Gamma = \hat{\Gamma}\sqrt{\frac{\gamma a}{\rho}}, \tag{13.5.45}$$

and letting

$$\alpha = \frac{K_0(\hat{k})}{K_1(\hat{k})}, \tag{13.5.46}$$

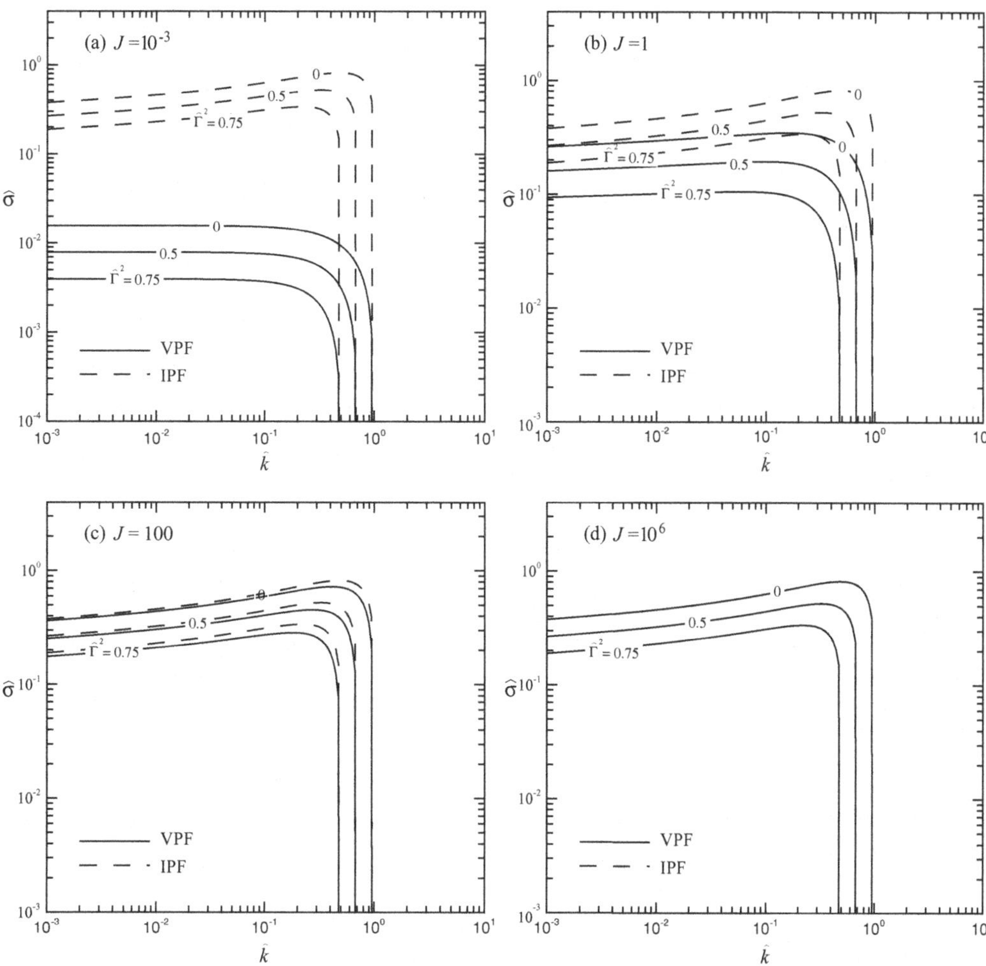

Figure 13.8. Growth rate $\hat{\sigma}$ as a function of the wavenumber $\hat{k}$ for a cylindrical bubble with circulation $\hat{\Gamma}^2$. Instability takes place for the results shown. Four values of the parameter J are selected. The definitions of the dimensionless quantities $\hat{\sigma}$, $\hat{k}$, and $\hat{\Gamma}$ are given in (13.5.45) and $J = \rho\gamma a/\mu^2$.

we can write expression (13.5.44) as

$$\alpha\hat{\sigma}^2 + \frac{2}{\sqrt{J}}\hat{k}\left(1 + \hat{k}\alpha\right)\hat{\sigma} = \hat{k}\left(1 - \hat{k}^2 - \hat{\Gamma}^2\right), \tag{13.5.47}$$

where $J = Oh^2 = \rho\gamma a/\mu^2$ and Oh is the Ohnesorge number. Relation (13.5.47) is a quadratic equation for the eigenvalue $\hat{\sigma}$ with roots

$$\hat{\sigma} = -\frac{\hat{k}\left(1 + \hat{k}\alpha\right)}{\alpha\sqrt{J}} \pm \sqrt{\left[\frac{\hat{k}\left(1 + \hat{k}\alpha\right)}{\alpha\sqrt{J}}\right]^2 + \frac{\hat{k}}{\alpha}\left(1 - \hat{k}^2 - \hat{\Gamma}^2\right)}. \tag{13.5.48}$$

The motion is unstable to axisymmetric disturbances when $Re[\hat{\sigma}] > 0$, which occurs if, and only if, $\hat{\Gamma}^2 < 1 - \hat{k}^2$ for which $\hat{\sigma}$ is real. Hence, when $\hat{\Gamma}^2 \geq 1$ the motion is stable for all $\hat{k}$ and when $\hat{k} \geq 1$ it is stable for all $\hat{\Gamma}$. When $\sqrt{J} \to \infty$, (13.5.47) goes to the inviscid result. Figure 13.8 shows the growth rate $\hat{\sigma}$ versus the wavenumber $\hat{k}$ according to VPF relation

(13.5.48) compared with the inviscid result for four values of $J = 10^{-3}$, 1, 100, and 10^6 and three values of the circulation $\hat{\Gamma}^2 = 0, 0.5$, and 0.75. For small J (e.g., large viscosity) the difference between VPF and the inviscid theory is significant and viscosity diminishes the growth rate. As J increases, VPF results tend to the inviscid ones. For instance, for $J = 10^6$ no difference is discernible. Moreover, for J fixed, the growth rate decreases and the minimum wavelength of the unstable waves increases with decreasing $\hat{\Gamma}^2$ until instabilities totally vanish ($\hat{\Gamma}^2 \geq 1$).

13.5.6 Boundary-integral study of vortex ring bubbles in a viscous liquid

Lundgren and Mansour (1991) studied toroidal bubbles with circulation by a boundary-integral method and by a physically motivated model equation, (13.5.49). The numerical calculations were the first to reveal the deformations of the bubble shape in a fully non-linear context. Two series of computations were performed:

> . . . one set shows the starting motion of an initially spherical bubble as a gravitationally driven liquid jet penetrates through the bubble from below causing a toroidal geometry to develop. The jet becomes broader as surface tension increases and fails to penetrate if surface tension is too large. The dimensionless circulation that develops is not very dependent on the surface tension. The second series of computations starts from a toroidal geometry, with circulation determined from the earlier series, and follows the motion of the rising and spreading vortex ring. Some modifications to the boundary-integral formulation were devised to handle the multiply connected geometry.

The irrotational effects of viscosity were not studied, but they can be obtained by boundary-integral methods following the work of Miksis *et al.* (1982), Georgescu *et al.* (2002), and Canot *et al.* (2003).

The physically motivated model of vortex ring bubbles proposed by Lundgren and Mansour is based on the force-momentum balance on a section of a slender ring, treated as locally 2D, given by

$$\rho A \frac{d\mathbf{u}}{dt} = \rho \Gamma_* \mathbf{t} \times \mathbf{u} + \rho A g \mathbf{e}_z. \tag{13.5.49}$$

The term on the left-hand side is the apparent mass per unit length times acceleration. The first term on the right-hand side is the Kutta–Joukowski lift per unit length on a vortex in cross flow. It acts perpendicularly to the relative velocity $\mathbf{u}$. The unit vector $\mathbf{t}$ is along the centerline of the vortex, in the direction of the vorticity. The last term is the buoyancy force per unit length acting in the upward direction.

Lundgren and Mansour consider the case in which the acceleration $(d\mathbf{u}/dt) = 0$ but the dimensionless ring radius $R = a/r_0$ changes with time. After expressing

$$\mathbf{u} = \mathbf{e}_r u + \mathbf{e}_z v$$

in dimensionless form with $\dot{u},\ \dot{v} = 0$, they find that $v = 0$,

$$u = \frac{2}{3\Gamma R}, \quad R = \left\{ R_0^2 + \frac{4t}{3\Gamma} \right\}^{1/2}, \tag{13.5.50}$$

where Γ is a dimensionless circulation. They note that equation (13.5.50) "is Turner's result for a constant volume buoyant vortex ring. Our interpretation is that the ring spreads radially at a velocity that gives just enough downward cross flow lift to balance the upward buoyancy force. This is the direct physical reason for the radial growth of the ring."

As in the case of impulse, the addition of an irrotational viscous drag per unit length

$$-\mu \tilde{D}\mathbf{e}_z$$

to equation (13.5.49) alters the dynamics in a fundamental way. In this case the system evolves to steady state in which the viscous drag balances the buoyant lift,

$$\mu \tilde{D} = \rho g A,$$

with no further increase of the radius of the core or ring.

13.5.7 Irrotational motion of a massless cylinder under the combined action of Kutta–Joukowski lift, acceleration of added mass, and viscous drag

Consider the irrotational motion of a cylinder of radius a in a viscous fluid with circulation Γ. Our analysis follows that given by Lundgren and Mansour (1991) for the same problem in an inviscid fluid. Let

$$\mathbf{R}(t) = X(t)\mathbf{i} + Y(t)\mathbf{j} \tag{13.5.51}$$

be the instantaneous position of the center of the cylinder. The velocity is then

$$\dot{\mathbf{R}} = U\mathbf{i} + V\mathbf{j}. \tag{13.5.52}$$

The direction of the velocity $\mathbf{n}$ is

$$\mathbf{n} = \frac{U}{W}\mathbf{i} + \frac{V}{W}\mathbf{j} \text{ with } W = \sqrt{U^2 + V^2}. \tag{13.5.53}$$

The motion is approximated by a potential flow. The dissipation per unit length for a cylinder moving with a speed W and a circulation Γ evaluated with the potential flow is (Wang and Joseph, 2006b)

$$\mathcal{D} = 8\pi\mu W^2 + \frac{\mu\Gamma^2}{\pi a^2}. \tag{13.5.54}$$

The dissipation should be equal to the power of the drag and torque on the cylinder. Ackeret (1952) did not consider the torque and used this dissipation to compute the drag on the cylinder:

$$D = \mathcal{D}/W = 8\pi\mu W + \frac{\mu\Gamma^2}{\pi a^2 W}. \tag{13.5.55}$$

Here we use this drag for the purpose of illustration of the viscous effect on the dynamics of the cylinder.

Lundgren and Mansour (1991) derived an equation (their Equation A7) governing the motion of the cylinder based on the assumption that the cylinder is a massless bubble and has no applied force. If drag force (13.5.55) is added, the governing equation for the motion becomes

$$\rho\pi a^2 \ddot{\mathbf{R}} = \rho\Gamma\mathbf{k} \times \mathbf{R} - D\mathbf{n}, \tag{13.5.56}$$

that is, the apparent mass times acceleration is balanced by the Kutta–Joukowski lift and the viscous drag. To make (13.5.56) dimensionless, we introduce the following scales:

$$[\text{length, velocity, time}] \sim \left[a,\ \frac{\Gamma}{a\pi},\ \frac{a^2\pi}{\Gamma}\right]. \tag{13.5.57}$$

In this analysis, we assume that the circulation Γ is a constant and does not depend on time. The dimensionless equations are written in the scalar form, as follows:

$$\dot{\tilde{U}} = -\tilde{V} - \frac{1}{Re}\left(8\tilde{W} + \frac{1}{\tilde{W}}\right)\frac{\tilde{U}}{\tilde{W}}, \tag{13.5.58}$$

$$\dot{\tilde{V}} = \tilde{U} - \frac{1}{Re}\left(8\tilde{W} + \frac{1}{\tilde{W}}\right)\frac{\tilde{V}}{\tilde{W}}, \tag{13.5.59}$$

where "$\sim$" indicates dimensionless parameters and the Reynolds number is defined as

$$Re = \frac{\rho\Gamma}{\mu\pi}. \tag{13.5.60}$$

We set the initial conditions for (13.5.58) and (13.5.59) arbitrarily to be

$$\tilde{U}(t=0) = 10, \quad \tilde{V}(t=0) = 0. \tag{13.5.61}$$

The set of equations (13.5.58), (13.5.59), and (13.5.61) can be solved analytically. First we assume that

$$\tilde{U} = \tilde{W}\cos\theta, \ \ \tilde{V} = \tilde{W}\sin\theta; \tag{13.5.62}$$

thus we have a set of equations:

$$\begin{aligned}
\frac{d\tilde{U}}{dt} &= \frac{d\tilde{W}}{dt}\cos\theta - \tilde{W}\sin\theta\frac{d\theta}{dt} = -\tilde{V} - \frac{1}{Re}\left(8\tilde{W} + \frac{1}{\tilde{W}}\right)\frac{\tilde{U}}{\tilde{W}} \\
&= -\tilde{W}\sin\theta - \frac{1}{Re}\left(8\tilde{W} + \frac{1}{\tilde{W}}\right)\cos\theta, \\
\frac{d\tilde{V}}{dt} &= \frac{d\tilde{W}}{dt}\sin\theta + \tilde{W}\cos\theta\frac{d\theta}{dt} = \tilde{U} - \frac{1}{Re}\left(8\tilde{W} + \frac{1}{\tilde{W}}\right)\frac{\tilde{V}}{\tilde{W}} \\
&= \tilde{W}\cos\theta - \frac{1}{Re}\left(8\tilde{W} + \frac{1}{\tilde{W}}\right)\sin\theta.
\end{aligned} \tag{13.5.63}$$

Multiply the first of (13.5.63) by $\cos\theta$ and the second by $\sin\theta$; then the sum gives

$$\frac{d\tilde{W}}{dt} = -\frac{1}{Re}\left(8\tilde{W} + \frac{1}{\tilde{W}}\right) \to \frac{d}{dt}\left(\frac{1}{2}\tilde{W}^2\right) = -\frac{1}{Re}\left(8\tilde{W}^2 + 1\right). \tag{13.5.64}$$

$$\tilde{W}^2 + \frac{1}{8} = C_1 \exp\left(-16t/Re\right). \tag{13.5.65}$$

Multiply the first by $\sin\theta$ and the second by $\cos\theta$; then the subtraction gives

$$\tilde{W}\frac{d\theta}{dt} = \tilde{W} \to \frac{d\theta}{dt} = 1 \to \theta = t + C_2. \tag{13.5.66}$$

The integration constants C_1 and C_2 are to be determined by initial conditions: $C_1 = 10^2 + \frac{1}{8}$ and $C_2 = 0$. When $Re \to \infty$, $\tilde{W} =$ constant and $\theta = t + C_2$. $\tilde{U}$ and $\tilde{V}$ are integrated to obtain the position of the cylinder. The calculation results are shown in figure 13.9.

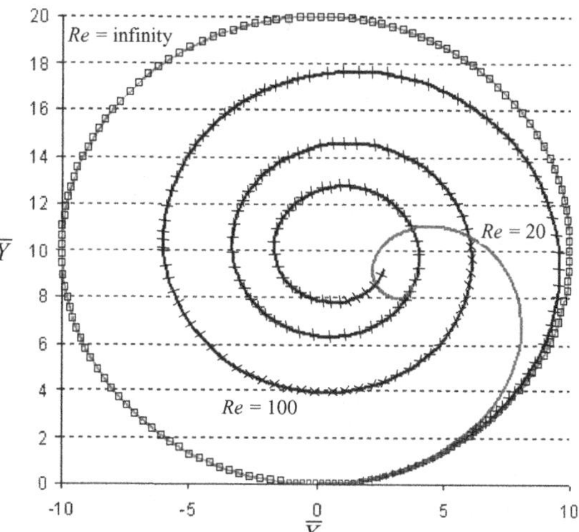

Figure 13.9. The path of the cylinder at different Reynolds numbers: $-\diamond-$, $Re \to \infty$; $\times$, $Re = 100$, $-$, $Re = 20$.

When $Re \to \infty$, the results tend to be the same as the inviscid solution given by Lundgren and Mansour (1991); the cylinder moves with a constant speed along a circular path. When the Reynolds number is finite, the cylinder moves in a spiral fashion. The speed decreases continuously because of the viscous effect and the cylinder eventually stops. The dimensionless stopping time is about 8.36 for $Re = 20$ and 41.79 for $Re = 100$. The paths from the start to the end of the motion are shown for $Re = 20$ and 100 in figure 13.9.

13.6 The motion of a spherical gas bubble in viscous potential flow

A spherical gas bubble accelerates to steady motion in an irrotational flow of a viscous liquid induced by a balance of the acceleration of the added mass of the liquid with the Levich drag. The equation of rectilinear motion is linear and may be integrated, giving rise to exponential decay with a decay constant $18\nu t/a^2$, where ν is the kinematic viscosity of the liquid and a is the bubble radius. The problem of decay to rest of a bubble moving initially when the forces maintaining motion are inactivated and the acceleration of a bubble initially at rest to terminal velocity are considered. The equation of motion follows from the assumption that the motion of the viscous liquid is irrotational. It is an elementary example of how potential flows can be used to study the unsteady motions of a viscous liquid suitable for the instruction of undergraduate students.

We consider a body moving with the velocity U in an unbounded VPF. Let M be the mass of the body and M' be the added mass; then the total kinetic energy of the fluid and body is

$$T = \frac{1}{2}(M + M')U^2. \tag{13.6.1}$$

Let D be the drag and F be the external force in the direction of motion; then the power of D and F should be equal to the rate of the total kinetic energy:

$$(F + D)U = \frac{\mathrm{d}T}{\mathrm{d}t} = (M + M')U\frac{\mathrm{d}U}{\mathrm{d}t}. \tag{13.6.2}$$

We next consider a spherical gas bubble, for which $M = 0$ and $M' = \frac{2}{3}\pi a^3 \rho_f$. The drag can be obtained by direct integration by use of the irrotational viscous normal stress and a viscous pressure correction: $D = -12\pi\mu a U$ (see Joseph and Wang, 2004). Suppose the external force just balances the drag; then the bubble moves with a constant velocity $U = U_0$. Imagine that the external force suddenly disappears; then (13.6.2) gives rise to

$$-12\pi\mu a U = \frac{2}{3}\pi a^3 \rho_f \frac{\mathrm{d}U}{\mathrm{d}t}. \tag{13.6.3}$$

The solution is

$$U = U_0 e^{-\frac{18\nu}{a^2}t}, \tag{13.6.4}$$

which shows that the velocity of the bubble approaches zero exponentially.

If gravity is considered, then $F = \frac{4}{3}\pi a^3 \rho_f g$. Suppose the bubble is at rest at $t = 0$ and starts to move because of the buoyant force. Equation (13.6.2) can be written as

$$\frac{4}{3}\pi a^3 \rho_f g - 12\pi\mu a U = \frac{2}{3}\pi a^3 \rho_f \frac{\mathrm{d}U}{\mathrm{d}t}. \tag{13.6.5}$$

The solution is

$$U = \frac{a^2 g}{9\nu}\left(1 - e^{-\frac{18\nu}{a^2}t}\right), \tag{13.6.6}$$

which indicates the bubble velocity approaches the steady-state velocity,

$$U = \frac{a^2 g}{9\nu}, \tag{13.6.7}$$

exponentially.

Another way to obtain the equation of motion is to argue following Lamb (1932) and Levich (1949) that the work done by the external force F is equal to the rate of the total kinetic energy and the dissipation:

$$FU = (M + M')U\frac{\mathrm{d}U}{\mathrm{d}t} + \mathcal{D}. \tag{13.6.8}$$

Because $\mathcal{D} = -DU$, (13.6.8) is the same as (13.6.2).

The motion of a single spherical gas bubble in a viscous liquid has been considered by some authors. Typically, these authors assemble terms arising in various situations, like Stokes flow (Hadamard–Rybczynski drag, Basset memory integral), high-Reynolds-number flow (Levich drag, boundary-layer drag, induced mass), and other terms, into a single equation. Such general equations have been presented by Yang and Leal (1991) and by Park, Klausner, and Mei (1995), and they have been discussed in the review paper of Magnaudet and Eames (2000; see their section 4). The equation of Yang and Leal has Stokes drag and no Levich drag. Our equation is not embedded in their equation. Park *et al.* listed five terms for the force on a gas bubble; our equation may be obtained from theirs if the free-stream velocity U is put to zero, the memory term is dropped, and the boundary-layer contribution to the drag given by Moore (1963) is neglected. Park *et al.* did not write the same equation as our equation (13.6.5) and did not obtain the exponential decay.

It is generally believed that the added-mass contribution, derived for potential flow, is independent of viscosity. Magnaudet and Eames say that "... results all indicate that the added mass coefficient is independent of the Reynolds number, strength of acceleration and ... boundary conditions." This independence of added mass on viscosity follows from the assumption that the motion of viscous fluids can be irrotational. The results cited by Magnaudet and Eames seem to suggest that induced mass is also independent of vorticity.

Chen (1974) did a boundary-layer analysis of the impulsive motion of a spherical gas bubble that shows that the Levich drag $48/Re$ at short times evolves to the drag $\frac{48}{Re}(1 - \frac{2.21}{\sqrt{Re}})$ obtained in a boundary-layer analysis by Moore (1963). The Moore drag cannot be distinguished from the Levich drag when Re is large. The boundary-layer contribution is vortical and is neglected in our potential flow analysis.

13.7 Steady motion of a deforming gas bubble in a viscous potential flow

Miksis, Vanden-Broeck, and Keller (1982) computed the shape of an axisymmetric rising bubble, or a falling drop, in an incompressible fluid assuming that the flow in the liquid is irrotational but viscous. The boundary condition for the normal stress including surface tension is satisfied but, as in other problems of VPF, the tangential stress is neglected. The shape function is obtained from the gravitational potential evaluated on the free surface; two shape functions are computed, one on the top and one on the bottom of the bubble. The shape is single valued on each function. The potential function is obtained from the values of the potential on the free surface by use of a Green's function approach following ideas introduced by Longuet-Higgins and Cokelet (1976), Vanden-Broeck and Keller (1980), and Miksis, Vanden-Broeck, and Keller (1981). The system of differential and integral equations is solved in a frame in which the bubble is stationary and the velocity at infinity is U, which is calculated by a drag balance in two ways. The first calculation is like that of Moore (1959), in which the drag comes from the normal irrotational viscous stress leading to $32/Re$. This direct method should not be used because of the additional contribution that is due to the irrotational viscous pressure.

This pressure is not easy to calculate in general, but the correct drag leading to $48/Re$ can be obtained, and was obtained by Miksis, Vanden-Broeck, and Keller in a second calculation, using the DM.

The solution of the system of governing equations was obtained as a power series in the Weber number and Re^{-1} and is therefore restricted to low Weber numbers (large surface tension) and high Reynolds numbers (small viscosity).

13.8 Dynamic simulations of the rise of many bubbles in a viscous potential flow

This problem was considered by Sangani and Didwania (1993). They placed N bubbles initially randomly within a unit cell and assumed that the entire space was filled with copies of this cell. They determined the potential flow around many bubbles exactly by using a multiple expansion. They computed a drag force on the bubbles by two different methods: (1) computing the gradient of the total viscous energy based on the relative velocity of individual bubbles, and (2) computing the bounce of colliding bubbles assuming that the collision time is short compared with the time scale for the inertial motion and that the momentum and kinetic energies are conserved in the motion (Sangani, 1991). These are apparently quite different mechanisms: One depends on viscosity, the other on inertia.

They performed simulations in which they included buoyancy and viscous forces. They found that the state of uniform bubbly liquids is unstable under the aforementioned conditions and that the bubbles form large aggregates by arranging themselves in planes perpendicular to gravity. These aggregates form even when a swarm of gas bubbles rises through a liquid at rest.

We think that the aggregates that form are due to the same inertial forces that turn long bodies broadside-on and cause aircraft to stall. Colliding bubbles are unstable long bodies that look for a stable configuration with lines of centers across the stream. The dynamic case underway is called "drafting, kissing, and tumbling." This case is discussed in §20.7.1 and §20.7.2.

14

Purely irrotational theories of the effect of viscosity on the decay of waves

14.1 Decay of free-gravity waves

It is generally believed that the major effects of viscosity are associated with vorticity. This belief is not always well founded; major effects of viscosity can be obtained from purely irrotational analysis of flows of viscous fluids. Here we illustrate this point by comparing irrotational solutions with Lamb's (1932) exact solution of the problem of the decay of free-gravity waves. Excellent agreements, even in fluids 10^7 more viscous than water, are achieved for the decay rates $n(k)$ for all wavenumbers k, excluding a small interval around a critical value k_c, where progressive waves change to monotonic decay.

14.1.1 Introduction

Lamb (1932, §348, §349) performed an analysis of the effect of viscosity on free-gravity waves. He computed the decay rate by a DM, using the irrotational flow only. He also constructed an ES for this problem, which satisfies both the normal and shear stress conditions at the interface.

Joseph and Wang (2004) studied Lamb's problem by using the theory of VPF and obtained a dispersion relation that gives rise to both the decay rate and the wave velocity. They also used VCVPF to obtain another dispersion relation. Because VCVPF is an irrotational theory, the shear stress cannot be made to vanish. However, the shear stress in the energy balance can be eliminated in the mean by the selection of an irrotational pressure that depends on viscosity.

Here we find that the viscous pressure correction gives rise to a higher-order irrotational correction to the velocity that is proportional to the viscosity and does not have a boundary-layer structure. The corrected velocity depends strongly on viscosity and is not related to vorticity. The corrected irrotational flow gives rise to a dispersion relation that is in splendid agreement with Lamb's exact solution, which has no explicit viscous pressure. The agreement with the exact solution holds for fluids even 10^7 times more viscous than water and for all wavenumbers away from the cutoff wavenumber k_c, which marks the place where progressive waves change to monotonic decay. We find that VCVPF gives rise to the same decay rate as in Lamb's exact solution and in his dissipation calculation when $k < k_c$. The exact solution agrees with VPF when $k > k_c$. The effects of vorticity are evident in only a small interval centered on the cutoff wavenumber. We present a comprehensive comparison for the decay rate and the wave velocity given by Lamb's ES and Joseph and Wang's VPF and VCVPF theories.

14.1.2 Irrotational viscous corrections for the potential flow solution

The gravity wave problem is governed by the linearized Navier–Stokes equation and the continuity equation

$$\frac{\partial \mathbf{u}}{\partial t} = -\frac{1}{\rho}\nabla p - g\mathbf{e}_y + \nu\nabla^2\mathbf{u}, \tag{14.1.1}$$

$$\nabla \cdot \mathbf{u} = 0, \tag{14.1.2}$$

subject to the boundary conditions at the free surface ($y \approx 0$),

$$T_{xy} = 0, \quad T_{yy} = 0, \tag{14.1.3}$$

where T_{xy} and T_{yy} are components of the stress tensor and the surface tension is neglected. Surface tension is important at high wavenumbers but, for simplicity, is neglected in the analyses given here. We divide the velocity and pressure field into two parts,

$$\mathbf{u} = \mathbf{u}_p + \mathbf{u}_v, \quad p = p_p + p_v, \tag{14.1.4}$$

where the subscript p denotes potential solutions and v denotes viscous corrections. The potential solutions satisfy

$$\mathbf{u}_p = \nabla\phi, \quad \nabla^2\phi = 0, \tag{14.1.5}$$

$$\frac{\partial \mathbf{u}_p}{\partial t} = -\frac{1}{\rho}\nabla p_p - g\mathbf{e}_y. \tag{14.1.6}$$

The viscous corrections are governed by

$$\nabla \cdot \mathbf{u}_v = 0, \tag{14.1.7}$$

$$\frac{\partial \mathbf{u}_v}{\partial t} = -\frac{1}{\rho}\nabla p_v + \nu\nabla^2\mathbf{u}_v. \tag{14.1.8}$$

We take the divergence of (14.1.8) and obtain

$$\nabla^2 p_v = 0, \tag{14.1.9}$$

which shows that the pressure correction must be harmonic. Next we introduce a stream function ψ so that (14.1.7) is satisfied identically:

$$u_v = -\frac{\partial\psi}{\partial y}, \quad v_v = \frac{\partial\psi}{\partial x}. \tag{14.1.10}$$

We eliminate p_v from (14.1.8) by cross differentiation and obtain the following equation for the stream function:

$$\frac{\partial}{\partial t}\nabla^2\psi = \nu\nabla^4\psi. \tag{14.1.11}$$

To determine the normal modes that are periodic in respect to x with a prescribed wavelength $\lambda = 2\pi/k$, we assume that

$$\psi = Be^{nt+ikx}e^{my}, \tag{14.1.12}$$

where m is to be determined from (14.1.11). Inserting (14.1.12) into (14.1.11), we obtain

$$(m^2 - k^2)\left[n - \nu(m^2 - k^2)\right] = 0. \tag{14.1.13}$$

The root $m^2 = k^2$ gives rise to irrotational flow; the root $m^2 = k^2 + n/\nu$ leads to the rotational component of the flow. The rotational component cannot give rise to a harmonic pressure satisfying (14.1.9) because

$$\nabla^2 e^{nt+ikx} e^{my} = (m^2 - k^2) e^{nt+ikx} e^{my} \tag{14.1.14}$$

does not vanish if $m^2 \neq k^2$. Thus the governing equation for the rotational part of the flow can be written as

$$\frac{\partial \psi}{\partial t} = \nu \nabla^2 \psi. \tag{14.1.15}$$

This is the equation used by Lamb (1932) for the rotational part of his ES.

The effect of viscosity on the decay of a free-gravity wave can be approximated by a purely irrotational theory in which the explicit appearance of the irrotational shear stress in the mechanical-energy equation is eliminated by a viscous contribution p_v to the irrotational pressure. In this theory $\mathbf{u} = \nabla \phi$ and a stream function, which is associated with vorticity, is not introduced. The kinetic energy, potential energy, and dissipation of the flow can be computed with the potential flow solution,

$$\phi = A e^{nt+ky+ikx}. \tag{14.1.16}$$

We insert the potential flow solution into mechanical-energy equation (12.5.14),

$$\int_{S_f} u_n \left[\rho \left(\frac{\partial \phi}{\partial t} + \frac{|\nabla \phi|^2}{2} + g\eta \right) + 2\mu \frac{\partial^2 \phi}{\partial n^2} + \gamma \nabla_{\text{II}} \cdot \mathbf{n} \right] \mathrm{d}S_f = - \int_{S_f} \tau_s u_s \mathrm{d}S_f, \tag{14.1.17}$$

where η is the elevation of the surface and $\mathbf{D}$ is the rate-of-strain tensor. The pressure correction p_v satisfies

$$\int_0^\lambda v(-p_v) \mathrm{d}x = \int_0^\lambda u \tau_{xy} \mathrm{d}x. \tag{14.1.18}$$

However, in our problem here, there is no explicit viscous pressure function in the ES [see (14.1.23) and (14.1.24)]. It turns out that the pressure correction defined here in the purely irrotational flow is related to quantities in the exact solution in a complicated way that requires further analysis [see (14.1.30)].

Joseph and Wang (2004) solved for the harmonic pressure correction from (14.1.9), then determined the constant in the expression of p_v by using (14.1.18), and obtained

$$p_v = -2\mu k^2 A e^{nt+ky+ikx}. \tag{14.1.19}$$

The velocity correction associated with this pressure correction can be obtained from (14.1.8). We seek normal-mode solution $\mathbf{u}_v \sim e^{nt+ky+ikx}$, and equation (14.1.8) becomes

$$\rho n \mathbf{u}_v = -\nabla p_v. \tag{14.1.20}$$

Hence, $\text{curl}(\mathbf{u}_v) = 0$ and $\mathbf{u}_v$ is irrotational. After assuming $\mathbf{u}_v = \nabla \phi_1$ and $\phi_1 = A_1 e^{nt+ky+ikx}$, we obtain

$$\rho n \phi_1 = -p_v \quad \Rightarrow \quad \phi_1 = \frac{2\mu k^2}{\rho n} A e^{nt+ky+ikx}. \tag{14.1.21}$$

Given ϕ_1, we can compute the correction η_1 of η from the equation $n\eta = \partial \phi_1 / \partial y$.

This calculation shows that the velocity $\mathbf{u}_v$ associated with the pressure correction is irrotational. Pressure correction (14.1.19) is proportional to μ, and it induces a correction ϕ_1 given by (14.1.21), which is also proportional to μ. The shear stress computed from $\mathbf{u}_v = \nabla\phi_1$ is then proportional to μ^2. To balance this nonphysical shear stress, we can add a pressure correction proportional to μ^2, which will in turn induce a correction for the velocity potential proportional to μ^2. We can continue to build higher-order corrections, and they will all be irrotational. The final velocity potential has the following form:

$$\phi = (A + A_1 + A_2 + \cdots +)e^{nt+ky+ikx}, \tag{14.1.22}$$

where $A_1 \sim \mu$, $A_2 \sim \mu^2 \ldots$. Thus the VCVPF theory is an approximation to the ES based on solely potential flow solutions, but the normal stress condition and n are not corrected.

Prosperetti (1976) considered viscous effects on standing free-gravity waves by using the same governing equations, (14.1.7) and (14.1.8), for the viscous correction terms. If we adapt our VCPVF method to treat standing waves represented by the potential $\phi = k^{-1}(\mathrm{d}a/\mathrm{d}t)e^{ky}\cos kx$, we can obtain $-p_v = 2\mu k(\mathrm{d}a/\mathrm{d}t)e^{ky}\cos kx$, which is exactly the same pressure correction obtained by Prosperetti (1976), who used a different method.

14.1.3 Relation between the pressure correction and Lamb's exact solution

It has been conjectured and is widely believed (Moore, 1963; Harper and Moore, 1968; Joseph and Wang, 2004) that a viscous pressure correction arises in the vortical boundary layer at the free surface that is neglected in the irrotational analysis. However, no viscous pressure correction arises in Lamb's ES. His solution is given by a potential ϕ and a stream function ψ:

$$u = \frac{\partial\phi}{\partial x} - \frac{\partial\psi}{\partial y}, \quad v = \frac{\partial\phi}{\partial y} + \frac{\partial\psi}{\partial x}, \quad \frac{p}{\rho} = -\frac{\partial\phi}{\partial t} - gy, \tag{14.1.23}$$

satisfying

$$\nabla^2\phi = 0, \quad \partial\psi/\partial t = \nu\nabla^2\psi. \tag{14.1.24}$$

The stream function gives rise to the rotational part of the flow. No pressure term enters into the stream function equation, as we have shown in the previous section that the only harmonic pressure for the rotational part is zero. The pressure p comes from Bernoulli's equation in (14.1.23) and no explicit viscous pressure exists, though p depends on the viscosity through the velocity potential. Lamb shows that (14.1.24) can be solved with normal modes:

$$\phi = Ae^{ky}e^{ikx+nt}, \quad \psi = Ce^{my}e^{ikx+nt}, \quad m^2 = k^2 + n/\nu, \tag{14.1.25}$$

where A and C are constants.

It is therefore of interest to derive the connection between the viscous pressure correction p_v in our VCVPF theory and Lamb's exact solution; the superscript E represents Lamb's exact solution and J represents the VCVPF theory of Joseph and Wang. The irrotational pressures in the two solutions are

$$p^E = -\rho\frac{\partial\phi^E}{\partial t} - \rho g\eta^E, \quad p_i^J = -\rho\frac{\partial\phi^J}{\partial t} - \rho g\eta^J. \tag{14.1.26}$$

Table 14.1. *The value of each term in (14.1.30) normalized by A^E for SO10000 oil at different wavenumbers: term 1* $= \partial(\phi^J - \phi^E)/\partial t$, *term 2* $= g(\eta^J - \eta^E)$, *term 3* $= 2\nu\partial^2(\phi^J - \phi^E)/\partial y^2$, *and term 4* $= 2\nu\partial^2\psi^E/\partial x\partial y$

k	p_v/ρ	Term 1	Term 2	Term 3	Term 4
0.01	-2.063×10^{-6}	$-1.325 \times 10^{-9} + i2.01 \times 10^{-4}$	$-2.063 \times 10^{-6} - i2.01 \times 10^{-4}$	-1.325×10^{-9}	$5.300 \times 10^{-9} + i5.300 \times 10^{-9}$
0.1	-2.057×10^{-4}	$-7.441 \times 10^{-7} + i0.00358$	$-2.071 \times 10^{-4} - i0.00358$	-7.461×10^{-7}	$2.980 \times 10^{-6} + i2.980 \times 10^{-6}$
1	-0.02022	$-4.207 \times 10^{-4} + i0.06272$	$-0.02106 - i0.06440$	-4.186×10^{-4}	$0.001679 + i0.001679$
10	-1.881	$-0.3131 + i0.6303$	$-2.423 - i1.513$	-0.1829	$1.038 + i0.8830$

The elevation η is obtained from the kinematic condition at $y \approx 0$:

$$\frac{\partial \eta^E}{\partial t} = \frac{\partial \phi^E}{\partial y} + \frac{\partial \psi^E}{\partial x}, \quad \frac{\partial \eta^J}{\partial t} = \frac{\partial \phi^J}{\partial y}. \tag{14.1.27}$$

The normal stress balance for the two solutions is

$$T_{yy}^E = -p^E + 2\mu\frac{\partial^2 \phi^E}{\partial^2 y} + 2\mu\frac{\partial^2 \psi^E}{\partial x \partial y} = 0, \tag{14.1.28}$$

$$T_{yy}^J = -p_i^J - p_v + 2\mu\frac{\partial^2 \phi^J}{\partial^2 y} = 0. \tag{14.1.29}$$

Therefore $T_{yy}^E - T_{yy}^J = 0$, and we can obtain

$$\frac{p_v}{\rho} = \frac{\partial\left(\phi^J - \phi^E\right)}{\partial t} + g(\eta^J - \eta^E) + 2\nu\frac{\partial^2\left(\phi^J - \phi^E\right)}{\partial y^2} - 2\nu\frac{\partial^2 \psi^E}{\partial x \partial y}. \tag{14.1.30}$$

The amplitude A for the potential is different in Lamb's exact solution and in VCPVF:

$$\phi^E = A^E e^{nt+ky+ikx}, \quad \phi^J = A^J e^{nt+ky+ikx}, \quad A^E \neq A^J. \tag{14.1.31}$$

To make the two solutions comparable, we compute the relation between A^E and A^J by equating the dissipation evaluated with Lamb's ES and evaluated with VCVPF. In table 14.1 we list the values of each term in (14.1.30) normalized by A^E. It seems that the term $g(\eta^J - \eta^E)$ gives the most important contribution to p_v, but the other terms are not negligible.

14.1.4 Comparison of the decay rate and the wave velocity given by the exact solution, VPF, and VCVPF

When the surface tension is ignored, Lamb's ES gives rise to the following dispersion relation:

$$n^2 + 4\nu k^2 n + 4\nu^2 k^4 + gk = 4\nu^2 k^3\sqrt{k^2 + n/\nu}. \tag{14.1.32}$$

Lamb considered the solution of (14.1.32) in the limits of small k and large k. When $k \ll k_c = (g/\nu^2)^{1/3}$, he obtained approximately

$$n = -2\nu k^2 \pm ik\sqrt{g/k}, \tag{14.1.33}$$

which gives rise to the decay rate $-2\nu k^2$, in agreement with the dissipation result, and the wave velocity $\sqrt{g/k}$, which is the same as the wave velocity for inviscid potential flow. When $k \gg k_c = (g/\nu^2)^{1/3}$, Lamb noted that the two roots of (14.1.32) are both real. One of them is

$$n_1 = -\frac{g}{2\nu k}, \tag{14.1.34}$$

and the other one is

$$n_2 = -0.91\nu k^2. \tag{14.1.35}$$

Lamb pointed out that n_1 is the more important root because the motion corresponding to n_2 dies out very rapidly.

14.1.4.1 *VPF results*

Joseph and Wang (2004) treated this problem by using VPF and obtained the following dispersion relation:

$$n^2 + 2\nu k^2 n + gk = 0. \tag{14.1.36}$$

When $k < k_c = (g/\nu^2)^{1/3}$, the solution of (14.1.36) is

$$n = -\nu k^2 \pm ik\sqrt{g/k - \nu^2 k^2}. \tag{14.1.37}$$

We note that the decay rate $-\nu k^2$ is half of that in (14.1.33) and the wave velocity $\sqrt{g/k - \nu^2 k^2}$ is slower than the inviscid wave velocity. When $k > k_c = (g/\nu^2)^{1/3}$, the two roots of (14.1.36) are both real and they are

$$n = -\nu k^2 \pm \sqrt{\nu^2 k^4 - gk}. \tag{14.1.38}$$

If $k \gg k_c = (g/\nu^2)^{1/3}$, the preceding two roots are approximately

$$n_1 = -\frac{g}{2\nu k}, \tag{14.1.39}$$

$$n_2 = -2\nu k^2 + \frac{g}{2\nu k}. \tag{14.1.40}$$

We note that (14.1.39) is the same as (14.1.34), and the magnitude of (14.1.40) is approximately twice that of (14.1.35).

14.1.4.2 *VCVPF results*

Joseph and Wang (2004) computed a pressure correction and added it to the normal stress balance to obtain

$$n^2 + 4\nu k^2 n + gk = 0, \tag{14.1.41}$$

which is the dispersion relation for VCVPF theory. When $k < k_c' = (g/4\nu^2)^{1/3}$, the solution of (14.1.41) is

$$n = -2\nu k^2 \pm ik\sqrt{g/k - 4\nu^2 k^2}. \tag{14.1.42}$$

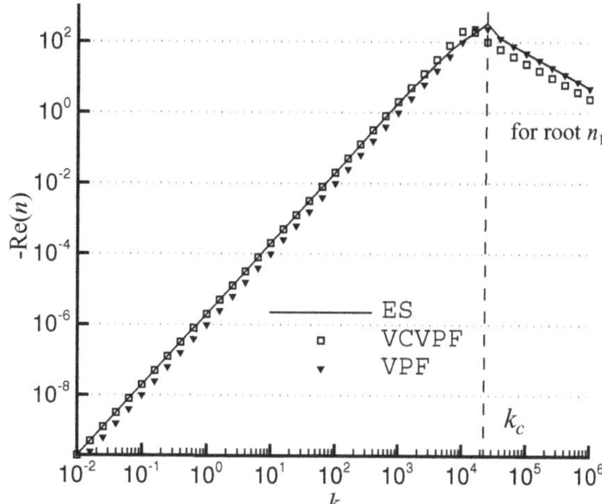

Figure 14.1. Decay rate $-\mathrm{Re}(n)$ vs. wavenumber k for water, $\nu = 10^{-6}$ m^2/s. Re(n) is computed for the exact solution (ES) from (14.1.32), for VPF from (14.1.36), and for VCVPF from (14.1.41). When $k < k_c$, the decay rate $-2\nu k^2$ for VCVPF is in good agreement with the ES, whereas the decay rate $-\nu k^2$ for VPF is only half of the ES. When $k > k_c$, n has two real solutions in each theory. In this figure, we plot the decay rate n_1 corresponding to (14.1.34), (14.1.39), and (14.1.44). The ES can be approximated by $-g/(2\nu k)$; the decay rate $-g/(2\nu k)$ for VPF is in agreement with the ES, whereas the decay rate $-g/(4\nu k)$ for VCVPF is only half of the ES.

We note that the decay rate $-2\nu k^2$ is the same as in (14.1.33) and the wave velocity $\sqrt{g/k - 4\nu^2 k^2}$ is slower than the inviscid wave velocity. When $k > k_c' = (g/4\nu^2)^{1/3}$, the two roots of (14.1.41) are both real and they are

$$n = -2\nu k^2 \pm \sqrt{4\nu^2 k^4 - gk}. \tag{14.1.43}$$

If $k \gg k_c' = (g/4\nu^2)^{1/3}$, the preceding two roots are approximately

$$n_1 = -\frac{g}{4\nu k}, \tag{14.1.44}$$

$$n_2 = -4\nu k^2 + \frac{g}{4\nu k}. \tag{14.1.45}$$

We note that (14.1.44) is half of (14.1.34), and the magnitude of (14.1.45) is approximately four times that of (14.1.35).

14.1.4.3 *Comparison of the exact and purely irrotational solutions*

We compute the solution of (14.1.32) and compare the real and imaginary parts of n with those we obtain by solving (14.1.36) and (14.1.41). Water, glycerin, and SO10000 oil, for which the kinematic viscosity is 10^{-6}, 6.21×10^{-4}, and 1.03×10^{-2} m^2/sec, respectively, are chosen as examples. Figures 14.1 and 14.2 show the decay rate $-\mathrm{Re}(n)$ for water; the root n_1 when $k > k_c$ is shown in figure 14.1 and the root n_2 in figure 14.2. The imaginary part of n, i.e., the wave velocity multiplied by k, is plotted in figure 14.3 for water. For glycerin and SO10000 oil (figures 14.4 and 14.5), we plot the decay rate corresponding to only the more important root n_1; the plots for the root n_2 and the wave velocity are omitted. Figure 14.6 shows the decay rate corresponding to the root n_1 for $\nu = 10$ m^2/sec, which is 1000 times more viscous than SO10000 oil; the comparison between the exact solution and VPF, VCVPF is still excellent. The cutoff wavenumber $k_c = (g/\nu^2)^{1/3}$ decreases as the viscosity increases. In Table 14.2 we list the values of k_c for water, glycerin, SO10000 oil, and the liquid with $\nu = 10$ m^2/s. In practice, waves associated with different wavenumbers may exist simultaneously. For very viscous fluids, k_c is small and the majority of the wavenumbers

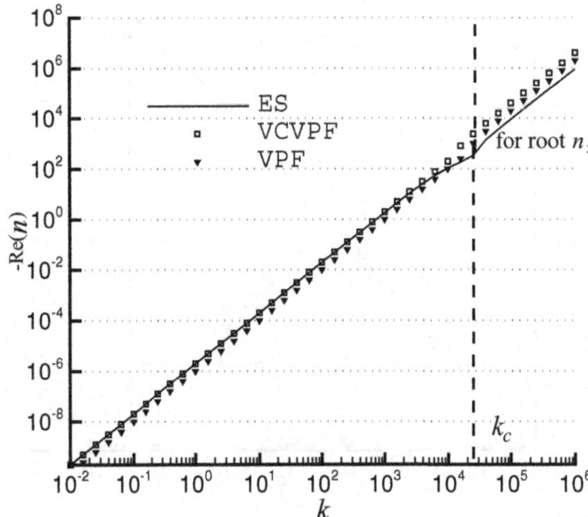

Figure 14.2. Decay rate $-\text{Re}(n)$ vs. wavenumber k for water, $\nu = 10^{-6}$ m^2/sec. Re(n) is computed for the exact solution (ES) from (14.1.32), for VPF from (14.1.36), and for VCVPF from (14.1.41). When $k > k_c$, n has two real solutions in each theory. In this figure, we plot the decay rate n_2 corresponding to (14.1.35), (14.1.40), and (14.1.45). The decay rate for the ES can be approximated by $-0.91\nu k^2$; the decay rate $\approx -2\nu k^2$ for VPF is closer to the ES than the decay rate $\approx -4\nu k^2$ for VCVPF.

are above k_c; therefore the motion of monotonic decay dominates; for less viscous fluids, the motion of progressive waves may dominate.

14.1.5 Why does the exact solution agree with VCVPF when $k < k_c$ and with VPF when $k > k_c$?

Our VCVPF solution and Lamb's dissipation calculation are based on the assumption that energy equation (14.1.17) for the ES is well approximated by an irrotational solution. To verify this, we computed and compared the rates of change of the kinetic energy, the potential energy, and dissipation terms in (14.1.17) for Lamb's exact solution and for the purely irrotational part of his solution. The agreement is excellent when $k < k_c$. This shows that the vorticity may be neglected in the computation of terms in the energy balance when $k < k_c$ and is consistent with results given in §14.1.4 that demonstrate that the decay rates from VCVPF and the dissipation calculation agree with the ES when

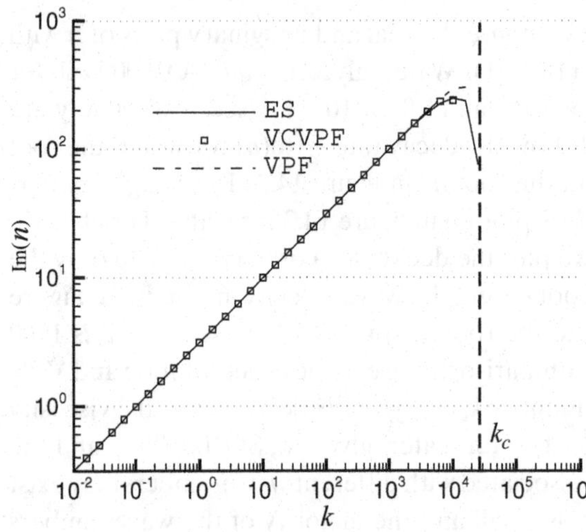

Figure 14.3. Im(n), i.e., the wave velocity multiplied by k, vs. wavenumber k for water, $\nu = 10^{-6}$m^2/sec. Im(n) is computed for the exact solution (ES) from (14.1.32), for VPF from (14.1.36), and for VCVPF from (14.1.41). When $k < k_c$, the three theories give almost the same wave velocity. When $k > k_c$, all the three theories give zero imaginary part of n.

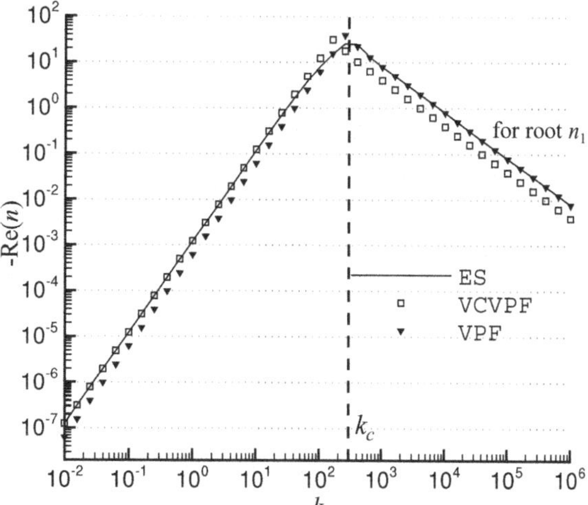

Figure 14.4. Decay rate $-\text{Re}(n)$ vs. wavenumber k for glycerin, $\nu = 6.21 \times 10^{-4}\text{m}^2/\text{sec}$. $\text{Re}(n)$ is computed for the exact solution (ES) from (14.1.32), for VPF from (14.1.36), and for VCVPF from (14.1.41). When $k < k_c$, the decay rate $-2\nu k^2$ for VCVPF is in good agreement with the ES, whereas the decay rate $-\nu k^2$ for VPF is only half of the ES. When $k > k_c$, n has two real solutions in each theory. In this figure, we plot the decay rate n_1 corresponding to (14.1.34), (14.1.39), and (14.1.44). The decay rate for the ES can be approximated by $-g/(2\nu k)$; the decay rate $-g/(2\nu k)$ for VPF is in agreement with the ES, whereas the decay rate $-g/(4\nu k)$ for VCVPF is only half of the ES.

k is small. The agreement is poor for k in the vicinity of k_c; therefore the decay rates from VCVPF deviate from the ES near k_c, as shown in figures 14.1–14.6.

When k is much larger than k_c, the energy equation is not well approximated by the irrotational part of the ES. However, this result does not mean that the vorticity is important. Lamb pointed out that $m \approx k$ when k is large, which is confirmed in our calculation. It follows that the vorticity of the ES is

$$\nabla^2\psi = \left(m^2 - k^2\right) Ce^{my+ikx+nt} \approx 0; \tag{14.1.46}$$

the vorticity is negligible when k is large. The result that the wave is nearly irrotational for large k was also pointed out by Tait (1890). Consequently, the decay rate $-g/(2\nu k)$ from VPF is in good agreement with the ES and no pressure correction is needed.

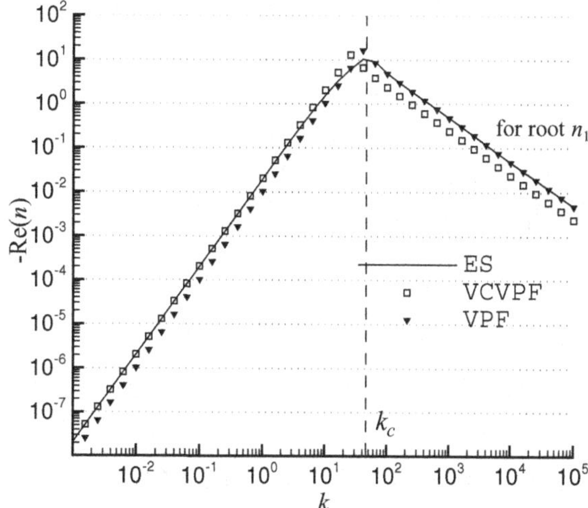

Figure 14.5. Decay rate $-\text{Re}(n)$ vs. wavenumber k for SO10000 oil, $\nu = 1.03 \times 10^{-2}$ m^2/sec. $\text{Re}(n)$ is computed for the exact solution (ES) from (14.1.32), for VPF from (14.1.36), and for VCVPF from (14.1.41). When $k < k_c$, the decay rate $-2\nu k^2$ for VCVPF is in good agreement with the exact solution, whereas the decay rate $-\nu k^2$ for VPF is only half of the ES. When $k > k_c$, n has two real solutions in each theory. In this figure, we plot the decay rate n_1 corresponding to (14.1.34), (14.1.39), and (14.1.44). The decay rate for the ES can be approximated by $-g/(2\nu k)$; the decay rate $-g/(2\nu k)$ for VPF is in agreement with the ES, whereas the decay rate $-g/(4\nu k)$ for VCVPF is only half of the ES.

Table 14.2. *The values for the cutoff wavenumber k_c for water, glycerin, SO10000 oil, and the liquid with $\nu = 10\ m^2/sec$*

k_c decreases as the viscosity increases

Fluid	Water	Glycerin	SO10000	–
ν (m²/sec)	10^{-6}	6.21×10^{-4}	1.03×10^{-2}	10
k_c (1/m)	21399.7	294.1	45.2	0.461

14.1.6 Conclusion and discussion

The problem of decay of free-gravity waves that is due to viscosity was analyzed with two different theories of viscous potential flow, VPF and VCVPF. The pressure correction leads to a hierarchy of potential flows in powers of viscosity. These higher-order contributions vanish more rapidly than the principal correction, which is proportional to μ. The higher-order corrections do not have a boundary-layer structure and may not have a physical significance.

The irrotational theory is in splendid agreement with Lamb's ES for all wavenumbers k except for those in a small interval around k_c, where progressive waves change to monotonic decay. VCVPF agrees with Lamb's solution when $k < k_c$ (progressive waves) and VPF agrees with Lamb's ES when $k > k_c$ (monotonic decay). The cutoff wavenumber $k_c = (g/\nu^2)^{1/3}$ decreases as the viscosity increases. In practice, waves associated with different wavenumbers may exist simultaneously. For very viscous fluids, k_c is small and the majority of the wavenumbers are above k_c; therefore the motion of monotonic decay dominates; for less viscous fluids, the motion of progressive waves may dominate.

There is a boundary layer of vorticity associated with the back-and-forth motion of the progressive waves. The confined vorticity layer has almost no effect on the solution except for k near k_c. There is no explicit pressure correction in the ES. The vortical part of the ES does not generate a pressure correction; the pressure depends on the viscosity through

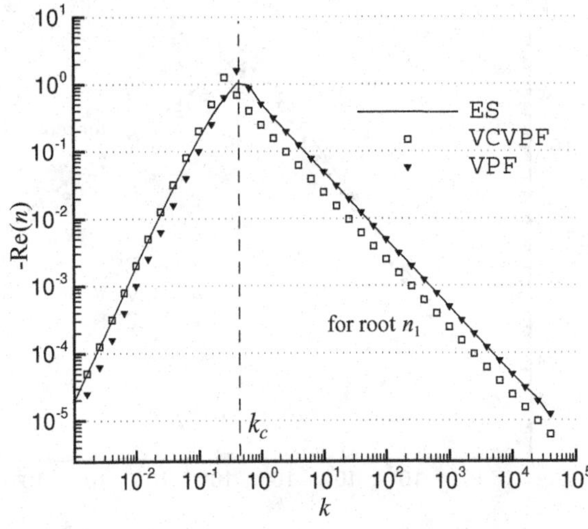

Figure 14.6. Decay rate $-\mathrm{Re}(n)$ vs. wavenumber k for $\nu = 10$ m²/sec. $\mathrm{Re}(n)$ is computed for the exact solution (ES) from (14.1.32), for VPF from (14.1.36), and for VCVPF from (14.1.41). When $k < k_c$, the decay rate $-2\nu k^2$ for VCVPF is in good agreement with the ES, whereas the decay rate $-\nu k^2$ for VPF is only half of the ES. When $k > k_c$, n has two real solutions in each theory. In this figure, we plot the decay rate n_1 corresponding to (14.1.34), (14.1.39), and (14.1.44). The decay rate for the ES can be approximated by $-g/(2\nu k)$; the decay rate $-g/(2\nu k)$ for VPF is in agreement with the ES, whereas the decay rate $-g/(4\nu k)$ for VCVPF is only half of the ES.

the velocity potential and the surface elevation; it is related to the potential and vortical parts of the ES in a complicated way. The vortical part is not dominant and the pressure correction is not primarily associated with a boundary layer [see (14.1.30) and table 14.1].

The analysis of capillary instability of liquid in gas (Wang, Joseph, and Funada, 2005a) is very much like the analysis of the decay of free-gravity waves. The purely potential flow analysis is in splendid agreement with Tomotika's (1935) exact solution, which has no explicit dependence on viscous pressure. In the case of capillary instability, the best result is based on VCVPF because the short waves that give rise to a sluggish decay in Lamb's problem are stabilized by surface tension.

14.1.7 Quasi-potential approximation – vorticity layers

A tractable theory for weakly damped, nonlinear Stokes waves was formulated by Ruvinsky and Freidman (1985a, 1985b, 1987). This theory has come to be known as the quasi-potential approximation. The approximation is based on a Helmholtz decomposition:

$$\mathbf{u} = \nabla\phi + \mathbf{v}, \tag{14.1.47}$$

where $\mathbf{v}$ is the rotational part. They apply a boundary-layer approximation to $\mathbf{v}$ and obtain the following system of coupled equations:

$$\nabla^2\phi = 0, \quad \frac{\partial\phi}{\partial z} \to 0, \quad \text{as } z \to \infty \tag{14.1.48}$$

and, on $z = \eta$,

$$\frac{\partial\phi}{\partial t} + \frac{1}{2}|\nabla\phi|^2 + g\eta - \frac{\gamma}{\rho}\kappa + 2\nu\frac{\partial^2\phi}{\partial z^2} = 0, \tag{14.1.49}$$

$$\frac{\partial\eta}{\partial t} + \frac{\partial\eta}{\partial x}\frac{\partial\phi}{\partial x} = \frac{\partial\phi}{\partial z} + v_z, \tag{14.1.50}$$

where $\kappa = (\partial^2\eta/\partial x^2)/[1 + (\partial\eta/\partial x)^2]^{3/2}$ and v_z is the vertical component of the rotational velocity satisfying

$$\frac{\partial v_z}{\partial t} = 2\nu\frac{\partial^3\phi}{\partial x^2\partial z}. \tag{14.1.51}$$

These are three equations in ϕ, η, and v_z. Their derivation is such that v_z is assumed small. This is one reason why the quasi-potential approximation is said to model weak effects of viscosity.

Spivak *et al.* (2002) applied the quasi-potential methodology for a slightly viscous fluid and small-surface elasticity to numerically compute typical free-surface profiles induced by a moving pressure distribution under the combined effects of gravity, viscosity, surface tension, and film elasticity.

Ruvinsky, Feldstein, and Freidman (1991) carried out simulations of ripple excitation by steep capillary-gravity waves. They note that, when one is describing the capillary-gravity excitation phenomenon, dissipative processes must be taken into account: "In an exact formulation, this problem has not yet been solved even with a computer." To solve their problem they used the quasi-potential methodology.

Longuet-Higgins (1992) gave a simplified derivation of the quasi-potential methodology in an application to the theory of weakly damped Stokes waves. He gave a simpler form of the equations by applying the boundary conditions on a slightly displaced surface.

Are the principal effects of viscosity associated with vorticity or are they purely irrotational? In his 1992 paper, Longuet-Higgins says that "... Lamb 1932... showed that for most wavelengths of interest the effects of viscosity on linear, deep-water waves are confined to a thin vortex layer near the free surface, of thickness $D_0 = (2\nu/\gamma)^{1/2}$ (where ν denotes the kinematic viscosity and σ the radian frequency). When $kD_0 \ll 1$ (k the wavenumber) we may say that the waves are weakly damped."

The foregoing statement by Longuet-Higgins is not correct. The effects of viscosity are not confined to a thin vortex layer; the main effects of viscosity on the decay of waves are irrotational. Typically the effects of viscosity arising from the boundary layer at a gas–liquid surface are small because the rates of strain in these layers are no larger than in the irrotational flow and the layer thickness is small. In a later work, Longuet-Higgins (1997, §14.3.2) calculated the decay of nonlinear capillary-gravity waves by computing the viscous dissipation of the irrotational flow without vorticity or vorticity layers.

14.2 Viscous decay of capillary waves on drops and bubbles

Now we consider the problem of decay of capillary waves on free surfaces of spherical form. Free surfaces are gas–liquid surfaces in which the dynamical effects of the gas are neglected. A drop is a liquid sphere with gas outside; a bubble is a gas sphere with liquid outside.

We may consider two types of waves on a spherical surface, capillary waves and gravity waves. Capillary waves are driven by surface tension. Gravity waves are driven by gravity; in the case of spheres some form of central gravity law must be adopted.

In the limit of a large radius of the spherical surface we obtain a plane problem such as the problem of free-gravity waves considered in the previous section. The problem of decay of capillary waves or capillary-gravity waves on plane or spherical surfaces can be considered.

In the last section we studied the decay of gravity waves on a plane free surface. The short waves decay monotonically and are well approximated by VPF. The long waves are progressive, and the decay of these progressive waves is well approximated by VCVPF. Here we show that the same result holds for capillary waves and hence may be expected for capillary-gravity waves.

From now on, we shall confine our attention to capillary waves on a spherical surface. The problem of the viscous decay of gravity waves on a globe was considered by Lamb (1932).

The following citation from Lamb (§355) makes the point about viscosity and irrotationality in oscillations of a liquid globe. He gives the formula for decay constant τ for $\exp(-t/\tau)$ from the dissipation analysis $\tau = a^2/(n-1)(2n+1)\nu$ and says that

> the most remarkable feature of this result is the excessively minute extent to which the oscillations of a globe of moderate dimensions are affected by such a degree of viscosity as is ordinarily met in nature. For a globe of the size of the earth, and of the same kinematic

viscosity as water, we have, on the cgs system, $a = 6.37 \times 10^8$, $\nu = 0.0178$ and the value of τ for the gravitational oscillation of longest period ($n = 2$) is therefore

$$\tau = 1.44 \times 10^{11} \text{ years.}$$

Even with the value found by Darwin (1878) for the viscosity of pitch near the freezing temperature, viz $\mu = 1.3 \times 10^8 \times g$, we find, taking $g = 980$, the value

$$\tau = 180 \text{ h}$$

for the modules of decay of the slowest oscillation of a globe the size of the earth, having the density of water and the viscosity of pitch. Since this is still large compared with the period of 1 h 34 min found in Art 262, it appears that such a globe would oscillate almost like a perfect fluid.

Padrino and Joseph (2007) have shown how to improve Lamb's (1932) analysis of dissipation (DM). Lamb's analysis predicts the effects of viscosity on the rate of decay of the waves but not on the frequency. The effects of viscosity on the frequency of irrotational waves were obtained by use of VCVPF. Furthermore, it was shown that the DM and VCVPF give the same dispersion relation. This result goes farther than Lamb's, as the effects of viscosity on the wave frequency can be obtained from the DM.

The analysis in this chapter is developed for the problem of decay of irrotational waves on an interface between two liquids. This kind of problem is not perfectly suited to analysis based on purely irrotational flow; the analysis in §16.1 of capillary instability of one liquid in another shows that vorticity is important for long waves but less important for short waves. The results we need for the free-surface problems in which purely irrotational theories give excellent results will be obtained as limits of a two-fluid case.

14.2.1 Introduction

A viscous liquid drop surrounded by a quiescent gas or a gas bubble immersed in a viscous liquid tends to an equilibrium spherical shape if the effects of surface tension are significantly large in comparison with gravitational effects. When the spherical interface of the bubble or drop is slightly perturbed by an external agent, the bubble or drop will recover their original spherical configuration through an oscillatory motion of decreasing amplitude. In the case of the drop, depending on its size and physical properties, the return to the spherical shape may consist of monotonically decaying waves. For a drop immersed in another viscous liquid, decaying oscillatory waves always occur at the liquid–liquid interface.

Early studies on the subject for inviscid liquids are due to Kelvin (1890) and Rayleigh (1896). The former obtained an expression for the frequencies of small oscillations of an inviscid liquid globe that tends to the spherical shape by the existence of an internal gravitational potential in the absence of surface tension. The latter included surface tension instead of gravitation. Lamb (1881, 1932) considered the effect of viscosity on the decay rate of the oscillations of a liquid globe, assuming an irrotational velocity field by using the DM. Lamb's result is independent of the nature of the forces that drive the interface to the spherical shape. Chandrasekhar (1961) studied fully viscous effects on the small oscillations of a liquid globe with self-gravitation forces, neglecting surface tension. The same form of the solution was also obtained by Reid (1960) when surface tension instead of self-gravitation is the force that tends to maintain the spherical shape. A good account of this solution is also presented in the treatise by Chandrasekhar (1961). Following Lamb's reasoning, Valentine, Sather, and Heideger (1965) applied the DM to the case of a drop surrounded by another viscous liquid.

A comprehensive analysis of viscous effects in a drop embedded in liquid was presented by Miller and Scriven (1968), who also included rheological properties of the interface. The limiting cases of a drop in a vacuum or a bubble surrounded by a liquid were considered. However, no numerical results were given. Prosperetti (1980a) presented numerical results for the eigenvalue problem for modes $\ell = 2$ to $\ell = 6$. He showed that the spectrum is continuous when the liquid outside is unbounded. Previous studies relied on normal modes. Prosperetti (1977, 1980b) considered the initial-value fully viscous problem posed by small perturbations about the spherical shape of a drop or a bubble in which no *a priori* form of the time dependence is assumed. The solution showed that the normal-mode results are recovered for large times.

Finite-size disturbances have received some attention. Tsamopoulus and Brown (1983) considered the small-to-moderate-amplitude inviscid oscillations by using perturbations methods. Lundgren and Mansour (1988) and Patzek *et al.* (1991) studied the inviscid problem posed by large oscillations, applying the boundary-integral and the finite-element methods, respectively. Lundgren and Mansour also investigated the effect of a "small" viscosity on drop oscillations. Basaran (1992) carried out the numerical analysis of moderate-to-large-amplitude axisymmetric oscillations of a viscous liquid drop.

Following Padrino, Funada, and Joseph (2007a), we seek approximate solutions of the linearized problem for small departures from the spherical shape for a drop surrounded by a gas of negligible density and viscosity or a bubble embedded in a liquid. These solutions, from VPF and VCVPF, are compared with each other, with IPF, and with ES.

14.2.2 VPF analysis of a single spherical drop immersed in another fluid

Consider a single spherical drop of radius a filled with a fluid with density ρ_l and viscosity μ_l immersed in another fluid with density ρ_a and viscosity μ_a. The coefficient of interfacial tension is denoted as γ. Both fluids are incompressible and Newtonian with gravity neglected. At the basic or undisturbed state both fluids are at rest and the pressure jump across the spherical interface is balanced by surface tension.

When the basic state is disturbed with small irrotational perturbations, the resulting velocity field can be written as the gradient of a potential. The disturbance of the spherical interface is denoted by $\zeta \equiv \zeta(t, \theta, \varphi)$; the interface position is $r = a + \zeta$.

For irrotational flow the incompressible Navier–Stokes equations reduce to the Bernoulli equation. The resulting pressure field can be decomposed into the undisturbed pressure plus a "small" disturbance.

After subtracting the basic state from the disturbed fluid motion and performing standard linearization of the resulting expressions by neglecting products of the small fluctuations and products of their derivatives, we obtain, for the interior motion ($0 \leq r < a$),

$$\nabla^2 \phi_l = 0, \tag{14.2.1}$$

$$p_l = -\rho_l \frac{\partial \phi_l}{\partial t}, \tag{14.2.2}$$

and, for the exterior motion ($a < r < \infty$),

$$\nabla^2 \phi_a = 0, \tag{14.2.3}$$

$$p_a = -\rho_a \frac{\partial \phi_a}{\partial t}, \tag{14.2.4}$$

where ϕ is the velocity potential and p is the pressure disturbance. For irrotational motion, the boundary conditions at the interface require the continuity of the radial velocity and the balance of the normal stresses by interfacial tension. For small departures about the spherical shape, $a \gg \zeta$, and the boundary conditions can be written as

$$u_r^l = u_r^a \tag{14.2.5}$$

for the continuity of the radial velocity at $r = a$ and

$$\left[\!\left[-p + 2\mu \frac{\partial u_r}{\partial r} \right]\!\right] = \frac{\gamma}{a^2}(L^2 - 2)\zeta \tag{14.2.6}$$

for the balance of normal stresses across the interface $r = a$, written in linearized form, accounting for the pressure balance in the undisturbed state. The notation $[\![\cdot]\!] = (\cdot)_{r=b^+} - (\cdot)_{r=b^-}$ is used to denote the jump across the interface located at $r = a$. The linearized kinematic condition is

$$u_r = \frac{\partial \zeta}{\partial t}, \tag{14.2.7}$$

at $r = a$ with $u_r = \partial\phi/\partial r$. The right-hand side of (14.2.6) is obtained from the linearized form of the divergence of the outward unit normal vector to the disturbed interface for the interior fluid. The operator L^2 is also known as the spherical Laplacian and emerges, for instance, in the solution of the Laplace equation obtained with spherical coordinates by application of the method of separation of variables. It is defined as

$$-L^2\zeta = \frac{1}{\sin\theta}\frac{\partial}{\partial\theta}\left(\sin\theta\frac{\partial\zeta}{\partial\theta}\right) + \frac{1}{\sin^2\theta}\frac{\partial^2\zeta}{\partial\varphi^2}. \tag{14.2.8}$$

Solutions of (14.2.1) and (14.2.3) for the interior and exterior of a sphere, respectively, can be sought in the form

$$\phi_l(r,\theta,\varphi,t) = \sum_{\ell=0}^{\infty} A_\ell \left(\frac{r}{a}\right)^{\ell} e^{-\sigma_\ell t} S_\ell(\theta,\varphi) + \text{c.c.}, \qquad 0 \le r < a, \tag{14.2.9}$$

$$\phi_a(r,\theta,\varphi,t) = \sum_{\ell=0}^{\infty} C_\ell \left(\frac{r}{a}\right)^{-\ell-1} e^{-\sigma_\ell t} S_\ell(\theta,\varphi) + \text{c.c.}, \quad a < r < \infty, \tag{14.2.10}$$

such that ϕ_l is finite at $r = 0$ and ϕ_a remains bounded as $r \to \infty$; σ_ℓ is an eigenvalue to be determined. The symbol c.c. designates the complex conjugate of the previous term. The functions S_ℓ are the surface harmonics of integral order,

$$S_\ell(\theta,\varphi) = \sum_{m=-\ell}^{\ell} B_{\ell m} Y_\ell^m(\theta,\varphi), \tag{14.2.11}$$

which, with the choice $\bar{B}_{\ell m} = B_{\ell,-m}$, are real functions. The functions $Y_\ell^m(\theta, \varphi)$ are known as the spherical harmonics (Strauss, 1992),

$$Y_\ell^m(\theta, \varphi) = P_\ell^{|m|}(\cos\theta)e^{im\varphi}, \tag{14.2.12}$$

where $P_\ell^{|m|}$ are the associated Legendre functions. The spherical harmonics satisfy

$$L^2 Y_\ell^m(\theta, \varphi) = \ell(\ell+1)Y_\ell^m(\theta, \varphi), \tag{14.2.13}$$

for $\ell = 0, 1, 2, \ldots,$ and $m = -\ell, \ldots, -1, 0, 1, \ldots, \ell$. The operator L^2 was defined in (14.2.8). Expressions for the radial components of the velocity can be obtained from (14.2.9) and (14.2.10) by application of $u_r = \partial\phi/\partial r$. Then the pressure disturbances p_l and p_a can be obtained from (14.2.2) and (14.2.4). It will be shown that σ_ℓ does not depend on the index m.

Let us write the disturbance of the spherical shape of the interface as a series expansion:

$$\zeta(\theta, \varphi, t) = \sum_{\ell=0}^{\infty} \zeta_\ell(\theta, \varphi)e^{-\sigma_\ell t} + \text{c.c.} \tag{14.2.14}$$

By considering $\zeta_\ell(\theta, \varphi) = \zeta_{0\ell} S_\ell(\theta, \varphi)$, where $\zeta_{0\ell}$ is a constant, and using conditions (14.2.5) and (14.2.7), we obtain

$$-\sigma_\ell\zeta_{0\ell} = \left(\frac{\ell}{a}\right)A_\ell, \qquad \sigma_\ell\zeta_{0\ell} = \left(\frac{\ell+1}{a}\right)C_\ell. \tag{14.2.15}$$

In addition, we have

$$(L^2-2)\zeta = \sum_{\ell=0}^{\infty}[\ell(\ell+1)-2]\zeta_{0\ell}e^{-\sigma_\ell t}S_\ell + \text{c.c.} = \sum_{\ell=0}^{\infty}(\ell+2)(\ell-1)\zeta_{0\ell}e^{-\sigma_\ell t}S_\ell + \text{c.c.}, \tag{14.2.16}$$

by virtue of (14.2.13) and (14.2.14). Substituting normal-mode expressions for u_r and p, obtained with (14.2.9) and (14.2.10), into the left-hand-side of (14.2.6), applying the result (14.2.16), and replacing A_ℓ and C_ℓ with (14.2.15) yields the dispersion relation for the eigenvalue σ_ℓ, which, after some manipulation, may be written as

$$\begin{aligned}\Big[\rho_l(\ell+1)+\rho_a\ell\Big]\sigma^2 - \left[\frac{2\mu_l}{a^2}(\ell+1)\ell(\ell-1) + \frac{2\mu_a}{a^2}(\ell+2)(\ell+1)\ell\right]\sigma \\ + \frac{\gamma}{a^3}(\ell+2)(\ell+1)\ell(\ell-1) = 0,\end{aligned} \tag{14.2.17}$$

for $\ell = 0, 1, 2, \ldots,$ where the subscript ℓ has been dropped from σ for convenience. Expression (14.2.17) may be written in dimensionless form with the following choices of dimensionless parameters (Funada and Joseph, 2002);

$$\hat{l} = \frac{\rho_a}{\rho_l}, \quad \hat{m} = \frac{\mu_a}{\mu_l}, \quad \hat{\sigma} = \sigma\frac{a}{U}, \quad \text{with} \quad U = \sqrt{\frac{\gamma}{\rho_l a}}. \tag{14.2.18}$$

In dimensionless form, expression (14.2.17) becomes

$$\begin{aligned}[(\ell+1)+\hat{l}\ell]\hat{\sigma}^2 - \frac{2}{\sqrt{J}}[(\ell+1)\ell(\ell-1)+\hat{m}(\ell+2)(\ell+1)\ell]\hat{\sigma} \\ + (\ell+2)(\ell+1)\ell(\ell-1) = 0,\end{aligned} \tag{14.2.19}$$

with a Reynolds number

$$J = \frac{\rho_l V a}{\mu_l} = Oh^2, \qquad \text{with} \qquad V = \frac{\gamma}{\mu_l}, \tag{14.2.20}$$

where Oh is the Ohnesorge number. Therefore the eigenvalue $\hat{\sigma}$ for VPF can be computed from

$$\hat{\sigma} = \frac{(\ell+1)\ell(\ell-1) + \hat{m}(\ell+2)(\ell+1)\ell}{\sqrt{J}\left[(\ell+1)+\hat{l}\ell\right]} \pm \sqrt{\left[\frac{(\ell+1)\ell(\ell-1) + \hat{m}(\ell+2)(\ell+1)\ell}{\sqrt{J}[(\ell+1)+\hat{l}\ell]}\right]^2 - \frac{(\ell+2)(\ell+1)\ell(\ell-1)}{(\ell+1)+\hat{l}\ell}}, \tag{14.2.21}$$

which has two different real roots or two complex roots. In the former case, the interface does not oscillate and the disturbances are damped. In the latter case, $\hat{\sigma} = \hat{\sigma}_R \pm i\hat{\sigma}_I$, where the real part represents the damping coefficient and the imaginary part corresponds to the frequency of the damped oscillations.

When both fluids are considered inviscid (IPF), expression (14.2.21) simplifies to ($\hat{m} \to 0$ and $\sqrt{J} \to \infty$)

$$\hat{\sigma} = \pm i\sqrt{\frac{(\ell+2)(\ell+1)\ell(\ell-1)}{(\ell+1)+\hat{l}\ell}}. \tag{14.2.22}$$

The same expression was found by Lamb (1932).

Drop: If the external fluid has negligible density and viscosity ($\hat{l} \to 0$ and $\hat{m} \to 0$), a drop surrounded by a dynamically inactive ambient fluid is obtained, in which case expression (14.2.21) becomes

$$\hat{\sigma} = \frac{\ell(\ell-1)}{\sqrt{J}} \pm \sqrt{\left[\frac{\ell(\ell-1)}{\sqrt{J}}\right]^2 - (\ell+2)\ell(\ell-1)}. \tag{14.2.23}$$

Moreover, for an inviscid drop $\sqrt{J} \to \infty$ and (14.2.23) reduces to

$$\hat{\sigma}_D = \pm i\sqrt{(\ell+2)\ell(\ell-1)} = \pm i\hat{\sigma}_D^*, \tag{14.2.24}$$

and the drop oscillates about the spherical form. This result was also obtained by Lamb (1932, 475).

Using the expression obtained from VPF in (14.2.23), we can readily find two roots for the rate of decay as $J \to 0$ in the drop (e.g., high viscosity). The relevant root on physical grounds is given by

$$\hat{\sigma} = \hat{\sigma}_D^{*2}\sqrt{J}\frac{1}{2\ell(\ell-1)}. \tag{14.2.25}$$

In the case $J \to \infty$ (low viscosity, say), the eigenvalues are complex, and then we encounter oscillatory decaying waves. These eigenvalues behave as

$$\hat{\sigma} = \frac{\ell(\ell-1)}{\sqrt{J}} \pm i\hat{\sigma}_D^*. \tag{14.2.26}$$

Bubble: By taking $\rho_l \to 0$ and $\mu_l \to 0$ in (14.2.17), we obtain for VPF the eigenvalue relation for a bubble of negligible density and viscosity embedded in a liquid:

$$\hat{\sigma} = \frac{(\ell+2)(\ell+1)}{\sqrt{J}} \pm \sqrt{\left[\frac{(\ell+2)(\ell+1)}{\sqrt{J}}\right]^2 - (\ell+2)(\ell+1)(\ell-1)}, \quad (14.2.27)$$

where J is defined in terms of the liquid properties. In the limit of an inviscid external fluid $\sqrt{J} \to \infty$ in (14.2.27), and we obtain

$$\hat{\sigma}_B = \pm i\sqrt{(\ell+2)(\ell+1)(\ell-1)} = \pm i\hat{\sigma}_B^*, \quad (14.2.28)$$

and the bubble oscillates about the spherical shape without damping. This expression was obtained by Lamb (1932).

The dispersion relation obtained from VPF in (14.2.27) can be used to study the trend followed by the disturbances when $J \to 0$ in the case of the bubble. For this case, monotonically decaying waves are predicted with decay rate

$$\hat{\sigma} = \hat{\sigma}_B^{*2}\sqrt{J}\frac{1}{2(\ell+2)(\ell+1)}. \quad (14.2.29)$$

In the case of $J \to \infty$, VPF analysis for the bubble yields

$$\hat{\sigma} = \frac{(\ell+2)(\ell+1)}{\sqrt{J}} \pm i\hat{\sigma}_B^*. \quad (14.2.30)$$

Therefore decaying oscillations are found.

14.2.3 VCVPF analysis of a single spherical drop immersed in another fluid

For irrotational motion, continuity of shear stress at the drop's interface is not satisfied. From Wang *et al.* (2005c), a pressure correction p^v can be added to the irrotational pressure p^i in order to compensate for this discontinuity. They showed that the power of the viscous pressure correction p^v equals the power of the irrotational shear stress:

$$\int_A (-p_l^v + p_a^v)u_n \mathrm{d}A = \int_A (\tau_l^s u_l^s - \tau_a^s u_a^s)\mathrm{d}A, \quad (14.2.31)$$

where the continuity of the normal velocity $u_n = \mathbf{u}\cdot\mathbf{n}_1$ at interface A is considered. We use the notation

$$\tau^S u^S \equiv \mathbf{n}_1 \cdot \mathbf{T} \cdot (\mathbf{u} - u_n \mathbf{n}_1),$$

where $\mathbf{n}_1$ is the outward unit normal vector at the interface for the interior fluid and $\mathbf{T}$ is the stress tensor.

For small disturbances, after linearization, the pressure corrections p_l^v and p_a^v are harmonic functions, such that

$$\nabla^2 p_l^v = 0, \quad \nabla^2 p_a^v = 0. \quad (14.2.32)$$

For the interior and exterior of a sphere of radius a, respectively, the following expressions solve (14.2.32):

$$-p_l^v = \sum_{k=0}^{\infty} D_k \left(\frac{r}{a}\right)^k e^{-\sigma_{kl}t} S_k(\theta,\varphi) + \text{c.c.}, \tag{14.2.33}$$

$$-p_a^v = \sum_{k=0}^{\infty} G_k \left(\frac{r}{a}\right)^{-k-1} e^{-\sigma_{kl}t} S_k(\theta,\varphi) + \text{c.c.}, \tag{14.2.34}$$

where D_k and G_k are constants. The balance of the normal stress at the sphere's interface $r = a$ is modified by the addition of the extra pressures p_l^v and p_a^v:

$$-p_a^i - p_a^v + p_l^i + p_l^v + 2\mu_a \frac{\partial^2 \phi_a}{\partial r^2} - 2\mu_l \frac{\partial^2 \phi_l}{\partial r^2} = \frac{\gamma}{a^2}\left(L^2 - 2\right)\zeta. \tag{14.2.35}$$

Substitution into (14.2.35) of the solutions for p_l^v (14.2.33) and p_a^v (14.2.34) and using (14.2.9), (14.2.10), (14.2.16), and the normal-mode expression for the irrotational pressure disturbance given in 14.2.2 yields

$$\begin{aligned}&\left[\rho_l(\ell+1) + \rho_a \ell\right] A_\ell \sigma_\ell^2 - (\ell+1)(D_\ell - G_\ell)\sigma_\ell \\ &\quad - \left[\frac{2\mu_l}{a^2}(\ell+1)\ell(\ell-1) + \frac{2\mu_a}{a^2}(\ell+2)\ell(\ell+1)\right] A_\ell \sigma_\ell + \frac{\gamma}{a^3}(\ell+2)(\ell+1)\ell(\ell-1)A_\ell = 0,\end{aligned} \tag{14.2.36}$$

where the relation $\ell A_\ell = -(\ell+1)C_\ell$, readily obtained from (14.2.15), has been used. For a sphere of radius a, the integrals in (14.2.31) can be written in terms of the velocity and shear stress components in spherical coordinates:

$$\int_A (-p_l^v + p_a^v) u_r \mathrm{d}A = \int_A \left(\tau_{r\theta}^l u_\theta^l + \tau_{r\varphi}^l u_\varphi^l - \tau_{r\theta}^a u_\theta^a - \tau_{r\varphi}^a u_\varphi^a\right) \mathrm{d}A, \tag{14.2.37}$$

with $u_r^a = u_r^l = u_r$ by continuity of normal velocities at $r = a$. The velocity components in spherical coordinates can be computed from well-known formulas for the gradient of a scalar field. Then the shear stress components in spherical coordinates are found from standard expressions for a Newtonian fluid. When this procedure is applied, the left-hand side of (14.2.37) becomes

$$\begin{aligned}\int_A (-p_l^v + p_a^v) u_r \mathrm{d}A = 2\,\mathrm{Re}\Bigg[&\sum_{\ell=0}^{\infty}\sum_{k=0}^{\infty} a^2 (D_\ell - G_\ell) e^{-\sigma_\ell t} \left(\frac{k}{a}\right)\left(A_k e^{-\sigma_{kl}t} + \bar{A}_k e^{-\bar{\sigma}_{kl}t}\right) \\ &\times \int_0^{2\pi}\int_0^{\pi} S_\ell S_k \sin\theta \mathrm{d}\theta \mathrm{d}\varphi\Bigg],\end{aligned} \tag{14.2.38}$$

where Re[z] delivers the real part of a complex number z. The equality in (14.2.38) follows from the formula

$$\int_{\tilde{A}} (B + \bar{B})(C + \bar{C}) \mathrm{d}\tilde{A} = 2\int_{\tilde{A}} \mathrm{Re}\left(BC + B\bar{C}\right) \mathrm{d}\tilde{A} = 2\mathrm{Re}\left[\int_{\tilde{A}} \left(BC + B\bar{C}\right) \mathrm{d}\tilde{A}\right], \tag{14.2.39}$$

where $\tilde{A}$ represents the domain of integration (e.g., volume) and B and C are complex fields. This formula is also used in the integration of the right-hand side of (14.2.37). Using the definition of S_ℓ in (14.2.11), we can write the double integral in (14.2.38) as

$$\int_0^{2\pi}\int_0^{\pi} S_\ell S_k \sin\theta \, d\theta d\varphi = \sum_{m=-\ell}^{\ell}\sum_{j=-k}^{k} B_{\ell m}\bar{B}_{kj}\int_0^{2\pi}\int_0^{\pi} P_\ell^{|m|}(\cos\theta)P_k^{|j|}(\cos\theta)e^{i(m-j)\varphi}\sin\theta \, d\theta d\varphi, \tag{14.2.40}$$

because S_k is real (i.e., $S_k = \bar{S}_k$). This integral is zero for all the different duplets $(\ell, m) \neq (k, j)$. Then we have

$$\begin{aligned}\int_0^{2\pi}\int_0^{\pi}[S_\ell(\theta,\varphi)]^2 \sin\theta \, d\theta d\varphi &= \sum_{m=0}^{\ell} 2\pi F_{\ell m}\int_0^{\pi}[P_\ell^m(\cos\theta)]^2\sin\theta \, d\theta \\ &= \sum_{m=0}^{\ell} F_{\ell m}\left[\frac{4\pi}{(2\ell+1)}\frac{(\ell+m)!}{(\ell-m)!}\right],\end{aligned} \tag{14.2.41}$$

using a standard result for the integral in the second term. Here, $F_{\ell m}$ represent real constants. Then, from (14.2.38), the power of the pressure correction can be expressed as

$$\begin{aligned}\int_A (-p_l^v + p_a^v)u_r \, dA = 2\,\mathrm{Re}\Bigg[&\sum_{\ell=0}^{\infty}\sum_{m=0}^{\ell} F_{\ell m}4\pi a(D_\ell - G_\ell)e^{-\sigma_\ell t}(A_\ell e^{-\sigma_\ell t} \\ &+ \bar{A}_\ell e^{-\bar{\sigma}_\ell t})\frac{\ell}{(2\ell+1)}\frac{(\ell+m)!}{(\ell-m)!}\Bigg].\end{aligned} \tag{14.2.42}$$

Similarly, the right-hand side of (14.2.37) yields

$$\begin{aligned}&\int_A \left(\tau_{r\theta}^l u_\theta^l + \tau_{r\varphi}^l u_\varphi^l - \tau_{r\theta}^a u_\theta^a - \tau_{r\varphi}^a u_\varphi^a\right) dA \\ &= 2\mathrm{Re}\Bigg\{\sum_{\ell=0}^{\infty}\sum_{k=0}^{\infty}\frac{2}{a}A_\ell e^{-\sigma_\ell t}\left[\mu_l(\ell-1) + \mu_a\frac{\ell(\ell+2)}{(\ell+1)}\frac{k}{(k+1)}\right](A_k e^{-\sigma_k t} + \bar{A}_k e^{-\bar{\sigma}_k t}) \\ &\quad \times \int_0^{2\pi}\int_0^{\pi}\left(\sin\theta\frac{\partial S_\ell}{\partial\theta}\frac{\partial S_k}{\partial\theta} + \frac{1}{\sin\theta}\frac{\partial S_\ell}{\partial\varphi}\frac{\partial S_k}{\partial\varphi}\right) d\theta d\varphi\Bigg\}.\end{aligned} \tag{14.2.43}$$

Using a standard formula (Bowman, Senior, and Uslenghi, 1987) for the double integral in the right-hand side of (14.2.43), we find that the power of the shear stress can be expressed as

$$\begin{aligned}&\int_A (\tau_{r\theta}^l u_\theta^l + \tau_{r\varphi}^l u_\varphi^l - \tau_{r\theta}^a u_\theta^a - \tau_{r\varphi}^a u_\varphi^a) dA \\ &= 2\,\mathrm{Re}\Bigg\{\sum_{\ell=0}^{\infty}\sum_{m=0}^{\ell} F_{\ell m}\frac{8\pi}{a}A_\ell e^{-\sigma_\ell t}(A_\ell e^{-\sigma_\ell t} + \bar{A}_\ell e^{-\bar{\sigma}_\ell t}) \\ &\quad \times\left[\mu_l(\ell+1)(\ell-1) + \mu_a\frac{\ell^2(\ell+2)}{\ell+1}\right]\frac{\ell}{(2\ell+1)}\frac{(\ell+m)!}{(\ell-m)!}\Bigg\}.\end{aligned} \tag{14.2.44}$$

By virtue of (14.2.37), equating (14.2.42) and (14.2.44) yields the relation

$$D_\ell - G_\ell = 2\frac{A_\ell}{a^2}\left[\mu_l(\ell+1)(\ell-1) + \mu_a\frac{\ell^2(\ell+2)}{\ell+1}\right]. \tag{14.2.45}$$

Replacing $D_\ell - G_\ell$ from (14.2.45) in (14.2.36) yields the dispersion relation for the complex eigenvalue σ_ℓ arising from VCVPF (again we drop the subscript ℓ for convenience):

$$\begin{aligned}\left[\rho_l(\ell+1) + \rho_a\ell\right]\sigma^2 &- \frac{2\mu_l}{a^2}(2\ell+1)(\ell+1)(\ell-1)\sigma \\ &- \frac{2\mu_a}{a^2}(2\ell+1)(\ell+2)\ell\sigma + \frac{\gamma}{a^3}(\ell+2)(\ell+1)\ell(\ell-1) = 0.\end{aligned} \tag{14.2.46}$$

In dimensionless form, expression (14.2.46) becomes

$$\begin{aligned}\left((\ell+1) + \hat{\ell}\ell\right)\hat{\sigma}^2 &- \frac{2}{\sqrt{J}}(2\ell+1)(\ell+1)(\ell-1)\hat{\sigma} \\ &- \frac{2\hat{m}}{\sqrt{J}}(2\ell+1)(\ell+2)\ell\hat{\sigma} + (\ell+2)(\ell+1)\ell(\ell-1) = 0,\end{aligned} \tag{14.2.47}$$

with eigenvalues

$$\begin{aligned}\hat{\sigma} = &\frac{(2\ell+1)(\ell+1)(\ell-1) + \hat{m}(2\ell+1)(\ell+2)\ell}{\sqrt{J}\left[(\ell+1)+\hat{\ell}\ell\right]} \\ &\pm\sqrt{\left\{\frac{(2\ell+1)(\ell+1)(\ell-1) + \hat{m}(2\ell+1)(\ell+2)\ell}{\sqrt{J}[(\ell+1)+\hat{\ell}\ell]}\right\}^2 - \frac{(\ell+2)(\ell+1)\ell(\ell-1)}{(\ell+1)+\hat{\ell}\ell}},\end{aligned} \tag{14.2.48}$$

which has two different real roots or a complex-conjugate pair of roots. The former case corresponds to monotonically decaying waves whereas the latter implies oscillatory decaying waves.

Drop: In the case of a drop surrounded by a gas of negligible density and viscosity, we have $\hat{\ell} \to 0$ and $\hat{m} \to 0$ in (14.2.48) and the eigenvalues become

$$\hat{\sigma} = \frac{(2\ell+1)(\ell-1)}{\sqrt{J}} \pm \sqrt{\left[\frac{(2\ell+1)(\ell-1)}{\sqrt{J}}\right]^2 - (\ell+2)\ell(\ell-1)}, \tag{14.2.49}$$

where the dimensionless parameter $\hat{\sigma}$ and the Reynolds number J have been defined in (14.2.18) and (14.2.20), respectively. In the case of decaying oscillations, the decay rate $(2\ell+1)(\ell-1)/\sqrt{J}$ was obtained by Lamb (1932) in §355 through the DM. In the present calculations, relation (14.2.49) gives the decay rate of the oscillatory waves $[\mathrm{Im}(\hat{\sigma}) \neq 0]$ as well as the wave velocity. For $\mathrm{Im}(\hat{\sigma}) = 0$, the decay rate for monotonically decaying waves, including the effect of surface tension, is obtained. Expression (14.2.49) can be compared with (14.2.23) from VPF and (14.2.24) from IPF.

As $J \to 0$, VCVPF produces two roots for the decay rate from (14.2.49); the following equation gives the lowest decay rate:

$$\hat{\sigma} = \hat{\sigma}_D^{*2}\sqrt{J}\frac{1}{2(2\ell+1)(\ell-1)}. \tag{14.2.50}$$

In the case of $J \to \infty$ the eigenvalues are complex:

$$\hat{\sigma} = \frac{(2\ell+1)(\ell-1)}{\sqrt{J}} \pm i\hat{\sigma}_D^*. \tag{14.2.51}$$

Hence we find oscillatory decaying waves. The definition of $\hat{\sigma}_D^*$ is given in (14.2.24).

Prosperetti (1977) studied the initial-value problem posed by small departures from the spherical shape of a viscous drop surrounded by another viscous liquid. In the limiting case of a drop in a vacuum, he found (14.2.49) in the limit $t \to 0$ if an irrotational initial condition is assumed. We remark that (14.2.49) was obtained here by a different method.

Bubble: We can determine the dispersion relation from VCVPF for a bubble of negligible density and viscosity immersed in a viscous liquid from (14.2.46) by taking $\rho_l \to 0$ and $\mu_l \to 0$, which gives rise to the following expression for the eigenvalues:

$$\hat{\sigma} = \frac{(2\ell+1)(\ell+2)}{\sqrt{J}} \pm \sqrt{\left[\frac{(2\ell+1)(\ell+2)}{\sqrt{J}}\right]^2 - (\ell+2)(\ell+1)(\ell-1)}, \tag{14.2.52}$$

with J determined from the liquid properties. In the case of small oscillations, the decay rate given in (14.2.52) as $(2\ell+1)(\ell+2)/\sqrt{J}$ is the same as the rate computed by Lamb (1932), who used the DM without the explicit inclusion of the surface-tension effects in the formulation.

VCVPF expression (14.2.52) yields the following result as $J \to 0$ ($\nu \to \infty$, say) for the bubble,

$$\hat{\sigma} = \hat{\sigma}_B^{*2}\sqrt{J}\frac{1}{2(2\ell+1)(\ell+2)}, \tag{14.2.53}$$

and thus monotonically decaying waves with $\hat{\sigma}_B^*$ defined in (14.2.28).

In the case of $J \to \infty$ ($\nu \to 0$, say), VCVPF predicts oscillatory decaying waves with eigenvalues

$$\hat{\sigma} = \frac{(2\ell+1)(\ell+2)}{\sqrt{J}} \pm i\hat{\sigma}_B^*. \tag{14.2.54}$$

14.2.4 Dissipation approximation (DM)

Lamb (1932) applied a dissipation approximation to compute the decay rate for small oscillatory waves in the free surface of a liquid spherical drop immersed in gas and a spherical bubble surrounded by liquid. Lamb's approach was extended to a drop surrounded by another viscous liquid by Valentine, Sather, and Heideger (1965). We showed in chapter 12 that in gas–liquid flows VCVPF gives rise to the same results as those of the DM. The construction of VCVPF theories for oscillatory decaying waves gives rise to a frequency that depends on viscosity. This dependence of frequency on viscosity was not obtained by Lamb, who assumed the frequency of an inviscid fluid in his dissipation calculation, or by Valentine *et al.* In §14.3.1 the correction of the frequency that is due to viscosity is calculated for capillary-gravity waves by use of the DM. The same kind of analysis can be applied in the case of two fluids.

In the case of a viscous liquid drop surrounded by another viscous liquid, addition of the mechanical-energy equations for the interior and exterior domains gives rise to the expression

$$\frac{d}{dt}\int_{V_a}\rho_a\frac{|\mathbf{u}_a|^2}{2}\mathrm{d}V+\frac{d}{dt}\int_{V_l}\rho_l\frac{|\mathbf{u}_l|^2}{2}\mathrm{d}V$$
$$=-\int_A\mathbf{n}_1\cdot\mathbf{T}_a\cdot\mathbf{u}_a\mathrm{d}A+\int_A\mathbf{n}_1\cdot\mathbf{T}_l\cdot\mathbf{u}_l\mathrm{d}A-\int_{V_a}2\mu_a\mathbf{D}_a:\mathbf{D}_a\mathrm{d}V-\int_{V_l}2\mu_l\mathbf{D}_l:\mathbf{D}_l\mathrm{d}V, \tag{14.2.55}$$

where V_a and V_l are the volumes occupied by the exterior and interior fluid, respectively, $\mathbf{D}$ is the strain-rate tensor, and $\mathbf{n}_1$ denotes the outward unit normal for the interior fluid. The last two terms in (14.2.55) represent the viscous dissipation. The first two integrals on the right-hand side of (14.2.55) can be expanded in terms of the pressure and viscous stress components. Then we assume that continuity of normal velocity and stress and continuity of tangential velocity and stress are satisfied at the interface $r=a$ and (14.2.55) reduces to

$$\frac{d}{dt}\int_{V_a}\rho_a\frac{|\mathbf{u}_a|^2}{2}\mathrm{d}V+\frac{d}{dt}\int_{V_l}\rho_l\frac{|\mathbf{u}_l|^2}{2}\mathrm{d}V$$
$$=-\int_A\frac{\gamma}{a^2}(L^2-2)\zeta u_r\mathrm{d}A+\int_A\mathbf{n}_1\cdot 2\mu_a\mathbf{D}_a\cdot\mathbf{u}_a\mathrm{d}A-\int_A\mathbf{n}_1\cdot 2\mu_l\mathbf{D}_l\cdot\mathbf{u}_l\mathrm{d}A, \tag{14.2.56}$$

where potential flow is assumed for the entire fluid domain, such that the dissipation volume integrals give rise to the last two terms in (14.2.56). With $|\mathbf{u}|^2=u_r^2+u_\theta^2+u_\varphi^2$ and the components of $\mathbf{D}$ expressed in spherical coordinates, such that $\mathbf{n}_1\cdot 2\mu\mathbf{D}\cdot\mathbf{u}=\tau_{rr}u_r+\tau_{r\theta}u_\theta+\tau_{r\varphi}u_\varphi$ computed from potential flow, the integrals in (14.2.56) can be evaluated (we use the standard results for the double integrals as presented in the VCVPF calculation). This calculation, which is omitted for brevity, yields the same dispersion relation, (14.2.46), obtained from VCVPF. This result differs from the one obtained by Valentine *et al.* (1965).

14.2.5 Exact solution of the linearized free-surface problem

In this subsection we present the dispersion relation for the effect of viscosity on small oscillations of a drop immersed in a vacuum and a bubble of negligible density and viscosity embedded in a viscous liquid obtained from the solution of the linearized equations of motion without the assumption of irrotational flow. The result for the drop was presented by Reid (1960), and we can obtain the solution for the bubble by following a similar path. In both cases, the dispersion relation coincides with the corresponding limiting results presented by Miller and Scriven (1968) and Prosperetti (1980a), who posed and solved the most general two-fluid problem.

The linearized Navier–Stokes equations govern this problem:

$$\rho\frac{\partial\mathbf{u}}{\partial t}=-\nabla p+\mu\nabla^2\mathbf{u}, \tag{14.2.57}$$

with $\nabla\cdot\mathbf{u}=0$ in $0\le r<a$ for the drop and in $a<r<\infty$ for the bubble. Continuity of velocity and shear stresses at $r=a$ is satisfied.

14.2.5.1 *Spherical drop*

Reid (1960) obtained the dispersion relation for the eigenvalue σ,

$$\alpha^4 = 2q^2 (\ell - 1) \left[\ell + (\ell + 1) \frac{q - 2\ell Q^J_{\ell+1/2}}{q - 2Q^J_{\ell+1/2}} \right] - q^4, \tag{14.2.58}$$

with

$$Q^J_{\ell+1/2}(q) = J_{\ell+3/2}(q)/J_{\ell+1/2}(q), \quad \alpha^2 = \frac{\sigma^*_D a^2}{\nu} = \hat{\sigma}^*_D \sqrt{J}, \quad \frac{q^2}{\alpha^2} = \frac{\sigma}{\sigma^*_D} = \frac{\hat{\sigma}}{\hat{\sigma}^*_D}, \tag{14.2.59}$$

where σ^*_D is the frequency of oscillations from IPF given in (14.2.24). A thorough discussion on the solution of (14.2.58) is presented by Chandrasekhar (1961) when q is real. With ℓ considered fixed, the right-hand side of (14.2.58) is a function of q, $\Phi(q)$ say. The graph of this function on the axis of positive q reveals that there are an infinite number of intervals where $\Phi(q)$ is positive. The first of these intervals, which contains $q = 0$, encloses a maximum (α^2_{max}). For $\alpha^2 < \alpha^2_{max}$ this first interval gives two real roots of (14.2.58), which determine the slowest decay rates. Because $\alpha^2 = \hat{\sigma}^*_D\sqrt{J}$, for every mode ℓ, the magnitude of the Reynolds number J defines the roots. In the other intervals, we have $0 \leq \Phi(q) < \infty$. When $\alpha^2 > \alpha^2_{max}$, (14.2.58) admits complex-conjugate eigenvalues with positive real parts that give the lowest decay rate; these waves oscillate as they decay. Results from this dispersion relation are presented and discussed in §14.2.7.

As $J \to 0$, the decay rate from the ES of a drop surrounded by gas behaves as

$$\hat{\sigma} = \hat{\sigma}^{*2}_D \sqrt{J} \frac{2\ell + 1}{2(\ell - 1)(2\ell^2 + 4\ell + 3)}. \tag{14.2.60}$$

From the ES previously given, the behavior of the complex eigenvalue σ for the drop as $J \to \infty$ is

$$\hat{\sigma} = \frac{(2\ell + 1)(\ell - 1)}{\sqrt{J}} \pm i\hat{\sigma}^*_D. \tag{14.2.61}$$

These expressions were also obtained by Chandrasekhar (1961) and Miller and Scriven (1968). The result given in (14.2.51) from VCVPF is the same as the expression obtained in (14.2.61) from the ES.

14.2.5.2 *Spherical bubble*

A procedure similar to the one applied to the drop gives rise to the following dispersion relation for the bubble:

$$\alpha^4 = (\ell + 2)\, q^2 \frac{(2\ell + 1)\, q^2 - 2\,(\ell + 1)\,(\ell - 1) \left[(2\ell + 1) - q Q^H_{\ell+1/2} \right]}{(2\ell + 1) + q^2/2 - q Q^H_{\ell+1/2}} - q^4, \tag{14.2.62}$$

with

$$Q^H_{\ell+1/2} = H^{(1)}_{\ell+3/2}(q)/H^{(1)}_{\ell+1/2}(q), \tag{14.2.63}$$

and by use of relations (14.2.59). In these relations, σ^*_B from IPF given in (14.2.28) for a bubble is used instead of σ^*_D.

Expression (14.2.62) is the same dispersion relation found by Miller and Scriven (1968). Prosperetti (1980a) indicates that this dispersion relation admits only complex roots as a consequence of the character of the Hankel functions. Therefore, for a bubble, only oscillatory decaying waves are predicted. For a drop, we recall that real eigenvalues can be found. Another feature of the dispersion relation is that the solutions occur in conjugate pairs, as in the drop case, because, for $\mathrm{Im}(\sigma) < 0$, one can choose the Hankel function of the second kind and (14.2.62) is satisfied.

For a gas bubble in a viscous liquid, a real σ can be approximated as $J \to 0$:

$$\hat{\sigma} = \hat{\sigma}_B^{*2}\sqrt{J}\frac{2\ell+1}{2(2\ell^2+1)(\ell+2)}. \tag{14.2.64}$$

For $J \to \infty$, damped oscillations take place with eigenvalues

$$\hat{\sigma} = \frac{(2\ell+1)(\ell+2)}{\sqrt{J}} \pm i\hat{\sigma}_B^{*}. \tag{14.2.65}$$

Both results are presented by Miller and Scriven (1968). As in the case of the drop, the result obtained from VCVPF in (14.2.54) is the same as the expression given in (14.2.65) from the ES.

14.2.6 VPF and VCVPF analyses for waves acting on a plane interface considering surface tension – comparison with Lamb's solution

The problem of the viscous effects on free-gravity waves was solved exactly by Lamb (1932). In his analysis, surface-tension effects were included. Joseph and Wang (2004) gave a solution to this problem by using VPF. They also added a viscous correction to the irrotational pressure and obtained another dispersion relation for the complex eigenvalue σ. In their analysis they neglected the surface-tension effects and considered gravity effects. Here we find the decay rate and wave velocity for free waves on an otherwise plane interface by neglecting gravity effects and including surface tension. These expressions are found as limiting results from our VPF and VCVPF analyses of waves on the surface of a liquid drop surrounded by gas presented at the end of §14.2.3 and §14.2.4, respectively. This procedure considers taking the limits $\ell \to \infty$ and $a \to \infty$ such that $\ell/a \to k$ in the expressions for the complex eigenvalue σ (Miller and Scriven, 1968).

For VPF, in the case of $k > k_c = \gamma/\rho\nu^2$, the decay rate is

$$\sigma = \nu k^2 \pm \sqrt{(\nu k^2)^2 - \gamma k^3/\rho}. \tag{14.2.66}$$

If $k < k_c = \gamma/\rho\nu^2$ then we have

$$\sigma = \nu k^2 \pm ik\sqrt{\gamma k/\rho - \nu^2 k^2}, \tag{14.2.67}$$

such that the wave velocity is $c = \sqrt{\gamma k/\rho - \nu^2 k^2}$ and the decay rate is νk^2.

Studying the asymptotic behavior for $k \gg k_c = \gamma/\rho\nu^2$, we find that the decay rate goes as

$$\sigma_1 = \frac{\gamma k}{2\rho\nu}, \quad \sigma_2 = 2\nu k^2. \tag{14.2.68}$$

Then, σ_1 gives the slowest decay rate for large k.

On the other hand, the asymptotic behavior as $k \ll k_c = \gamma/\rho\nu^2$ can be expressed as

$$\sigma = \nu k^2 \pm ik\sqrt{\gamma k/\rho}, \tag{14.2.69}$$

such that the wave velocity is given by $c = \sqrt{\gamma k/\rho}$; this is the same result from IPF, as subsequently shown. The decay rate goes as νk^2.

For VCVPF, in the case of $k > k_c = \gamma/4\rho\nu^2$, the decay rate is

$$\sigma = 2\nu k^2 \pm \sqrt{(2\nu k^2)^2 - \gamma k^3/\rho}. \tag{14.2.70}$$

If $k < k_c = \gamma/4\rho\nu^2$ then we have

$$\sigma = 2\nu k^2 \pm ik\sqrt{\gamma k/\rho - 4\nu^2 k^2}, \tag{14.2.71}$$

such that the wave velocity is $c = \sqrt{\gamma k/\rho - 4\nu^2 k^2}$ and the decay rate is $2\nu k^2$.

Studying the asymptotic behavior for $k \gg k_c = \gamma/4\rho\nu^2$, we find that the decay rate goes as

$$\sigma_1 = \frac{\gamma k}{4\rho\nu}, \qquad \sigma_2 = 4\nu k^2. \tag{14.2.72}$$

Hence σ_1 gives the slowest decay rate for large k.

On the other hand, the asymptotic behavior as $k \ll k_c = \gamma/4\rho\nu^2$ can be expressed as

$$\sigma = 2\nu k^2 \pm ik\sqrt{\gamma k/\rho}, \tag{14.2.73}$$

such that the wave velocity is given by $c = \sqrt{\gamma k/\rho}$, which is the same result from IPF, as subsequently shown. In this case the decay rate goes as $2\nu k^2$. It can be verified that the same results are found if we start from the VPF and VCVPF solutions obtained for a bubble surrounded by a viscous liquid.

Lamb (1932) solved the problem of free-gravity waves exactly, considering the effects of surface tension (§349, p. 625). Lamb's dispersion relation, neglecting gravitational effects, is

$$\left(\sigma - 2\nu k^2\right)^2 + \frac{\gamma}{\rho}k^3 = 4\nu^2 k^3 m, \tag{14.2.74}$$

with

$$m^2 = k^2 - \sigma/\nu.$$

Therefore the dispersion relation is biquadratic. As the viscosity goes to zero, Lamb found that the complex eigenvalue σ behaves as

$$\sigma = 2\nu k^2 \pm ik\sqrt{\gamma k/\rho}, \tag{14.2.75}$$

which is exactly the result from VCVPF in the case of $k \ll k_c = \gamma/4\rho\nu^2$, as shown in (14.2.73). In the limit $\nu \to 0$ Lamb's ES reduces to

$$\sigma_P^2 = \gamma k^3/\rho, \tag{14.2.76}$$

corresponding to IPF analysis. Then the wave velocity for inviscid flow is $c = \sqrt{\gamma k/\rho}$, as previously mentioned.

For the case of $\nu \to \infty$, Lamb (1932) found two real roots from his ES, neglecting surface tension. Following the same procedure but neglecting gravity effects and keeping surface tension, we find that the relevant root on physical grounds is

$$\sigma = \frac{\gamma k}{2\rho \nu}, \tag{14.2.77}$$

which is the same result obtained from VPF in (14.2.68). Using asymptotic formulas for Bessel functions of large order and large variable such that $a \to \infty$ and $\ell \to \infty$ for fixed k, Prosperetti (1980a) showed that the dispersion relation for the bubble, (14.2.62), reduces to Lamb's result (14.2.74). Prosperetti (1976) posed and solved the initial-value problem for small disturbances of the infinite plane without assuming an exponential decay with time. For an irrotational initial condition and with gravity neglected, in the limit $t \to 0$, his dispersion relation is the same as (14.2.71), obtained here by a different method.

14.2.7 Results and discussion

In this subsection, a detailed comparison of the results for the decay rate and frequency of the waves according to VPF and VCVPF with the ES of the fully viscous linear problem is presented for a drop and a bubble. A wide interval is selected for the mode number ℓ, ranging from $\ell = 2$ up to $\ell = 100$. The smallest value of $\ell = 2$ is chosen because lower values yield compressive or expansive motions of the drop interface that are not compatible with the incompressibility assumption or a nonphysical static disturbed interface. For higher values of ℓ, the exact fully viscous solution for the drop predicts oscillations that decay faster (Miller and Scriven, 1968). The same lowest value of $\ell = 2$ is selected for the bubble case.

Figure 14.7(a) shows the critical Reynolds number J_c as a function of ℓ for a drop. The critical Reynolds number is defined as the value of J at a given ℓ for which transition from monotonically decaying waves to oscillatory decaying waves occurs. For $J \leq J_c$ the eigenvalues $\hat{\sigma}$ are real and monotonically decaying waves take place, whereas for $J > J_c$ the eigenvalues are complex and the waves decay through oscillations. Figure 14.7(b) presents the trends of J_c with ℓ for VPF and VCVPF for a bubble. Recall that the ES always predicts decaying oscillations (i.e., complex eigenvalues) in the bubble case. Therefore the ES does not give rise to a critical J. If VCVPF predicts oscillatory decaying waves, then VPF gives the same prediction. If VPF predicts monotonically decaying waves, then the same behavior is obtained from VCVPF.

For a drop, the decay rate and wave frequency as functions of the Reynolds number J are presented in figures 14.8(a) and 14.8(b), respectively, for $\ell = 2$ as predicted by VPF, VCVPF, and the ES. The wave frequency given by IPF is also included for comparison. For large J, VCVPF and the ES show excellent agreement, whereas VPF is off the mark. The wave frequencies from the three viscous theories tend to the inviscid solution for large J. As J decreases (below $J = 10$, say), transition from the oscillatory-decaying-wave regime to the monotonically-decaying-wave regime occurs. In the point of transition, the frequency becomes identically zero and the curve of $\text{Re}(\hat{\sigma})$ bifurcates, yielding two real and different roots. In figure 14.8(a), the lowest root, representing the least-damped mode of decay, is presented.

For a bubble, the decay rate and wave frequency are presented in figures 14.9(a) and 14.9(b), respectively for $\ell = 2$. Note that the results follow similar trends as those

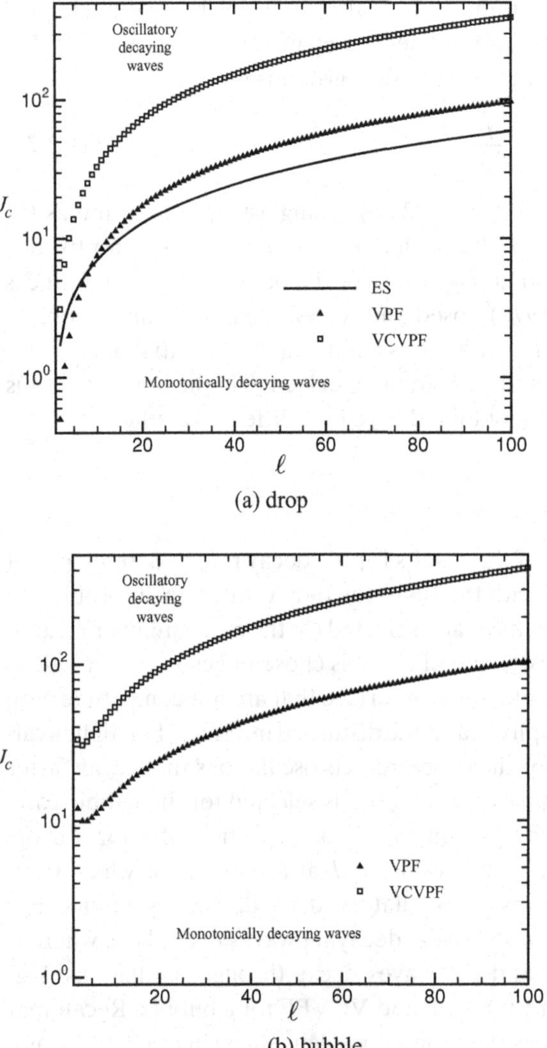

Figure 14.7. Critical Reynolds number J_c as a function of the mode number ℓ for a *drop* and a *bubble*. At a given ℓ, for $J > J_c$ oscillatory decaying waves are predicted (i.e., the eigenvalues $\hat{\sigma}$ are complex), whereas for $J \leq J_c$ monotonically decaying waves are obtained (i.e., the eigenvalues $\hat{\sigma}$ are real). The results are presented for VPF, VCVPF, and the exact solution (ES). Note that, for the bubble, the ES does not provide a crossover value, because the imaginary part is never identically zero.

just described for the drop. An important difference, however, is that the ES does not predict transition to the monotonically-decaying-wave regime, but the wave frequency tends smoothly to zero as J decreases (the viscosity increases, say). On the other hand, the irrotational viscous theories, VPF and VCVPF, do render a crossover J_c for which transition to monotonically decaying waves occurs.

For large values of J, if one of the fluids has negligible density and viscosity, a thin boundary layer results (Miller and Scriven, 1968). Thus an irrotational velocity field works as a good approximation with a pressure correction arising within the boundary layer and resolving the discrepancy between the nonzero irrotational shear stress and the actual zero shear stress at the interface. These considerations explain the excellent agreement of VCVPF and the ES for large J. In terms of the dissipation approximation, such a thin boundary layer yields a negligible contribution to the total viscous dissipation, which is thus determined by the irrotational flow. By contrast, as J decreases (e.g., the liquid

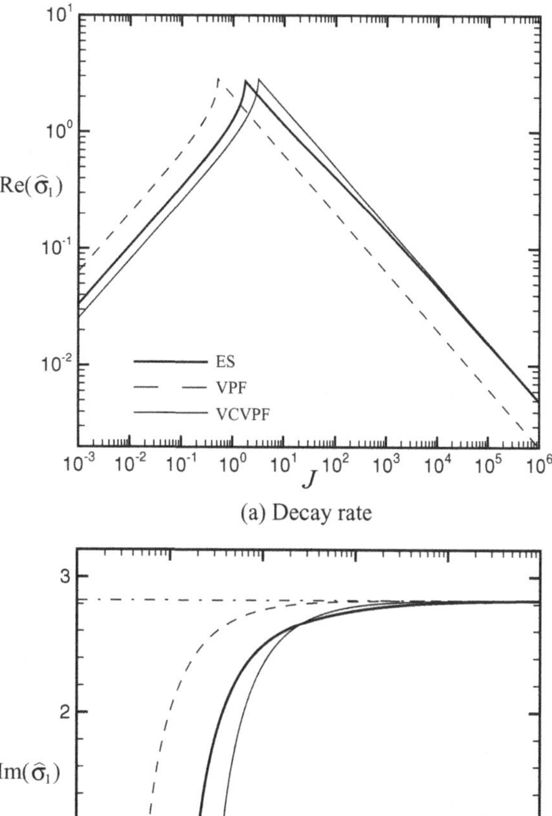

Figure 14.8. Decay rate and wave frequency for the fundamental mode $\ell = 2$ as functions of the Reynolds number J for a *drop* from the exact solution (ES), VPF, VCVPF, and IPF. The decay rate predicted by IPF is identically zero for all ℓ.

viscosity increases), the boundary layer becomes thicker and the performance of the irrotational approximations (VPF and VCVPF) deteriorates. A nonnegligible boundary-layer flow contributes substantially to the rate of viscous dissipation. The motion for small J, as discussed by Prosperetti (1980a), is restrained in such a drastic way that the energy dissipation per unit time and thus the decay rate decrease with the Reynolds number J.

The sharp crossover from the oscillatory-decaying-wave regime to the monotonically-decaying-wave regime can be readily obtained from the dispersion relations (figure 14.8). For a bubble, oscillatory-decaying waves are always predicted by the exact linearized theory yet the smooth region where the decay-rate graph reaches a maximum as a function of the Reynolds number J suggests a transition in the structure of the flow (figure 14.9). This was associated by Prosperetti (1980a) with the two length scales characteristic of the problem, namely, the sphere's radius a and the vorticity diffusion length of order $(\nu/\omega_B)^{1/2}$, where ω_B is the frequency of the oscillations for the inviscid case given in (14.2.28).

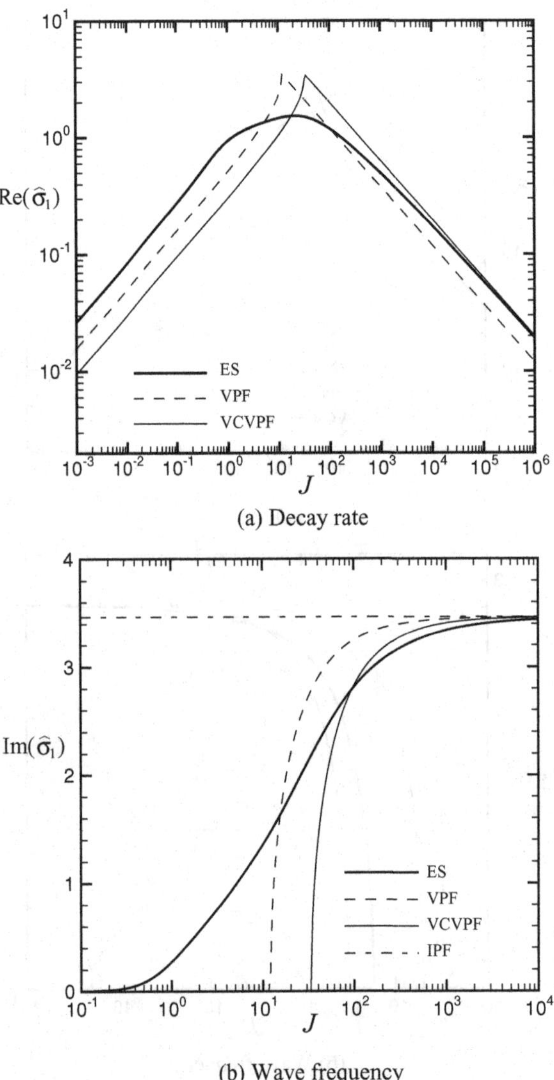

Figure 14.9. Decay rate and wave frequency for the fundamental mode $\ell = 2$ as functions of the Reynolds number J for a *bubble* from the exact solution (ES), VPF, VCVPF, and IPF. The decay rate predicted by IPF is identically zero for all ℓ.

When $a \gg (\nu/\omega_B)^{1/2}$, which represents the right-hand branch of the exact solution in figure 14.9(a), a thin vorticity layer at the interface occurs, having a minute effect on the rate of viscous dissipation. On the other hand, the left-hand branch in figure 14.9(a), for which the decay rate is reduced with decreasing J, as previously explained, is associated with a relatively thicker vorticity layer such that $a \ll (\nu/\omega_B)^{1/2}$.

Computations carried out for several higher modes ($\ell = 3$, 4, and 10) have shown that the features commented on for the fundamental mode are also observed for these other modes. The general trend is that the decay rate increases with increasing ℓ. The analysis of the predictions from the ES indicates that the changeover from oscillatory decaying waves to monotonically decaying waves takes place for a larger critical J as ℓ increases for a drop. Even though no transition to monotonically decaying waves occurs for the bubble, low frequencies are obtained in a wider J interval as ℓ increases. The viscous irrotational theories follow these tendencies. These results show that viscosity damps the motion more effectively for shorter waves.

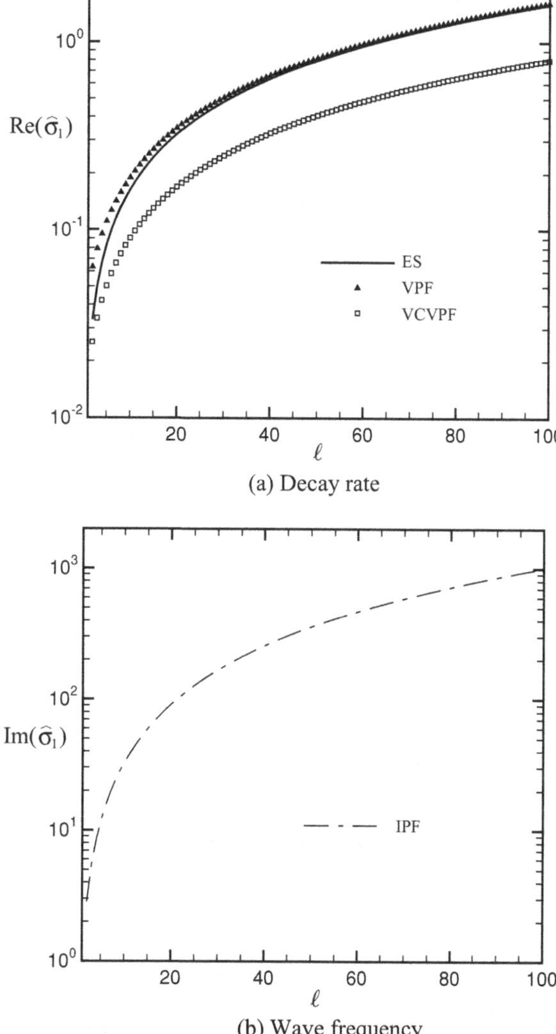

Figure 14.10. Decay rate $\mathrm{Re}(\hat{\sigma}_1)$ and wave frequency $\mathrm{Im}(\hat{\sigma}_1)$ for $J = 10^{-3}$ versus the mode number ℓ for a *drop* from the exact solution (ES), VPF, VCVPF, and IPF. For the interval of ℓ shown, the eigenvalues from the viscous theories are real and different. The lowest decay rate is plotted in (a) and the trend exhibited by $\mathrm{Im}(\hat{\sigma}_1)$ is shown in (b) for IPF, which predicts oscillatory waves with constant amplitude.

For a drop, figures 14.10–14.12 show predictions from VPF, VCVPF, and the ES for the decay rate and frequency of the oscillations for several values of the Reynolds number J, namely, $J = 10^{-3}$, 40, and 10^6 as a function of ℓ. The frequency from IPF is also presented. When J is small, in the monotonically-decaying-wave regime, VCVPF gives a better approximation to the exact solution than VPF for the fundamental mode $\ell = 2$, as can be observed in figure 14.10 for $J = 10^{-3}$. By contrast, for larger ℓ (i.e., shorter waves), VPF shows a better performance. As J increases, the waves start to oscillate. For instance, in the case of $J = 40$, the crossover values of ℓ according to every viscous theory enters the analysis (see figure 14.11). In this case, oscillatory decaying waves exist according to the viscous theories for $\ell < \ell_{\mathrm{VCVPF}}$, whereas these theories agree and predict monotonically decaying waves for $\ell > \ell_{\mathrm{ES}}$. According to figure 14.7(a), this trend seems to hold for $J \geq 6$. In figure 14.12 for $J = 10^6$, the crossover ℓ_c obtained from each viscous theory is larger than 100 and lies out of the figure. For large J, in the

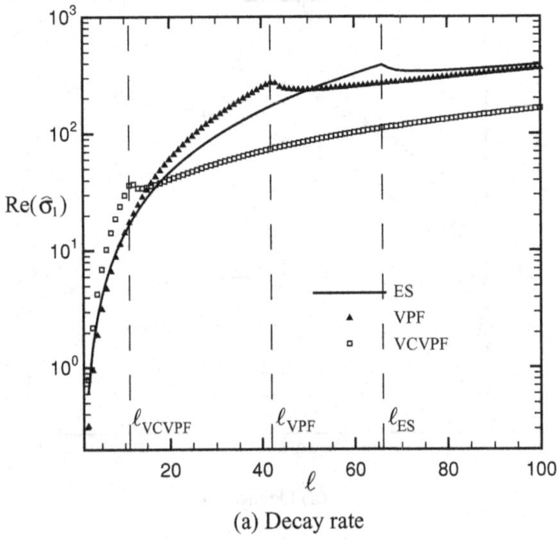

(a) Decay rate

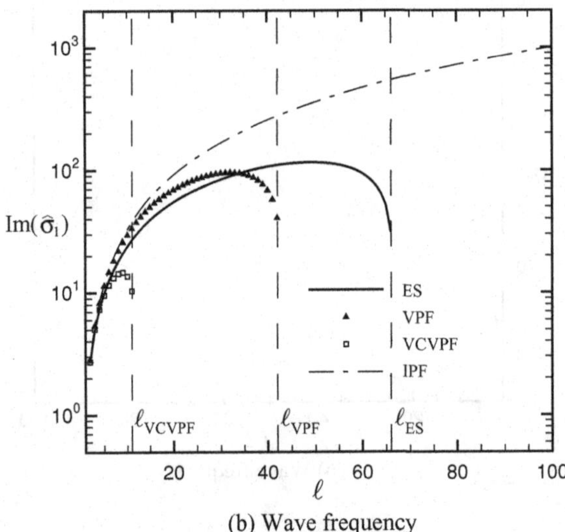

(b) Wave frequency

Figure 14.11. Decay rate Re($\hat{\sigma}_1$) and wave frequency Im($\hat{\sigma}_1$) for $J = 40$ versus the mode number ℓ for a *drop* from the exact solution (ES), VPF, VCVPF, and IPF. In this case, the eigenvalues are a pair of complex conjugates for the interval of $\ell < \ell_c$ and they are real and different for $\ell \geq \ell_c$. For the latter case, the lowest decay rate is plotted in (a). The symbol ℓ_c stands for the highest value of ℓ for which a nonzero imaginary part is obtained, i.e., decaying oscillations occur. For instance, $\ell_c = \ell_{ES}$ from the ES; analogous definitions can be set for VPF (ℓ_{VPF}) and VCVPF (ℓ_{VCVPF}).

regime of oscillatory decaying waves, the region of good agreement between VCVPF and the ES extends to higher values of ℓ (e.g., $2 \leq \ell \leq 100$ for $J = 10^6$ in figure 14.12), whereas VPF shows poor agreement in comparison. At least for values of $\ell \ll \ell_{ES}$ in the neighborhood of $\ell = 2$, VCVPF provides the better approximation of the decay rate.

For very small values of J, as in the case of $J = 10^{-3}$, the ES, VPF, and VCVPF predict zero frequency, whereas IPF is off the mark, giving rise to a nonzero frequency for nondecaying waves [see figure 14.10(b)]. For moderate to large values of J, the oscillatory regime appears for a certain ℓ interval starting with $\ell = 2$. This interval becomes wider with increasing J. The crossover ℓ_c moves to the right as J increases and the agreement in terms of the frequency becomes much better, in particular for low values of ℓ [figure 14.11(b)]. On the other hand, because the crossover value ℓ_{VPF} is larger than ℓ_{VCVPF}, VPF follows

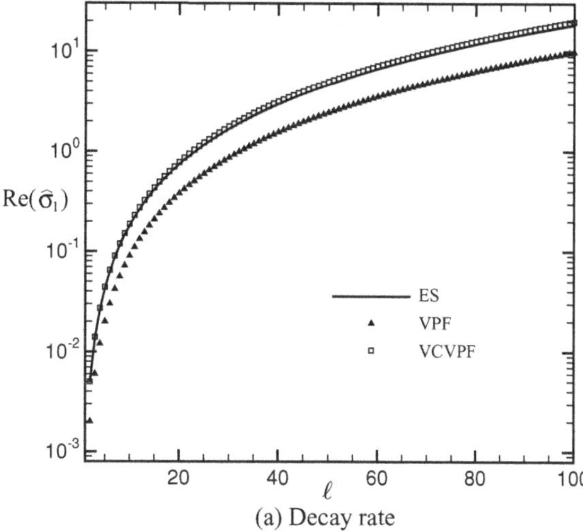

(a) Decay rate

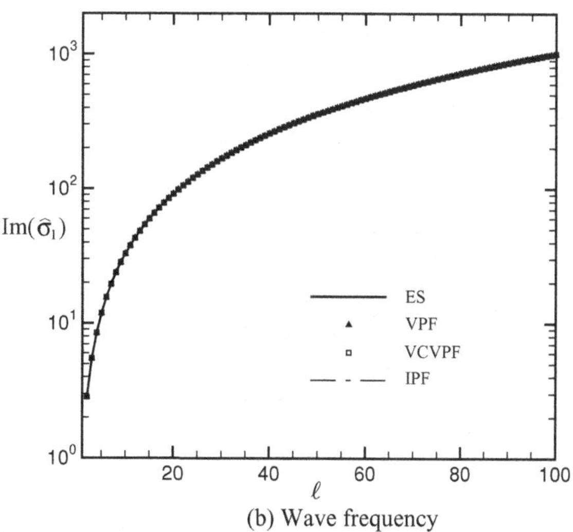

(b) Wave frequency

Figure 14.12. Decay rate $\text{Re}(\hat{\sigma}_l)$ and wave frequency $\text{Im}(\hat{\sigma}_l)$ for $J = 10^6$ versus the mode number ℓ for a *drop* from the exact solution (ES), VPF, VCVPF, and IPF. In this case, the eigenvalues are a pair of complex conjugates for the interval of ℓ considered in this study.

the trend of the ES in the oscillatory regime for a wider interval. For very large values of J, the three approximations for the frequency cannot be distinguished from the predictions given by the exact solution for a wide ℓ interval [e.g., $2 \leq \ell \leq 100$ for $J = 10^6$ in figure 14.12(b)].

To some extent, similar trends as those described for the drop are observed for the bubble. However, a noteworthy difference is that, for low to moderate J (from 10^{-3} to 500, say); the decay rates computed from VPF are closer to the exact solution ES than rates computed from VCVPF. This trend is reversed for large J.

Kojo and Ueno (2006) present experimental results for oscillatory bubbles under ultrasonic vibration. Figure 14.13 shows the time series of shape oscillations for an air bubble in water under ultrasonic vibration of 20 kHz and the predicted shape with spherical harmonics. Several other figures showing the trends previously described are included in the work by Padrino *et al.* (2007a), available from the URL: http://www.aem.umn.edu/people/faculty/joseph/ViscousPotentialFlow/

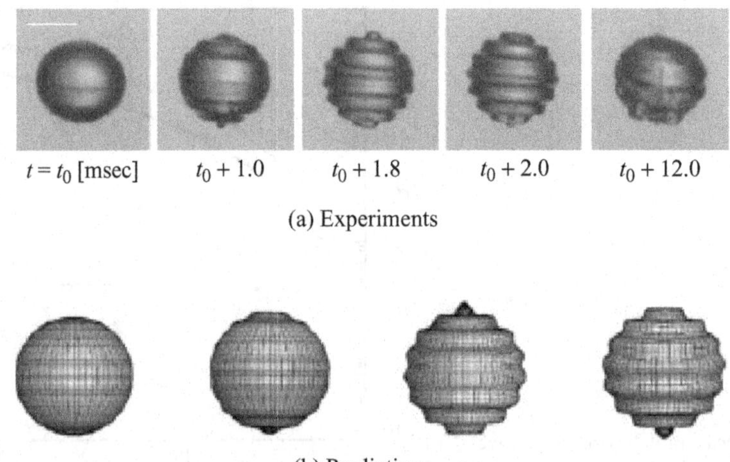

Figure 14.13. Experimental and theoretical time evolutions of shape oscillations of an air bubble in water with oscillation mode $(\ell, m) = (15,0)$ under an ultrasonic vibration of 20 kHz. The bar at the top of the left-hand frame in (a) corresponds to 1.0 mm (Kojo and Ueno, 2006).

14.2.8 Concluding remarks

The results obtained from the viscous purely irrotational approximations for the decay rate and frequency of the oscillations for a drop and a bubble follow the trends described by the ES. The damping role of viscosity in the dynamics of the waves is adequately described by the viscous irrotational theories through the modeling of the decay rate and frequency of the oscillations, whereas the classical inviscid theory predicts undamped oscillations, and thus an identically zero decay rate. Quantitative agreement is also demonstrated for certain intervals of modes and J. Some features are noteworthy from the comparison carried out in this study for the drop and the bubble:

- In the case of short waves (i.e., large mode number ℓ), VPF gives a very good approximation of the decay rate for monotonically decaying waves for both the drop and the bubble. On the other hand, VCVPF gives rise to values of the decay rate in closer agreement with the exact solution within a certain ℓ interval, including $\ell = 2$, in the oscillatory-decaying-wave regime (i.e. long waves) for large values of the Reynolds number J. This trend resembles the tendencies for free-gravity waves perturbing a plane interface obtained by Wang and Joseph (2006a), presented in §14.1.4. Nonetheless, a notable difference between their results and those given here is that surface tension has a stronger regularizing effect on short waves than gravity does.
- VPF also follows the trend described by the ES for the frequency of the oscillations, because the transition from oscillatory to monotonically decaying waves predicted by this irrotational theory occurs at a higher critical value of ℓ than the threshold given by VCVPF.
- The viscous irrotational approximations predict effects of viscosity on the frequency of the oscillations. For every mode, there is a Reynolds number J for which transition from oscillatory decaying waves to monotonically decaying waves occurs for either the drop or the bubble. Whereas a transitional value of J is predicted by the ES for a drop,

only oscillatory decaying waves are found by this theory for a bubble. In this case, very small frequencies are obtained as $J \to 0$ (e.g., $\nu \to \infty$).

- The viscous irrotational theories do not give rise to a continuous spectrum of eigenvalues for the bubble as has been found for the ES by Prosperetti (1980a).
- After the calculations are carried out, it is verified that the dissipation approximation yields the same dispersion relation as VCVPF. A general proof of this equivalence was shown in chapter 12.

14.3 Irrotational dissipation of capillary-gravity waves

In §14.1 we considered irrotational theories of the effect of viscosity on the decay of free-gravity waves. In §14.2 we considered the same problem for capillary waves on a spherical surface. Capillary-gravity waves combine these two mechanisms of propagation and decay. The dissipation of irrotational waves was considered by Stokes (1851) and applied extensively by Lamb (1932). Here we consider extensions of Lamb's DM to obtain the effect of viscosity on frequency and to nonlinear effects.

14.3.1 Correction of the wave frequency assumed by Lamb

The dissipation method stems from the mechanical-energy equation. To apply the dissipation approximation to capillary-gravity waves, we obtain the working equation after subtracting the basic state of rest from the incompressible Navier–Stokes equation and then taking the scalar ("dot") product with the velocity vector. Integration over the region of interest yields the mechanical-energy equation for the flow disturbances in integral form. Next, it is assumed that the zero-shear-stress condition at the interface is satisfied. After some manipulation, this equation can be written as

$$\frac{\mathrm{d}}{\mathrm{d}t}\left(\int_V \rho\,|\mathbf{u}|^2/2\,\mathrm{d}V + \int_0^\lambda \rho g\eta^2/2\,\mathrm{d}x\right) = \int_0^\lambda \upsilon\gamma\frac{\partial^2\eta}{\partial x^2}\,\mathrm{d}x - \int_V 2\mu\mathbf{D}:\mathbf{D}\,\mathrm{d}V, \tag{14.3.1}$$

where γ is the surface tension. The last term in (14.3.1) gives the viscous dissipation. This term is computed assuming that the fluid motion is irrotational, neglecting the thickness of the vorticity layer at the free surface. For irrotational flow, the following identity holds,

$$\int_V 2\mu\mathbf{D}:\mathbf{D}\mathrm{d}V = \int_A \mathbf{n}\cdot 2\mu\mathbf{D}\cdot\mathbf{u}\mathrm{d}A, \tag{14.3.2}$$

which is used in the calculation of the viscous dissipation; in (14.3.2), $\mathbf{n}$ is the outward normal. The integration of (14.3.1) is carried out with the velocity potential (14.1.16) over the region defined by $0 \le x \le \lambda$ and $-\infty < y \le 0$. Periodic boundary conditions at $x = 0$ and $x = \lambda$ and vanishingly small disturbances (both velocity and pressure) as $y \to -\infty$ are taken into account in (14.3.1). Therefore, the surface integrals are evaluated at $y = 0$. As a result, the following expression is obtained for the eigenvalues:

$$n = -2\nu k^2 \pm ik\sqrt{(g/k + \gamma' k) - 4\nu^2 k^2}, \tag{14.3.3}$$

with $\gamma' = \gamma/\rho$. For $2\nu k^2 \ge \sqrt{gk + \gamma' k^3}$, the eigenvalues are real and monotonically decaying waves occur. On the other hand, for $2\nu k^2 \le \sqrt{gk + \gamma' k^3}$ we find progressive-decaying waves with rate of decay $-2\nu k^2$. This is the same value computed by Lamb (1932) by the

DM. However, Lamb's approach did not account for the effects of viscosity in the wave speed of traveling decaying waves. For this type of waves, the wave speed is extracted from (14.3.3) as

$$c = \sqrt{(g/k + \gamma' k) - 4\nu^2 k^2}, \tag{14.3.4}$$

which is slower than the inviscid result $\sqrt{g/k + \gamma' k}$ used by Lamb (1932, §348). Therefore, these calculations indicate that effects of viscosity, considered through the viscous dissipation of the mechanical energy, decrease the traveling speed of capillary-gravity waves. This trend also holds for either zero gravity or zero surface tension, as can be deduced from (14.3.4). According to the analysis of §12.5 and §12.6, the result given in (14.3.3) is also obtained from VCVPF.

The dispersion relations for the viscous irrotational theories (VPF and DM) and for Lamb's (1932) exact solution of the linearized Navier–Stokes equations for capillary-gravity waves (see also §14.1.3) can be conveniently written in dimensionless form as follows:

$$\text{VPF}: \qquad \tilde{n} = -\theta \pm i\sqrt{1-\theta^2}; \tag{14.3.5}$$

$$\text{DM}: \qquad \tilde{n} = -2\theta \pm i\sqrt{1-4\theta^2}; \tag{14.3.6}$$

$$\text{ES}: [(\tilde{n} + 2\theta)^2 + 1]^2 = 16\theta^3(\tilde{n} + \theta). \tag{14.3.7}$$

In these expressions we have defined $\tilde{n} = n/n_0$ and $\theta = \nu k^2/n_0$, a factor introduced by Lamb in his ES, with the inviscid frequency $n_0 = \sqrt{gk + \gamma' k^3}$. Thus, for IPF, $\tilde{n} = i$. The analysis of (14.3.5), (14.3.6), and (14.3.7) reveals that a threshold θ_c can be obtained that separates progressive waves ($\theta < \theta_c$, $\text{Im}[\tilde{n}] \neq 0$) from standing waves ($\theta \geq \theta_c$, $\text{Im}[\tilde{n}] = 0$) from each theory. We obtain $\theta_c = 1$ for VPF and $\theta_c = 0.5$ for the DM. For the ES, (14.3.7) gives rise to, nearly, $\theta_c = 1.3115$ (also reported by Prosperetti, 1976). We notice that the first-order approximation in θ of (14.3.6) for the DM is equivalent to the first-order approximation in this parameter of the ES presented by Lamb (1932) and Basset (1888).

From the definition of θ, we have that the respective cutoff wavenumber k_c can be obtained for each theory using the corresponding value of θ_c given previously. When $k < k_c$, progressive waves decay, whereas for $k \geq k_c$, the waves decay monotonically.

Figure 14.14 shows the dimensionless decay rate $-\text{Re}[\tilde{n}]$ and frequency of the oscillations $\text{Im}[\tilde{n}]$ as functions of the dimensionless parameter θ from (14.3.5), (14.3.6), and (14.3.7) for VPF, DM, and the ES, respectively. IPF predictions for the frequency, $\text{Im}[\tilde{n}] = 1$, are also included. For $\theta > \theta_c$ only the slowest decay rate, given by the smallest real eigenvalue, is plotted. In these figures, only the cutoff θ_c given by the ES is presented, which hereinafter is referred to as θ_c'.

Both viscous irrotational theories follow the trend described by the ES as shown in figure 14.14. With respect to the decay rate, this figure indicates that the DM approaches the ES in the progressive wave regime ($\theta \leq \theta_c'$) for $\theta \ll 1$. In particular, for $\theta \leq 0.02$ we found that the relative error for the DM in absolute value remains below 10% and the agreement becomes outstanding as θ decreases following $\text{Re}[\tilde{n}] = -2\theta$. On the other hand, VPF is off the mark by 50%. In the standing wave regime ($\theta \geq \theta_c'$), VPF shows excellent agreement with the ES; for $\theta \geq 2$ this irrotational theory predicts values of the

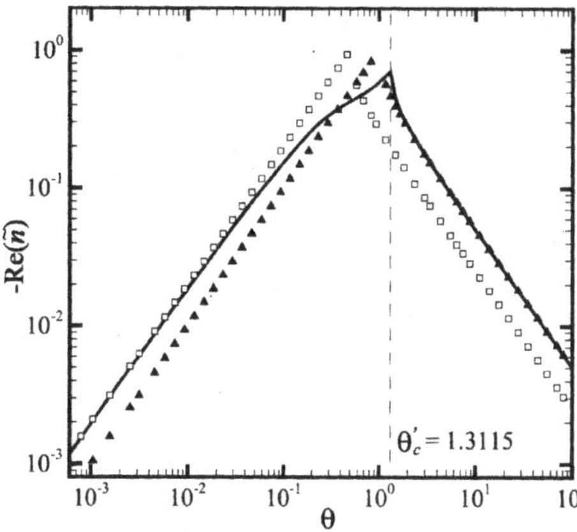

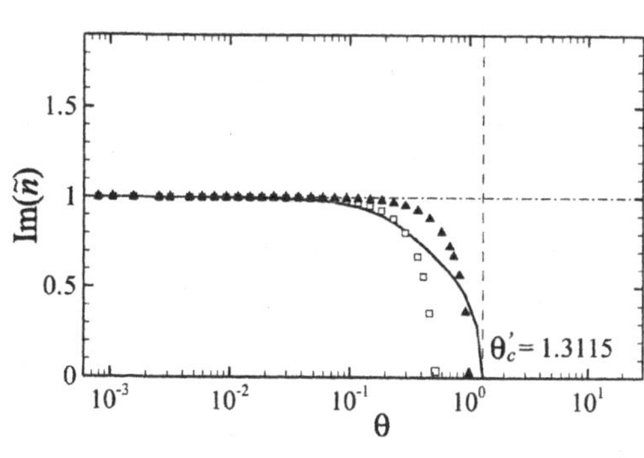

Figure 14.14. Dimensionless decay rate $-\mathrm{Re}[\tilde{n}]$ and frequency of the oscillations $\mathrm{Im}[\tilde{n}]$ as a function of the dimensionless parameter $\theta = \nu k^2/n_0$ from the exact solution (ES) and the irrotational theories VPF, DM, and IPF: Solid curve, ES; ▲, VPF; □, DM; dashed–dotted line, IPF. In the latter case, the eigenvalues $\tilde{n}$ are purely imaginary. The dimensionless eigenvalue is $\tilde{n} = n/n_0$, where n_0 is the inviscid frequency. For $\theta > \theta_c$, the solutions are purely real; in this case, only the slowest decay rate, given by the lowest real eigenvalue, is presented. The cutoff θ_c' corresponds to the ES.

decay rate with relative errors within 5% in absolute value and the agreement with the ES improves substantially following $\tilde{n} = -1/(2\theta)$ as θ increases. By contrast, the DM underpredict the decay rate by 50% in this regime. In the transition region ($0.02 \leq \theta \leq 2$, say), both viscous irrotational theories compute values of the decay rate that poorly approach the ES. However, each of these approximations gives rise to a critical value θ_c that qualitatively resembles the crossover from progressive to standing waves depicted by the ES.

Regarding the frequency of the oscillations, $\mathrm{Im}[\tilde{n}]$, figure 14.14 reveals that viscous effects are significant when $\theta \geq 0.1$, for which the ES deviates from the inviscid result. The frequency becomes damped and, for $\theta \geq \theta_c'$, the oscillations are suppressed. These features in the dynamics of the waves are also described, on qualitative grounds, by the viscous irrotational approximations. By taking into account viscous effects in the irrotational theories, the crossover from the traveling wave regime to the standing wave regime is predicted by these approximations, in a fashion similar to that of the ES.

We have shown that the effect of viscosity on the frequency of capillary-gravity waves, which Lamb (1932) assumed to be the same as in an inviscid fluid, can be obtained from the dissipation integral in the mechanical-energy balance. In addition, results from this DM are not restricted to oscillatory waves, as is the case with Lamb's dissipation calculation, but they also predict values for the decay rate of standing waves that follow the trend described by the ES. More details on the various aspects of the VPF and DM analyses for capillary-gravity waves are presented in Padrino and Joseph (2007).

14.3.2 Irrotational dissipation of nonlinear capillary-gravity waves

Many studies of nonlinear irrotational waves can be found in the literature, but the only study of the effects of viscosity on the decay of these waves known to us is due to

Longuet-Higgins (1997), who used the DM to determine the decay that is due to viscosity of irrotational steep capillary-gravity waves in deep water. He finds that the limiting rate of decay for small-amplitude solitary waves is twice that for linear periodic waves computed by the DM. The dissipation of very steep waves can be more than 10 times that of linear waves because of the sharply increased curvature in wave troughs. He assumes that the nonlinear wave maintains its steady form while decaying under the action of viscosity. The wave shape could change radically from its steady shape in very steep waves.

15

Irrotational Faraday waves on a viscous fluid

When a vessel containing liquid is made to vibrate vertically with constant frequency and amplitude, a pattern of standing waves on the gas–liquid surface can appear. For some combinations of frequency and amplitude, waves appear; for other combinations the free surface remains flat. These waves were first studied in the experiments of Faraday (1831), who noticed that the frequency of the liquid vibrations was only half that of the vessel. Nowadays, this would be described as a symmetry-breaking vibration of a type that characterized the motion of a simple pendulum subjected to a vertical oscillation of its purpose.

The first mathematical study of Faraday waves are due to Rayleigh (1883a, 1883b) but the first definitive study is due to Benjamin and Ursell (1954; hereafter BU) who remark that "The present work has been made possible by the development of the theory of Mathieu functions."

Faraday's problem is a rich source of problems in pattern formation, bifurcation, chaos, and other topics within the framework of fluid mechanics applications in the modern theory of dynamical system. Under the excitation of different parameters governing the Faraday system, different patterns, stripes, squares, hexagons, and time-dependent states can be observed. These features have spawned a large recent literature on Faraday waves. The experiments of Ciliberto and Gollub (1985) and Simonelli and Gollub (1989) on chaos, symmetry, and mode interactions are often cited. Theoretical approaches are restricted to "weak nonlinearities"; practically, some forms of expansion truncated at low order are used. The theoretical studies of Miles (1967), Ockendon and Ockendon (1973), Gu and Sethna (1987), and Feng and Sethna (1989) are weakly nonlinear studies of the Faraday instability for an inviscid fluid. Vega *et al.* (2001) and Higuera and co-workers (2005, 2006) studied weakly nonlinear and nearly inviscid Faraday waves with weak viscous effect, driven from boundary layers at the bottom of the containers and at the free surface.

Here we study the effects of viscosity of Faraday waves. There are two kinds of effects, those associated with vorticity and those that are purely irrotational. The purely irrotational effects have not been considered before, and they are studied here.

Effects associated with boundary layers at which vorticity is generated have been discussed by Miles (1967), who notes that

> ... the damping of surface waves in closed basins appears to be due to (a) viscous dissipation at the boundary of the surrounding basin, (b) viscous dissipation at the surface in consequence of surface contamination, and (c) capillary hysteresis associated with the meniscus surrounding the free surface.

Free surfaces are a source for the generation of vorticity because the shear stress vanishes there but the shear stress computed on irrotational flow does not vanish. Vorticity generation does not have a strong effect on damping even for progressive waves (§14.1), much less for RT waves that are closely related to the Faraday waves studied here.

Miles (1984) used a Lagrangian–Hamiltonian approach and included weak linear damping in treating the Faraday wave problem while retaining the inviscid approximation. Miles (1984) and Miles and Henderson (1990) advocate an empirical approach to these damping effects. Miles and Henderson note that

> Weak linear damping may be incorporated... by introducing the dissipation function $D = \frac{1}{2}\alpha_n(p_n^2 + q_n^2)$ where $\alpha_n \equiv \delta_n/\epsilon$ and δ_n is the ratio of actual to critical damping for free oscillations in the n-th mode [δ_n is best determined experimentally, although theoretical estimates are available (Miles 1967)].

The possibility that viscous damping can be rigorously calculated from purely irrotational theories without boundary layers has not been considered. Here we show that this possibility can be realized in the linearized case. The same kind of techniques can be used to calculate damping in nonlinear problems based on the equations of potential flow of an inviscid fluid. Of particular interest are the bifurcation studies on nonlinear Faraday waves on an inviscid fluid by Feng and Sethna (1989) and Gu and Sethna (1987), which are discussed in §15.8.

15.1 Introduction

The seminal paper of BU (1954) on the stability of the plane free surface of a liquid in vertical periodic motion has spawned a huge literature that extends their analysis to include effects of viscosity, two liquids, sidewall, bottom, and free-surface boundary layers, and nonlinear effects associated with bifurcation and pattern formation. Review papers emphasizing different aspects of this problem have been prepared by Miles and Henderson (1990), Dias and Kharif (1999), and Perlin and Schultz (2000). Here we focus on the effects of viscosity by use of VPF for exactly the same problem, in exactly the same formulation as that of BU (1954).

We derived the following two damped Mathieu equations from systematic analysis of the irrotational motion of viscous fluids in the formulation and using the notations introduced by BU. Thus

$$\ddot{a}_m + N\nu\dot{a}_m k_m^2 + k_m \tanh(k_m h)\left\{\frac{\gamma}{\rho}k_m^2 + [g - f\cos(\omega t)]\right\} a_m = 0, \qquad (15.1.1)$$

where k_m is an eigenvalue of the vibrating membrane equation,

$$\frac{\partial^2 S_m}{\partial x^2} + \frac{\partial^2 S_m}{\partial y^2} + k_m^2 S_m = 0, \qquad (15.1.2)$$

$$N = 2\ (\text{VPF}), \quad N = 4\ (\text{VCVPF}).$$

The number $N = 4$ also appears in the approximation called phenomenological by Kumar and Tuckerman (1994; hereafter KT), who incorporated damping by using a damping coefficient given by Landau and Lifshitz (1987, §25). KT say that "... for small damping (*i.e.* for $\lambda^2\omega \gg \nu$) and for small deformation of the interface..., the flow can be considered to be irrotational except for a thin layer around the interface." The restriction to small deformations of the interface is required for linearization, but the restriction to small damping, to small viscosities, is not implied by any mathematical argument and, in fact, is erroneous. The damping coefficient of Landau and Lifshitz (1987) is correct for gas–liquid flow but the application of this damping coefficient by KT (1994) to waves on the interface between two liquids is less justified and will lead to large errors for the long waves (§16.1.8) that occur in the Faraday problem. We find that the damped Mathieu equation, (15.1.1), with $N = 2$ gives a better approximation to small-amplitude Faraday dynamics than $N = 4$ for which the waves are "overdamped."

15.2 Energy equation

The energy equation is the basis of our VCVPF theory. We derive the mechanical-energy equation from Navier–Stokes equations,

$$\rho \frac{\mathrm{d}\boldsymbol{u}}{\mathrm{d}t} = \nabla \cdot \mathbf{T} + \rho \left[g - f\cos(\omega t) \right] \mathbf{e}_z, \tag{15.2.1}$$

in the usual way; we scalar multiply (15.2.1) by $\mathbf{u}$, integrate over the fluid domain V, and apply the Reynolds transport theorem and the Gauss theorem to find

$$\begin{aligned} \frac{\mathrm{d}}{\mathrm{d}t} \int_V \frac{1}{2}\rho \left|\mathbf{u}\right|^2 \mathrm{d}V = & \int_{A_f} \mathbf{u}\cdot\mathbf{T}\cdot\mathbf{n}\mathrm{d}A + \int_{A_w} \mathbf{u}\cdot\mathbf{T}\cdot\mathbf{n}\mathrm{d}A \\ & - \int_V 2\mu\mathbf{D} : \mathbf{D}\mathrm{d}V + \left[g - f\cos(\omega t) \right] \int_V \rho \frac{\mathrm{d}z}{\mathrm{d}t}\mathrm{d}V, \end{aligned} \tag{15.2.2}$$

where A_f is the free surface, A_w represents both the sidewalls and the bottom wall, and $\mathbf{n}$ is the outward normal of V on A. The integrals $\int_{A_f} \mathbf{u}\cdot\mathbf{T}\cdot\mathbf{n}\mathrm{d}A$ and $\int_{A_w} \mathbf{u}\cdot\mathbf{T}\cdot\mathbf{n}\mathrm{d}A$ are the power of traction. On the free surface $\mathbf{n} \approx -\mathbf{e}_z$, and we can show readily that

$$\mathbf{u}\cdot\mathbf{T}\cdot\mathbf{n} = -\left(u_x T_{xz} + u_y T_{yz} + u_z T_{zz}\right). \tag{15.2.3}$$

We shall be considering the potential flow of viscous fluids called VPF. For these flows, the no-slip conditions usually cannot be satisfied and they are replaced with (15.3.1), given in §15.3. The stresses on the free surface are evaluated by use of potential flow:

$$\begin{aligned} T_{xz} &= \tau^i_{xz} = \mu\left(\frac{\partial u_x}{\partial z} + \frac{\partial u_z}{\partial x}\right), \\ T_{yz} &= \tau^i_{yz} = \mu\left(\frac{\partial u_y}{\partial z} + \frac{\partial u_z}{\partial y}\right), \\ T_{zz} &= -p_i + \tau^i_{zz} = -p_i + 2\mu\frac{\partial u_z}{\partial z}, \end{aligned} \tag{15.2.4}$$

where p_i is the irrotational pressure computed from the Bernoulli equation. Then the mechanical-energy equation for VPF may be written as

$$\begin{aligned} \frac{\mathrm{d}}{\mathrm{d}t} \int_V \frac{1}{2}\rho \left|\mathbf{u}\right|^2 \mathrm{d}V = & \int_{A_f} -\left[u_x\tau^i_{xz} + u_y\tau^i_{yz} + u_z\left(-p_i + \tau^i_{zz}\right)\right]\mathrm{d}A + \int_{A_w} \mathbf{u}\cdot\mathbf{T}\cdot\mathbf{n}\mathrm{d}A \\ & - \int_V 2\mu\mathbf{D} : \mathbf{D}\mathrm{d}V + \left[g - f\cos(\omega t) \right] \int_V \rho \frac{\mathrm{d}z}{\mathrm{d}t}\mathrm{d}V. \end{aligned} \tag{15.2.5}$$

The shear stresses τ^i_{xz} and τ^i_{yz} from the potential flow are not zero at the free surface. However, the shear stresses should be zero physically. T_{xz} and T_{yz} cannot be made zero in irrotational flows, but we can remove the power by the shear stress $\int_{A_f}(u_x\tau^i_{xz}+u_y\tau^i_{yz})\mathrm{d}A$ from the mechanical-energy equation. At the same time, a pressure correction p_v is added to p_i to compensate for the shear stresses. The mechanical-energy equation for VCVPF is then written as

$$\begin{aligned}\frac{\mathrm{d}}{\mathrm{d}t}\int_V \frac{1}{2}\rho\,|\mathbf{u}|^2\,\mathrm{d}V &= \int_{A_f} -\left[u_z\left(-p_i - p_v + \tau^i_{zz}\right)\right]\mathrm{d}A + \int_{A_w} \mathbf{u}\cdot\mathbf{T}\cdot\mathbf{n}\mathrm{d}A \\ &\quad - \int_V 2\mu\mathbf{D}:\mathbf{D}\mathrm{d}V + \left[g - f\cos(\omega t)\right]\int_V \rho\frac{\mathrm{d}z}{\mathrm{d}t}\mathrm{d}V.\end{aligned} \tag{15.2.6}$$

A comparison of (15.2.5) and (15.2.6) gives rise to the relation between the pressure correction and the irrotational shear stresses:

$$\int_{A_f}\left(u_x\tau^i_{xz} + u_y\tau^i_{yz}\right)\mathrm{d}A = \int_{A_f}\left(-p_v u_z\right)\mathrm{d}A. \tag{15.2.7}$$

15.3 VPF and VCVPF

The two theories, VPF and VCVPF, give rise to different results. Our experience with other problems is such as to suggest that VCVPF is closer to exact results for progressive waves and VPF is closer to exact results when waves do not propagate; more precisely, in this case, the eigenvalues are real. This second case, in which VPF is better, applies here to irrotational Faraday waves on viscous fluids.

15.3.1 Potential flow

The velocity $\mathbf{u} = \nabla\phi = (u_x, u_y, u_z)$ is expressed in terms of a harmonic potential $\nabla^2\phi = 0$ in a coordinate system moving with the container. Boundary conditions at the container walls are given by

$$\begin{aligned}\frac{\partial\phi}{\partial n} &= 0 \text{ on the sidewalls,} \\ \frac{\partial\phi}{\partial z} &= 0 \text{ on the bottom wall at } z = h.\end{aligned} \tag{15.3.1}$$

A harmonic solution satisfying (15.3.1) can be written as

$$\phi(x, y, z, t) = \sum_{m=0}^{\infty} f_m(t)\cosh\left[k_m(h - z)\right]S_m(x, y), \tag{15.3.2}$$

where the eigenfunctions $S_m(x, y)$ satisfy

$$\frac{\partial S_m}{\partial n} = 0 \tag{15.3.3}$$

on the sidewall. The condition at the bottom wall gives

$$\left(\frac{\partial\phi}{\partial z}\right)_{z=h} = \sum_{m=0}^{\infty} f_m(t)\left(-k_m\right)\sinh\left[k_m(h - z)\right]S_m(x, y)\bigg|_{z=h} = 0. \tag{15.3.4}$$

The normal stress balance at the free surface,

$$z = \zeta(x, y, t), \tag{15.3.5}$$

in the linearized approximation is

$$\left(p - 2\mu \frac{\partial u_z}{\partial z}\right)_{z=0} = \gamma \left(\frac{\partial^2 \zeta}{\partial x^2} + \frac{\partial^2 \zeta}{\partial y^2}\right). \tag{15.3.6}$$

For VPF, $p = p_i$, where p_i is given by the Bernoulli equation. For VCVPF, $p = p_i + p_v$, where p_v is a viscous correction of the irrotational pressure p_i.

15.3.2 Amplitude equations for the elevation of the free surface

Now consider the kinematic condition at $z = 0$:

$$\frac{\partial \zeta}{\partial t} = u_z = \frac{\partial \phi}{\partial z} \quad \text{at} \quad z = 0, \tag{15.3.7}$$

where

$$\left(\frac{\partial \phi}{\partial z}\right)_{z=0} = \sum_{m=0}^{\infty} f_m(t)\,(-k_m) \sinh(k_m h)\, S_m(x, y). \tag{15.3.8}$$

If we write the surface elevation as

$$\zeta = \sum_{m=0}^{\infty} a_m(t) S_m(x, y), \tag{15.3.9}$$

then

$$\frac{\partial \zeta}{\partial t} = \sum_{m=0}^{\infty} \frac{\mathrm{d}a_m}{\mathrm{d}t} S_m(x, y). \tag{15.3.10}$$

Because the total volume of fluid is constant, $a_0(t)$ is constant:

$$\frac{\mathrm{d}a_0}{\mathrm{d}t} = 0,$$

$$\frac{\gamma}{\rho}\left(\frac{\partial^2 \zeta}{\partial x^2} + \frac{\partial^2 \zeta}{\partial y^2}\right) = -\frac{\gamma}{\rho} \sum_{m=1}^{\infty} k_m^2 a_m(t) S_m(x, y). \tag{15.3.11}$$

Because $k_0 = 0$, (15.3.11) and (15.3.8) show that $f_0(t)$ is undetermined. BU showed that $a_0(t)$ can be put to zero.

For $m \geq 1$, (15.3.7), (15.3.8), and (15.3.10) give

$$\frac{\mathrm{d}a_m}{\mathrm{d}t} S_m(x, y) = f_m(t)\,(-k_m) \sinh(k_m h)\, S_m(x, y); \quad \text{hence } f_m(t) = -\frac{\mathrm{d}a_m}{\mathrm{d}t} \frac{1}{k_m \sinh(k_m h)},$$

so that the potential is given by

$$\phi(x, y, z, t) = -\sum_{m=1}^{\infty} \frac{\mathrm{d}a_m}{\mathrm{d}t} \frac{\cosh[k_m(h - z)]}{k_m \sinh(k_m h)} S_m(x, y). \tag{15.3.12}$$

Bernoulli's equation is

$$\frac{p_i}{\rho} + \frac{\partial \phi}{\partial t} - [g - f \cos(\omega t)]\, z = 0. \tag{15.3.13}$$

Normal stress balance (15.3.6) is

$$\left(p_i + p_v - 2\mu\frac{\partial u_z}{\partial z}\right)_{z=0} = \gamma\left(\frac{\partial^2\zeta}{\partial x^2} + \frac{\partial^2\zeta}{\partial y^2}\right). \tag{15.3.14}$$

Linearized governing equations for the viscous corrections are

$$\rho\frac{\partial \mathbf{u}_v}{\partial t} = -\nabla p_v + \mu\nabla^2\mathbf{u}_v, \quad \nabla\cdot\mathbf{u}_v = 0. \tag{15.3.15}$$

Hence,

$$\nabla^2 p_v = 0. \tag{15.3.16}$$

The solution of (15.3.16) may be written as

$$-p_v = \sum_{m=0}^{\infty} C_m\hat{r}_m(t)\theta_m(z)S_m(x, y), \quad \theta_m = c_{m1}e^{k_m z} + c_{m2}e^{-k_m z}. \tag{15.3.17}$$

At $z = 0$,

$$-p_v(z=0) = \sum_{m=0}^{\infty} C_m r_m(t)S_m(x, y), \tag{15.3.18}$$

where $r_m = \hat{r}_m(t)\,(c_{m1} + c_{m2})$.

We may eliminate p_i from (15.3.14) by using (15.3.13):

$$\left\{p_v + \rho\,[g - f\cos(\omega t)]\,\zeta - \rho\frac{\partial\phi}{\partial t} - 2\mu\frac{\partial^2\phi}{\partial z^2}\right\}_{x=0} = \gamma\left(\frac{\partial^2\zeta}{\partial x^2} + \frac{\partial^2\zeta}{\partial y^2}\right). \tag{15.3.19}$$

Equation (15.3.19) may be evaluated on modal functions by use of (15.3.9) and (15.3.11):

$$2\mu\frac{\partial^2\phi}{\partial z^2} = -2\mu\sum_{m=1}^{\infty}\frac{\mathrm{d}a_m}{\mathrm{d}t}k_m\frac{\cosh\,[k_m(h-z)]}{\sinh\,(k_m h)}S_m(x, y), \tag{15.3.20}$$

$$\frac{\partial\phi}{\partial t} = -\sum_{m=1}^{\infty}\frac{\mathrm{d}^2 a_m}{\mathrm{d}t^2}\frac{\coth\,[k_m(h-z)]}{k_m\sinh\,(k_m h)}S_m(x, y). \tag{15.3.21}$$

Hence

$$\begin{aligned}\sum_{m=1}^{\infty}\Bigg\{&-C_m r_m(t) + \rho\,[g - f\cos(\omega t)]\,a_m(t) + \rho\frac{\mathrm{d}^2 a_m}{\mathrm{d}t^2}\frac{\coth\,(k_m h)}{k_m}\\ &+ 2\mu\frac{\mathrm{d}a_m}{\mathrm{d}t}k_m\coth\,(k_m h) + \gamma k_m^2 a_m(t)\Bigg\}\,S_m(x, y) = 0.\end{aligned} \tag{15.3.22}$$

The coefficients of the linearly independent functions $S_m(x, y)$ vanish. Hence the amplitude equation for VCVPF is

$$\begin{aligned}&\frac{\mathrm{d}^2 a_m}{\mathrm{d}t^2} + 2\nu k_m^2\frac{\mathrm{d}a_m}{\mathrm{d}t} + k_m\tanh\,(k_m h)\left[\frac{\gamma}{\rho}k_m^2 + g - f\cos(\omega t)\right]a_m\\ &-\frac{k_m}{\rho}\tanh\,(k_m h)\,C_m r_m(t) = 0.\end{aligned} \tag{15.3.23}$$

To evaluate $C_m r_m(t)$ in (15.3.23) we need to work with only mode m. To simplify the writing we suppress the subscript m and write

$$u_x = -\frac{\mathrm{d}a}{\mathrm{d}t}\frac{\coth(kh)}{k}\frac{\partial S}{\partial x}, \tag{15.3.24}$$

$$\tau_{xz} = 2\mu\frac{\mathrm{d}a}{\mathrm{d}t}\frac{\partial S}{\partial x}, \tag{15.3.25}$$

$$u_y = -\frac{\mathrm{d}a}{\mathrm{d}t}\frac{\coth(kh)}{k}\frac{\partial S}{\partial y}, \tag{15.3.26}$$

$$\tau_{yz} = 2\mu\frac{\mathrm{d}a}{\mathrm{d}t}\frac{\partial S}{\partial y}, \tag{15.3.27}$$

$$u_z = \frac{\mathrm{d}a}{\mathrm{d}t}S, \tag{15.3.28}$$

$$p_v = -Cr(t)S, \tag{15.3.29}$$

$$\tau_{zz} = -2\mu\frac{\mathrm{d}a}{\mathrm{d}t}k\coth(kh)\,S, \tag{15.3.30}$$

$$\int_A (u_x\tau_{xz} + u_y\tau_{yz} + p_v u_z)\,\mathrm{d}A$$
$$= \int_A \left\{-2\mu\left(\frac{\mathrm{d}a}{\mathrm{d}t}\right)^2 \frac{\coth(kh)}{k}\left[\left(\frac{\partial S}{\partial x}\right)^2 + \left(\frac{\partial S}{\partial y}\right)^2\right] - Cr(t)\frac{\mathrm{d}a}{\mathrm{d}t}S^2\right\}\mathrm{d}A = 0.$$

Using Gauss' theorem and the boundary condition on the sidewall (15.3.3), we obtain

$$\int_A \left[\frac{\partial}{\partial x}\left(S\frac{\partial S}{\partial x}\right) + \frac{\partial}{\partial y}\left(S\frac{\partial S}{\partial y}\right)\right]\mathrm{d}A = \int_L \left[S\frac{\partial S}{\partial n}\right]\mathrm{d}L = 0,$$

where L is the boundary of the free surface A and L is on the sidewall. With the condition $\nabla_2^2 S = -k^2 S$, we can show that

$$\int_A \left[\left(\frac{\partial S}{\partial x}\right)^2 + \left(\frac{\partial S}{\partial y}\right)^2\right]\mathrm{d}A = -\int_A S\nabla_2^2 S\,\mathrm{d}A = k^2\int_A S^2\mathrm{d}A. \tag{15.3.31}$$

We find that

$$\left[2\mu\left(\frac{\mathrm{d}a}{\mathrm{d}t}\right)^2 k\coth(kh) + Cr(t)\frac{\mathrm{d}a}{\mathrm{d}t}\right]\int S^2\mathrm{d}A = 0, \tag{15.3.32}$$

$$C_m r_m(t) = Cr(t) = -2\mu k\frac{\mathrm{d}a}{\mathrm{d}t}\coth(kh). \tag{15.3.33}$$

Inserting (15.3.33) into (15.3.23), we find the amplitude equation for VCVPF:

$$\frac{\mathrm{d}^2 a}{\mathrm{d}t^2} + 4\nu k^2\frac{\mathrm{d}a}{\mathrm{d}t} + k\tanh(kh)\left[\frac{\gamma}{\rho}k^2 + g - f\cos(\omega t)\right]a = 0. \tag{15.3.34}$$

VPF is the same as VCVPF without the pressure correction p_v. If p_v is set to zero, we find that

$$\frac{\mathrm{d}^2 a}{\mathrm{d}t^2} + 2\nu k^2\frac{\mathrm{d}a}{\mathrm{d}t} + k\tanh(kh)\left[\frac{\gamma}{\rho}k^2 + g - f\cos(\omega t)\right]a = 0. \tag{15.3.35}$$

The damping term can be written as

$$N\nu k^2 \frac{\mathrm{d}a}{\mathrm{d}t}, \tag{15.3.36}$$

with $N = 2$ for VPF and $N = 4$ for VCVPF.

15.4 Dissipation method

The DM leads to the same amplitude equation, (15.3.34), that we derived for VCVPF. The two theories are equivalent, but no pressure whatsoever is required for implementing the DM. In the dissipation calculation, the power of the pressure correction is replaced with the power of shear tractions [see (12.6.1)].

The dissipation calculation gives the same results as VCVPF. (15.3.34) is also in agreement with (4.21) of KT (1994).

15.5 Stability analysis

The amplitude equations are

$$\ddot{a} + N\nu\dot{a}k^2 + k\tanh(kh)\left\{\frac{\gamma}{\rho}k^2 + [g - f\cos(\omega t)]\right\}a = 0, \tag{15.5.1}$$

where

$$N = 2\ \text{(VPF)}, \quad N = 4\ \text{(VCVPF)}. \tag{15.5.2}$$

In the fourth-order Runge–Kutta integration, we may take time difference $\Delta t = \pi/2^{12} = \pi/4096$ for which time at n steps is given by $t = n \times \Delta t$ and periodic time T may be defined as

$$T = \left[\frac{t}{2\pi}\right], \tag{15.5.3}$$

with Gauss symbol [ ]. According to Floquet theory, we may represent the solutions of (15.5.1) in the unstable region as

$$\ln[a(t)] = \sigma t + \beta = \sigma 2\pi T + \beta, \tag{15.5.4}$$

where $\exp[\beta(t)] = b(t)$ is periodic in t but constant in T and the growth rate σ is positive; $\sigma = 0$ at the marginal state. To check $\exp[\beta(t)] = b(t)$, we may use a Fourier series expressed as

$$b(t) = \sum_{n=-\infty}^{\infty} A_n \exp(int), \tag{15.5.5}$$

where the Fourier coefficient A_{-n} is the complex conjugate of A_n. The coefficient is evaluated as

$$\begin{aligned} A_m &= \frac{1}{2\pi}\int_t^{t+2\pi} b(t)\exp(-imt)\mathrm{d}t \\ &= \frac{1}{2\pi}\sum_{j=1}^{8192}[b(t_j)\exp(-imt_j) + b(t_{j-1})\exp(-imt_{j-1})]\frac{\Delta t}{2} \end{aligned} \tag{15.5.6}$$

with the trapezoidal rule; $t_j = t + j \times \Delta t$.

The solution of (15.5.1) can be written in Floquet form:

$$a(t) = e^{\sigma t} b(t) \tag{15.5.7}$$

where σ is the growth rate and $b(t)$ is a bounded oscillatory function that is periodic when

$$\sigma = 0 \quad \text{(marginal state)} \tag{15.5.8}$$

or

$$\sigma > 0 \quad \text{(unstable state).} \tag{15.5.9}$$

A growth-rate curve is given by

$$\sigma = \sigma(k) = \sigma(-k), \tag{15.5.10}$$

which is an even function of k.

The maximum growth rate is

$$\sigma_m = \max_{k>0} [\sigma(k)] = \sigma(k_m). \tag{15.5.11}$$

The flow is stable when $k > k_c$:

$$\sigma(k) < 0, \quad k > k_c, \tag{15.5.12}$$

where k_c is called the cutoff wavenumber. This says that short waves are stable.

Governing equation (15.5.1) for the oscillation amplitude in a deep liquid $h \to \infty$ is

$$\ddot{a} + N\nu \dot{a} k^2 + k \left\{ \frac{\gamma}{\rho} k^2 + [g - f \cos(\omega t)] \right\} a = 0, \tag{15.5.13}$$

where $N = 2$ (VPF) or $N = 4$ (VCVPF). The main goal of this calculation is to show that, for any value of the kinematic viscosity ν, the other parameters being constant, $N = 2$ has a larger maximum growth rate than $N = 4$. Hence the irrotational theory associated with the direct effects of the viscous normal stress (VPF) on the motion of the waves will give a better description of the effects of viscosity on the waves than the value $N = 4$ (VCVPF), which has been universally used by researchers in this subject since the study of KT (1994).

Cerda and Tirapegui (1998) considered Faraday's instability in a viscous fluid and found a Mathieu equation based on lubrication theory rather than on potential flow. They interpret the irrotational theory leading to $N = 4$ as appropriate to weak dissipation. Our irrotational theories are not restricted to small viscosity, but they do not account for vorticity generated by the no-slip condition on the container sidewalls or bottom. For periodic disturbances on deep water, the irrotational theories are valid for all values of ν.

Our results apply to silicon oils with kinematic viscoucities ranging from zero to 10 cm^2/sec, a density of 0.97 g/cm^3, and a surface tension of $\gamma = 21$ dyn/cm. The frequency $\omega = 15.87 \times 2\pi$ sec^{-1} is fixed and $(g, f) = (981, 981)$ cm/sec^2.

In figures 15.1 and 15.2 we present graphs and tables for the Floquet representation $a = e^{\sigma t} b(t)$ for the stability of Faraday waves on an inviscid and a viscous fluid.

In figure 15.3 we plot σ_m vs. ν for Faraday waves on an inviscid fluid ($N = 0$) and on a viscous fluid by using VPF ($N = 2$) and VCVPF ($N = 4$). These theories are all irrotational. VCVPF gives the same results as the DM, as shown here in §15.4. Previously, the dissipation theory with $N = 4$ was proposed by KT (1994) from heuristic considerations. In figure 15.4 we plot the critical wavenumber k_m vs. ν. The growth rates and critical wavenumber are largest for $N = 0$ and are larger for VPF than for VCVPF at each fixed ν.

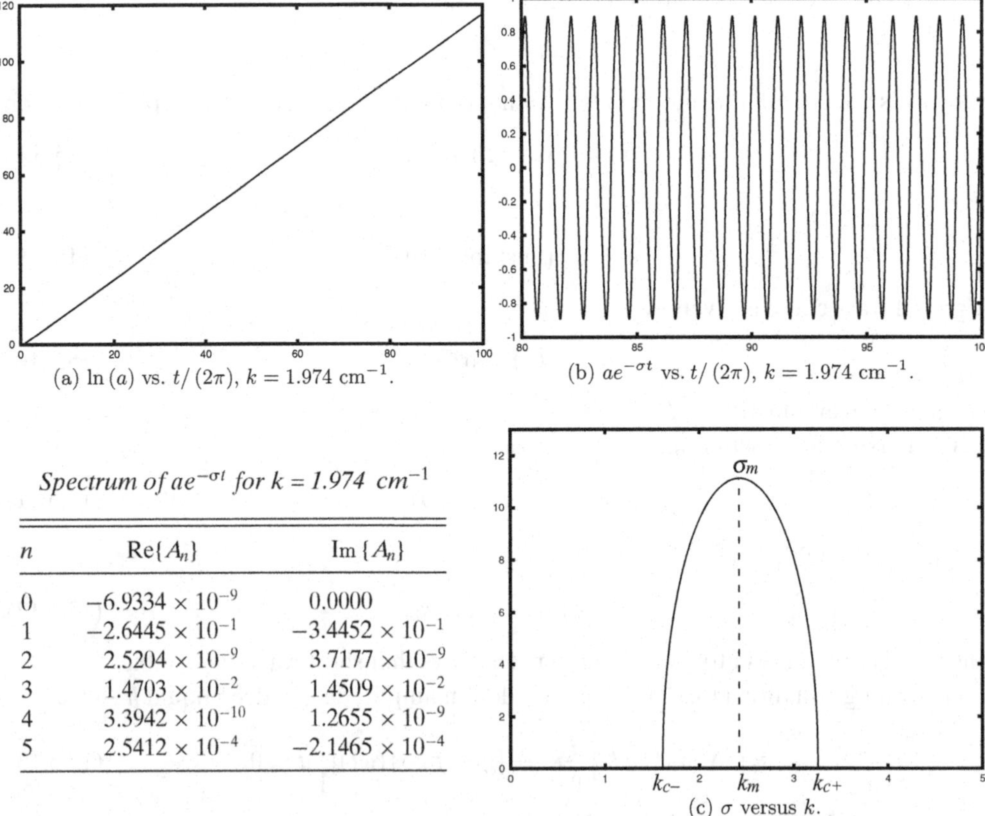

(a) $\ln(a)$ vs. $t/(2\pi)$, $k = 1.974$ cm^{-1}. (b) $ae^{-\sigma t}$ vs. $t/(2\pi)$, $k = 1.974$ cm^{-1}.

Spectrum of $ae^{-\sigma t}$ for $k = 1.974$ cm^{-1}

n	Re$\{A_n\}$	Im $\{A_n\}$
0	-6.9334×10^{-9}	0.0000
1	-2.6445×10^{-1}	-3.4452×10^{-1}
2	2.5204×10^{-9}	3.7177×10^{-9}
3	1.4703×10^{-2}	1.4509×10^{-2}
4	3.3942×10^{-10}	1.2655×10^{-9}
5	2.5412×10^{-4}	-2.1465×10^{-4}

(c) σ versus k.

Figure 15.1. Floquet theory $a = e^{\sigma t}b(t)$ for Faraday waves on an inviscid fluid ($N = 0$): $\rho = 0.97$ g/cm^3, $\gamma = 21$ dyn/cm, $g = 981$ cm/sec^2, $\omega = 2\pi \times 15.87$ sec^{-1}, $f = g$ cm/sec^2. (a) $\ln(a)$ vs. $t/(2\pi)$, (b) $ae^{-\sigma t} = b(t) = b(t + 2\pi)$ vs. $t/(2\pi)$, (c) $\sigma(N\nu)$ vs. k. $\sigma_m = 11.1274$ sec^{-1} at $k_m = 2.4246$ cm^{-1}.

15.6 Rayleigh–Taylor instability and Faraday waves

This section follows the work of Kumar (2000) who compared wavenumber selection in RT instability and Faraday instability on deep and highly viscous liquids. He used the dissipation theory (VCVPF) of Faraday instability with $N = 4$, first proposed by KT (1994). Our main goal is to introduce VPF with $N = 2$ into this comparison. We also revise slightly the comparisons made by Kumar (2000) so that a direct comparison of maximum growth rates in the two problems can be made.

Kumar (2000) compared a critical $k_c = k_m$ at $f = f_c$ [where $\sigma(k_m) = 0$, $\sigma < 0$ for $k \neq k_m = k_c$] for Faraday waves with the maximum growth rate of RT waves when the gravitational acceleration is replaced with

$$\bar{a}_c = \frac{\omega}{\pi} \int_{3\pi/2\omega}^{5\pi/2\omega} [f_c \cos(\omega t) - g]\, \mathrm{d}t = \frac{2}{\pi} f_c - g. \tag{15.6.1}$$

This value of $\bar{a}_c$ is an average upward acceleration in the Faraday problem. Kumar used (15.6.1) with $N = 4$ in his calculation.

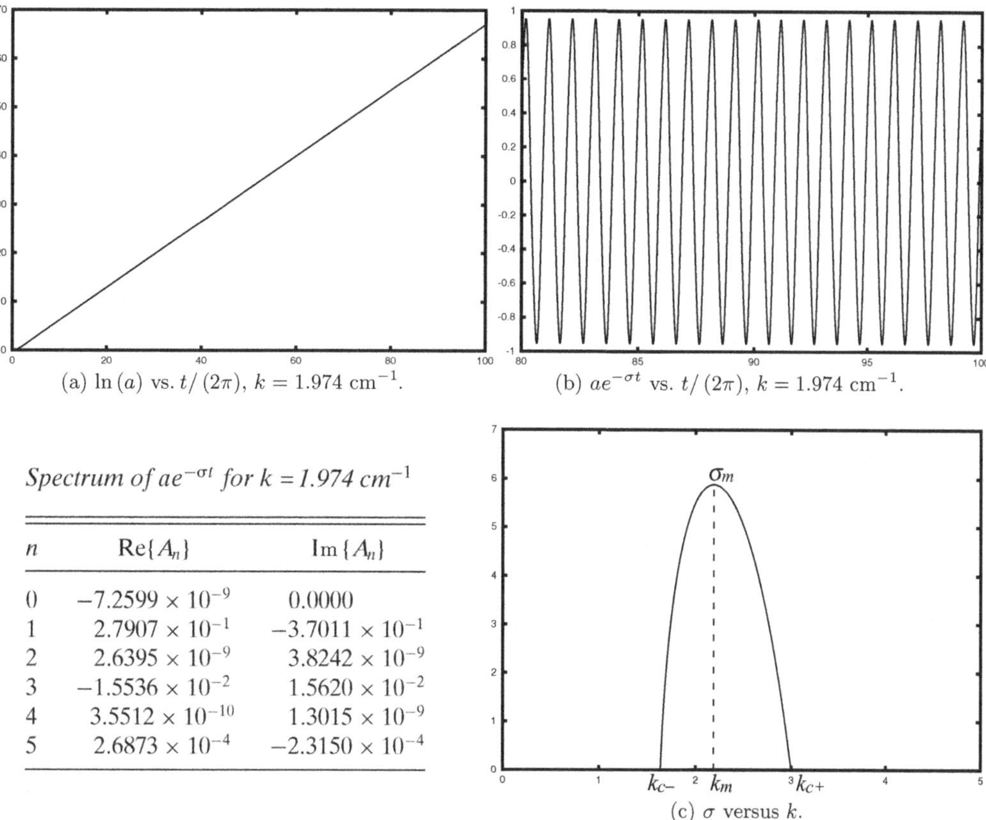

(a) $\ln(a)$ vs. $t/(2\pi)$, $k = 1.974\ \text{cm}^{-1}$.

(b) $ae^{-\sigma t}$ vs. $t/(2\pi)$, $k = 1.974\ \text{cm}^{-1}$.

Spectrum of $ae^{-\sigma t}$ *for* $k = 1.974\ cm^{-1}$

n	$\text{Re}\{A_n\}$	$\text{Im}\{A_n\}$
0	-7.2599×10^{-9}	0.0000
1	2.7907×10^{-1}	-3.7011×10^{-1}
2	2.6395×10^{-9}	3.8242×10^{-9}
3	-1.5536×10^{-2}	1.5620×10^{-2}
4	3.5512×10^{-10}	1.3015×10^{-9}
5	2.6873×10^{-4}	-2.3150×10^{-4}

(c) σ versus k.

Figure 15.2. Floquet theory $a = e^{\sigma t} b(t)$ for Faraday waves on a viscous liquid $\nu = 1\ \text{cm}^2/\text{sec}$ with $N = 2$ (VPF): $\rho = 0.97\ \text{g/cm}^3$, $\gamma = 21\ \text{dyn/cm}$, $g = 981\ \text{cm/sec}^2$, $\omega = 2\pi \times 15.87\ \text{sec}^{-1}$, $f = g\ \text{cm/sec}^2$. (a) $\ln(a)$ vs. $t/(2\pi)$, (b) $ae^{-\sigma t} = b(t) = b(t + 2\pi)$, (c) σ vs. k. $\sigma_m = 5.8790\ \text{sec}^{-1}$ at $k_m = 2.1874\ \text{cm}^{-1}$. The dissipation theory with $N = 4$ (VCVPF) is stable, $\text{Re}\{\sigma\} < 0$.

The maximum growth rate for RT instability can be computed from the exact linear theory given by equation (18) in Joseph, Belanger, and Beavers (1999) or more easily and with good accuracy by the purely irrotational theory by equation (28) with $\nu = \mu_2/\rho_2$ and g replaced with $\bar{a}_c$. In this case, their (28) gives

$$\sigma = -k^2\nu \pm \sqrt{k\bar{a}_c - \frac{k^3\gamma}{\rho} + k^4\nu^2}. \tag{15.6.2}$$

The function $k_m(\bar{a}_c)$ is given by maximizing σ given by (15.6.2) with respect to k:

$$\frac{d\sigma}{dk} = -2\nu k + \frac{1}{2}\frac{\bar{a} - 3\frac{\gamma}{\rho}k^2 + 4k^3\nu^2}{\sqrt{\bar{a}k - \frac{\gamma}{\rho}k^3 + k^4\nu^2}} = 0, \tag{15.6.3}$$

which can be arranged as

$$8\nu^2 k^3\left(\bar{a} + \frac{\gamma}{\rho}k^2\right) = \left(\bar{a} - 3\frac{\gamma}{\rho}k^2\right)^2. \tag{15.6.4}$$

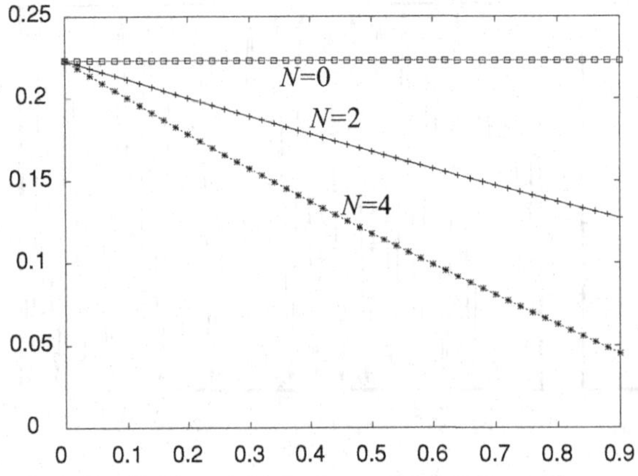

Figure 15.3. σ_m vs. ν cm^2/sec.

This is a fifth-order algebraic equation for $k = k_m$. When $\bar{a}$ is large, we have

$$8\nu^2 k^3 \bar{a} = (\bar{a})^2 . \tag{15.6.5}$$

Hence

$$k_m = \left(\frac{\bar{a}}{8\nu^2} \right)^{1/3} \tag{15.6.6}$$

gives the maximum growth rate for large $\bar{a}$.

The maximum growth rate $\sigma_m = \sigma(k_m)$ and the wavenumber $k = k_m$ of maximum growth rate for RT instability and dissipative Faraday waves are compared in figures 15.5, 15.6, and 15.7. These figures show that the dissipative theory with $N = 4$ introduced by KT (1994) and used by Kumar (2000) are more damped than the dissipative potential flow solution VPF with $N = 2$; damped solutions with $\sigma_m < 0$ at small values of f/g are shown in figures 15.6 and 15.7. We can say that the demonstration that damped Faraday waves at large viscosities are driven by the same acceleration mechanism that produces RT waves is better demonstrated by VPF with $N = 2$ than by the dissipative theory, which is equivalent to our irrotational VCVPF with $N = 4$.

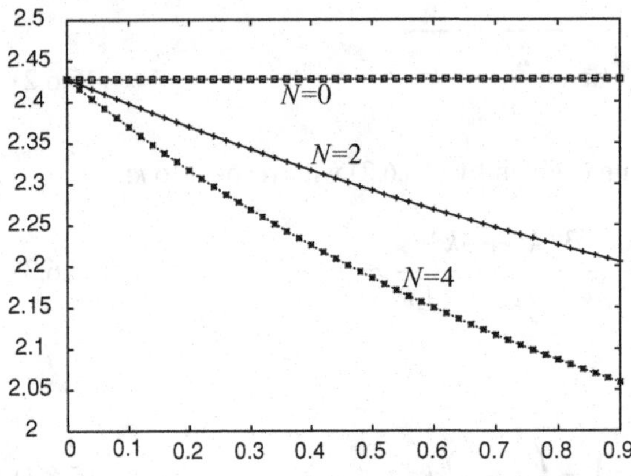

Figure 15.4. k_m vs. ν cm^2/sec.

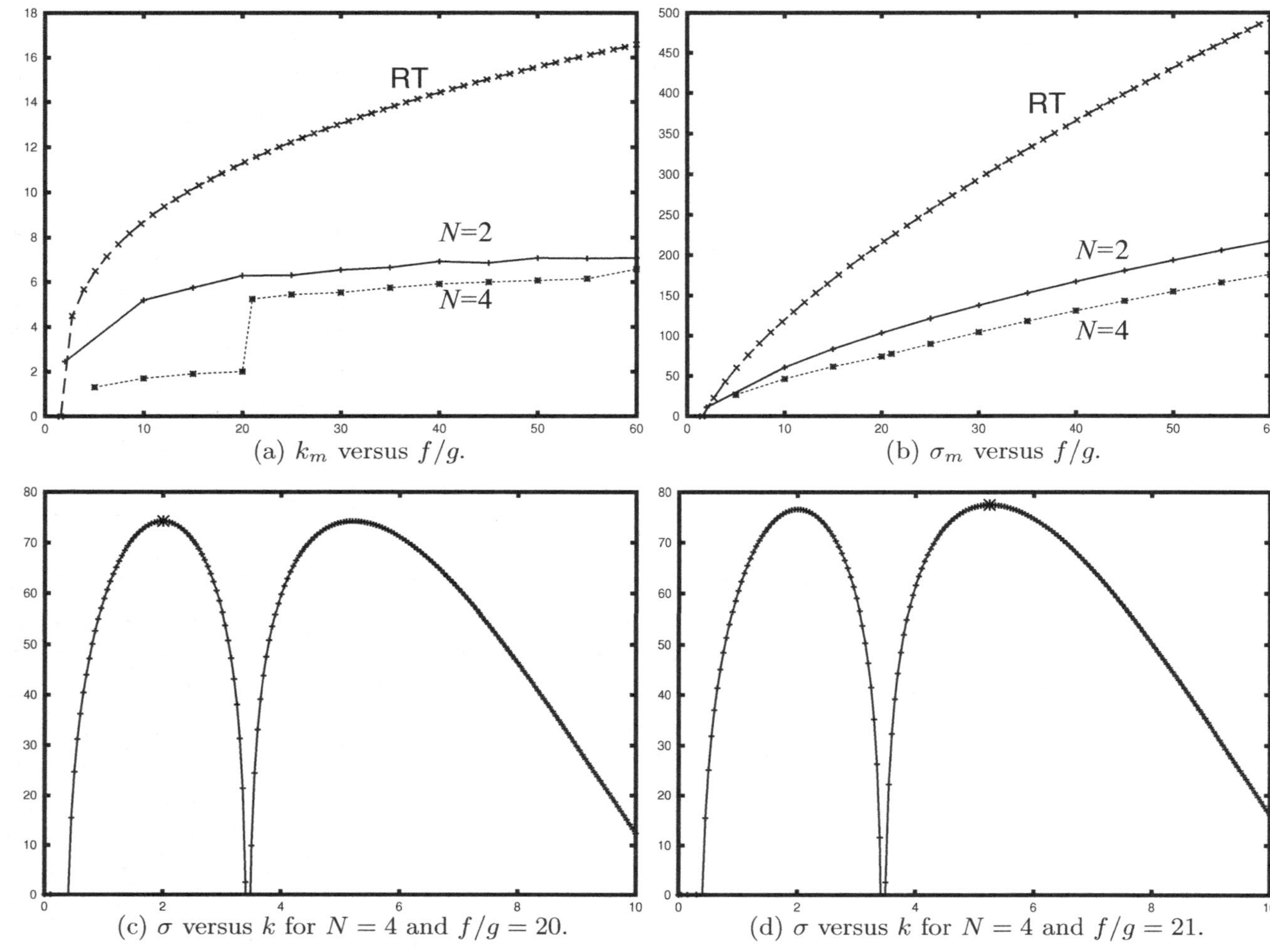

Figure 15.5. (a) k_m vs. f/g, (b) σ_m vs. f/g, for $\nu = 1$ cm^2/sec. (c), (d) σ vs. k for $N = 4$ in a transition region, in which $*$ denotes the maximum growth rate.

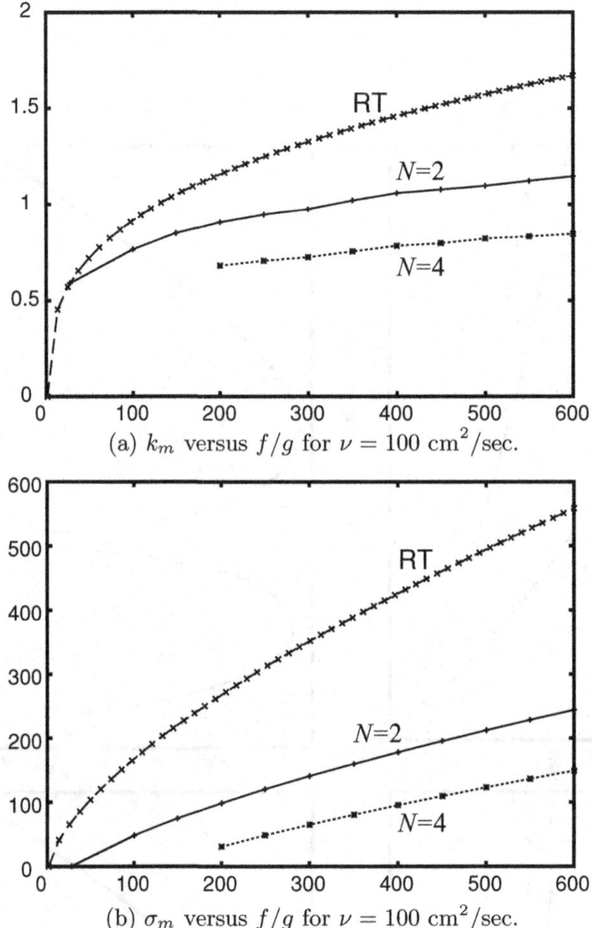

Figure 15.6. (a) k_m vs. f/g and (b) σ_m vs. f/g, for $\nu = 100$ cm^2/sec.

15.7 Comparison of purely irrotational solutions with exact solutions

KT (1994) presented a linear stability analysis of the interface between two viscous fluids. Starting from the Navier–Stokes equations, they derived the relevant equations describing the hydrodynamic system in the presence of parametric forcing and carried out a Floquet analysis to solve the stability problem. The viscous problem does not reduce to a system of Mathieu equations with a linear damping term, which is traditionally considered to represent the effect of viscosity. The traditional approach ignores the viscous boundary conditions at the interface of two fluids. To determine the effect of neglecting these, they compared their exact viscous fluid results with those derived from the traditional phenomenological approach. They call the exact theory a fully hydrodynamic system (FHS). The traditional phenomenological approach is an application of the dissipation method; it is called a model. When applied to an air–liquid system the model is the same as the DM, which is the same as our irrotational theory VCVPF with damping proportional to 4ν.

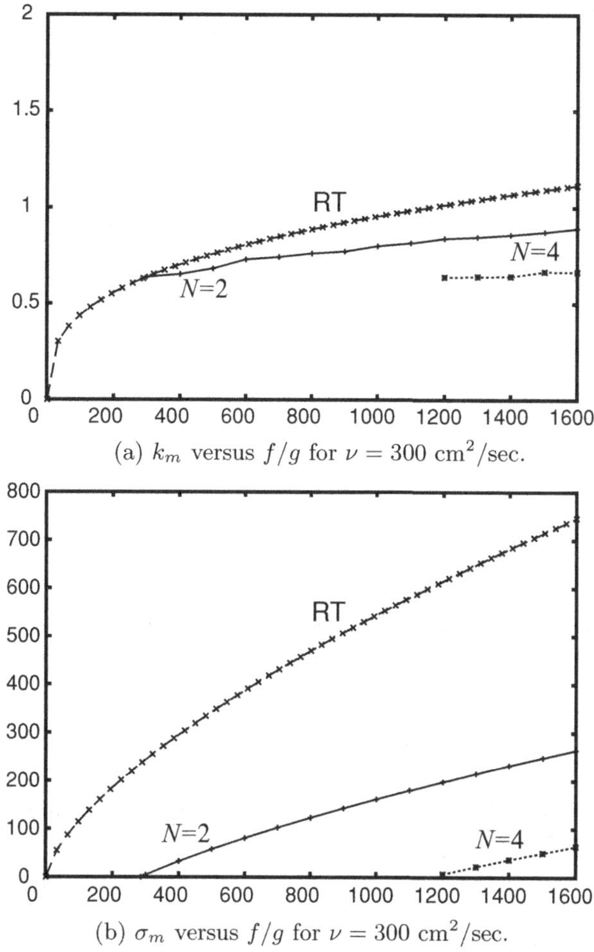

Figure 15.7. (a) k_m vs. f/g and (b) σ_m vs. f/g, for $\nu = 300$ cm^2/sec. For small values of f/g the potential flow solutions for Faraday waves are stable, $\sigma_m < 0$ but $N = 2$ is less stable and more like RT waves than $N = 4$.

They compared the results of the FHS and of the model with experimental results obtained in a viscous glycerin–water mixture (Edwards and Fauve, 1993) in contact with air. They considered the glycerin–water mixture to be a layer of finite height $h = 0.29$ cm, in contact with a layer of air of infinite height. In their figure 3 (our figure 15.8) they plotted the experimental data for the critical wavelength λ_c and amplitude f_c as functions of forcing frequency. The solid and dashed curves are obtained from the FHS and from the model with finite depth corrections, respectively. They noted, however, that the values for the surface tension γ and the viscosity ν were chosen so as to best fit the FHS to the experimental data. This led to values $\gamma = 67.6 \times 10^{-3}$ N/m and $\nu = 1.02 \times 10^{-4}$ m^2/sec, which are in good agreement with the corresponding values given in the literature for the mixture composed of 88% (by weight) glycerol and 12% water, at a temperature of 23 °C. With these values, both the model and the FHS agree reasonably well with the experimentally measured wavelengths. They noted that "... the experimentally measured amplitudes agree quite well with the FHS over the entire frequency range, and not at all

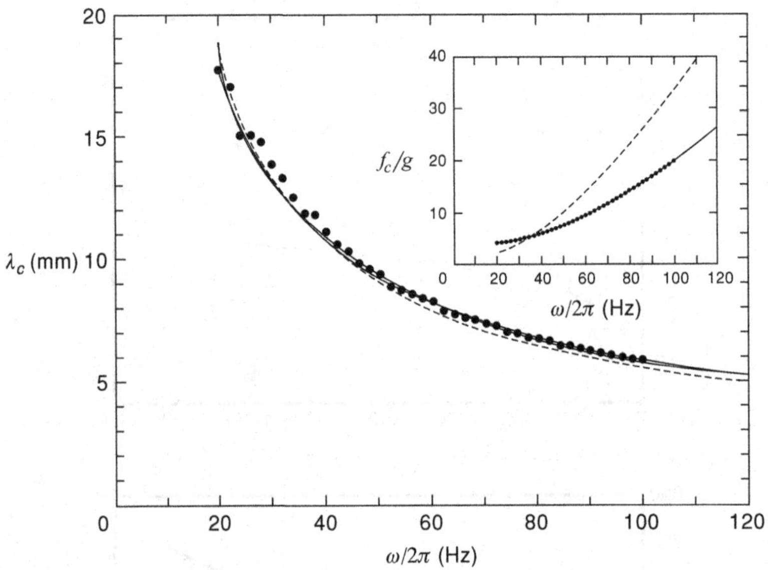

Figure 15.8. Dispersion relation for glycerin–water mixture in contact with air at atmospheric pressure. Fitting the experimental data (Edwards and Fauve, 1993) with the results of the FHS (solid curves) leads to $\gamma = 67.6 \times 10^{-3}$ N/m. Inset: Fitting of the experimental data for the stability threshold leads to $\nu = 1.02 \times 10^{-4}$ m^2/sec.

with the model. It is impossible to improve the fit of the critical amplitudes to the model by varying γ and ν."

The model results shown in the inset for f_c/g in figure 15.8 are for the dissipative approximation VCVPF ($N = 4$).

In figure 15.9 we compare the ES with the irrotational approximation for $N = 4$ and $N = 2$. It is apparent that the fit of the critical amplitudes to the model with $N = 2$ is rather good.

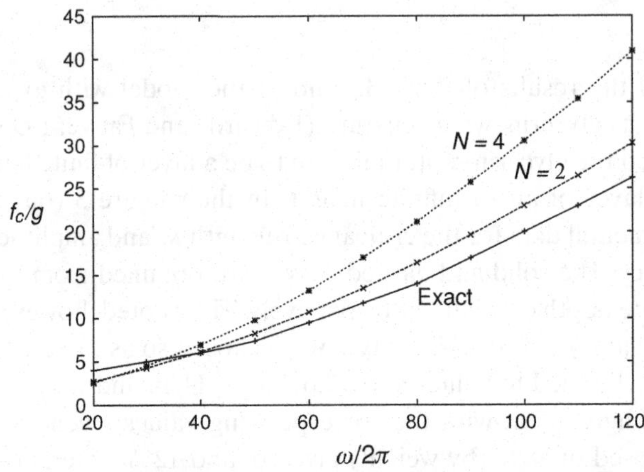

Figure 15.9. The values of f_c/g vs. $\omega/2\pi$ for VPF ($N = 2$) and VCVPF ($N = 4$) are compared with the "exact solution," FHS of KT (1994), using the values of λ_c given in figure 15.8. VPF is closer to the ES than VCVPF. $\rho = 1.1848$ g/cm^3, $h = 0.29$ cm, $\nu = 1.02$ cm^2/sec, $\gamma = 67.6$ dyn/cm.

15.8 Bifurcation of Faraday waves in a nearly square container

This problem has been studied by Feng and Sethna (1989), who used techniques of the theory of dynamical systems. Their study is of interest here because it is based on the nonlinear equations governing the potential flow of an inviscid fluid and can be updated to include the irrotational effects of viscosity. Their equations are the same as those given by BU. They look for a harmonic potential ϕ satisfying $\mathbf{n} \cdot \nabla\phi = 0$ on sidewalls and the container bottom, such that

$$\frac{\partial \phi}{\partial z} = \frac{\partial \eta}{\partial t} + \frac{\partial \phi}{\partial x}\frac{\partial \eta}{\partial x} + \frac{\partial \phi}{\partial y}\frac{\partial \eta}{\partial y}, \tag{15.8.1}$$

$$\frac{\gamma}{\rho}(\kappa_1 + \kappa_2) + \frac{\partial \phi}{\partial t} + \frac{1}{2}|\nabla\phi|^2 - (g - f\cos\omega t)\,\eta = 0. \tag{15.8.2}$$

Surface tension $\gamma = 0$ in the analysis of Feng and Sethna (1989). In our studies of purely irrotational effects of viscosity in the linear problem, we found that the less damped of the irrotational studies gave better results; VPF was closer to experiments and exact analysis than were VCVPF and the DM. To implement VPF in the nonlinear problem, we add $2\mu\mathbf{n} \cdot (\nabla \otimes \nabla\phi) \cdot \mathbf{n}$ to the left-hand side of (15.8.2)

It would be of interest to see how the addition of surface tension and viscosity alter the agreements between theory and experiment reported by Miles and Henderson (1990):

> Feng & Sethna (1989) and Simonelli & Gollub (1989) have measured the stability boundaries for square and almost square cylinders. Feng & Sethna's experiments confirm some, and do not qualitatively contradict any, of their theoretical predictions. Simonelli & Gollub's experiments in square containers are in qualitative accord with Feng & Sethna's theory for $|\beta_1 - \beta_2| \ll 1$. However, their experiments reveal a much richer structure than those of either Ciliberto & Gollub (1985) or Feng & Sethna. In particular, they find that the transition between the various states may be hysteretic and associate this with a subcritical bifurcation. For an almost square container, their experiments appear to be compatible with Feng & Sethna's theory, although direct comparison is difficult owing to different choices of control parameters. In particular, they find that the parameter space includes a region of time-dependent behavior but no region of a stable superposition of the two modes. Feng & Sethna do find the stable coexistence of the pure modes. In addition, they report slow rotational (about a vertical axis) motions not observed by Simonelli & Gollub. Experiments on standing waves in square cylinders also have been reported by Douady & Fauve (1988).

15.9 Conclusion

We developed two purely irrotational theories for the effects of viscosity on Faraday waves. In both theories the velocity is computed from the potential, and the viscous term in the normal stress balance at the free surface is evaluated on potential flow. In one theory, called VPF, the pressure is given by the Bernoulli equation; it is the same pressure as would be computed for an inviscid fluid. The second irrotational theory, called VCVPF, is the same as the first except for the introduction of an additional pressure generated to remove the unphysical irrotational shear stress from the energy balance. The first theory leads to an amplitude equation with a damping coefficient proportional to 2ν. The second theory leads to the same amplitude equation except that the damping coefficient is proportional to 4ν. We show that the VCVPF theory with damping coefficcient 4ν is identical to the well-known dissipation theory in which no pressure, inviscid or viscous, appears. We show then that the dissipation theory is identical to the damped theory that was introduced by KT (1994), who followed a heuristic argument. This theory, equivalent to VCVPF, with damping 4ν has been universally regarded as the correct irrotational approximation for viscous damping for small ν. Here, we show that both these ideas are not correct; the VPF theory with damping equal to 2ν is a better approximation, and the approximation is not restricted to small viscosities.

16

Stability of a liquid jet into incompressible gases and liquids

In this chapter we carry out an analysis of the stability of a liquid jet into a gas or another fluid by using VPF. This instability may be driven by KH instability that is due to a velocity difference and a neck-down that is due to capillary instability. KH instabilities are driven by pressures generated by a dynamically active ambient flow, gas or liquid. On the other hand, capillary instability can occur in a vacuum; the ambient can be neglected. KH instability is included by a discontinuity of the velocity at a two-fluid interface. This discontinuity is inconsistent with the no-slip condition for Navier–Stokes studies of viscous fluids, but is consistent with the theory of potential flow of a viscous fluid. We start our study with an analysis of capillary instability.

16.1 Capillary instability of a liquid cylinder in another fluid

The study of this problem is especially valuable because it can be solved exactly and was solved by Tomotika (1935). This solution allows one to compute the effects of vorticity generated by the no-slip condition. The ES can be compared with irrotational solutions of the same problem. One effect of viscosity on the irrotational motion may be introduced by evaluation of the viscous normal stress at the liquid–liquid interface on the irrotational motions. In a second approximation, the explicit effects of the discontinuity of the shear stress and tangential component of velocity that cannot be resolved pointwise in irrotational flows can be removed in the mean from the energy equation by the selection of two viscous corrections of the irrotational pressure. We include the irrotational stress and pressure correction in the normal stress balance and compare the computed growth rates with the growth rates of the exact viscous flow solution. The agreement is excellent when one of the liquids is a gas; for two viscous liquids, the agreement is good to reasonable for the maximum growth rates but poor for long waves. Calculations show that good agreement is obtained when the vorticity is relatively small or the irrotational part is dominant in the exact viscous solution. We show that the irrotational viscous flow with pressure corrections gives rise to exactly the same dispersion relation as that of the DM in which no pressure at all is required and the viscous effect is accounted for by evaluation of the viscous dissipation by use of the irrotational flow.

16.1.1 Introduction

A liquid thread of mean radius R immersed in another liquid is subject to capillary instability. The capillary collapse can be described as a neck-down that is due to the

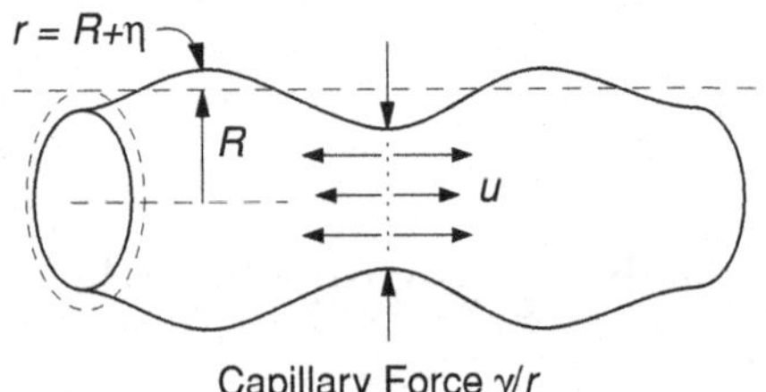

Figure 16.1. Capillary instability. The force γ/r drives fluid away from the throat, leading to collapse.

surface tension γ in which the liquid is ejected from the throat of the neck, as seen in figure 16.1. Capillary instability is responsible for drop formation in applications such as ink-jet printing, fiber spinning, and silicon chip technology.

The dynamical theory of instability of a long cylindrical column of liquid of radius R under the action of capillary force was given by Rayleigh (1879) following earlier work by Plateau (1873), who showed that a long cylinder of liquid is unstable to disturbances with wavelengths greater than $2\pi R$. Rayleigh showed that the effect of inertia is such that the wavelength λ corresponding to the mode of maximum instability is $\lambda = 4.51 \times 2R$, exceeding considerably the circumference of the cylinder. The idea that the wavelength associated with the fastest-growing growth rate would become dominant and be observed in practice was first put forward by Rayleigh (1879). The analysis of Rayleigh is based on potential flow of an inviscid liquid, neglecting the effect of the outside fluid. An attempt to account for viscous effects was made by Rayleigh (1892), again neglecting the effect of the surrounding fluid. One of the effects considered is meant to account for the forward motion of an inviscid fluid with a resistance proportional to velocity. The effect of viscosity is treated in the special case in which the viscosity is so great that inertia may be neglected. He showed that the wavelength for maximum growth is very large, strictly infinite. Weber (1931) extended Rayleigh's theory by considering an effect of viscosity and that of surrounding air on the stability of a columnar jet. The effect of viscosity on the stability of a liquid cylinder when the surrounding fluid is neglected and on a hollow (dynamically passive) cylinder in a viscous liquid was treated briefly by Chandrasekhar (1961). Eggers (1997) has given a comprehensive review of nonlinear dynamics and breakup of free-surface flows.

Tomotika (1935) studied capillary instability and gave an exact normal-mode solution of the linearized Navier–Stokes equations. Funada and Joseph (2002) analyzed the same problem by using VPF. Wang, Joseph, and Funada (2005a) considered capillary instability in cases in which one liquid is viscous and the other is a gas of negligible density and viscosity. They included the pressure correction in the normal stress balance at the free surface and showed that the growth rates computed with VCVPF are almost indistinguishable from the exact solution.

Here, we extend the VCVPF analysis to cases involving the interface of two viscous fluids. The formulation for the pressure correction is derived and used to compute growth rates for capillary instability of two viscous fluids. The computed values of the maximum growth rate and the associated wavenumber computed from VCVPF are close to those from the ES; but the growth rates at small wavenumbers are not in good agreement.

16.1.2 Linearized equations governing capillary instability

Consider the stability of a liquid cylinder of radius R with viscosity μ_l and density ρ_l surrounded by another fluid with viscosity μ_a and density ρ_a under capillary forces generated by interfacial tension γ. Note that we use the subscript "l" for the inside fluid and "a" for the outside fluid. The analysis is done in cylindrical coordinates (r, θ, z), and only axisymmetric disturbances independent of θ are considered. The linearized Navier–Stokes equations and interfacial conditions are made dimensionless with the following scales:

$$[\text{length, velocity, time, pressure}] = [D, U, D/U, p_0],$$

where D is the diameter of the liquid cylinder, $U = \sqrt{\gamma/(\rho_l D)}$, $p_0 = \rho_l U^2 = \gamma/D$. The three dimensionless parameters controlling the solution are $m = \mu_a/\mu_l$, $l = \rho_a/\rho_l$, and a Reynolds number $J = VD\rho_l/\mu_l = Oh^2$, where $V = \gamma/\mu_l$ and Oh is the Ohnesorge number. The governing equations are

$$\frac{\partial u_l}{\partial r} + \frac{u_l}{r} + \frac{\partial w_l}{\partial z} = 0, \tag{16.1.1}$$

$$\frac{\partial u_l}{\partial t} = -\frac{\partial p_l}{\partial r} + \frac{1}{\sqrt{J}}\left(\nabla^2 u_l - \frac{u_l}{r^2}\right), \quad \frac{\partial w_l}{\partial t} = -\frac{\partial p_l}{\partial z} + \frac{1}{\sqrt{J}}\nabla^2 w_l, \tag{16.1.2}$$

$$\frac{\partial u_a}{\partial r} + \frac{u_a}{r} + \frac{\partial w_a}{\partial z} = 0, \tag{16.1.3}$$

$$l\frac{\partial u_a}{\partial t} = -\frac{\partial p_a}{\partial r} + \frac{m}{\sqrt{J}}\left(\nabla^2 u_a - \frac{u_a}{r^2}\right), \quad l\frac{\partial w_a}{\partial t} = -\frac{\partial p_a}{\partial z} + \frac{m}{\sqrt{J}}\nabla^2 w_a, \tag{16.1.4}$$

with $\nabla^2 = \frac{\partial^2}{\partial r^2} + \frac{1}{r}\frac{\partial}{\partial r} + \frac{\partial^2}{\partial z^2}$. The kinematic condition at the interface $r = 1/2 + \eta$ (where η is the varicose displacement) is given by

$$\frac{\partial \eta}{\partial t} = u_l, \quad \frac{\partial \eta}{\partial t} = u_a. \tag{16.1.5}$$

The normal stress balance at the interface is given by

$$p_a - p_l + \frac{2}{\sqrt{J}}\frac{\partial u_l}{\partial r} - \frac{2m}{\sqrt{J}}\frac{\partial u_a}{\partial r} = \frac{\partial^2 \eta}{\partial z^2} + \frac{\eta}{R^2}. \tag{16.1.6}$$

The tangential stress balance at the interface is given by

$$\left(\frac{\partial u_l}{\partial z} + \frac{\partial w_l}{\partial r}\right) = m\left(\frac{\partial u_a}{\partial z} + \frac{\partial w_a}{\partial r}\right). \tag{16.1.7}$$

The continuity of the velocity at the interface requires

$$u_l = u_a, \tag{16.1.8}$$

$$w_l = w_a. \tag{16.1.9}$$

16.1.3 Fully viscous flow analysis

Tomotika (1935) gave a normal-mode solution to the linearized governing equations. This is an ES that satisfies all the four interfacial conditions in (16.1.6)–(16.1.9). He expressed the velocities with a stream function $\psi(r, z, t)$,

$$u = \frac{1}{r}\frac{\partial\psi}{\partial z}, \quad w = -\frac{1}{r}\frac{\partial\psi}{\partial r}, \tag{16.1.10}$$

and the basic variables in normal modes:

$$\psi_l = [A_1 r I_1(kr) + A_2 r I_1(k_l r)]\exp(\sigma t + ikz), \tag{16.1.11}$$

$$\psi_a = [B_1 r K_1(kr) + B_2 r K_1(k_a r)]\exp(\sigma t + ikz), \tag{16.1.12}$$

$$\eta = H\exp(\sigma t + ikz), \tag{16.1.13}$$

where σ is the complex growth rate and k is the wavenumber; the modified Bessel functions of the first order are denoted by I_1 for the first kind and K_1 for the second kind. Substitution of (16.1.11)–(16.1.13) into (16.1.6)–(16.1.9) leads to the solvability condition, which is given as the dispersion relation:

$$\begin{vmatrix} I_1(kR) & I_1(k_l R) & K_1(kR) & K_1(k_a R) \\ kI_0(kR) & k_l I_0(k_l R) & -kK_0(kR) & -k_a K_0(k_a R) \\ 2k^2 I_1(kR) & (k^2+k_l^2)\, I_1(k_l R) & 2mk^2 K_1(kR) & m\,(k^2+k_a^2)\, K_1(k_a R) \\ F_1 & F_2 & F_3 & F_4 \end{vmatrix} = 0, \tag{16.1.14}$$

where

$$F_1 = i\sigma I_0(kR) + 2i\frac{k^2}{\sqrt{J}}\left[\frac{\mathrm{d}I_1(kR)}{\mathrm{d}(kR)}\right] - \left(\frac{1}{R^2} - k^2\right) i\frac{k}{\sigma} I_1(kR), \tag{16.1.15}$$

$$F_2 = 2i\frac{kk_l}{\sqrt{J}}\left[\frac{\mathrm{d}I_1(k_l R)}{\mathrm{d}(k_l R)}\right] - \left(\frac{1}{R^2} - k^2\right) i\frac{k}{\sigma} I_1(k_l R), \tag{16.1.16}$$

$$F_3 = -il\sigma K_0(kR) + 2i\frac{mk^2}{\sqrt{J}}\left[\frac{\mathrm{d}K_1(kR)}{\mathrm{d}(kR)}\right], \quad F_4 = 2i\frac{mkk_a}{\sqrt{J}}\left[\frac{\mathrm{d}K_1(k_a R)}{\mathrm{d}(k_a R)}\right], \tag{16.1.17}$$

with

$$k_l = \sqrt{k^2 + \sqrt{J}\sigma}, \quad k_a = \sqrt{k^2 + \frac{l}{m}\sqrt{J}\sigma}. \tag{16.1.18}$$

16.1.4 Viscous potential flow analysis

The potential flow solution is given by $\mathbf{u} = \nabla\phi$, $\nabla^2\phi = 0$, where ϕ is the velocity potential. Normal stress balance (16.1.6) and normal velocity continuity (16.1.8) are satisfied; shear stress and tangential velocity conditions (16.1.7) and (16.1.9) cannot be enforced. This problem was first considered by Funada and Joseph (2002). Here, we follow Wang *et al.* (2005a). The potential solution can be expressed as

$$\psi_l = A_1 r I_1(kr)\exp(\sigma t + ikz), \tag{16.1.19}$$

$$\psi_a = B_1 r K_1(kr)\exp(\sigma t + ikz), \tag{16.1.20}$$

$$\eta = H\exp(\sigma t + ikz), \tag{16.1.21}$$

for which the dispersion relation is given by

$$(\alpha_l + l\alpha_a)\sigma^2 + \frac{2k^2}{\sqrt{J}}(\beta_l + m\beta_a)\sigma = \left(\frac{1}{R^2} - k^2\right)k, \tag{16.1.22}$$

with

$$\alpha_l = \frac{I_0(kR)}{I_1(kR)}, \quad \alpha_a = \frac{K_0(kR)}{K_1(kR)}, \quad \beta_l = \alpha_l - \frac{1}{kR}, \quad \beta_a = \alpha_a + \frac{1}{kR}. \tag{16.1.23}$$

Solving (16.1.22), we obtain

$$\sigma = -\frac{k^2(\beta_l + m\beta_a)}{\sqrt{J}(\alpha_l + l\alpha_a)} \pm \sqrt{\left[\frac{k^2(\beta_l + m\beta_a)}{\sqrt{J}(\alpha_l + l\alpha_a)}\right]^2 + \left(\frac{1}{R^2} - k^2\right)\frac{k}{\alpha_l + l\alpha_a}}. \tag{16.1.24}$$

Thus instability arises in $0 < kR < 1$, for which the dimensionless critical wavenumber $k_c = 1/R = 2$. When $\sqrt{J} \to \infty$, (16.1.24) reduces to

$$\sigma = \pm\sqrt{\left(\frac{1}{R^2} - k^2\right)\frac{k}{\alpha_l + l\alpha_a}}, \tag{16.1.25}$$

which is just the dispersion relation in IPF; the same dispersion relation was obtained by Christiansen and Hixson (1957).

16.1.5 Pressure correction for viscous potential flow

Joseph and Wang (2004) derived a viscous correction for the irrotational pressure at free surfaces of steady flows, which is induced by the discrepancy between the nonzero irrotational shear stress and the zero-shear-stress condition at free surfaces. In the VPF analysis of capillary instability, the interface between two viscous fluids is involved and the two potential flows are unsteady. We derive the pressure correction for capillary instability from the basic mechanical-energy equation. The energy analysis for the two-fluid problem was given first by Wang, Joseph, and Funada (2005c).

If we ignore the small deformation η in the linear problem, we have $\mathbf{n}_1 = \mathbf{e}_r$ as the outward normal at the interface for the inside fluid; $\mathbf{n}_2 = -\mathbf{n}_1$ is the outward normal for the outside fluid; $\mathbf{t} = \mathbf{e}_z$ is the unit tangential vector. We use the superscripts i for irrotational and v for viscous. The normal and shear parts of the viscous stress are represented by τ^n and τ^s, respectively.

The velocities and stresses are evaluated with the potentials, which are expressed by stream functions (16.1.19) and (16.1.20). The mechanical-energy equations for the outside and inside fluids are, respectively,

$$\begin{aligned}\frac{\mathrm{d}}{\mathrm{d}t}\int_{V_a}\frac{\rho_a}{2}|\mathbf{u}_a|^2\,\mathrm{d}V &= \int_A[\mathbf{u}_a\cdot\mathbf{T}_a\cdot\mathbf{n}_2]\,\mathrm{d}A - \int_{V_a}2\mu_a\mathbf{D}_a:\mathbf{D}_a\mathrm{d}V \\ &= -\int_A\left[\mathbf{u}_a\cdot\mathbf{n}_1\left(-p_a^i + \tau_a^n\right) + \mathbf{u}_a\cdot\mathbf{t}\tau_a^s\right]\mathrm{d}A - \int_{V_a}2\mu_a\mathbf{D}_a:\mathbf{D}_a\mathrm{d}V,\end{aligned} \tag{16.1.26}$$

$$\begin{aligned}\frac{\mathrm{d}}{\mathrm{d}t}\int_{V_l}\frac{\rho_l}{2}|\mathbf{u}_l|^2\,\mathrm{d}V &= \int_A[\mathbf{u}_l\cdot\mathbf{T}_l\cdot\mathbf{n}_1]\,\mathrm{d}A - \int_{V_l}2\mu_l\mathbf{D}_l:\mathbf{D}_l\mathrm{d}V \\ &= \int_A\left[\mathbf{u}_l\cdot\mathbf{n}_1\left(-p_l^i + \tau_l^n\right) + \mathbf{u}_l\cdot\mathbf{t}\tau_l^s\right]\mathrm{d}A - \int_{V_l}2\mu_l\mathbf{D}_l:\mathbf{D}_l\mathrm{d}V.\end{aligned} \tag{16.1.27}$$

With the continuity of the normal velocity,

$$\mathbf{u}_a \cdot \mathbf{n}_1 = \mathbf{u}_l \cdot \mathbf{n}_1 = u_n, \tag{16.1.28}$$

the sum of (16.1.26) and (16.1.27) can be written as

$$\begin{aligned} \frac{\mathrm{d}}{\mathrm{d}t}\int_{V_a}\frac{\rho_a}{2}|\mathbf{u}_a|^2\,\mathrm{d}V + \frac{\mathrm{d}}{\mathrm{d}t}\int_{V_l}\frac{\rho_l}{2}|\mathbf{u}_l|^2\,\mathrm{d}V = \int_A \big[u_n\left(-p_l^i + \tau_l^n + p_a^i - \tau_a^n\right) \\ + \mathbf{u}_l \cdot \mathbf{t}\tau_l^s - \mathbf{u}_a \cdot \mathbf{t}\tau_a^s\big]\,\mathrm{d}A - \int_{V_a} 2\mu_a \mathbf{D}_a : \mathbf{D}_a \mathrm{d}V - \int_{V_l} 2\mu_l \mathbf{D}_l : \mathbf{D}_l \mathrm{d}V. \end{aligned} \tag{16.1.29}$$

Now consider the boundary-layer approximation of VPF. We propose two pressure corrections, p_l^v and p_a^v, for the inside and outside potential flows, respectively, together with the continuity conditions

$$\tau_a^s = \tau_l^s = \tau^s, \quad \mathbf{u}_a \cdot \mathbf{t} = \mathbf{u}_l \cdot \mathbf{t} = u_s. \tag{16.1.30}$$

We assume that the boundary-layer approximation has a negligible effect on the flow in the bulk liquid but it changes the pressure and continuity conditions at the interface. Hence the mechanical-energy equations become

$$\frac{\mathrm{d}}{\mathrm{d}t}\int_{V_l}\frac{\rho_l}{2}|\mathbf{u}_l|^2\,\mathrm{d}V = \int_A \left[u_n\left(-p_l^i - p_l^v + \tau_l^n\right) + u_s\tau^s\right]\mathrm{d}A - \int_{V_l} 2\mu_l \mathbf{D}_l : \mathbf{D}_l \mathrm{d}V, \tag{16.1.31}$$

$$\frac{\mathrm{d}}{\mathrm{d}t}\int_{V_a}\frac{\rho_a}{2}|\mathbf{u}_a|^2\,\mathrm{d}V = -\int_A \left[u_n\left(-p_a^i - p_a^v + \tau_a^n\right) + u_s\tau^s\right]\mathrm{d}A - \int_{V_a} 2\mu_a \mathbf{D}_a : \mathbf{D}_a \mathrm{d}V. \tag{16.1.32}$$

The sum of (16.1.31) and (16.1.32) can be written as

$$\begin{aligned} \frac{\mathrm{d}}{\mathrm{d}t}\int_{V_a}\frac{\rho_a}{2}|\mathbf{u}_a|^2\,\mathrm{d}V + \frac{\mathrm{d}}{\mathrm{d}t}\int_{V_l}\frac{\rho_l}{2}|\mathbf{u}_l|^2\,\mathrm{d}V = \int_A \left[u_n\left(-p_l^i - p_l^v + \tau_l^n + p_a^i + p_a^v - \tau_a^n\right)\right]\mathrm{d}A \\ - \int_{V_a} 2\mu_a \mathbf{D}_a : \mathbf{D}_a \mathrm{d}V - \int_{V_l} 2\mu_l \mathbf{D}_l : \mathbf{D}_l \mathrm{d}V. \end{aligned} \tag{16.1.33}$$

Comparing (16.1.29) and (16.1.33), we obtain an equation that relates the pressure corrections to the uncompensated irrotational shear stresses:

$$\int_A u_n\left(-p_l^v + p_a^v\right)\mathrm{d}A = \int_A \left(\mathbf{u}_l \cdot \mathbf{t}\tau_l^s - \mathbf{u}_a \cdot \mathbf{t}\tau_a^s\right)\mathrm{d}A. \tag{16.1.34}$$

Joseph and Wang (2004) showed that, in linearized problems, the governing equation for the pressure corrections is

$$\nabla^2 p^v = 0. \tag{16.1.35}$$

Solving equation (16.1.35), we obtain the two pressure corrections:

$$-p_l^v = \sum_{j=0}^{\infty} C_j' i I_0\left(\frac{2\pi}{\lambda} jr\right)\exp\left(\sigma t + i\frac{2\pi}{\lambda} jz\right), \tag{16.1.36}$$

$$-p_a^v = \sum_{j=0}^{\infty} D_j' i K_0\left(\frac{2\pi}{\lambda} jr\right)\exp\left(\sigma t + i\frac{2\pi}{\lambda} jz\right), \tag{16.1.37}$$

where C_j' and D_j' are constants to be determined, j is an integer, and λ is the period in the z direction. Suppose $2\pi j_0/\lambda = k$, $C_{j_0}' = C_k$, and $D_{j_0}' = D_k$; then the two pressure corrections can be written as

$$-p_l^v = C_k i I_0(kr)\exp(\sigma t + ikz) + \sum_{j\neq j_0} C_j' i I_0\left(\frac{2\pi}{\lambda} jr\right)\exp\left(\sigma t + i\frac{2\pi}{\lambda} jz\right), \tag{16.1.38}$$

$$-p_a^v = D_k i K_0(kr)\exp(\sigma t + ikz) + \sum_{j\neq j_0} D_j' i K_0\left(\frac{2\pi}{\lambda} jr\right)\exp\left(\sigma t + i\frac{2\pi}{\lambda} jz\right). \tag{16.1.39}$$

With the pressure corrections, the normal stress balance has the following form:

$$p_a^i + p_a^v - p_l^i - p_l^v + \frac{2}{\sqrt{J}}\frac{\partial u_l}{\partial r} - \frac{2m}{\sqrt{J}}\frac{\partial u_a}{\partial r} = \frac{\partial^2 \eta}{\partial z^2} + \frac{\eta}{R^2}, \tag{16.1.40}$$

which gives rise to

$$\begin{aligned}
&\left\{\begin{aligned} &l B_1 K_0(kR)\sigma - D_k K_0(kR) + A_1\sigma I_0(kR) + C_k I_0(kR) \\ &+\frac{2k^2}{\sqrt{J}} A_1\left[I_0(kR) - \frac{I_1(kR)}{kR}\right] + \frac{2mk^2}{\sqrt{J}} B_1\left[K_0(kR) + \frac{K_1(kR)}{kR}\right]\end{aligned}\right\}\exp(\sigma t + ikz) \\
&+\sum_{j\neq j_0}\left[C_j' I_0\left(\frac{2\pi}{\lambda} jR\right) - D_j' K_0\left(\frac{2\pi}{\lambda} jR\right)\right]\exp\left(\sigma t + i\frac{2\pi}{\lambda} jz\right) \\
&= A_1\frac{k}{\sigma} I_1(kR)\left(\frac{1}{R^2} - k^2\right)\exp(\sigma t + ikz).
\end{aligned} \tag{16.1.41}$$

By orthogonality of the Fourier series, we obtain

$$\begin{aligned}
&l B_1 K_0(kR)\sigma - D_k K_0(kR) + A_1\sigma I_0(kR) + C_k I_0(kR) + \frac{2k^2}{\sqrt{J}} A_1\left[I_0(kR) - \frac{I_1(kR)}{kR}\right] \\
&+\frac{2mk^2}{\sqrt{J}} B_1\left[K_0(kR) + \frac{K_1(kR)}{kR}\right] = A_1\frac{k}{\sigma} I_1(kR)\left(\frac{1}{R^2} - k^2\right),
\end{aligned} \tag{16.1.42}$$

$$C_j' I_0\left(\frac{2\pi}{\lambda} jR\right) - D_j' K_0\left(\frac{2\pi}{\lambda} jR\right) = 0 \ \text{when}\ j \neq j_0. \tag{16.1.43}$$

Equation (16.1.42) replaces the normal stress balance and can be solved for the growth rate σ. However, the undetermined part, $C_k I_0(kR) - D_k K_0(kR)$, has to be computed from (16.1.34) before we can solve (16.1.42). Substitution of (16.1.38), (16.1.39), and (16.1.43) into the left-hand side of (16.1.34) gives rise to

$$\begin{aligned}
\int_A \bar{u}_n(-p_l^v + p_a^v)\,\mathrm{d}A &= \int_0^{2\pi} R\mathrm{d}\theta \int_z^{z+\lambda} \bar{u}_n(-p_l^v + p_a^v)\,\mathrm{d}z \\
&= 2\pi R\lambda\left[\bar{A}_1 C_k I_0(kR) I_1(kR) - \bar{B}_1 D_k K_0(kR) K_1(kR)\right] k\exp(\sigma + \bar{\sigma})t,
\end{aligned} \tag{16.1.44}$$

where $\bar{u}_n$ is the conjugate of u_n. The right-hand side of (16.1.34) can be evaluated:

$$\begin{aligned}
&\int_A [(\overline{\mathbf{u}_l \cdot \mathbf{t}})\tau_l^s - (\overline{\mathbf{u}_a \cdot \mathbf{t}})\tau_a^s]\,\mathrm{d}A \\
&= \frac{4\pi R\lambda}{\sqrt{J}}\left[\bar{A}_1 A_1 I_0(kR) I_1(kR) + m\bar{B}_1 B_1 K_0(kR) K_1(kR)\right] k^3\exp(\sigma + \bar{\sigma})t.
\end{aligned} \tag{16.1.45}$$

Table 16.1. *The properties of the five pairs of fluids used to study capillary instability and the controlling dimensionless parameters l, m, and J*

Case	1	2	3	4	5
Fluids	Mercury–water	Water–benzene	Glycerin–mercury	Goldensyrup–paraffin	Goldensyrup–BBoil
ρ_l (kg m^{-3})	13500	1000	1257	1400	1400
ρ_a (kg m^{-3})	1000	0.001	13500	1600	900
μ_l (kg/m sec)	0.00156	860	0.782	11.0	11.0
μ_a (kg/m sec)	0.001	0.00065	0.00156	0.0034	6.0
γ (N/m)	0.375	0.0328	0.375	0.023	0.017
$l = \rho_a/\rho_l$	0.07407	0.86	10.74	1.143	0.6429
$m = \mu_a/\mu_l$	0.6410	0.65	1.995×10^{-3}	3.091×10^{-4}	0.5455
$J = \rho_l \gamma D/\mu_l^2$	2.080×10^7	3.280×10^5	7.708	2.661×10^{-3}	1.967×10^{-3}

Combining (16.1.44) and (16.1.45), we obtain

$$\begin{aligned}&\bar{A}_1 C_k I_0(kR) I_1(kR) - \bar{B}_1 D_k K_0(kR) K_1(kR)\\&= \frac{2}{\sqrt{J}} \bar{A}_1 A_1 k^2 I_0(kR) I_1(kR) + \frac{2m}{\sqrt{J}} \bar{B}_1 B_1 k^2 K_0(kR) K_1(kR).\end{aligned} \tag{16.1.46}$$

Normal velocity continuity condition (16.1.28) leads to

$$B_1 = A_1 \frac{I_1(kR)}{K_1(kR)}. \tag{16.1.47}$$

Substitution of (16.1.47) into (16.1.46) leads to

$$C_k I_0(kR) - D_k K_0(kR) = \frac{2}{\sqrt{J}} A_1 k^2 I_0(kR) + \frac{2m}{\sqrt{J}} A_1 k^2 I_1(kR) K_0(kR)/K_1(kR). \tag{16.1.48}$$

Inserting (16.1.47) and (16.1.48) into (16.1.42), we obtain the dispersion relation:

$$(\alpha_l + l\alpha_a)\sigma^2 + \frac{2k^2}{\sqrt{J}}\left[(\alpha_l + \beta_l) + m(\alpha_a + \beta_a)\right]\sigma = \left(\frac{1}{R^2} - k^2\right)k, \tag{16.1.49}$$

where α_l, α_a, β_a and β_l are defined in (16.1.23). Solving (16.1.49), we obtain the growth rate:

$$\begin{aligned}\sigma = &-\frac{k^2\left[(\alpha_l + \beta_l) + m(\alpha_a + \beta_a)\right]}{\sqrt{J}(\alpha_l + l\alpha_a)}\\&\pm \sqrt{\left\{\frac{k^2\left[(\alpha_l + \beta_l) + m(\alpha_a + \beta_a)\right]}{\sqrt{J}(\alpha_l + l\alpha_a)}\right\}^2 + \left(\frac{1}{R^2} - k^2\right)\frac{k}{\alpha_l + l\alpha_a}}.\end{aligned} \tag{16.1.50}$$

16.1.6 Comparison of growth rates

We calculate the growth rate σ by using IPF (16.1.25), VPF (16.1.24), and VCVPF (16.1.50), and compare these results with the ES (16.1.14). We choose five pairs of fluids to study capillary instability; the properties of the fluids and controlling dimensionless parameters are listed in table 16.1.

We are essentially comparing solutions assuming irrotational flows with the ES. To better understand the potential flow approximation to the fully viscous flow, we may evaluate the vorticity,

$$\omega = \frac{\partial u_r}{\partial z} - \frac{\partial u_z}{\partial r} = \frac{\partial u}{\partial z} - \frac{\partial w}{\partial r}, \tag{16.1.51}$$

in the interior and exterior fluids from the ES. When the vorticity is great, the potential flow cannot give a satisfactory approximation.

The vorticity is

$$\omega_l = A_2 I_1(k_l r)\,(k_l^2 - k^2)\exp(\sigma t + ikz) = A_2 I_1(k_l r)\sqrt{J}\sigma \exp(\sigma t + ikz) \tag{16.1.52}$$

in the interior fluid and is

$$\omega_a = B_2 K_1(k_a r)\,(k_a^2 - k^2)\exp(\sigma t + ikz) = B_2 K_1(k_a r)\frac{l}{m}\sqrt{J}\sigma \exp(\sigma t + ikz) \tag{16.1.53}$$

in the exterior fluid. The magnitudes of ω_l and ω_a are proportional to A_2 and B_2, respectively. We normalize them by B_2 so that the two magnitudes are measured by the same scale and can be compared. The normalized magnitudes of the vorticities at the interface ($r \approx R$) are

$$\omega_l^* = -\frac{A_2}{B_2} I_1(k_l R)\sqrt{J}\sigma, \quad \omega_a^* = K_1(k_a R)\frac{l}{m}\sqrt{J}\sigma. \tag{16.1.54}$$

Note that we add a minus sign for ω_l^*. The reason is that the vorticity vectors in the interior and exterior fluids are in opposite directions, leading to vorticities of opposite signs. By adding a minus sign for ω_l^*, we obtain the absolute value of the vorticity in the interior fluid. To compute (16.1.54), we need to know the value of A_2/B_2. This can be achieved by manipulation of dispersion relation (16.1.14):

$$\begin{bmatrix} I_1(kR) & I_1(k_l R) & K_1(kR) \\ kI_0(kR) & k_l I_0(k_l R) & -kK_0(kR) \\ 2k^2 I_1(kR) & (k^2+k_l^2) I_1(k_l R) & 2mk^2 K_1(kR) \end{bmatrix} \begin{bmatrix} A_1/B_2 \\ A_2/B_2 \\ B_1/B_2 \end{bmatrix} = \begin{bmatrix} -K_1(k_a R) \\ k_a K_0(k_a R) \\ -m(k^2+k_a^2) K_1(k_a R) \end{bmatrix}, \tag{16.1.55}$$

from which we can solve for A_1/B_2, A_2/B_2, and B_1/B_2.

We note that stream functions (16.1.11) and (16.1.12) in the ES can be divided into an irrotational part and a rotational part:

$$\psi_l^i = A_1 r I_1(kr)\exp(\sigma t + ikz), \quad \psi_l^r = A_2 r I_1(k_l r)\exp(\sigma t + ikz); \tag{16.1.56}$$

$$\psi_a^i = B_1 r K_1(kr)\exp(\sigma t + ikz), \quad \psi_a^r = B_2 r K_1(k_a r)\exp(\sigma t + ikz). \tag{16.1.57}$$

The irrotational parts are exactly the potential flow solution, whereas the vorticities are solely determined by the rotational parts. When the irrotational parts dominate, potential flows can give a good approximation to the ES; when the rotational parts are important, the approximation cannot be satisfactory. We define two ratios of the irrotational part to the rotational part:

$$f_l = \left|\frac{\psi_l^i(r=R)}{\psi_l^r(r=R)}\right| = \left|\frac{A_1 I_1(kR)}{A_2 I_1(k_l R)}\right|, \tag{16.1.58}$$

$$f_a = \left|\frac{\psi_a^i(r=R)}{\psi_a^r(r=R)}\right| = \left|\frac{B_1 K_1(kR)}{B_2 K_1(k_a R)}\right|. \tag{16.1.59}$$

These two ratios characterize the relative importance of the irrotational and rotational parts at the interface. When the Reynolds number is large, we expect the values of the ratios to be high.

We present the comparison of the growth rates computed from IPF, VPF, VCVPF, and the ES in figures 16.2–16.6 for the five pairs of fluids listed in table 16.1. In each case, a growth rate vs. wavenumber plot, a vorticity vs. wavenumber plot, and a plot for the two ratios f_l and f_a vs. wavenumber are shown. The vorticity plot and the plot for the two ratios can help us to understand the agreement and disagreement between the growth rates from potential-based solutions and from the ES.

When the Reynolds number is high (figures 16.2 and 16.3), the three potential-flow-based solutions are essentially the same; they are in good agreement with the ES in the maximum growth region (see table 16.2) but deviate from the ES when $k \ll 1$. When the Reynolds number is lower (figures 16.4 and 16.5), IPF and VPF deviate from the exact solution in the maximum growth region whereas VCVPF can give almost the same maximum growth rate σ_m and the associated wavenumber k_m as the ES. However, VCVPF does not differ greatly from IPF or VPF when $k \ll 1$. Figure 16.6 shows the results for case 5 in which the Reynolds number is low and the viscosity ratio $m \sim O(1)$; VCVPF does not give the correct values of σ_m and k_m in this case.

The vorticity as a function of the wavenumber k helps us to understand the nonuniform agreement between the VCVPF results and the exact results. In cases 1, 2, and 3 (figures 16.2, 16.3, and 16.4), the magnitude of the vorticity is large when $k \ll 1$ and small in the maximum growth region. This could explain the good agreement in the maximum growth region and poor agreement when $k \ll 1$. In cases 4 and 5 (figures 16.5 and 16.6), the magnitude of the vorticity is large at almost all the values of k except when k is very close to $k_c = 2$. The distribution of the vorticity is helpful for understanding the growth-rate calculation, but a clear explanation for the nonuniform agreement is not obtained in cases 4 and 5.

We find that the values of the two ratios f_l and f_a are close to 1 when $k \ll 1$ in all the cases, indicating that the rotational parts are important for long waves even at high Reynolds number. This is consistent with our growth-rate calculation, which shows that the agreement between VCVPF and the ES is poor for long waves. In the maximum growth region (k close to 1), the values of f_l and f_a are larger than for long waves. When the Reynolds number is large, the maximum value of f_l and f_a is large, indicating that the irrotational parts dominate the solution; at the same time, we observe good agreement between VCVPF and the ES for wavenumbers near k_m. When the Reynolds number is small, the values of f_l and f_a are close to 1 in the whole range of k; at the same time, we observe that the agreement between VCVPF and the ES is poor at almost all the values of k.

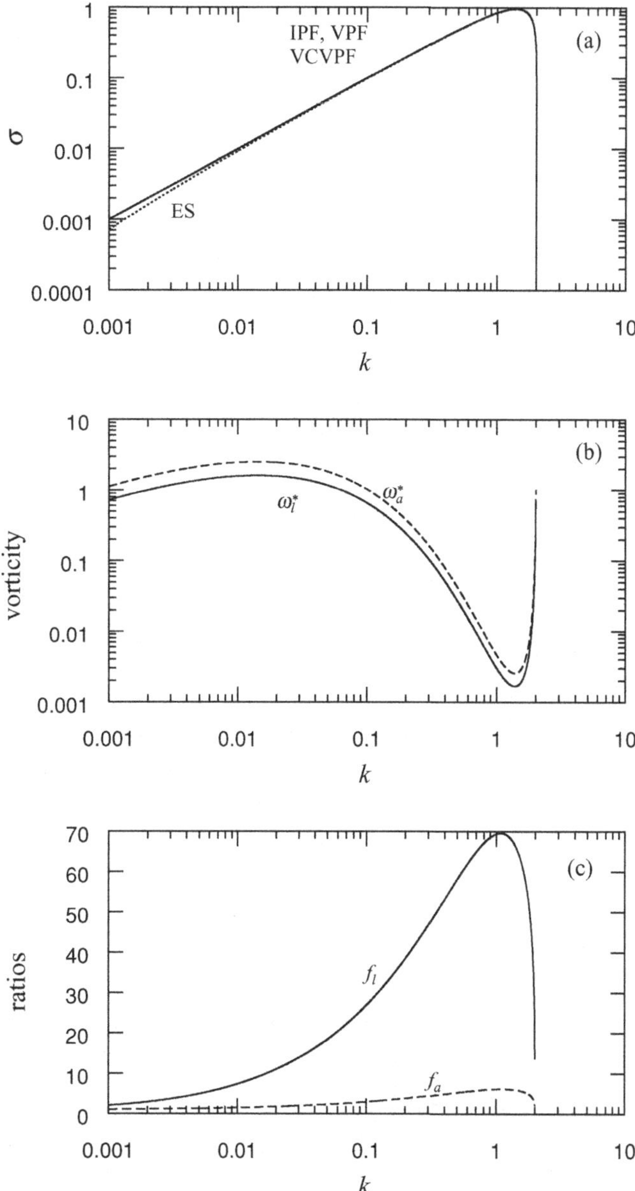

Figure 16.2. (a) The growth rate σ vs. k for case 1, mercury in water. The three potential-flow-based analyses agree with the exact solution (ES) well but deviate from it slightly when $k \ll 1$. (b) The vorticities ω_l^* and ω_a^* vs. k for case 1. The magnitude of the vorticity is large when $k \ll 1$ and small when k is about 1. (c) The two ratios f_l and f_a vs. k for case 1. The irrotational parts dominate when k is close to 1; the irrotational and rotational parts are comparable when $k \ll 1$ or $k \approx 2$. The dominance of the irrotational part in the maximum growth region is understandable because the Reynolds number is very high, 2.080×10^7. Both the vorticities and the two ratios could help us to understand the deviation of the potential-based analyses from the ES when $k \ll 1$.

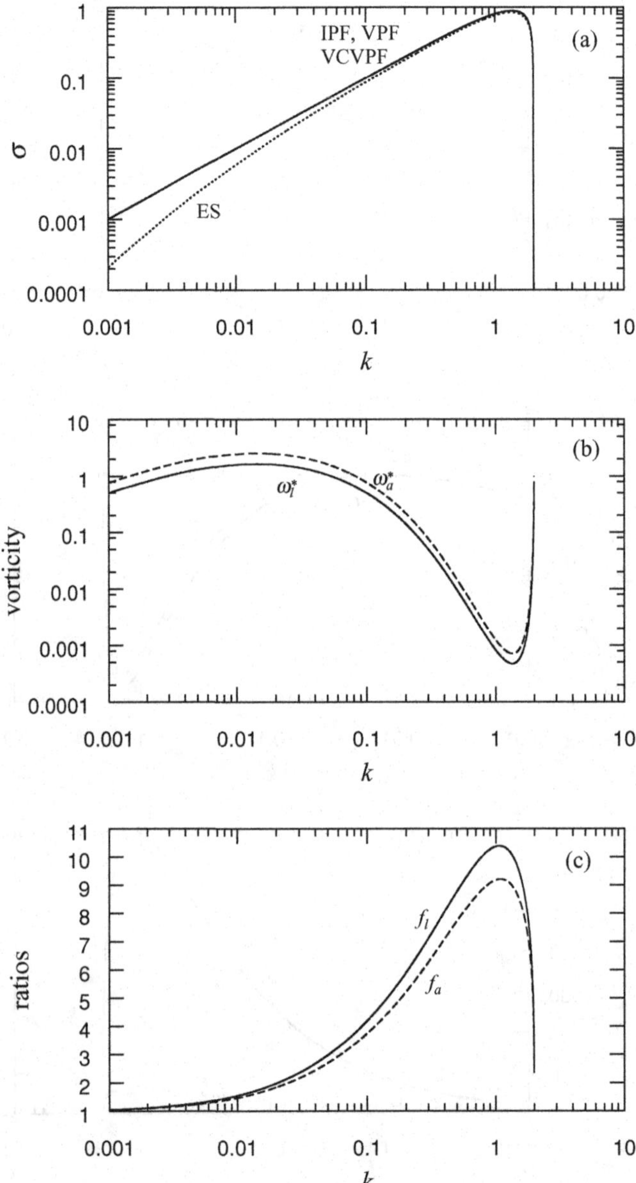

Figure 16.3. (a) The growth rate σ vs. k for case 2, water in benzene. The three potential-flow-based analyses agree with the exact solution (ES) in the maximum growth region but deviate from it considerably when k is small. (b) The vorticities ω_l^* and ω_a^* vs. k for case 2. The magnitude of the vorticity is large when $k \ll 1$ and small in the maximum growth region. (c) The two ratios f_l and f_a vs. k for case 2. The ratios are high when k is close to 1 but close to 1 when $k \ll 1$ or $k \approx 2$. The maximum value of f_l is 10.37 here, smaller than the value 66.82 in case 1. The reason is that the Reynolds number in case 2 is smaller than in case 1. Both the vorticities and the two ratios could help us to understand the good agreement in the maximum growth region and poor agreement when $k \ll 1$, as shown in (a).

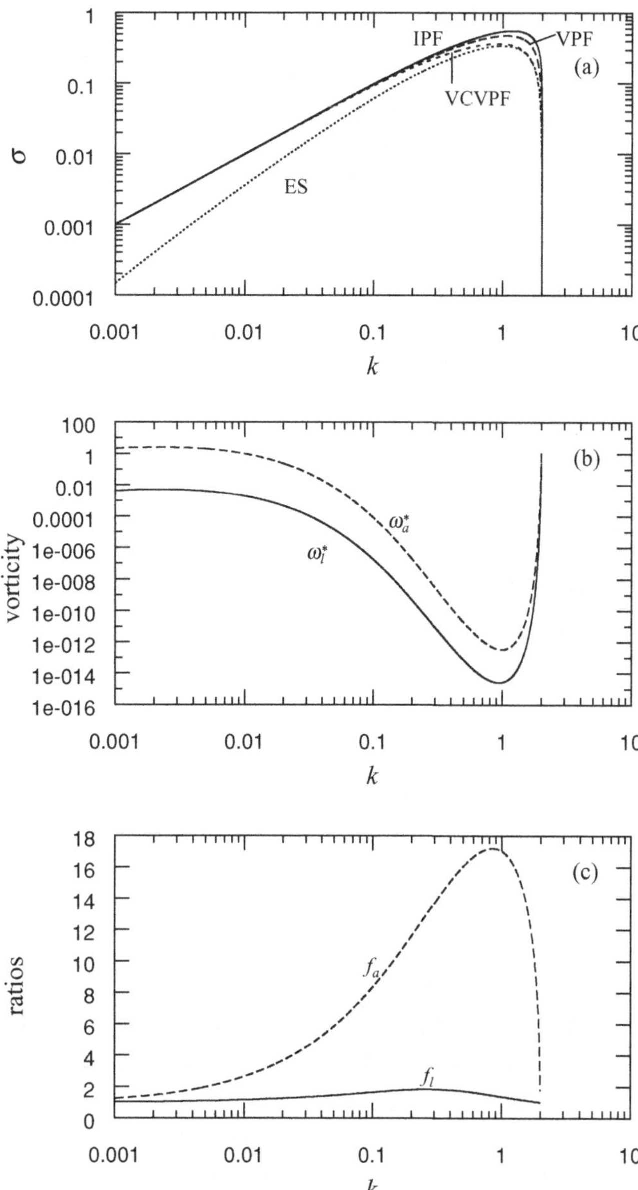

Figure 16.4. (a) The growth rate σ vs. k for case 3, glycerin in mercury. In the maximum region, IPF and VPF overestimate the growth rate whereas VCVPF gives almost the same value as the exact solution (ES). IPF, VPF, and VCVPF deviate from the exact solution considerably when $k \ll 1$. (b) The vorticities ω_l^* and ω_a^* vs. k for case 3. The magnitude of the vorticity is large when $k \ll 1$ and small in the maximum growth region. (c) The two ratios f_l and f_a vs. k for case 3. The ratios are high when k is close to 1 but close to 1 when $k \ll 1$ or $k \approx 2$. The maximum value of f_l is 1.83, much smaller than in case 1 and case 2. At the same time, the Reynolds number is also much smaller than in case 1 and case 2. It is noted that the maximum value of f_a is 17.19, much larger than f_l. The reason is that the value of f_a should correspond to the Reynolds number based on ρ_a and μ_a, which is 2.08×10^7 in case 3. Both the vorticities and the two ratios could help us to understand the good agreement in the maximum growth region and poor agreement when $k \ll 1$, as shown in (a).

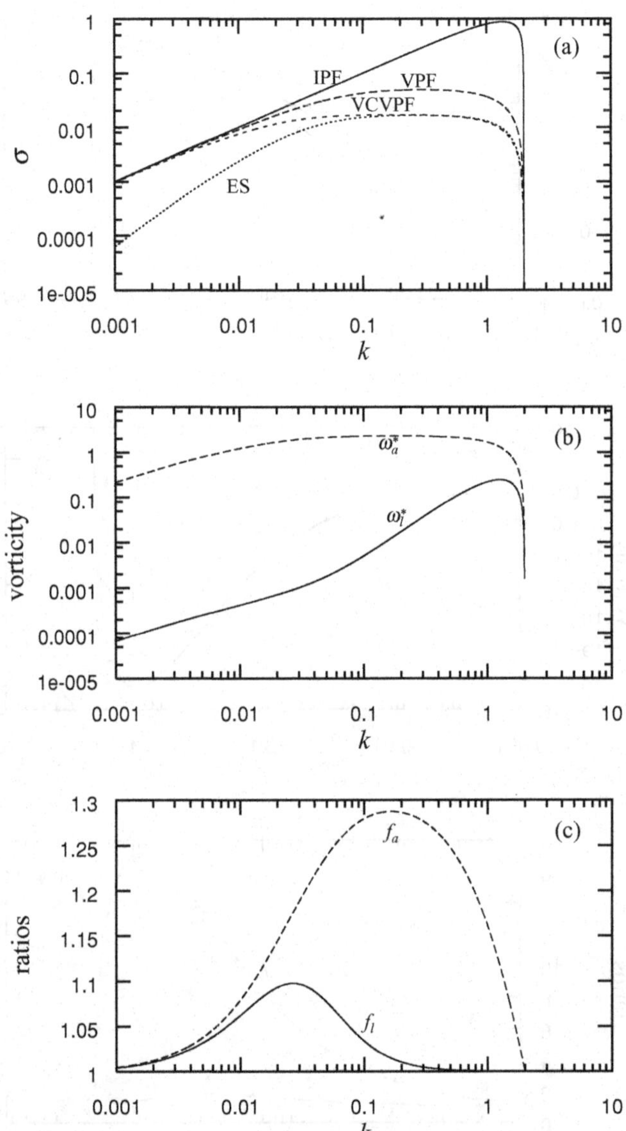

Figure 16.5. (a) The growth rate σ vs. k for case 4, goldensyrup in paraffin. IPF and VPF deviate from the exact solution (ES) considerably in the whole range of $k \leq k_c = 2$. VCVPF is still in good agreement with the exact solution in the maximum growth region. (b) The vorticities ω_l^* and ω_a^* vs. k for case 4. The magnitude of the vorticity is large at almost all the values of k except when k is very close to $k_c = 2$. (c) The two ratios f_l and f_a vs. k for case 4. The maximum value of f_l and f_a does not exceed 1.3, indicating that the rotational parts are important in the whole range of k. This could explain the deviation of IPF and VPF from the ES in the whole range. At the same time, the curve for f_a shows that the ratio is higher in the maximum growth region than in the region where $k \ll 1$ or $k \approx 2$. This may help us to understand the good agreement between VCVPF and the ES in the neighborhood of the maximum growth rate.

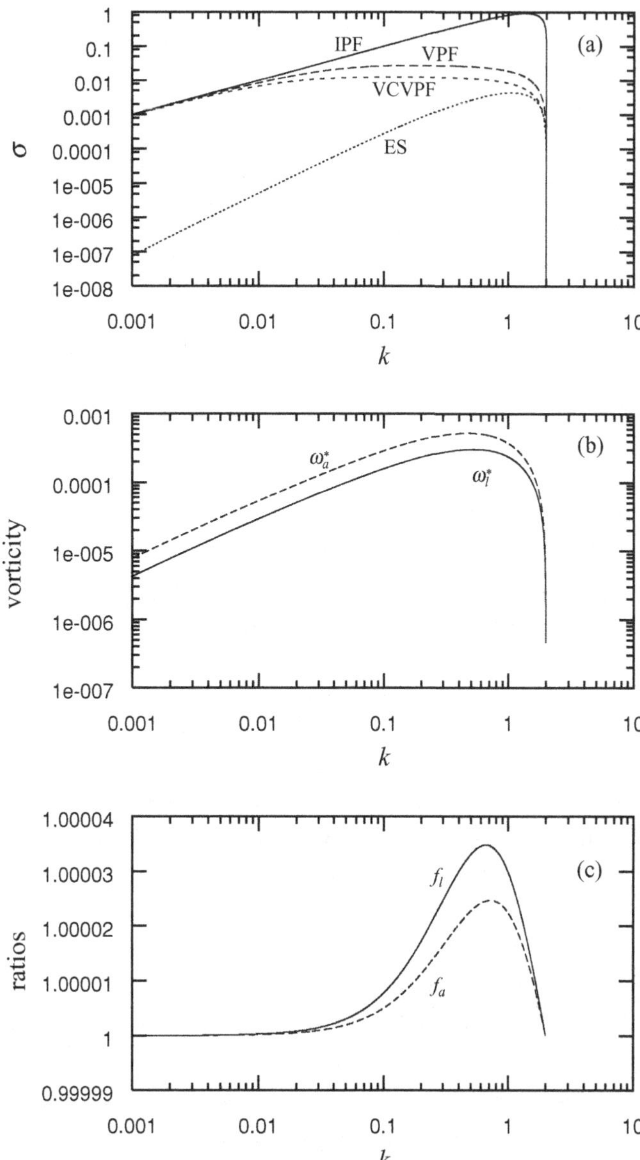

Figure 16.6. (a) The growth rate σ vs. k for case 5, goldensyrup in BBoil. The agreement among IPF, VPF, and VCVPF with the exact solution (ES) is poor at almost all the values of k. (b) The vorticities ω_l^* and ω_a^* vs. k for case 5. The magnitude of the vorticity is relatively large at almost all the values of k and becomes small only when k is very close to $k_c = 2$. (c) The two ratios f_l and f_a vs. k for case 5. The maximum value of f_l and f_a does not exceed 1.000035, indicating that the irrotational and rotational parts are almost equally important in the whole range of k. Both the vorticities and the two ratios could help to understand the poor agreement among IPF, VPF, and VCVPF with the ES shown in (a).

Table 16.2. *Data of the maximum growth rate and the associated wavenumber*

	IPF		ES	
Case	k_m	σ_m	k_m	σ_m
1	1.3909616	9.6377400×10^{-1}	1.3893986	9.5517399×10^{-1}
2	1.3585050	8.9589722×10^{-1}	1.3508895	8.4918585×10^{-1}
3	1.1897181	5.5734616×10^{-1}	9.9496394×10^{-1}	3.4140194×10^{-1}
4	1.3493716	8.7507633×10^{-1}	2.7032697×10^{-1}	1.6552208×10^{-2}
5	1.3661634	9.1303869×10^{-1}	1.0632024	4.3484116×10^{-3}

	VPF		VCVPF	
Case	k_m	σ_m	k_m	σ_m
1	1.3909616	9.6337186×10^{-1}	1.3893986	9.6289992×10^{-1}
2	1.3569785	8.9325293×10^{-1}	1.3539306	8.9017625×10^{-1}
3	1.0716030	4.7575270×10^{-1}	9.1862735×10^{-1}	3.6760629×10^{-1}
4	2.9643429×10^{-1}	4.8733702×10^{-2}	1.7693239×10^{-1}	1.6878141×10^{-2}
5	2.0203403×10^{-1}	2.7575187×10^{-2}	1.3433168×10^{-1}	1.2291103×10^{-2}

16.1.7 Dissipation calculation for capillary instability

We have shown in chapter 12 that the DM gives the same result as that of VCVPF for gas–liquid flows. This is also true for the two-fluid problem considered in §16.1.5, as the following calculation shows. Here we extend the dissipation calculation to capillary instability of two viscous fluids.

The sum of the mechanical-energy equations of the interior and exterior fluids can be written as

$$\begin{aligned}\frac{\mathrm{d}}{\mathrm{d}t}\int_{V_a}\frac{\rho_a}{2}|\mathbf{u}_a|^2\,\mathrm{d}V + \frac{\mathrm{d}}{\mathrm{d}t}\int_{V_l}\frac{\rho_l}{2}|\mathbf{u}_l|^2\,\mathrm{d}V = \int_A [u_n(-p_l+\tau_l^n+p_a-\tau_a^n) \\ +\mathbf{u}_l\cdot\mathbf{t}\tau_l^s-\mathbf{u}_a\cdot\mathbf{t}\tau_a^s]\,\mathrm{d}A - \int_{V_a}2\mu_a\mathbf{D}_a:\mathbf{D}_a\mathrm{d}V - \int_{V_l}2\mu_l\mathbf{D}_l:\mathbf{D}_l\mathrm{d}V.\end{aligned} \tag{16.1.60}$$

We assume that the normal stress balance,

$$p_a-\tau_a^n-p_l+\tau_l^n = \frac{\partial^2\eta}{\partial z^2}+\frac{\eta}{R^2}, \tag{16.1.61}$$

and the continuity of the tangential velocity and stress,

$$\tau_a^s=\tau_l^s=\tau^s, \quad \mathbf{u}_a\cdot\mathbf{t}=\mathbf{u}_l\cdot\mathbf{t}=u_s, \tag{16.1.62}$$

are all satisfied at the interface. At the same time, the flow in the bulk of the fluids is approximated by potential flow, for which the following identity can be easily proved:

$$\int_V 2\mu\mathbf{D}:\mathbf{D}\mathrm{d}V = \int_A \mathbf{n}\cdot 2\mu\mathbf{D}\cdot\mathbf{u}\mathrm{d}A, \tag{16.1.63}$$

where A is the surface of V and $\mathbf{n}$ is the unit normal pointing outward. Inserting (16.1.61), (16.1.62), and (16.1.63), into (16.1.60), we obtain

$$\begin{aligned}\frac{\mathrm{d}}{\mathrm{d}t}\int_{V_a}\frac{\rho_a}{2}|\mathbf{u}_a|^2\,\mathrm{d}V+\frac{\mathrm{d}}{\mathrm{d}t}\int_{V_l}\frac{\rho_l}{2}|\mathbf{u}_l|^2\,\mathrm{d}V &= \int_A u_n\gamma\left(\frac{\partial^2\eta}{\partial z^2}+\frac{\eta}{R^2}\right)\mathrm{d}A\\ &+\int_A \mathbf{n}_1\cdot 2\mu_a\mathbf{D}_a\cdot\mathbf{u}_a\mathrm{d}V-\int_A \mathbf{n}_1\cdot 2\mu_l\mathbf{D}_l\cdot\mathbf{u}_l\mathrm{d}V.\end{aligned} \tag{16.1.64}$$

The dimensionless form of (16.1.64) is

$$\begin{aligned}l\frac{\mathrm{d}}{\mathrm{d}t}\int_{V_a}\frac{1}{2}|\mathbf{u}_a|^2\,\mathrm{d}V+\frac{\mathrm{d}}{\mathrm{d}t}\int_{V_l}\frac{1}{2}|\mathbf{u}_l|^2\,\mathrm{d}V &= \int_A u_n\left(\frac{\partial^2\eta}{\partial z^2}+\frac{\eta}{R^2}\right)\mathrm{d}A\\ &+\frac{m}{\sqrt{J}}\int_A \mathbf{n}_1\cdot 2\mathbf{D}_a\cdot\mathbf{u}_a\mathrm{d}A-\frac{1}{\sqrt{J}}\int_A \mathbf{n}_1\cdot 2\mathbf{D}_l\cdot\mathbf{u}_l\mathrm{d}A.\end{aligned} \tag{16.1.65}$$

The integrals in (16.1.65) are evaluated as

$$\begin{aligned}\frac{\mathrm{d}}{\mathrm{d}t}\int_{V_l}\frac{|\mathbf{u}_l|^2}{2}\mathrm{d}V &= \frac{\mathrm{d}}{\mathrm{d}t}\int_0^{2\pi}\mathrm{d}\theta\int_z^{z+\lambda}\int_0^R\frac{|\mathbf{u}_l|^2}{2}r\,\mathrm{d}r\mathrm{d}z\\ &= |A_1|^2\,\pi\lambda R^2kI_0(kR)I_1(kR)(\sigma+\bar{\sigma})\exp(\sigma+\bar{\sigma})t,\end{aligned} \tag{16.1.66}$$

$$\begin{aligned}l\frac{\mathrm{d}}{\mathrm{d}t}\int_{V_a}\frac{|\mathbf{u}_a|^2}{2}\mathrm{d}V &= l\frac{\mathrm{d}}{\mathrm{d}t}\int_0^{2\pi}\mathrm{d}\theta\int_z^{z+\lambda}\int_R^{\infty}\frac{|\mathbf{u}_a|^2}{2}r\,\mathrm{d}r\mathrm{d}z\\ &= l\,|B_1|^2\,\pi\lambda R^2kK_0(kR)K_1(kR)(\sigma+\bar{\sigma})\exp(\sigma+\bar{\sigma})t,\end{aligned} \tag{16.1.67}$$

$$\int_A\left(\frac{\partial^2\eta}{\partial z^2}+\frac{\eta}{R^2}\right)u_l\mathrm{d}A=2\,|A_1|^2\,\pi\lambda R\frac{k^2}{\sigma}I_1^2(kR)\left(\frac{1}{R^2}-k^2\right)\exp(\sigma+\bar{\sigma})t, \tag{16.1.68}$$

$$\frac{1}{\sqrt{J}}\int_A \mathbf{n}_1\cdot 2\mathbf{D}_l\cdot\mathbf{u}_l\mathrm{d}A=\frac{4}{\sqrt{J}}\,|A_1|^2\,\pi\lambda Rk^3I_1(kR)\left[2I_0(kR)-\frac{I_1(kR)}{kR}\right]\exp(\sigma+\bar{\sigma})t, \tag{16.1.69}$$

$$\frac{m}{\sqrt{J}}\int_A \mathbf{n}_1\cdot 2\mathbf{D}_a\cdot\mathbf{u}_a\mathrm{d}A=\frac{4m}{\sqrt{J}}\,|A_1|^2\,\pi\lambda Rk^3K_1(kR)\left[2K_0(kR)+\frac{K_1(kR)}{kR}\right]\exp(\sigma+\bar{\sigma})t, \tag{16.1.70}$$

where u_l is the radial component of velocity in the liquid. Inserting (16.1.66)–(16.1.70) into (16.1.65), we obtain

$$(\alpha_l+l\alpha_a)\frac{\sigma+\bar{\sigma}}{2}+\frac{2k^2}{\sqrt{J}}\left[(\alpha_l+\beta_l)+m(\alpha_a+\beta_a)\right]=\left(\frac{1}{R^2}-k^2\right)\frac{k}{\sigma}, \tag{16.1.71}$$

where α_l, α_a, β_a, and β_l are defined in (16.1.23). If we assume that σ is real, (16.1.71) is the same as dispersion relation (16.1.49) from the VCVPF solution. The solution of the dispersion relation is given in (16.1.50). In the range $0<k\leq 1/R=2$, σ is real and our assumption is satisfied. Therefore the growth rate computed by the dissipation calculation is the same as that computed by the VCVPF.

16.1.8 Discussion of the pressure corrections at the interface of two viscous fluids

Our pressure corrections arise in a boundary layer induced by the discontinuity of the tangential velocity and shear stress at the interface evaluated with the potential solution. We assume that the boundary layer is thin, and when the boundary layer is considered, the tangential velocity and shear stress are continuous at the interface. This assumption leads to good agreements between the ES and VCVPF for the liquid–gas cases and less good agreements, but better than what might be expected, in the cases of two fluids for which the boundary-layer assumptions do not hold uniformly; this is discussed in this subsection. In the boundary layer near the interface, we divide the velocity and pressure into an irrotational part and a viscous correction part:

$$u = u^i + u^v, \quad w = w^i + w^v, \quad p = p^i + p^v. \tag{16.1.72}$$

The irrotational tangential velocities at the interface are

$$w_l^i = -A_1 k I_0(kR) \exp(\sigma t + ikz), \tag{16.1.73}$$

$$w_a^i = A_1 k \frac{I_1(kR)}{K_1(kR)} K_0(kR) \exp(\sigma t + ikz). \tag{16.1.74}$$

The tangential stresses at the interface evaluated with the potential flows are

$$\tau_l^s = -\left(2/\sqrt{J}\right) A_1 k^2 I_1(kR) \exp(\sigma t + ikz), \tag{16.1.75}$$

$$\tau_a^s = -\left(2m/\sqrt{J}\right) A_1 k^2 I_1(kR) \exp(\sigma t + ikz). \tag{16.1.76}$$

The continuity of the tangential velocity requires that

$$w_l^i + w_l^v = w_a^i + w_a^v, \tag{16.1.77}$$

which gives rise to

$$w_l^v - A_1 k I_0(kR) \exp(\sigma t + ikz) = w_a^v + A_1 k \frac{I_1(kR)}{K_1(kR)} K_0(kR) \exp(\sigma t + ikz). \tag{16.1.78}$$

If we assume $w_a^v = q w_l^v$, (16.1.78) can be written as

$$w_l^v (1 - q) = A_1 k \left[I_0(kR) + \frac{I_1(kR)}{K_1(kR)} K_0(kR) \right] \exp(\sigma t + ikz). \tag{16.1.79}$$

The continuity of the shear stress requires that

$$\begin{aligned} &\frac{1}{\sqrt{J}} \left(\frac{\partial u_l^v}{\partial z} + \frac{\partial w_l^v}{\partial r} \right) - \frac{2}{\sqrt{J}} A_1 k^2 I_1(kR) \exp(\sigma t + ikz) \\ &= \frac{m}{\sqrt{J}} \left(\frac{\partial u_a^v}{\partial z} + \frac{\partial w_a^v}{\partial r} \right) - 2 \frac{m}{\sqrt{J}} A_1 k^2 I_1(kR) \exp(\sigma t + ikz). \end{aligned} \tag{16.1.80}$$

Because the potential flow solution satisfies the continuity of the normal velocity $u_l^i = u_a^i$, the viscous corrections to the normal velocity in the boundary layer should be very small. Thus $\partial u_l^v / \partial z$ and $\partial u_a^v / \partial z$ can be ignored and (16.1.80) becomes

$$\frac{\partial}{\partial r} \left(w_l^v - m w_a^v \right) = 2(1 - m) A_1 k^2 I_1(kR) \exp(\sigma t + ikz). \tag{16.1.81}$$

Assuming that the boundary-layer thickness δ is small, we can write (16.1.81) approximately as

$$w_l^v\,(1-mq)\,/\delta = 2(1-m)A_1k^2I_1(kR)\exp(\sigma t+ikz). \tag{16.1.82}$$

Comparing (16.1.79) and (16.1.82), we obtain

$$\frac{1-mq}{(1-q)(1-m)}\frac{1}{\delta} = \frac{2kI_1(kR)}{I_0(kR)+I_1(kR)\frac{K_0(kR)}{K_1(kR)}}. \tag{16.1.83}$$

Therefore (16.1.83) needs to be satisfied if the continuities of the tangential velocity and stress are to be enforced. In other words, the assumptions on which VCVPF is based are valid only if (16.1.83) is satisfied. Because we do not solve the boundary-layer problem, q and δ are unknown and we are not able to determine if (16.1.83) is satisfied. However, we may assume that δ is very small and estimate the possibility of satisfying (16.1.83) under different conditions, i.e., different values of m and k. We have the following observations regarding (16.1.83).

(1) Because δ is supposed to be small, (16.1.83) is easier to satisfy when the right-hand side is larger. Calculation shows that the right-hand side of (16.1.83) is 5×10^{-7} when $k=0.001$ and is 1.36 when $k=2$. Therefore (16.1.83) can be satisfied for k close to $k_c=2$ but is very difficult to satisfy for small k. This observation could help us to understand the results of the growth-rate calculation, i.e., the agreement between VCVPF and the ES is good for k close to k_c but is poor when $k\ll1$.

(2) The term on the left-hand side of (16.1.83),

$$\frac{1-mq}{(1-q)(1-m)}, \tag{16.1.84}$$

should be comparable with or even smaller than δ, so that (16.1.83) can be satisfied. We find that, for certain values of q, (16.1.84) is small when m is much smaller than 1 and is large when m is close to 1. For example, if we fix q at 10, the value of (16.1.84) is 0.0125 when $m=0.11$ and is 8.89 when $m=0.9$. This observation may help us to understand our growth-rate calculation at low Reynolds numbers. In case 4, $m=3.091\times10^{-4}$ and VCVPF is in good agreement with the ES in the region of maximum growth; in case 5, $m=0.5455$ and VCVPF does not give the correct maximum growth rate σ_m and the associated wavenumber k_m. The value of m does not seem to be important at high Reynolds numbers. In cases 1 and 2, IPF, VPF, and VCVPF all agree well with the ES in the region of maximum growth even when the value of m is relatively close to 1. This is because the boundary layer and the viscous correction are not important at high Reynolds numbers.

We have shown that condition (16.1.83) can be satisfied in some cases (k close to $k_c=2$ and $m\ll1$) and is very difficult to satisfy in other cases ($k\ll1$ or m close to 1). When (16.1.83) is satisfied, the assumption of continuous tangential velocity and stress is realized approximately and our calculation does show good agreement between VCVPF and VPF. When (16.1.83) is difficult to satisfy, VCVPF may not give a good approximation to the ES, especially at low Reynolds numbers. Thus viscous potential flows with pressure corrections can be used to approximate viscous flows in problems involving the interface of two viscous fluids, but this approximation is not uniformly valid.

When the viscous corrections for the velocity and pressure are added to the irrotational flow, the complete form of the normal stress balance is

$$p_a^i + p_a^v - p_l^i - p_l^v + \frac{2}{\sqrt{J}}\left(\frac{\partial u_l^i}{\partial r} + \frac{\partial u_l^v}{\partial r}\right) - \frac{2m}{\sqrt{J}}\left(\frac{\partial u_a^i}{\partial r} + \frac{\partial u_a^v}{\partial r}\right) = \frac{\partial^2 \eta}{\partial z^2} + \frac{\eta}{R^2}. \quad (16.1.85)$$

The viscous corrections of the velocity satisfy the continuity equation

$$\frac{\partial u_l^v}{\partial r} + \frac{u_l^v}{r} + \frac{\partial w_l^v}{\partial z} = 0. \quad (16.1.86)$$

We may estimate the order of the terms in (16.1.86):

$$2\frac{u_l^v}{\delta} \sim \frac{w_l^v}{\lambda}. \quad (16.1.87)$$

Combining (16.1.82) and (16.1.87), we obtain

$$\frac{\partial u_l^v}{\partial r} \sim \frac{u_l^v}{\delta} \sim \frac{\delta}{\lambda}\frac{1-m}{1-mq}A_1 k^2 I_1(kR). \quad (16.1.88)$$

At the same time, we have

$$\frac{\partial u_l^i}{\partial r} \sim A_1 k^2\left[I_0(kR) - \frac{I_1(kR)}{kR}\right]. \quad (16.1.89)$$

If $(1-m)/(1-mq)$ is not a very big value, $\partial u_l^v/\partial r$ may be significantly smaller than $\partial u_l^i/\partial r$. A similar argument can show that $\partial u_a^v/\partial r$ could be significantly smaller than $\partial u_a^i/\partial r$. Therefore $\partial u_l^v/\partial r$ and $\partial u_a^v/\partial r$ can be ignored in (16.1.85), and we obtain the normal stress balance equation, (16.1.40), used in the VCVPF calculation. Admittedly, omission of $\partial u_l^v/\partial r$ and $\partial u_a^v/\partial r$ is not justified for certain values of m and q, which may be partially responsible for the poor agreement between VCVPF and the ES in some cases.

16.1.9 Capillary instability when one fluid is a dynamically inactive gas

There are two cases; (i) the exterior fluid is a gas or (ii) the interior fluid is a gas. The gas is dynamically inactive when $m = \ell = 0$. We can obtain dispersion relations for the cases in which one fluid is a dynamically inactive gas by putting $m = \ell = 0$ in the fluid formulas

VPF	[equation (16.1.24)],
IPF	[equation (16.1.25)],
VCVPF	[equation (16.1.50)],
ES	[equation (16.1.14)],

where ES stands for the exact solution of Tomotika (1935).

Table 16.3. *(Wang, et al. 2005a) The properties of five fluids surrounded by air used to study capillary instability and the Reynolds number $J = Oh^2$, where Oh is the Ohnesorge number*

Fluids	(1) Mercury–air	(2) Water–air	(3) SO100–air	(4) Glycerin–air	(5) SO10000–air
ρ_l (kg m^{-3})	1.35×10^4	1.00×10^3	9.69×10^2	1.26×10^3	9.69×10^2
μ_l (kg/m sec)	1.56×10^{-3}	1.0×10^{-3}	0.1	0.782	10.0
γ (N/m)	0.482	7.28×10^{-2}	2.1×10^{-2}	6.34×10^{-2}	2.1×10^{-2}
$J = \rho_l \gamma D/\mu_l^2$	2.67×10^7	7.28×10^5	20.4	1.30	2.04×10^{-3}

(i) Exterior fluid is a gas:

$$\text{VPF}: \sigma = -\frac{k^2\beta_l}{\sqrt{J}\alpha_l} \pm \sqrt{\left[\frac{k^2\beta_l}{\sqrt{J}\alpha_l}\right]^2 + \left(\frac{1}{R^2} - k^2\right)\frac{k}{\alpha_l}}; \tag{16.1.90}$$

$$\text{IPF}: \sigma = \pm\sqrt{\left(\frac{1}{R^2} - k^2\right)\frac{k}{\alpha_l}}; \tag{16.1.91}$$

$$\text{VCVPF}: \sigma = -\frac{k^2(\alpha_l + \beta_l)}{\sqrt{J}\alpha_l} \pm \sqrt{\left[\frac{k^2(\alpha_l + \beta_l)}{\sqrt{J}\alpha_l}\right]^2 + \left(\frac{1}{R^2} - k^2\right)\frac{k}{\alpha_l}}; \tag{16.1.92}$$

ES: σ is the solution of

$$\begin{vmatrix} 2k^2 I_1(kR) & (k^2 + k_l^2) I_1(k_l R) \\ F_1 & F_2 \end{vmatrix} = 0, \tag{16.1.93}$$

where F_1 and F_2 are defined in (16.1.15) and (16.1.16).

Calculations made with these formulas were carried out for the five liquids listed in table 16.3.

At high Reynolds numbers, the results obtained with IPF, VPF, VCVPF, and ES are essentially the same (all the theories give rise to essentially the same result in the case of water or mercury; readers are referred to Funada and Joseph, 2002). At lower Reynolds numbers, IPF overestimates the growth rate considerably in the maximum growth region; the growth rate by VPF is in better agreement with the ES solution; the curves for VCVPF are almost indistinguishable from the ES curves (figures 16.7 and 16.8). The remarkably good agreement between VCVPF and ES seems universal; it is observed for $10^{-3} < J < 10^7$ and for any $k < k_c = 2$ (see table 16.4).

(ii) Inverse cases; the interior fluid is a gas: Next we consider cases in which the interior fluid is a gas with negligible viscosity and density; these are the inverse cases of those in table 16.3. We omit the details of the VCVPF calculation because they are similar to

Table 16.4. *Maximum growth rate σ_m and the associated wavenumber k_m for VCVPF and ES in the 5 cases shown in table 16.3 (A) and in the 5 inverse cases (B) [see subsection 16.1.9 (ii)], e.g., air–mercury*

	VCVPF (A)		ES (A)		VCVPF (B)		ES (B)	
Case	k_m	σ_m	k_m	σ_m	k_m	σ_m	k_m	σ_m
1	1.39	0.97	1.39	0.97	0.97	2.32	0.97	2.31
2	1.39	0.97	1.39	0.97	0.96	2.31	0.96	2.31
3	1.06	0.58	1.09	0.59	0.53	1.58	0.60	1.61
4	0.72	0.27	0.74	0.28	0.21	0.84	0.28	0.86
5	0.17	0.015	0.17	0.015	0.0066	0.045	0.013	0.045

those presented in §16.1.9 (i). We directly present the growth rates computed by IPF, VPF, VCVPF, and ES as follows:

$$\text{IPF}: \sigma = \pm\sqrt{\left(\frac{1}{R^2} - k^2\right)\frac{k}{\alpha_a}}; \tag{16.1.94}$$

$$\text{VPF}: \sigma = -\frac{k^2\beta_a}{\sqrt{J'}\alpha_a} \pm \sqrt{\left[\frac{k^2\beta_a}{\sqrt{J'}\alpha_a}\right]^2 + \left(\frac{1}{R^2} - k^2\right)\frac{k}{\alpha_a}}; \tag{16.1.95}$$

$$\text{VCVPF}: \sigma = -\frac{k^2(\alpha_a + \beta_a)}{\sqrt{J'}\alpha_a} \pm \sqrt{\left[\frac{k^2(\alpha_a + \beta_a)}{\sqrt{J'}\alpha_a}\right]^2 + \left(\frac{1}{R^2} - k^2\right)\frac{k}{\alpha_a}}; \tag{16.1.96}$$

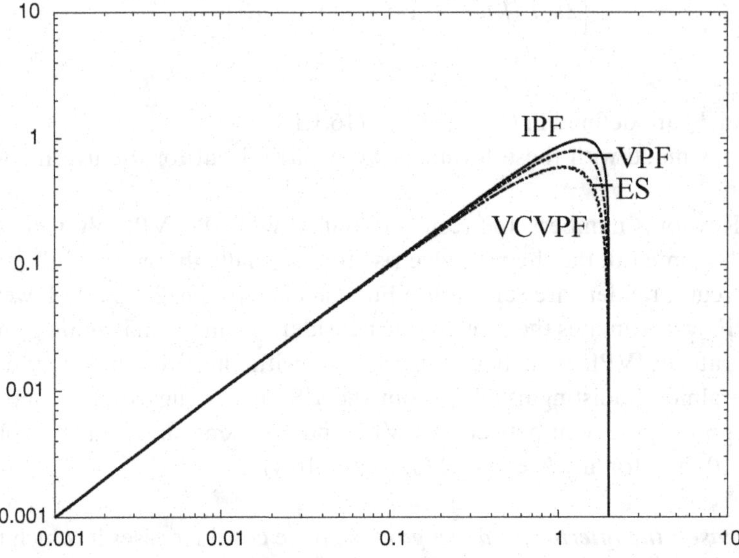

Figure 16.7. The growth rate σ vs. k for case 3, SO100 in air. IPF and VPF slightly overestimate the growth rate in the region near the peak; the curve for the corrected solution (VCPVF) is almost indistinguishable from the exact solution (ES).

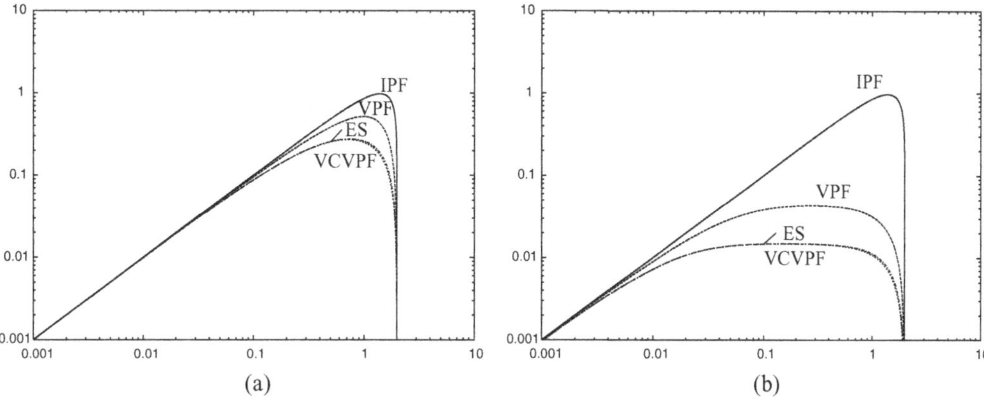

Figure 16.8. The growth rate σ vs. k for case 4, glycerin in air (a) and for case 5, SO10000 in air (b). The growth rates computed from IPF and VPF deviate considerably from the exact solution (ES), but the growth rates from the corrected solution (VCVPF) are nearly the same as those of the ES (see table 16.4).

ES: σ is the solution of

$$\begin{vmatrix} 2k^2 K_1(kR) & (k^2+k_a'^2)\,K_1(k_a'R) \\ F_5 & F_6 \end{vmatrix} = 0, \tag{16.1.97}$$

where α_a, β_a are defined in (16.1.23) and

$$F_5 = -\sigma K_0(kR) - \frac{2k^2}{\sqrt{J'}}\left[\frac{\mathrm{d}K_1(kR)}{\mathrm{d}(kR)}\right] + \left(\frac{1}{R^2}-k^2\right)\frac{k}{\sigma}K_1(kR), \tag{16.1.98}$$

$$F_6 = -\frac{2kk_a'}{\sqrt{J'}}\left[\frac{\mathrm{d}K_1(k_a'R)}{\mathrm{d}(k_a'R)}\right] + \left(\frac{1}{R^2}-k^2\right)\frac{k}{\sigma}K_1(k_a'R), \tag{16.1.99}$$

with

$$k_a' = \sqrt{k^2+\sqrt{J'}\sigma}, \quad J' = \frac{\rho_a D\gamma}{\mu_a^2}. \tag{16.1.100}$$

We calculate the growth-rate curves for the inverse cases of those listed in table 16.3. Example curves are plotted in figure 16.9 for air in SO10000. At low Reynolds numbers (figure 16.9), the growth rate by IPF is significantly larger than that by ES for any $k < k_c = 2$; VPF and VCVPF both give good approximations to the ES. At higher Reynolds numbers, the growth-rate curves computed with IPF, VPF, VCVPF, and ES are almost the same.

16.1.10 Conclusions

Purely irrotational theories of capillary instability of one viscous fluid cylinder in another viscous fluid were derived and compared with Tomotika's exact theory. In the case of two liquids the continuity condition, like no-slip and continuity of shear stresses, generates regions of vorticity that have an important effect on viscous decay. The ES, however, is surprisingly well approximated by VCVPF for waves near the peak value of the growth-rate curve but is not well approximated by VCVPF for long waves.

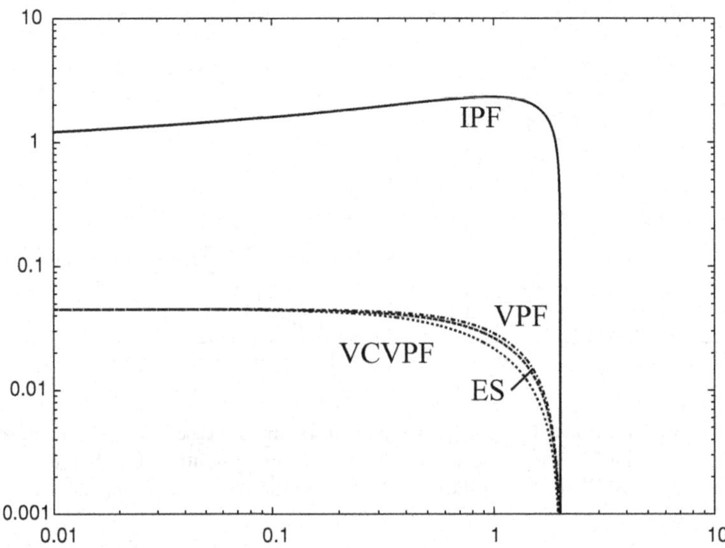

Figure 16.9. The growth rate σ vs. k for case 5 (inverse), air in SO10000. The growth rate by IPF is significantly larger than that by the exact solution (ES); the results by VPF, VCVPF, and ES are almost the same (see table 16.4).

The extra pressure derived for VCVPF is generally thought to arise in a vorticity layer at the gas–liquid boundary. This idea still needs clarification, but in any case it is not used or needed in our purely irrotational theories.

In the cases in which one of the fluids is a dynamically passive gas (or vacuum), the purely irrotational theory of VCVPF with extra pressure gives rise to a dispersion relation that is almost indistinguishable from the exact one, uniformly in the wavenumber. The same dispersion relation as given by VCVPF can be derived from the DM. The approximation to the ES generated by VCVPF is much better than that of VPF. As a rule of thumb, we can say that VPF gives a good approximation of the viscous effects in the case of short waves and VCVPF gives a good approximation for long waves. In the capillary instability problems studied here, the short waves are stabilized by surface tension and do enter into the stability analysis.

16.2 Stability of a liquid jet into incompressible gases: Temporal, convective, and absolute instabilities

We carry out an analysis of the stability of a liquid jet into a gas or another liquid by using VPF. The instability may be driven by KH instability that is due to a velocity difference and a neck-down that is due to capillary instability. An explicit dispersion relation is derived and analyzed for temporal and convective/absolute (C/A) instability. We find that, for all values of the relevant parameters, there are wavenumbers for which the liquid jet is temporally unstable. The cutoff wavenumber and the wavenumber of maximum growth are most important; the variation of these quantities with the density and viscosity ratios, the Weber number, and the Reynolds number is computed and displayed as graphs and asymptotic formulas. The instabilities of a liquid jet are due to capillary and KH instabilities. We show that KH instability cannot occur in a vacuum but capillary instability (and RT instability) can occur in vacuum. We present comprehensive results, based on VPF, of the effects of the ambient.

Temporally unstable liquid jet flows can be analyzed for spatial instabilities by C/A theory; they are either convectively unstable or absolutely unstable, depending on the sign of the temporal growth rate at a singularity of the dispersion relation. The study of such singularities is greatly simplified by the analysis here, which leads to an explicit dispersion relation: an algebraic function of a complex frequency and complex wavenumber. Analysis of this function gives rise to an accurate Weber–Reynolds criterion for the border between absolute and convective instabilities. Some problems of the applicability to physics of C/A analysis of stability of spatially uniform and nearly uniform flows are discussed.

16.2.1 Introduction

This chapter is also allied to the analysis of temporal instability of the capillary jet given by Funada and Joseph (2002). Their analysis generalizes the inviscid analysis of Rayleigh (1878) in two ways: by accounting (i) for the viscosity of the jet and (ii) for the viscosity and density of the ambient media. The liquid jet is prey to capillary instability and KH instability that are due to the difference of the velocity of the jet and of the ambient media. The effects of discontinuous velocity can be obtained by a Galilean transformation to a fixed coordinate system relative to which the jet is moving with a velocity U. As a practical matter, the transformation $z \to z + Ut$ transforms disturbances of the form $\exp[i(kz - \omega t)]$ to $\exp[i(kz - \tilde{\omega} t)]$, where $\tilde{\omega} = \omega - kU$. A KH instability that is due to velocity discontinuity cannot occur when the no-slip condition is applied.

We compute in a coordinate system fixed on the ambient fluid relative to which the jet velocity is U. We consider temporal, convective, and absolute instabilities. The analysis of these instabilities is greatly simplified by VPF, which leads to an explicit dispersion relation $D(k, \omega) = 0$, in which the ambient media with viscosity μ_a and density ρ_a are fully represented. Moreover, the analysis applies equally to jets into liquid and jet into gas; ρ_a and μ_a stand for "air" for jets into air and for "ambient" for jets into liquid. This chapter is organized so as to emphasize the case of jets into air that can be compared with the prior literature, which is subsequently reviewed. The case of jets of viscous liquids into viscous liquids can be analyzed because VPF is consistent with a discontinuous velocity across the jet boundary.

The liquid jet is subject to KH instability that is due to the discontinuous velocity and to capillary instability that is due to surface tension. We show that KH instabilities cannot occur in a vacuum. The only other paper to treat the case of combined KH and capillary instability is that of Lin and Lian (1989), who analyzed the viscous problem by using the Navier–Stokes equations but neglecting the effects of shear from the viscous gas.

Other papers relevant to the stability of a liquid jet neglect viscosity altogether or neglect the effects of the ambient. Rayleigh's (1878) study of temporal instability, the study of spatial instability of Keller, Rubinow, and Tu (1973), and Leib and Goldstein's (1986a) study of C/A instability neglect viscosity and the effects of the ambient. Leib and Goldstein (1986b) and Dizès (1997) account for the jet viscosity but neglect the ambient. Many examples of transition from convective to absolute instability arising in the breakup of sheets and jets are presented in the monograph of S. P. Lin (2003). In particular, Lin has an interesting discussion of the relevance of this transition to the transition from dripping to jetting under gravity. The comparisons are suggestive, but the agreement between C/A theory and experiments is not definitive.

16.2.2 Problem formulation

A long liquid cylinder of density ρ, viscosity μ, and mean radius a moves with a uniform axial velocity U relative to an ambient gas (air) of ρ_a, μ_a. With a cylindrical frame (r, θ, z) fixed on the gas, the liquid cylinder is put in the region of $0 \le r < a + \eta$ and $-\infty < z < \infty$, where $\eta = \eta(z, t)$ is the varicose interface displacement. The governing Navier–Stokes equations and interface conditions for disturbances of the cylinder and gas are made dimensionless with the following scales:

$$[\text{length, velocity, time, pressure}] = \left[2a,\ U,\ \frac{2a}{U},\ \rho U^2\right]. \tag{16.2.1}$$

In terms of this normalization, we may define Weber number We, Reynolds number Re, density ratio ℓ, and viscosity ratio m:

$$We = \frac{\gamma}{\rho 2aU^2}, \quad Re = \frac{U2a}{\nu}, \quad \ell = \frac{\rho_a}{\rho}, \quad m = \frac{\mu_a}{\mu}, \tag{16.2.2}$$

where γ is the surface-tension coefficient, $\nu = \mu/\rho$, $\nu_a = \mu_a/\rho_a$, and $m/\ell = \nu_a/\nu$.

This problem is a combination of capillary instability and KH instability. When $We = 0$ $(\gamma = 0)$ the instability is generated by the velocity difference. An interesting feature of this instability is that even though the density and viscosity of the gas are much smaller than those of the liquid, the dynamical effects of the gas cannot be neglected. The relevant physical quantity is the kinematic viscosity $\nu = \mu/\rho$; Funada and Joseph (2001) found that the stability limit for VPF is nearly independent of the viscosity when $\nu_\ell > \nu_a$, with a sensible dependence when $\nu_\ell < \nu_a$, for small viscosities, the opposite of what intuition would suggest. Essentially the same result holds for the KH instability of liquid jet, studied here. The other limit, $We \to \infty$ or $U \to 0$, leads to capillary instability that was studied with VPF by Funada and Joseph (2002). Our scaling fails when U tends to zero in the case in which the scale velocity is γ/μ, which is the characteristic velocity for capillary collapse and the relevant Reynolds number is $J = \rho\gamma 2a/\mu^2$. The basic flow in dimensionless coordinates is $(\partial\Phi/\partial z, \partial\Phi_a/\partial z) = (1, 0)$ in terms of the velocity potential Φ and Φ_a.

For the liquid cylinder in a disturbed state $(0 \le r < 1/2 + \eta$ and $-\infty < z < \infty)$, the velocity potential $\phi \equiv \phi(r, z, t)$ of an axisymmetric disturbance satisfies the Laplace equation,

$$\left(\frac{\partial^2}{\partial r^2} + \frac{1}{r}\frac{\partial}{\partial r} + \frac{\partial^2}{\partial z^2}\right)\phi = 0, \tag{16.2.3}$$

and the Bernoulli equation,

$$\frac{\partial\phi}{\partial t} + \frac{\partial\phi}{\partial z} + \frac{1}{2}\left(\frac{\partial\phi}{\partial r}\right)^2 + \frac{1}{2}\left(\frac{\partial\phi}{\partial z}\right)^2 + p = f(t), \tag{16.2.4}$$

where $p \equiv p(r, z, t)$ is the pressure and $f(t)$ is an arbitrary function of time t that may be put to zero. For the gas disturbance of infinite extent ($1/2 + \eta < r < \infty$ and $-\infty < z < \infty$), the velocity potential $\phi_a \equiv \phi_a(r, z, t)$ satisfies

$$\left(\frac{\partial^2}{\partial r^2} + \frac{1}{r}\frac{\partial}{\partial r} + \frac{\partial^2}{\partial z^2} \right) \phi_a = 0, \tag{16.2.5}$$

$$\ell \left[\frac{\partial \phi_a}{\partial t} + \frac{1}{2}\left(\frac{\partial \phi_a}{\partial r} \right)^2 + \frac{1}{2}\left(\frac{\partial \phi_a}{\partial z} \right)^2 \right] + p_a = f_a(t). \tag{16.2.6}$$

The kinematic condition at the interface $r = 1/2 + \eta$ is given for each fluid by

$$\frac{\partial \eta}{\partial t} + \frac{\partial \eta}{\partial z} + \frac{\partial \phi}{\partial z}\frac{\partial \eta}{\partial z} = \frac{\partial \phi}{\partial r}, \quad \frac{\partial \eta}{\partial t} + \frac{\partial \phi_a}{\partial z}\frac{\partial \eta}{\partial z} = \frac{\partial \phi_a}{\partial r}, \tag{16.2.7}$$

and the normal stress balance at $r = 1/2 + \eta$ is given by

$$\begin{aligned} &p - p_a - \frac{1}{Re}\tau + \frac{m}{Re}\tau_a \\ &= -We \left\{ \frac{\partial^2 \eta}{\partial z^2} \left[1 + \left(\frac{\partial \eta}{\partial z} \right)^2 \right]^{-3/2} - (1/2 + \eta)^{-1} \left[1 + \left(\frac{\partial \eta}{\partial z} \right)^2 \right]^{-1/2} + 2 \right\}, \end{aligned} \tag{16.2.8}$$

where the pressures at the interface are expressed by (16.2.4) and (16.2.6) and τ and τ_a denote the normal viscous stresses acting on the interface:

$$\tau = 2 \left[\frac{\partial^2 \phi}{\partial r^2} - 2\frac{\partial^2 \phi}{\partial r \partial z}\frac{\partial \eta}{\partial z} + \frac{\partial^2 \phi}{\partial z^2}\left(\frac{\partial \eta}{\partial z} \right)^2 \right] \left[1 + \left(\frac{\partial \eta}{\partial z} \right)^2 \right]^{-1}, \tag{16.2.9}$$

$$\tau_a = 2 \left[\frac{\partial^2 \phi_a}{\partial r^2} - 2\frac{\partial^2 \phi_a}{\partial r \partial z}\frac{\partial \eta}{\partial z} + \frac{\partial^2 \phi_a}{\partial z^2}\left(\frac{\partial \eta}{\partial z} \right)^2 \right] \left[1 + \left(\frac{\partial \eta}{\partial z} \right)^2 \right]^{-1}. \tag{16.2.10}$$

For a case of an interface displacement that is small compared with the mean radius, (16.2.7)–(16.2.10) may be expanded around $r = 1/2$ to give a linear system of boundary conditions for small disturbances. We do not require the continuity of tangential velocity and shear stress. The other conditions are that the liquid velocity is finite at the center $r = 0$ and that the gas velocity should vanish as $r \to \infty$.

16.2.3 Dispersion relation

The potentials ϕ and ϕ_a are determined by (16.2.3) and (16.2.5). At the interface approximated by $r = 1/2$, the kinematic conditions are given by

$$\frac{\partial \eta}{\partial t} + \frac{\partial \eta}{\partial z} = \frac{\partial \phi}{\partial r}, \quad \frac{\partial \eta}{\partial t} = \frac{\partial \phi_a}{\partial r}, \tag{16.2.11}$$

and the normal stress balance is given by

$$-\left(\frac{\partial \phi}{\partial t} + \frac{\partial \phi}{\partial z} \right) + \ell\frac{\partial \phi_a}{\partial t} - \frac{2}{Re}\frac{\partial^2 \phi}{\partial r^2} + \frac{2m}{Re}\frac{\partial^2 \phi_a}{\partial r^2} = -We\left(\frac{\partial^2 \eta}{\partial z^2} + 4\eta \right). \tag{16.2.12}$$

Thus, we may have the solutions of the form

$$\eta = AE + \text{c.c.}, \quad \phi = A_1 I_0(kr)E + \text{c.c.}, \quad \phi_a = A_2 K_0(kr)E + \text{c.c.}, \tag{16.2.13}$$

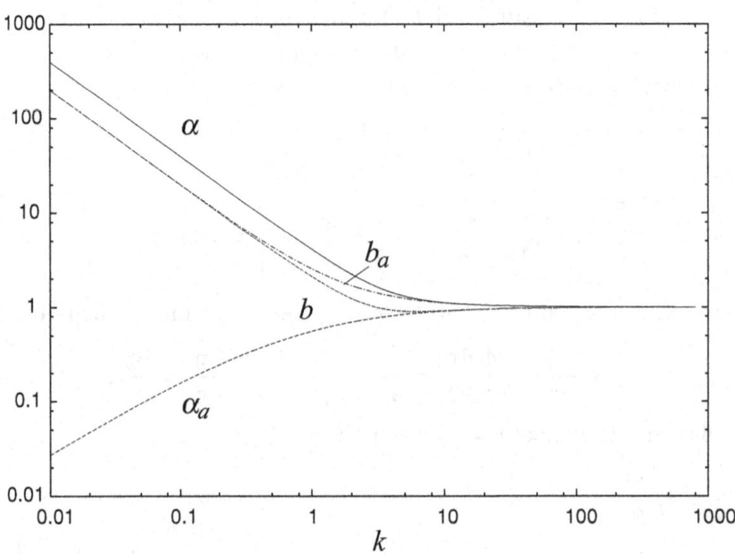

Figure 16.10. Functions α, b_a, b, and α_a versus real k; these functions tend to one for $k > 10$. The neutral curves of IPF and VPF for $\ell = m$ are identical when $k > 10$; this is seen in (16.2.33). The functions for $k > 10$ lead to the asymptotic forms (16.2.23), (16.2.24), and (16.2.39)–(16.2.46).

where A, A_1, and A_2 are the complex amplitudes, $E \equiv \exp(ikz - i\omega t)$, $\omega \equiv \omega_R + i\omega_I$ denotes the complex angular frequency, $k \equiv k_R + ik_I$ is the complex wavenumber, $i = \sqrt{-1}$, and c.c. stands for the complex conjugate of the preceding expression; $I_0(kr)$ and $K_0(kr)$ denote the zeroth order of modified Bessel functions of the first and second kind. Then ϕ gives the finite velocity at $r = 0$ and ϕ_a gives the velocity that vanishes as $r \to \infty$.

Substitution of (16.2.13) into (16.2.11) and (16.2.12) gives the dispersion relation,

$$D(k,\omega) = (\omega - k)^2\alpha + \ell\omega^2\alpha_a + i\frac{2k^2}{Re}(\omega - k)b + i\frac{2mk^2}{Re}\omega b_a - We\left(k^3 - 4k\right) = 0,$$

which is a quadratic equation in ω:

$$c_2\omega^2 + 2c_1\omega + c_0 = 0, \tag{16.2.14}$$

with the coefficients $c_2 \equiv c_2(k)$, $c_1 \equiv c_1(k)$, and $c_0 \equiv c_0(k)$:

$$\begin{aligned} c_2 &= \alpha + \ell\alpha_a, \quad c_1 = -k\alpha + i\frac{k^2}{Re}(b + mb_a) = c_{1R} + ic_{1I}, \\ c_0 &= k^2\alpha - i\frac{2k^3}{Re}b - We\left(k^3 - 4k\right) = c_{0R} + ic_{0I}, \end{aligned} \tag{16.2.15}$$

where α, α_a, b, and b_a are defined as

$$\alpha = \frac{I_0(k/2)}{I_1(k/2)}, \quad \alpha_a = \frac{K_0(k/2)}{K_1(k/2)}, \quad b = \alpha - \frac{2}{k}, \quad b_a = \alpha_a + \frac{2}{k}. \tag{16.2.16}$$

It is noted for real k that $k\alpha \to 4$ and $\alpha_a \to 0$ as $k \to 0$, whereas $\alpha \to 1$ and $\alpha_a \to 1$ as $k \to \infty$; this is shown in figure 16.10. Apart from the Bessel functions, (16.2.14) is a cubic equation in k; numerical calculations show that, for each and every fixed set of parameters studied here, (16.2.14) gives rise to three complex roots.

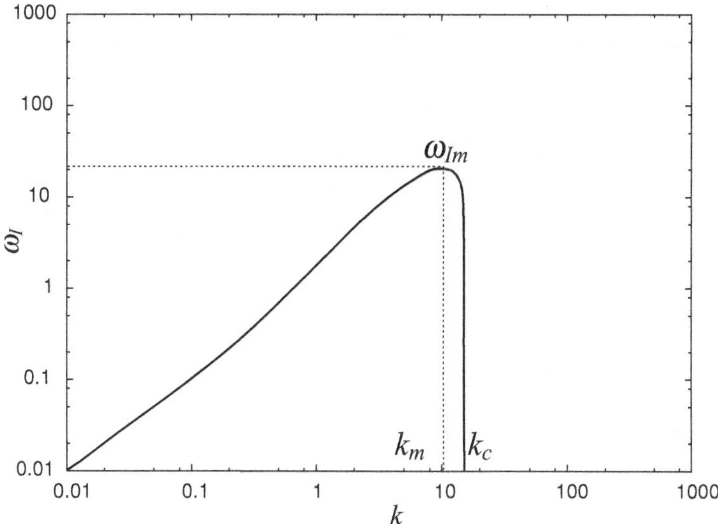

Figure 16.11. Example of a growth-rate curve defined in subsection 16.2.4, showing the main features: the shape, maximum ω_{Im}, k_m, and the cutoff wavenumber k_c.

16.2.4 Temporal instability

For this case k is real and $\omega = \omega_R + i\omega_I$. Recalling that $We = \gamma/(\rho 2aU^2)$, we obtain pure KH instability with $We = 0$ and pure capillary instability with $We \to \infty$. A temporal growth-rate curve is shown in figure 16.11, by which the maximum growth rate ω_{Im}, the associated wavenumber k_m, and the cutoff wavenumber k_c are defined.

16.2.4.1 *Inviscid fluids*

When the fluids are both inviscid as $Re \to \infty$, (16.2.14)–(16.3.10) reduce to

$$(\alpha + \ell\alpha_a)\,\omega^2 - 2k\alpha\omega + k^2\alpha - We\left(k^3 - 4k\right) = 0. \tag{16.2.17}$$

The complex angular frequency $\omega = \omega_R + i\omega_I$ is given by

$$\omega_R = \frac{k\alpha}{\alpha + \ell\alpha_a}, \qquad \omega_I = \pm\sqrt{\frac{k^2\ell\alpha\alpha_a}{(\alpha + \ell\alpha_a)^2} - \frac{We\,(k^3 - 4k)}{\alpha + \ell\alpha_a}} \tag{16.2.18}$$

in the unstable case, and

$$\omega_R = \frac{k\alpha}{\alpha + \ell\alpha_a} \pm \sqrt{-\frac{k^2\ell\alpha\alpha_a}{(\alpha + \ell\alpha_a)^2} + \frac{We\,(k^3 - 4k)}{\alpha + \ell\alpha_a}}, \qquad \omega_I = 0, \tag{16.2.19}$$

in the stable case. The neutral state is defined by $\omega_I = 0$; either $k = 0$ or $We = We_c$ where

$$We_c^{-1} = \left[\frac{\alpha + \ell\alpha_a}{\ell\alpha\alpha_a}\left(k - \frac{4}{k}\right)\right]_{k=k_c}. \tag{16.2.20}$$

Instability may arise in $0 < k < k_c$, where the *cutoff wavenumber* k_c ($k_c \geq 2$ for which $We_c^{-1} \geq 0$) is evaluated by (16.2.20) for given values of ℓ and We_c. For k large for which α and α_a approach 1 (see figure 16.10), (16.2.20) is approximated as $We_c^{-1} = \left(\ell^{-1} + 1\right) k_c$.

The effects of surface tension, leading to capillary instability, are absent when $We = 0$; hence $We = 0$ is pure KH instability. Inspection of (16.2.18) shows that *KH instability cannot occur when* $\ell = 0$; the viscosity and density of the ambient vanish so the *KH instability cannot occur in vacuum* (no pressure can be generated in vacuum). Pure KH instability is Hadamard unstable[1] with a growth rate proportional to k; the short waves grow exponentially with k at fixed t. The regularizing effect of surface tension is to stabilize short waves with $k > k_c$, given by (16.2.20). The maximum growth rate,

$$\omega_{Im} = \max_k \omega_I(k) = \omega_I(k_m), \tag{16.2.21}$$

may be obtained from (16.2.18) at an interior maximum for which

$$\partial\omega_I/\partial k = 0. \tag{16.2.22}$$

This computation is slightly complicated by the fact that $\alpha(k)$ and $\alpha_a(k)$ depend on k weakly. The values ω_{Im} and k_m depend on We and are plotted in figure 16.12. For large k, $\alpha(k) = \alpha_a(k) = 1$, and we find that

$$\omega_{Im} = \frac{2\ell\sqrt{\ell}}{3\sqrt{3}\,(1+\ell)^2\, We}, \tag{16.2.23}$$

$$k_m = \frac{2\ell}{3\,(1+\ell)\, We}. \tag{16.2.24}$$

Equations (16.2.23) and (16.2.24) show that the maximum growth rate and the associated wavenumber tend to infinity for small We like $1/We$; the wavelength $\lambda_m = 2\pi/k_m$ tends to zero with We. Viscosity regularizes the growth rate but the wavelength tends to zero with We, as in the inviscid case (see figure 16.13).

For pure capillary instability with $We \to \infty$, the neutral boundaries are given by $k = 0$ and $k_c = 2$. In this limiting case, we may rescale as $\omega = \hat{\omega}\sqrt{We}$, by which (16.2.17) is expressed as

$$(\alpha + \ell\alpha_a)\,\omega^2 - We\left(k^3 - 4k\right) = 0 \;\rightarrow\; (\alpha + \ell\alpha_a)\,\hat{\omega}^2 - \left(k^3 - 4k\right) = 0, \tag{16.2.25}$$

so that the solution $\hat{\omega}$ is given by

$$\hat{\omega} = \hat{\omega}_R + i\hat{\omega}_I = \pm\sqrt{\frac{k^3 - 4k}{\alpha + \ell\alpha_a}}; \tag{16.2.26}$$

hence instability ($\hat{\omega} = i\hat{\omega}_I$) may arise in $0 < k < 2$. Disturbances with $k > 2$ are stable and have an angular frequency $\hat{\omega} = \hat{\omega}_R$. The single column with $\ell = 0$ is the case that Lord Rayleigh analyzed in 1878.

[1] Hadamard instability is defined differently by different authors. For stability studies the growth rates $\sigma(k)$ go to infinity with k; the growth rates are not bounded for short waves. Say, for example $\sigma = k$; the disturbance amplitude is proportional to $\exp(kt)$. This is a very bad instability; the amplitude tends to infinity with k for any fixed t no matter how small t; the more you refine the mesh, the worse is the result; they are very unstable to short waves. See Joseph (1990) and Joseph and Saut (1990).

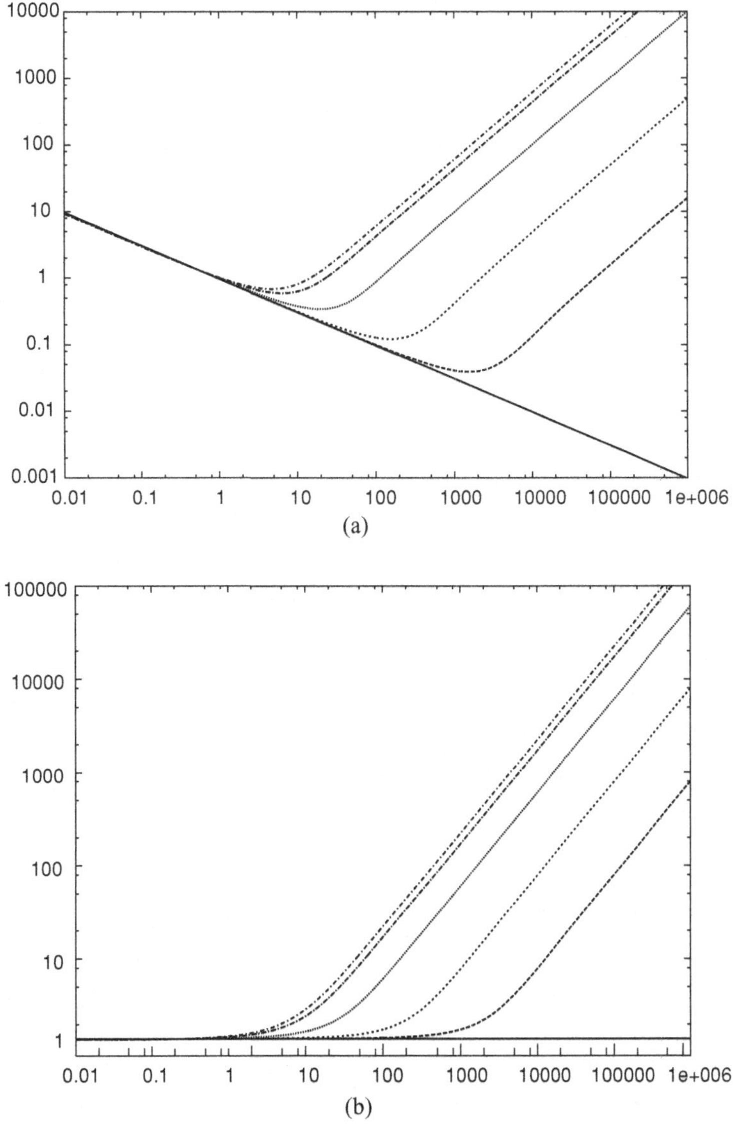

Figure 16.12. (a) ω_{Im} versus We^{-1} and (b) k_m versus We^{-1}, for $\ell = 0$ (solid lines), 0.0012 (broken curves), 0.012 (dashed curves), 0.1 (dotted curves), 0.3455 (broken dotted curves), 0.5 (dash–dotted curves). For large We^{-1} and $\ell \neq 0$, the curves approach the asymptotic form given respectively by (16.2.23) and (16.2.24).

16.2.4.2 *Viscous fluids*

For $\omega = \omega_R + i\omega_I$, quadratic equation (16.2.14) is separated into real and imaginary parts,

$$c_2\left(\omega_R^2 - \omega_I^2\right) + 2\left(c_{1R}\omega_R - c_{1I}\omega_I\right) + c_{0R} = 0, \tag{16.2.27}$$

$$2c_2\omega_R\omega_I + 2\left(c_{1R}\omega_I + c_{1I}\omega_R\right) + c_{0I} = 0 \rightarrow \omega_R = -\frac{2c_{1R}\omega_I + c_{0I}}{2c_2\omega_I + 2c_{1I}}, \tag{16.2.28}$$

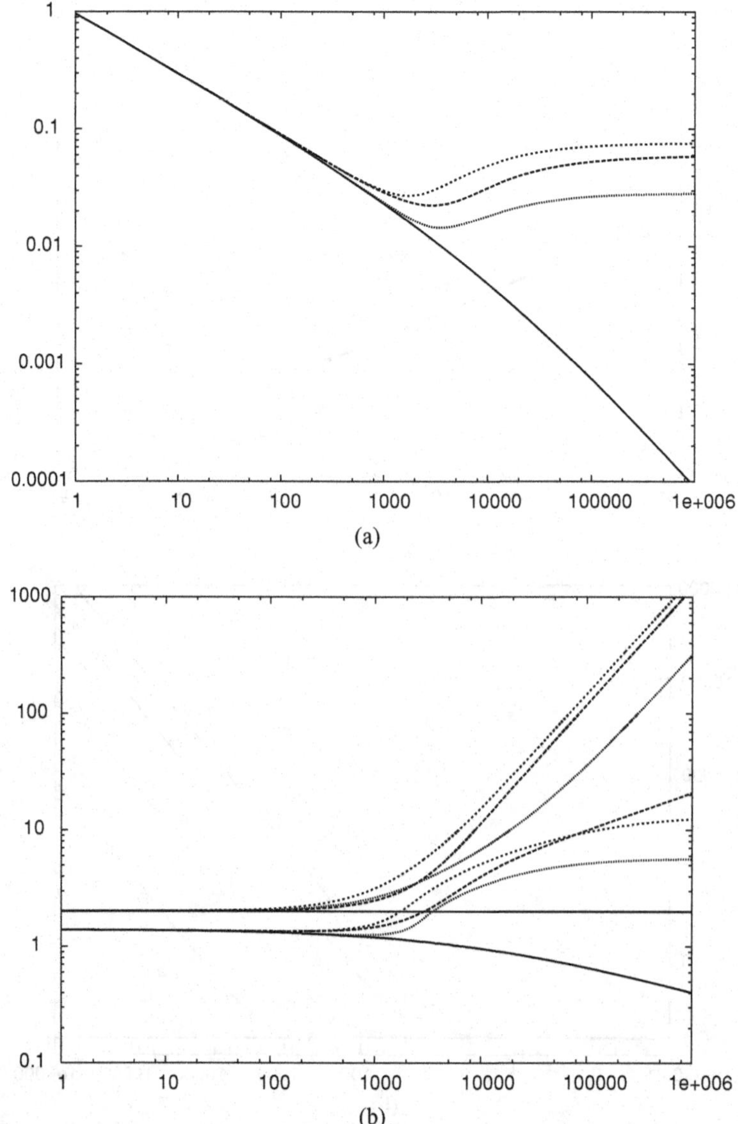

Figure 16.13. (a) ω_{Im} versus We^{-1} $(= \rho 2aU^2/\gamma)$ and (b) k_c (the upper curves) and k_m (the lower curves) versus We^{-1} for $Re = 100$; the solid curves are for $\ell = 0$ and $m = 0$, the broken curves for $\ell = 0.0012$ and $m = 0$, the dashed curves for $\ell = 0.0012$ and $m = 0.018$, the dotted curves for $\ell = 0$ and $m = 0.018$. KH instability for the liquid jet corresponds to $\gamma \to 0$ or $We^{-1} \to \infty$. The neutral curve is independent of the Reynolds number Re. If surface tension and gravity are zero, KH flows are unstable for all k [see equation (2.27) in Funada and Joseph, 2001]. When $U \to 0$, we get capillary instability that is unstable to all waves with $0 < k < 2$. The interval of unstable wave $0 < k < k_c$ increases as the Weber number decreases (larger U, smaller γ). In general, the neutral curve for VPF lies above that for IPF with equality for a given k when $m\alpha b_a = \ell b\alpha_a$ and for large $k > 10$, when $\ell = m\,(\nu = \nu_a)$ [see equations (16.2.36)–(16.2.38)]. The values $k_m(We^{-1})$ for which the growth is maximum depends on Re. The maximum growth rates ω_{Im} are finite for $We \to 0$ but the associated wavenumbers are proportional to $1/We$ for small We.

to give the quartic equation in ω_I alone:

$$a_4\omega_I^4 + a_3\omega_I^3 + a_2\omega_I^2 + a_1\omega_I + a_0 = 0, \tag{16.2.29}$$

with

$$\begin{aligned} &a_4 = c_2^3, \quad a_3 = 4c_2^2 c_{1I}, \quad a_2 = c_2 c_{1R}^2 + 5c_2 c_{1I}^2 - c_2^2 c_{0R}, \\ &a_1 = 2c_{1R}^2 c_{1I} + 2c_{1I}^3 - 2c_2 c_{1I} c_{0R}, \quad a_0 = c_{1R} c_{1I} c_{0I} - c_{1I}^2 c_{0R} - \frac{1}{4} c_2 c_{0I}^2. \end{aligned} \tag{16.2.30}$$

Neutral curves: Neutral curves, $\omega_I = 0$ in (16.2.29), are generated by the condition $a_0 = 0$:

$$a_0 = -\frac{k^6}{Re^2}\left(m^2\alpha b_a^2 + \ell b^2\alpha_a\right) + \frac{k^4}{Re^2}\left(b + mb_a\right)^2 We\left(k^3 - 4k\right) = 0. \tag{16.2.31}$$

One root of (16.2.31) is $k = 0$, and it is the only root when $We = 0$ (KH instability). The other roots are given by

$$We^{-1} = \frac{(b + mb_a)^2}{m^2\alpha b_a^2 + \ell b^2\alpha_a}\left(k - \frac{4}{k}\right), \tag{16.2.32}$$

which has the general form shown in figure 16.13. Equations (16.2.29) and (16.2.31) show that, when $\ell = 0$, $m = 0$, the only instability is due to capillarity. KH instability is not possible in vacuum. An identical conclusion for KH instability of stratified gas–liquid flow in a horizontal rectangular channel follows from equation (3.4) in the paper by Funada and Joseph (2001). For large values of k, b, b_a, α and α_a tend to 1 and (16.2.32) reduces to

$$We^{-1} = \frac{(1 + m)^2}{m^2 + \ell}k. \tag{16.2.33}$$

When $\ell = m$ $(\nu = \nu_a)$, this reduces to $(\ell + 1)\,k/\ell$, which is the same as the inviscid case given by (16.2.20).

A striking conclusion that follows from (16.2.31) and (16.2.32) is that the cutoff wavenumber $k = k_c$ satisfying (16.2.32) is independent of the Reynolds number Re; when $k > k_c$, the liquid jet is stable.

A further comparison, (16.2.32) and (16.2.20), of IPF and VPF shows that the neutral curves are identical under the condition that $m\alpha b_a = \ell b\alpha_a$. Figure 16.10 shows that $\alpha b_a = b\alpha_a$ for $k > 10$; in this case the neutral curves are identical when $\ell = m$ (or $\nu = \nu_a$). It is of interest that for jets of liquid into air $\ell \ll 1$ and $m \ll 1$. In this limit both (16.2.32) and (16.2.20) reduce to

$$We^{-1} = \frac{1}{\ell\alpha_a}\left(k - \frac{4}{k}\right). \tag{16.2.34}$$

This surprising and anti-intuitive result says that the neutral condition for a highly viscous liquid $m \to 0$ is the same as for two inviscid fluids provided that $\ell \ll 1$.

The ratio (16.2.32)/(16.2.20) may be written as

$$\frac{We_{\text{VPF}}^{-1}}{We_{\text{IPF}}^{-1}} = \frac{(b + mb_a)^2\,\ell\alpha\alpha_a}{(m^2\alpha b_a^2 + \ell b^2\alpha_a)(\alpha + \ell\alpha_a)}. \tag{16.2.35}$$

For large $k > 10$, this reduces to

$$\frac{We_{\text{VPF}}^{-1}}{We_{\text{IPF}}^{-1}} = \frac{(1+m)^2 \ell}{(m^2+\ell)(1+\ell)}. \tag{16.2.36}$$

For $m = 0$ (the viscosity of the jet is much larger than the ambient),

$$\frac{We_{\text{VPF}}^{-1}}{We_{\text{IPF}}^{-1}} = \frac{1}{1+\ell} \leq 1. \tag{16.2.37}$$

For $m \to \infty$ (the viscosity of the jet is much smaller than the ambient),

$$\frac{We_{\text{VPF}}^{-1}}{We_{\text{IPF}}^{-1}} = \frac{\ell}{1+\ell} \leq 1. \tag{16.2.38}$$

In general, the neutral curve for VPF is below (or at least not above) that of IPF.

For all values of ℓ, Re, and We, there are wavenumbers for which $\omega_I > 0$; the liquid jet is always unstable to temporal disturbances in analysis based on VPF.

Growth-rate curves: An example of a growth-rate curve is shown in figure 16.11. All of the growth-rate curves have this same form and may be characterized by three parameters: the maximum growth rate ω_{Im} and wavenumber k_m, $\omega_{Im} = \omega_I(k_m)$, and the cutoff wavenumber k_c, as shown in figure 16.11. ω_{Im} and k_m depend on Re, but k_c is independent of Re.

The variation of ω_{Im} and k_m with We^{-1} is shown in figure 16.13. The effect of viscosity is to regularize the Hadamard instability, ω_{Im} tends to a finite value as $We \to \infty$ (cf. figures 16.13 and 16.14). For large values of We^{-1}, KH instability dominates. The great difference between stability in vacuum $(\ell, m) = (0, 0)$ and inviscid gas $m = 0$, $\ell \neq 0$ is apparent for large values of We^{-1}.

For large k (>10) for which $\alpha = \alpha_a = b = b_a = 1$, the imaginary part of dispersion relation (16.2.28) gives ω_R:

$$\omega_R = \frac{k\omega_I + \frac{k^3}{Re}}{(1+\ell)\,\omega_I + \frac{k^2}{Re}(1+m)} = \frac{k}{(1+\ell)\,X}\left[X + \frac{k^2}{Re}(\ell - m)\right], \tag{16.2.39}$$

where X is defined as

$$X = (1+\ell)\,\omega_I + \frac{k^2}{Re}(1+m). \tag{16.2.40}$$

The real part of dispersion relation (16.2.27) leads to the quadratic equation of X^2:

$$\frac{k^2}{X^2}\frac{k^4}{Re^2}(\ell - m)^2 - X^2 + 2Y = 0, \tag{16.2.41}$$

with

$$Y = \frac{1}{2}\left[\frac{k^4}{Re^2}(1+m)^2 + \ell k^2 - (1+\ell)\,We\left(k^3 - 4k\right)\right], \tag{16.2.42}$$

whence the solution to (16.2.41) is expressed as

$$X = (1+\ell)\,\omega_I + \frac{k^2}{Re}(1+m) = \left[Y + \sqrt{Y^2 + \frac{k^6}{Re^2}(\ell - m)^2}\right]^{1/2}. \tag{16.2.43}$$

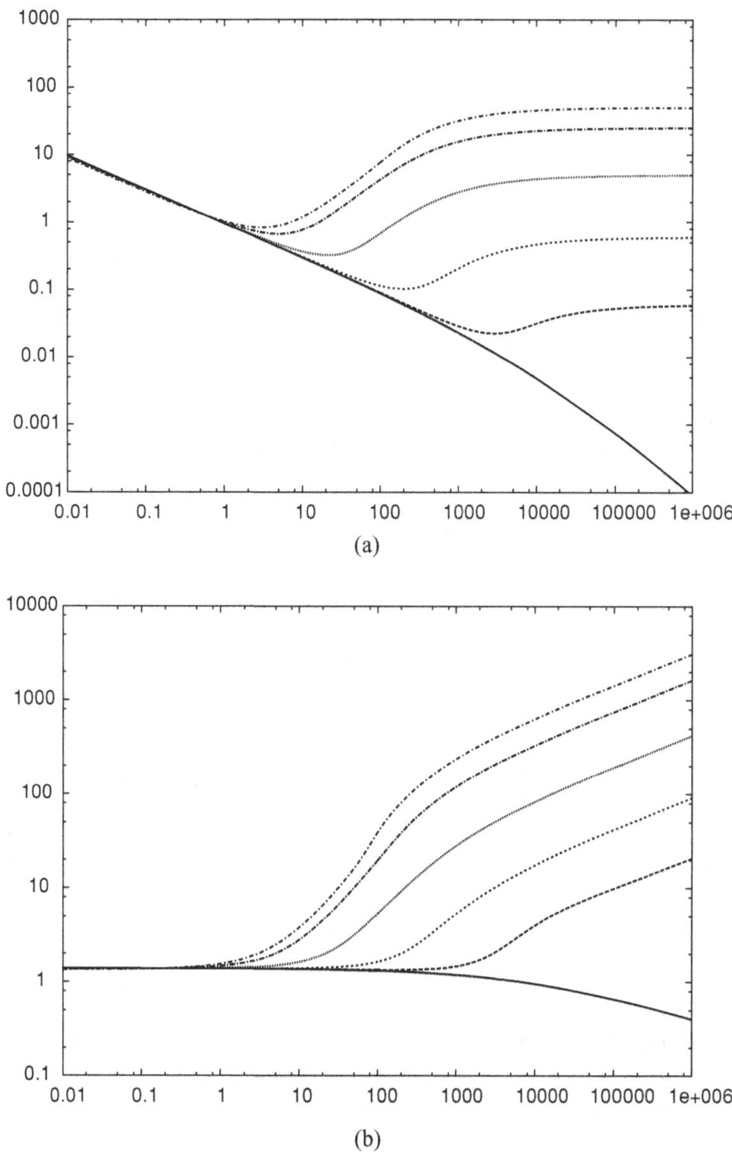

Figure 16.14. (a) ω_{Im} versus We^{-1} and (b) k_m versus We^{-1} for $Re = 100$, $m = 0$, and various ℓ; $\ell = 0$ (solid curves), 0.0012 (broken curves), 0.012 (dashed curves), 0.1 (dotted curves), 0.5 (broken dotted curves), 1 (dash–dotted curves). For large We^{-1} and $\ell \neq 0$, the curves approach the asymptotic form given by (16.2.46).

When $\ell = m$, solution (16.2.43) reduces to

$$X = (1+\ell)\,\omega_I + \frac{k^2}{Re}(1+m) = \sqrt{2Y}. \tag{16.2.44}$$

When $Re \to \infty$, solution (16.2.43) reduces to the inviscid case:

$$X = (1+\ell)\,\omega_I = \left[\ell k^2 - (1+\ell)\,We\left(k^3 - 4k\right)\right]^{1/2}. \tag{16.2.45}$$

Solution (16.2.43) is available to have the maximum growth rate and the cutoff wavenumber when those exist in large k under the condition that $\ell \neq 0$ or $m \neq 0$, for which We^{-1} is large.

We find that the extremum value of ω_I is given by differentiating (16.2.43) and imposing $\partial\omega_I/\partial k = 0$ at $k = k_m$:

$$\omega_{Im} = -\frac{k_m^2}{Re}\frac{(1+m)}{(1+\ell)} + \frac{Re/(1+m)}{4k_m(1+\ell)} \times \left\{ \frac{\partial Y}{\partial k} + \left[Y\frac{\partial Y}{\partial k} + \frac{3k^5}{Re^2}(\ell - m)^2 \right] \left[Y^2 + \frac{k^6}{Re^2}(\ell - m)^2 \right]^{-1/2} \right\}_{k=k_m}. \quad (16.2.46)$$

The expression for k_m is rather cumbersome; it shows that $k_m \propto 1/We$ for small We even though $\omega_I(k_m)$ is bounded as $We \to 0$ (figures 16.13 and 16.14). It follows that the wavelength $\lambda_m = 2\pi/k_m$ tends to zero with We. If this KH disturbance leads to breakup, we would find small liquid fragments even to fine mist.

16.2.4.3 *Nonaxisymmetric disturbances*

The authors of papers on spatial, temporal, and C/A theory cited at the end of subsection 16.2.1 restrict their attention to axisymmetric disturbances. Yang (1992) studied the stability of an inviscid liquid jet to axisymmetric and nonaxisymmetric temporal disturbances. He found wavenumber ranges for which the nonaxisymmetric disturbances grow faster, but the greatest peak values of $\omega_I(k_m)$ are for axisymmetric disturbances. A preliminary study of axisymmetric disturbances proportional to $\exp(in\theta)$ with $n = 0$ and asymmetric disturbances with $n > 0$, especially with $n = 1$, yielded results similar to those found by Yang (1992) for the inviscid jet. The peak growth rates are always attained for $n = 0$ in flows with capillary numbers We larger than small value, say 1/10; for KH instability $We \to \infty$, the peak values for $n = 0$ and $n = 1$ are nearly identical (see figure 16.15).

Li and Kelly (1992) did an analysis of an inviscid liquid jet in a compressible high-speed airstream. They found that $n = 1$ is the most dangerous mode when the Mach number is near to one. The case of nonaxisymmetric disturbances needs further study.

16.2.5 Numerical results of temporal instability

Here we present neutral curves and growth rates for the stability of viscous liquids into air, comparing VPF with IPF.

From the data of Funada and Joseph (2002), various liquid–gas cases are shown in table 16.5. The parameters of the growth-rate curves for the 10 cases are defined in figure 16.11 and given in table 16.6 for typical values of U. The neutral curves $We^{-1}(k)$ for all 10 cases start at $k = 2$. For large k (>10), they may be computed exactly from (16.2.33) and compared with IPF by use of (16.2.36). The differences between VPF and IPF vanish for $k > 10$ when $m\alpha b_a = \ell b\alpha_a$ and for all k when $\rho_a \ll \rho$, $\mu_a \ll \mu$ when μ is very large, as shown previously.

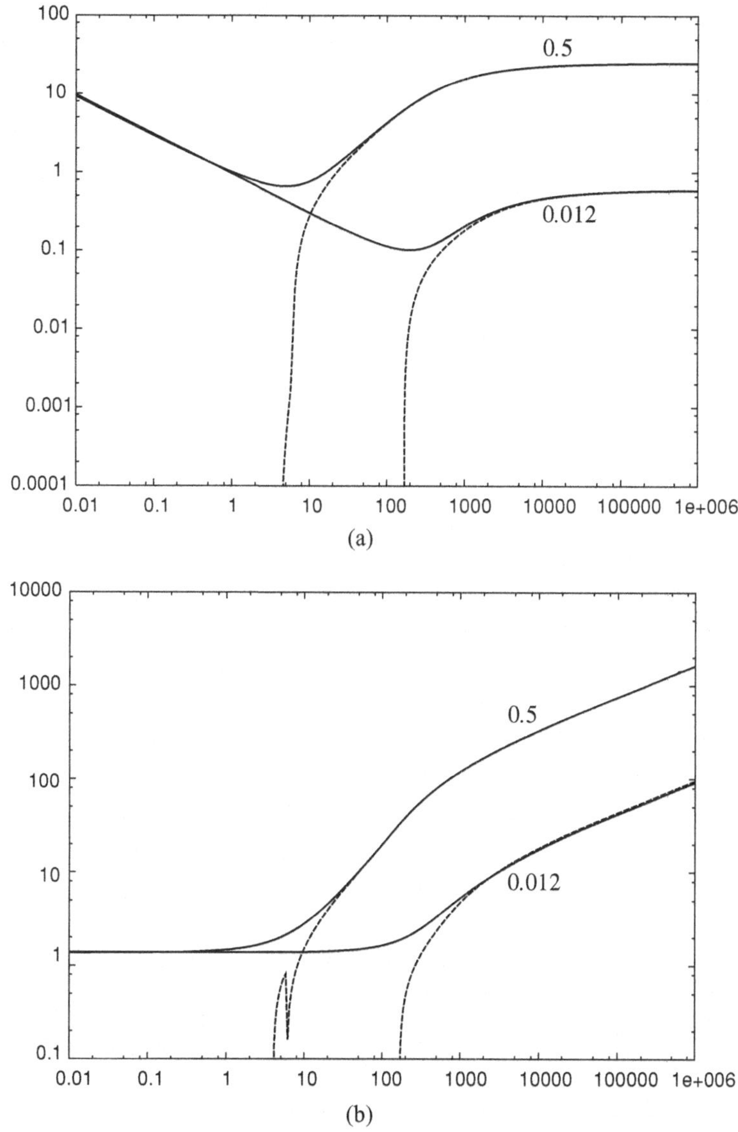

Figure 16.15. (a) ω_{Im} versus We^{-1} and (b) k_m versus We^{-1} for $Re = 100$, and $m = 0$; $\ell = 0.012$ and $n = 0$ (solid curve), $\ell = 0.012$ and $n = 1$ (broken curve), $\ell = 0.5$ and $n = 0$ (solid curve), $\ell = 0.5$ and $n = 1$ (broken curve).

16.2.6 Spatial, absolute, and convective instability

The motivation for considering spatial instability of a liquid jet was very clearly expressed by Keller, Rubinow, and Tu (1973), who noted that the disturbance initiating from the nozzle tip actually grows in space as it is swept downstream, where it is observed to break into drops, leaving a section of jet intact near the nozzle tip as would occur for a disturbance that is convectively unstable. Such disturbances proportional to $\exp(ikz - i\omega t)$ can be described by allowing k to be complex and ω real so that disturbances can grow in space but not in time. They found that Rayleigh's results are relevant only when the Weber

Table 16.5. *Data of various liquid–gas cases*

Case	Liquid cylinder–gas	$\ell = \rho_a/\rho$	$m = \mu_a/\mu$	$m/\ell = \nu_a/\nu$	$\sqrt{\gamma/(\rho 2a)}$ (m/sec)
1	Mercury–air	8.889×10^{-5}	1.154×10^{-2}	1.298×10^{2}	5.976×10^{-2}
2	Water–air	1.200×10^{-3}	1.800×10^{-2}	1.500×10	8.532×10^{-2}
3	Benzene–air	1.395×10^{-3}	2.769×10^{-2}	1.985×10	5.793×10^{-2}
4	SO100–air	1.238×10^{-3}	1.800×10^{-4}	1.454×10^{-1}	4.655×10^{-2}
5	Glycerin–air	9.547×10^{-4}	2.302×10^{-5}	2.411×10^{-1}	7.102×10^{-2}
6	Oil–air	1.478×10^{-3}	3.830×10^{-5}	2.592×10^{-2}	5.984×10^{-2}
7	SO10000–air	1.238×10^{-3}	1.800×10^{-6}	1.454×10^{-3}	4.655×10^{-2}
8	SO10–air	1.238×10^{-3}	1.800×10^{-3}	1.454	4.655×10^{-2}
9	Silicon oil–nitrogen	1.345×10^{-3}	8.750×10^{-4}	6.508×10	4.697×10^{-2}
10	Silicon oil–nitrogen	1.326×10^{-3}	1.750×10^{-3}	1.319	4.620×10^{-2}

Table 16.6. *Parameters of the growth-rate curves identified in figure 16.11 for the 10 cases of liquid–gas flow in Table 16.5*

Viscous (VPF) and (IPF) are compared

Case	U (m/sec)	We	We^{-1}	Re	ω_{Im}^{VPF}	k_m^{VPF}	k_c^{VPF}	ω_{Im}^{IPF}	k_m^{IPF}	k_c^{IPF}
1	0.02	8.928	0.112	0.17308×10^4	0.9709	1.394	2.000	0.9711	1.394	2.000
	0.06	0.992	1.008	0.51923×10^4	0.9709	1.394	2.000	0.9711	1.394	2.000
	0.10	0.357	2.800	0.86538×10^4	0.9709	1.394	2.000	0.9712	1.394	2.000
2	0.02	18.200	0.055	0.20000×10^3	0.9698	1.393	2.000	0.9710	1.394	2.000
	0.06	2.022	0.495	0.60000×10^3	0.9699	1.393	2.000	0.9711	1.394	2.000
	0.10	0.728	1.374	0.10000×10^4	0.9701	1.394	2.000	0.9713	1.394	2.000
3	0.02	8.390	0.119	0.26462×10^3	0.9696	1.393	2.000	0.9710	1.394	2.000
	0.06	0.932	1.073	0.79385×10^3	0.9699	1.394	2.000	0.9713	1.394	2.000
	0.10	0.336	2.980	0.13231×10^4	0.9704	1.394	2.001	0.9718	1.395	2.001
4	0.02	5.418	0.185	0.19380×10	0.7890	1.244	2.000	0.9710	1.394	2.000
	0.06	0.602	1.661	0.58140×10	0.7893	1.244	2.000	0.9714	1.395	2.000
	0.10	0.217	4.614	0.96900×10	0.7898	1.245	2.001	0.9722	1.396	2.001
5	0.02	12.609	0.079	0.32148	0.5138	0.988	2.000	0.9710	1.394	2.000
	0.06	1.401	0.714	0.96445	0.5139	0.988	2.000	0.9712	1.394	2.000
	0.10	0.504	1.983	0.16074×10	0.5140	0.988	2.000	0.9714	1.395	2.000
6	0.02	8.951	0.112	0.36431	0.5028	0.977	2.000	0.9710	1.395	2.000
	0.06	0.995	1.005	0.10929×10	0.5029	0.977	2.000	0.9713	1.395	2.000
	0.10	0.358	2.793	0.18215×10	0.5031	0.977	2.001	0.9718	1.396	2.001
7	0.02	5.418	0.185	0.19380×10^{-1}	0.0430	0.278	2.000	0.9710	1.394	2.000
	0.06	0.602	1.661	0.58140×10^{-1}	0.0430	0.279	2.000	0.9714	1.395	2.000
	0.10	0.217	4.614	0.96900×10^{-1}	0.0430	0.278	2.002	0.9722	1.396	2.001
8	0.02	5.418	0.185	0.19380×10^2	0.9488	1.376	2.000	0.9710	1.394	2.000
	0.06	0.602	1.661	0.58140×10^2	0.9492	1.377	2.000	0.9714	1.395	2.000
	0.10	0.217	4.614	0.96900×10^2	0.9499	1.378	2.002	0.9722	1.396	2.001
9	0.02	5.515	0.181	0.95200×10	0.9273	1.359	2.000	0.9710	1.394	2.000
	0.06	0.613	1.632	0.28560×10^2	0.9277	1.359	2.000	0.9714	1.395	2.000
	0.10	0.221	4.533	0.47600×10^2	0.9284	1.361	2.002	0.9722	1.396	2.002
10	0.02	5.337	0.187	0.19300×10^2	0.9485	1.376	2.000	0.9710	1.394	2.000
	0.06	0.593	1.686	0.57900×10^2	0.9489	1.377	2.000	0.9714	1.395	2.000
	0.10	0.213	4.684	0.96500×10^2	0.9497	1.378	2.002	0.9723	1.396	2.002

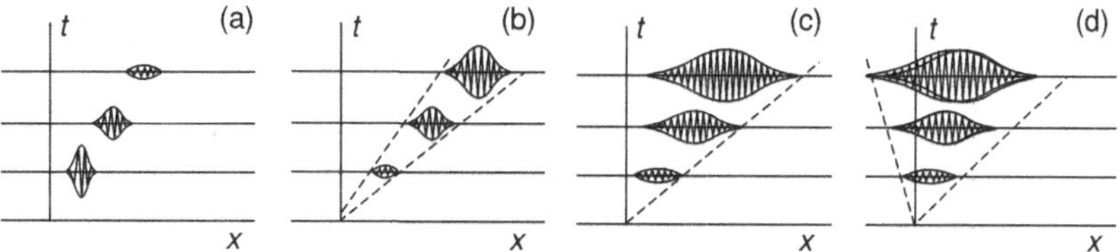

Figure 16.16. Linear impulse response: (a) linearly stable flow, (b) linearly convectively unstable flow, (c) marginally C/A unstable flow, (d) absolutely unstable flow (after Huerre, 2000, figure 8). The pictures here are for a response to a linear impulses; stability cannot be determined from the evolution of impulses alone. Convectively unstable flows in (b) are also temporally unstable; at any x, real wavenumbers exist for which disturbances outside the unstable wedge in (b) will grow.

number $We = \gamma/(\rho 2aU^2)$ is small, the spatial growth rate k_I is related to the temporal growth rate ω_I by the relation $k_I = \pm\omega_I + O(We)$, and the disturbance travels at the jet velocity. For large values of We, they found a new mode of faster-growing disturbances whose wavelengths are perhaps too long to be observable.

Leib and Goldstein (1986b) showed that the new mode corresponds to an absolute instability that arises from a pinch-point singularity in the dispersion relation. An absolutely unstable wave packet propagates upstream and downstream, and hence such disturbances spread over the whole z domain of flow; the flow is unstable at any z as t increases. This is not the picture advanced by Keller, Rubinow, and Tu (1973), in which the region close to the discharge is never corrupted by growing disturbances.

One of the aims of the theory of absolute and convective instability is to provide a framework for the problem of spatial development of disturbances. The spatial development of controlled disturbances such as are generated by a vibrating ribbon at the start of a growing boundary layer or at the inlet of a plane Poiseuille flow calculated from spatial theory yielded good results with experiments in which disturbances were suppressed. However, the spatial theory has no rigorous foundation; for example, a spatial mode, when it can be defined, is inadmissible when it is unbounded at infinity, though it may describe the spatial evolution of disturbances of a given frequency for a long time (see Drazin and Reid, 1981, §32 for experiments, §47.1 for theoretical problems).

To deal with the problem of propagation of impulses, the concept of convective and absolutely unstable solutions (C/A for short) has been introduced. The definitions of instability in the C/A context are formulated in terms of the evolution of impulses at the origin in the (x, t) plane proportional initially to the product $\delta(x)\delta(t)$. A flow is called *linearly stable* if this disturbance decays to zero along all rays $x/t =$ const. It is *linearly unstable* if the impulse tends to infinity along at least one ray $x/t =$ const. An unstable flow is linearly *convectively unstable* if the impulse tends to zero along the ray $x/t = 0$ and is *absolutely unstable* if the impulse tends to infinity along the ray $x/t = 0$. Obviously an absolutely unstable impulse is linearly unstable.

A line drawing for wave packets propagating from the origin in these four cases is shown in figure 16.16.

The C/A concepts are not straightforward and need explanation. Huerre (2000) notes that

> A parallel shear flow of given velocity profile is said to be *convectively unstable* if the growing wavepacket produced in response to an impulsive source localized in space and time is

advected away. It is *absolutely unstable* if the growing wavepacket expands around the source to contaminate the entire medium. In the case of parallel flows that are invariant under Galilean transformations, this distinction appears at first sight to be preposterous: a simple change of reference frame transforms a flow from convectively unstable to absolutely unstable and vice versa, and the "laboratory frame" is not properly defined. However, when Galilean invariance is broken, e.g. in spatially developing flows, in flows with a definite origin, or in flows forced at a specific streamwise station, the laboratory frame is singled out and it is precisely in these instances that the distinction between convective and absolute instability becomes of interest. It should be emphasized that, in order for these concepts to be relevant, one must enforce a *scale-separation* assumption: the flow under consideration must be slowly evolving along the stream over a typical instability wavelength. This strong hypothesis is made throughout the ensuing theoretical developments in order to recover the locally parallel flow instability properties as a leading-order approximation at each streamwise station.

Parallel flows that are temporally stable are also stable in the C/A theory. Temporally unstable flows can be absolutely or convectively unstable. Disturbances from a source, like a vibrating ribbon, can propagate without corrupting the source only if the flow is convectively unstable and only if random disturbances at fixed points that are temporally unstable are suppressed.

In this section we follow Funada, Joseph, and Yamashita (2004) and others in considering the open basic flow that has no spatial variation in the axial direction that is Galilean invariant. The point of novelty of our analyses is that we carry out the stability analysis by using the equations of VPF that allow a discontinuous velocity at the jet surface but accommodate effects of viscosity, viscosity ratios, density ratios, etc., and still lead to an explicit dispersion relation.

Huerre and Monkewitz (1985) point out that the spatial stability theory is applicable when the flow is convectively unstable and not when the flow is absolutely unstable. They describe a methodology based on Bers (1975) criterion to determine when a free shear layer is convectively unstable. Their problem is difficult and does not give rise to an explicit dispersion; numerical computations are required. The search for the border between absolute and convective instability is a function of prescribed parameters and requires knowledge of the dispersion relation for complex frequencies and wavenumbers. This search is greatly simplified in the case of the stability of the liquid jet based on VPF because the search for singularities of the dispersion relation is reduced to the study of algebraic equations of two complex variables.

We use the criterion of Bers (1975), which says that an unstable flow is convectively unstable if the modes proportional to $\exp(ikz - i\omega t)$ of complex frequency ω and k, which have a zero group velocity

$$G_V = \partial\omega_R/\partial k_R = 0, \tag{16.2.47}$$

are all temporally damped, $\omega_I < 0$. Otherwise the system is absolutely unstable. If ω_R does not change with k_R and $\omega_I < 0$, then a disturbance with excitation frequency ω_R will decay in time but can grow in space. On the other hand, we do not use the method of Briggs (1964) and Bers (1975). Instead, we implement an algebraic study of the dispersion relation.

Following Schmid and Henningson (2001), we may characterize the singularities of $D(k, \omega)$ at k_0, ω_0 by

$$D(k_0, \omega_0) = 0, \quad \frac{\partial D}{\partial k}(k_0, \omega_0) = 0, \quad \frac{\partial^2 D}{\partial k^2}(k_0, \omega_0) \neq 0. \tag{16.2.48}$$

In the neighborhood of k_0, ω_0 a Taylor series expansion of $D(k, \omega)$ leads to

$$0 = \left.\frac{\partial D}{\partial \omega}\right|_0 (\omega - \omega_0) + \frac{1}{2}\left.\frac{\partial^2 D}{\partial k^2}\right|_0 (k - k_0)^2 + HO, \tag{16.2.49}$$

where HO are terms that go to zero faster than the terms retained. This results in a square-root singularity for the local map between the k and ω planes.

If we imagine $D(k, \omega) = 0$ to be solved for $\omega = \omega(k)$, then $D[k, \omega(k)] = 0$ is an identity in k and

$$0 = \frac{\mathrm{d}D}{\mathrm{d}k} = \frac{\partial D}{\partial k} + \frac{\partial \omega}{\partial k}\frac{\partial D}{\partial \omega}, \tag{16.2.50}$$

$$\hat{c} \equiv \frac{\partial \omega}{\partial k} = -\frac{\partial D}{\partial k} \Big/ \frac{\partial D}{\partial \omega} = \hat{c}_R + i\hat{c}_I \tag{16.2.51}$$

can be said to be a complex-valued "generalized" group velocity that must be zero at the singularity. This is not the ordinary group velocity. If $\partial D/\partial \omega \neq 0$, and it is not equal to zero in this study, then

$$\hat{c} = \frac{\partial \omega}{\partial k} = 0 \quad \text{when} \quad \frac{\partial D}{\partial k} = 0. \tag{16.2.52}$$

Moreover,

$$\frac{\partial \omega}{\partial k} = \frac{1}{2}\left(\frac{\partial \omega_R}{\partial k_R} + \frac{\partial \omega_I}{\partial k_I}\right) + \frac{i}{2}\left(\frac{\partial \omega_I}{\partial k_R} - \frac{\partial \omega_R}{\partial k_I}\right) \tag{16.2.53}$$

and $\hat{c} = 0$ does not imply that the group velocity $G_V = \partial \omega_R/\partial k_R = 0$. However, because the Cauchy–Riemann condition for a function $\omega(k)$ holds, then $\partial \omega/\partial \bar{k} = 0$ and

$$\frac{\partial \omega_R}{\partial k_R} = \frac{\partial \omega_I}{\partial k_I}, \quad \frac{\partial \omega_I}{\partial k_R} = -\frac{\partial \omega_R}{\partial k_I}. \tag{16.2.54}$$

Hence, if (16.2.52) holds, then (16.2.54) implies that

$$G_V = \frac{\partial \omega_R}{\partial k_R} = \frac{\partial \omega_I}{\partial k_I} = 0, \quad \frac{\partial \omega_R}{\partial k_I} = -\frac{\partial \omega_I}{\partial k_R} = 0. \tag{16.2.55}$$

16.2.7 Algebraic equations at a singular point

A *singular point* satisfies (16.2.48); alternatively, $D = \hat{c} = 0$. These are complex equations, four real equations for k_R, k_I, ω_R, and ω_I when the other parameters are prescribed. If $\omega_I < 0$ at a singular point, the flow is convectively unstable. A *critical singular point* is a singular point such that $\omega_I = 0$.

For given values of the parameters (ℓ, We, Re, and $m = 0$), the solution $k = k_R + ik_I$ $[\omega(k) = \omega_R + i\omega_I]$ of the dispersion relation (16.2.14, 15 and 16) is obtained implicitly. Equation (16.2.14) is quadratic in ω and has two roots, ω_1 and ω_2:

$$\omega_1 = -\frac{c_1}{c_2} + \sqrt{\left(\frac{c_1}{c_2}\right)^2 - \frac{c_0}{c_2}}, \quad \omega_2 \equiv \omega = -\frac{c_1}{c_2} - \sqrt{\left(\frac{c_1}{c_2}\right)^2 - \frac{c_0}{c_2}}, \tag{16.2.56}$$

where c_0, c_1, and c_2 are defined in (16.2.15). The second root can be singular, thus here and henceforth, we drop the subscript $_2$ to simplify notation. A singular point $(k, \omega) = (k_0, \omega_0)$ can now be defined relative to ω

$$\omega_0 = \omega(k_0), \quad \omega(k) = \omega_0 + \left.\frac{\partial \omega}{\partial k}\right|_0 (k - k_0) + \frac{1}{2}\left.\frac{\partial^2 \omega}{\partial k^2}\right|_0 (k - k_0)^2 + \cdots + . \tag{16.2.57}$$

If

$$\left.\frac{\partial \omega}{\partial k}\right|_0 = 0 \quad \text{at} \quad k = k_0 \quad \text{and} \quad \left.\frac{\partial^2 \omega}{\partial k^2}\right|_0 \neq 0 \quad \text{at} \quad k = k_0, \tag{16.2.58}$$

a pinch in the k plane is a square-root branch point in the ω plane. We identify k_0 as the roots of

$$\frac{\partial \omega}{\partial k}(k_0) = 0, \quad (k - k_0)^2 = (\omega - \omega_0)/\left[\frac{1}{2}\left.\frac{\partial^2 \omega}{\partial k^2}\right|_0\right]. \tag{16.2.59}$$

The critical singular point satisfies (16.2.58) and $\omega_I = 0$.

Our solution procedure is as follows: the root $\omega = \omega_R + i\omega_I$, where $\omega_R \equiv \omega_R(k_R, k_I)$ and $\omega_I \equiv \omega_I(k_R, k_I)$, could be inverted implicitly for

$$k_R = k_R(\omega_R, \omega_I), \quad k_I = k_I(\omega_R, \omega_I). \tag{16.2.60}$$

The singular point is determined from conditions (16.2.58). The solutions must be implicit because of the Bessel function. The two real equations in (16.2.59) may be solved for k_R and k_I.

To seek a singular point k_0 at which $\hat{c} = \mathrm{d}\omega/\mathrm{d}k = 0$, we make the computation by means of Newton's method:

$$\hat{c}(k) - \hat{c}(k_s) = \left(\frac{\mathrm{d}\hat{c}}{\mathrm{d}k}\right)_s (k - k_s), \tag{16.2.61}$$

where k_s is a starting value that may be close to the singular point k_0. Because the solution k_0 is to satisfy $\hat{c}(k_0) = 0$, we may rewrite (16.2.61) as the following iteration algorithm:

$$k = k_s - \hat{c}(k_s)/\left(\frac{\mathrm{d}\hat{c}}{\mathrm{d}k}\right)_s. \tag{16.2.62}$$

For given k_s, the right-hand side of this equation is calculated to give a next approximate solution k. The iteration is repeated until $|k - k_s| < \epsilon$ ($\epsilon < 10^{-6}$) or until the iteration is made over 30 times. The solution k_0 also gives $\omega(k_0)$; then we can find the critical singular point when $\omega_I = 0$.

16.2.8 Subcritical, critical, and supercritical singular points

The formation and properties of singular points are similar for all cases. Here we look at some typical cases for the formation of pinch points in the (k_R, k_I) plane and cusp points in the (ω_R, ω_I) plane. We use the Weber number parameter $\beta = \frac{\rho 2aU^2}{\gamma} = We^{-1}$, where We is the Weber number defined in (16.2.2).

First we fix the parameters $(\ell, Re, m) = (0, 100, 0)$ and plot curves of constant ω_I and ω_R in the k_R, k_I plane in the subcritical ($\beta = 4.934$, figure 16.17), critical ($\beta = 5.134$,

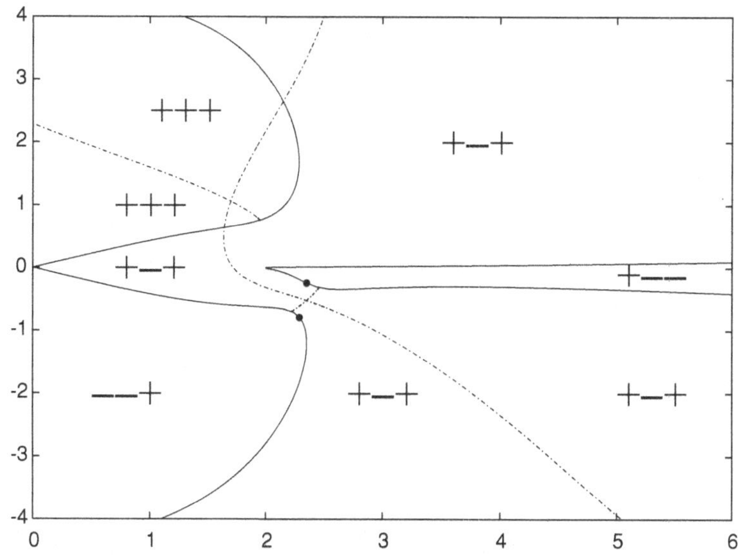

Figure 16.17. k_I vs. k_R for $\ell = 0$, $Re = 100$, $\beta = 4.934$, and $m = 0$. Equation (16.2.14) gives rise to three complex roots k for each prescribed set of parameters; for each of the three k's there is one value of ω_I whose sign is marked on the figure. The value $\beta = 4.934 < \beta_c = 5.134$ is subcritical. The singular point $D = 0$ and $\hat{c} = 0$ (or $c = \partial\omega_R/\partial k_R = 0$) has $\omega_I > 0$ in the subcritical case and the flow is absolutely unstable; this point is not shown but the points • that will merge into a pinch point • in figure 16.18 are identified. The solid curves are given by $D = 0$ and $\omega_I = 0$. The dashed curves are for $D = 0$, $\omega_R = 1.7178$, and $\omega_I \geq 0$.

figure 16.18), and supercritical ($\beta = 5.334$, figure 16.19) cases. The cusp singularity, with $\omega_I < 0$ at the cusp point, is shown in figure 16.20. The pinch point in the subcritical case is in the region $\omega_I > 0$; this is in the region of absolute instability. The pinch point in the supercritical case is in the region $\omega_I < 0$; this is in the region of convective instability.

We draw the reader's attention to the fact the curves $k_I(k_R)$ on which $\omega_R =$ constant typically pass through regions in which ω_I is positive and negative. The only curves for which

$$\omega\left[k_R, k_I(k_R)\right] = \text{const} \tag{16.2.63}$$

that lie entirely in regions in which ω_I is of one sign pass through the pinch point. In the supercritical case shown in figure 16.19, $\omega_R = 1.743$ is the only frequency for $\omega_I < 0$ for all k_R. Of course, we may interpret ω_R as an excitation frequency.

Huerre and Monkewitz (1985) describe the evolution of a wave packet from an impulsive source in the convectively unstable case. They note that

> ... among all the wavenumbers contained in the impulsive source, the flow selects, along each ray $x/t =$ const. one particular complex wavenumber k^* given by
>
> $$\frac{\mathrm{d}\omega}{\mathrm{d}k}(k^*) = \frac{x}{t}. \tag{16.2.64}$$
>
> The group velocity then is real and the temporal amplification rate of the wave reduces to
>
> $$\sigma = \omega_I(k^*) - k_I^*(\mathrm{d}\omega/\mathrm{d}k)(k^*). \tag{16.2.65}$$

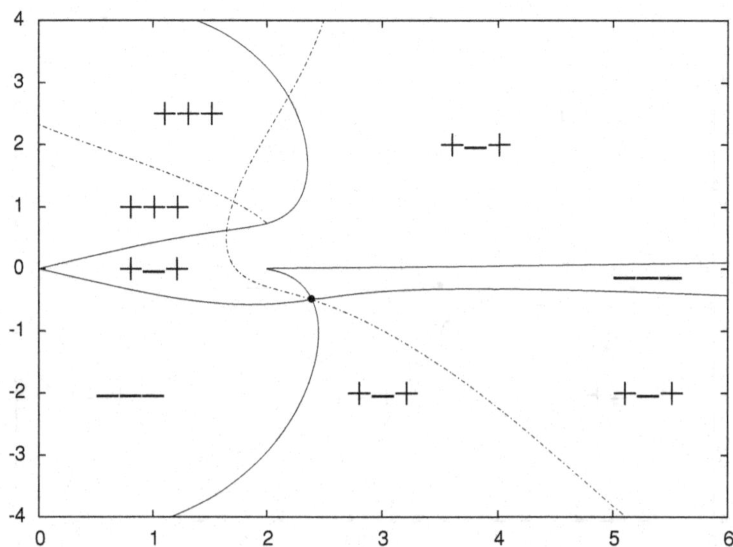

Figure 16.18. k_I vs. k_R for $\ell = 0$, $Re = 100$, $m = 0$, and $\beta = \beta_c = 5.134$ is critical and identified by •. At this point $D = 0$, $\hat{c} = 0$, $(\omega_R, \omega_I) = (1.7304, 0)$ and $(k_R, k_I) = (2.392, -0.496)$. The dashed curve $D = 0$, and $\omega_R = 1.7304$ passes through the critical point and has $\omega_I \leq 0$.

The growth function then becomes

$$\exp(-k_I^* x + \omega_I t) = \exp[\omega_I - k_I^* (\mathrm{d}\omega/\mathrm{d}k)]\, t. \tag{16.2.66}$$

Using (16.2.53) and (16.2.54), we find that the real part of the group velocity is

$$\frac{\partial \omega}{\partial k}(k^*) = \frac{\partial \omega_R}{\partial k_R}. \tag{16.2.67}$$

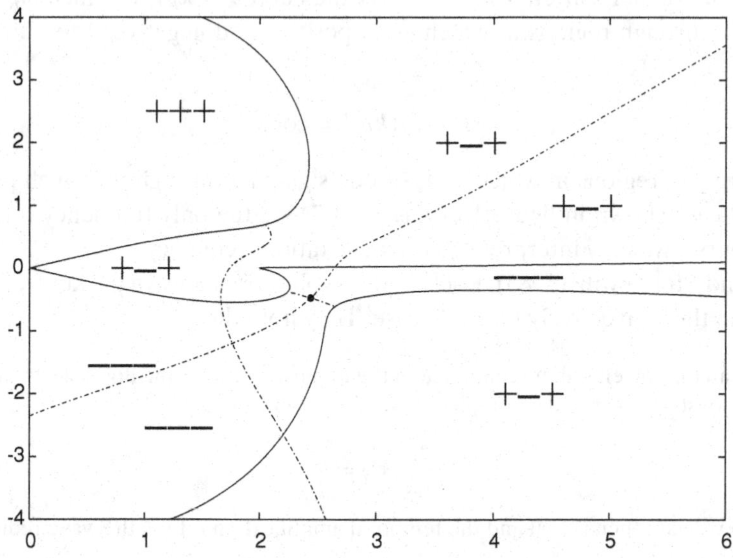

Figure 16.19. k_I vs. k_R for $\ell = 0$, $Re = 100$, $m = 0$, and $\beta = 5.334 > \beta_c$ is supercritical. The singular point is shown as a dot • and $\omega_I < 0$ there. On dashed curves $\omega_R = 1.743$.

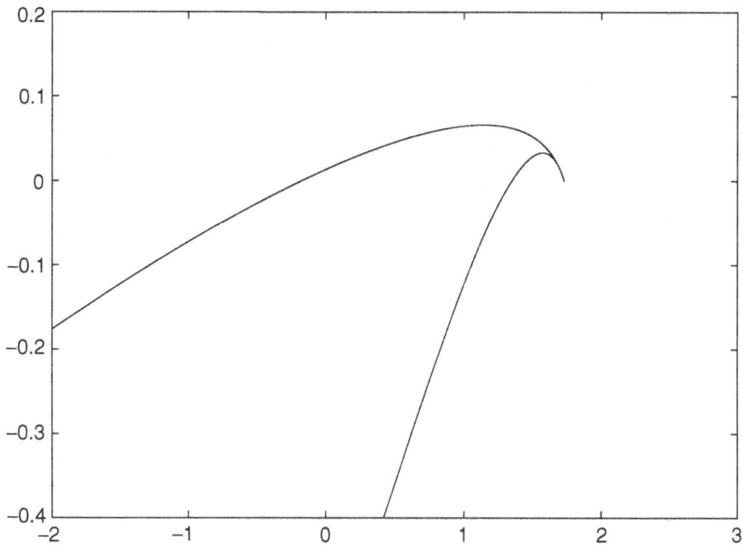

Figure 16.20. Cusp point $(\omega_R, \omega_I) = (1.7304, 0)$. ω_I vs. ω_R for $\ell = 0$, $\beta = 5.134$, $Re = 100$, and $m = 0$; the solid curves are for $D = 0$ and $\hat{c} = 0$, which passes through the pinch point $(k_R, k_I) = (2.392, -0.496)$ in the (k_R, k_I) plane (see figure 16.18).

Focusing now on the lines $\omega_R[k_R, k_I(k_R)] = \text{const}$ in the k plane (figures 16.17–16.19) we find that

$$\frac{\mathrm{d}\omega_R}{\mathrm{d}k_R}[k_R, k_I(k_R)] = \frac{\partial \omega_R}{\partial k_R} + \frac{\partial \omega_R}{\partial k_I}\frac{\mathrm{d}k_I}{\mathrm{d}k_R} = \frac{\partial \omega_R}{\partial k_R} = 0. \qquad (16.2.68)$$

The only curve $\omega_R = \text{const}$ that has $\omega_I \leq 0$ for all $k_R > 0$ is the one that passes through the pinch point. The harmonic content of the impulsive source produced by this frequency is not restricted. In general, the harmonic content of impulsive sources that are convectively unstable depends on the frequency ω_R of the excitation.

From now on, we describe singular points as points at which the group velocity $G_V = \partial\omega_R/\partial k_R = 0$. This is a shorthand for the conditions $D = 0$ and $\hat{c} = 0$, which are four equations for k_R, k_I, ω_R, and ω_I. All of the graphs shown in the figures that follow satisfy the dispersion relation $D = 0$. We have shown in subsection 16.2.6 that the condition $\hat{c} = 0$ implies that $G_V = 0$. This leads to the condition of Bers (1975), that is, a flow is convectively unstable when $\omega_I < 0$ and when $G_V = 0$.

The qualitative properties of singularities are the same in all cases. At a critical singular point $\omega_I = 0$. The values of parameters at critical singular points are given for $Re = 100$, $Re = 200$, and $Re = 2000$ (table 16.7) for three or four different values of m and $0 \leq \ell \leq 0.4$. We found that when $m = 0$, so that the ambient viscosity is zero, one and only one critical β is found for given values of Re and ℓ. When the ambient viscosity is finite ($m = 0.5, 1$) two critical values are found, but only one is spatially unstable, $k_I < 0$.

The inviscid case $Re \to \infty$ is degenerate and is treated in subsection 16.2.9.

Figures 16.17–16.19 look at the (k_R, k_I) plane in the subcritical, critical, and supercritical cases for $(\ell, Re, m) = (0, 100, 0)$. In these figures we plot curves in k_I vs. k_R that arise from the real and imaginary part of the dispersion relation $D(k_R + ik_I, \omega_R + i\omega_I) = 0$ when ω_R or ω_I is fixed. The singular points are points on the curves that satisfy (16.2.52).

Table 16.7. *Critical values of* $\beta = \beta_c(\ell, Re, m)$ *at a generic singular point* $(D = 0, \hat{c} = 0)$

At such a point the group velocity $\partial\omega_R/\partial k_R = 0$, (k_R, k_I) is a pinch point; (ω_R, ω_I) is a cusp point. A critical singular point also has $\omega_I = 0$. When $\beta < \beta_c$ the flow is subcritical (absolutely unstable) and disturbances with zero group velocity are amplified ($\omega_I > 0$, $k_I < 0$). When $\beta > \beta_c$ (convectively unstable) these disturbances decay temporally ($\omega_I < 0$, $k_I < 0$).

Re	m	ℓ	β_c	k_R	k_I	ω_R
100	0	0.00	5.134	2.392	−0.496	1.730
		0.08	5.714	2.550	−0.565	1.823
		0.16	6.495	2.762	−0.668	1.945
		0.24	7.596	3.065	−0.834	2.116
		0.32	9.192	3.510	−1.127	2.360
		0.40	11.358	4.139	−1.643	2.689
100	1	0.00	13.690	5.899	−1.024	2.585
		0.08	15.296	6.377	−1.340	2.832
		0.16	17.452	7.004	−1.830	3.158
		0.24	20.396	7.837	−2.590	3.589
		0.32	24.320	8.889	−3.692	4.135
		0.40	29.182	10.044	−5.129	4.760
100	0.5	0.00	7.206	3.225	−0.309	1.882
		0.08	8.158	3.478	−0.424	2.026
		0.16	9.484	3.840	−0.630	2.226
		0.24	11.352	4.374	−1.010	2.508
		0.32	13.814	5.129	−1.647	2.879
		0.40	16.664	6.058	−2.505	3.308
200	0	0.00	5.493	2.454	−0.403	1.759
		0.08	6.198	2.646	−0.467	1.868
		0.16	7.193	2.920	−0.569	2.020
		0.24	8.696	3.350	−0.753	2.249
		0.32	11.042	4.074	−1.126	2.612
		0.40	14.276	5.203	−1.804	3.131
200	1	0.00	13.638	6.046	−0.534	2.612
		0.08	15.306	6.606	−0.716	2.885
		0.16	17.720	7.411	−1.036	3.279
		0.24	21.594	8.698	−1.679	3.906
		0.32	28.344	10.943	−3.081	4.975
		0.40	38.728	14.247	−5.688	6.539
200	0.5	0.00	7.244	3.238	−0.162	1.891
		0.08	8.242	3.502	−0.230	2.044
		0.16	9.694	3.894	−0.374	2.266
		0.24	11.956	4.565	−0.706	2.616
		0.32	15.388	5.740	−1.368	3.161
		0.40	20.154	7.497	−2.412	3.943
2000	0	0.00	6.065	2.539	−0.192	1.799
		0.08	7.011	2.788	−0.229	1.937
		0.16	8.488	3.191	−0.296	2.151
		0.24	11.154	3.978	−0.447	2.542
		0.32	16.744	5.865	−0.865	3.417
		0.40	35.516	12.573	−2.650	6.620

Re	m	ℓ	β_c	k_R	k_I	ω_R
2000	1	0.00	13.622	6.097	−0.054	2.621
		0.08	15.320	6.690	−0.073	2.905
		0.16	17.868	7.574	−0.109	3.333
		0.24	22.550	9.181	−0.201	4.120
		0.32	38.966	14.774	−0.847	6.873
		0.40	196.642	72.962	−21.449	33.075
2000	0.5	0.00	7.260	3.242	−0.016	1.894
		0.08	8.274	3.506	−0.024	2.050
		0.16	9.806	3.899	−0.042	2.285
		0.24	12.584	4.610	−0.132	2.713
		0.32	19.602	6.916	−0.665	3.830
		0.40	63.200	22.614	−4.872	11.276

The cusp point in (ω_R, ω_I) plane at criticality is shown in figure 16.20. These graphs are representative for all the nondegenerate cases $Re < \infty$. The explanations of the figures are given in the captions.

In figures 16.21 and 16.22 we have plotted β_c versus ℓ for $Re = 2000, 200$, and 100 when $m = 0$ and $m = 0.5$, respectively.

Figure 16.23 gives a summary of the behavior of singular points for inviscid $Re \to \infty$ and viscous fluids $Re = 100, 200$. A detailed explanation of this summary is given in the figure caption.

16.2.9 Inviscid jet in inviscid fluid ($Re \to \infty, m = 0$)

This problem was studied for the case $\ell = 0$ by Keller, Rubinow, and Tu (1973); they did not look at the problem of C/A instability, which was treated later by Leib and Goldstein (1986a). Here we look at the inviscid problem for all density ratios and find the border, the

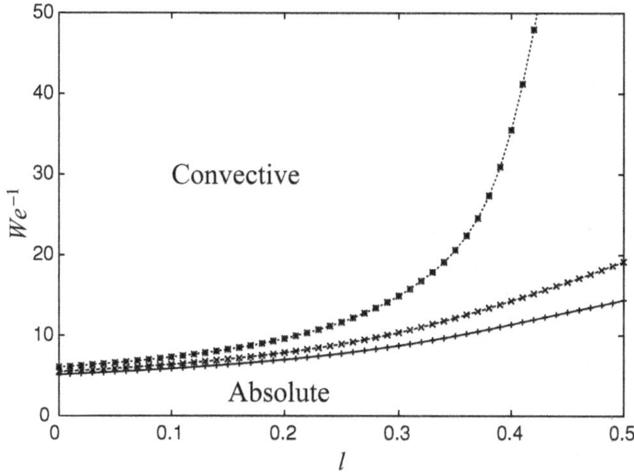

Figure 16.21. β_c versus ℓ when $m = 0$; $Re = 2000\ (*)$, $200\ (\times)$, $100\ (+)$.

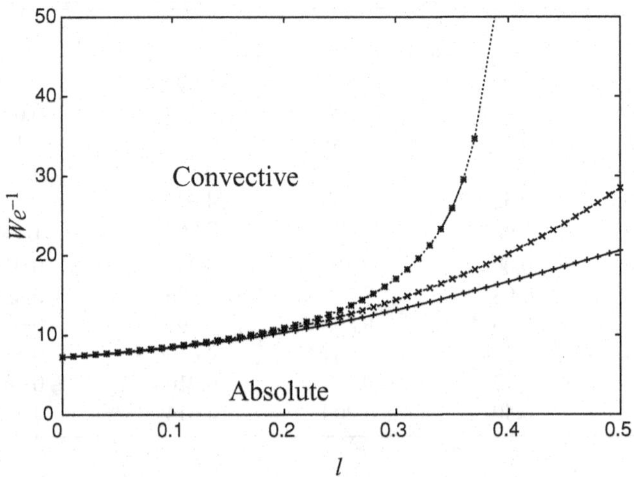

Figure 16.22. β_c versus ℓ when $m = 0.5$; $Re = 2000$ (∗), 200 (×), 100 (+).

critical $\beta = \beta_c$ at the singular point (16.2.48). We find that the singular point is degenerate, because the imaginary part of $\partial^2 D/\partial k^2$ at the pinch point vanishes for $0 < \ell < 0.3455$. The degeneracy appears in the collapse of the region in which $\omega_I < 0$ for large k_R into a line; this region of convective instability collapses onto a neutral region for which $\omega_I = 0$.

Table 16.8 lists values of parameter at the pinch point; figure 16.24 gives the critical curve $\beta_c = \beta(\ell)$ in the β versus ℓ plane. $\ell = 0.3455$ is an asymptote; when $\ell > 0.3455$, there is no pinch point and the flow is absolutely unstable.

16.2.10 Exact solution; comparison with previous results

Using equations (2.17)–(2.21) in Funada and Joseph (2002) for capillary instability of both viscous fluids, we can modify dispersion relation (2.17) so as to make the ambient fluid inviscid and the viscous column moving with uniform velocity. The resultant dispersion relation D for a viscous jet in an inviscid fluid is given by

$$D = \begin{vmatrix} I_1(k/2) & I_1(k_\ell/2) & K_1(k/2) \\ 2k^2 I_1(k/2) & \left(k^2 + k_\ell^2\right) I_1(k_\ell/2) & 0 \\ F_1 & F_2 & F_3 \end{vmatrix} = 0, \tag{16.2.69}$$

where

$$F_1 = (\omega - k)^2 I_0(k/2) + 2i(\omega - k)\frac{k^2}{Re}\left[\frac{dI_1(k/2)}{d(k/2)}\right] + We\left(4 - k^2\right) kI_1(k/2), \tag{16.2.70}$$

$$F_2 = 2i(\omega - k)\frac{kk_\ell}{Re}\left[\frac{dI_1(k_\ell/2)}{d(k_\ell/2)}\right] + We\left(4 - k^2\right) kI_1(k_\ell/2), \tag{16.2.71}$$

$$F_3 = -\ell\omega^2 K_0(k/2), \tag{16.2.72}$$

with k_ℓ defined as

$$k_\ell = \sqrt{k^2 - iRe(\omega - k)}. \tag{16.2.73}$$

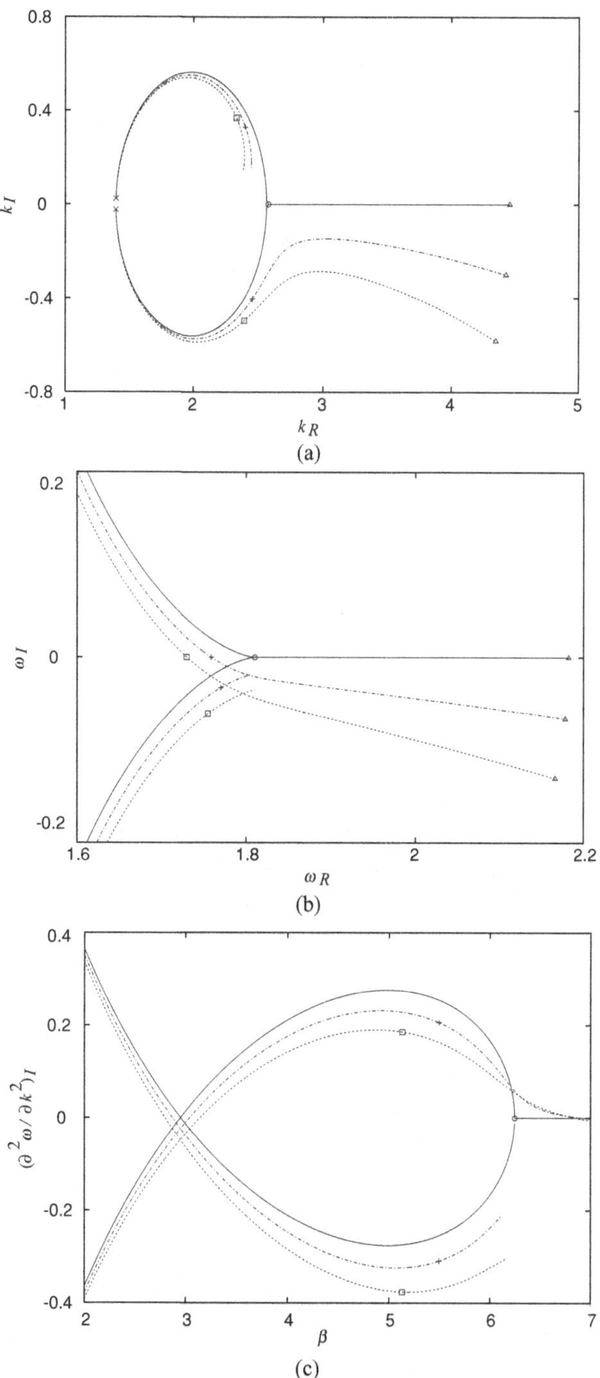

Figure 16.23. Locus of singular points $D = 0, \hat{c} = 0$ for $m = 0, \ell = 0, Re = 100$ (dashed curves), $Re = 200$ (dash–dotted curves), $Re \to \infty$ (solid curves) for $10^{-3} \leq \beta \leq 10$. Critical singular points are those for which $\omega_I = 0$. $\beta = 10^{-3}$, ×; $\beta_c = 6.246$, ○; $\beta_c = 5.134$, □; $\beta_c = 5.493$, +; and $\beta = 10$, △. (a) k_I vs. k_R, (b) ω_I vs. ω_R, (c) $\left(\partial^2\omega/\partial k^2\right)_I$ vs. β. $\beta < \beta_c$ is subcritical, $\beta > \beta_c > 0$ is supercritical. The supercritical inviscid branch $Re \to \infty$, $\beta > 6.246$, is degenerate $\left(\partial^2\omega/\partial k^2\right)_I = \omega_I = k_I = 0$ there. Disturbances with zero group velocity are neutrally stable. The values of β on the upper branches $k_I > 0$ of the (k_R, k_I) plane are less than $\beta_c = 6.246$ and $\omega_I < 0$. These branches are subcritical and spatially and temporally damped. The lower branches $k_I < 0$ go from subcritical values $10^{-3} < \beta < \beta_c$ for $\omega_I > 0$ (for which the disturbances are spatially and temporally amplified) to supercritical values $\beta > \beta_c$: $\beta_c = 6.246$ for $Re \to \infty$ and $\omega_I = 0$, $k_I = 0$ for $\beta > 6.246$, $\beta_c = 5.134$ for $Re = 200$ and $\omega_I < 0$ there, $\beta_c = 5.493$ for $Re = 100$ and $\omega_I < 0$ there. Disturbances with zero group velocity are temporally damped but spatially amplified ($\omega_I < 0$, $k_I < 0$) for supercritical values of β.

Table 16.8. *Inviscid fluids ($Re \to \infty$, $m = 0$)*

Values of (k_R, k_I) at pinch point singularity (see figure 16.24) are indexed by the density ratio ℓ. The values of the frequency ω_R and the Weber number parameter β $(= We^{-1})$ at the pinch point are also listed. Pinch-point singularities do not exist when $\ell > 0.3455$; in this case all flows are absolutely unstable.

Case	ℓ	β_c	k_R	k_I	ω_R	ω_I
1	0.00	6.246	2.576	0	1.810	0
2	0.04	6.710	2.696	0	1.877	0
3	0.08	7.280	2.840	0	1.958	0
4	0.12	8.004	3.025	0	2.060	0
5	0.16	8.962	3.290	0	2.196	0
6	0.20	10.298	3.681	0	2.387	0
7	0.24	12.282	4.274	0	2.677	0
8	0.28	15.496	5.317	0	3.162	0
9	0.32	22.038	7.514	0	4.199	0
10	0.3455	52.350	17.647	0	9.189	0
11	0.36	—	—	—	—	—
12	0.40	—	—	—	—	—

The parameters ℓ, We, and Re are the ones defined in (16.2.2). The top row of (16.2.69) arises from the continuity of normal velocity, of the tangential stress, and of the normal stress. Solving (16.2.69) implicitly by using Newton's method for given values of (ω_R, ω_I) and the parameters, we have (k_R, k_I) and $\hat{c}$ numerically.

In figure 16.25 we plot the critical value β_c $(= We_c^{-1})$ versus Re, giving the border between absolute and convective instability for viscous jets in an inviscid fluid as computed by Leib and Goldstein (1986b) for $\ell = 0$ and by Lin and Lian (1989) for $\ell = 0.0013$ and $\ell = 0.03$. The value $\beta_c = 6.3$ for an inviscid jet in an inviscid fluid was calculated by Leib

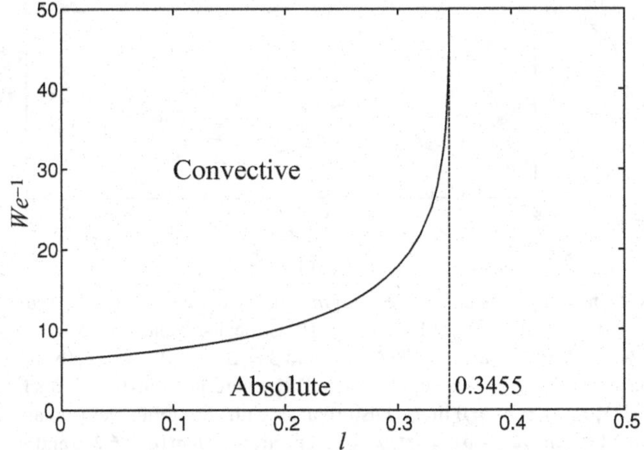

Figure 16.24. Border between absolute and convective instability in the inviscid case $Re \to \infty, m = 0$. The value of $\ell = 0.3455$ is asymptotic. The inviscid case is degenerate because the imaginary part of $\partial^2\omega/\partial k^2 = 0$ at the singular point. The consequence of this degeneracy is that at criticality $(\omega_I, k_I) = (0, 0)$ for all $\ell < 0.3455$. The condition $\omega_I < 0$ at the pinch point cannot be realized; ω_I at $\partial\omega_R/\partial k_R = 0$.

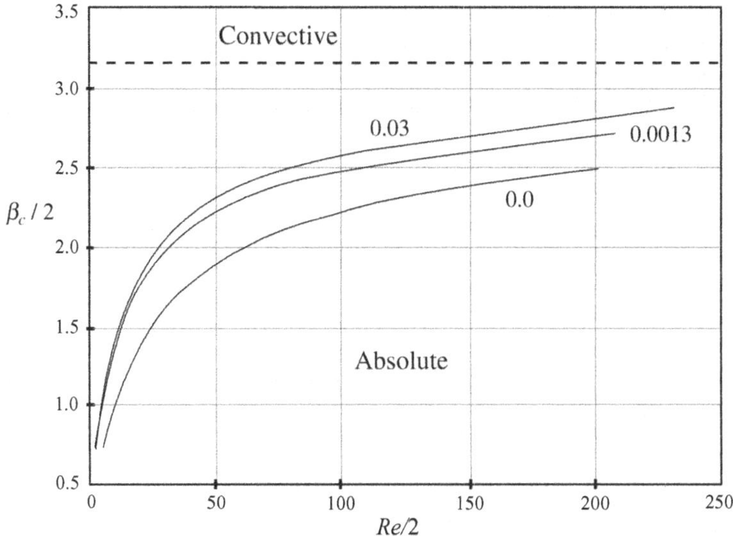

Figure 16.25. Critical Weber number β_c ($= We_c^{-1}$) versus Reynolds number Re from the literature: Leib and Goldstein (1986a) for inviscid jet in an inviscid fluid, $\beta_c = 6.3$ for $\ell = 0$, denoted by dashed line; solid curves are for viscous jets in an inviscid fluid ($m = 0$) for $\ell = 0$ (Leib and Goldstein, 1986b) and $\ell = 0.0013$ and $\ell = 0.03$ (Lin and Lian, 1989).

and Goldstein (1986a). In figure 16.26 we compare the results from the theory of VPF given in subsection 16.2.3 with the results for viscous flow according to (16.2.69). The stability limits from the two theories are close.

The analysis for fully viscous flow (FVF) neglects the dynamical effects of the viscous gas; these effects require the imposition of continuity of the tangential component of

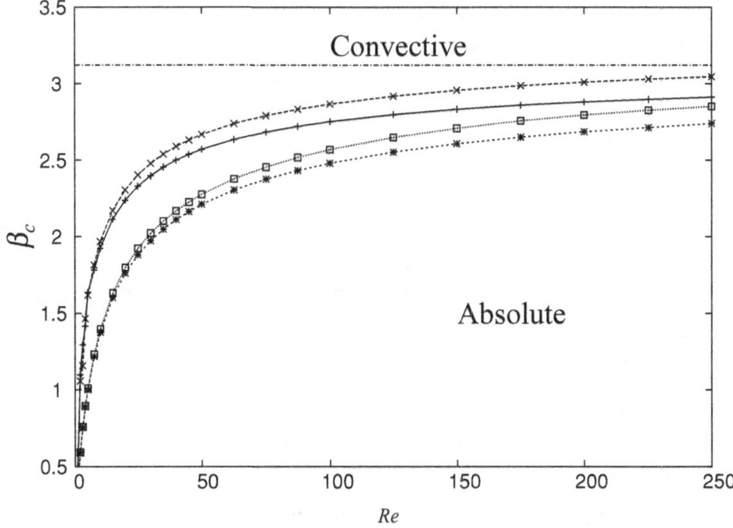

Figure 16.26. β_c versus Re comparing viscous flow (lower two curves) computed from the theory in subsection 16.2.10 for $m = 0$, $\ell = 0.0013$, $*$, and $\ell = 0.03$, $\square$, with VPF (upper two curves) for the same values ($\ell = 0.0013$, $+$; $\ell = 0.03$, $\times$; $m = 0$).

velocity and stress. The discontinuous velocity that induces KH instability is then inconsistent with the aforementioned continuity requirements. S. P. Lin (2003) attempted to address the effects of shear by a study of core-annular flow in a vertical pipe. This is a very different problem. He found a transition from convective to absolute instability, but comparisons with experiments are not available.

16.2.11 Summary and discussion

A summary of results and a brief discussion of results are presented below:

(1) The computation of temporal and C/A instability of the liquid jet by use of VPF.

(2) Extensive computation of the effects of the ambient density and viscosity and the viscosity of the liquid.

(3) A demonstration that KH instability cannot occur in vacuum (but capillary instability and RT instability can occur in vacuum).

(4) The derivation of a dispersion relation in the form of a polynomial in the complex frequency that can be used to study temporal, spatial, and C/A instabilities.

(5) A comprehensive analysis of temporal instability; for all values of the parameters there are wavenumbers for which the liquid jet is unstable.

(6) An analysis of the relative importance of KH and capillary instability on the maximum growth, on the wavelength of maximum growth, and the cutoff wavenumber for inviscid and viscous liquids. KH instability is the dominant mechanism for small Weber numbers, and capillary instability is dominant for large Weber numbers; the variation of the growth rates and wavenumbers in the two regimes is sharply different.

(7) When viscosity and surface tension are zero, the liquid jet is Hadamard unstable to KH instability with growth rates proportional to k. Surface tension stabilizes the short waves. The maximum growth rate and the associated wavenumber are proportional to $1/We$ for small We when the fluid is inviscid. For viscous fluids the maximum growth rate is finite as $We \to 0$ (pure KH instability) but the wavelength $\lambda_m = 2\pi/k_m$ tends to zero with We, as in the inviscid case. It can be said that the wavelengths for breakup that is due to KH instability are exceedingly short and that breakup that is due to KH instability leads to surpassingly small drops, essentially mist.

(8) The critical wavenumber k_c for marginal instability (stability for $k > k_c$) is independent of the Reynolds number of the liquid jet but the maximum growth rate and the wavenumber of maximum growth depend on the Reynolds number.

(9) Under the realizable condition that $m\alpha b_a = \ell b \alpha_a$ the neutral curves for VPF are the same as for IPF; this is a remarkable result. When $k > 10$ this criterion reduces to a statement that the neutral curves are the same when the kinematic viscosity of the liquid equals the kinematic viscosity of the gas; this result was proved for KH instability of a plane layer by Funada and Joseph (2001).

(10) A comprehensive study of the transition between convective and absolute stability of the combined KH stability and capillary instability of the liquid jet under a wide range of ambient condition for different liquid was carried out. Precise results were obtained for full ranges of the Weber and Reynolds numbers and density and viscosity ratios.

(11) The study of pinch-point and cusp-point singularities was greatly simplified by the fact that an explicit dispersion relation in the complex frequency and wavenumber planes could be studied by an algebraic method rather than by the geometric method of Bers (1975) and Briggs (1964).

(12) We show that the singular point for inviscid fluids when the density ratio ℓ is in $0 \leq \ell < 0.3455$ is degenerate; when $\ell > 0.3455$ the inviscid jet is absolutely unstable. The singular points for inviscid solutions in the supercritical case $\beta > \beta_c$ are degenerate because $\left(\partial^2\omega/\partial k^2\right)_I = 0$. In this case $k_I = 0$ and $\omega_I = 0$, so that the transition from absolute instability is to a neutral rather than to a convectively unstable state. All other cases are not degenerate, and cusp points in the ω plane are associated with pinch points in the k plane.

(13) The transition between convective and absolute stability computed by VPF is in reasonably good agreement with the transition computed by FVF in which the flow is not assumed to be irrotational. It must be understood that KH instability cannot be studied exactly in the frame of the Navier–Stokes equations because the basic flow has a discontinuous velocity.

The liquid jet with no spatial development studied here is always unstable to temporal disturbances, convectively and absolutely unstable jets are temporally unstable; flows like Hagen–Poiseuille flow that are temporally stable will not admit C/A instability. Instability for jets with very slow spatial development should not be very different for jets with no spatial development. The predictions of temporal and C/A instabilities achieved here have a somewhat tentative relation to actual experiments that emanate from nozzles in which the spatial development of the basic flow could have an important effect.

16.3 Viscous potential flow of the Kelvin–Helmholtz instability of a cylindrical jet of one fluid into the same fluid

Batchelor and Gill (1962) constructed a simple analysis of a cylindrical jet of one fluid into the same fluid. Their analysis is based on potential flow of an inviscid fluid. The interface in their problem is defined by a discontinuity of velocity. The fluid on either side of the discontinuity is the same. Here we show how to generalize the analysis to irrotational flow of a viscous fluid. Our analysis is a paradigm for generalizing problems of potential flow with discontinuous velocity profiles to include the effects of viscosity.

16.3.1 Mathematical formulation

Laplace's equations are given by

$$\nabla^2\phi_0 = 0, \quad \nabla^2\phi_1 = 0, \tag{16.3.1}$$

and Bernoulli equations are

$$\frac{\partial\phi_0}{\partial t} + U\frac{\partial\phi_0}{\partial x} + \frac{p_0}{\rho} = f_0(t), \quad \frac{\partial\phi_1}{\partial t} + \frac{p_1}{\rho} = f_1(t). \tag{16.3.2}$$

The boundary conditions at the cylindrical surface $r = a + \eta \approx a$ are the kinematic conditions,

$$\frac{\partial\eta}{\partial t} + U\frac{\partial\eta}{\partial x} = \frac{\partial\phi_0}{\partial r}, \quad \frac{\partial\eta}{\partial t} = \frac{\partial\phi_1}{\partial r}, \tag{16.3.3}$$

and the normal stress balance,

$$p_0 - 2\mu\frac{\partial^2\phi_0}{\partial r^2} = p_1 - 2\mu\frac{\partial^2\phi_1}{\partial r^2}. \tag{16.3.4}$$

16.3.2 Normal modes; dispersion relation

The normal-mode solution of the problem by (16.3.1)–(16.3.4) is in the form

$$\eta = Ae^{ikx+in\theta-ikct}, \quad \phi_0 = CI_n(kr)e^{ikx+in\theta-ikct}, \quad \phi_1 = DK_n(kr)e^{ikx+in\theta-ikct}. \tag{16.3.5}$$

The kinematic conditions give

$$ik(U_0-c)\,A = kCI_n'(ka), \quad -ikcA = kDK_n'(ka). \tag{16.3.6}$$

The normal stress balance gives

$$\frac{\partial\phi_0}{\partial t} + U\frac{\partial\phi_0}{\partial x} + 2\nu\frac{\partial^2\phi_0}{\partial r^2} = \frac{\partial\phi_1}{\partial t} + 2\nu\frac{\partial^2\phi_1}{\partial r^2}, \tag{16.3.7}$$

which is then written as

$$ik(U_0-c)\,CI_n(ka) + 2\nu Ck^2 I_n''(ka) = -ikcDK_n(ka) + 2\nu Dk^2 K_n''(ka), \tag{16.3.8}$$

where $\nu = \mu/\rho$.

For $n = 0$, α, α_a, b, and b_a are defined as

$$\alpha = \frac{I_0(k/2)}{I_1(k/2)} = \frac{I_0(k/2)}{I_0'(k/2)}, \quad \alpha_a = \frac{K_0(k/2)}{K_1(k/2)} = -\frac{K_0(k/2)}{K_0'(k/2)}, \tag{16.3.9}$$

$$b = \alpha - \frac{2}{k} = \frac{I_0''(k/2)}{I_0'(k/2)}, \quad b_a = \alpha_a + \frac{2}{k} = -\frac{K_0''(k/2)}{K_0'(k/2)}. \tag{16.3.10}$$

It is noted for real k that $k\alpha \to 4$ and $\alpha_a \to 0$ as $k \to 0$, whereas $\alpha \to 1$ and $\alpha_a \to 1$ as $k \to \infty$, as shown in figure 16.10.

In the limit $ka \to \infty$,

$$\frac{I_n(ka)}{I_n'(ka)} \to 1, \quad \frac{I_n''(ka)}{I_n'(ka)} \to 1, \quad -\frac{K_n(ka)}{K_n'(ka)} \to 1, \quad -\frac{K_n''(ka)}{K_n'(ka)} \to 1. \tag{16.3.11}$$

If $\nu = 0$, then

$$(c-U_0)^2\frac{I_n(ka)}{I_n'(ka)} = c^2\frac{K_n(ka)}{K_n'(ka)}, \tag{16.3.12}$$

$ka \to \infty$,

$$(c-U_0)^2\frac{I_n(ka)}{I_n'(ka)} = c^2\frac{K_n(ka)}{K_n'(ka)} \to (c-U_0)^2 + c^2 = 0 \to c^2 - U_0c + \frac{U_0^2}{2} = 0, \tag{16.3.13}$$

$$c = \frac{U_0}{2} \pm \sqrt{\frac{U_0^2}{4} - \frac{U_0^2}{2}} = \frac{U_0}{2}[1 \pm i], \to \frac{c}{U_0} = \frac{1}{2}[1 \pm i]. \tag{16.3.14}$$

When $\nu \neq 0$ and $ka \to \infty$, we have

$$(c - U_0)^2 + 2i\nu k(c - U_0) + c^2 + 2i\nu kc = 0, \tag{16.3.15}$$

$$c = \frac{U_0 - 2i\nu k}{2} \pm \sqrt{\left(\frac{U_0 - 2i\nu k}{2}\right)^2 - \frac{U_0^2 - 2i\nu kU_0}{2}} = \frac{U_0 - 2i\nu k}{2} \pm i\sqrt{\frac{U_0^2}{4} + \nu^2k^2}, \tag{16.3.16}$$

$$c_R = \frac{U_0}{2}, \quad c_I = -\nu k \pm \sqrt{\frac{U_0^2}{4} + \nu^2k^2} = -\nu k + \sqrt{\frac{U_0^2}{4} + \nu^2k^2}. \tag{16.3.17}$$

When $c_I = 0$, the neutral state is given by

$$0 = -\nu k + \sqrt{\frac{U_0^2}{4} + \nu^2k^2} \to \frac{U_0^2}{4} = 0. \tag{16.3.18}$$

The flow is unstable for all values of n and ka and all viscosities ν.

16.3.3 Growth rates and frequencies

The evolution of disturbances of the cylindrical jet is governed by the function

$$e^{-ikct}, \quad c = c_R + ic_I. \tag{16.3.19}$$

Batchelor and Gill (1962) plotted c_R vs. k and c_I vs. k for an inviscid fluid. It is more revealing to compute the real and imaginary parts of the complex frequency $\omega = c/k$, with

$$e^{-ikct} = e^{\omega_I(k,\nu)t}e^{-i\omega_R(k,\nu)t}, \tag{16.3.20}$$

because the analytical character of the evolution is not disguised by the extraneous multiplicative factor k.

In figure 16.27 we present graphs of ω_I vs. k and ω_R vs. k for $\nu = 0$ and $\nu = 0.1$.

16.3.4 Hadamard instabilities for piecewise discontinuous profiles

The study of problems of irrotational flow of an inviscid fluid with a discontinuous velocity profile is widespread. There are two prototypical problems. In one problem, the velocity itself is discontinuous, as in the jet problem considered here, or in the problem of instability of the Rankine vortex with axial flow studied by Loiseleux, Chomaz, and Huerre (1998), and related problems used to study absolute and convective instability. In another problem, the velocity is continuous but the slope of the profile is discontinuous. Such discontinuities are compatible with irrotational flows in which the effects of shear stresses are ignored but the effects of viscous normal stresses are not ignored. In fact, the effects of shear in any real fluid would quickly smooth discontinuities, though the effects of a thin shear layer might persist for a time.

One of the troublesome difficulties that arises in the study of instability of vortex sheets defined by discontinuities of velocity of the type just mentioned is Hadamard instability. In all these problems the growth rate increases without bound at any finite time as the wavelength $\lambda = 2\pi/k$ tends to zero. This kind of disaster cannot occur; as they say in

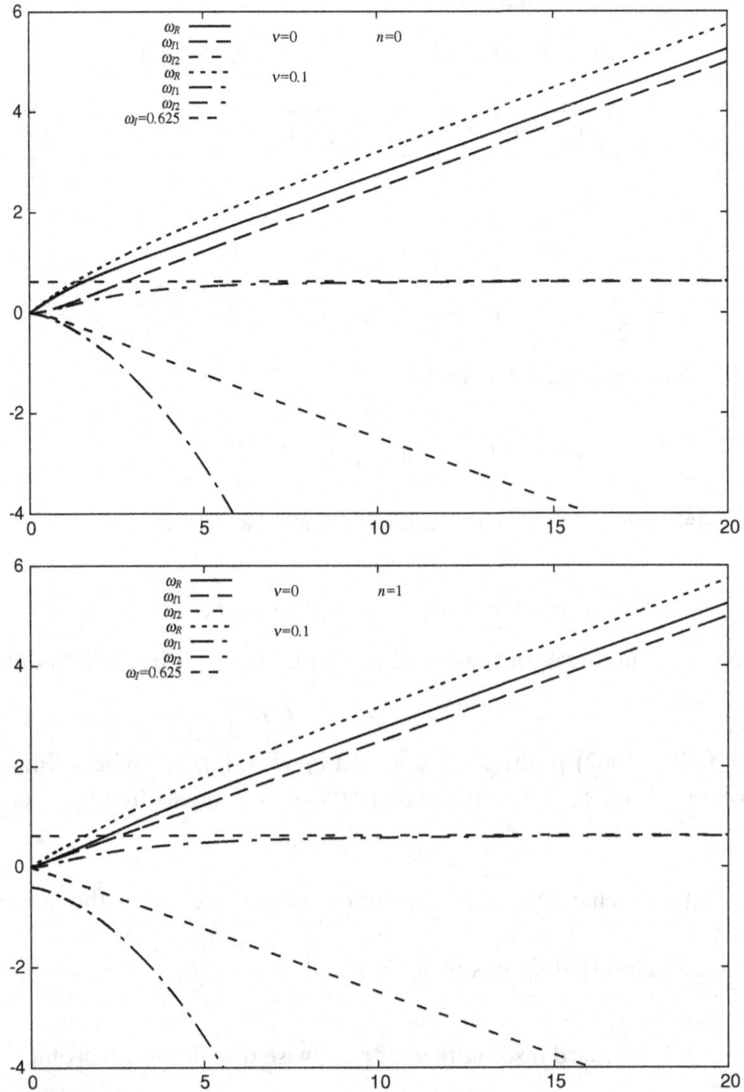

Figure 16.27. ω_I vs. k and ω_R vs. k for $\nu = 0$ and $\nu = 0.1$. The problem considered by Batchelor and Gill (1962) is Hadamard unstable, $e^{\omega_I t} \to \infty$ as $k \to \infty$, no matter how small t is. $n = 0$ above.

medicine, "the bleeding always stops." The way in which short waves are regularized ought to describe the physical mechanisms at work.

The instability of the viscous jet described by equation (16.3.17) is given by

$$\omega_I = -\nu k^2 + k\sqrt{\frac{U_0^2}{4} + \nu^2 k^2}. \tag{16.3.21}$$

When $\nu = 0$, $\omega_I = kU_0/2$ tends to infinity with k. If $\nu > 0$, then

$$\omega_I \to \frac{U_0^2}{8\nu}, \tag{16.3.22}$$

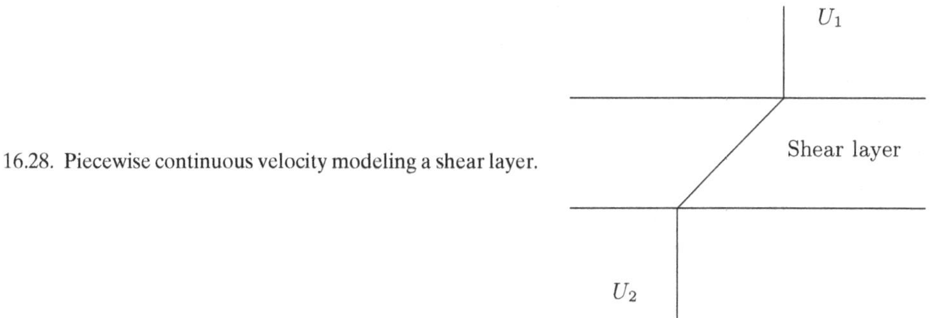

Figure 16.28. Piecewise continuous velocity modeling a shear layer.

and the small waves are unstable but not Hadamard unstable. The growth rate for short waves increases as ν decreases and it is large when ν is small.

Hadamard instabilities may also be regularized by weakening the discontinuity as in the second problem in which the vorticity is discontinuous but the velocity is continuous. Shear layers may be modeled by such vorticity discontinuities as in figure 16.28.

The flow in figure 16.28 is not Hadamard unstable even when $\nu = 0$ (Joseph and Saut, 1990, §7).

17

Stress-induced cavitation

The usual criterion for cavitation is that cavities will form in a liquid when and where the pressure falls below a critical value. In the ideal case, the cavitation threshold is the vapor pressure. The pressure in an incompressible viscous liquid is not a thermodynamic or material property; it is the average stress (actually the negative of the average stress, which is positive in tension). The viscous part of the stress is proportional to the rate of strain, which has a zero average with positive and negative values on the leading diagonal in the principal coordinates. It follows that in motion the liquid will develop stresses that are both larger and smaller than the average value. The theory of stress-induced cavitation seeks to relate the fracture or cavitation of a liquid to its state of stress rather than to its average stress. This kind of theory requires that the state of stress be monitored in the evolving field of motion to determine when and where the liquid will fracture. The theory can be thought of as an application area for Navier–Stokes fluid dynamics that can be studied by VPF when the flows are irrotational or nearly irrotational. The link between the theory of stress-induced cavitation and VPF is the fact that viscous stresses can be computed on irrotational motions. In §17.1 we give a comprehensive presentation of the physical foundations and mathematical formulation of the theory of stress-induced cavitation. This presentation is taken from the 2007b paper of Padrino *et al.*; it is new, comprehensive, and otherwise unavailable. In §17.2 we consider cavitation of a liquid in aperture flow by computing stresses on the well-known potential flow solution for this problem in complex variables. This is a simple example of how the extension of potential flow solutions to viscous fluids leads to new results at small cost. Padrino *et al.* (2007b) studied stress-induced cavitation in streaming flow past a sphere. This solution is discussed in §17.3; there are potential flow solutions there, and they are compared with numerical solutions of the Navier–Stokes equation. In §17.4 we construct a nonlinear model of capillary collapse and rupture of a liquid thread under the assumption that the collapse is symmetric with respect to a stagnation point at the throat. Stagnation-point flow is irrotational. The line drawing for the model in figure 17.13 is defined by a periodic array of stagnation points between points of liquid depletion and liquid accumulation. The passage of liquid from points of depletion to points of accumulation around which a liquid ball expands cannot be done without the generation of vorticity in an axisymmetric analog to hyperbolic vortices $xy = \text{const}$ nested in the four quadrants at each stagnation point in the vortex array described by (4.6.33). The analysis can be said to generate a Helmholtz decomposition for rotational flows near points of stagnation. This model gives rise to a rupture criterion based on the generation of large viscous extensional stresses at the throat in the final stage of collapse.

17.1 Theory of stress-induced cavitation

Winer and Bair (1987) and, independently, Joseph (1995, 1998) proposed the maximum tension criterion for cavitation that states that the flowing liquid will cavitate if the maximum tensile stress exceeds a critical value. Because this maximum stress is associated with a principal direction, this criterion is not isotropic. Winer and Bair (1987) introduced the idea that stress-induced cavitation may enter into the apparent shear thinning of liquid lubricants. They remarked that shear thinning may be the result of a yielding or cavitation event that takes place at a critical value of the liquid's tensile stress. They further note that "for some high shear rate viscosity data at atmospheric pressure the principal normal stress may approach quite low values relative to one atmosphere suggesting the possibility of cavitation or fracture of the material resulting in a reduced shear stress." In a private communication, Bair noted that "There was little interest from tribologists, so we dropped it until recently. In the original work we were able to see to the voids by eye using a clear plastic outer cylinder."

The maximum tension criterion is embedded as one possibility for liquid failure presented by analysis of the state of stress in Joseph's theory. A comparison of these two theories can be found in the study of cavitation in creeping shear flows by Kottke, Bair, and Winer (2005). Numerous examples of cavitation in shear flow by other researchers are discussed by Kottke *et al.* (2005). Examples of stress-induced cavitation in extensional flow and shear flow were discussed by Joseph (1998). Pereira *et al.* (2001) made a theoretical study of cavitation in a journal bearing with axial throughput. They found that the inception of cavitation in a moving fluid is always stress induced. Funada *et al.* (2006) carried out an analysis of stress-induced cavitation in a 2D aperture flow modeling atomizers in which cavitation is well documented. The aperture flow was expressed by a complex potential and the stress calculated with VPF. They found that the viscous stress was huge near the tips of the aperture; thus cavitation could be induced. The region at risk to cavitation is larger for a fixed cavitation number when the Reynolds number is smaller.

17.1.1 Mathematical formulation

The stress in an incompressible Newtonian fluid is given by

$$\mathbf{T} = -p\mathbf{1} + 2\mu\mathbf{D}[\mathbf{u}], \tag{17.1.1}$$

where $\mathbf{D}[\mathbf{u}]$ is the symmetric part of the velocity gradient and $\operatorname{tr}\mathbf{D}[\mathbf{u}] = 0$, such that

$$\operatorname{tr}\mathbf{T} = T_{11} + T_{22} + T_{33} = -3p. \tag{17.1.2}$$

We define the stress at the cavitation threshold as p_c. It is positive when compressive and negative when tensile. Classically, the vapor pressure is taken as the threshold stress; however, in the next section, we discuss examples for which different values, including tensile values, should be used for the cavitation threshold.

In the pressure criterion, the viscous part of the stress tensor is not considered and the liquid will cavitate when

$$-p + p_c > 0. \tag{17.1.3}$$

The pressure criterion assumes that cavitation inception is determined by the average stress, called the pressure. The fluid cannot average its stresses; it sees only principal stresses, and when the actual state of stress is considered there is at least one stress that is more compressive and another that is more tensile than the average stress. The most conservative criterion is the one that requires that the most compressive stress be larger than the cavitation threshold; suppose T_{22} is the most compressive and T_{11} is the most tensile (or least compressive) stress; then, if

$$T_{22} + p_c > 0 \tag{17.1.4}$$

for cavitation, it will surely be true that

$$-p + p_c > 0, \quad T_{11} + p_c > 0. \tag{17.1.5}$$

The maximum tension theory, which perhaps embodies the statement that liquids that are not specially prepared will cavitate when they are subject to tension, can be expressed by the condition that, if T_{11} is the maximum of the three principal stresses,

$$T_{11} + p_c > 0. \tag{17.1.6}$$

The cavitation number K compares the cavitation threshold p_c with a typical pressure. For example, in our sphere problem, the typical pressure is the pressure p_∞ at infinity. We define

$$K = \frac{p_\infty - p_c}{N_R}, \tag{17.1.7}$$

where $N_R = \mu U/L$ for Stokes flow and $N_R = \rho U^2/2$ when inertia acts; L is a characteristic length scale. Later, the analysis will show that the stress difference between the free-stream pressure and the cavitation threshold $p_\infty - p_c$ is the critical value rather than the cavitation threshold by itself.

Foteinopoulou *et al.* (2004) have recently studied maximum tension criterion (17.1.6) in a numerical simulation of bubble growth in Newtonian and viscoelastic filaments undergoing stretching. They base their analysis on the Navier–Stokes equations for Newtonian fluids and the Phan-Thien–Tanner model for viscoelastic fluids. They compute the principal stresses and evaluate the cavitation threshold for maximum tension criterion (17.1.6), pressure theory (17.1.3), and minimum principal theory (17.1.4). They find that the capillary number at inception is smallest for (17.1.6). As remarked by Kottke *et al.* (2005), the cavitation threshold p_c could be negative or positive. In the case of $p_c < 0$, the liquid shows tensile strength; for $p_c = 0$, a cavity will open if the maximum principal stress becomes positive (i.e., tensile), and, if $p_c > 0$, the cavitation threshold is given by a positive pressure (i.e., compressive stress). The latter case is typified by the pressure theory of cavitation determined by the local pressure dropping below the vapor pressure.

To each principal stress there corresponds a principal direction, which plays a role in the physics of cavity inception. Joseph (1998) asserts that "if a cavitation bubble opens up, it will open in the direction of maximum tension. Since this tension is found in the particular coordinate system in which the stress is diagonal, the opening direction is in the direction of maximum extension, even if the motion is a pure shear. It may open initially as an ellipsoid before flow vorticity rotates the major axis of ellipsoid away from the principal tension axis of stress, or it may open abruptly into a 'slit' vacuum cavity perpendicular to the tension axis before vapour fills the cavity as in the experiments of Kuhl *et al.* (1994)". These ideas are illustrated in §17.3 in line drawings showing the orientation of the principal directions on the surface of the sphere for each approach considered in this study.

Consider the expression for the stress tensor for a Newtonian fluid given in (17.1.1). Adding the diagonal tensor $p_c\mathbf{1}$ to both sides and decomposing $p = p_\infty + p^*$ yields

$$\mathbf{T} + p_c\mathbf{1} = -(p_\infty - p_c)\mathbf{1} - p^*\mathbf{1} + 2\mu\mathbf{D}. \tag{17.1.8}$$

Dividing through by the normalizing factor N_R, we find that (17.1.8) becomes

$$\frac{\mathbf{T} + p_c\mathbf{1}}{N_R} = -\left(K + \frac{p^*}{N_R}\right)\mathbf{1} + \frac{2\mu}{N_R}\mathbf{D}. \tag{17.1.9}$$

The strain–rate tensor can be readily diagonalized. Thus the principal stresses and directions can be determined. Suppose now that $K = K_c$ at the marginal state separating cavitation from no cavitation. For brevity, let us call K_c the incipient cavitation number. This marginal state is defined by an equality in one of the three criteria, (17.1.3), (17.1.4), or (17.1.6). For the maximum tension theory $K = K_c$ when $T_{11} + p_c = 0$. In particular, for the maximum principal stress T_{11}, (17.1.9) yields

$$(T_{11} + p_c)/N_R = -(K + p^*/N_R) + 2\mu D_{11}/N_R, \tag{17.1.10}$$

where D_{11} denotes the maximum principal rate of strain. Then $K_c = (-p^* + 2\mu D_{11})/N_R$ is, in general, a scalar function of the position in the fluid domain. For a positive cavitation number, consider $K = K_c + K^*$ such that $(T_{11} + p_c)/N_R = -K^*$. It is thus clear from (17.1.10) that

$$T_{11} + p_c < 0 \text{ when } K > K_c, \tag{17.1.11}$$

$$T_{11} + p_c > 0 \text{ when } K < K_c. \tag{17.1.12}$$

The latter condition implies that the liquid is at the most risk to cavitation in regions where $K < K_c$. For instance, for a fixed cavitation number K, no cavity will open if $K > K_{c,\max}$, the maximum value that K_c takes in the entire fluid domain. On the other hand, the cavitation number K based on the actual hydrodynamics may vary in the fluid domain, as the cavitation threshold p_c may also change with position. For example, Singhal *et al.* (2002) included in their cavitation model the effect of the local turbulence pressure fluctuations on the phase-change threshold pressure.

17.1.2 Cavitation threshold

Cavitation can be defined as the formation, growth, and collapse of a cavity in a liquid. In general, the "formation" of a cavity implies both the appearance of a new void and the extension of a preexisting nucleus beyond a critical size large enough to be observed with the unaided eye (Young, 1989). The idea of the opening of a cavity in the liquid continuum brings into consideration the concept of liquid tensile strength. The pressure criterion for cavitation states that the liquid cavitates when the local pressure reaches the vapor pressure somewhere in the domain. Knapp, Daily, and Hammit (1970) discuss that, although the inception of a cavity has been observed in experiments when local pressure is near the vapor pressure, deviations of various degrees have been reported for different liquids such that the results do not agree with the vapor pressure criterion. Knapp *et al.* (1970) define the vapor pressure as "the equilibrium pressure, at a specified temperature, of the liquid's vapor which is in contact with an *existing* free surface." They argue that the stress required for rupturing the continuum in a homogeneous liquid is determined by the tensile strength, not by the vapor pressure. The literature on the tensile strength of liquids is vast, and a good account of experimental results is given in the book by Knapp *et al.* (1970) for various liquids. In particular for water, values ranging from 13 to 200 atm are listed. Briggs (1950) reports the inception of cavities in water induced by centrifugal force for pressures between vapor presssure and −300 atm (tension). Recently, Kottke *et al.* (2003) measured the tensile strength of nine liquids, including water, lubricants, and polymeric liquids. Theoretical estimates of the tensile strength of water render large negative values in the interval −500 to −10,000 atm, which, however, have never been reported from experiments (Strasberg, 1959). Both observed phenomena, the wide scatter of the experimental results, and the inception of cavitation at pressures much higher than the theoretical tensile strength reported in the literature, indicate the existence of weak spots in the fluid that allows breaking of the continuum. Plesset (1969) comments that bubbles can grow to macroscopic dimensions starting from voids of sizes already beyond the molecular level under tensile stresses much lower than the theoretical values predicted for pure liquids.

Fisher (1948) reasons that, in a similar manner as very greatly subcooled liquids (such as glass) may fail by the nucleation and growth of a crack, a fluid may fail under tension by the growth of a cavity starting from very small holes. By applying methods of nucleation theory, Fisher predicts fracture tensions for several liquids with values, however, 1 order of magnitude higher than the experimental evidence. Some mechanism is required for stabilizing preexisting nuclei in the liquid. For a very small bubble suspended in the liquid, the pressure inside the bubble is much higher than the pressure in the surrounded liquid because of surface–tension forces. This pressure difference diffuses the gas out of the gas void until it vanishes. On the other hand, bubbles not so small will rise and escape through the surface. Harvey and collaborators (1944a) introduced the idea of stabilized gas pockets attached to submicroscopic and hydrophobic crevices in the surface of the liquid container or in solid impurities. The size of these nuclei can be of the order of micrometers. Harvey *et al.* (1944b) supported their theory with results from a series of experiments in which previously pressurized and unpressurized samples of water were boiled at atmospheric pressure such that the saturation pressure corresponding to the boiling temperature was taken as a rough measure of the effective tensile strength. Although quite broad scatter was observed in the results for the pressurized samples, they all boiled at temperatures much higher than the saturation temperature for atmospheric pressure, which was the boiling temperature showed by the unpressurized samples. Tensile strength of 16 atm was reported in some samples previously pressurized.

Harvey *et al.* (1947) performed a different type of experiment to investigate the tensile strength of water by high-speed removal of a square-ended glass rod from a narrow glass tube containing the liquid. Meticulous cleaning of the glass surfaces and pressurization of the sample with the rod in position were done to remove hydrophobic spots and gas nuclei. In terms of the rod-withdrawal speed, they found that "if the rod surface contained glass nuclei, or was hydrophobic and free of gas nuclei, cavitation ocurred at the rear end when the velocity was less than 3 m/s, but if completely hydrophilic and free of gas nuclei, the velocity could be 37 m/s... without cavitation." Knapp (1958) confirmed the results of Harvey *et al.* (1947), performing experiments on a rather larger scale. Strasberg (1959) explored the onset of acoustically induced cavitation in tap water, finding that microscopic undissolved air cavities, which show a slow motion toward the surface, play as important a role as nuclei. Apfel (1970) extended the theory of Harvey *et al.* (1947) to consider the condition required in a liquid for the inception of a vapor cavity from a solid impurity in the liquid. Crum (1982) examined the crevice model of Harvey *et al.*, comparing its predictions with experimental evidence.

From the standpoint of hydrodynamic cavitation, stream nuclei carried by the moving liquid as particulates or microbubbles have a greater contribution as sites for the onset of cavitation than do the surface nuclei that originate in crevices or cracks on the solid boundaries (Billet, 1985, and references therein). Turbulence has been shown to influence cavitation inception, and its effect has been accounted for in models through the phase-change threshold (Singhal *et al.*, 2002).

The inception of a cavity can be an abrupt event, during which the liquid must rupture, instead of a continuous one. Chen and Israelachvili (1991) and Kuhl *et al.* (1994) monitored the elastohydrodynamic deformations of two curved surfaces that move relative to each other separated by a thin-liquid film of nanoscopic dimensions. A low-molecular-weight polymer liquid of polybutadiene and bare mica smooth surfaces having strong adhesion to the liquid were utilized in the tests. When the surfaces move normally with a slow separation speed they bulge outward, becoming pointed at the location of the shortest surface separation. This shape indicates the existence of a tensile stress acting on the surface. If the separation speed is increased beyond a critical value, a vapor cavity opens in the liquid at the position of the shortest separation, reducing the tensile stress, while the pointed surfaces suddenly recover their original shape. Chen and Israelachvili (1991) also used surfactant-coated mica surfaces, which have weak adhesion to the liquid, resulting in cavity formation at the liquid–solid interface. Kuhl *et al.* (1994) considered lateral sliding of a curved surface over a mostly plane surface with a thin-liquid film in between. Describing the shape of the sliding element, they observed that "the leading edge becomes more rounded and lifts off while the trailing edge becomes more pointed." For a sliding speed larger than some critical value, the pointed trailing edge snaps back, while a small cavity opens in the wake.

Cavitation inception has been observed in liquids undergoing shearing, suddenly changing the rheological response of the samples. Bair and Winer (1990) inferred cavitation inception by detecting yielding of a synthetic oil during rheological tests using a rotating concentric cylinder rheometer for a shear stress near the hydrostatic pressure (1.73 MPa). A similar phenomenon was noted by Bair and Winer (1992) for polybutene in simple shear at low pressures (0.1–1 MPa) for a shear stress of 0.075 MPa in excess of the internal absolute pressure. This magnitude may represent the amount of tension that this liquid can resist without the opening of a cavity. Archer, Ternet, and Larson (1997) visualized the opening of bubbles within a sample of low-molecular-weight polysterene subjected to start-up of steady shearing flow. They noted that bubbles seemed to appear near dust particles. As a consequence of cavitation, the shear stress abruptly drops after it reaches a maximum of 0.1 MPa. Kottke *et al.* (2005) observed the inception of cavities in polybutene undergoing shearing tests using a Couette viscometer. Cavities become visible when the measured shear stress matches the ambient pressure. According to their principal normal stress cavitation criterion, this result implies that the sample liquid is not able to withstand tension. They suggest that cavitation grows from preexisting nuclei stabilized in some cracks or crevices on the solid boundaries.

The previous survey has shown that the idea of minute gas and vapor pockets in the liquid acting as nucleation sites is plausible and generally accepted. Nevertheless, a precise definition of the cavitation threshold and a clear description of the wide gamut of factors that influence this critical value are yet to be accomplished. We use the words "cavitation threshold" and "breaking strength" as synonymous with the threshold at which the liquid continuum will fracture. This threshold can vary from place to place in a sample. The threshold need not be a material parameter. In the case of heterogeneous nucleation, the cavitation threshold depends on the sample preparation, the density and nature of nucleation sites. In the case of homogeneous nucleation, the threshold may be taken as the vapor pressure. The vapor pressure is a thermodynamic quantity that is defined for uniform isotropic samples for which the stress tensor is isotropic; for static samples, bubble nucleation is a function of pressure and nothing else. In this chapter, the cavitation threshold, p_c, is not necessarily the vapor pressure; this value is regarded as given and is not a subject for study here. For liquid that cannot withstand tension, $p_c = 0$.

17.2 Viscous potential flow analysis of stress-induced cavitation in an aperture flow

It is well known that cavitation may be induced at sharp edges of the inlet of nozzles such as those used in atomizers. It is at just such edges that the pressure of an inviscid fluid into a nozzle is minimum. At higher-pressure drops (larger cavitation number), the liquid in the nozzle may break away from the nozzle wall; the flow then attaches to the sharp edge of the nozzle and is surrounded by atmospheric gas. The term "incipient cavitation" is used to define the situation in which cavitation first appears. The term "supercavitation" describes the situation in which there is a strong cavitation flow near the nozzle exit, which is very beneficial to atomization. Total hydraulic flip describes the situation in which the liquid jet completely separates from the nozzle wall. Hydraulic flip occurs in a variety of nozzles of different cross sections, provided that the edge at the inlet is sharp and not round. The aperture flow in a flat plate considered here (figure 17.1) is a nearly perfect 2D model of total hydraulic flip. Experiments documenting the transition to hydraulic flip from cavitating have been presented by Bergwerk (1959), Soteriou *et al.* (1995), Chaves *et al.* (1995), Laoonual *et al.* (2001), and a few others.

The outstanding property of the hydraulic flip is the disappearance of any sign of the cavitation that was there before the flow detached. To our knowledge, reports of the observations of the disappearance of cavitation are for very low-viscosity liquids, such as water and diesel oil. In the analysis of aperture flow that follows, we find cavitation at the sharp edge for all fluids with viscosity larger than zero, but for low-viscosity liquids it would be very hard to observe.

Our analysis is based on the theory of stress-induced cavitation; the flow will cavitate at places where the principal tensile stress $T_{11} > -p_v$, where p_v is the vapor pressure.

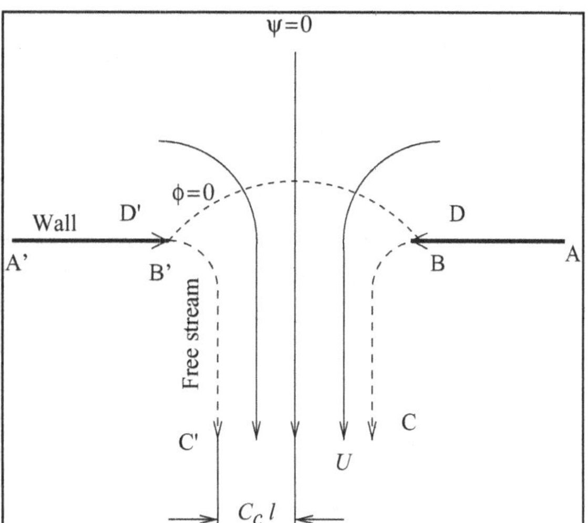

Figure 17.1. Flow through an aperture in a flat plate.

The theory of VPF allows us to compute these stresses directly and easily from the classical potential flow solution for aperture flow. Liquid samples that are not specially prepared ordinarily do not cavitate at the vapor pressure; various impurities can reduce the cavitation threshold and degassing followed by massive pressurization can increase the cavitation threshold. It is more realistic to think of p_v as the breaking strength of the liquid that depends on the history of the liquid sample. Readers may find it convenient to consider the case $p_v = 0$; in this case the sample will fail under tension.

17.2.1 Analysis of stress-induced cavitation

The aperture flow in a flat plate is shown in figure 17.1. The magnitude in the resulting jet will reach some uniform value U downstream of the edges. The half-width of the jet is $C_c\ell$, where C_c is the contraction coefficient and ℓ is the half-width of the aperture. The complex potential for this flow is given implicitly by (Currie, 1974, 129)

$$f(z) = \phi + i\psi = -\frac{2C_c\ell U}{\pi}\ln\left\{\cosh\left[\ln\left(U\frac{dz}{df}\right)\right]\right\} - iC_c\ell U. \tag{17.2.1}$$

The stress is calculated by

$$\mathbf{T} = -p\mathbf{1} + 2\mu\nabla\otimes\nabla\phi. \tag{17.2.2}$$

The pressure can be calculated with Bernoulli's equation:

$$p + \frac{\rho}{2}(u^2 + v^2) = p_d + \frac{\rho}{2}U^2 = p_u, \tag{17.2.3}$$

where p_u is the upstream pressure and p_d is the downstream pressure at a position where the velocity reaches the uniform velocity U. The velocities are evaluated from the potential:

$$u = \frac{1}{2}\left(\frac{df}{dz} + \frac{d\bar{f}}{d\bar{z}}\right), \qquad v = \frac{i}{2}\left(\frac{df}{dz} - \frac{d\bar{f}}{d\bar{z}}\right). \tag{17.2.4}$$

It follows that the rate-of-strain tensor is

$$2\mathbf{D} = \begin{bmatrix} \left(\frac{d^2 f}{dz^2} + \frac{d^2 \bar{f}}{d\bar{z}^2}\right) & i\left(\frac{d^2 f}{dz^2} - \frac{d^2 \bar{f}}{d\bar{z}^2}\right) \\ i\left(\frac{d^2 f}{dz^2} - \frac{d^2 \bar{f}}{d\bar{z}^2}\right) & -\left(\frac{d^2 f}{dz^2} + \frac{d^2 \bar{f}}{d\bar{z}^2}\right) \end{bmatrix}. \tag{17.2.5}$$

To use the maximum tension criterion for cavitation, the principal axes coordinates in which $2\mathbf{D}$ is diagonalized need to be found. In the two-dimensional case under consideration here, the diagonalized rate-of-strain tensor is

$$2\mathbf{D} = \begin{bmatrix} \lambda & 0 \\ 0 & -\lambda \end{bmatrix}, \quad \text{where} \quad \lambda = 2\left|\frac{d^2 f}{dz^2}\right|. \tag{17.2.6}$$

Thus the maximum tension T_{11} is given by

$$T_{11} = -p + \mu\lambda = -p_u + \frac{\rho}{2}(u^2 + v^2) + \mu\lambda, \tag{17.2.7}$$

and the cavitation threshold is given by

$$T_{11} = -p_v. \tag{17.2.8}$$

Combining (17.2.7) and (17.2.8), we obtain

$$T_{11} + p_v = -p_u + p_v + \frac{\rho}{2}(u^2 + v^2) + \mu\lambda = 0. \tag{17.2.9}$$

We use $\frac{\rho}{2}U^2$ to render (17.2.9) dimensionless:

$$\begin{aligned} \frac{T_{11} + p_v}{\frac{\rho}{2}U^2} &= \frac{-p_u + p_v}{p_u - p_d} + \frac{u^2 + v^2}{U^2} + \frac{\mu\lambda}{\frac{\rho}{2}U^2} \\ &= -\frac{1+K}{K} + \left|\alpha\left(\frac{\phi}{\ell U}, \frac{\psi}{\ell U}\right)\right|^2 + \frac{1}{Re}\frac{2\pi}{C_c}\left|\beta\left(\frac{\phi}{\ell U}, \frac{\psi}{\ell U}\right)\right| = 0, \end{aligned} \tag{17.2.10}$$

where the dimensionless parameters are defined as

$$\text{cavitation number} \quad K = \frac{p_u - p_d}{p_d - p_v}, \tag{17.2.11}$$

$$\text{Reynolds number} \quad Re = \frac{\rho\ell U}{\mu}, \tag{17.2.12}$$

and the complex functions α and β are given by

$$\alpha = \left(e^\gamma \pm \sqrt{e^{2\gamma} - 1}\right)^{-1}, \quad \beta = \left(e^\gamma \pm \frac{e^{2\gamma}}{\sqrt{e^{2\gamma} - 1}}\right)\alpha^3, \tag{17.2.13}$$

with

$$\gamma = -\frac{f + iC_c\ell U}{2C_c\ell U}\pi = -\frac{\pi}{2C_c}\frac{\phi}{\ell U} - \frac{i\pi}{2C_c}\frac{\psi}{\ell U} - \frac{i\pi}{2}. \tag{17.2.14}$$

Definition (17.2.11) for the cavitation number follows Bergwerk (1959) and Soteriou *et al.* (1995) and is somewhat different from the definition by Brennen (1995) in which the denominator is the dynamic stagnation pressure.

For a flow with given cavitation number K and Reynolds number, equation (17.2.10) gives the positions where cavitation inception occurs in terms of $\phi/\ell U$ and $\psi/\ell U$. The dimensionless description of (17.2.10) is independent of the vapor pressure or any dimensional parameters entering into the definition of K and Re.

17.2.2 Stream function, potential function, and velocity

The complex potential of flow (17.2.1) is implicit and not convenient to use. Therefore we invert the potential to obtain a function in the form $z = z(f)$. First we transform the variables as

$$z' = \frac{\pi z}{2C_c\ell},$$

which gives

$$U\frac{dz}{df} = U\frac{dz}{dz'}\frac{dz'}{d\gamma}\frac{d\gamma}{df} = -\frac{dz'}{d\gamma}. \tag{17.2.15}$$

By virtue of (17.2.14) and (17.2.15), we have

$$\gamma = \ln\left\{\cosh\left[\ln\left(-\frac{dz'}{d\gamma}\right)\right]\right\},$$

which can be written as

$$e^{\gamma} = -\frac{1}{2}\left[\frac{dz'}{d\gamma} + \left(\frac{dz'}{d\gamma}\right)^{-1}\right].$$

Thus we obtain

$$\frac{dz'}{d\gamma} = -e^{\gamma} - \sqrt{e^{2\gamma} - 1}. \tag{17.2.16}$$

Integration of (17.2.16) gives

$$z' = -e^{\gamma} - \sqrt{e^{2\gamma} - 1} + \frac{1}{2i}\ln\left(\frac{1 + i\sqrt{e^{2\gamma} - 1}}{1 - i\sqrt{e^{2\gamma} - 1}}\right) - \pi/2. \tag{17.2.17}$$

We prescribe the value of the complex potential and compute the corresponding position z by (17.2.17). The computational results are shown in figure 17.2(a) where the stream and potential functions are plotted in the z plane. The velocity is then obtained with (17.2.4), shown as a vector plot in figure 17.2(b).

Nearly all the flow through the aperture emanates from regions of irrotationality; vorticity generated by no-slip at the wall of the aperture is confined to a boundary layer, and its effects are neglected here. Stress-induced cavitation can arise where the irrotational stresses are very large, but the maximum stress, hence the point of inception, is probably at the edge of the aperture where the vorticity is greatest. We show in the next subsection that VPF predicts cavitation at the tip whenever the viscosity is larger than zero. Therefore the effects of vorticity could be important only if they were such as to suppress cavitation at the tip. Even in this case, the fluid could cavitate in the irrotational region outside the boundary layer. We note that the boundary layer is shrinking near the tip of the aperture because of a favorable pressure gradient; the aperture flow is analogous to the flow with suctions on the wall that suppress the boundary-layer thickness. Therefore the boundary layer could be very thin near the tip of the aperture, and our theory can be applied to the

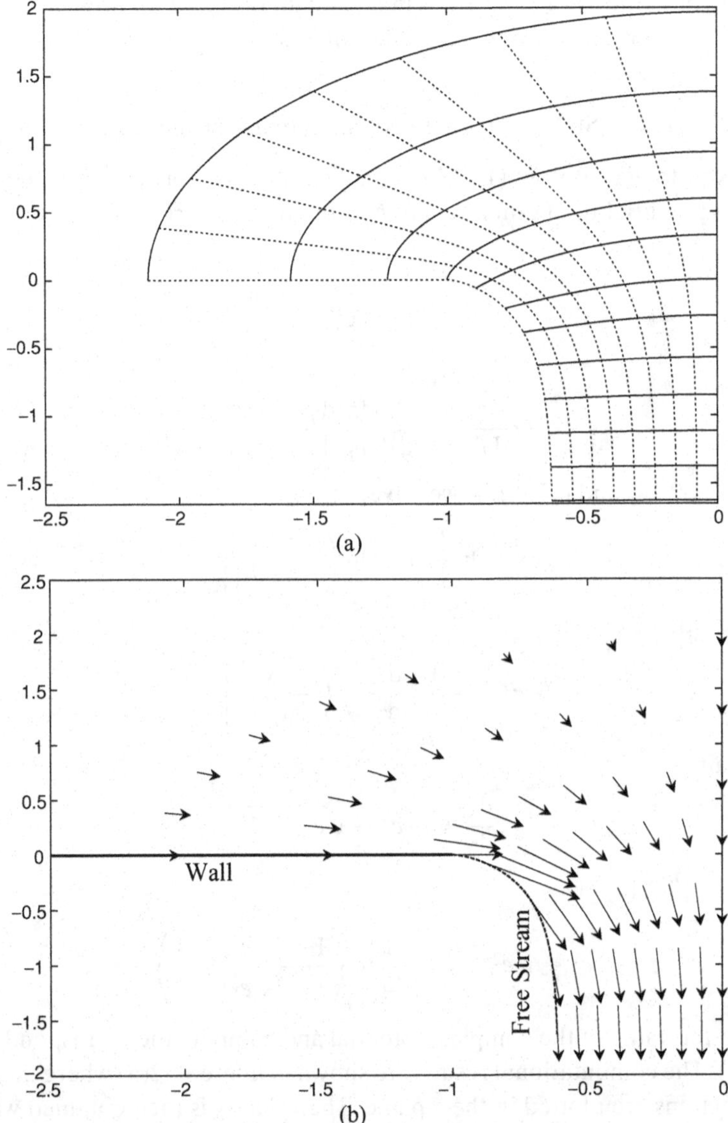

Figure 17.2. (a). The stream and potential functions in the z plane. The x and y coordinates are normalized as $(x/\ell, y/\ell)$. The stream and potential functions are in the range $-C_c < \psi/(\ell U) < 0$ and $-0.375 < \phi/(\ell U) < 1.75$, respectively. The contraction coefficient $C_c = \pi/(2+\pi) = 0.611$, and the edge of the nozzle is at $(x/\ell, y/\ell) = (-1, 0)$. (b) The velocity in the z plane. Only half of the flow field is shown because of the symmetry.

region near the tip but outside the boundary layer. Our results indicate that cavitation may occur in such regions. The main unsolved question is the extent to which vorticity generated at the boundary of the aperture corrupts the main flow when the Reynolds number is not large.

17.2.3 Cavitation threshold

We use water as an example to illustrate stress-induced cavitation. The vapor pressure of water at 20 °C is 2339 Pa, and we assume that the downstream pressure is the atmospheric

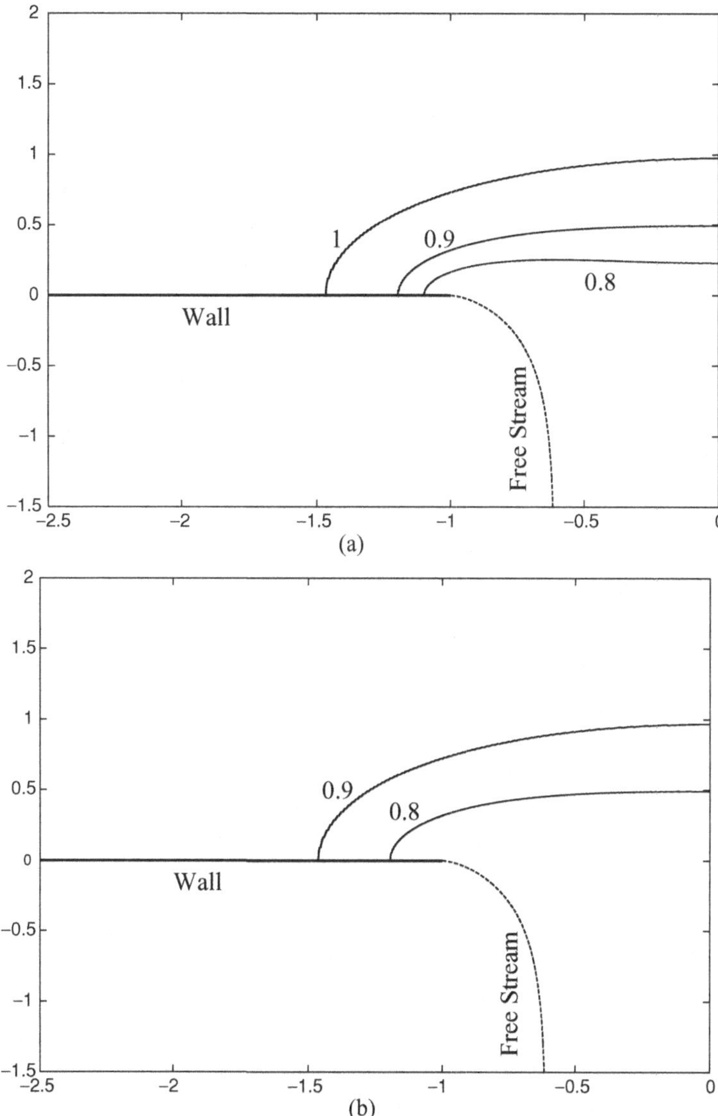

Figure 17.3. Contour plot for $(p - p_v)/(\rho U^2/2)$ in the $(x/\ell, y/\ell)$ plane: (a) $K = 10$; (b) $K = 1000$. In the flow field, $p - p_v > 0$ everywhere. Thus there is no cavitation according to the pressure criterion.

pressure: $p_d = p_a = 10^5$ Pa. First we calculate the pressure by using Bernoulli's equation (17.2.3). The pressure does not depend on the Reynolds number, and we show the pressure distribution for different cavitation numbers in figure 17.3.

The pressure criterion for cavitation is that cavitation occurs when the pressure is lower than the vapor pressure. The minimum pressure in the aperture flow is the downstream pressure and $p_d = 10^5$ Pa $> p_v = 2339$ Pa. Thus the pressure criterion predicts no cavitation for the case under consideration. However, we will show that cavitation occurs in the aperture flow according to the tensile stress criterion (Joseph, 1995, 1998).

Next we include the viscous part of the stress and consider the maximum tension T_{11}. The cavitation criterion is that cavitation occurs when $T_{11} + p_v \geq 0$. T_{11} depends on both the Reynolds number and the cavitation number. We show the contour plot for

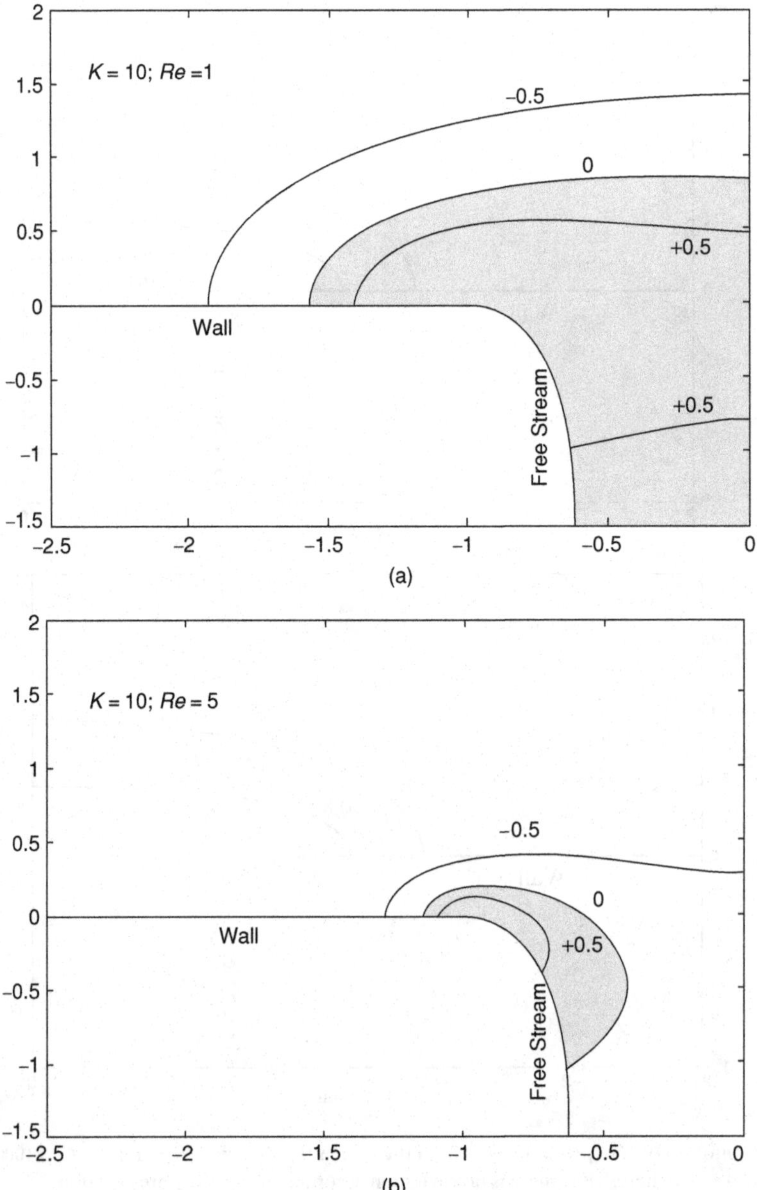

Figure 17.4. Contour plot for $(T_{11} + p_v)/(\rho U^2/2)$ in the $(x/\ell,\ y/\ell)$ plane: (a) $K = 10$ and $Re = 1$; (b) $K = 10$ and $Re = 5$. Cavitation occurs inside the curve on which $T_{11} + p_v = 0$. Figures 17.3(a) and 17.4(a) can be compared to show the effect of the viscous stress cavitation.

$(T_{11} + p_v)/(\rho U^2/2)$ with different *Re* and *K* in figure 17.4. More graphs, with different values of *Re* and *K*, can be found in Funada, Wang and Joseph (2006).

Although the velocity is continuous everywhere in the aperture flow, its derivative and therefore the viscous stress are singular at the sharp edge. Thus at the sharp edge for all fluids with viscosity larger than zero, $T_{11} + p_v$ is always larger than zero and cavitation occurs. In our analysis here, we avoid the singular points and calculate the stresses at points very close to the edges. This is partially justified by the fact that in reality the edges

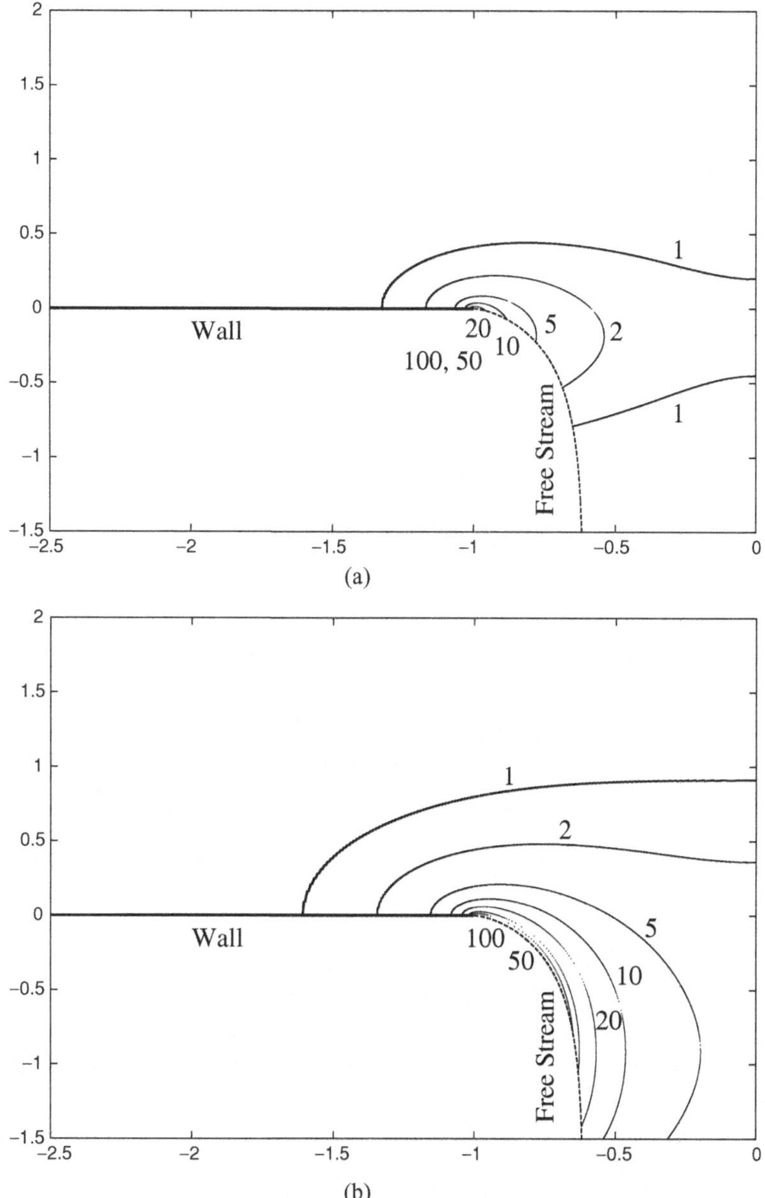

Figure 17.5. The cavitation threshold curves on which $T_{11} + p_v = 0$ in different flows with $Re = 1, 2, 5, 10, 20, 50$, and 100. The cavitation number is fixed at $K = 1$ in (a) and at $K = 100$ in (b). Cavitation occurs inside the curve on which $T_{11} + p_v = 0$.

are not perfectly sharp. As Chaves *et al.* (1995) noted, "Microscopic pictures of the nozzle inlet still show however small indentations of the corner, i.e. less than 5 micrometers." (The diameter of the nozzle in their experiments was 0.2 or 0.4 mm.)

In figure 17.4, the curves on which $T_{11} + p_v = 0$ are the thresholds for cavitation. On the side of a $T_{11} + p_v = 0$ curve that is closer to the sharp edge, $T_{11} + p_v > 0$ and cavitation appears; on the other side of the $T_{11} + p_v = 0$ curve, $T_{11} + p_v < 0$ and there is no cavitation.

In figure 17.5, we single out the threshold curves on which $T_{11} + p_v = 0$ and plot these curves corresponding to different Re at $K = 1$ and $K = 100$. The cavitation region is

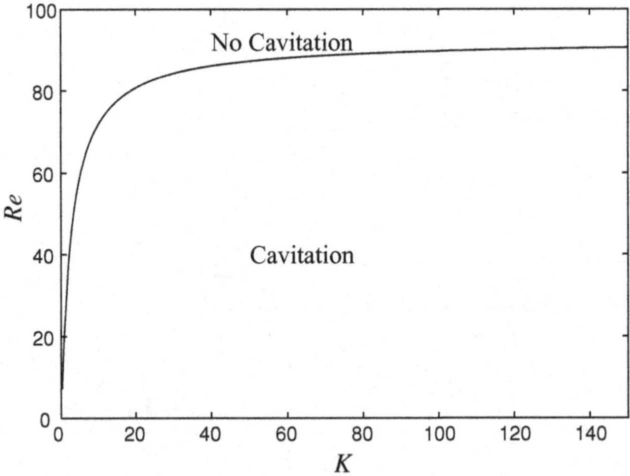

Figure 17.6. The curve on which $T_{11} + p_v = 0$ at the point $(x/\ell = -1.01,\ y/\ell = 0)$ in the *Re* vs. *K* plane. Below the curve, $T_{11} + p_v > 0$ and cavitation occurs at the point; above the curve, $T_{11} + p_v < 0$ and there is no cavitation at the point.

larger when *Re* is smaller; when *Re* is larger, cavitation is confined to a very small region near the edge of the aperture. The cavitation region is larger and is stretched to far downstream when *K* is larger. This shows that supercavitation, i.e., cavitation extending to the nozzle exit, can be achieved when *K* is large. This predicted supercavitation occurs under the hydraulic flip condition and is not the same as the supercavitation observed in experiments.

Now we focus on the point $(x/\ell = -1.01,\ y/\ell = 0)$, which is upstream to the left sharp edge and very close to it. We can identify values of *Re* and *K* that give rise to $T_{11} + p_v = 0$ at this point. These values can be plotted on the *Re* and *K* plane, as shown in figure 17.6. On one side of the plotted curve, $T_{11} + p_v > 0$ and cavitation occurs at this point; on the other side, there is no cavitation at this point. Figure 17.6 shows that small *Re* and large *K* favor cavitation. This can be understood readily because small *Re* leads to large viscous stress and large *K* leads to small pressure. Both of these effects contribute to a large value of $T_{11} = -p + \mu\lambda$.

17.2.4 Conclusions

We use the potential flow through an aperture as a 2D model to study the hydraulic flip observed in injection flows at a nozzle. The pressure in the flow field is computed with Bernoulli's equation and the viscous stress is evaluated on the potential. The stress tensor is transformed to the principal axes coordinates and the principal stress T_{11} is obtained. If T_{11} is larger than the negative value of the vapor pressure p_v, the flow will cavitate. We find that cavitation occurs for all fluids with viscosity larger than zero at the sharp edges of the aperture. The region in which cavitation occurs depends on the Reynolds number *Re* and the cavitation number *K*. The cavitation region is larger if *Re* is smaller and *K* is larger. The cavitation is confined to very small regions near the edges of the aperture when *Re* is larger and *K* is smaller.

Researchers do not observe cavitation in a hydraulic flip. The reason may be that the Reynolds numbers in nozzle flows are usually very high (of the order of thousands and tens of thousand). Thus, even if cavitation occurred at the edge of the nozzle, the cavities would collapse quickly outside the small-cavitation region (the time for cavities to collapse is of the order of microseconds according to Chaves *et al.*, 1995) and would be very difficult to observe. The effects of liquid viscosity on cavitation are apparently not known; we could not find an evaluation of these effects in the literature. The results obtained here and in Joseph (1998) suggest that an increase in viscosity lowers the threshold to stress-induced cavitation.

17.2.5 Navier–Stokes simulation

A full Navier–Stokes simulation of stress-induced cavitation in aperture flow was carried out by Dabiri, Sirignano, and Joseph (2007) to evaluate the conclusions stated in 17.2.4. The computations were done with the finite-volume package SIMPLER. The free surface was generated with level sets. The effects of surface tension were not computed. To secure convergence when this numerical package is used, it is necessary to have an active ambient. In these calculations the gas–liquid density ratio is 10^{-4} and the viscosity ratio is 1.5×10^{-4}. The computations are not sensitive to changes of these ratios for small ratios. The Reynolds number in these calculations is based on Bernoulli's velocity of the jet, which is larger than the average velocity of a jet for low Reynolds number. For example, for flow with $Re = 1$ the jet velocity is about 20% of Bernoulli velocity, and this causes the strain rates and therefore the stresses to be scaled down with the same ratio. This could be the reason that the regions with large stresses are smaller in the Navier–Stokes solution.

The results of the computations are exhibited in figures 17.7 and 17.8. Isolines for tensile stresses are marked on the figures; high numbers identify high stresses. The liquid is most at risk to stress-induced cavitation when the stress level is positive and high. The regions of high stress are near the corner in the Navier–Stokes and VPF calculations. We could say that the computations of regions vulnerable to stress-induced cavitation are given with reasonable accuracy by VPF, even for low Reynolds numbers.

It is of interest that the free surface in the VPF calculation done here is a streamline associated with potential flow of an inviscid fluid. The pressure is constant on the free streamline, so that the speed there is constant, as shown by Bernoulli's equation. Dabiri, Sirignano, and Joseph (2007) have shown that the normal strain rate $\partial U_n/\partial n$ vanishes on the free streamline of the potential flow used by Funada, Wang and Joseph (2006). The irrotational flow satisfies the constant normal stress condition on the free surface, but does not satisfy the zero shear stress on the free streamline.

17.3 Streaming motion past a sphere

Padrino *et al.* (2007b) solved the problem of stress-induced cavitation in the axisymmetric streaming flow past a sphere. This is a canonical problem for cavitation behind blunt bodies; the application of the maximum tension theory to this problem is new. Padrino *et al.* derived explicit analytic formulas for this theory in the Stokes flow limit and for the irrotational flow of a viscous fluid. The cavitation thresholds for these analytic approaches were compared with exact results from a direct numerical simulation. The direct simulation

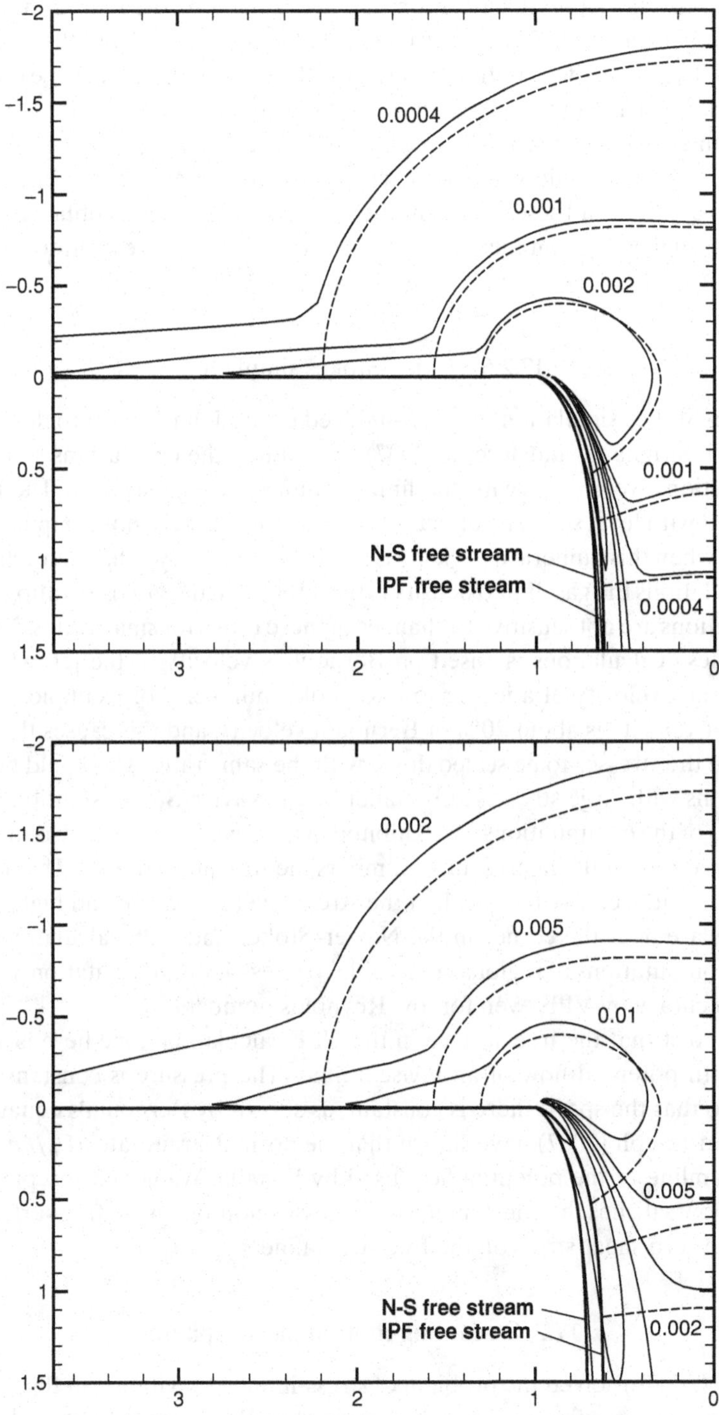

Figure 17.7. Viscous stress contours for Navier–Stokes (N-S) solution (solid curves) compared with VPF solution (dashed curves) for ρ ratio = 10^{-4} and μ ratio = 1.5×10^{-4}. Top, *Re* = 500; bottom, *Re* = 100.

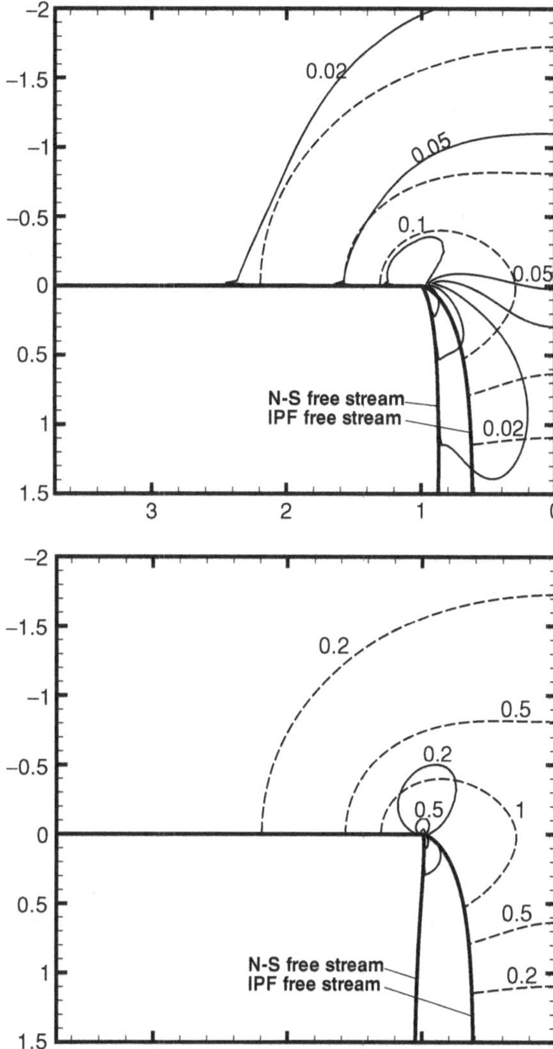

Figure 17.8. Viscous stress contours for Navier–Stokes (N-S) solution (solid curves) compared with VPF solution (dashed curves) for ρ ratio $= 10^{-4}$ and μ ratio $= 1.5 \times 10^{-4}$. Top, $Re = 10$; bottom, $Re = 1$.

allows us to compare cavitation thresholds between the stress-induced maximum tension theory with pressure alone theory. (This is different from the calculation in aperture flow, in which the flow is irrotational and the fluid viscous. In the irrotational flow of a viscoius fluid the pressure is given by Bernoulli's equation and does not depend on viscosity.) Both theories computed with the same Navier–Stokes package depend on the Reynolds number. The liquid is always at greater risk of cavitation from viscous stresses than from the pressure alone theory. The difference between the approaches of the two theories increases as the Reynolds number decreases and is greatest at the boundary of the sphere where the motion is pure shear; if the shear stress is large enough, the liquid will fracture even when the pressure is much higher than the critical pressure. Readers with an interest in the theory of stress-induced cavitation will profit from reading the paper of Padrino *et al.* For our purpose here we can show how the cavitation threshold formula is formed from the irrotational theory.

17.3.1 Irrotational flow of a viscous fluid

De Chizelle, Ceccio, and Brennen (1995) studied the interactions between a traveling cavity and the potential flow exterior to the thin boundary layer around an axisymmetric headform. Liu and Brennen (1998) presented a mechanistic model for a hydrodynamic cavitation event rate for flow over a headform that utilizes the pressure distribution given by potential flow modified to accomodate boundary-layer effects. There is no literature other than the paper of Funada, Wang, and Joseph (2006) on analysis of stress-induced cavitation by use of potential flow. The analysis of this problem, subsequently discussed, is completely transparent; the effects of vorticity on cavitation on a solid sphere are mainly associated with the formation of wakes and the displacement of the region of irrotational flow. Considering the irrotational flow of a viscous fluid, we can determine the principal strain rates and stresses and their corresponding principal directions. Then the maximum tension criterion is applied to evaluate the cavitation threshold. The theory of VPF considered here includes the viscous components in the definition of the state of stress in the flowing liquid.

Irrotational flows of incompressible viscous fluids satisfy the Navier–Stokes equations and give rise to the usual Bernoulli's equation because

$$\mu\nabla^2\mathbf{u} = \mu\nabla\nabla^2\phi = 0, \tag{17.3.1}$$

no matter what the value of μ. The stresses are given by

$$\mathbf{T} = -p\mathbf{1} + 2\mu\nabla\otimes\nabla\phi = \frac{1}{3}\mathrm{Tr}\,(\mathbf{T})\,\mathbf{1} + 2\mu\nabla\otimes\nabla\phi, \tag{17.3.2}$$

where p is the average stress given by (17.1.2).

The flow is axisymmetric and steady, and the potential $\phi(r, \theta)$ satisfies $\nabla^2\phi = 0$. In this analysis, spherical-polar coordinates are utilized.

The potential for this flow is

$$\phi = U\left(r + \frac{1}{2}\frac{a^3}{r^2}\right)\cos\theta. \tag{17.3.3}$$

The velocity $\mathbf{u} = \mathbf{e}_r u_r + \mathbf{e}_\theta u_\theta$ is given by

$$u_r = \frac{\partial\phi}{\partial r} = U\left(1 - \frac{a^3}{r^3}\right)\cos\theta, \tag{17.3.4}$$

$$u_\theta = \frac{1}{r}\frac{\partial\phi}{\partial\theta} = -U\left(1 + \frac{1}{2}\frac{a^3}{r^3}\right)\sin\theta. \tag{17.3.5}$$

Note that the no-slip condition must be relaxed for VPF, so the boundary layer is not resolved. The pressure is given by

$$\begin{aligned} p &= p_\infty + \rho\frac{U^2}{2} - \frac{\rho}{2}\left(u_r^2 + u_\theta^2\right) \\ &= p_\infty + \rho\frac{U^2}{2}\left[1 - \left(1 - \frac{a^3}{r^3}\right)^2\cos^2\theta - \left(1 + \frac{1}{2}\frac{a^3}{r^3}\right)^2\sin^2\theta\right], \end{aligned} \tag{17.3.6}$$

where p_∞ is the constant value of the pressure at infinity. The nonzero components of the viscous stress,

$$2\mu\mathbf{D}\left[\nabla\phi\right], \tag{17.3.7}$$

are

$$2\mu D_{rr} = 2\mu\frac{\partial u_r}{\partial r} = 6\mu U\frac{a^3}{r^4}\cos\theta, \tag{17.3.8}$$

$$2\mu D_{\theta\theta} = 2\mu\left(\frac{1}{r}\frac{\partial u_\theta}{\partial\theta} + \frac{u_r}{r}\right) = -3\mu U\frac{a^3}{r^4}\cos\theta, \tag{17.3.9}$$

$$2\mu D_{\varphi\varphi} = 2\mu\left(\frac{u_r}{r} + \frac{u_\theta}{r}\cot\theta\right) = -3\mu U\frac{a^3}{r^4}\cos\theta, \tag{17.3.10}$$

$$2\mu D_{r\theta} = \mu\left[r\frac{\partial}{\partial r}\left(\frac{u_\theta}{r}\right) + \frac{1}{r}\frac{\partial u_r}{\partial\theta}\right] = 3\mu U\frac{a^3}{r^4}\sin\theta. \tag{17.3.11}$$

The matrix of components,

$$2\mu\begin{bmatrix} D_{rr} & D_{r\theta} & 0 \\ D_{r\theta} & D_{\theta\theta} & 0 \\ 0 & 0 & D_{\varphi\varphi}\end{bmatrix} = 3\mu U\frac{a^3}{r^4}\begin{bmatrix} 2\cos\theta & \sin\theta & 0 \\ \sin\theta & -\cos\theta & 0 \\ 0 & 0 & -\cos\theta\end{bmatrix}, \tag{17.3.12}$$

can be rotated into diagonal form through an angle α satisfying

$$\tan 2\alpha = \frac{2}{3}\tan\theta. \tag{17.3.13}$$

From (17.3.13) we look for the angle α that puts by rotation with axis $\mathbf{e}_\varphi$ the direction given by the unit vector $\mathbf{e}_r$ into the principal direction corresponding to the most tensile (or the least compressive) principal stress in the plane of the motion. Without lack of generality, we consider this angle α to be in the interval $-0.5 \leq \alpha/\pi \leq 0.5$.

The diagonal form of $2\mu\nabla\otimes\nabla\phi$ is given by

$$3\mu U\frac{a^3}{r^4}\begin{bmatrix} \frac{1}{2}\cos\theta + \frac{\sin\theta}{\sin 2\alpha} & 0 & 0 \\ 0 & \frac{1}{2}\cos\theta - \frac{\sin\theta}{\sin 2\alpha} & 0 \\ 0 & 0 & -\cos\theta\end{bmatrix}. \tag{17.3.14}$$

At $\theta = \pi/2$, where the pressure is smallest, $\alpha = \pi/4$ and the diagonal form is

$$3\mu U\frac{a^3}{r^4}\begin{bmatrix} 1 & 0 & 0 \\ 0 & -1 & 0 \\ 0 & 0 & 0\end{bmatrix}, \tag{17.3.15}$$

giving rise to tension and compression.

We next consider the whole stress by using (17.3.2), which may be written as

$$\mathbf{T} + p_c\mathbf{1} = (-p + p_c)\mathbf{1} + 2\mu\nabla\otimes\nabla\phi \tag{17.3.16}$$

with the addition of the cavitation threshold p_c. After arranging, expression (17.3.16) becomes

$$\frac{\mathbf{T}+p_c\mathbf{1}}{\frac{1}{2}\rho U^2} = -\left[K+1-\left(1-\frac{a^3}{r^3}\right)^2\cos^2\theta-\left(1+\frac{a^3}{2r^3}\right)^2\sin^2\theta\right]\begin{bmatrix}1&0&0\\0&1&0\\0&0&1\end{bmatrix}$$

$$+\frac{3}{Re}\left(\frac{a}{r}\right)^4\begin{bmatrix}\cos\theta+\frac{2\sin\theta}{\sin 2\alpha}&0&0\\0&\cos\theta-\frac{2\sin\theta}{\sin 2\alpha}&0\\0&0&-2\cos\theta\end{bmatrix}, \tag{17.3.17}$$

where

$$K=\frac{p_\infty-p_c}{\frac{1}{2}\rho U^2} \tag{17.3.18}$$

is the cavitation number and

$$Re=\frac{\rho U a}{\mu} \tag{17.3.19}$$

is the Reynolds number.

For VPF solution (17.3.17), we can show that at any point the most tensile (least compressive) and most compressive (least tensile) principal stresses lie in the plane of motion (i.e., the plane where the velocity vector is contained). Suppose now that T_{11} is the largest of the three principal values of stress. Then, according to the maximum tension theory, the locus of the cavitation threshold is given by

$$T_{11}+p_c=0, \tag{17.3.20}$$

giving rise to isolines $(a/r,\theta)=f(K_c, Re)$ for the cavitation threshold. The largest values of the viscous irrotational stress are at the boundary $r=a$ where the neglected vorticity is largest. In §17.3.2, it is shown that $\theta=0$ for very low Re and $\theta=\pi/2$ for very high Re are the points most vulnerable for cavitation under the respective conditions.

Equation (17.3.17) gives the form of the diagonalized stress tensor at each point (r,θ) in the axially symmetric flow. T_{11}, T_{22}, and $T_{33}=T_{\varphi\varphi}$ are principal stresses in the principal axes coordinates with bases $\mathbf{e}_1$,$\mathbf{e}_2$, and $\mathbf{e}_\varphi$. In the present case, the angle α changes with θ, $\tan 2\alpha=\frac{2}{3}\tan\theta$. A representation of the orientation of the principal axes in the plane of motion at the surface of the sphere $r=a$ as predicted by (17.3.13) is presented in figure 17.9.

It is apparent from (17.3.17) that the largest stresses are at the boundary of the sphere where $r=a$. Certainly the liquid will cavitate when $K=(p_\infty-p_c)/(\frac{\rho}{2}U^2)<0$; only $K>0$ is of interest. Using now the maximum tension criterion, we see that cavitation occurs for $0<K<K_c$ and that the fluid is most at risk to cavitation for θ at which $K_c(\theta)$ is greatest. For VPF, this most dangerous θ is at $\theta=0$ when Re is small and at $\theta=\pi/2$ when Re is large. It follows that the place most at risk to cavitation runs from the rear stagnation point at $\theta=0$ when Re is small to $\theta=\pi/2$ when Re is large.

Next, contour plots with lines of constant K representing the cavitation threshold from (17.3.17) for various Re are presented and compared with the results from the numerical simulation and the Stokes solution for the lowest Re (see figures 17.10, 17.11 and 17.12).

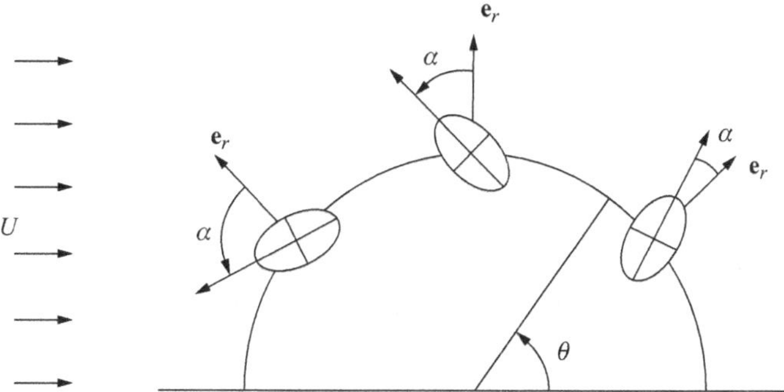

Figure 17.9. Schematic view of the orientation of the principal directions in the plane of motion for irrotational flow of a viscous fluid according to (17.3.13) on the surface of the sphere. The major axis in the ellipse represents the maximum tensile stress. The angle α puts the direction defined by the unit outward normal vector $\mathbf{e}_r$ into the principal direction of the maximum tensile stress.

17.3.2 An analysis for maximum K

Here we look for the angular position on the surface of the sphere at which the maximum value of the cavitation number K occurs. This position is the most vulnerable to cavitation.

For the potential flow solution, the stress tensor is given in (17.3.17). Suppose that T_{11} is the maximum tensile stress such that $T_{11} \geq T_{33} \geq T_{22}$. If we consider the surface of the sphere $r = a$ and use the cavitation criterion $T_{11} + p_c = 0$, we obtain from (17.3.17)

$$K = \frac{9}{4}\sin^2\theta - 1 + \frac{3}{Re}\left(\cos\theta + \frac{2\sin\theta}{\sin 2\alpha}\right). \tag{17.3.21}$$

Considering the expression for α given in (17.3.13),

$$\tan 2\alpha = \frac{2}{3}\tan\theta, \tag{17.3.22}$$

we can write (17.3.21) as

$$K = \frac{9}{4}\sin^2\theta - 1 + \frac{3}{Re}\left(\cos\theta + 3\cos\theta\sqrt{1 + \frac{4}{9}\tan^2\theta}\right) \tag{17.3.23}$$

for $0 \leq \theta \leq \pi/2$, whereas

$$K = \frac{9}{4}\sin^2\theta - 1 + \frac{3}{Re}\left(\cos\theta - 3\cos\theta\sqrt{1 + \frac{4}{9}\tan^2\theta}\right) \tag{17.3.24}$$

for $\pi/2 < \theta \leq \pi$. Taking the derivative of K in (17.3.23), we find

$$\frac{\partial K}{\partial\theta} = \frac{9}{2}\sin\theta\cos\theta - \frac{3}{Re}\sin\theta\left(1 + \frac{5}{\sqrt{9 + 4\tan^2\theta}}\right). \tag{17.3.25}$$

It is obvious that $\theta = 0$ is a solution of $\partial K/\partial\theta = 0$ at any Reynolds number. We compute $\partial^2 K/\partial\theta^2$ to determine whether K at $\theta = 0$ is a local maximum or minimum. From (17.3.25), the second derivative of K is

$$\frac{\partial^2 K}{\partial\theta^2} = \frac{9}{2}\cos 2\theta + \frac{60\sec\theta\tan^2\theta}{Re\,(9 + 4\tan^2\theta)^{3/2}} - \frac{\cos\theta}{Re}\left(3 + \frac{15}{\sqrt{9 + 4\tan^2\theta}}\right). \tag{17.3.26}$$

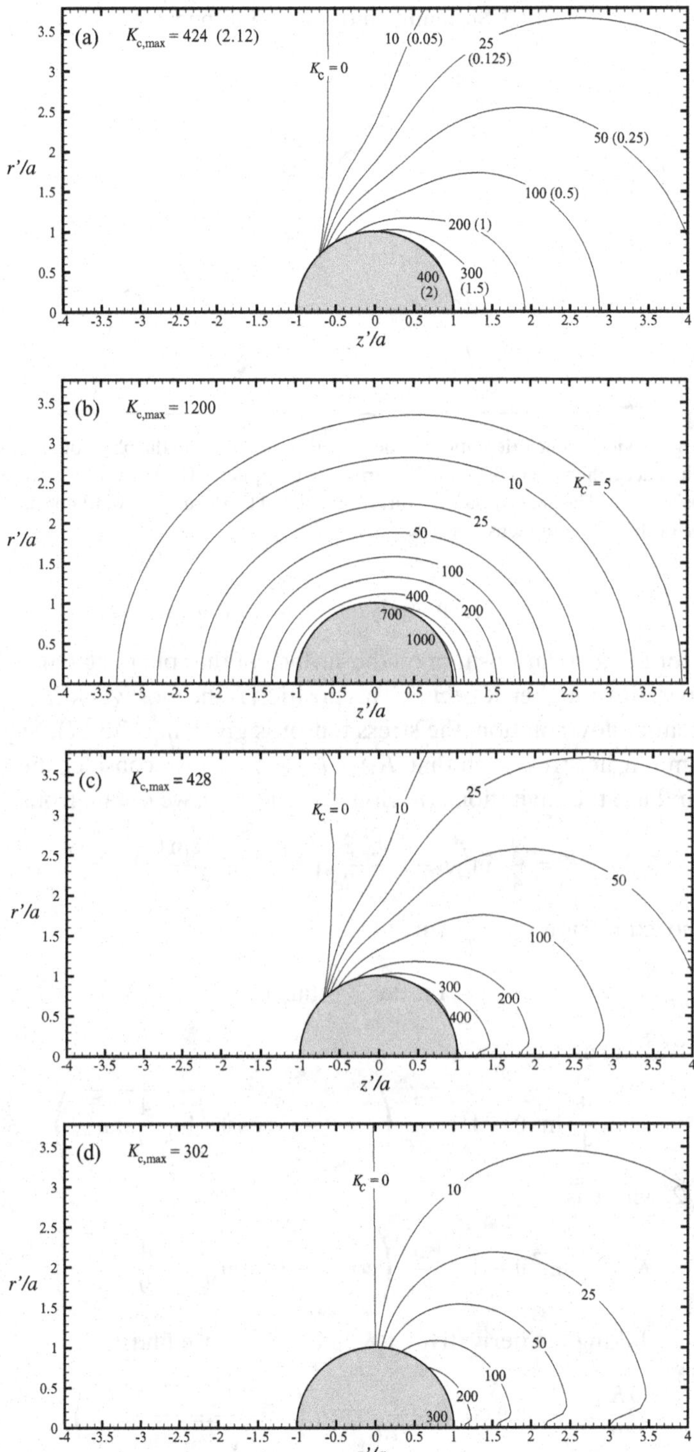

Figure 17.10. Contours of critical cavitation number K_c given by the condition $T_{11} + p_c = 0$ according to the maximum tension criterion for a Reynolds number $Re = 0.01$ from (a) Stokes flow, (b) the irrotational flow of a viscous fluid (17.3.17), and (c) numerical solution; the pressure criterion given by $K_c = -c_p$ is shown in (d) with the numerical pressure field. The cavitation number K is defined in terms of the dynamic pressure $\rho U^2/2$. For a given cavitation number K, cavitation occurs in the region where $K < K_c$. A different normalization of the cavitation number and of the critical cavitation number is used for Stokes flow rather than the normalization used for the other cases (17.3.18). The contour lines for the normalization of $p_\infty - p_c$ with the viscous stress scale $\mu U/a$ are presented in parentheses in (a). The ratio of the normalization factors is $Re/2$.

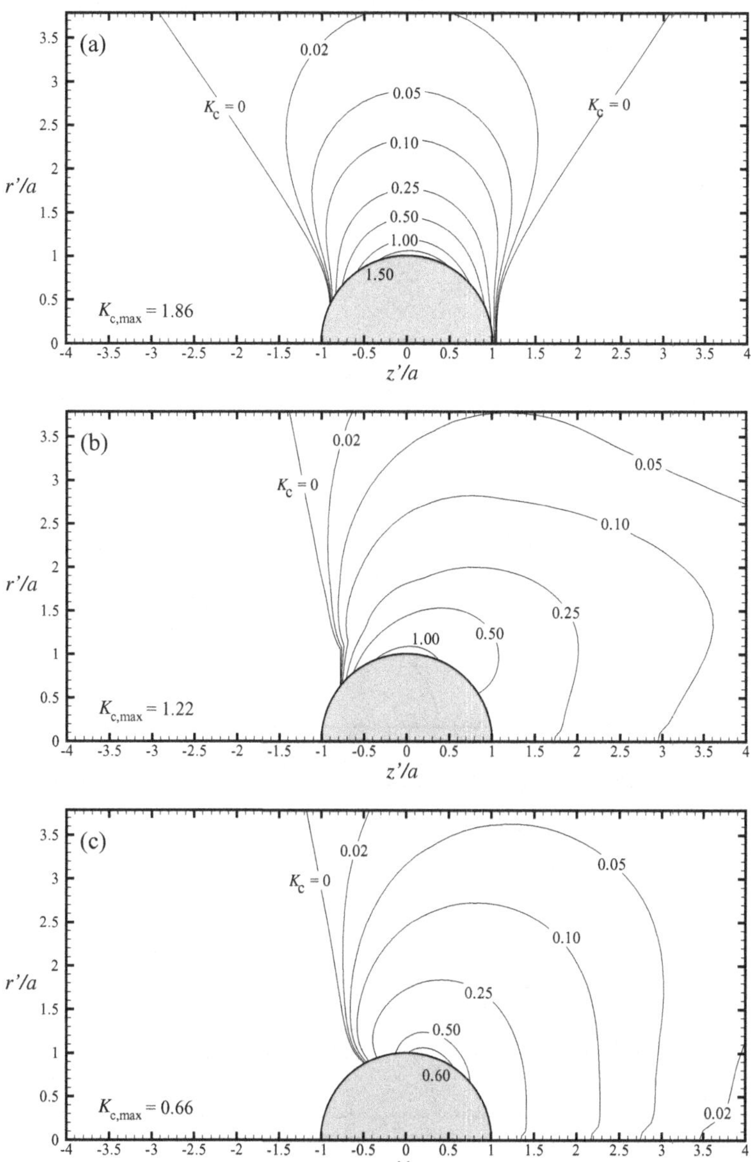

Figure 17.11. Contours of critical cavitation number K_c given by the condition $T_{11} + p_c = 0$ according to the maximum tension criterion for a Reynolds number $Re = 10$ from (a) the irrotational flow of a viscous fluid (17.3.17), and (b) numerical solution; the pressure criterion given by $K_c = -c_p$ is shown in (c) with the numerical pressure field. The cavitation number K is defined in terms of the dynamic pressure $\rho U^2/2$. For a given cavitation number K, cavitation occurs in the region where $K < K_c$.

When $\theta = 0$, we have

$$\frac{\partial^2 K}{\partial \theta^2}(\theta = 0) = \frac{9}{2} - \frac{8}{Re}. \tag{17.3.27}$$

Thus $\partial^2 K/\partial \theta^2$ at $\theta = 0$ is negative when $Re < 16/9$ and positive when $Re > 16/9$. This result indicates that K at $\theta = 0$ is a local maximum when $Re < 16/9$ and is a local minimum

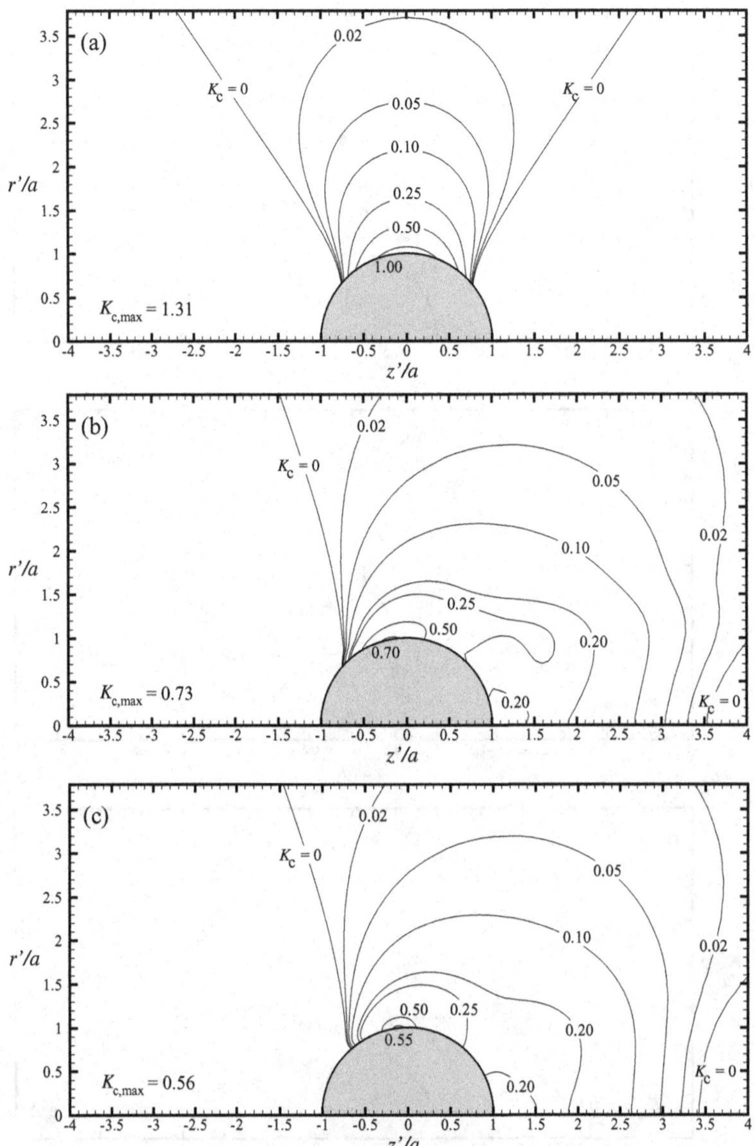

Figure 17.12. Contours of critical cavitation number K_c given by the condition $T_{11} + p_c = 0$ according to the maximum tension criterion for a Reynolds number $Re = 100$ from (a) the irrotational flow of a viscous fluid (17.3.17), and (b) numerical solution; the pressure criterion given by $K_c = -c_p$ is shown in (c) with the numerical pressure field. The cavitation number K is defined in terms of the dynamic pressure $\rho U^2/2$. For a given cavitation number K, cavitation occurs in the region where $K < K_c$.

when $Re > 16/9$. Substitution of $\theta = 0$ into (17.3.23) yields

$$K(\theta = 0) = \frac{12}{Re} - 1, \qquad (17.3.28)$$

which is a local maximum of K for $Re < 16/9$ in the interval $0 \leq \theta \leq \pi/2$.

Besides $\theta = 0$, there is a second solution for $\partial K/\partial\theta = 0$ that depends on Re and has a very complicated expression. If the value of Re is very high, then $\partial K/\partial\theta \approx 9\sin(2\theta)/4$.

The second solution is close to $\theta = \pi/2$. We also note that $\partial^2 K/\partial\theta^2 < 0$ at $\theta = \pi/2$ when *Re* is high. Therefore the maximum value of K occurs near $\theta = \pi/2$ when *Re* is high in the interval $0 \leq \theta \leq \pi/2$.

Similarly, from (17.3.24), K has a local maximum at $\theta = \pi$ when $Re < 4/9$ and has a local minimum at this position when $Re > 4/9$ in the interval $\pi/2 < \theta \leq \pi$. Substitution of $\theta = \pi$ into (17.3.24) gives this local maximum for $Re < 4/9$ in the interval $\pi/2 < \theta \leq \pi$:

$$K(\theta = \pi) = \frac{6}{Re} - 1. \tag{17.3.29}$$

A second solution, as a complicated function of *Re*, can be written for $\partial K/\partial\theta = 0$. For high *Re*, this solution gives that K goes to a maximum when θ approaches $\pi/2$.

A comparison of (17.3.28) and (17.3.29) allows us to discard the position $\theta = \pi$. Summarizing our findings for VPF, we find that K reaches a maximum at $\theta = 0$ when $Re < 16/9$ in the interval of interest $0 \leq \theta \leq \pi$. In addition, in the limit of high *Re*, K is maximum at $\theta = \pi/2$ in this interval. These results are verified in figure 17.10(b) for low *Re* and figure 17.12(a) for high *Re*.

17.4 Symmetric model of capillary collapse and rupture

The breakup of a liquid capillary filament is analyzed as a VPF near a stagnation point on the centerline of the filament toward which the surface collapses under the action of surface-tension forces.[1] The analysis given here is restricted to cases in which the neck-down is symmetric around the stagnation point. The line drawing for the model in figure 17.13 is defined by a periodic array of stagnation points between points of liquid depletion and liquid accumulation. The passage of liquid from points of depletion to points of accumulation around which a liquid ball expands cannot be done without the generation of vorticity in an axisymmetric analog to the hyperbolic vortices $xy = \text{const}$ nested in the four quadrants at each stagnation point in the vortex array described by (4.6.33). The analysis can be said to generate a Helmholtz decomposition for rotational flow near a point of stagnation. We find that the neck is of parabolic shape and its radius collapses to zero in a finite time; the curvature at the throat tends to zero much faster than does the radius, leading ultimately to a microthread of nearly uniform radius. During the collapse the tensile stress that is due to viscosity increases in value until at a certain finite radius (which may be estimated as $2\gamma/p_a$, where γ is surface tension and p_a is atmospheric pressure, about 1.5 μm for water in air), the stress in the throat passes into tension, presumably inducing cavitation there. The Reynolds number at which the stress passes into tension decreases as $1/\nu^2$, where ν is the kinematic viscosity. Viscous threads rupture at very low Reynolds numbers. Numerical studies and experiments show that low-viscosity liquids generate satellite drops, but highly viscous liquids may fail in a symmetric mode before satellite drops form.

17.4.1 Introduction

The breakup of liquid jets is generally framed in terms of the capillary pressure $\gamma/R(z,t)$ because of surface tension γ acting at the neck of radius $R(z,t)$. The capillary pressure

[1] T. S. Lundgren and D. D. Joseph, 2007.

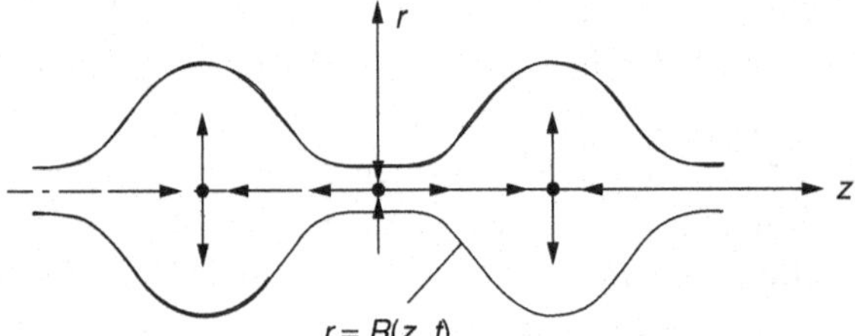

Figure 17.13. Periodic structure of stagnation points as a line drawing of the dynamics of capillary collapse. The collapse will give rise to a periodic string of liquid drops. The analysis here is local focusing on dynamics of collapse at $z = 0$.

is greatest at the position z where R is smallest, an unstable situation in which liquid is squeezed out of the neck, further reducing R and increasing the capillary pressure there. This picture leads to an inevitable collapse of the radius to zero. The conventional view is that the capillary instability just described leads to "pinchoff," but the physics required for actually rupturing the thread is not revealed. Here we examine the idea that the filaments can rupture under the action of tensile stresses which are induced by the extrusion of fluid out of the neck. In our model, the viscous extensional stresses in the neck are balanced by surface tension. Low viscosity jets do not break up in a symmetric mode; they develop satellite drops driven by high capillary pressure in the neck. The idea that liquids can break in tension is not new, but the idea that cavities will open when the principal tensile stress exceeds the local breaking stress is new; in a pure shear flow, cavities are opened by a tensile stress 45° from the direction of shearing.

Capillary collapse is the final stage of dynamics that may be framed as starting from the capillary instability of a liquid cylinder. The initial instability of the liquid cylinder was studied by Funada and Joseph (2002) and Wang, Joseph, and Funada (2005a). Funada and Joseph (2002) assumed that the motion of the viscous cylinder is irrotational; the velocity is given by $\mathbf{u} = \nabla\phi$, $\nabla^2\phi = 0$, and the viscous terms in the normal stress balances are evaluated from the potential. They derived a dispersion relation, σ vs. k, where σ is the growth rate and k is the wavenumber. They compared growth-rate curves for potential flows of inviscid and viscous fluids in which the conditions on the tangential components of the velocity and stress are neglected with the growth rates from the normal-mode reduction of the Navier–Stokes equations (called exact) in which the vorticity and continuity of the tangential velocity and stress are not neglected. Many liquids with viscosities differing by many orders of magnitude were studied. In all cases there is a strict separation of the growth-rate curves computed by the three theories. The growth rates for IPF are largest and those for the Navier–Stokes theory are smallest, with VPF in between. The curves are crowded when the viscosity is small and are widely separated when the viscosity is large. The potential flow solution for viscous fluids is in a modest agreement with the exact results, but the result for inviscid fluids are well off the mark.

Wang, Joseph, and Funada (2005a) implemented a method proposed by Joseph and Wang (2004) for computing a viscous correction of the irrotational pressure induced by the discrepancy between the nonzero irrotational shear stress and the zero-shear-stress boundary condition at a free surface. The theory with an additional viscous pressure correction added to the irrotational pressure is called the viscous correction of VPF (VCVPF). The corrected theory leads to a connection formula between the irrotational shear stress and the added viscous pressure that arises in a very thin boundary layer, which is not analyzed and not needed. The linearized equations in this layer are used to show that the added pressure is harmonic and the additional contributions of the viscosity to the normal stress are small compared with the viscous irrotational contribution.

The analysis of capillary instability by use of the added pressure is in remarkable agreement with the results of exact analysis for cases in which a liquid and a gas are involved. The growth-rate curves for VCVPF are nearly identical to those computed from the exact theory, uniformly in k. The two theories differ at most by a few percent whereas, for the case of highly viscous liquids, the analysis for inviscid liquids gives large, unrealistic growth rates. The popular idea that VPF should be a small perturbation of IPF is wrong.

The reader may think that the calculation of an added viscous pressure correction takes the theory away from purely irrotational flow, even though the velocities are obtained from the potential. However, exactly the same results that arise from VCVPF also arise from the DM, in which the pressure never enters; all the quantities needed are obtained from solutions of the Laplace equation for the potential.

The problem of capillary collapse considered here is rather different than the problem of capillary instability of a liquid cylinder. One obvious difference is that the instability problem is linear but the collapse problem is very nonlinear. Less obvious is the role of the pressure correction, which arises in the linear problem from the need to compensate for the unbalanced irrotational shear stress. The normal and shear stress in our model are balanced on the free surface at the neck $z = 0$, but the normal stress is not balanced near the neck at order z^2. A harmonic correction p_v is required for balancing the normal stresses away from the neck. This additional contribution (17.4.18) to the pressure generates a vortical contribution (17.4.22) to the velocity and the shear stress

$$\tau_{rz} = -Crz, \tag{17.4.1}$$

where C is given under (17.4.18), does not vanish on the free surface when $z \neq 0$.

17.4.2 Analysis

In this subsection we study the collapse of a capillary filament under surface-tension forces that squeeze liquid symmetrically from the throat at $z = 0$ in figure 17.13. The analysis is local; terms of the order of z^4 are neglected, but the local stagnation flow can be thought to be embedded in a global periodic structure of stagnation points with depletion at throats and accumulation at crests, as shown in figure 17.13.

we assume that the flow in the neighborhood of the throat is an axially symmetric straining flow, or stagnation-point flow, with velocity components

$$\begin{aligned} u_z &= a\,(t)\,z, \\ u_r &= -\frac{1}{2}a\,(t)\,r, \end{aligned} \tag{17.4.2}$$

and determine the strain rate $a(t)$ and the capillary shape, $r = R(z, t)$, by satisfying the appropriate boundary conditions at the capillary surface. The velocity field described by equations (17.4.1) and (17.4.2) is incompressible and irrotational; therefore, despite being a viscous flow, it may be described by a velocity potential ($u_z = \frac{\partial\phi}{\partial z}$, $u_r = \frac{\partial\phi}{\partial r}$) of the form

$$\phi = \frac{1}{2}az^2 - \frac{1}{4}ar^2. \tag{17.4.3}$$

The pressure p_i in the flow is determined from the unsteady version of Bernoulli's equation,

$$\frac{\partial\phi}{\partial t} + \frac{1}{2}\left(u_r^2 + u_z^2\right) + \frac{p_i}{\rho} = \frac{p_0}{\rho}, \tag{17.4.4}$$

in the form

$$\frac{p - p_0}{\rho} = -\left(\frac{1}{2}\dot{a} + \frac{1}{2}a^2\right)z^2 + \left(\frac{1}{4}\dot{a} - \frac{1}{8}a^2\right)r^2. \tag{17.4.5}$$

The stagnation pressure p_0 may be related to a distant state of rest at the center of the inflating balloon into which liquid is driven from the collapsing throat (see figure 17.13); p_0 is a global reference in an otherwise local solution. The overdot denotes a time derivative.

In this flow the state of stress is given by two principal stresses:

$$T_{zz} = -p + 2\mu\frac{\partial u_z}{\partial z} = -p + 2\mu a, \tag{17.4.6}$$

$$T_{rr} = -p + 2\mu\frac{\partial u_r}{\partial r} = -p_i - \mu a. \tag{17.4.7}$$

The normal traction at a point on the free surface, the force per unit area that the surface exerts on the fluid, is

$$T_{nn} = n_r^2 T_{rr} + n_z^2 T_{zz}, \tag{17.4.8}$$

where n_r and n_z are components of the unit outward normal. A force balance at the free surface gives the boundary condition

$$-T_{nn} - p_a = \gamma\kappa, \tag{17.4.9}$$

where p_a is atmospheric pressure, p_0 is the surface-tension force per unit length, and p_0 is the mean curvature, given by

$$\kappa = -\frac{\dfrac{\partial^2 R}{\partial z^2}}{\left[1 + \left(\dfrac{\partial R}{\partial z}\right)^2\right]^{3/2}} + \frac{1}{R\left[1 + \left(\dfrac{\partial R}{\partial z}\right)^2\right]^{1/2}}. \tag{17.4.10}$$

Equation (17.4.9) is the condition that drives the capillary collapse. It should be pointed out that the condition of zero shear stress at the boundary is satisfied exactly at the throat because u_z is independent of r.

Because the free surface must move with the fluid, we also have the kinematic condition

$$u_r = \frac{\partial R}{\partial t} + u_z\frac{\partial R}{\partial z} \tag{17.4.11}$$

at $r = R(z, t)$. This may be written as

$$-\frac{1}{2}aR = \frac{\partial R}{\partial t} + az\frac{\partial R}{\partial z}. \tag{17.4.12}$$

The mathematical problem is to find a function $R(z,t)$ that satisfies the conditions expressed by equations (17.4.9) and (17.4.12). We show that a function of form

$$R(z,t) = R_0(t) + R_2(t)\,z^2 + O\left(z^4\right) \tag{17.4.13}$$

is suitable, and we determine $R_0(t)$, $R_2(t)$, and the strain rate $a(t)$ by expanding these conditions for small z. To the lowest order in z^2,

$$\begin{aligned}\frac{T_{nn}}{\rho} = \frac{T_{rr}}{\rho} &= -\frac{p_i}{\rho} - \nu a \\ &= -\frac{p_0}{\rho} - \nu a + \frac{1}{2}\left(\dot{a} + a^2\right) z^2 - \left(\frac{1}{4}\dot{a} - \frac{1}{8}a^2\right) R^2 \\ &= -\frac{p_0}{\rho} - \nu a + \frac{1}{2}\left(\dot{a} + a^2\right) z^2 - \frac{1}{4}\left(\dot{a} - \frac{a^2}{2}\right)\left(R_0^2 + 2R_2 z^2\right),\end{aligned} \tag{17.4.14}$$

and, to the same order,

$$\kappa = \frac{1}{R_0} - 2R_2. \tag{17.4.15}$$

Equation (17.4.12) gives the two equations

$$-\frac{1}{2}a R_0 = \dot{R}_0, \tag{17.4.16}$$

$$-\frac{5}{2}a R_2 = \dot{R}_2. \tag{17.4.17}$$

From these, we see that $R_2 = \hat{C} R_0^5$, where $\hat{C}$ is a constant depending on starting conditions. This result implies that R_2 tends to zero faster than R_0, which means that the parabola flattens out during collapse. It follows then that the R_2 term in (17.4.14) is of lower order and the term proportional to z^2 cannot be balanced.

To balance the terms proportional to z^2 in (17.4.14), we introduce a pressure correction p_v, where $p = p_i + p_v$. We find this correction among harmonic functions $\nabla^2 p_v = 0$ so that

$$\nabla^2 p = \nabla^2 p_i = -\rho\,\mathrm{div}\left(\mathbf{u}\cdot\nabla\mathbf{u}\right),$$

where $\mathbf{u} = \nabla\phi$ and p_i is given by (17.4.5). The required harmonic function is found in the form

$$p_v = C\left(-\frac{r^2}{2} + z^2\right), \tag{17.4.18}$$

and, after adding p_v to (17.4.5), we get

$$\frac{p - p_0}{\rho} = -\left(\frac{1}{2}\dot{a} + \frac{1}{2}a^2 - \frac{C}{\rho}\right) z^2 + \left(\frac{1}{4}\dot{a} - \frac{1}{8}a^2 - \frac{C}{2\rho}\right) r^2.$$

To balance the terms proportional to z^2 in (17.4.14), we choose

$$C = \frac{\rho}{2}\left(\dot{a} + a^2\right).$$

Then

$$\frac{p - p_0}{\rho} = -\frac{3}{8}a^2 r^2. \tag{17.4.19}$$

This pressure difference is negative and is most negative at the boundary $r = R$.

The pressure correction induces a vortical velocity $\mathbf{v}$ that vanishes at the throat. The velocity $\mathbf{u} = \mathbf{u}_i + \mathbf{v}$, where the components of $\mathbf{u}_i = \nabla\phi$ are given by (17.4.1), and

$$\frac{\partial \mathbf{v}}{\partial t} + \mathbf{v}\cdot\nabla\mathbf{v} + \mathbf{u}_i\cdot\nabla\mathbf{v} + \mathbf{v}\cdot\nabla\mathbf{u}_i = -\nabla\frac{p_v}{\rho} + \nu\nabla^2\mathbf{v}. \tag{17.4.20}$$

We show that the left-hand side of (17.4.20) is of lower order and may be neglected near the stagnation point. Writing $(v_r,\ v_z) = (u,\ w)$ and using (17.4.18), we have

$$\begin{aligned} 2Cz &= \mu\left[\frac{\partial^2 w}{\partial z^2} + \frac{1}{r}\frac{\partial}{\partial r}\left(r\frac{\partial w}{\partial r}\right)\right], \\ -Cr &= \mu\left[\frac{\partial^2 u}{\partial r^2} + \frac{1}{r}\frac{\partial}{\partial r}\left(r\frac{\partial u}{\partial r}\right) - \frac{u}{r^2}\right], \\ \frac{\partial w}{\partial z} &+ \frac{1}{r}\frac{\partial}{\partial r}(ru) = 0. \end{aligned} \tag{17.4.21}$$

The solution of (17.4.21) that vanishes at the origin is

$$\begin{aligned} w &= \frac{C}{3\mu}z^3, \\ u &= -\frac{C}{2\mu}rz^2. \end{aligned} \tag{17.4.22}$$

The vorticity for this axisymmetric solution is given by

$$\frac{\partial u}{\partial z} - \frac{\partial w}{\partial r} = -\frac{C}{\mu}rz. \tag{17.4.23}$$

The largest terms on the right-hand side of (17.4.20) for values of r and z are $O(z^3)$ and $O(rz^2)$. The vortical velocity $\mathbf{v}$ does not enter into any leading balance, as subsequently discussed.

To leading order, $T_{nn} = T_{rr} = -p + \mu a$ and $-T_{rr} - p_a = \dfrac{\gamma}{R_0}$, which is in the form

$$-\frac{3\rho}{8}R_0^2 - \mu a + p_0 - p_a = \frac{\gamma}{R_0}. \tag{17.4.24}$$

Because we are most interested in small R_0, we can pick out the dominant terms in (17.4.24) as R_0 tends to zero. These are

$$-2\mu\frac{\dot{R}_0}{R_0} = \frac{\gamma}{R_0}, \tag{17.4.25}$$

which is a balance between the *viscous* part of the normal force (which resists the collapse) and the surface-tension force (which drives it). The large R_0^{-1} term cancels from each side, giving

$$\dot{R}_0 = -\frac{1}{2}\frac{\gamma}{\mu} \tag{17.4.26}$$

with the solution

$$R_0 = \frac{\gamma}{2\mu}(t_* - t), \tag{17.4.27}$$

where t_* is a constant of integration. Therefore we have a solution in which R_0 tends to zero in a finite time. It is easy to see that the neglected terms in equation (17.4.24) give a correction to R_0 of the order of $(t_* - t)^2$. With the additional term the solution becomes

$$R_0 = \frac{\gamma}{2\mu}(t_* - t) - \frac{\gamma}{2\mu}\frac{p_0 - p_a}{4\mu}(t_* - t)^2 + \cdots + . \tag{17.4.28}$$

The strain rate is

$$a = -2\frac{\dot{R}_0}{R_0} = \frac{2}{t_* - t} - \frac{p_0 - p_a}{2\mu} + \cdots + . \tag{17.4.29}$$

The axial stress at leading order is given by

$$\begin{aligned} T_{zz} &= -p + 2\mu a = -p_0 + \frac{3}{8}\rho a^2 r^2 + 2\mu a \\ &= -p_0 + \frac{3}{2}\dot{R}_0^2\left(\frac{r}{R_0}\right)^2 + 2\mu a \\ &= \frac{3}{8}\frac{\rho\gamma^2}{\mu^2}\left(\frac{r}{R_0}\right)^2 + \frac{4\mu}{t - t_*} - (2p_0 - p_a), \end{aligned} \tag{17.4.30}$$

where $\dot{R}^2 = \gamma^2/4\mu^2$, from (17.4.26). The stress-induced cavitation will occur when and where the axial stress passes into tension. This will always occur first at the boundary of the capillary where $r/R_0 = 1$.

Consider next the axial stress at the stagnation point:

$$T_{zz} = -(2p_0 - p_a) + \frac{2\gamma}{R_0(t)}. \tag{17.4.31}$$

The thread will pass into tension over the whole cross section at $z = 0$ when T_{zz}, given by (17.4.31), becomes positive and passes into tension.

We see that when R_0 is sufficiently small, T_{zz} can become positive. This means that the axial stress becomes tension instead of compression. Liquids cannot support much tension without rupturing. This would occur here when R_0 is somewhat less than the critical value

$$R_{0\text{cr}} = \frac{2\gamma}{(2p_0 - p_a)}, \tag{17.4.32}$$

which, it should be noted, is independent of the viscosity. This value is fairly large; for water with $\gamma = 75$ dyn/cm and estimating $p_0 - p_a$ in the static balloon to be approximately $2\gamma/B$, which is nearly zero for a large radius B, we get $p_0 = p_a = 10^6$ dyn/cm^2 and $R_{0\text{cr}} = 1.5\ \mu$m.

A Reynolds number for the collapsing capillary may be defined by

$$Re = \frac{R_0\dot{R}_0}{\nu}, \tag{17.4.33}$$

based on the throat radius and the velocity of collapse; using equation (17.4.26) for the latter quantity gives

$$Re = \frac{R_0\gamma}{2\rho\nu^2} \tag{17.4.34}$$

which is the ratio of R_0 to a viscous length (Peregrine *et al.*, 1990), $2\rho\nu^2/\gamma$, which is very small for water, about 0.027 μm. Therefore with $R_{0\text{cr}}(=1.5\ \mu\text{m})$ used for R_0, the Reynolds number at collapse is about 55 for water (the collapse velocity is about 37 m/s). For more viscous liquids the Reynolds number at collapse could be very small. For the solution presented here there is no restriction on the magnitude of the Reynolds number.

17.4.3 Conclusions and discussion

The neck-down of a liquid capillary thread was studied in a local analysis based on VPF. One objective of this study was to show that, during collapse, the thread will enter into tension because of viscosity and can be expected to fracture, or cavitate, at a finite radius.

The flow in the throat of the collapsing capillary is locally a uniaxial extensional flow, linear in z and r, with a time-dependent strain rate $a(t)$. This VPF satisfies the Navier–Stokes equation and all the relevant interfacial conditions, including continuity of the shear stress at $z = 0$ (see (17.4.1)).

The principal dynamic balance is between the surface-tension forces, which are trying to collapse the capillary, and the radial viscous stress, which is resisting the collapse. Because mass must be conserved, a large axial flow results from squeezing liquid out of the neck, and this results in a large viscous extensional stress. The extensional stress passes into tension at $R_0 = 1.5$ μm (for water and air) long before R_0 actually collapses to zero.

The solution is symmetric about $z = 0$, the position of the smallest radius; the axial velocity is odd and the radial velocity, pressure, and interface shape,

$$R(z,t) = R_0(t) + R_2(t)z^2 + O\left(z^4\right),$$

are even in z. At lowest order the interface is a parabola in which $R_2(t)$ is proportional to R_0^5; hence in the limit of collapsing radius $R_2 \to 0$ much more rapidly then R_0 and the shape approaches that of a straight cylinder. The radius tends to zero linearly, like $(t_* - t)$, collapsing to zero in a finite time. At the same time the strain rate $a(t)$ tends to infinity like $(t_* - t)^{-1}$.

An authorative review of the capillary breakup literature before 1997 was given by Eggers (1997). A part of this literature review describes theoretical studies of IPF. These studies can be easily extended to accommodate the irrotational effects of viscosity that are in fact hugely important in the extensional flows under consideration here. Eggers notes that the most extensive numerical study of breakup before 1997 is due to Ashgriz and Mashayek (1995). They used periodic boundary conditions in the axial direction and considered initial sinusoidal disturbances of different wavelengths of wavenumber k. Their computations lead to the symmetric configurations shown in figure 17.14. They note that "The linear theory predicts that the breakup point is always at the neck (trough) of the initial disturbance. However, figure [17.14] clearly shows that the breakup point is closer to the swell points." When $Re > 0.1$, the breakup point moves to the swell and the center stagnation point is a point of fluid accumulation leading to satellite drops. When $Re \leq 0.1$, the formation of satellite drops is not evident; it is not known if the symmetric form of breakup persists at later times. A symmetric form of breakup of a liquid bridge at a low Reynolds number (3.7×10^{-3}) is shown in figure 17.15, taken from the paper of Spiegelberg, Gaudet, and McKinley (1994b). The outer fluid eliminates buoyancy and has a viscosity 1000 times smaller than that of the inner fluid. Brenner *et al.* (1997) discuss the breakdown of the similarity solution of Eggers at high Reynolds number. In their figure 1(c) they show the shape of a pure glycerol interface just before rupture that appears to be symmetric.

In the literature on capillary collapse and rupture, the focus is on collapse that is universally framed as a pinchoff and the fundamental physics governing the rupture of the thread is not considered. A pinchoff is a squeezing flow; the radius of the jet at the pinch

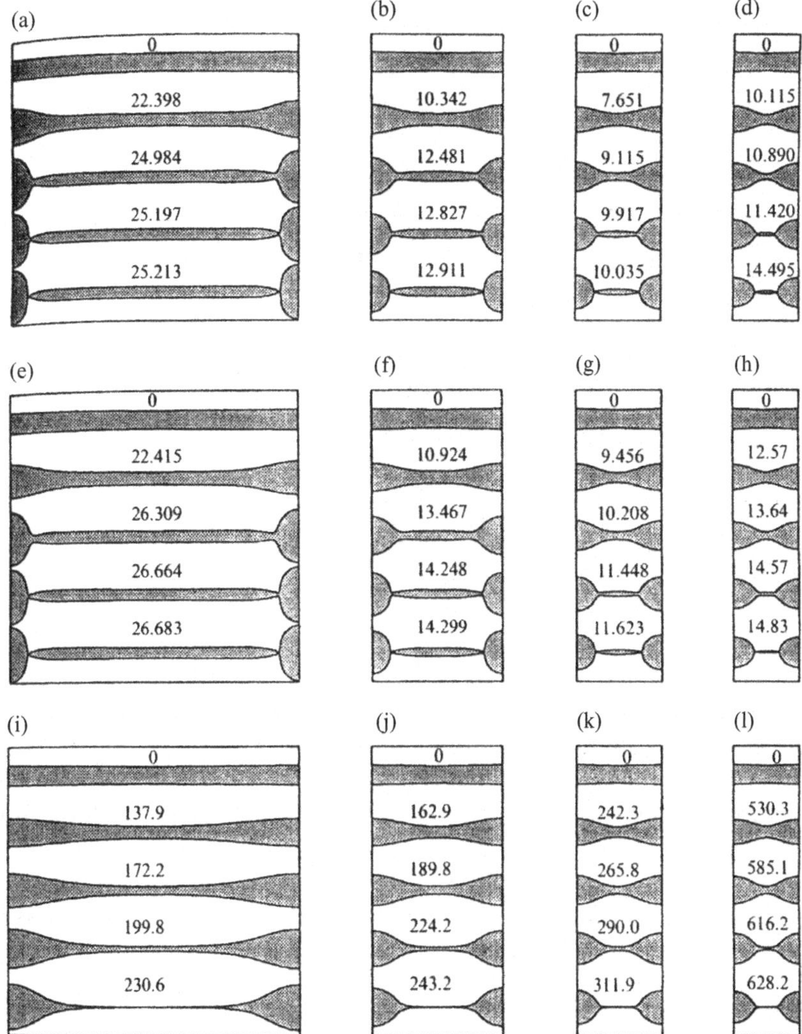

Figure 17.14. (Ashgritz and Mashayek, 1995) Time evolution of the instability of a capillary liquid jet, $\epsilon_0 = 0.05$: (a) $k = 0.2$, $Re = 200$; (b) $k = 0.45$, $Re = 200$; (c) $k = 0.7$, $Re = 200$; (d) $k = 0.9$, $Re = 200$; (e) $k = 0.2$, $Re = 10$; (f) $k = 0.45$, $Re = 10$; (g) $k = 0.7$, $Re = 10$; (h) $k = 0.9$, $Re = 10$; (i) $k = 0.2$, $Re = 0.1$; (j) $k = 0.45$, $Re = 0.1$; (k) $k = 0.7$, $Re = 0.1$; (l) $k = 0.9$, $Re = 0.1$. The numbers on the figures indicate the corresponding times.

point collapses, squeezing fluid out as the filament collapses. Here one finds a stagnation point; stagnation-point flow, is a potential flow, and the effects of viscosity in such a flow may be huge. Certainly potential flow of an inviscid fluid is not the right tool here. We can get results in which viscosity acts strongly by using VPF. The question is not whether viscosity is important, which it is, but whether vorticity is important.

Pinchoff at swell points (figure 17.16) is greatly different than the pinchoff at a neck (figure 17.15). It is generally believed the breakup at swell points in long, skinny threads is very well described by Eggers (1993) one-dimensional universal solution. The final stage of breakup at the neck in the Lundgren–Joseph model developed here leads also to a long skinny thread, but the dynamics of the breakup of the long skinny threads at the final stage

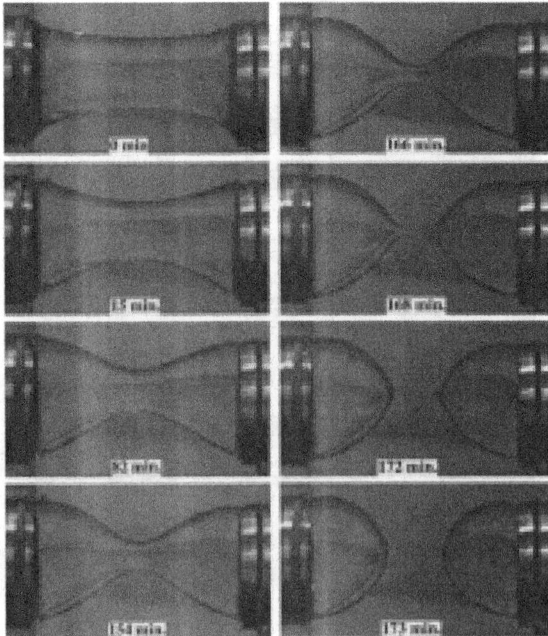

Figure 17.15. Liquid-bridge evolution starting from an unstable configuration. The disk diameter is 3.8 cm, the Reynolds number is 3.7×10^{-3}. The outer fluid, which eliminates buoyancy forces, has a viscosity approximately 1000 times smaller than that of the inner fluid. (Spiegelberg, Gaudet, and McKinley, 1994b.)

is greatly different (the details are given in the appendix to this section). It is of interest that the low-Reynolds-number similarity solution of Papageriou that falls into the Eggers class of similarity solutions is symmetric. Rothert *et al.* (2001) discuss the transition from a symmetric to asymmetric scaling function before pinchoff; symmetric forms are shown in their figure 2.

Chen *et al.* (2002) have studied pinchoff and scaling during drop formation by using high-accuracy computation and ultrafast high-resolution imaging. They discuss dynamic

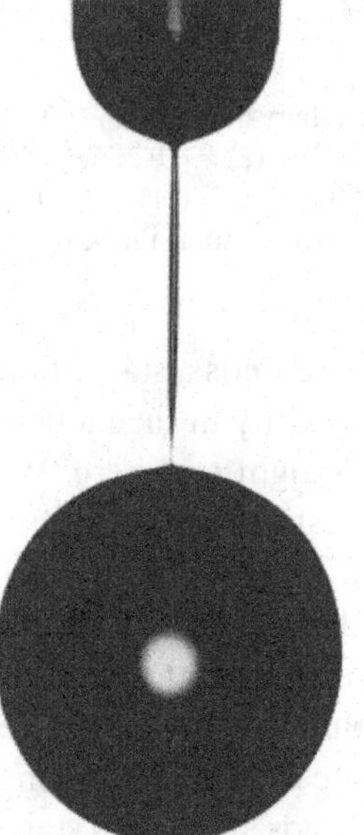

Figure 17.16. A drop of a glycerol and water mixture, 100 times as viscous as water, falling from a nozzle 1.5 mm in diameter. Here, as opposed to the case of water, a long neck is produced. (Shi, Brenner, and Nagel, 1994.)

transition from potential flow with a 2/3 scaling, which is due to Keller and Miksis (1983), to an inertial-viscous regime, described by Eggers' (1993) universal solution. They find overturn before breakup in experiments in water (1 cP) well before the dynamic transition from the potential flow to the inertial viscous regime. On the other hand, an 85-cP glycerol–water solution is said to exhibit this transition. The potential flow solutions discussed by Chen *et al.* (2002) are for inviscid solutions. Of course, water and glycerol are not inviscid. The scaling of Keller and Miksis (1983), which gives rise to the 2/3 power collapse law, does not work for VPF. The spoiler is their equation (3.3), expressing the normal stress balance. To this equation we must add the viscous component $2\mu\partial U_n/\partial n$. The term $(\nabla\phi)^2$ in (3.3) scales like ϕ^2/L^2 whereas the viscous component scales like ϕ/L^2, so that the similarity transformation does not factor through. Analogies have been put forward between capillary pinchoff of a viscous fluid thread and van der Waals-driven ruptures of a free thin viscous sheet (Vaynblat *et al.*, 2001). The observation that a filament under capillary collapse ruptures in a pinchoff does not come to grips with the physics that leads to a loss of the continuum. One idea is that thread breaks under the action of disjoining pressures. Unfortunately, a mathematical theory for disjoining pressures for thin threads is not available. The thread may disjoin when the capillary force increases to intermolecular values.

The recent literature on capillary collapse is currently dominated by the discovery of self-similar, finite-time singularity formation. These solutions are discussed in the recent papers by Chang *et al.* (1999), McKinley and Tripathi (2000), and Chen *et al.* (2002). This literature does not treat the physics of rupture or breakup by cavitation and does not compute stresses. All of the just-mentioned authors find that the capillary radius decreases to zero linearly in time, but the rate of collapse differs from author to author. McKinley and Tripathi (2000) give the formula

$$R_{\text{mid}}^{(t)} = R_1 - \frac{2X-1}{6}\frac{\gamma}{\mu}t$$

for the neck radius of the collapsing capillary in the stage of final decay as t increases to t^* when $R_{\text{mid}}^{(t^*)} = 0$. They give the X obtained by different authors in their table 1, but without the value $X = 2$ obtained here for VPF, giving the fastest decay. Eggers (1993, 1997) obtained $X = 0.5912$ and Papageoriou (1995) obtained $X = 0.7127$. The solutions of the two authors last named have vorticity; Papageriou's solution has no inertia. McKinley and Tripathi (2000) note that very close to breakup the solution of Papageriou crosses over to Egger's similarity solution. The transitions between different similarity solutions are less well understood than the similarity solutions themselves. These transitions can be regarded as a form of instability.

The solution of Eggers (1993) gives rise to a universal scaling law that has been observed for viscous liquids but not in water (see Chen *et al.*, 2002). The long-wave approximation used to derive universal scalings may prevent it from resolving the dynamics of rupture. The similarity solution of Eggers does not lead to a cavitation threshold; in his solution the tension that is due to extension increases but not fast enough to overcome the compression that is due to the capillary pressure of the thinning filament.

Our criterion for the termination of the continuum is probably not a finite-time singularity; the thread radius does not go to zero. It comes apart before then.

17.4.4 Appendix

A comparison of the symmetric solution of Lundgren and Joseph (2007) (hereafter LJ) with the similarity solution of Eggers (1993) (called E) is made.

We refer to equations (1)–(6) in the 1993 paper of E. The variables used by E are here designated with a caret:

$$\hat{\mathbf{v}} = (\hat{v}_z, \hat{v}_r),$$

$$\frac{\partial \hat{v}_r}{\partial r} + \frac{\partial \hat{v}_z}{\partial z} + \frac{\hat{v}_r}{r} = 0,$$

$$\hat{v}_z = \hat{v}_0(z,t) + \hat{v}_z(z,t)r^2,$$

$$\hat{v}_r(z,r,t) = -\hat{v}_0'(z,t)\frac{r}{2} - \hat{v}_z'(z,t)\frac{r^3}{4},$$

$$\hat{p}(z,r,t) = \hat{p}_0(z,t) + \hat{p}_z(z,t)r^2.$$

Using these expressions, E reduces normal stress balance (17.4.9) to a pressure equation,

$$p = \hat{p}_0(z,t) = \gamma\left(\frac{1}{R_1} + \frac{1}{R_2}\right) - \mu\hat{v}_0'(z,t),$$

at leading order.

In E's analysis, the lowest-order balance is for the r independent part of the foregoing representations:

$$\hat{v}_r = 0,$$

$$\hat{v}_z = \hat{v}_0(z,t),$$

$$p = \hat{p}_0(z,t) = \gamma\left(\frac{1}{R_1} + \frac{1}{R_2}\right) - \mu\hat{v}_0'.$$

The expressions for $\hat{v}_r = -\hat{v}_0'(z,t)\frac{r}{2}$ are needed to compute $\hat{v}_r$. The radial component of momentum at lowest order has no r independent terms and therefore is dropped as lower order and does not enter into the analysis leading to his similarity solution. According to E, the radial momentum is not relevant at leading order.

E neglects the radial momentum equation [(1) in E's paper]. LJ solve this problem at $O(r)$; it is not neglected.

Recalling now the analysis of LJ just given, we have the velocity decomposed in Helmholtz form, where $\phi = \frac{1}{2}az^2 - \frac{1}{4}ar^2$, $\mathbf{v} = \nabla\phi + \mathbf{u}$, where $\nabla\phi$ is the irrotational velocity and $\mathbf{u}$ is rotational velocity:

$$\mathbf{u} = (u_z, u_r) = (w, u),$$

$$\mathbf{v} = (v_z, v_r) = \left(\frac{\partial\phi}{\partial z} + w, \frac{\partial\phi}{\partial r} + u\right),$$

$$w = \frac{c}{3\mu}z^3,$$

$$u = -\frac{c}{2\mu}rz^2,$$

$$c = \frac{\rho}{2}\left(\dot{a} + a^2\right).$$

The value of c is what it takes to balance the normal stress at order z^2. We get

$$\begin{aligned} v_z &= az + \frac{c}{3\mu} z^3, \\ v_r &= -\frac{a}{2} r - \frac{c}{2\mu} z^2 r, \\ \frac{p - p_0}{\rho} &= -\frac{3}{8} a^2 r^2. \end{aligned}$$

Now we verify that E's equation (1) of radial momentum E is satisfied at leading order. The left-hand side of (1) in E is evaluated with the expansion of LJ just given.

$$\frac{\partial \hat{v}_r}{\partial t} + \hat{v}_r \frac{\partial \hat{v}_r}{\partial r} + \hat{v}_z \frac{\partial \hat{v}_z}{\partial z} \quad \to \quad -\frac{\dot{a}}{2} + \frac{a^2}{4}.$$

The right-hand side reduces as follows:

$$\begin{aligned} &-\frac{1}{\rho}\frac{\partial \hat{p}}{\partial r} + \nu \left(\frac{\partial^2 \hat{v}_r}{\partial r^2} + \frac{\partial^2 \hat{v}_r}{\partial z^2} + \frac{1}{r}\frac{\partial \hat{v}_r}{\partial r} - \frac{\hat{v}_r}{r^2} \right) \\ &\to -\frac{3}{4} a^2 - \frac{\nu c}{\mu} = \frac{3}{4} a^2 - \frac{1}{2} \left(\dot{a} + a^2 \right). \end{aligned}$$

The two sides are in balance.

A summary of the comparison just derived follows:

$$\text{LJ} \begin{cases} v_z = az, \\ v_r = -\dfrac{a}{2} r, \\ \dfrac{p - p_0}{\rho} = -\dfrac{3}{8} a^2 r, \end{cases}$$

where p_0 is a reference pressure equal to pressure at the stagnation point in the expanding ball (figure 17.13).

$$\text{E} \begin{cases} v_z = \hat{v}_0(z, t), \\ \hat{v}_r = 0, \\ \hat{p} = \gamma \left(\dfrac{1}{R_1} + \dfrac{1}{R_2} \right) + \mu \hat{v}_0'. \end{cases}$$

The normal stress balance for E is the expansion for $\hat{p}$. The normal stress balance for LJ is (17.4.24). The pressure in the two solutions is radically different. The pressure does not enter into the leading balance of LJ; their analysis balances capillarity against the viscous part of the normal stress.

18

Viscous effects of the irrotational flow outside boundary layers on rigid solids

High-Reynolds-number flows may be approximated by an outer irrotational flow and small layers on the boundary and narrow wakes where vorticity is important. The irrotational flow gives rise to an extra viscous dissipation over and above the dissipation in the boundary layer. At high Reynolds numbers the viscous dissipation in the irrotational flow outside is a very small fraction of the total that vanishes asymptotically as the Reynolds number tends to infinity.

Prandtl's boundary-layer theory is asymptotic and does not account for the viscous effects of the outer irrotational flow. Viscous effects on the normal stresses at the boundary of a solid cannot be obtained from Prandtl's theory. It is very well known and easily demonstrated that, as a consequence of the continuity equation, the viscous normal stress must vanish on a rigid solid. The only way that viscous effects can act on a boundary is through the pressure, but the pressure in Prandtl's theory is not viscous. It is determined by Bernoulli's equation in the irrotational flow and is imposed unchanged on the wall through the thin boundary layer. Therefore the important pressure drag cannot be calculated from Prandtl's theory. In addition, the mismatch between the irrotational shear stress and the shear stress at the outer edge of the boundary layer given by Prandtl's theory is not resolved.

Our work here (Wang and Joseph (2006b, 2006c); Padrino and Joseph (2006)) is motivated by the desire to understand the dynamical effect of the fact that the viscous dissipation of the outer irrotational flow is not zero and that the viscous effects on the normal stress on a solid are due only to the pressure, and, at a finite Reynolds number, no matter how large, there will be a viscous effect on the pressure, not given in Prandtl's theory. The viscous dissipation of the irrotational flow outside the boundary layer was considered by Romberg (1967). We take into account the viscous effects of the outer irrotational flow; it can be said that this work gives rise to a boundary-layer theory at a finite Reynolds number.

From the previous studies of gas–liquid flows, we have seen two closely related methods to account for the viscous effects of irrotational flows, DM and VCVPF. The dissipation of the outer potential flow increases the drag calculated from the boundary layer alone. Our calculation in this chapter shows that the drag increase is proportional to $1/Re$. The pressure correction is the kernel of VCVPF. At the outer edge of the boundary layer, the shear stress evaluated on the boundary-layer solution by use of Prandtl's theory does not necessarily equal the irrotational shear stress; this is analogous to the discrepancy between the zero-shear-stress condition and the nonzero irrotational shear stress at a

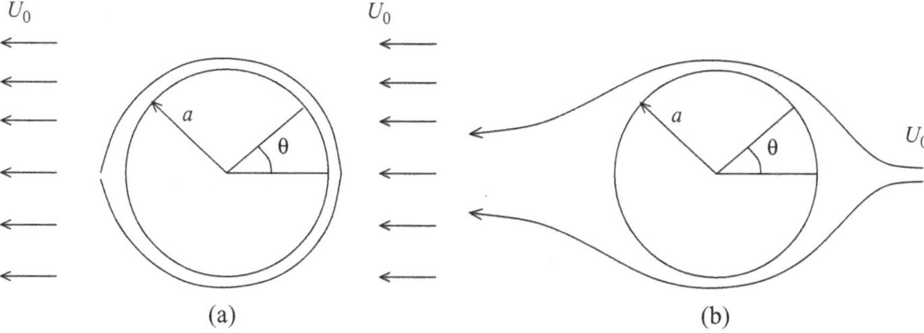

Figure 18.1. The flow past a circular cylinder (a) without separation of the boundary layer, (b) with separation of the boundary layer.

gas–liquid interface. This discrepancy induces vorticity, and a pressure correction can be calculated.

The method of VCVPF can determine only the pressure correction at the outer edge of the boundary layer, not the variation through the layer. We develop a new boundary analysis, in which the pressure inside the boundary layer is solved and the continuity of the shear stress at the outer edge of the boundary layer is imposed. This analysis is applied to the problem of the flow past a rapidly rotating cylinder. Inside the boundary layer, the velocity field is decomposed into an irrotational purely rotary flow and a boundary-layer flow. Inserting this decomposition into the Navier–Stokes equations, we obtain a new set of equations for the unknown boundary-layer flow. We can solve this new set of equations by expanding the solution into a power series. Our solution is in reasonable to excellent agreement with the numerical simulation of Padrino and Joseph (2006). The decomposition for the velocity inside the boundary layer can be regarded as a Helmholtz decomposition. However, the flow outside is approximated by a potential.

18.1 Extra drag due to viscous dissipation of the irrotational flow outside the boundary layer

For a body moving with a constant velocity in an otherwise quiescent liquid, the nonzero viscous dissipation of the outer potential flow gives rise to an additional drag, increasing the drag calculated from the boundary layer alone. The drag is considered in three cases here, on a 2D circular gas bubble in a streaming flow, at the edge of the boundary layer around a rapidly rotating cylinder in a uniform flow, and on an airfoil in a streaming flow. The drag may be computed with the DM or the viscous pressure correction of the irrotational pressure.

In subsection 18.1.1 we compute the drag on a 2D circular gas bubble by using the DM and VCVPF. This problem sets the frame for considerations of the additional drag on the boundary layer around a solid. Figure 18.1 shows the flow past a circular cylinder. Suppose that there is no separation of the boundary layer [figure 18.1(a)]; the flow is like a uniform flow past a circular gas bubble. The additional drag at the edge of the boundary layer can be computed just like the drag on a gas bubble. Practically, boundary-layer separation occurs [figure 18.1(b)] and the potential flow solution for the outer flow is not known. One of the methods to suppress separation is to rotate the cylinder rapidly. We compute the additional drag at the edge of the boundary layer of a rapidly rotating cylinder in a uniform flow in subsection 18.1.2. The flow past an airfoil, which can be obtained by conformal transformation from the flow over a rotating cylinder, is the subject in subsection 18.1.3.

18.1.1 Pressure corrections for the drag on a circular gas bubble

The drag D per unit length on a stationary circular gas bubble of radius a in a uniform stream $-U_0$ may be obtained by use of the DM introduced by Levich (1949) to compute the drag on a spherical gas bubble.[1] The steady rise velocity U_0 of the circular gas bubble in the irrotational flow of a viscous liquid can be obtained from the stationary bubble in a uniform stream by a Galilean transformation. This problem is a good frame to set the considerations that lead to viscous effects on boundary layers around solid bodies because of extra pressure generated by the unphysical shear stress as the outer edge of the boundary layer. The solid and its entrained boundary layer can be regarded as a boundary-layer bubble.

The irrotational flow of a viscous liquid over a stationary gas bubble is given by VPF $\mathbf{u} = \nabla\phi$, $\nabla^2\phi = 0$. $p = p_i$ is the pressure according to Bernoulli's equation; the stress in the liquid is $\mathbf{T} = -p_i\mathbf{1} + 2\mu\nabla\otimes\nabla\phi$, where μ is the viscosity.

The velocity potential for the stationary gas bubble is

$$\phi = -U_0 r\left(1 + \frac{a^2}{r^2}\right)\cos\theta, \tag{18.1.1}$$

and, at $r = a$, we get

$$u_r = 0, \quad u_\theta = 2U_0\sin\theta. \tag{18.1.2}$$

$$(\tau_{rr},\ \tau_{r\theta}) = -\frac{4\mu U_0}{a}(\cos\theta,\ \sin\theta) \tag{18.1.3}$$

are the normal and shear stresses, respectively, and p_i is determined by Bernoulli's equation:

$$p_i = p_\infty + \frac{\rho}{2}U_0^2(1 - 4\sin^2\theta). \tag{18.1.4}$$

The dissipation $\mathcal{D}$ per unit length of the potential flow may be evaluated by use of the identity

$$\begin{aligned} \mathcal{D} \equiv \int_V 2\mu\mathbf{D}:\mathbf{D}\,\mathrm{d}V &= \int_A \mathbf{u}\cdot 2\mu\mathbf{D}\cdot\mathbf{n}\,\mathrm{d}A \\ &= \int_A -(u_r\tau_{rr} + u_\theta\tau_{r\theta})\,\mathrm{d}A = 8\pi\mu U_0^2, \end{aligned} \tag{18.1.5}$$

where $\mathbf{D}$ is the rate-of-strain tensor, V is the volume occupied by the fluid, and A is the boundary of V. The drag that is due to the dissipation of the potential flow can then be calculated:

$$D = \mathcal{D}/(-U_0) = -8\pi\mu U_0. \tag{18.1.6}$$

[1] We call the readers' attention to the fact that, in our problem, the uniform flow is from right to left (see figure 18.1). The drag on the bubble is in the uniform flow direction and is negative.

A direct calculation of the drag on the bubble, with VPF used to calculate the stress traction at $r = a$, yields a different result:

$$D = \int_A \mathbf{e}_x \cdot \mathbf{T} \cdot (-\mathbf{n})\, \mathrm{d}A = \int_A [(-p_i + \tau_{rr})\mathbf{e}_x \cdot \mathbf{e}_r + \tau_{r\theta}\mathbf{e}_x \cdot \mathbf{e}_\theta]\, \mathrm{d}A = 0. \quad (18.1.7)$$

This is because the integral of p_i vanishes and

$$\int_A \tau_{rr}\mathbf{e}_x \cdot \mathbf{e}_r\, \mathrm{d}A = -\int_A \tau_{r\theta}\mathbf{e}_x \cdot \mathbf{e}_\theta\, \mathrm{d}A. \quad (18.1.8)$$

This result, $D = 0$ with a nonzero dissipation $8\pi\mu U_0^2$, is a paradox that is even more paradoxical than D'Alembert's.

In an exact formulation of the flow past a circular bubble, without assuming potential flow and with $\tau_{r\theta} = 0$ at $r = a$, we have

$$D = \int_A (-p + \tau_{rr})\mathbf{e}_x \cdot \mathbf{e}_r\, \mathrm{d}A. \quad (18.1.9)$$

The effects of viscosity can enter this integral through p or τ_{rr}. We next assume that the nonphysical irrotational shear stress $\tau_{r\theta}$ is removed in a boundary layer in which the vorticity is not zero. The thickness δ of the vortical layer is very small at high Reynolds number. The rate of strain in the vortical layer is of the order of U_0/a in order for the shear stress to be zero; the volume of the vortical layer is of the order of $a\delta$ per unit length. Therefore the dissipation per unit length in the vortical layer is of the order of $\mu U_0^2\delta/a$, which is negligible compared with the dissipation in bulk volume (18.1.5). It is further assumed that the boundary-layer contribution to τ_{rr} is also negligible. It follows then that the direct calculation of drag can agree with the dissipation calculation only if

$$p = p_i + p_v, \quad (18.1.10)$$

where p_v is the additional contribution to pressure in the vorticity boundary layer. The mechanical-energy equation at steady state gives rise to

$$\mathcal{D} \equiv \int_V 2\mu\mathbf{D} : \mathbf{D}\, \mathrm{d}V = \int_A \mathbf{u} \cdot \mathbf{T} \cdot \mathbf{n}\, \mathrm{d}A. \quad (18.1.11)$$

Given the structure previously described, we have

$$\mathcal{D} = -\int_A u_r(-p_v + \tau_{rr})\, \mathrm{d}A. \quad (18.1.12)$$

Comparing (18.1.12) with (18.1.5), we can see that (12.6.1) holds with $\mathbf{n} = -\mathbf{e}_r$, $\mathbf{t} = -\mathbf{e}_\theta$, and $\tau_s = \tau_{r\theta}$ in the case of the circular gas bubble.

The extra pressure must be a 2π periodic solution on the circle and can be represented by a Fourier series:

$$-p_v = \sum_{k=0}^{\infty}(C_k \cos k\theta + D_k \sin k\theta). \quad (18.1.13)$$

Inserting now (18.1.13) and (18.1.3) into (12.6.1), we find that

$$-\int_0^{2\pi} U_0 \cos\theta \left[C_1 \cos\theta + D_1 \sin\theta + \sum_{k\neq 1} (C_k \cos k\theta + D_k \sin k\theta) \right] a\, d\theta = 4\pi\mu U_0^2. \tag{18.1.14}$$

The preceding integration is performed on the surface of the bubble, and the vortical layer is not considered. Evaluation of (18.1.14) by use of orthogonality gives

$$C_1 = -4\mu U_0/a. \tag{18.1.15}$$

The other coefficients are undetermined. The only term in Fourier series (18.1.13) entering into the direct calculation of the drag is proportional to $\cos\theta$. Hence

$$D = \int_0^{2\pi} (-p_v + \tau_{rr})\mathbf{e}_x \cdot \mathbf{e}_r\, a d\theta = \int_0^{2\pi} (-p_v + \tau_{rr}) a \cos\theta\, d\theta = -8\pi\mu U_0 \tag{18.1.16}$$

is the same D as calculated by the DM in (18.1.6).

It is of interest to consider the separate contribution to the drag of $-p_v$ and τ_{rr} in (18.1.16):

$$D = D_{p_v} + D_{\tau_{rr}} = -4\pi\mu U_0 - 4\pi\mu U_0 = -8\pi\mu U_0. \tag{18.1.17}$$

If somehow the surface of the bubble were made rigid so that the no-slip condition could be realized, then the continuity equation would imply that $D_{\tau_{rr}} = 0$ and D, would be the pressure drag on the rigid solid. Moore (1959) calculated the drag on a spherical gas bubble by using the viscous normal stress alone and got $D = -8\pi\mu U_0 a$. The Levich drag is $-12\pi\mu U_0 a$, and the difference is the drag $D_{p_v} = -4\pi\mu U_0 a$, which is, in the present mode of imagination, the viscous drag on a rigid sphere that is due to the viscous irrotational flow.

The existence and asymptotic validity of a boundary layer of the type assumed here and elsewhere have not been established. The details of the size of the layer, the boundary-layer equations, and the variation of velocity, vorticity, and pressure in the layer have not been given. Kang and Leal (1988a) did calculations from the vorticity equation in the case of the drag on a spherical gas bubble. Results indicating a boundary-layer structure of the type described here were obtained, but their results are partial and do not give the details previously listed.

The nature of the boundary layer may be determined in the appropriate asymptotic limit more easily in two dimensions than in three. In the 2D problem we may obtain an exact solution of the stream-function equation

$$\frac{1}{r}\frac{\partial\psi}{\partial\theta}\frac{\partial}{\partial r}\nabla^2\psi - \frac{1}{r}\frac{\partial\psi}{\partial r}\frac{\partial}{\partial\theta}\nabla^2\psi = \nu\nabla^4\psi, \tag{18.1.18}$$

where

$$\nabla^2\psi = \frac{1}{r}\frac{\partial}{\partial r}\left(r\frac{\partial\psi}{\partial r}\right) + \frac{1}{r^2}\frac{\partial^2\psi}{\partial\theta^2}$$

in the region outside the circle subject to the conditions that

$$\mathbf{u} = -\mathbf{e}_x U_0 \quad \text{at } \infty, \tag{18.1.19}$$

$$u_r = 0, \quad \tau_{r\theta} = 0 \quad \text{at } r = a. \tag{18.1.20}$$

This problem is well posed; it is like the flow over a stationary solid cylinder except that the no-slip condition on the tangential velocity on the stationary solid circle is replaced with a zero-shear-stress condition on a circular bubble.

The solution of (18.1.18), (18.1.19), and (18.1.20) determines a stream function $\psi(r, \theta)$. Once this function is determined, the pressure may be determined from the equations of motion and the pressure correction can be obtained.

18.1.2 A rotating cylinder in a uniform stream

The potential flow over a rotating cylinder in a uniform stream plays an important role in classical airfoil theory in which the flow and airfoil shape are obtained by conformal transformation, and the Kutta condition suppressing separation at the trailing edge is obtained by adjustment of the ratio of the rotational speed to the streaming speed.[2]

We study the extra pressure contribution to the drag at the outer edge of Prandtl's boundary layer on a solid cylinder rotating so fast that the separation of the boundary layer is suppressed. We compare the analysis of the extra pressure associated with the viscous dissipation of the irrotational flow outside the boundary layer with a numerical solution of the unapproximated equations for values as close to the appropriate asymptotic values as the numerical solution will allow.

Readers should remember that our work here is motivated by the desire to understand the dynamical effect of the fact that the viscous dissipation of the irrotational flow outside Prandtl's boundary layer is not zero and that the viscous effects on the normal stress on a solid are due only to the pressure, and at finite Reynolds number, no matter how large, there will be a viscous effect on the pressure not given in Prandtl's theory.

18.1.2.1 *Dissipation calculation*

We consider the uniform flow $-U_0$ past a fixed circular cylinder with circulation Γ. Suppose no separation of the boundary layer occurs; the flow outside the boundary layer is given by the potential

$$\phi = -U_0 r\left(1 + \frac{a^2}{r^2}\right)\cos\theta + \frac{\Gamma\theta}{2\pi}. \tag{18.1.21}$$

The velocity and stress at the surface of the cylinder can be evaluated by use of (18.1.21):

$$u_r = 0, \quad u_\theta = 2U_0 \sin\theta + \frac{\Gamma}{2\pi a}; \tag{18.1.22}$$

$$\tau_{rr} = -4\mu U_0 \cos\theta/a, \quad \tau_{r\theta} = -4\mu U_0 \sin\theta/a - \mu\Gamma/(\pi a^2). \tag{18.1.23}$$

[2] We call the readers' attention to the fact that in our problem, the uniform flow is from right to left (see figure 18.1) and the cylinder rotates counterclockwise. The lift on the cylinder points upward and is positive. The drag on the cylinder is negative if it is in the uniform flow direction; the drag is positive if it is opposite to the uniform flow direction.

The dissipation $\mathcal{D}$ of the potential flow can be evaluated:

$$\mathcal{D} = -\int_A (u_r\tau_{rr} + u_\theta\tau_{r\theta})\,\mathrm{d}A = 8\pi\mu U_0^2 + \frac{\mu\Gamma^2}{\pi a^2}. \tag{18.1.24}$$

The dissipation is equal to the sum of the dissipation of an irrotational purely rotary flow and a streaming flow past a fixed cylinder; the cross terms in $u_\theta\tau_{r\theta}$ do not appear in the dissipation expression because they integrate to zero. The dissipation of the potential flow should be equal to the power of the drag D and the torque T:

$$D(-U_0) + T\frac{\Gamma}{2\pi a^2} = \mathcal{D}. \tag{18.1.25}$$

Ackeret (1952) computed the same dissipation for the problem under consideration. He did not consider the torque and equated the dissipation to the power of the drag alone and obtained

$$D = \mathcal{D}/(-U_0) = -8\pi\mu U_0 - \frac{\mu\Gamma^2}{\pi a^2 U_0}. \tag{18.1.26}$$

Ackeret argued that it is worthwhile to consider the potential flow solution if the viscous liquid is allowed to slip at solid boundaries. He did not mention gas bubbles, liquid–gas flows, the additional drag, or the relation of his solution to the unphysical irrotational shear stress at the edge of the boundary layer.

We argue that the additional drag cannot be computed from (18.1.25) with the torque T undetermined. We will obtain the additional drag in a later section by computing the pressure correction p_v, following the method laid down in our calculation of p_v in the case of a circular gas bubble.

18.1.2.2 *Boundary-layer analysis*

Glauert (1957) carried out a boundary-layer analysis of the flow past a rotating cylinder. He assumed that the ratio

$$\alpha = 2U_0/Q, \tag{18.1.27}$$

where U_0 is magnitude of the uniform stream velocity and Q is the circulatory velocity of the flow at the outer edge of the boundary layer, is smaller than unity and separation is suppressed. He obtained a solution of the boundary-layer equations in the form of a power series in α and deduced the ratio Q/q, where q is the cylinder's peripheral velocity that is related to the angular velocity Ω of the cylinder by $q = \Omega a$. Glauert's solution suggests that Q is approximately equal to q for large values of q; it follows that

$$\alpha \to 2U_0/q = \frac{2}{q/U_0} \quad \text{as} \quad q \to \infty. \tag{18.1.28}$$

Glauert used Prandtl's boundary-layer theory in which the irrotational pressure of the outer flow is imposed on the solid wall through the boundary layer. Assuming that the boundary-layer thickness is negligible compared with the cylinder radius, Glauert used the boundary-layer equations for steady 2D flows:

$$\frac{\partial u}{\partial x} + \frac{\partial v}{\partial y} = 0, \tag{18.1.29}$$

$$u\frac{\partial u}{\partial x} + v\frac{\partial u}{\partial y} = U\frac{\mathrm{d}U}{\mathrm{d}x} + \nu\frac{\partial^2 u}{\partial y^2}, \tag{18.1.30}$$

where U is the irrotational velocity at the edge of the boundary layer; x is measured around the cylinder circumference and y is normal to it. Glauert chose $x = 0$ to be the point at which the surface moves in the same direction as the uniform stream (the top of the cylinder). We follow his choice here. Let φ represent the polar angle measured from the point $x = 0$; then $\varphi = x/a$. Glauert obtained the following solutions:

$$u = Q\left\{1 + \alpha f_1'(y)e^{i\varphi} + \alpha^2\left[f_2'(y)e^{2i\varphi} + g_2'(y)\right] + \cdots +\right\}, \tag{18.1.31}$$

$$v = -Q\left[\frac{i}{a}\alpha f_1(y)e^{i\varphi} + \frac{2i}{a}\alpha^2 f_2(y)e^{2i\varphi} + \cdots +\right], \tag{18.1.32}$$

$$\frac{\partial u}{\partial y} = Q\left\{\alpha f_1''e^{i\varphi} + \alpha^2\left[f_2''(y)e^{2i\varphi} + g_2''(y)\right] + \cdots +\right\}, \tag{18.1.33}$$

where $f_1(y)$, $f_2(y)$, and $g_2(y)$ are functions of y and were determined by Glauert. Because $f_1'(0) = f_2'(0) = 0$ and $g_2'(0) > 0$ given by Glauert's solution, the velocity at the surface of the cylinder can be obtained from (18.1.31):

$$q = Q\left[1 + \alpha^2 g_2'(0) + \cdots +\right], \tag{18.1.34}$$

which shows that $Q < q$. Because the shear stress at the cylinder surface is given by $\mu(\partial u/\partial y)_{y=0}$, it can be inferred from (18.1.33) that the shear stress is zero at the cylinder surface when α is zero. In other words, when there is no streaming flow but only viscous irrotational rotary flow, Glauert's solution suggests that the shear stress at the cylinder surface is zero. However, the real shear stress is $-2\mu q/a$.

The reason for this discrepancy is that the irrotational rotary flow component is not considered in Glauert's solution, which is an approximation consistent with the assumption that δ/a is negligible compared with 1. Thus the shear stress induced by the rotary flow is ignored. The irrotational rotary component of the velocity inside the boundary layer can be written as

$$u_{p\varphi} = Q\frac{a+\delta}{r}, \tag{18.1.35}$$

where δ is the thickness of the boundary layer. We propose a simple modification of Glauert's solution:

$$u_\varphi = u_{p\varphi} + u_b = Q\left\{\frac{a+\delta}{r} + \alpha f_1'(y)e^{i\varphi} + \alpha^2\left[f_2'(y)e^{2i\varphi} + g_2'(y)\right] + \cdots +\right\}, \tag{18.1.36}$$

$$u_r = v_b = -Q\left[\frac{i}{a}\alpha f_1(y)e^{i\varphi} + \frac{2i}{a}\alpha^2 f_2(y)e^{2i\varphi} + \cdots +\right]. \tag{18.1.37}$$

$f_1(y)$, $f_2(y)$, $g_2(y)$, ..., are solutions of boundary-layer equations (18.1.29) and (18.1.30), which are based on the assumption that δ/a is negligible compared with 1. Under the same assumption, $(a+\delta)/r \approx 1$ inside the boundary layer and (18.1.36) reduces to Glauert's solution, (18.1.31). Thus it appears that the $(a+\delta)/r$ term is not consistent with the solutions of $f_1(y)$, $f_2(y)$, and $g_2(y)$. However, (18.1.36) is a simple modification to address the defect of ignoring the irrotational rotary component of the flow inside the boundary layer. We will show that the modified Glauert's solution is in better agreement with numerical simulation data than Glauert's solution. In section 18.2, we carry out a new boundary-layer analysis for the flow past a rotating cylinder, in which the inconsistency previously mentioned is resolved.

A key problem in boundary-layer analysis is to determine the circulatory velocity Q when given the cylinder rotational speed q. At $y=0$ $(r=a)$, (18.1.36) gives rise to

$$\begin{aligned} q &= Q\left[\frac{a+\delta}{a} + \alpha^2 g_2'(0) + \alpha^4 h_4'(0)\right] \\ &= Q\left[1+\frac{\delta}{a}+3\left(\frac{U_0}{Q}\right)^2 - 5.76\left(\frac{U_0}{Q}\right)^4\right], \end{aligned} \tag{18.1.38}$$

where Glauert's solutions for g_2 and h_4 have been used and the terms of the order of α^5 or higher are ignored. We invert (18.1.38) to obtain the expression for Q in terms of q:

$$\frac{Q}{q} = \frac{1}{1+\delta/a} - 3\left(\frac{U_0}{q}\right)^2 - 3.23\left(1-0.803\frac{\delta}{a}\right)\left(\frac{U_0}{q}\right)^4. \tag{18.1.39}$$

If δ/a is ignored, (18.1.39) reduces to

$$\frac{Q}{q} = 1 - 3\left(\frac{U_0}{q}\right)^2 - 3.23\left(\frac{U_0}{q}\right)^4, \tag{18.1.40}$$

which is the same as Glauert's result. When $U_0=0$, there is only irrotational purely rotary flow and the boundary layer does not exist. Thus $\delta=0$ and (18.1.39) indicates that $Q=q$.

We calculate the shear stress at the cylinder surface. The contribution from the irrotational purely rotary flow is

$$\mu\left(\frac{\partial u_{p\varphi}}{\partial r} - \frac{u_{p\varphi}}{r}\right) = -2\mu Q\frac{a+\delta}{a^2} \quad \text{at } r=a,$$

which is added to Glauert's shear stress to obtain the total shear stress:

$$\tau_{r\varphi} = \mu Q\left\{-2\frac{a+\delta}{a^2} + \alpha f_1''(0)e^{i\varphi} + \alpha^2\left[f_2''(0)e^{2i\varphi} + g_2''(0)\right] + \cdots +\right\}. \tag{18.1.41}$$

The torque T on the cylinder is given by

$$T = -a^2\int_0^{2\pi} \tau_{r\varphi}\,\mathrm{d}\varphi. \tag{18.1.42}$$

Only terms independent of φ in (18.1.41) contribute to (18.1.42), and we obtain

$$T = 8\pi\rho U_0^2\frac{a(a+\delta)}{Re}\frac{Q}{U_0} + 4\pi\rho U_0^2\frac{a^2}{\sqrt{Re}}\left[\left(\frac{U_0}{Q}\right)^{\frac{1}{2}} - 2.022\left(\frac{U_0}{Q}\right)^{\frac{5}{2}} + \cdots +\right], \tag{18.1.43}$$

$$C_T = \frac{T}{2\rho U_0^2 a^2} = 4\pi\left(1+\frac{\delta}{a}\right)\frac{1}{Re}\frac{Q}{U_0} + \frac{2\pi}{\sqrt{Re}}\left[\left(\frac{U_0}{Q}\right)^{\frac{1}{2}} - 2.022\left(\frac{U_0}{Q}\right)^{\frac{5}{2}} + \cdots +\right], \tag{18.1.44}$$

where C_T is the torque coefficient and

$$Re = 2U_0 a/\nu \tag{18.1.45}$$

is the Reynolds number based on U_0. The first term on the right-hand side of (18.1.43) is the torque induced by the rotary flow component and is of the order of $1/Re$; the second term is the torque given by Glauert and is of the order of $1/\sqrt{Re}$. When $Re \to \infty$, the term by Glauert is the dominant one. However, when Re is finite and Q/U_0 is large, the first term can be more significant than the second one. When $U_0 = 0$, the torque is equal to $T = 4\pi\mu q a$, which is the torque on the cylinder when there is only the viscous irrotational purely rotary flow.

Glauert cited Reid's (1924) experimental result about the torque,

$$T = 20\pi\rho U_0^2 \frac{a^2}{\sqrt{Re}}, \tag{18.1.46}$$

which was measured for $q = U_0$. Glauert noted that Reid's torque was far above the value given by him and remarked on this discrepancy: "... but it is doubtful if it has much accuracy or relevance, in view of the experimental imperfections and also the separation occurring at this low rotational speed." We compare our torque expression with the results of numerical simulation in which the rotational speed is high and separation is suppressed.

The lift and drag on the cylinder are given by the pressure and shear stress at the wall. The pressure in Glauert's solution is a constant across the boundary layer and is equal to the irrotational pressure at the outer edge of the boundary layer; it does not give drag and the pressure lift can be computed by use of the classical lift coefficient formula in aerodynamics:

$$C_{L_p} = \frac{\rho U_0 \Gamma}{\rho U_0^2 a} = \frac{2\pi Q}{U_0}. \tag{18.1.47}$$

In our simple modification of Glauert's solution, we add the irrotational rotary flow component $u_{p\varphi}$ to the velocity, and the pressure induced by $u_{p\varphi}$ is

$$p_p = p_{pc} - \frac{\rho}{2}u_{p\varphi}^2 = p_{pc} - \frac{\rho}{2}\frac{(a+\delta)^2}{r^2}Q^2, \tag{18.1.48}$$

where p_{pc} is a constant for the pressure. As an approximation, we assume that the total pressure is obtained by a simple addition of p_p and the pressure given by Glauert. On the cylinder surface $r = a$, p_p is independent of θ and does not contribute to the lift. Therefore pressure lift expression (18.1.47) still holds. After inserting (18.1.39) into (18.1.47), we obtain

$$C_{L_p} = 2\pi\frac{q}{U_0}\left[\frac{1}{1+\delta/a} - 3\left(\frac{U_0}{q}\right)^2 - 3.23\left(1 - 0.803\frac{\delta}{a}\right)\left(\frac{U_0}{q}\right)^4\right]. \tag{18.1.49}$$

Because our Q (18.1.39) is smaller than Glauert's result (18.1.40), our pressure lift is smaller than Glauert's. Glauert did not consider the friction drag and lift, but they can be computed from his solution easily:

$$C_{D_f} = \frac{D_f}{\rho U_0^2 a} = -\frac{2\pi}{\sqrt{Re}}\sqrt{\frac{Q}{U_0}}, \qquad C_{L_f} = \frac{L_f}{\rho U_0^2 a} = \frac{2\pi}{\sqrt{Re}}\sqrt{\frac{Q}{U_0}}. \tag{18.1.50}$$

Table 18.1. *Comparison of the coefficients for the pressure lift and torque on the cylinder obtained from Glauert's solution, the simple modification of Glauert's solution, and numerical simulation*

In the simple modification of Glauert's solution, we use an effective boundary-layer thickness δ_L/a, which is determined by matching C_{L_p} computed from our simple modification (18.1.49) to the results of numerical simulation. For $(Re, q/U_0) = (1000, 3)$, $\alpha > 1$ for both our simple modification and Glauert's solution. These solutions are not expected to converge to the true results.

Solution	Re	q/U_0	δ_L/a	α	C_{L_p}	C_T
Glauert's solution	200	4	–	0.625	20.102	0.215
Modified Glauert's solution	200	4	0.145	0.741	16.961	0.390
Numerical simulation	200	4	–	–	16.961	0.453
Glauert's solution	200	5	–	0.457	27.483	0.195
Modified Glauert's solution	200	5	0.0434	0.480	26.183	0.465
Numerical simulation	200	5	–	–	26.183	0.514
Glauert's solution	400	4	–	0.625	20.102	0.152
Modified Glauert's solution	400	4	0.112	0.714	17.609	0.237
Numerical simulation	400	4	–	-	17.609	0.275
Glauert's solution	400	5	–	0.457	27.483	0.138
Modified Glauert's solution	400	5	0.0354	0.476	26.415	0.272
Numerical simulation	400	5	–	–	26.415	0.297
Glauert's solution	400	6	–	0.365	34.463	0.126
Modified Glauert's solution	400	6	0.0380	0.380	33.087	0.299
Numerical simulation	400	6	–	–	33.087	0.316
Glauert's solution	1000	3	–	1.064	11.812	0.108
Modified Glauert's solution	1000	3	0.0837	1.207	10.409	0.0632
Numerical simulation	1000	3	–	–	10.409	0.118

Our simple modification changes the shear stress at the wall by only a constant; thus the expressions for the friction drag and lift do not change, but their values change because of Q.

We compare our simple modification of Glauert's solution, the results of numerical simulation from Padrino and Joseph (2006), and Glauert's solution in table 18.1 (see section 18.3). Six cases, $(Re, q/U_0) = (200, 4)$, $(200, 5)$, $(400, 4)$, $(400, 5)$, $(400, 6)$, and $(1000, 3)$, are considered. Although the boundary-layer thickness δ/a is not needed in Glauert's solution, it must be prescribed in our simple modification. We choose an effective boundary-layer thickness $\delta/a = \delta_L/a$ that we determine by matching C_{L_p} computed from our simple modification (18.1.49) to the results of numerical simulation. Table 18.1 shows that $\delta_L/a \ll 1$, δ_L/a decreases with increasing Re at a fixed q/U_0, and δ_L/a generally decreases with increasing q/U_0, because the rotary flow suppresses the boundary layer. The torque coefficient is not sensitive to the choice of δ/a as long as $\delta/a \ll 1$, which can be seen from (18.1.44) and (18.1.39). The pressure lift and torque in numerical simulation are obtained by integration at the cylinder surface. The values of $\alpha = 2U_0/Q$ are listed for Glauert's solution and our simple modification for each pair of Re and q/U_0. For $(Re, q/U_0) = (1000, 3)$, $\alpha > 1$ for both our simple modification and Glauert's solution. These solutions are not expected to converge to the true results. A comparison of the solutions with $\alpha < 1$ shows that the values of C_{L_p} from our simple modification and numerical simulation are smaller than those from Glauert's solution; the values of the torque from our simple modification are much closer to the numerical results than those from Glauert's solution.

If the results of numerical simulation are not available, our analysis cannot provide the value of δ/a. Then Glauert's pressure lift may be taken as a reasonable approximation, and the torque coefficient may be computed from (18.1.44) with $\delta/a = 0$. Because the torque coefficient is not sensitive to the choice of δ/a, it still improves Glauert's solution of the torque substantially.

18.1.2.3 *Pressure correction and the additional drag*

We consider the pressure correction at the outer edge of the boundary layer and the additional drag induced by it. The shear stress at the outer edge of the boundary layer can be computed in two ways: from the outside potential flow or from the boundary-layer solution. If we consider a rotating cylinder with its entrained boundary layer moving with U_0 in a liquid, the potential flow outside $r = a + \delta$ has the following velocity:

$$u_\theta = U_0 \frac{(a+\delta)^2}{r^2} \sin\theta + Q\frac{a+\delta}{r}, \quad u_r = U_0 \frac{(a+\delta)^2}{r^2} \cos\theta. \tag{18.1.51}$$

The irrotational shear stress at $r = a + \delta$ is

$$\tau_{r\theta} = -\mu \left(\frac{4U_0 \sin\theta}{a+\delta} + \frac{2Q}{a+\delta} \right). \tag{18.1.52}$$

The shear stress from the boundary-layer analysis is

$$\begin{aligned} \tau_{r\theta} &= \mu \left(\frac{\partial u_{p\varphi}}{\partial r} - \frac{u_{p\varphi}}{r} + \frac{\partial u_b}{\partial y} + \frac{\partial v_b}{\partial x} \right) \\ &= \mu \left[-2Q\frac{a+\delta}{r^2} + Q\alpha f_1''(y)\sin\theta + \cdots + Q\frac{\alpha}{a^2} f_1 \sin\theta + \cdots + \right]. \end{aligned} \tag{18.1.53}$$

Glauert's solution gives $f_1''(\delta) \approx 0$ and $f_1(\delta) \sim O(\delta)$, which is negligible. Thus the shear stress at $r = a + \delta$ from the boundary-layer solution is approximately

$$\tau_{r\theta} \approx -\mu \frac{2Q}{a+\delta}. \tag{18.1.54}$$

Comparing (18.1.52) with (18.1.54), we can see that the shear stress is not continuous and the discrepancy is

$$\tau_{r\theta}^d = -\mu \frac{4U_0 \sin\theta}{a+\delta}. \tag{18.1.55}$$

This shear stress discrepancy induces extra vorticity at the outer edge of the boundary layer and a pressure correction. The power of the pressure correction is equal to the power of the shear stress discrepancy:

$$-\int_0^{2\pi} u_r(-p_v)(a+\delta)\,d\theta = -\int_0^{2\pi} u_\theta \tau_{r\theta}^d (a+\delta)\,d\theta = 4\pi\mu U_0^2. \tag{18.1.56}$$

Again we expand the pressure correction as Fourier series (18.1.13) and insert it into (18.1.56):

$$-\int_0^{2\pi} U_0 \cos\theta \left[C_1 \cos\theta + D_1 \sin\theta + \sum_{k\neq 1}(C_k \cos k\theta + D_k \sin k\theta) \right] (a+\delta)\, d\theta = 4\pi\mu U_0^2, \tag{18.1.57}$$

which gives rise to

$$\begin{aligned} -C_1 &= \frac{4\mu U_0}{a+\delta}, \\ p_v &= \frac{4\mu U_0}{a+\delta}\cos\theta - D_1 \sin\theta - \sum_{k\neq 1}(C_k \cos k\theta + D_k \sin k\theta). \end{aligned} \tag{18.1.58}$$

We evaluate the additional drag by direct integration of the traction vector at the outer edge of the boundary layer:

$$D = \int_0^{2\pi} [(-p_i - p_v + \tau_{rr})\mathbf{e}_x \cdot \mathbf{e}_r + \tau_{r\theta}\mathbf{e}_x \cdot \mathbf{e}_\theta](a+\delta)\, d\theta, \tag{18.1.59}$$

where τ_{rr} is the viscous normal stress evaluated on potential flow velocity (18.1.51) and $\tau_{r\theta}$ is shear stress (18.1.54) evaluated by use of the boundary-layer solution. The preceding choices are made because τ_{rr} is essentially continuous at the outer edge of the boundary layer but $\tau_{r\theta}$ is not; we choose $\tau_{r\theta}$ from the boundary-layer solution, and this is analogous to using zero shear stress at a gas–liquid interface. The irrotational pressure p_i does not contribute to the drag, and we may write (18.1.59) as

$$\begin{aligned} D &= \int_0^{2\pi} (-p_v)\cos\theta(a+\delta)\, d\theta + \int_0^{2\pi} \tau_{rr}\cos\theta(a+\delta)\, d\theta - \int_0^{2\pi} \tau_{r\theta}\sin\theta(a+\delta)\, d\theta \\ &= -4\pi\mu U_0 - 4\pi\mu U_0 - 0 = -8\pi\mu U_0, \end{aligned} \tag{18.1.60}$$

which is the same as the drag on a circular gas bubble, (18.1.16). Our additional drag, (18.1.60), is much smaller than the one, (18.1.26), computed by Ackeret (1952) when the rotational velocity is much larger than the streaming velocity. Equation (18.1.60) indicates that the additional drag depends on only the forward speed U_0 and not on the spinning speed q. The additional drag should be the drag evaluated at the outer edge of the boundary layer, but the boundary-layer thickness does not affect the additional drag because δ/a does not appear in (18.1.60).

If we consider only the additional drag that is due to the pressure, we obtain

$$D_p = -4\pi\mu U_0, \quad C_{D_p} = \frac{D_p}{\rho U_0^2 a} = -\frac{8\pi}{Re}, \tag{18.1.61}$$

which should be compared with C_{D_p} computed from numerical simulation at the outer edge of the boundary layer. However, in practice the vorticity extends to infinity and a clear-cut boundary-layer edge does not exist. To address this difficulty, we present C_{D_p} computed from numerical simulation at different values of r,

$$C_{D_p}(r) = \frac{1}{\rho U_0^2 a}\int_0^{2\pi} (-p)\mathbf{e}_x \cdot \mathbf{e}_r\, r\, d\theta, \tag{18.1.62}$$

Table 18.2. *The values of δ_{D1}/a and δ_{D2}/a at which C_{D_p} given by (18.1.61) is equal to C_{D_p} computed from numerical simulation (18.1.62)*

The magnitude of the vorticity on the circle with the radius $r = a + \delta_{D1}$ or $r = a + \delta_{D2}$ was estimated from the numerical data and expressed as a certain percentage of the maximum magnitude of the vorticity field. This percentage is between 12.6% and 20.4% at $r = a + \delta_{D1}$ and is between 0.003% and 0.913% at $r = a + \delta_{D2}$.

Re	q/U_0	C_{D_p}	δ_{D1}/a	Vorticity (%)	δ_{D2}/a	Vorticity (%)
200	4	−0.126	0.161	13.1	0.594	0.913
200	5	−0.126	0.0838	14.6	2.03	0.00455
400	4	−0.0628	0.0835	18.1	0.811	0.418
400	5	−0.0628	0.0553	20.8	2.46	0.00340
400	6	−0.0628	0.0473	20.7	2.65	0.00365
1000	3	−0.0251	0.0552	20.4	1.56	0.0139

and compare with C_{D_p} from (18.1.61). As an example, we plot $C_{D_p}(r)$ from numerical simulation for $Re = 400$ and $q/U_0 = 4$, 5, and 6 in figure 18.2; the straight line gives C_{D_p} computed from (18.1.61) for $Re = 400$. Note that the results of numerical simulation depend on q/U_0 but equation (18.1.61) does not. Each curve for $C_{D_p}(r)$ has two intersections with the straight line, the one close to the wall denoted by δ_{D1}/a and the other one far way from the wall denoted by δ_{D2}/a. In table 18.2 we list the values of δ_{D1}/a and δ_{D2}/a for (Re, q/U_0) = (200, 4), (200, 5), (400, 4), (400, 5), (400, 6), and (1000, 3). The vorticity field in the whole domain was computed in numerical simulation. The magnitude of the vorticity on the circle with the radius $r = a + \delta_{D1}$ or $r = a + \delta_{D2}$ was estimated from the numerical data and expressed as a certain percentage of the maximum magnitude of the vorticity field. This percentage is between 12.6% and 20.4% at $r = a + \delta_{D1}$ and is between 0.003% and 0.913% at $r = a + \delta_{D2}$. This percentage is 20.6% at $r = a + \delta_{D1}$ for $(Re, q/U_0) = (400, 6)$, which indicates that, roughly speaking, the vorticity magnitude at a radial position

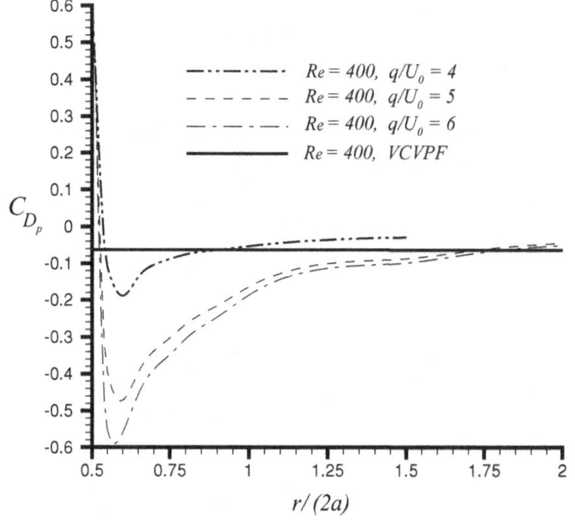

Figure 18.2. The pressure drag coefficient C_{D_p} at different radial positions $r/(2a)$ computed from numerical simulation (18.1.62) for $Re = 400$: dash–double-dotted curve, $q/U_0 = 4$; dashed curve, $q/U_0 = 5$; dash–dotted curve, $q/U_0 = 6$. The solid straight line gives C_{D_p} computed from (18.1.61) for $Re = 400$. Each curve for $C_{D_p}(r)$ has two intersections with the straight line, at which C_{D_p} given by (18.1.61) is equal to C_{D_p} computed from numerical simulation at $r = a + \delta$.

Table 18.3. *The calculation of C_{L_p} (18.1.49) and C_T (18.1.44) on the cylinder with δ_{D1}/a determined by matching C_{D_p} as an effective boundary-layer thickness*

The results are in fair agreement with the numerical data shown in table 18.1. This demonstrates that δ_{D1}/a can be used not only as an effective boundary-layer thickness for C_{D_p}, but also for C_{L_p} and C_T.

Re	q/U_0	δ_{D1}/a	α	C_{L_p}	C_T
200	4	0.161	0.754	16.659	0.388
200	5	0.0838	0.501	25.065	0.466
400	4	0.0835	0.691	18.188	0.239
400	5	0.0553	0.486	25.845	0.273
400	6	0.0473	0.384	32.765	0.299
1000	3	0.0552	1.157	10.862	0.0718

$r > a + \delta_{D1}$ is less than 20.6% of the maximum vorticity magnitude. The reason is that the vorticity magnitude generally decreases as r increases. When $r > a + \delta_{D2}$, the vorticity is almost negligible.

A comparison of tables 18.1 and 18.2 shows that δ_{D1}/a is close to the effective boundary-layer thickness δ_L/a we determine by matching C_{L_p}. When we insert δ_{D1}/a into the expressions for C_{L_p} and C_T on the cylinder, (18.1.49) and (18.1.44), respectively, the results are in fair agreement with the numerical simulation and are better than Glauert's solutions (see tables 18.3 and 18.1). Thus δ_{D1}/a can be used not only as an effective boundary-layer thickness for C_{D_p}, but also for C_{L_p} and C_T. This result shows that one effective boundary-layer thickness for both the VCVPF calculation and the simple modification of Glauert's solution exists.

Figure 18.2 shows that C_{D_p} changes significantly with r near the wall; C_{D_p} reaches its minimum and then increases; the magnitude of C_{D_p} approaches zero as r increases to infinity. In the region near the second intersection $r = a + \delta_{D2}$, the C_{D_p} curve is rather flat and the straight line given by (18.1.61) is a reasonable approximation to the numerical results. This region may be viewed as a transition region from the inner flow, where the vorticity is important, to the outer flow, where the vorticity is negligible. The VCVPF calculation cannot predict variation of C_{D_p} near the wall.

The term $D_1 \sin\theta$ in the pressure correction should give rise to an extra lift force in addition to the contribution from the irrotational pressure. However, D_1 is not determined in the VCVPF calculation. In the new boundary-layer analysis in section 18.2 the pressure is not assumed to be a constant across the boundary layer and it is solved from the governing equations. We determine the terms proportional to $\sin\theta$, $\cos\theta$, $\sin 2\theta$, and $\cos 2\theta$ up to $O(\alpha^2)$.

18.1.3 The additional drag on an airfoil by the dissipation method

We consider a symmetrical Joukowski airfoil moving in a liquid at an angle of attack β with a constant velocity U_0 (figure 18.3). The airfoil is obtained by the Joukowski transformation

$$z = \zeta + \frac{c^2}{\zeta}, \tag{18.1.63}$$

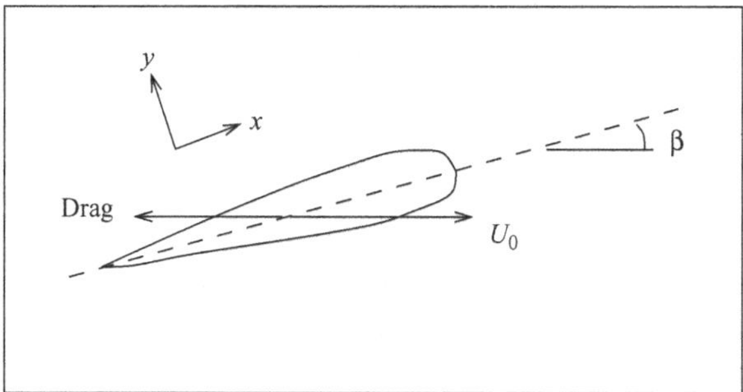

Figure 18.3. A symmetrical airfoil moving in a liquid at an angle of attack β with a constant velocity U_0. The additional drag on the airfoil computed with the DM is opposite to the moving direction of the airfoil and is defined as negative.

in conjunction with a circle in the ζ plane. The radius a of the circle is slightly larger than the transformation coefficient c,

$$a = c + m = c(1 + \varepsilon), \tag{18.1.64}$$

where $\varepsilon = m/c$ is assumed to be small compared with unity. The center of the circle is displaced from the origin to $(-m, 0)$, so that the circle passes through one of the critical points of the Joukowski transformation, $\zeta = c$, which gives rise to the cusp of the airfoil in the z plane.

In the ζ plane, a generic point (r, θ) on the circle satisfies

$$(c + m)^2 = r^2 + m^2 + 2rm\cos\theta, \tag{18.1.65}$$

which leads to

$$r = -m\cos\theta + \sqrt{c^2 + 2cm + m^2\cos^2\theta}. \tag{18.1.66}$$

The airfoil surface is then given by

$$z = re^{i\theta} + \frac{c^2}{r}e^{-i\theta} \tag{18.1.67}$$

or

$$x = \left(r + \frac{c^2}{r}\right)\cos\theta, \quad y = \left(r - \frac{c^2}{r}\right)\sin\theta. \tag{18.1.68}$$

The complex potential for a uniform flow past a circle with circulation is

$$f(\zeta) = U_0\left[(\zeta + m)e^{-i\beta} + \frac{a^2}{\zeta + m}e^{i\beta}\right] + \frac{i\Gamma}{2\pi}\log\left(\frac{\zeta + m}{a}\right). \tag{18.1.69}$$

Equation (18.1.69), along with the inverse Joukowski transformation

$$\zeta = \frac{z}{2} \pm \sqrt{\frac{z^2}{4} - c^2}, \tag{18.1.70}$$

gives the potential for the flow past an airfoil in the z plane. The Kutta condition requires the circulation to be

$$\Gamma = 4\pi U_0 a \sin\beta. \tag{18.1.71}$$

The dissipation calculation will be carried out in dimensionless form. We choose U_0 and c to be the scales for velocity and length, respectively. The dimensionless form of the potential is

$$\frac{f(\zeta)}{U_0 c} = (\zeta+\varepsilon)e^{-i\beta} + \frac{(1+\varepsilon)^2}{\zeta+\varepsilon}e^{i\beta} + 2i\sin\beta(1+\varepsilon)\log\left(\frac{\zeta+\varepsilon}{1+\varepsilon}\right). \qquad (18.1.72)$$

Note that we use the same symbols for the dimensional and dimensionless variables. The inverse Joukowski transformation in dimensionless form is

$$\zeta = \frac{z}{2} \pm \sqrt{\frac{z^2}{4} - 1}. \qquad (18.1.73)$$

The velocities can be evaluated on the potential,

$$u = \frac{1}{2}\left(\frac{\mathrm{d}f}{\mathrm{d}z} + \frac{\mathrm{d}\bar{f}}{\mathrm{d}\bar{z}}\right), \quad v = \frac{i}{2}\left(\frac{\mathrm{d}f}{\mathrm{d}z} - \frac{\mathrm{d}\bar{f}}{\mathrm{d}\bar{z}}\right), \qquad (18.1.74)$$

and the rate-of-strain tensor is

$$2\mathbf{D} = \begin{bmatrix} \frac{\mathrm{d}^2 f}{\mathrm{d}z^2} + \frac{\mathrm{d}^2 \bar{f}}{\mathrm{d}\bar{z}^2} & i\left(\frac{\mathrm{d}^2 f}{\mathrm{d}z^2} - \frac{\mathrm{d}^2 \bar{f}}{\mathrm{d}\bar{z}^2}\right) \\ i\left(\frac{\mathrm{d}^2 f}{\mathrm{d}z^2} - \frac{\mathrm{d}^2 \bar{f}}{\mathrm{d}\bar{z}^2}\right) & -\left(\frac{\mathrm{d}^2 f}{\mathrm{d}z^2} + \frac{\mathrm{d}^2 \bar{f}}{\mathrm{d}\bar{z}^2}\right) \end{bmatrix}. \qquad (18.1.75)$$

The surface of the airfoil is given by

$$x = \left(r + \frac{1}{r}\right)\cos\theta, \quad y = \left(r - \frac{1}{r}\right)\sin\theta, \qquad (18.1.76)$$

where

$$r = -\varepsilon\cos\theta + \sqrt{1 + 2\varepsilon + \varepsilon^2\cos^2\theta}. \qquad (18.1.77)$$

Let $\dot{x} = \mathrm{d}x/\mathrm{d}\theta$ and $\dot{y} = \mathrm{d}y/\mathrm{d}\theta$; then the norm on the surface can be written as

$$\mathbf{n} = n_x\mathbf{e}_x + n_y\mathbf{e}_y = \frac{-\dot{y}\mathbf{e}_x + \dot{x}\mathbf{e}_y}{\sqrt{\dot{x}^2 + \dot{y}^2}}, \qquad (18.1.78)$$

$$\mathrm{d}s = \sqrt{\dot{x}^2 + \dot{y}^2}\,\mathrm{d}\theta. \qquad (18.1.79)$$

Now we calculate the dissipation

$$\mathcal{D} = \mu U_0^2 \int_A \mathbf{u}\cdot 2\mathbf{D}\cdot\mathbf{n}\,\mathrm{d}A = \mu U_0^2 I, \qquad (18.1.80)$$

where

$$I = \int_0^{2\pi} [n_x(2\mathbf{D})_{xx}u + n_x(2\mathbf{D})_{xy}v + n_y(2\mathbf{D})_{yx}u + n_y(2\mathbf{D})_{yy}v]\sqrt{\dot{x}^2 + \dot{y}^2}\,\mathrm{d}\theta. \qquad (18.1.81)$$

The integral I is computed numerically. The additional drag is then obtained, $D = \mathcal{D}/(-U_0)$, and the drag coefficient

$$C_D = \frac{D}{\frac{1}{2}\rho U_0^2 4c} = -\frac{\mu U_0 I}{\frac{1}{2}\rho U_0^2 4c} = -\frac{1}{Re'}\frac{I}{2}, \qquad (18.1.82)$$

Table 18.4. *The integral I as a function of the attack angle β and the nose sharpness parameter ε (the smaller the ε, the sharper the nose)*

The drag coefficient can be obtained by $C_D = -I/(2Re')$, where the Reynolds number $Re' = \frac{\rho U_0 c}{\mu}$. Here $c \approx l/4$, where l is the length of the airfoil

β	ε = 0.3	ε = 0.2	ε = 0.1	ε = 0.05	ε = 0.01
0	4.34	3.48	2.53	2.05	1.67
π/20	6.24	6.30	9.39	22.7	410
π/10	11.8	14.5	29.3	82.8	1.59×10^3
π/6	23.7	32.3	72.6	213	4.17×10^3
π/4	43.2	61.1	142	424	8.34×10^3

where the Reynolds number is

$$Re' = \frac{\rho U_0 c}{\mu}. \tag{18.1.83}$$

The drag coefficient depends on the parameter ε and the angle of attack β. The parameter ε determines the maximum thickness of the airfoil and the roundness of the leading nose. The smaller the ε, the thinner the airfoil and the sharper the leading nose. In table 18.4 we present the magnitude of the drag coefficient multiplied by the Reynolds number as a function of the parameter ε and the angle of attack β. When ε is fixed, the magnitude of the drag coefficient increases with β. The reason is that, when β is not zero, the stream must turn around the leading nose and a large amount of dissipation is generated near the leading nose. When β is fixed at zero, the dissipation decreases as ε decreases. This is because a slimmer airfoil leads to less disturbance to the uniform flow and smaller dissipation. At the limit when ε is zero, the flow becomes a uniform flow past a flat plate at a zero attack angle, in which the drag is zero. However, when β is fixed at nonzero values, the magnitude of the drag coefficient increases as ε decreases. The reason is that the major contribution to the dissipation is from the flow that turns around the leading nose. A smaller value of ε leads to a sharper leading nose and larger dissipation. At the limit when ε is zero, the leading edge coincides with one of the critical points of the Joukowski transformation, the velocity at the leading edge is singular, and the dissipation calculation breaks down.

If the Reynolds number is of the order of hundreds or thousands, the additional drag is negligible when $\beta = 0$ or when β is small and the airfoil is not very sharp at the leading edge; the additional drag is evident when β is nonzero and the airfoil has a very sharp leading edge.

18.1.4 Discussion and conclusion

This work concerns the drag on a body moving at a constant velocity $U_0\mathbf{e}_i$ in an otherwise quiescent viscous liquid. The Reynolds number is high, and the flow can be approximated by an outer potential flow and a boundary layer adjacent to the surface of the body. The drag is defined as

$$D = \int_A \mathbf{e}_i \cdot \mathbf{T} \cdot (-\mathbf{n}) \, \mathrm{d}A. \tag{18.1.84}$$

The dissipation calculation is one of the methods to compute the drag, and it is based on the mechanical-energy equation:

$$\frac{\mathrm{d}}{\mathrm{d}t}\int_V \frac{\rho|\mathbf{u}|^2}{2}\mathrm{d}V = \int_A \mathbf{u}\cdot\mathbf{T}\cdot\mathbf{n}\,\mathrm{d}A - \int_V 2\mu\mathbf{D}:\mathbf{D}\,\mathrm{d}V. \tag{18.1.85}$$

At steady state, (18.1.85) becomes

$$\int_A \mathbf{u}\cdot\mathbf{T}\cdot\mathbf{n}\,\mathrm{d}A = \int_A [\mathbf{u}\cdot\mathbf{n}(-p+\tau_n) + \mathbf{u}\cdot\mathbf{t}\tau_s]\,\mathrm{d}A = \mathcal{D}. \tag{18.1.86}$$

If the body is a gas of negligible density and viscosity, the shear stress τ_s is zero at the interface. The continuity of the normal velocity at the gas–liquid interface gives $\mathbf{u}\cdot\mathbf{n} = U_0\mathbf{e}_i\cdot\mathbf{n}$. Thus (18.1.86) can be written as

$$\int_A U_0\mathbf{e}_\mathrm{i}\cdot\mathbf{n}(-p+\tau_n)\,\mathrm{d}A = \mathcal{D} \quad \Rightarrow \quad U_0(-D) = \mathcal{D} = \mathcal{D}_{\mathrm{BL}} + \mathcal{D}_P, \tag{18.1.87}$$

where $\mathcal{D}_{\mathrm{BL}}$ is the dissipation inside the boundary layer and $\mathcal{D}_P$ is the dissipation of the outer potential flow. In gas–liquid flows, the boundary layer is assumed to be very weak and $\mathcal{D}_{\mathrm{BL}}$ is negligible to the first-order approximation. Thus we have the drag on a gas body,

$$D \approx \mathcal{D}_P/(-U_0), \tag{18.1.88}$$

which is used in our calculation of the drag on a circular gas bubble in subsection 18.1.1.

If the body is solid, the no-slip condition at the wall gives $\mathbf{u} = U_0\mathbf{e}_i$. Thus (18.1.86) can be written as

$$\int_A U_0\mathbf{e}_\mathrm{i}\cdot\mathbf{T}\cdot\mathbf{n}\,\mathrm{d}A = \mathcal{D}_{\mathrm{BL}} + \mathcal{D}_P \quad \Rightarrow \quad U_0(-D) = \mathcal{D}_{\mathrm{BL}} + \mathcal{D}_P. \tag{18.1.89}$$

The boundary layer near a solid wall is usually strong and accounts for the major part of the total dissipation. However, $\mathcal{D}_P$ is not zero and does contribute to the drag. We call $\mathcal{D}_P/U_0$ an additional drag, and it is computed for an airfoil in subsection 18.1.3. The dissipation of the outer potential flow increases the drag calculated from the boundary-layer flow alone. Our calculation shows that the coefficient of the additional drag is proportional to $1/Re$. Thus the additional drag is small when the Reynolds number is high.

The situation is different for a rotating cylinder moving in a liquid. The no-slip condition at the wall gives $\mathbf{u} = U_0\mathbf{e}_i + \Omega a\mathbf{e}_\theta$, where Ω is the angular speed of the cylinder. Equation (18.1.86) can be written as

$$\begin{aligned}\int_A (U_0\mathbf{e}_\mathrm{i} + \Omega a\mathbf{e}_\theta)\cdot\mathbf{T}\cdot\mathbf{n}\,\mathrm{d}A &= U_0\int_A \mathbf{e}_\mathrm{i}\cdot\mathbf{T}\cdot\mathbf{n}\,\mathrm{d}A + \Omega\int_A a\tau_{\theta n}\,\mathrm{d}A,\\ \Rightarrow \quad U_0(-D) + \Omega T &= \mathcal{D}.\end{aligned} \tag{18.1.90}$$

Equation (18.1.90) is not enough to determine two unknowns, the drag D and the torque T. Thus the DM alone cannot give the drag or the torque in this case.

Padrino and Joseph (2006) numerically simulated the flow past a rapidly rotating cylinder. When the Reynolds number is $Re = 400$ and the ratio between the cylinder rotating speed and the streaming flow speed $q/U_0 = 4$, they obtain $\mathcal{D}_{\mathrm{BL}}:\mathcal{D}_P = 1.72:1$. Although in this case the dissipation cannot be used to compute the drag or the torque

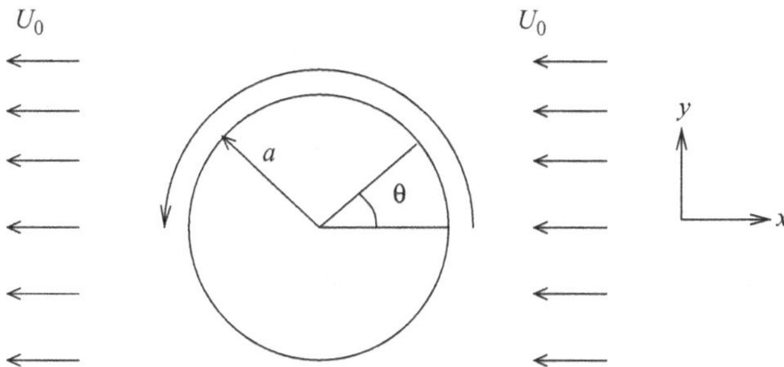

Figure 18.4. The uniform streaming flow past a rotating cylinder.

independently, the data show that the viscous dissipation of the outer potential flow can be significant.

We extend the idea of the viscous pressure correction from gas–liquid flows to Prandtl's boundary layer outside a solid. At the outer edge of the boundary layer, the shear stress evaluated on the boundary-layer solution by use of Prandtl's theory does not necessarily equal the shear stress evaluated on the outside potential flow; this is analogous to the discrepancy between zero shear stress and nonzero irrotational shear stress at a gas–liquid interface. The shear stress discrepancy at the outer edge of the boundary layer induces extra vorticity and a viscous pressure correction. The power of the pressure correction is equal to the power of this shear stress discrepancy.

We apply the method of VCVPF to the boundary layer around a rapidly rotating cylinder in a uniform flow in subsection 18.1.2. The pressure correction is expanded as a Fourier series, and we determine the coefficient for the $\cos\theta$ term, which is the only term in the Fourier series contributing to the drag. We integrate the pressure correction and viscous stresses to obtain the additional drag at the outer edge of the boundary layer, which is not obtained by the dissipation calculation for this problem. Numerical simulations confirm that the pressure in the region near the cylinder surface gives rise to a noticeable drag. After choosing an effective boundary-layer thickness, we are able to fit the pressure drag computed from VCVPF theory to the pressure drag from numerical simulation. We note that this pressure drag at the outer edge of the boundary layer is different from the pressure drag on the cylinder. Actually, the simulations show that the pressure drag changes sign across the boundary layer. The method of VCVPF can determine only the pressure correction at the outer edge of the boundary layer, not the variation inside the boundary layer. A boundary-layer analysis for the flow past a rapidly rotating cylinder is presented in the next section, in which the pressure inside the boundary layer is solved.

18.2 Glauert's solution of the boundary layer on a rapidly rotating cylinder in a uniform stream revisited

We perform a boundary-layer analysis for the streaming flow past a rapidly rotating circular cylinder (figure 18.4).[3] The starting point is the full continuity and momentum

[3] We call the readers' attention to the fact that in our problem, the uniform flow is from right to left and the cylinder rotates counterclockwise (see figure 18.4) The lift on the cylinder points upward and is defined as positive. The drag on the cylinder is negative if it is in the uniform flow direction; the drag is positive if it is opposite to the uniform flow direction.

equations without any approximations. We assume that the solution is a boundary-layer flow near the cylinder surface with the potential flow outside the boundary layer. The order of magnitude of the terms in the continuity and momentum equations can be estimated inside the boundary layer. When terms of the order of δ/a and higher are dropped, where δ is the boundary-layer thickness and a is the radius of the cylinder, the equations used by Glauert (1957) are recovered. Glauert's solution ignores the irrotational rotary component of the flow inside the boundary layer, which is consistent with dropping δ/a terms in the governing equations.

We propose a new solution to this problem, in which the velocity field is decomposed into two parts. Outside the boundary layer, the flow is irrotational and can be decomposed into a purely rotary flow and a potential flow past a fixed cylinder. Inside the boundary layer, the velocity is decomposed into an irrotational purely rotary flow and a boundary-layer flow. Inserting this decomposition of the velocity field inside the boundary layer into the governing equations, we obtain a new set of equations for the boundary-layer flow, in which we do not drop the terms of the order of δ/a or higher. The pressure can no longer be assumed to be a constant across the boundary layer, and the continuity of shear stress at the outer edge of the boundary layer is enforced. We solve this new set of equations by using Glauert's method, i.e., to expand the solutions as a power series of $\alpha = 2U_0/Q$, where U_0 is the uniform stream velocity and Q is the circulatory velocity at the outer edge of the boundary layer. The pressure from this boundary-layer solution has two parts, an inertial part and a viscous part. The inertial part comes from the inertia terms in the momentum equations and is in agreement with the irrotational pressure; the viscous part comes from the viscous stress terms in the momentum equations and may be viewed as a viscous pressure correction, which contributes to both drag and lift. Our boundary-layer solution is in reasonable to excellent agreement with the numerical simulation that will be presented in §18.3.

18.2.1 Introduction

The flow pattern depends critically on the ratio between the uniform stream velocity U_0 and the fluid circulatory velocity Q. Potential flow theory shows that, when $2U_0/Q < 1$, there is no stagnation point on the cylinder and a region of closed streamlines exists near the cylinder. The fluid circulatory velocity Q is closely related to the peripheral velocity of the cylinder $q = \Omega a$, where Ω is the angular velocity of the cylinder; Q and q are approximately equal for large values of q. Experiments (Prandtl and Tietjens, 1934) and simulations (Padrino and Joseph, 2006) confirm that separation is largely suppressed and a closed boundary layer around the cylinder may be expected when q is much larger than U_0.

A number of theoretical studies have been dedicated to this problem based on the assumption that the ratio q/U_0 is high, separation is suppressed, and a steady-state solution of the problem exists. Glauert (1957) solved the steady-state, 2D boundary-layer equations and obtained the solution in the form of a power series in $\alpha = 2U_0/Q$, which is related to the speed ratio q/U_0 by

$$\alpha \to 2U_0/q = \frac{2}{q/U_0} \quad \text{as} \quad q \to \infty. \tag{18.2.1}$$

He carried out the analysis up to and including boundary-layer functions associated with α^4 and obtained

$$\frac{Q}{q} = 1 - 3\left(\frac{U_0}{q}\right)^2 - 3.24\left(\frac{U_0}{q}\right)^4 + \cdots + . \tag{18.2.2}$$

In his boundary-layer equations, the pressure was assumed to be a constant across the boundary layer; thus the irrotational pressure is the only component in the normal stress acting on the cylinder. The pressure does not contribute to the drag, and its contribution to the lift is the same as in the classical aerodynamics equation $L_p = \rho U_0 \Gamma$, where $\Gamma = 2\pi a Q$ is the circulation. The coefficient for the pressure lift is

$$C_{L_p} = \frac{L_p}{\rho U_0^2 a} = 2\pi \frac{Q}{U_0} = 2\pi \frac{q}{U_0}\left[1 - 3\left(\frac{U_0}{q}\right)^2 - 3.24\left(\frac{U_0}{q}\right)^4 + \cdots +\right]. \tag{18.2.3}$$

The effect of the boundary-layer analysis on the pressure lift is through the value of Q, and it should be noted that C_{L_p} is independent of the Reynolds number. Glauert did not consider the friction drag and lift, but they can be readily obtained from his solution:

$$C_{D_f} = \frac{D_f}{\rho U_0^2 a} = -\frac{2\pi}{\sqrt{Re}}\sqrt{\frac{Q}{U_0}}, \qquad C_{L_f} = \frac{L_f}{\rho U_0^2 a} = \frac{2\pi}{\sqrt{Re}}\sqrt{\frac{Q}{U_0}}, \tag{18.2.4}$$

where

$$Re = \frac{2a\rho U_0}{\mu} \tag{18.2.5}$$

is the Reynolds number based on the uniform streaming velocity. Glauert computed the torque needed to maintain the rotation,

$$T = 2\sqrt{2}\pi\rho U_0^{\frac{3}{2}} a^{\frac{3}{2}} \sqrt{\nu}\left[\left(\frac{U_0}{q}\right)^{\frac{1}{2}} - 0.522\left(\frac{U_0}{q}\right)^{\frac{5}{2}} + \cdots +\right], \tag{18.2.6}$$

and the torque coefficient is

$$C_T = \frac{T}{2\rho U_0^2 a^2} = \frac{2\pi}{\sqrt{Re}}\left[\left(\frac{U_0}{q}\right)^{\frac{1}{2}} - 0.522\left(\frac{U_0}{q}\right)^{\frac{5}{2}} + \cdots +\right]. \tag{18.2.7}$$

Wood (1957) studied a class of 2D, laminar boundary-layer flows with closed streamlines. The velocity at the solid boundary was supposed uniform, and the velocity in the boundary layer was supposed to differ only slightly from that of the boundary. A formal solution of the boundary layer was then derived by expansion of the velocity in a power series in a small parameter representative of the small differences of the speed through the boundary layer. He applied the theory to the uniform streaming flow past a rotating cylinder and obtained a circulation that was equivalent to the first two terms of equation (18.2.2).

Moore (1957) also considered this problem, assuming that the cylinder rotation velocity was much greater than that of the uniform stream. He argued that the effect of the uniform streaming flow could be regarded as a perturbation of the viscous irrotational rotary flow induced by a rotating cylinder and obtained a uniformly valid first approximation to the flow field by solving the stream function equation. In the limit of large Reynolds number, Moore also obtained a circulation that was equivalent to the first two terms of equation

(18.2.2). Moore showed that the drag was small, of the order of $(U_0/q)^3$, and the lift coefficient was

$$C_L = 2\pi\frac{q}{U_0}\left[1 + O\left(\frac{U_0}{q}\right)^2\right], \tag{18.2.8}$$

which is comparable with Glauert's pressure lift, (18.2.3), but Moore did not give the coefficient of $(U_0/q)^2$. In the limit of large Reynolds number, Moore showed that the torque was

$$T = 4\pi\mu aq\left[1 + \frac{\sqrt{2}}{2}\left(\frac{U_0}{q}\right)^2\sqrt{Re}\right], \tag{18.2.9}$$

where the first term on the right-hand side, $4\pi\mu aq$, is the torque when there is only viscous irrotational rotary flow but no streaming flow. The torque coefficient is

$$C_T = \frac{4\pi}{Re}\frac{q}{U_0} + \frac{2\sqrt{2}\pi}{\sqrt{Re}}\frac{U_0}{q}, \tag{18.2.10}$$

where the second term on the right-hand side is similar to Glauert's torque coefficient, (18.2.7), but the powers of U_0/q are different.

Numerical simulations have been widely used to study the flow past a rotating cylinder. A review of the numerical studies can be found in §18.3. The numerical results in §18.3 show that separation is largely suppressed and a steady-state solution can be obtained. The numerical simulation will serve as the benchmark for the analysis in this work.

We investigate the uniform streaming flow past a rotating cylinder adopting the same assumption as in Glauert (1957), Moore (1957), and Wood (1957), i.e., the cylinder rotation velocity is much greater than that of the uniform stream, separation is suppressed, and a steady-state solution of the problem exists. Our work here is intended to be an improvement of Glauert's boundary-layer solution. The boundary-layer equations used by Glauert can be recovered when terms of the order of δ/a and higher are dropped from the unapproximated continuity and momentum equations. To be consistent, one should have

$$Q\frac{a+\delta}{r} \sim Q \quad \text{for} \quad a \le r \le a + \delta,$$

which means that the irrotational rotary component of the flow inside the boundary layer is ignored. The tangential velocity given by Glauert is

$$u = Q\left[1 + \alpha f_1' e^{i\theta} + \alpha^2\left(f_2' e^{2i\theta} + g_2'\right) + \cdots +\right], \tag{18.2.11}$$

which is a perturbation about a uniform flow, not a perturbation about the viscous irrotational rotary flow. It can be inferred from (18.2.11) that the shear stress at the cylinder surface approaches zero when α approaches zero. In other words, Glauert's solution suggests that the shear stress at the cylinder surface approaches zero when the streaming flow is extremely weak compared with the rotation of the cylinder. However, the real limiting value for the shear stress is $-2\mu q/a$. This discrepancy carries on to the computation of the torque. Glauert's torque expression, (18.2.6), indicates that the torque is zero when $U_0 = 0$; however, the actual torque to maintain the rotation of the cylinder in a viscous irrotational purely rotary flow is $4\pi\mu a q$, which is also shown in Moore's torque expression, (18.2.9).

We propose a new solution to this problem, in which the velocity is decomposed into two parts. Outside the boundary layer, the flow is irrotational and can be decomposed into a viscous irrotational purely rotary flow and a potential flow past a fixed cylinder. Inside the boundary layer, the velocity field is decomposed into a viscous irrotational purely rotary flow and a boundary-layer flow that is expanded as a power series of $\alpha = 2U_0/Q$. This decomposition of the velocity field in the boundary layer is actually a perturbation of the purely rotary flow with α being the perturbation parameter, which is similar to Moore's approach. The difference is that Moore tried to obtain a uniformly valid solution for the flow, whereas we are seeking the solution valid in the boundary layer. Inserting this decomposition of the velocity field inside the boundary layer into the governing equations, we obtain a new set of equations for the boundary-layer flow, in which we do not drop the terms of the order of δ/a or higher. There are significant differences between our new equations and Glauert's boundary-layer equations. In Glauert's study the pressure is assumed to be a constant across the boundary layer, and the momentum equation in the radial direction is not used. The direct result of this approximation by Glauert (and Prandtl) is that the normal stress on a solid is imposed by the irrotational pressure, independent of the Reynolds number. Viscous effects on the normal stress on a solid wall, which always exist at finite Reynolds number, no matter how large, are not available. In our new equations, the pressure is an unknown and the momentum equation in the radial direction does appear. Because we have an extra unknown, an extra boundary condition is needed, and we choose to enforce the continuity of the shear stress at the outer edge of the boundary layer. The technique to solve this new set of equations is almost the same as that used by Glauert. The power-series expansions are inserted into the new set of equations and the coefficients of different powers of α are compared; then ODEs for the functions in the power series are obtained and solved. The inertia terms in the momentum equations give rise to the irrotational pressure, and the viscous terms lead to a viscous pressure correction, which contributes to both drag and lift.

One of the key differences between our new boundary-layer analysis and the classical boundary-layer theory of Prandtl is about the calculation of the pressure drag. In the classical boundary-layer approximation, the pressure is constant across the layer and the irrotational pressure of the outer flow is imposed on the surface of the body. This approximation is not good enough for the purpose of the drag calculation and leads to zero pressure drag. Lighthill (1963) remarked "Errors, due to neglecting either the pressure gradient across the layer, or the displacement-thickness effect on U, produce a resultant pressure force ('form drag') comparable with the whole viscous force on the body ('skin-friction drag'). Accordingly, such errors cannot be neglected, as often no drag is present from other causes, the pressure forces in pure irrotational flow having zero resultant." Various techniques were developed to calculate the pressure drag as a patch for Prandtl's theory. Lighthill (1963) described these methods: "To get round these difficulties, one does not in practice attempt to calculate surface pressure more precisely, but uses a combination of arguments (Chapter X) in which drag is inferred, from conservation of momentum for large masses of fluid, in terms of the state of the boundary layer at the trailing edge." Schlichting (1960) reviewed methods for the calculation of the profile drag (the sum of the friction drag and the pressure drag) devised by Pretsch (1938) and Squire and Young (1938). These methods are tied in with the boundary-layer calculation, and the drag is obtained based on the principle of momentum conservation. Schlichting remarked about these methods: "However, in order to be in a position to calculate pressure drag it is necessary in each case to make use of certain additional empirical relations." The method by which we treat the drag is totally different. We solve for the pressure on the body from governing equations. The pressure drag is computed by direct integration of the pressure over the surface of the body, not by arguments of conservation of the momentum.

In our new set of equations for the boundary-layer flow and its boundary conditions, we assume that the boundary-layer thickness δ/a is known; then we can compute the solution. This is different from the problems such as Blasius' solution, in which the boundary condition at the outer edge of the boundary layer can be stretched to infinity and the solution is obtained without knowledge of the boundary-layer thickness. We prescribe δ/a at different values, compute the solution, and then compare them with the results of numerical simulation; the value of δ/a that leads to the best agreement with the simulation results may be viewed as a proper boundary-layer thickness. The boundary-layer thickness determined in this way satisfies approximately $(\delta/a) \propto (1/\sqrt{Re})$ and decreases with increasing q/U_0. Comparison of our solution by use of the proper δ/a with the simulation results and Glauert's and Moore's solutions shows that our lift and torque are in reasonable to excellent agreement with the simulation results and the agreement for the drag is less good if the speed ratio q/U_0 is not high enough. It is also demonstrated our solution is indeed an improvement of Glauert's solution.

18.2.2 Unapproximated governing equations

When the polar coordinate system (r, θ) is used, the continuity equation is

$$\frac{\partial v_r}{\partial r} + \frac{v_r}{r} + \frac{1}{r}\frac{\partial v_\theta}{\partial \theta} = 0, \tag{18.2.12}$$

and the momentum equations for steady flows are

$$\left(v_r \frac{\partial}{\partial r} + \frac{v_\theta}{r}\frac{\partial}{\partial \theta}\right) v_\theta + \frac{v_r v_\theta}{r} = -\frac{1}{r\rho}\frac{\partial P}{\partial \theta} + \nu\left(\nabla^2 v_\theta - \frac{v_\theta}{r^2} + \frac{2}{r^2}\frac{\partial v_r}{\partial \theta}\right), \tag{18.2.13}$$

$$\left(v_r \frac{\partial}{\partial r} + \frac{v_\theta}{r}\frac{\partial}{\partial \theta}\right) v_r - \frac{v_\theta^2}{r} = -\frac{1}{\rho}\frac{\partial P}{\partial r} + \nu\left(\nabla^2 v_r - \frac{v_r}{r^2} - \frac{2}{r^2}\frac{\partial v_\theta}{\partial \theta}\right), \tag{18.2.14}$$

where

$$\nabla^2 = \frac{\partial^2}{\partial r^2} + \frac{1}{r}\frac{\partial}{\partial r} + \frac{1}{r^2}\frac{\partial^2}{\partial \theta^2}.$$

18.2.3 Boundary-layer approximation and Glauert's equations

The flow may be approximated by a boundary layer near the cylinder surface and a potential flow outside. Inside the boundary layer, we have the following estimations:

$$v_\theta \sim Q, \quad r \sim a, \quad \frac{\partial}{\partial r} \sim \frac{1}{\delta}, \quad \frac{\partial}{\partial \theta} \sim 1. \tag{18.2.15}$$

With these estimates, the magnitude of the terms in (18.2.12) can be written as

$$\frac{v_r}{\delta} + \frac{v_r}{a} + \frac{Q}{a} = 0.$$

If we consider v_r/a to be negligible compared with v_r/δ, we have

$$\frac{v_r}{\delta} + \frac{Q}{a} = 0 \quad \Rightarrow \quad v_r \sim Q\frac{\delta}{a}, \tag{18.2.16}$$

and the continuity equation may be written as

$$\frac{\partial v_r}{\partial r} + \frac{1}{r}\frac{\partial v_\theta}{\partial \theta} = 0. \tag{18.2.17}$$

We estimate the magnitude of terms in equation (18.2.13):

$$\frac{Q^2}{a}\left(1 + 1 + \frac{\delta}{a}\right) = -\frac{1}{r\rho}\frac{\partial P}{\partial \theta} + \nu\frac{Q}{\delta^2}\left(1 + \frac{\delta}{a} + \frac{\delta^2}{a^2} - \frac{\delta^2}{a^2} + \frac{\delta^3}{a^3}\right). \tag{18.2.18}$$

If we drop the terms of the order of δ/a and higher, equation (18.2.13) becomes

$$v_r\frac{\partial v_\theta}{\partial r} + \frac{v_\theta}{r}\frac{\partial v_\theta}{\partial \theta} = -\frac{1}{r\rho}\frac{\partial P}{\partial \theta} + \nu\frac{\partial^2 v_\theta}{\partial r^2}. \tag{18.2.19}$$

Now we estimate the magnitude of terms in (18.2.14):

$$\frac{Q^2}{a}\left(\frac{\delta}{a} + \frac{\delta}{a} - 1\right) = -\frac{1}{\rho}\frac{\partial P}{\partial r} + \nu\frac{Q}{\delta^2}\left(\frac{\delta}{a} + \frac{\delta^2}{a^2} + \frac{\delta^3}{a^3} - \frac{\delta^3}{a^3} + \frac{\delta^2}{a^2}\right). \tag{18.2.20}$$

Thus (18.2.14) becomes

$$\frac{v_\theta^2}{r} = \frac{1}{\rho}\frac{\partial P}{\partial r}, \tag{18.2.21}$$

which indicates that the change of the pressure across the boundary layer is of the order of δ and the pressure can still be assumed to be constant if δ/a is negligible (Schlichting, 1960).

Now if we use x for $r\theta$, y for r, u for v_θ, and v for v_r, equations (18.2.17) and (18.2.19) may be written as

$$\frac{\partial u}{\partial x} + \frac{\partial v}{\partial y} = 0, \tag{18.2.22}$$

$$u\frac{\partial u}{\partial x} + v\frac{\partial u}{\partial y} = -\frac{1}{\rho}\frac{\partial P}{\partial x} + \nu\frac{\partial^2 u}{\partial y^2}, \tag{18.2.23}$$

which are the 2D boundary-layer equations used by Glauert (1957). If δ/a terms are dropped, the irrotational rotary component of the velocity inside the boundary layer will be ignored. In reality, the boundary-layer thickness is never zero and is found to be rather large in numerical simulations. Therefore dropping δ/a terms can cause substantial error.

18.2.4 Decomposition of the velocity and pressure field

We propose a new solution, in which the total velocity and pressure are decomposed into two parts:

$$v_\theta = u_{p\theta} + u_\theta, \quad v_r = u_r, \quad P = p_p + p, \tag{18.2.24}$$

where

$$u_{p\theta} = Q\frac{a+\delta}{r}, \quad p_p = p_\infty - \frac{\rho}{2}\frac{(a+\delta)^2}{r^2}Q^2 \tag{18.2.25}$$

are the irrotational purely rotary velocity and the pressure induced by rotation. It is noted that $v_\theta = u_{p\theta}$, $v_r = 0$, and $P = p_p$ is a potential solution and is an exact solution for the unapproximated governing equations and no-slip boundary condition.

Outside the boundary layer, the flow is irrotational and can be decomposed into two potential flows: the irrotational purely rotary flow and the uniform flow past a circle with the radius $a+\delta$. The velocity from the second potential flow is

$$u_\theta = U_0\left[1+\frac{(a+\delta)^2}{r^2}\right]\sin\theta, \quad u_r = -U_0\left[1-\frac{(a+\delta)^2}{r^2}\right]\cos\theta. \tag{18.2.26}$$

At the outer edge of the boundary layer ($r = a+\delta$), the tangential velocity is

$$v_\theta = 2U_0\sin\theta + Q. \tag{18.2.27}$$

The total pressure at $r = a+\delta$ can be obtained from Bernoulli's equation:

$$P = p'_\infty + \frac{\rho}{2}U_0^2\left(1-4\sin^2\theta\right) - 2\rho U_0 Q\sin\theta - \frac{\rho}{2}Q^2. \tag{18.2.28}$$

After subtracting p_p from (18.2.28), we obtain the pressure p at $r = a+\delta$:

$$p = c + \frac{\rho}{2}U_0^2\left(1-4\sin^2\theta\right) - 2\rho U_0 Q\sin\theta = c - \frac{\rho}{2}U_0^2 + \rho U_0^2\cos 2\theta - 2\rho U_0 Q\sin\theta, \tag{18.2.29}$$

where c is a certain constant.

Inside the boundary layer, u_θ, u_r, and p need to be obtained from the governing equations. We insert (18.2.24) into governing equations (18.2.12), (18.2.13), and (18.2.14), subtract the equations satisfied by $u_{p\theta}$ and p_p, and obtain

$$\frac{\partial u_r}{\partial r} + \frac{u_r}{r} + \frac{1}{r}\frac{\partial u_\theta}{\partial \theta} = 0, \tag{18.2.30}$$

$$\begin{aligned} u_r\frac{\partial}{\partial r}(u_{p\theta}+u_\theta) + \frac{u_{p\theta}+u_\theta}{r}\frac{\partial u_\theta}{\partial\theta} + \frac{u_r(u_{p\theta}+u_\theta)}{r} &= -\frac{1}{r\rho}\frac{\partial p}{\partial\theta} \\ +\nu\left(\frac{\partial^2 u_\theta}{\partial r^2} + \frac{1}{r}\frac{\partial u_\theta}{\partial r} + \frac{1}{r^2}\frac{\partial^2 u_\theta}{\partial\theta^2} - \frac{u_\theta}{r^2} + \frac{2}{r^2}\frac{\partial u_r}{\partial\theta}\right)&, \end{aligned} \tag{18.2.31}$$

$$\begin{aligned} u_r\frac{\partial u_r}{\partial r} + \frac{u_{p\theta}+u_\theta}{r}\frac{\partial u_r}{\partial\theta} - \frac{2u_{p\theta}u_\theta + u_\theta^2}{r} &= -\frac{1}{\rho}\frac{\partial p}{\partial r} \\ +\nu\left(\frac{\partial^2 u_r}{\partial r^2} + \frac{1}{r}\frac{\partial u_r}{\partial r} + \frac{1}{r^2}\frac{\partial^2 u_r}{\partial\theta^2} - \frac{u_r}{r^2} - \frac{2}{r^2}\frac{\partial u_\theta}{\partial\theta}\right)&. \end{aligned} \tag{18.2.32}$$

18.2.5 Solution of the boundary-layer flow

We solve equations (18.2.30), (18.2.31), and (18.2.32) for u_θ, u_r, and p. Three boundary conditions are imposed on the velocities, u_θ, u_r at $r = a$ and u_θ at $r = a+\delta$; these are the same as in Glauert's analysis. The fourth boundary condition is that the shear stress $\tau_{r\theta}^{BL}$

evaluated with the boundary-layer solution is equal to the shear stress $\tau^I_{r\theta}$ evaluated with the outer irrotational flow at $r = a + \delta$. The four boundary conditions are as follows:

$$Q\frac{a+\delta}{a} + u_\theta = q \quad \text{at} \quad r = a, \tag{18.2.33}$$

$$u_r = 0 \quad \text{at} \quad r = a, \tag{18.2.34}$$

$$Q + u_\theta = Q + 2U_0 \sin\theta \quad \text{at} \quad r = a + \delta; \tag{18.2.35}$$

$$\tau^{\text{BL}}_{r\theta} = \tau^I_{r\theta} \quad \text{at} \quad r = a + \delta. \tag{18.2.36}$$

Because we are considering one single fluid, the viscosity is the same inside and outside the boundary layer. The continuity of shear stress (18.2.36) is equivalent to continuity of velocity gradients. Glauert's boundary-layer equations can give solutions with only continuous velocity, but our new equations can give solutions with continuous velocity and velocity gradients. We use complex variables to solve the equations, and (18.2.35) is written as

$$u_\theta = Q\alpha(-i)e^{i\theta} \quad \text{at} \quad r = a + \delta. \tag{18.2.37}$$

Note that only the real part of the equation has physical significance.

We follow Glauert and expand the solution as a power series of α. A stream function can be written as

$$\psi = Q\left\{\alpha f_1(r)e^{i\theta} + \alpha^2\left[f_2(r)e^{2i\theta} + g_2(r)\right]\ldots\right\}, \tag{18.2.38}$$

and the velocities are

$$u_\theta = \frac{\partial\psi}{\partial r} = Q\left\{\alpha f_1'(r)e^{i\theta} + \alpha^2\left[f_2'(r)e^{2i\theta} + g_2'(r)\right]\ldots\right\}, \tag{18.2.39}$$

$$u_r = -\frac{1}{r}\frac{\partial\psi}{\partial\theta} = -\frac{Q}{r}\left[\alpha i f_1(r)e^{i\theta} + \alpha^2 2i f_2(r)e^{2i\theta} + \cdots +\right]. \tag{18.2.40}$$

Continuity equation (18.2.30) is automatically satisfied. The pressure is assumed to be

$$p = p_c + \rho Q^2\left\{\alpha s_1(r)e^{i\theta} + \alpha^2\left[s_2(r)e^{2i\theta} + t_2(r)\right]\ldots\right\}, \tag{18.2.41}$$

where p_c is a constant.

We evaluate $\tau^I_{r\theta}$ by using potential flow (18.2.26) and the irrotational rotary flow,

$$\tau^I_{r\theta} = -\frac{2\mu Q}{a+\delta} + \mu Q\alpha\frac{2i}{a+\delta}e^{i\theta}. \tag{18.2.42}$$

From the boundary-layer solutions, we obtain

$$\begin{aligned}\tau^{\text{BL}}_{r\theta} = &-\frac{2\mu Q}{a+\delta} + \mu Q\alpha\left[f_1'' - \frac{f_1'}{a+\delta} + \frac{f_1}{(a+\delta)^2}\right]e^{i\theta} \\ &+ \mu Q\alpha^2\left[\left(f_2''e^{2i\theta} + g_2''\right) - \frac{1}{a+\delta}\left(f_2'e^{2i\theta} + g_2'\right) + \frac{4}{(a+\delta)^2}f_2e^{2i\theta}\right] + \cdots +.\end{aligned} \tag{18.2.43}$$

Comparing the terms in (18.2.42) and (18.2.43) linear in α, we obtain

$$\frac{2i}{a+\delta} = f_1'' - \frac{f_1'}{a+\delta} + \frac{f_1}{(a+\delta)^2} \quad \text{at} \quad r = a + \delta. \tag{18.2.44}$$

Consideration of the terms quadratic in α gives

$$f_2'' - \frac{1}{a+\delta} f_2' + \frac{4}{(a+\delta)^2} f_2 = 0, \quad g_2'' - \frac{1}{a+\delta} g_2' = 0 \quad \text{at} \quad r = a+\delta. \tag{18.2.45}$$

In this study, one of the major objects is to determine the relation between the cylinder velocity q and the fluid circulatory velocity Q. In the expansion of boundary-layer velocities (18.2.39) and (18.2.40), Q is used as the fundamental parameter rather than q. As noted by Glauert, this approach is convenient for the study of the boundary-layer equations because the velocity at the outer edge of the boundary layer is completely specified. Although Q is an unknown quantity and q is prescribed, the relationship between q and Q can be established by means of (18.2.33), giving Q in terms of q. We also note that the boundary-layer thickness δ appears in the boundary conditions and it must be prescribed to obtain the solution. It can be expected that the boundary-layer thickness is a function of θ, but we are not able to determine the shape of the boundary layer. We assume that δ is a constant for given Re and q/U_0; it may be viewed as the average boundary-layer thickness. The choice of δ has significant effects on the solution and will be discussed later.

We insert (18.2.39), (18.2.40), and (18.2.41) into (18.2.31) and (18.2.32) and compare the coefficients of different powers of α to obtain ODEs for $f_1(r)$, $f_2(r)$, $g_2(r)\ldots$. The terms linear in α in (18.2.31) and (18.2.32) satisfy, respectively,

$$\frac{Q(a+\delta)}{r} f_1' = -Qs_1 + \frac{\nu}{i}\left(r f_1''' + f_1'' - \frac{2f_1'}{r} + \frac{2f_1}{r^2}\right), \tag{18.2.46}$$

$$\frac{Q(a+\delta)}{r^3} f_1 - \frac{2Q(a+\delta)}{r^2} f_1' = -Qs_1' + \frac{\nu}{i}\left(\frac{f_1''}{r} + \frac{1}{r^2} f_1' - \frac{1}{r^3} f_1\right). \tag{18.2.47}$$

After eliminating s_1 from (18.2.46) and (18.2.47), we obtain a fourth-order ODE for f_1,

$$\frac{Q(a+\delta)}{\nu} i \left(\frac{f_1''}{r} + \frac{f_1'}{r^2} - \frac{f_1}{r^3}\right) = r f_1'''' + 2 f_1''' - \frac{3}{r} f_1'' + \frac{3}{r^2} f_1' - \frac{3}{r^3} f_1, \tag{18.2.48}$$

where $Q(a+\delta)/\nu$ is a Reynolds number, and we write

$$k = \frac{\nu}{Q(a+\delta)}. \tag{18.2.49}$$

The solution of (18.2.48) is

$$f_1(r) = \frac{c_1}{r} + c_2 r^{2-\beta} + c_3 r^{2+\beta} + c_4 r, \tag{18.2.50}$$

where c_1, c_2, c_3, and c_4 are constants to be determined by boundary conditions and

$$\beta = \sqrt{1 + \frac{i}{k}}. \tag{18.2.51}$$

Three boundary conditions for f_1 are obtained from (18.2.33), (18.2.34), and (18.2.37)

$$f_1(a) = 0, \quad f_1'(a) = 0, \quad f_1'(a+\delta) = -i. \tag{18.2.52}$$

The fourth condition is the continuity of the shear stress (18.2.44), which can be written as

$$\frac{i}{a+\delta} = f_1'' + \frac{f_1}{(a+\delta)^2} \quad \text{at} \quad r = a+\delta. \tag{18.2.53}$$

With these four boundary conditions, we can determine c_1, c_2, c_3, and c_4 and the function f_1. The expression for f_1 is long and is not shown here.

After we obtain $f_1(r)$, we can compute $s_1(r)$ by using equation (18.2.46):

$$s_1 = -\frac{a+\delta}{r} f_1' + \frac{\nu}{Q}\zeta, \tag{18.2.54}$$

where

$$\zeta = \frac{1}{i}\left(r f_1''' + f_1'' - \frac{2}{r} f_1' + \frac{2}{r^2} f_1\right). \tag{18.2.55}$$

The two parts of s_1, $(-\frac{a+\delta}{r} f_1')$ and $\frac{\nu}{Q}\zeta$, come from the inertia term and viscous stress term in the momentum equation, respectively.

Next we carry out the calculation for terms quadratic in α. As pointed out by Glauert, care should be taken when computing product of two complex numbers A and B:

$$\mathrm{Re}(A)\mathrm{Re}(B) = \mathrm{Re}[A(B+\bar{B})/2], \tag{18.2.56}$$

where the overbar denotes a complex conjugate. We collect terms quadratic in α from (18.2.31) and they can be divided into two groups, terms proportional to $e^{2i\theta}$ and terms independent of θ. The two groups of terms satisfy the following equations, respectively:

$$\frac{1}{r}\frac{Qi}{\nu}\left[\frac{1}{2} f_1'^2 - \frac{1}{2} f_1 f_1'' - \frac{1}{2r} f_1 f_1' + \frac{2(a+\delta)}{r} f_2'\right] = -\frac{Qi}{\nu}\frac{2s_2}{r} + \left(f_2''' + \frac{f_2''}{r} - \frac{5f_2'}{r^2} + \frac{8f_2}{r^3}\right), \tag{18.2.57}$$

$$-\frac{1}{2r}\frac{Qi}{\nu}\left(f_1\overline{f_1''} + f_1'\overline{f_1'} + \frac{f_1}{r}\overline{f_1'}\right) = g_2''' + \frac{g_2''}{r} - \frac{g_2'}{r^2}. \tag{18.2.58}$$

Consideration of terms in (18.2.32) quadratic in α also yields two equations:

$$\frac{1}{r}\frac{Qi}{\nu}\left[\frac{1}{2r^2} f_1^2 - \frac{1}{2} f_1'^2 + \frac{4(a+\delta)}{r^2} f_2 - \frac{2(a+\delta)}{r} f_2'\right] = -\frac{Qi}{\nu} s_2' - \left(\frac{8}{r^3} f_2 - \frac{2}{r^2} f_2' - \frac{2}{r} f_2''\right), \tag{18.2.59}$$

$$\frac{1}{r}\left[\frac{1}{2r} f_1\overline{f_1'} + \frac{1}{2r} f_1'\overline{f_1} - \frac{1}{2r^2} f_1\overline{f_1} - \frac{1}{2} f_1'\overline{f_1'} - \frac{2(a+\delta)}{r} g_2'\right] = -t_2'. \tag{18.2.60}$$

We can first solve $g_2(r)$ from (18.2.58), then eliminate $s_2(r)$ from (18.2.57) and (18.2.59), solve for $f_2(r)$, and finally obtain $t_2(r)$ from (18.2.60).

Equation (18.2.58) is a third-order ODE for g_2. We prescribe the stream function at $r = a$ to be zero, which gives the condition

$$g_2(a) = 0. \tag{18.2.61}$$

Boundary condition (18.2.37) leads to

$$g_2'(a+\delta) = 0. \tag{18.2.62}$$

The continuity of shear stress (18.2.45) leads to

$$g_2''(a+\delta) = 0. \tag{18.2.63}$$

No condition can be applied to $g_2'(r{=}a)$, because it is known only that the surface velocity is independent of θ. Thus we have three boundary conditions, (18.2.61), (18.2.62), and

(18.2.63), for third-order ODE (18.2.58). A closed-form solution for $g_2(r)$ can be obtained, but it is long and tedious and is not shown here.

We eliminate s_2 from (18.2.57) and (18.2.59) and obtain a fourth-order ODE for f_2:

$$\begin{aligned}&\frac{Qi}{\nu}\left(-\frac{1}{2}f_1 f_1''' + \frac{1}{2}f_1' f_1'' - \frac{1}{2r}f_1 f_1'' + \frac{1}{2r}f_1' f_1' + \frac{1}{2r^2}f_1 f_1' - \frac{1}{r^3}f_1^2\right)\\ &= r f_2'''' + 2 f_2''' - \frac{9}{r}f_2'' + \frac{9}{r^2}f_2' - \frac{Q(a+\delta)i}{\nu}\left(\frac{2}{r}f_2'' + \frac{2}{r^2}f_2' - \frac{8}{r^3}f_2\right).\end{aligned} \tag{18.2.64}$$

Three boundary conditions for $f_2(r)$ are obtained from (18.2.33), (18.2.34), and (18.2.37),

$$f_2(a) = 0, \quad f_2'(a) = 0, \quad f_2'(a+\delta) = 0, \tag{18.2.65}$$

and the fourth condition comes from the continuity of shear stress (18.2.45):

$$f_2'' + \frac{4}{(a+\delta)^2}f_2 = 0 \quad \text{at} \quad r = a+\delta. \tag{18.2.66}$$

Equation (18.2.64) and the boundary conditions are solved by numerical integration.

After $f_2(r)$ is obtained, we compute $s_2(r)$ by using (18.2.57):

$$s_2 = -\frac{1}{4}f_1'^2 + \frac{1}{4}f_1 f_1'' + \frac{1}{4r}f_1 f_1' - \frac{a+\delta}{r}f_2' + \frac{\nu}{2Q}\xi, \tag{18.2.67}$$

where

$$\xi = \frac{1}{i}\left(r f_2''' + f_2'' - \frac{5}{r}f_2' + \frac{8}{r^2}f_2\right). \tag{18.2.68}$$

The function s_2 can be divided into two parts, the term $\frac{\nu}{2Q}\xi$ comes from the viscous stress, and other terms in s_2 come from the inertia terms in the momentum equation.

The last step in the calculation of terms quadratic in α is to integrate (18.2.60) to obtain $t_2(r)$. There will be an undetermined constant in the process of integration, which can be absorbed into the pressure constant p_c.

With the functions s_1 and s_2, we can write the pressure as

$$\begin{aligned}p &= p_c + \rho Q^2\alpha s_1 e^{i\theta} + \rho Q^2\alpha^2\left(s_2 e^{2i\theta} + t_2\right) + O(\alpha^3)\\ &= p_c + 2\rho U_0 Q\left(-\frac{a+\delta}{r}f_1' + \frac{\nu}{Q}\zeta\right)e^{i\theta}\\ &\quad + 4\rho U_0^2\left[\left(-\frac{1}{4}f_1'^2 + \frac{1}{4}f_1 f_1'' + \frac{1}{4r}f_1 f_1' - \frac{a+\delta}{r}f_2' + \frac{\nu}{2Q}\xi\right)e^{2i\theta} + t_2\right] + O(\alpha^3).\end{aligned} \tag{18.2.69}$$

The pressure at the outer edge of the boundary layer is of interest because it can be compared with the irrotational pressure (18.2.29) at $r = a+\delta$, and the difference between them gives the pressure correction. From (18.2.52) $f_1'(a+\delta) = -i$ and from (18.2.65) $f_2'(a+\delta) = 0$, the pressure at $r = a+\delta$ is

$$p = p_c + 2\rho U_0 Q\left(i + \frac{\nu}{Q}\zeta\right)e^{i\theta} + \rho U_0^2\left[\left(1 + f_1 f_1'' - \frac{i}{a+\delta}f_1 + \frac{2\nu}{Q}\xi\right)e^{2i\theta} + 4t_2\right] + O(\alpha^3). \tag{18.2.70}$$

The real part of the preceding equation is

$$p = \mathrm{Re}(p_c) + 4\rho U_0^2 \,\mathrm{Re}(t_2) - 2\rho U_0 Q \sin\theta + \rho U_0^2 \cos 2\theta + \rho U_0^2 \,\mathrm{Re}\left[\left(f_1 f_1'' - \frac{i}{a+\delta} f_1\right) e^{2i\theta}\right]$$
$$+ 2\mu U_0 \left[\mathrm{Re}(\zeta)\cos\theta - \mathrm{Im}(\zeta)\sin\theta\right] + 2\mu \frac{U_0^2}{Q}\left[\mathrm{Re}(\xi)\cos 2\theta - \mathrm{Im}(\xi)\sin 2\theta\right] + O(\alpha^3). \tag{18.2.71}$$

Because the radial component u_r of the velocity is small in the boundary layer, we may neglect f_1, and then compare (18.2.71) with the irrotational pressure (18.2.29). The terms $-2\rho U_0 Q \sin\theta$ and $\rho U_0^2 \cos 2\theta$ are the same in the two pressure expressions, and the terms proportional to μ in (18.2.71) are the extra pressures arising in the boundary layer. This comparison demonstrates that the inertia terms in the momentum equations give rise to the irrotational pressure and the viscous stress terms give rise to a viscous pressure correction. In general, the pressure correction can be expanded as a Fourier series:

$$p_v = \sum_{m=0}^{\infty} \left[h_m(r)\cos m\theta + j_m(r)\sin m\theta\right].$$

Here we determine the coefficients of $\sin\theta$, $\cos\theta$, $\sin 2\theta$, and $\cos 2\theta$ up to $O(\alpha^2)$ terms. These coefficients may be modified, and more coefficients in the Fourier series can be obtained if calculations for $O(\alpha^3)$ terms are carried out. The $\cos\theta$ and $\sin\theta$ terms in the pressure correction contribute to the drag and lift, respectively.

Up to this point, our solutions are in terms of the fluid circulatory velocity Q. We solve for Q in terms of the prescribed quantities by using an iterative method. There are two prescribed dimensionless parameters in this problem, the Reynolds number Re and the speed ratio q/U_0. Our first guess of Q comes from the irrotational purely rotary flow,

$$Q^{(1)} = q\frac{a}{a+\delta} \quad \Rightarrow \quad \frac{Q^{(1)}}{q} = \frac{1}{1+\delta/a},$$

where the superscript (1) indicates the value for Q in the first iteration. By use of $Q^{(1)}/q$, the value of k is computed in equation (18.2.49) and $f_1(r)$ is subsequently obtained. Then we solve for $g_2(r)$ from equation (18.2.58) and obtain g_2'. The velocity u_θ at $r = a$ is then

$$u_\theta = Q^{(1)}\alpha^2 \,\mathrm{Re}\{g_2'[r=a,\ Q^{(1)}/q]\} + O(\alpha^3). \tag{18.2.72}$$

Inserting (18.2.72) into (18.2.33), we obtain

$$\frac{Q^{(2)}}{q}\frac{a+\delta}{a} + 4\frac{U_0^2}{q^2}\frac{q}{Q^{(1)}}\mathrm{Re}\{g_2'[r=a,\ Q^{(1)}/q]\} = 1, \tag{18.2.73}$$

from which we can solve for $Q^{(2)}$, which is the value for Q in the second iteration. We repeat the calculation, using $Q^{(2)}$ to obtain the value for Q in the next iteration, until the value of Q converges.

The functions s_1, f_2, s_2, and t_2 are computed from the procedure just described, and the solutions of the boundary-layer equations are determined up to $O(\alpha^2)$. We can compute the pressure and shear stress at the cylinder surface and integrate to obtain the drag, lift, and torque. The drag and lift by the pressure are

$$D_p = \int_A \mathbf{e}_x \cdot (-P\mathbf{1}) \cdot \mathbf{e}_r \, dA = \int_0^{2\pi} (-P)\cos\theta \, a d\theta = -\rho Q^2 \alpha \, \mathrm{Re}(s_1)\pi a, \tag{18.2.74}$$

$$L_p = \int_A \mathbf{e}_y \cdot (-P\mathbf{1}) \cdot \mathbf{e}_r \, dA = \int_0^{2\pi} (-P)\sin\theta \, a d\theta = \rho Q^2 \alpha \, \mathrm{Im}(s_1)\pi a. \tag{18.2.75}$$

The friction drag and lift by the shear stress are

$$D_f = \int_A \mathbf{e}_x \cdot (\tau_{\theta r}\mathbf{e}_\theta \mathbf{e}_r) \cdot \mathbf{e}_r \, dA = \int_0^{2\pi} \tau_{\theta r}(-\sin\theta) \, a d\theta = \mu Q \alpha \, \mathrm{Im}(f_1'')\pi a, \tag{18.2.76}$$

$$L_f = \int_A \mathbf{e}_y \cdot (\tau_{\theta r}\mathbf{e}_\theta \mathbf{e}_r) \cdot \mathbf{e}_r \, dA = \int_0^{2\pi} \tau_{\theta r}\cos\theta \, a d\theta = \mu Q \alpha \, \mathrm{Re}(f_1'')\pi a. \tag{18.2.77}$$

We call the readers' attention to the fact that, in our problem, the drag on the cylinder is negative if it is in the uniform flow direction; the drag is positive if it is opposite to the uniform flow direction (see figure 18.4). The drag and lift coefficients are defined as

$$C_{D_p} = \frac{D_p}{\rho U_0^2 a}, \qquad C_{D_f} = \frac{D_f}{\rho U_0^2 a}, \qquad C_D = \frac{D_p + D_f}{\rho U_0^2 a}; \tag{18.2.78}$$

$$C_{L_p} = \frac{L_p}{\rho U_0^2 a}, \qquad C_{L_f} = \frac{L_f}{\rho U_0^2 a}, \qquad C_L = \frac{L_p + L_f}{\rho U_0^2 a}. \tag{18.2.79}$$

The torque is

$$T = -a^2 \int_0^{2\pi} \tau_{\theta r} \, d\theta = 2\pi\mu a^2 \left(2Q\frac{a+\delta}{a^2} - Q\alpha^2 g_2'' + \frac{Q\alpha^2}{a} g_2' \right), \tag{18.2.80}$$

with the dimensionless torque defined as

$$C_T = \frac{T}{2\rho U_0^2 a^2}. \tag{18.2.81}$$

In tables 18.5–18.10, we list the drag, lift, and torque computed from our boundary-layer solutions and compare them with the results of numerical simulation for six cases: $(Re, q/U_0) = (200, 4), (200, 5), (400, 4), (400, 5), (400, 6)$, and $(1000, 3)$. The boundary-layer thickness δ/a is prescribed at different values; when the value of δ/a falls into a certain range (boldfaced in tables 18.5–18.10), our analysis gives rise to lift and torque in good agreement with the simulation results. The agreement for the drag is less good, which is partly because the absolute value of the drag is small and the relative error is apparent. Nevertheless, good agreement for the drag is obtained in the cases $(Re, q/U_0) = (400, 5)$ and $(400, 6)$, which are the ones with relatively large values of Re and q/U_0 in the six cases. This indicates that the agreement for the drag becomes better as the prescribed parameters move toward the range in which the theory is supposed to work better. Our solution for $(Re, q/U_0) = (400, 6)$ obtained with $\delta/a = 0.14$ (see table 18.9) is in excellent agreement with the results of numerical simulation.

We highlight the range of δ/a in which the lift and torque are in good agreement with the simulation results in tables 18.5–18.10. We choose one value from this range (typically the median) as a proper boundary-layer thickness: $\delta/a = 0.25, 0.21, 0.17, 0.15, 0.14$, and 0.12 for $(Re, q/U_0) = (200, 4), (200, 5), (400, 4), (400, 5), (400, 6)$, and $(1000, 3)$, respectively.

Table 18.5. *The comparison of the coefficients for the drag, lift, and torque with the simulation results for Re = 200 and $q/U_0 = 4$*

The lift and torque computed with $\delta/a = 0.24$, 0.25, or 0.26 are in reasonable agreement with the results of numerical simulation. The drag, especially the drag that is due to the pressure, does not agree well with the simulation results. When $\delta/a = 0.28$, the value of q/Q is such that $\alpha = 2U_0/Q > 1$, which makes the power-series expansions of the solutions in terms of α divergent. The calculation can be performed but cannot be expected to converge to the true result.

δ/a	q/Q	C_{D_p}	C_{D_f}	C_D	C_{L_p}	C_{L_f}	C_L	C_T
0.1	1.114	−13.551	−1.388	−14.939	21.563	0.414	21.978	0.277
0.15	1.221	−5.090	−1.010	−6.100	21.303	0.587	21.891	0.328
0.2	1.409	−1.669	−0.848	−2.517	20.153	0.697	20.850	0.400
0.23	1.580	−0.620	−0.788	−1.408	18.938	0.725	19.663	0.446
0.24	**1.649**	**−0.381**	**−0.770**	**−1.152**	**18.444**	**0.729**	**19.173**	**0.461**
0.25	**1.726**	**−0.188**	**−0.754**	**−0.942**	**17.907**	**0.729**	**18.636**	**0.476**
0.26	**1.812**	**−0.0351**	**−0.737**	**−0.772**	**17.329**	**0.727**	**18.056**	**0.490**
0.27	1.907	0.0815	−0.721	−0.639	16.710	0.722	17.432	0.504
0.28	2.012	0.166	−0.704	−0.538	16.055	0.714	16.769	0.517
Simulation results		0.728	−0.604	0.124	16.961	0.621	17.582	0.453

As expected, the boundary-layer thickness decreases with increasing Reynolds number and the relation $(\delta/a) \propto (1/\sqrt{Re})$ seems to hold when q/U_0 is fixed. The boundary-layer thickness also decreases with increasing q/U_0, because the rotary flow suppresses the boundary layer induced by the streaming flow.

The choice of δ/a is vital in our calculation. If δ/a is much smaller than the proper boundary-layer thickness, the flow there cannot match the potential flow outside, which breaks the assumptions of our calculation. If δ/a is much larger than the proper boundary-layer thickness, the value of Q is small and $\alpha = 2U_0/Q$ could be close to 1 or even larger

Table 18.6. *The comparison of the coefficients for the drag, lift, and torque with the simulation results for Re = 200 and $q/U_0 = 5$*

The lift and torque computed with $\delta/a = 0.2$, 0.21, or 0.22 are in excellent agreement with the results of numerical simulation. The agreement of drag, especially the drag that is due to the pressure, is not good. When $\delta/a = 0.35$, $\alpha = 2U_0/Q > 1$, and the power-series expansions of the solutions in terms of α are divergent.

δ/a	q/Q	C_{D_p}	C_{D_f}	C_D	C_{L_p}	C_{L_f}	C_L	C_T
0.1	1.114	−13.061	−1.400	−14.461	27.039	0.517	27.556	0.346
0.15	1.218	−4.157	−1.044	−5.201	27.007	0.730	27.736	0.407
0.19	1.350	−0.988	−0.928	−1.916	26.414	0.845	27.259	0.473
0.2	**1.392**	**−0.468**	**−0.909**	**−1.377**	**26.164**	**0.866**	**27.029**	**0.490**
0.21	**1.437**	**−0.0296**	**−0.893**	**−0.922**	**25.870**	**0.883**	**26.752**	**0.507**
0.22	**1.486**	**0.337**	**−0.878**	**−0.541**	**25.531**	**0.896**	**26.428**	**0.524**
0.25	1.659	1.089	−0.841	0.248	24.243	0.920	25.163	0.574
0.3	2.072	1.531	−0.779	0.752	21.111	0.903	22.014	0.650
0.35	2.804	1.318	−0.690	0.628	16.512	0.813	17.324	0.714
Simulation results		0.824	−0.835	−0.0107	26.183	0.846	27.029	0.514

Table 18.7. *The comparison of the coefficients for the drag, lift, and torque with the simulation results for Re = 400 and* $q/U_0 = 4$

The lift and torque computed with $\delta/a = 0.17$, or 0.18 are in excellent agreement with the results of numerical simulation. The agreement of drag, especially the drag that is due to the pressure, is not good. When $\delta/a = 0.23$, $\alpha = 2U_0/Q > 1$, and the power-series expansions of the solutions in terms of α are divergent.

δ/a	q/Q	C_{D_p}	C_{D_f}	C_D	C_{L_p}	C_{L_f}	C_L	C_T
0.1	1.155	−5.629	−0.724	−6.352	21.104	0.396	21.499	0.173
0.13	1.277	−2.249	−0.610	−2.859	20.292	0.470	20.762	0.215
0.15	1.398	−1.008	−0.565	−1.572	19.357	0.498	19.855	0.246
0.16	1.472	−0.582	−0.547	−1.129	18.766	0.506	19.272	0.261
0.17	**1.558**	**−0.258**	**−0.530**	**−0.788**	**18.094**	**0.509**	**18.603**	**0.276**
0.18	**1.657**	**−0.0178**	**−0.514**	**−0.532**	**17.343**	**0.509**	**17.852**	**0.291**
0.2	1.906	0.260	−0.483	−0.223	15.613	0.497	16.110	0.318
0.23	2.466	0.316	−0.432	−0.116	12.551	0.453	13.004	0.352
Simulation results		0.534	−0.451	−0.0836	17.609	0.447	18.057	0.275

Table 18.8. *The comparison of the coefficients for the drag, lift, and torque with the simulation results for Re = 400 and* $q/U_0 = 5$

The lift and torque computed with $\delta/a = 0.14$ or 0.15 and the drag computed with $\delta/a = 0.16$ are in good agreement with the results of numerical simulation. When $\delta/a = 0.25$, $\alpha = 2U_0/Q > 1$, and the power-series expansions of the solutions in terms of α divergent.

δ/a	q/Q	C_{D_p}	C_{D_f}	C_D	C_{L_p}	C_{L_f}	C_L	C_T
0.1	1.153	−4.803	−0.746	−5.550	26.670	0.491	27.187	0.215
0.13	1.266	−1.190	−0.650	−1.840	26.140	0.582	26.722	0.263
0.14	**1.315**	**−0.457**	**−0.631**	**−1.089**	**25.816**	**0.602**	**26.419**	**0.280**
0.15	**1.370**	**0.112**	**−0.617**	**−0.504**	**25.420**	**0.618**	**26.038**	**0.297**
0.16	**1.431**	**0.548**	**−0.604**	**−0.056**	**24.952**	**0.630**	**25.582**	**0.314**
0.2	1.750	1.386	−0.565	0.821	22.366	0.641	23.007	0.374
0.25	2.502	1.313	−0.493	0.820	16.923	0.577	17.500	0.440
Simulation results		0.591	−0.601	−0.010	26.415	0.597	27.011	0.297

Table 18.9. *The comparison of the coefficients for the drag, lift, and torque with the simulation results for Re = 400 and* $q/U_0 = 6$

The drag, lift, and torque computed with $\delta/a = 0.14$ are in excellent agreement with the results of numerical simulation. The calculation is reasonable accurate in the range $0.135 \leq \delta/a \leq 0.145$.

δ/a	q/Q	C_{D_p}	C_{D_f}	C_D	C_{L_p}	C_{L_f}	C_L	C_T
0.1	1.151	−3.860	−0.773	−4.633	32.48	0.585	33.06	0.256
0.12	1.216	−1.031	−0.714	−1.744	32.48	0.660	33.14	0.290
0.13	1.254	−0.0650	−0.697	−0.762	32.42	0.690	33.11	0.308
0.135	**1.274**	**0.331**	**−0.691**	**−0.361**	**32.37**	**0.703**	**33.07**	**0.316**
0.14	**1.296**	**0.676**	**−0.686**	**−0.0105**	**32.31**	**0.714**	**33.03**	**0.325**
0.145	**1.317**	**0.975**	**−0.682**	**0.293**	**32.24**	**0.725**	**32.97**	**0.333**
0.15	1.340	1.233	−0.679	0.554	32.16	0.734	32.90	0.342
0.2	1.577	2.318	−0.678	1.641	31.31	0.788	32.10	0.408
Simulation results		0.668	−0.681	−0.0136	33.09	0.682	33.77	0.316

Table 18.10. *The comparison of the coefficients for the drag, lift, and torque with the simulation results for Re = 1000 and* $q/U_0 = 3$

The lift and torque computed with $\delta/a = 0.1$ or 0.12 are close to the results of numerical simulation. However, it should be noted that $\alpha = 1.17 > 1$ when $\delta/a = 0.12$ and the power-series expansions of the solutions in terms of α divergent. This is caused by the relatively low value of the speed ratio $q/U_0 = 3$. If Glauert's solution is used for this case, $\alpha = 1.064$ and Glauert's solution also diverges.

δ/a	q/Q	C_{D_p}	C_{D_f}	C_D	C_{L_p}	C_{L_f}	C_L	C_T
0.08	1.217	−3.185	−0.362	−3.547	14.72	0.220	14.94	0.0801
0.1	1.418	−1.509	−0.310	−1.819	13.20	0.237	13.44	0.106
0.12	**1.755**	**−0.755**	**−0.273**	**−1.028**	**11.08**	**0.232**	**11.31**	**0.130**
Simulation results		0.213	−0.197	0.0155	10.41	0.192	10.60	0.118

than 1, which makes the power-series expansion of the solutions in terms of α slow to converge or even divergent. On the other hand, there is a range of δ/a values that can lead to lift and torque in good agreement with simulation results, because there is no clear-cut boundary-layer edge physically. The calculation is reasonably accurate when δ/a falls in this range.

We compare the drag, lift, and torque given by our solution we obtain by using the proper δ/a, by Glauert's solution, by Moore's solution, and by the numerical simulation in table 18.11. Equations (18.2.3), (18.2.4), and (18.2.7) are used to compute the lift, drag, and torque coefficients given by Glauert's solution. Equation (18.2.10) is used to compute the torque given by Moore's solution; the drag and lift are not computed because Moore did not give the necessary coefficients. The comparison demonstrates that Moore's torque is relatively close to the simulation results, and Glauert's solution gives reasonable approximations for the friction drag and lift but poor approximation for the torque. It also confirms that our solution is indeed an improvement of Glauert's solution, especially in the category of torque.

A key feature of this boundary-layer analysis is that the variation of the pressure across the boundary layer is obtained. We integrate the drag and lift components of the pressure over circles concentric with the cylinder but with different radii; then C_{D_p} and C_{L_p} become functions of r. We compare these functions computed from our boundary-layer analysis and from numerical simulation in figures 18.5 and 18.6.

The functions $C_{D_p}(r)$ for three cases, $Re = 400$ and $q/U_0 = 4$, 5, and 6 are shown in figures 18.5(a), 18.5(b), and 18.5(c), respectively. Two curves, computed from our boundary-layer analysis with different values of δ/a, are compared with the numerical simulation for $(Re, q/U_0) = (400, 4)$ in figure 18.5(a). The dashed curve gives the results with $\delta/a = 0.17$, which is the boundary-layer thickness leading to the best fit for the lift and torque (see table 18.7). The dashed curve correctly predicts that C_{D_p} decreases with increasing r, but the values of C_{D_p} are not close to the results of numerical simulation. The solid curve gives the results with $\delta/a = 0.2$, which are much closer to the simulation results and correctly predict that C_{D_p} changes sign across the boundary layer. Figure 18.5(b) shows the comparison for the case $(Re, q/U_0) = (400, 5)$. Again, the dashed curve gives the results from our boundary-layer analysis we obtained by using the value of δ/a

Table 18.11. *The comparison of the solution in this work, using $\delta/a = 0.25, 0.21, 0.17, 0.15, 0.14$ and 0.12 for $(Re, q/U_0) = (200, 4), (200, 5), (400, 4), (400, 5), (400, 6)$ and $(1000, 3)$, respectively, with the simulation results and Glauert's and Moore's solutions*

Note that, in our problem, the drag on the cylinder is negative if it is in the uniform flow direction; the drag is positive if it is opposite to the uniform flow direction (see figure 18.4). We call the readers' attention to the fact that $\alpha > 1$ in our solution and in Glauert's solution when $(Re, q/U_0) = (1000, 3)$; the solutions are not expected to converge to the true results.

Solution	Re	q/U_0	α	C_{D_p}	C_{D_f}	C_D	C_{L_p}	C_{L_f}	C_L	C_T
Numerical simulation	200	4	–	0.728	−0.604	0.124	16.961	0.621	17.582	0.453
This work	200	4	0.863	−0.188	−0.754	−0.942	17.907	0.729	18.636	0.476
Glauert's solution	200	4	0.625	0	−0.795	−0.795	20.102	0.795	20.897	0.215
Moore's solution	200	4	–	–	–	–	–	–	–	0.408
Numerical simulation	200	5	–	0.824	−0.835	−0.0107	26.183	0.846	27.029	0.514
This work	200	5	0.575	−0.0296	−0.893	−0.922	25.870	0.883	26.752	0.507
Glauert's solution	200	5	0.457	0	−0.929	−0.929	27.483	0.929	28.412	0.195
Moore's solution	200	5	–	–	–	–	–	–	–	0.440
Numerical simulation	400	4	–	0.534	−0.451	−0.0836	17.609	0.447	18.057	0.275
This work	400	4	0.779	−0.258	−0.530	−0.788	18.094	0.509	18.603	0.277
Glauert's solution	400	4	0.625	0	−0.562	−0.562	20.102	0.562	20.664	0.152
Moore's solution	400	4	–	–	–	–	–	–	–	0.237
Numerical simulation	400	5	–	0.591	−0.601	−0.010	26.415	0.597	27.011	0.297
This work	400	5	0.548	0.112	−0.617	−0.504	25.420	0.618	26.038	0.297
Glauert's solution	400	5	0.457	0	−0.657	−0.657	27.483	0.657	28.140	0.138
Moore's solution	400	5	–	–	–	–	–	–	–	0.246
Numerical simulation	400	6	–	0.668	−0.681	−0.0136	33.09	0.682	33.77	0.316
This work	400	6	0.432	0.676	−0.686	−0.0105	32.31	0.714	33.03	0.325
Glauert's solution	400	6	0.365	0	−0.736	−0.736	34.46	0.736	35.20	0.126
Moore's solution	400	6	–	–	–	–	–	–	–	0.263
Numerical simulation	1000	3	–	0.213	−0.197	0.0155	10.41	0.192	10.60	0.118
This work	1000	3	1.17	−0.755	−0.273	−1.028	11.08	0.232	11.31	0.130
Glauert's solution	1000	3	1.06	0	−0.273	−0.273	11.81	0.273	12.08	0.108
Moore's solution	1000	3	–	–	–	–	–	–	–	0.131

leading to the best fit for the lift and torque (see table 18.8). The solid curve gives the results obtained with a larger δ/a, which are in excellent agreement with the simulation results. Figure 18.5(c) shows the comparison for the case $(Re, q/U_0) = (400, 6)$. We plot only one curve from the boundary-layer analysis, using $\delta/a = 0.14$. This value leads to not only the best fit for the lift and torque (see table 18.9), but also excellent agreement for C_{D_p} in figure 18.5(c). This comparison demonstrates that our boundary-layer analysis can be used to compute the variation of the pressure drag across the boundary layer and the agreement with the numerical simulation becomes better as q/U_0 increases.

The functions $C_{L_p}(r)$ for three cases, $Re = 400$ and $q/U_0 = 4$, 5, and 6 are shown in figure 18.6. In all the three cases, C_{L_p} computed from our boundary-layer analysis, are in excellent agreement with the numerical simulation. The theory correctly predicts the variation of C_{L_p} with r inside the boundary layer, which is a significant improvement on

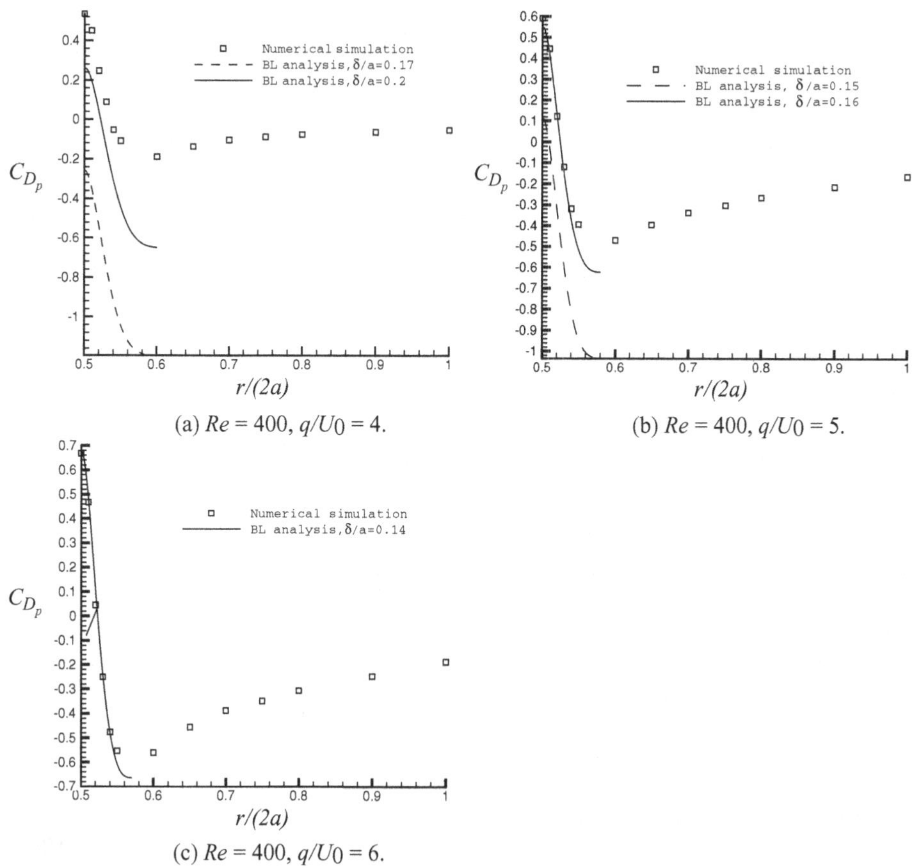

Figure 18.5. Comparison of the coefficient for the pressure drag C_{D_p} as a function of the radial position. (a) $Re = 400$, $q/U_0 = 4$. Our boundary-layer analysis, dashed curve, with $\delta/a = 0.17$; solid curve, with $\delta/a = 0.2$. The results of numerical simulation, □. (b) $Re = 400$, $q/U_0 = 5$. Our boundary-layer analysis, dashed curve, with $\delta/a = 0.15$; solid curve, with $\delta/a = 0.16$. The results of numerical simulation, □. (c) $Re = 400$, $q/U_0 = 6$. Our boundary-layer analysis with $\delta/a = 0.14$, solid curve. The results of numerical simulation, □. C_{D_p} from our boundary-layer analysis can be computed only inside the boundary layer: $a \leq r \leq a + \delta$; C_{D_p} from numerical simulation is plotted up to $r = 2a$.

the irrotational theory and the classical boundary-layer theory of Prandtl. The lift force $L = \rho U_0 \Gamma$ from the irrotational theory is a constant at any $r \geq a$ because the circulation is a constant. In the classical boundary-layer theory, the pressure is a constant across the boundary layer, and the variation of $C_{L_p}(r)$ shown in figure 18.6 cannot be obtained.

18.2.6 Higher-order boundary-layer theory

Glauert's analysis is a first-order boundary-layer approximation for the flow past a rotating cylinder. Our analysis here is intended to be an improvement of his boundary-layer solution. Another possible way to improve Glauert's solution is the higher-order boundary-layer theory based on the method of matched asymptotic expansions (Lagerstrom and

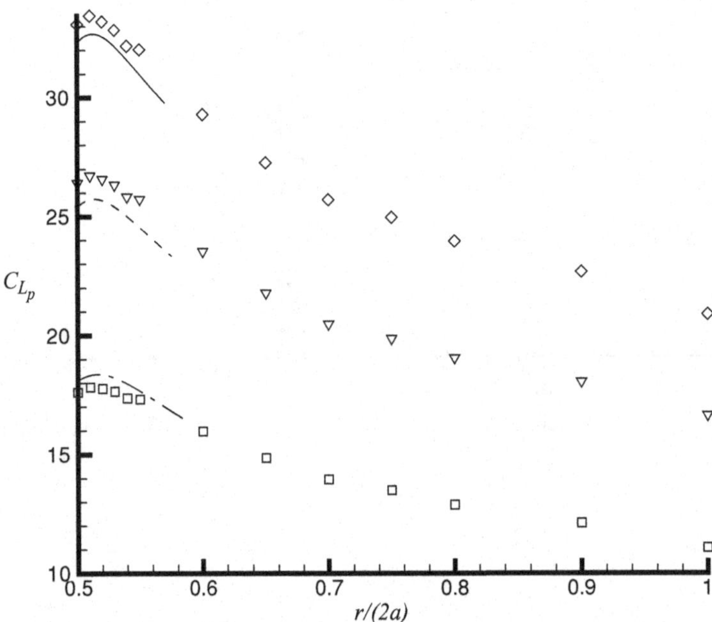

Figure 18.6. Comparison of the coefficient for the pressure lift C_{L_p} as a function of the radial position for $Re = 400$. Our boundary-layer analysis, dash–dotted line, $q/U_0 = 4$; dashed line, $q/U_0 = 5$; solid curve, $q/U_0 = 6$. Numerical simulation, $\square$, $q/U_0 = 4$; ∇, $q/U_0 = 5$; $\diamond$, $q/U_0 = 6$. C_{L_p} from our boundary-layer analysis can be computed only inside the boundary layer: $a \leq r \leq a + \delta$; C_{L_p} from numerical simulation is plotted up to $r = 2a$.

Cole, 1955; Van Dyke, 1962a, 1969; Maslen, 1963). We discuss the differences between our approach and the higher-order boundary-layer theory.

The basic idea of the higher-order boundary-layer theory is to construct outer and inner asymptotic expansions by iterating the Navier–Stokes equations about the outer solution and about the boundary-layer solution, respectively, and to match the two expansions in their overlap region of validity. Tani (1977) remarked "Higher approximations have thus been found only for flows without separation. In such cases the first term of the outer expansion is the inviscid irrotational flow, from which the first term of the inner expansion is determined by Prandtl's approximation. The second term of the outer expansion is the irrotational flow due to an apparent source distribution representing the displacement effect of Prandtl's boundary layer. This then determines a correction to the boundary-layer solution, yielding the second term of the inner expansion." The second-order corrections are terms proportional to $1/\sqrt{Re}$ and the third-order terms are proportional to $1/Re$. Because the viscous term in the Navier–Stokes equations for the outer flow is of the order of $1/Re$, the higher-order theory needs to compute the third-order corrections to account for the viscous effects of the outer flow. Van Dyke (1969) remarked "Definite results in the literature are restricted mostly to laminar boundary layer, to steady motion, to plane or axisymmetric flows, and to the second approximation." We are not aware of any third-order corrections in the literature. For incompressible fluids, the first- and second-order terms for the outer expansion are irrotational (Panton 1984, Tani 1977); it is not clear whether the third-order term is irrotational or not. Suppose the outer flow is irrotational at all orders and the fluid is incompressible; the viscous term $\mu\nabla^2\mathbf{u}$ disappears identically, which indicates that the viscous effects of the outer potential flow do not enter the higher-order boundary-layer theory if only velocity is matched but stress is not considered. Suppose the third-order term for the outer flow is rotational; the viscous term is then proportional to $1/Re$, which should give viscous effects to the inner solution at the third order.

Our new approach to boundary-layer flow is different from the higher-order boundary layer theory and is not based on the method of matched asymptotic expansions. The matching conditions at the outer edge of the boundary layer are for the velocity in higher-order boundary-layer theory; shear stress has not been considered. We enforce the continuity of the shear stress at the outer edge of the boundary layer. Because we are considering one single fluid, the viscosity is the same inside and outside the boundary layer. The continuity of the shear stress is equivalent to continuity of velocity gradients. Because the velocity gradients for the outer flow are of order 1, our approach is not the same as the third-order corrections of the higher-order boundary theory.

Glauert's analysis is a first-order boundary-layer approximation. He ignored the irrotational rotary flow component of the flow in the boundary layer, which is justifiable because the irrotational rotary flow is a second-order effect in the boundary layer. The torque coefficient by Glauert is of the order of $1/\sqrt{Re}$, and the torque coefficient in a purely irrotational rotary flow (without forward flow) is of the order of $1/Re$. However, numerical simulation shows that the higher-order correction is not negligible in this case. When $Re = 400$ and $q/U_0 = 6$, the torque coefficient by Glauert is only 40% of the result of numerical simulation. It is conceivable that higher-order boundary-layer theory using the method of matched asymptotic expansions can be applied to this problem and yield corrections for Glauert's solution; but this has not been done. Our analysis is compared with the numerical simulation and good to excellent agreement is observed. Admittedly, the outer flow in our analysis is a first-order approximation and can be improved by higher-order corrections.

The pressure across the boundary layer can no longer be taken as a constant in higher-order theory. On a curved wall, centrifugal force produces a pressure gradient across the boundary layer, which is a second-order effect. Van Dyke (1969) inserted the irrotational surface speed with a correction that is due to the surface curvature into Bernoulli's equation for the external flow to compute the pressure at the outer edge of the boundary layer and it has no viscous terms. The pressure inside the boundary layer can be computed with this condition and the equation

$$\frac{\partial \overline{p}_2}{\partial n} = \kappa \overline{u}_1^2, \tag{18.2.82}$$

where $\overline{p}_2$ is the second-order correction for the pressure, n is normal to the surface, κ is the surface curvature, and $\overline{u}_1$ is the first-order velocity from Prandtl's boundary-layer theory. Because $\overline{u}_1$ has viscous terms, the pressure at the wall computed from (18.2.82) should have viscous terms. However, in the applications of the higher-order theory to problems of leading edges and parabola in uniform stream by Van Dyke (1962b, 1964), the second-order correction for the pressure was not computed; the drag is computed only with skin friction and the pressure is not considered.

In summary, the higher-order boundary-layer theory has not yet been applied to determine (1) the effect of the viscous dissipation of the outer irrotational flow; (2) the effect generated by a mismatch between the shear stress at the effective edge of the boundary layer and the irrotational shear stress there; and (3) the drag and lift on the body that are due to normal stress associated with the viscous contribution to the pressure.

The numerical simulations show that the region in which the vortical effects are important is thick around the rotating cylinder. For example, the thickness of the vortical region determined with 1% of the maximum vorticity magnitude criterion is 26% of the cylinder radius for $Re = 400$ and $q/U_0 = 5$. The higher-order boundary-layer theory might encounter difficulty when treating such problems. Weinbaum *et al.* (1976) proposed an approximate method, which is not based on asymptotic analysis, to improve Prandtl's boundary-layer theory. They focused on flows with the Reynolds number range $O(1) < Re < O(10^2)$, where the boundary layer is thick and a steady laminar wake is present. They remarked "It is not surprising in view of the large changes in effective body shape which the external inviscid flow must experience at these Reynolds numbers that a theory of successive approximation which is based on the potential flow past the original body shape will converge very slowly. This would appear to be the basic difficulty encountered in extending the results of second-order boundary-layer theory (Van Dyke, 1962) to flows with Reynolds numbers less than about 10^3." The method of Weinbaum *et al.* is based on a pressure hypothesis that enables one to take into account the displacement interaction and centrifugal effects in thick boundary layers by using conventional first-order boundary-layer equations. Weinbaum *et al.* neglected the viscous term in the pressure that we mentioned in the discussion of equation (18.2.82). They solved the momentum integral of the boundary-layer equations by using the fourth-order Pohlhausen profile to obtain the displacement thickness. They treated the flows past parabolic and circular cylinders and obtained results in excellent agreement with numerical Navier–Stokes solutions. The method of Weinbaum *et al.* shares a common feature with ours in that the boundary-layer thickness has to be taken into account in the solution. However, like the higher-order boundary-layer theory, the method of Weinbaum *et al.* does not consider the shear stress discrepancy at the effective edge of the boundary layer or the viscous contribution to the pressure.

18.2.7 Discussion and conclusion

The dependence of the lift on Re and q/U_0 is a key problem in the study of the flow past a rotating cylinder. Our work here and numerical simulations (Mittal and Kumar, 2003; Padrino and Joseph 2006) show clearly that the lift force increases with increasing q/U_0; the major contribution to the lift is from the pressure and the friction lift is much smaller than the pressure lift. The numerical simulation results show that the influence of the Reynolds number on the pressure lift is small; the friction lift seems to decrease with increasing Reynolds number (table 18.11). Glauert's prediction that the pressure lift is independent of Re is a good approximation to the results of numerical simulation; our solution, which considers the viscous effects on the pressure, is in even better agreement with the results of numerical simulation. Kang, Choi, and Lee (1999) simulated the flows with $Re = 40, 60, 100,$ and 160 and q/U_0 between 0 and 2.5. The temporal-averaged values of pressure lift, pressure drag, friction lift, and friction drag, computed after the flow becomes fully developed, were presented in their paper. They showed that the friction lift decreases with increasing Re and the pressure lift is nearly independent of Re. These results are consistent with those of Padrino and Joseph (2006) and our work, despite the fact that most of the flows studied by Kang, Choi, and Lee (1999) do not satisfy the assumptions that the separation is suppressed and a steady-state solution exists.

The dependence of the drag on Re and q/U_0 is more complicated than the lift. Mittal and Kumar (2003) simulated the flows with $Re = 200$ and q/U_0 between 0 and 5; they presented the total drag coefficients C_D for the fully developed flows. The results show that when $q/U_0 < 1.91$, the flow is unsteady and the drag is oscillating; but the drag on the cylinder is always in the direction of the uniform flow. When $2 < q/U_0 < 4.34$ or $4.75 < q/U_0 < 5$, separation is suppressed and steady-state drag coefficients are obtained. The magnitude of C_D decreases with q/U_0 first, from about 0.3 at $q/U_0 = 2$ to about 0 at $q/U_0 = 3.25$. If q/U_0 is higher than 3.25, the magnitude of C_D is very close to zero; C_D could be slightly positive or negative. Kang, Choi, and Lee (1999) presented the temporal-averaged values for the total drag, pressure drag, and friction drag. They showed that the magnitude of the total drag decreases with increasing q/U_0 but the total drag is in the same direction as the uniform flow for all the flows they studied. The magnitude of the pressure drag also decreases with increasing q/U_0; when $0 < q/U_0 < 2$, the pressure drag is in the same direction as the uniform flow but when $q/U_0 = 2.5$, the pressure drag becomes opposite to the uniform flow. Similar results were obtained by Padrino and Joseph (2006), who showed that the pressure drag is opposite to the uniform flow for $q/U_0 = 3, 4, 5, 6$, and it is in competition with the friction drag, resulting in a total drag that is close to zero (see tables 18.5–18.10). The reason for the pressure drag becoming opposite to the uniform flow is not understood.

The pressure drag is a viscous effect. It cannot be studied with the classical boundary-layer theory, in which the irrotational pressure is imposed on the solid. Our boundary-layer solution is able to give a pressure drag. The agreement between this pressure drag and the result of numerical simulation depends on the choice of the boundary-layer thickness in our calculation. When q/U_0 is not high enough ($q/U_0 = 4$ or 5), it seems that the value of δ/a that gives rise to a good agreement for the pressure drag is larger than the value of δ/a, which leads to good agreements for the lift and torque (see tables 18.5 – 18.8 and figure 18.5). In the case $(Re, q/U_0) = (400, 6)$, we can find a single value of δ/a that leads to good agreement for all the three quantities, lift, torque and drag (see table 18.9), demonstrating that the agreement between our solution and numerical simulation becomes better as q/U_0 increases.

We presented a comprehensive comparison for the drag, lift, and torque on the cylinder given by our solution, by Glauert's (1957) solution, by Moore's solution (1957), and by the numerical simulation. The comparison demonstrates that Moore's torque is relatively close to the simulation results, and Glauert's solution gives reasonable approximations for the friction drag and lift but poor approximation for the torque. Our solution gives the best approximation to the numerical simulation when the value of δ/a is chosen to fit the numerical data. We also compared the profiles of the pressure drag and lift inside the boundary layer given by our solution and given by numerical simulation. The agreement of the lift profile is good (figure 18.6); the agreement of the drag profile is less good for small values of q/U_0 but improves as q/U_0 increases (figure 18.5). Such profiles are not available in Prandtl's boundary-layer theory, in which the pressure is equal to the irrotational pressure throughout the boundary layer.

The accuracy of our solution is mainly affected by the values of $\alpha = 2U_0/q$ and δ/a. Because we carried out the calculation only up to terms quadratic in α, the solution can be accurate only when α is very small. From table 18.11, one can see that the smallest value of α corresponds to $(Re, q/U_0) = (400, 6)$; other values of α in our work are all larger than 0.5. This is one of the reasons why the agreement between our solution and the numerical simulation is best for the case $(Re, q/U_0) = (400, 6)$. The boundary-layer thickness depends on the azimuthal angle θ, but we are not able to determine this dependence. We assume that δ is a constant for given Re and q/U_0 and the value of δ/a used in our calculation may be viewed as an average boundary-layer thickness. Numerical simulation will be used to determine δ/a at different azimuthal angles by use of the criterion that the vorticity magnitude at $r = a + \delta$ is approximately 1% of the maximum magnitude of the vorticity field. Plots for δ/a as a function of θ will be shown in figure 18.14 in §18.3. The figure shows that the deviation of δ/a from its average are large when Re or q/U_0 is small, and the deviation is small when Re and q/U_0 are large. This result may explain why the agreement between our solution and numerical simulation becomes better when Re and q/U_0 increases.

The problem confronted in this work is that there is no precise end to the boundary layer, although most of the vorticity is confined to a region near to the spinning cylinder when the ratio of cylinder rotating speed to uniform stream speed q/U_0 is large. We have addressed this problem by using the idea of an effective boundary-layer thickness, which is determined by matching with the results of numerical simulation. The thickness depends on the choice of the quantities for the matching. We are able to match lift, drag, and torque from our boundary-layer analysis for large values of q/U_0 and Re. In §18.1, an effective boundary-layer thickness was found that gave rise to reasonable matching for the pressure lift and torque on the cylinder computed from a simple modification of Glauert's solution (1957), and for the pressure drag computed from the method of viscous correction of VPF (VCVPF). The values of the effective thickness in §18.1 are about 1/2 or 1/3 of the values in this section. A method to determine the boundary-layer thickness without the aid of numerical simulation needs to be developed.

18.3 Numerical study of the steady-state uniform flow past a rotating cylinder

Results from the numerical simulation of the 2D incompsressible unsteady Navier–Stokes equations for streaming flow past a rotating circular cylinder are presented in this section. The numerical solution of the equations of motion is conducted with a commercial computational fluid dynamics package that discretizes the equations, applying the control volume method. The numerical setup is validated by comparison of results for a Reynolds number based on the free stream of $Re = 200$ and dimensionless peripheral speed of $\tilde{q} = 3, 4$, and 5 with results from the literature. After the validation stage, various pairs of Re and $\tilde{q}$ are specified in order to carry out the numerical experiments. These values are $Re = 200$ with $\tilde{q} = 4$ and 5; $Re = 400$ with $\tilde{q} = 4$, 5, and 6, and $Re = 1000$ with $\tilde{q} = 3$. In all these cases, gentle convergence to a fully developed steady state is reached. From the numerical vorticity distribution, the position of the outer edge of the vortical region is determined as a function of the angular coordinate. This position is found by means of a reasonable criterion set to define the outmost curve around the cylinder where the vorticity magnitude reaches a certain cutoff value. By consideration of the average value of this profile, a uniform vortical region thickness is specified for every pair of Re and $\tilde{q}$.

18.3.1 Introduction

Two aspects have drawn attention from researchers with respect to streaming flow past a rotating circular cylinder. The first aspect is the observation that the spinning action is able to suppress the separation of the boundary layer around the cylinder as well as to avoid vortex shedding from the surface of the cylinder while reaching steady state when a critical dimensionless velocity is achieved. This threshold has been reported to be a function of the Reynolds number of the free stream. The second aspect is the lift generated on the cylinder by the surrounding fluid, also called the Magnus effect. Prandtl's famous limiting value of the lift force generated by a rotating circular cylinder has encountered contradictory evidence from theoretical studies, experiments, and computations (e.g., Glauert, 1957; Tokumaru and Dimotakis, 1993; Mittal and Kumar, 2003), thus making this problem even more attractive as the subject for improved numerical methods and experimental techniques.

The literature reveals that two relevant parameters are usually specified to describe the problem, namely, the Reynolds number $Re = 2U_0a/\nu$, based on the free-stream velocity U_0, the diameter of the cylinder $2a$, fluid kinematic viscosity ν, and the dimensionless peripheral velocity $\tilde{q}$, defined as the ratio of the velocity magnitude at the surface of the cylinder to the free-stream velocity. From the point of view of the numerical simulations, setting the appropriate range for these parameters is a delicate task on which part of the success of the numerical work relies. Another feature is the choice of the form of the governing equations to solve. For a 2D problem, the vorticity and stream function form of the Navier–Stokes equations is the preferred option. However, some researchers have carried out their numerical work with the equations of motion written in terms of the primitive variables, velocity, and pressure. Ingham (1983) obtained numerical solutions of the 2D steady incompressible Navier–Stokes equations in terms of vorticity and stream function by using finite differences for flow past a rotating circular cylinder for Reynolds numbers $Re = 5$ and 20 and dimensionless peripheral velocity $\tilde{q}$ between 0 and 0.5. Solving the same form of the governing equations but expanding the range for $\tilde{q}$, Ingham and Tang (1990) showed numerical results for $Re = 5$ and 20 and $0 \leqslant \tilde{q} \leqslant 3$. With a substantial increase in Re, Badr *et al.* (1990) studied the unsteady 2D flow past a circular cylinder that translates and rotates starting impulsively from rest both numerically and experimentally for $10^3 \leqslant Re \leqslant 10^4$ and $0.5 \leqslant \tilde{q} \leqslant 3$. They solved the unsteady equations of motion in terms of vorticity and stream function. The agreement between numerical and experimental results was good except for the highest rotational velocity, where they observed 3D and turbulence effects. Choosing a moderate interval for Re, Tang and Ingham (1991) followed with numerical solutions of the steady 2D incompressible equations of motion for $Re =$ 60 and 100 and $0 \leqslant \tilde{q} \leqslant 1$. They used a scheme that avoids the difficulties regarding the boundary conditions far from the cylinder.

Considering a moderate constant $Re = 100$, Chew, Cheng, and Luo (1995) further expanded the interval for the dimensionless peripheral velocity $\tilde{q}$, such that $0 \leqslant \tilde{q} \leqslant 6$. They used a vorticity stream function formulation of the incompressible Navier–Stokes equations. The numerical method consisted of a hybrid vortex scheme, in which the time integration is split into two fractional steps, namely pure diffusion and convection. They separated the domain into two regions: the region close to the cylinder where viscous effects are important and the outer region where viscous effects are neglected and potential flow is assumed. Using the expression for the boundary-layer thickness for flow past a flat plate, they estimated the thickness of the inner region. Their results indicated a critical value for $\tilde{q}$ about 2 where vortex shedding ceases and the lift and the drag coefficients tend to asymptotic values. Nair, Sengupta, and Chauhan (1998) expanded their choices for the Reynolds number by selecting a moderate $Re = 200$ with $\tilde{q} = 0.5$ and 1 and two relatively high values of $Re = 1000$ and $Re = 3800$, with $\tilde{q} = 3$ and $\tilde{q} = 2$, respectively. They performed the numerical study of flow past a translating and rotating circular cylinder, solving the 2D unsteady Navier–Stokes equations in terms of vorticity and stream function using a third-order upwind scheme. Kang, Choi, and Lee (1999) followed with the numerical solution of the unsteady governing equations in the primitive variables velocity and pressure for flows with $Re = 60$, 100, and 160 with $0 \leqslant \tilde{q} \leqslant 2.5$. Their results showed that vortex shedding vanishes when $\tilde{q}$ increases beyond a critical value, which follows a logarithmic dependence on the Reynolds number (e.g., the critical dimensionless peripheral velocity $\tilde{q} = 1.9$ for $Re = 160$).

Chou (2000) worked on the ground of high Reynolds numbers by presenting a numerical study that included computations falling into two categories: $\tilde{q} \leqslant 3$ with $Re = 10^3$ and $\tilde{q} \leqslant 2$ with $Re = 10^4$. Chou solved the unsteady 2D incompressible Navier–Stokes equations written in terms of vorticity and stream function. In contrast, the recent work of Mittal and Kumar (2003) performed a comprehensive numerical investigation by fixing a moderate value of $Re = 200$ while considering a wide interval for the dimensionless peripheral velocity of $0 \leqslant \tilde{q} \leqslant 5$. They used the finite-element method to solve the unsteady incompressible Navier–Stokes equations in two dimensions for the primitive variable velocity and pressure. They observed vortex shedding for $\tilde{q} < 1.91$. Steady-state fully developed flow was achieved for higher rotation rates except for the narrow region $4.34 < \tilde{q} < 4.8$, where vortex shedding was again reported, perhaps, for the first time. This literature survey indicates that researchers have favored moderate Reynolds numbers $Re \leqslant 200$ in order to keep the turbulence effects away and to prevent the appearance of nonphysical features in their numerical results. For similar reasons, the peripheral speeds have been chosen such that $\tilde{q} \leqslant 3$ in the most of the cases, whereas few researchers have simulated beyond this value, with $\tilde{q} = 6$ as an upper bound.

Many studies have been devoted to the numerical simulation of this type of fluid motion to address the problem of presence or suppression of separation and vortex shedding. However, rather less attention has been paid to the application of numerical results to delimit and describe the region around the cylinder where vorticity effects are far from negligible. This fluid zone is called the vortical region. Once this region is delimited, the evaluation of theoretical boundary-layer-type solutions is feasible. When Re and $\tilde{q}$ are not so high, say $Re < 1000$ and $\tilde{q} < 5$, the vortical region can be relatively thick, so the classical thin boundary-layer analysis may not work with acceptable accuracy; nevertheless, there is still an identifiable region where the effects of vorticity are significant. Outside this region, the potential flow theory for flow past a circular cylinder with circulation may be applied.

This section concerns two main objectives: The first objective is to simulate numerically the steady-state limit of the flow past a rotating circular cylinder. The second objective is to bound the region around the rotating cylinder where the vorticity effects are mostly confined. The numerical results presented in this section are intended to test the validity of the theoretical approaches of §§18.1 and 18.2. A value for an effective, uniform thickness of the vortical region needs to be prescribed in these models. The flow field obtained from numerical analysis represents reliable data that can be utilized to estimate the limits of the vortical region. We perform numerical simulations by solving the 2D incompressible unsteady Navier–Stokes equations utilizing the commercial package Fluent® 6.1. Tests of mesh refinement are used to select the size of the computational domain and mesh structure. We validate the numerical setup by comparing our results with those from the literature for three cases.

Next, the velocity and pressure fields are computed for Reynolds numbers based on the free-stream velocity $Re = 200$ and 400, with dimensionless peripheral velocity $\tilde{q} = 4, 5$ and 6. Results for $Re = 1000$ with $\tilde{q} = 3$ are also considered. The drag and lift coefficients on the rotating cylinder are presented. From the numerical vorticity distribution in the fluid domain, the position of the outer edge of the vortical region is determined as a function of the angular coordinate. This position is found by means of a reasonable criterion set to define the outmost curve around the cylinder where the vorticity magnitude reaches a certain cutoff value. By considering the average value of this profile, a uniform vortical region thickness is specified for every pair of Re and $\tilde{q}$. The selection of this cutoff value is somewhat arbitrary, and moderate changes in this parameter may yield significant changes in the extension of the vortical region. This feature motivates the introduction of an effective vortical region thickness, which represents an alternative approach to define the position of the outer edge of the vortical region. The theoretical approach in §18.1 and the numerical results are utilized to determine two different values of the effective vortical region thickness. Exhaustive comparisons have been presented and discussed in §§18.1 and 18.2.

18.3.2 Numerical features

The 2D unsteady incompressible Navier–Stokes equations are the governing expressions for the problem at hand. In dimensionless form, these equations can be written as

$$\frac{\partial \tilde{\mathbf{u}}}{\partial \tilde{t}} + \tilde{\mathbf{u}} \cdot \nabla \tilde{\mathbf{u}} = -\nabla \tilde{p} + \frac{1}{Re} \nabla^2 \tilde{\mathbf{u}}, \qquad (18.3.1)$$

$$\nabla \cdot \tilde{\mathbf{u}} = 0, \qquad (18.3.2)$$

on a domain Φ with boundaries Λ and subject to appropriate boundary conditions. The symbol "$\sim$" designates dimensionless variables. Unless otherwise noted, the following scales are considered to make the equations dimensionless:

$$[\mathit{length},\ \mathit{velocity},\ \mathit{time},\ \mathit{pressure}] \equiv \left[2a,\ U_0,\ \frac{2a}{U_0},\ \rho U_0^2\right]. \qquad (18.3.3)$$

Three relevant parameters computed from the velocity and pressure fields are the drag, lift, and torque coefficients, which represent dimensionless expressions of the forces and torque that the fluid produces on the circular cylinder. These are defined, respectively, as follows:

$$C_D = \frac{D}{\rho U_0^2 a}, \qquad C_L = \frac{L}{\rho U_0^2 a}, \qquad C_T = \frac{T}{2\rho U_0^2 a^2}, \qquad (18.3.4)$$

where D is the drag force, L is the lift force, and T is the torque with respect to the center of the cylinder.

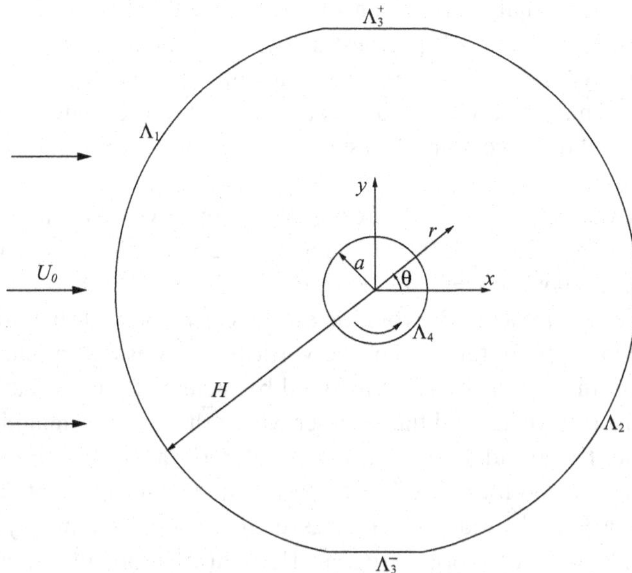

Figure 18.7. Scheme of the computational domain showing the Cartesian and polar reference coordinate systems. The boundary conditions correspond to Λ_1, inflow; Λ_2, outflow; Λ_3, zero-shear stress boundaries; and Λ_4, wall with prescribed velocity and no-slip condition.

The numerical solution of the governing system of PDEs is carried out through the computational fluid dynamics package Fluent® 6.1. This computer program applies a control-volume method to integrate the equations of motion, constructing a set of discrete algebraic equations with conservative properties. The segregated numerical scheme, which solves the discretized governing equations sequentially, is selected. An implicit scheme is applied to obtain the discretized system of equations. The sequence updates the velocity field through the solution of the momentum equations using known values for pressure and velocity. Then, it solves a "Poisson-type" pressure correction equation obtained by a combination of the continuity and momentum equations. A quadratic upwind interpolation for convective kinematics (QUICK) scheme is used to discretize the convective term in the momentum equations. Pressure implicit with splitting of operators (PISO) is selected as the pressure–velocity coupling scheme. Finally, the time integration of the unsteady momentum equations is performed with a second-order approximation.

Although from a mathematical point of view the problem setup imposes boundary conditions at infinity, the numerical approach necessarily considers a finite computational domain. Hence there is an outer boundary where inflow and outflow boundary conditions should be applied. Figure 18.7 shows the computational domain and the reference frames selected for this study. We use a modified O-type mesh similar to the one adopted by Kang *et al.* (1999). Other mesh types reported in the literature are the C-type mesh referred to by Kang *et al.* (1999) and the square-type mesh used by Mittal and Kumar (2003) in their finite-element computations. An O-type mesh is expected to save computational effort as compared with a C-type mesh or a square mesh with sides of length H/a. In this study, the domain is partially delimited by two arcs of a circle, one upstream of the cylinder and the other one downstream, and both have the same radius H. The dimensionless radius of the upstream and downstream arcs is determined as $\tilde{H} \equiv H/2a$.

The boundary conditions applied in this investigation can be described as follows: The left arc Λ_1 (figure 18.7) is the inflow section or upstream section, where a Dirichlet-type boundary condition for the Cartesian velocity components, $\tilde{u} = 1$ and $\tilde{v} = 0$, is prescribed, i.e., the free-stream velocity is imposed. The right arc Λ_2 represents the outflow boundary, where it is considered that the diffusion flux in the direction normal to the exit surface is zero for all variables. Therefore extrapolation from inside the computational domain is used to compute the flow variables at the outflow plane, which do not influence the upstream conditions.[4] On the straight horizontal segments Λ_3^+ and Λ_3^- a zero normal velocity and a zero normal gradient of all variables are prescribed. As a consequence, a zero-shear-stress condition is imposed at these two boundaries. These relatively short segments are two chords in a circle of radius H, parallel to the horizontal x axis, and are symmetric with respect to the vertical y axis. The sectors of the circle that contain these segments have a span of 10° each. The inclusion of these segments defines a transition region between the inlet and outlet sections and can be thought of as the adaptation to an O-type mesh of the zero-shear-stress upper and lower boundaries, parallel to the free stream, that Mittal and Kumar (2003) used in their domain. Finally, the dimensionless peripheral or tangential velocity $\tilde{q}$ is prescribed on the surface of the rotating cylinder along with a no-slip boundary condition. The cylinder rotates in the counterclockwise direction.

As initial conditions in this numerical investigation the values given to the velocity components at the inflow section are extended over the interior of the computational domain. Because we are focused on the fully developed flow, as Kang *et al.* (1999) pointed out, the simulations may be started with arbitrary initial conditions. They performed a numerical study with different initial conditions, including the impulsive startup, for $Re = 100$ and $\tilde{q} = 1.0$ and the same fully developed response of the flow motion was eventually reached in all the cases. In contrast, solving the steady version of the Navier–Stokes equations may yield multiple numerical solutions, depending on the given Re and $\tilde{q}$ and the initial guess used to start the computations, as was demonstrated from simulations carried out by Mittal and Kumar (2003). Keeping the unsteady term in the equations of motion prevents the occurrence of unrealistic predictions and permits us to acknowledge and describe the unsteady behavior, which is a major feature of the process of vortex shedding, when it occurs. For instance, Tang and Ingham (1991) dropped the unsteady term of the Navier–Stokes equations and found steady-state solutions for $Re = 60$ and 100 and $0 \leqslant \tilde{q} \leqslant 1$ for which experimental and numerical evidence indicates that unsteady periodic flow takes place (e.g., Kang *et al.*, 1999). The study of fully developed flows, in which a periodic unsteady state prevails, lies beyond the scope of this work. Here, the numerical experiments are focused on the steady-state (i.e., fully developed nonperiodic) flow motion.

To find a suitable position for the outer boundary, such that it appropriately approximates the real condition far from the surface of the rotating cylinder, different values of the $\tilde{H}$ parameter are considered ranging from 50 to 175 with increments of 25 units. For these grid sizes, a numerical study is performed for $Re = 400$ and $\tilde{q} = 5$ and $Re = 1000$ and $\tilde{q} = 3$ to determine the variation of the lift, drag, and torque coefficients with the

[4] The reader is referred to the Fluent® 6.1's User's Guide for details about the numerical schemes and boundary conditions used by the package.

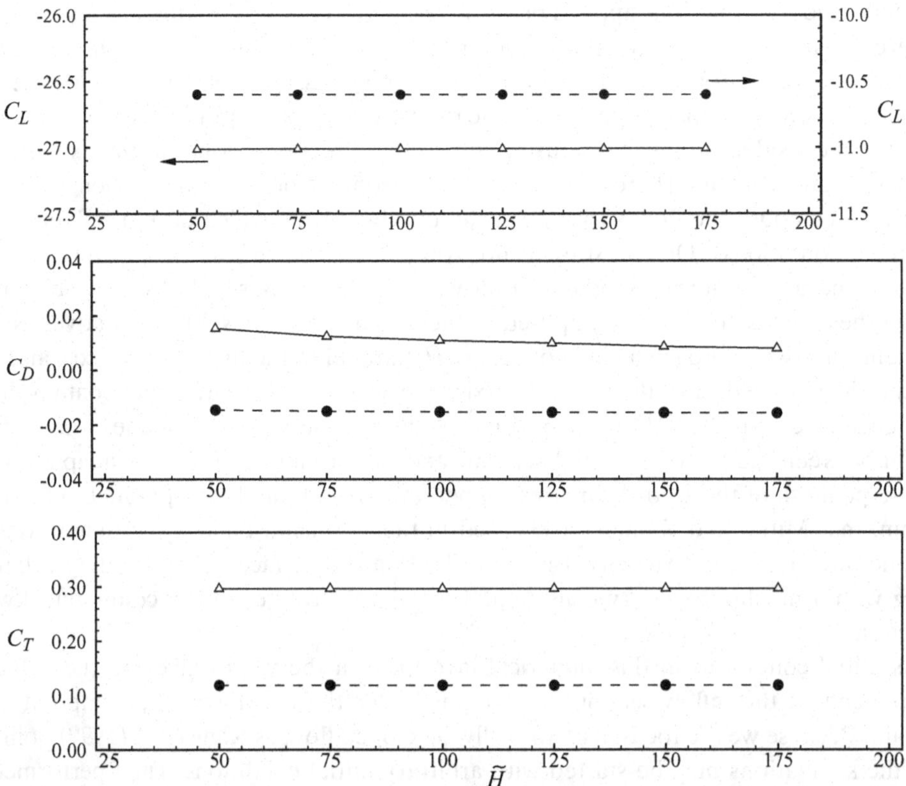

Figure 18.8. Influence of the dimensionless position of the outer boundaries (inlet and outlet) $\tilde{H}$ on the lift, drag, and torque coefficients for $Re = 400$ and $\tilde{q} = 5.0$ (solid line with $\triangle$) and $Re = 1000$ and $\tilde{q} = 3.0$ (dashed line with $\bullet$).

parameter $\tilde{H}$ (figure 18.8). From this figure it is clear that beyond $\tilde{H} = 75$ the coefficients show an asymptotic behavior; then, $\tilde{H} = 125$ is selected as the fixed radial position of the upstream and downstream arcs of circle of the outer boundary. This analysis is carried out with a dimensionless time step of $\Delta\tilde{t} = 0.02$. This time step was chosen in agreement with Kang *et al.* (1999), whereas Mittal and Kumar (2003) used a dimensionless time step of 0.0125. Because the segregated method selected from the solver is implicit, no dependency on the time step occurs in terms of numerical stability. With respect to the spatial step size, we assign the same value recommended by Mittal and Kumar (2003) for the thickness or radial step size of the first layer of cells (i.e., cells attached to the wall), $\tilde{h}_a = 0.0025$. A very fine mesh is used around the cylinder, with the size of the cells gradually increasing as the distance from the wall becomes larger. Following the approach of Chew *et al.* (1995), a rough calculation using the Blasius solution for the boundary-layer thickness $\tilde{\delta}$ for flow past a flat plate but with the Reynolds number based on the peripheral velocity of the cylinder [i.e., $\tilde{\delta} \sim (\pi/\tilde{q}\,Re)^{1/2}$] indicates that this choice of the radial-spatial step size provides a good resolution of the boundary-layer thickness. Table 18.12 gives the parameters defining the various meshes considered in this analysis; N_a is the number of nodes in the circumferential direction. Tests of the sensitivity of simulation results to mesh refinement

Table 18.12. *Properties of the meshes considered in the numerical simulations*

Mesh	Nodes	Cells	N_a	$\tilde{H}$	$\tilde{h}_a$ ($\times 10^{-3}$)	$\Delta\tilde{t}$
M50	22,080	21,920	160	50	2.50	0.02
M75	24,160	24,000	160	75	2.50	0.02
M100	25,760	25,600	160	100	2.50	0.02
M125	26,880	26,720	160	125	2.50	0.01/0.02
M150	28,000	27,840	160	150	2.50	0.02
M175	29,120	28,960	160	175	2.50	0.02
M125b	50,820	50,600	220	125	1.25	0.01

were carried out with the meshes designated as M125 (lower $\Delta\tilde{t}$) and M125b in table 18.12 for *Re* $= 400$ and $\tilde{q} = 5$ and *Re* $= 1000$ and $\tilde{q} = 3$. The lift and torque coefficients do not change much under mesh refinement but the drag coefficient does change. The changes in the drag coefficient seem relatively large because the coefficients are small, one or more orders of magnitude lower than the lift and torque coefficients (see table 18.15). This issue is also addressed in the next subsection. As a result of this systematic study, the mesh M125 from Table 18.12 with $\Delta\tilde{t} = 0.02$ was selected for our numerical experiments. All the results presented in the forthcoming sections are computed with this mesh and time step. Figure 18.9 shows the structure of a typical mesh (for $\tilde{H} = 125$, mesh M125), which is more refined near the wall.

18.3.3 Results and discussion

In this section we present the numerical results for streaming flow past a rotating circular cylinder for various *Re* and $\tilde{q}$. First, we validate our numerical setup by comparing results for selected cases with those from a previous publication. Then, the streamlines and vorticity contours are plotted and discussed. Next, the shape and extension of the vortical

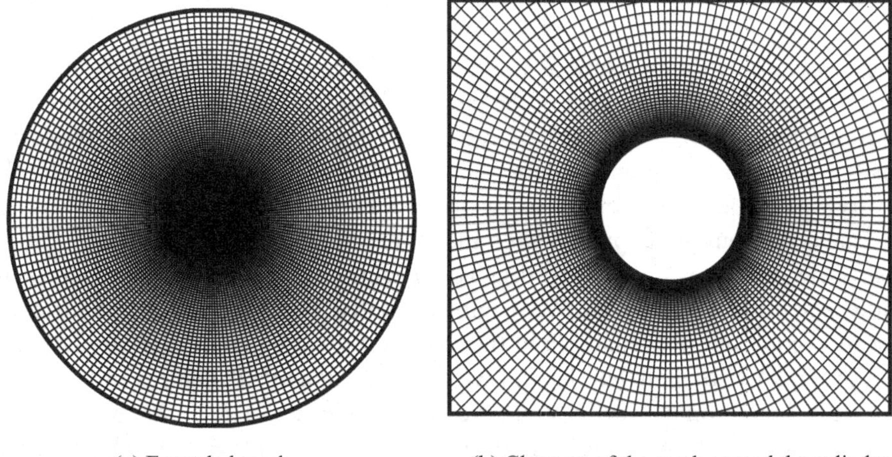

(a) Extended mesh. (b) Close-up of the mesh around the cylinder.

Figure 18.9. O-type mesh used in the numerical simulations (M125).

Table 18.13. *Comparison between the lift and drag coefficients acting on the surface of the rotating cylinder, C_L and C_D, computed in the present study with the results of Mittal and Kumar (2003)*

		Present study		Mittal and Kumar	
Re	$\tilde{q}$	C_L	C_D	C_L	C_D
200	3	−10.3400	0.0123	−10.3660	0.0350
200	4	−17.5820	−0.1240	−17.5980	−0.0550
200	5	−27.0287	0.0107	−27.0550	0.1680

region around the rotating cylinder are addressed based on the vorticity field obtained from the numerical experiments. The prediction of the outer edge of the vortical region is accomplished by the imposition of a cutoff value such that the magnitude of the vorticity at this position approximates this critical value. The outmost curve that satisfies this criterion is found. We also present the drag and lift coefficients on the rotating cylinder as computed from the numerical simulations.

18.3.3.1 *Validation of the numerical approach*

The first step is to validate the problem setup, the choice of numerical methods, and mesh attributes by comparing results from our numerical simulations with results obtained from the literature, provided the same conditions are imposed. This comparison is performed with the numerical results of Mittal and Kumar (2003) for $Re = 200$ with $\tilde{q} = 4$ and 5 under the steady-state condition. The outcomes included in the comparison are the lift and drag coefficients as defined in (18.3.4) as well as the dimensionless vorticity and pressure coefficients on the surface of the rotating circular cylinder. The dimensionless form of the vorticity is $\tilde{\omega} = 2a\omega/U_0$, and the pressure coefficient is defined as

$$c_p = \frac{p - p_\infty}{\frac{1}{2}\rho U_0^2}, \tag{18.3.5}$$

where p_∞ represents the pressure as the radial coordinate r goes to infinity and p represents the pressure on the surface of the circular cylinder. In the numerical simulations, the reference pressure p_∞ is taken to be zero at the point where the axis $y = 0$ intercepts the upstream boundary. In our simulations, it is verified that the pressure tends closely to zero everywhere along the outer boundary of the domain. This result prevails because the free-stream conditions are approached on the outer boundary.

Table 18.13 compares the lift and drag coefficients computed here with values given by Mittal and Kumar (2003). We have already noted that the agreement for the lift coefficient is good but discrepancies in the values of the drag coefficient are larger; the drag coefficients are so small that the relative errors are magnified. Figures 18.10 and 18.11 show that the dimensionless vorticity and pressure coefficient for $Re = 200$ with $\tilde{q} = 4$ and 5 are in good agreement with slightly less good agreement for $Re = 200$ and $\tilde{q} = 5$, where the pressure coefficient on the upper surface of the cylinder ($0^\circ \leqslant \theta \leqslant 180^\circ$) is slightly disturbed. This behavior may be related to the differences between our results and those of Mittal and Kumar for the drag coefficient. The pressure distribution around the surface

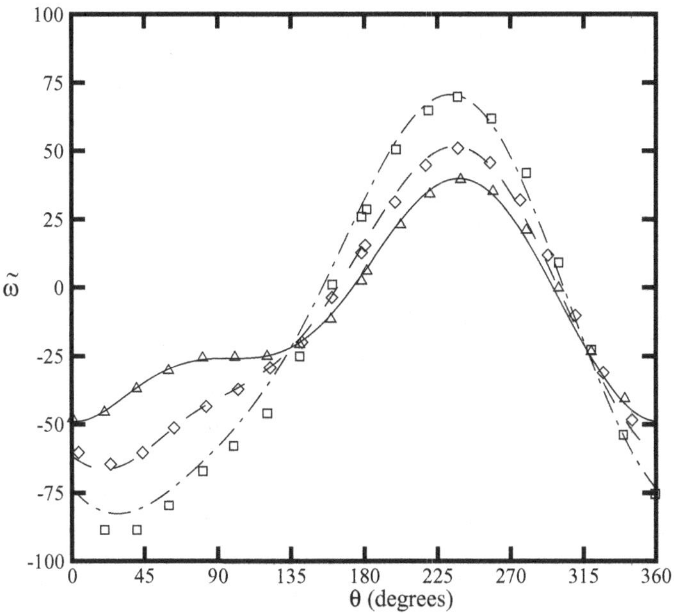

Figure 18.10. Dimensionless vorticity profiles on the surface of the rotating cylinder for $Re = 200$. Present computations: solid curve, $\tilde{q} = 3$; dashed curve, $\tilde{q} = 4$; dash–dotted curve, $\tilde{q} = 5$. Results of Mittal and Kumar (2003): $\triangle, \tilde{q} = 3$; $\diamond, \tilde{q} = 4$; $\square, \tilde{q} = 5$.

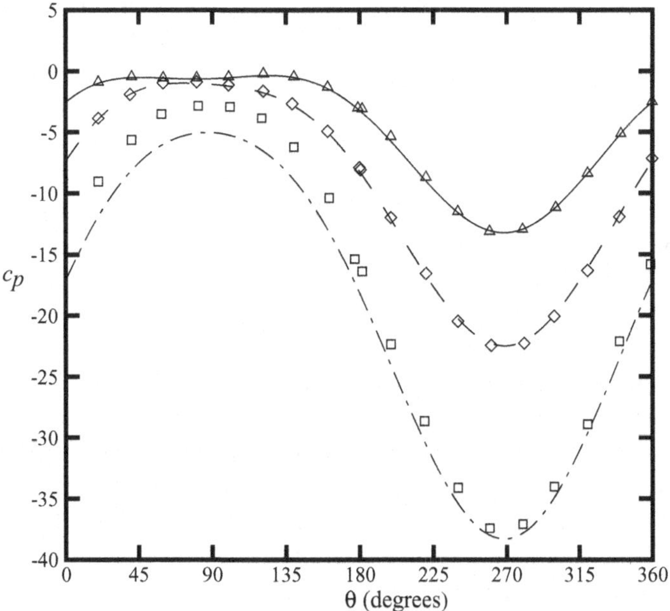

Figure 18.11. Pressure coefficient profiles on the surface of the rotating cylinder for $Re = 200$. Present computations: solid curve, $\tilde{q} = 3$; dashed curve, $\tilde{q} = 4$; dash–dotted curve, $\tilde{q} = 5$. Results of Mittal and Kumar (2003): $\triangle, \tilde{q} = 3$; $\diamond, \tilde{q} = 4$; $\square, \tilde{q} = 5$.

of the cylinder contributes to the drag and the larger discrepancies are evident at large values of $\tilde{q}$. For the largest value of $\tilde{q}$ $(=5)$ the differences in the computational strategies used here and by Mittal and Kumar (2003) in terms of mesh shape and numerical schemes are most evident in the values of the drag. We do not know which computational approach is more accurate.

18.3.3.2 *Vortical region thickness from the numerical flow field*

After the evaluation of the numerical setup with results from the literature have been accomplished, we compute the flow field for flow past a rotating circular cylinder for $Re = 400$ with $\tilde{q} = 4$, $\tilde{q} = 5$ and $\tilde{q} = 6$ and for $Re = 1000$ and $\tilde{q} = 3$. In addition, the computations for $Re = 200$ with $\tilde{q} = 4$ and $\tilde{q} = 5$, already used for comparison with results presented in a previous publication, are also included in this data set. The choice of this set of Reynolds number Re and peripheral speed $\tilde{q}$ for the numerical simulations renders a gentle convergence to steady-state fully developed flow when the unsteady incompressible Navier–Stokes equations are solved. For flow past a rotating circular cylinder, the boundary layer remains attached to the surface of the cylinder if the dimensionless peripheral velocity $\tilde{q}$ lies beyond a certain threshold so that separation is avoided and vortex shedding is suppressed. As previously mentioned, this critical value of $\tilde{q}$ is a function of Re. Furthermore, the selected range of parameters Re and $\tilde{q}$ is likely to avoid large inertia effects that yield transition to turbulence and 3D effects. A reasonable assumption, based on literature review and previous computations (Chew *et al.* 1995), indicates taking $\tilde{q} \leqslant 6$ as a limiting value. However, this upper limit may be expected to be also a function of Re.

Figure 18.12 shows the streamline patterns for the various pairs of Re and $\tilde{q}$ considered in this investigation. Note that the stagnation point lies above the cylinder, in the region where the direction of the free stream opposes the motion induced by the rotating cylinder. As the dimensionless peripheral speed at the surface of the cylinder increases, for a fixed Re, the region of close streamlines around the cylinder extends far from the wall and, as a consequence, the stagnation point moves upward. For the lowest $\tilde{q} = 3$, the region of close streamlines become narrow and the stagnation point lies near the upper surface of the cylinder. The contours of positive and negative vorticity are presented in figure 18.13. The positive vorticity is generated mostly in the lower half of the surface of the cylinder whereas the negative vorticity is generated mostly in the upper half. For the dimensionless peripheral speeds of $\tilde{q} = 3$ and 4, a zone of relative high vorticity stretches out beyond the region neighboring the rotating cylinder for $0^\circ \leqslant \theta \leqslant 90^\circ$, resembling "tongues" of vorticity. Increasing $\tilde{q}$, the rotating cylinder drags the vorticity so the "tongues" disappears and the contours of positive and negative vorticity appear wrapped around each other within a narrow region close to the surface. Based on the velocity and pressure fields obtained from the simulations for the various Re and $\tilde{q}$ considered, the next step in this numerical study is to identify the region where the vorticity effects are mostly confined. In a classical sense, the term boundary layer has been reserved for a narrow or thin region, attached to a solid surface, where the vorticity is nonnegligible. The concept of a boundary layer attached to a wall is linked to the idea of potential flow. Once the boundary layer has been delimited, the analysis follows by applying the relatively simple but still powerful theory of potential flow to approximate the external fluid motion. Nevertheless, this approximation may become inadequate when separation occurs. For flow past a

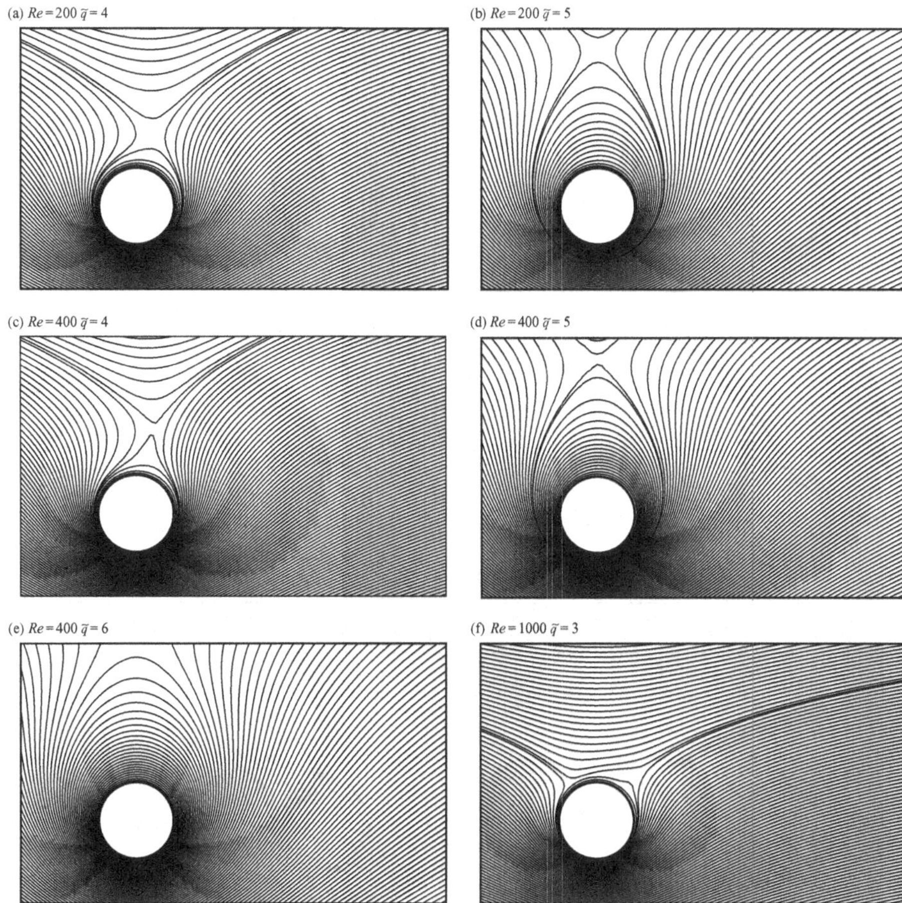

Figure 18.12. Streamlines for various pairs of *Re* and $\tilde{q}$. The rotation of the cylinder is counterclockwise, and the streaming flow is from left to right. The stagnation point lies above the cylinder. The stagnation point moves upward as the peripheral speed $\tilde{q}$ increases for a fixed *Re*.

stationary cylinder, separation is present near the downstream end of the cylinder even at *Re* as low as 5 (Panton, 1984). In contrast, for flow past a rapidly rotating cylinder the separation of the boundary-layer can be suppressed for a critical $\tilde{q}$ given *Re*. The Reynolds number based on the free-stream velocity also plays an important role in the boundary-layer analysis. In this investigation, a more general term, vortical region, is used to designate the region where the effects of viscosity are mostly restricted. The vortical region may extend relatively far from the surface where it is attached in opposition to the thinness implied in the classical boundary-layer concept. A boundary layer is certainly a vortical region; however, a relatively thick vortical region may or may not be regarded as a boundary layer.

For the type of fluid motion considered in this investigation, the vortical region lies in the fluid zone enclosed between the surface of the rotating circular cylinder and a contour surrounding this solid cylinder, called the outer edge of the vortical region. Beyond this surface, the effect of vorticity is regarded as negligible. As a first approximation, we propose that the radial position of the outer edge of the vortical region is determined such that, for

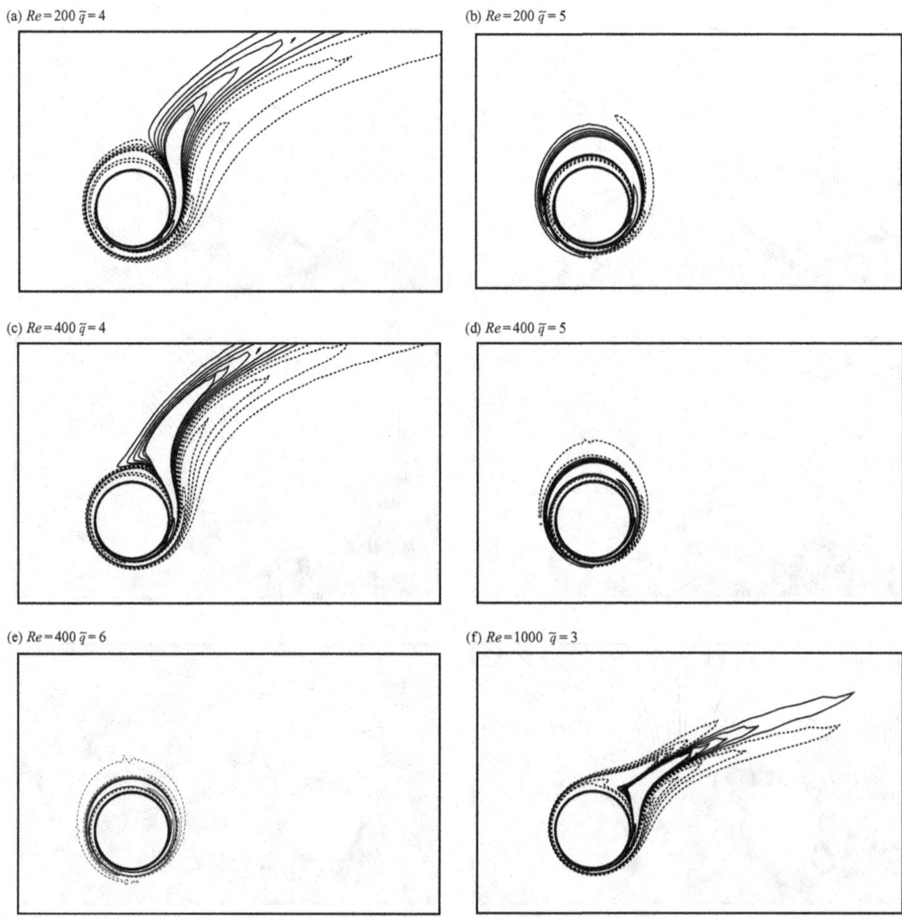

Figure 18.13. Vorticity contours for various pairs of *Re* and $\tilde{q}$. The negative vorticity is shown as dashed curves. The rotation of the cylinder is counterclockwise, and the streaming flow is from left to right.

a given angular position, the vorticity magnitude is approximately 1% of the maximum value of the vorticity magnitude field. This *ad hoc* 1% criterion is then applied to a set of discrete angular positions around the rotating cylinder ($0° \leqslant \theta \leqslant 360°$), with a constant incremental angular step, by use of the flow field obtained from the numerical simulations. Therefore the radial position of the outer edge of the vortical region as a discrete function of the azimuthal coordinate θ can be determined for every pair of *Re* and $\tilde{q}$. Because 1% of the maximum vorticity magnitude criterion may be satisfied at multiple radial positions, for any given azimuthal position, the point with the largest $\tilde{r}$ among them is chosen to determine the outer edge of the vortical region. The radial position of the outer edge of the vortical region as a function of θ determines a nonconstant thickness of the vortical region. Hence a profile of this thickness as a function of the angular position can be generated.

For practical reasons, it is convenient to deal with a constant value of the radial position of the outer edge of the vortical region and then with a constant vortical region thickness. To find this uniform value, $\tilde{r}_\delta$, a straightforward choice is the average of the discrete set of radial positions that define the edge of the vortical region as a function of θ from the previous methodology. Then, a uniform vortical region thickness, $\delta_{1\%}/a$, expressed in dimensionless fashion, is easily computed from the simple geometric formula:

$$\delta_{1\%}/a = 2\tilde{r}_\delta - 1. \tag{18.3.6}$$

Figure 18.14 shows the thickness of the vortical region as a function of the azimuthal coordinate as well as its average value for different pairs of *Re* and $\tilde{q}$. In all the cases considered, the vortical region is thick in the upper half of the cylinder ($0° \leqslant \theta \leqslant 180°$), where the fluid is retarded and the viscous effects are emphasized, whereas its thickness is decreased in the lower half ($180° \leqslant \theta \leqslant 360°$), where the fluid is accelerated. For $Re = 200$ and 400 and $\tilde{q} = 4$ and $Re = 1000$ and $\tilde{q} = 3$ the graphs show a prominent peak in the region $45° \leqslant \theta \leqslant 90°$. This trend indicates that a region of vorticity magnitude higher than 1% of the maximum vorticity magnitude in the whole domain lies relatively far from the wall. For $Re = 200$ and $\tilde{q} = 5$ and $Re = 400$ and $\tilde{q} = 5$ and 6 the peak is replaced with a hump that reaches its maximum by $\theta = 90°$. This result indicates that the region of high vorticity has been wrapped around the cylinder as a consequence of the higher rotational speed. This is verified in figure 18.13. In addition, the constant radial position of the vortical region edge $\tilde{r}_\delta$ and its corresponding vortical region thickness $\delta_{1\%}/a$ for various *Re* and $\tilde{q}$ are listed in table 18.14. These results reveal that the vortical region thickness is far from negligible for all the cases. Moreover, it is shown in table 18.14 that, as $\tilde{q}$ increases for a fixed *Re*, the average vortical region thickness $\delta_{1\%}/a$ decreases. When the rotational speed of the cylinder is increased, the local Reynolds number near its wall also increases and the inertia effects then become even more dominant than the viscosity effects, which turn out to be confined to a smaller region. A similar reasoning applies to the trend observed for a fixed $\tilde{q}$, where $\delta_{1\%}/a$ decreases as *Re* increases.

The outer edge of the vortical region from the 1% criterion as a function of the angular position θ for various pairs of *Re* and $\tilde{q}$ is presented in figure 18.15 along with the vorticity contours. The corresponding outer edge of the vortical region for a uniform thickness $\delta_{1\%}/a$ is also included. Only levels of vorticity whose magnitude is greater than or equal 1% of the maximum vorticity magnitude in the fluid domain are shown. Because the position of the outer edge of the vortical region is obtained from a discrete set of angular positions, short sections of some isovorticity lines may lie outside the nonconstant thickness vortical region. The large spikes presented in figure 18.14 are also represented here, corresponding to the regions of vorticity magnitude greater than 1% that extend far from the cylinder in the interval $0° \leqslant \theta \leqslant 90°$.

It is recognized that the criterion set to define the outer edge of the vortical region is somewhat arbitrary. A new reasonable cutoff value can be prescribed, and substantial differences may be found in terms of the position of the outer edge of the vortical region, its thickness and its shape. Here, this preliminary criterion is introduced to show that a region can be delimited where the effects of vorticity are circumscribed. This region is found to be attached to the rotating cylinder and its outer edge varies with the polar angle for a sufficiently large rotational speed.

18.3.3.3 *Drag and lift coefficients and pressure distribution from the numerical solution*

The analysis of the forces that the fluid motion produces on the rotating cylinder has been a topic of major importance in aerodynamics. For streaming flow past a rotating

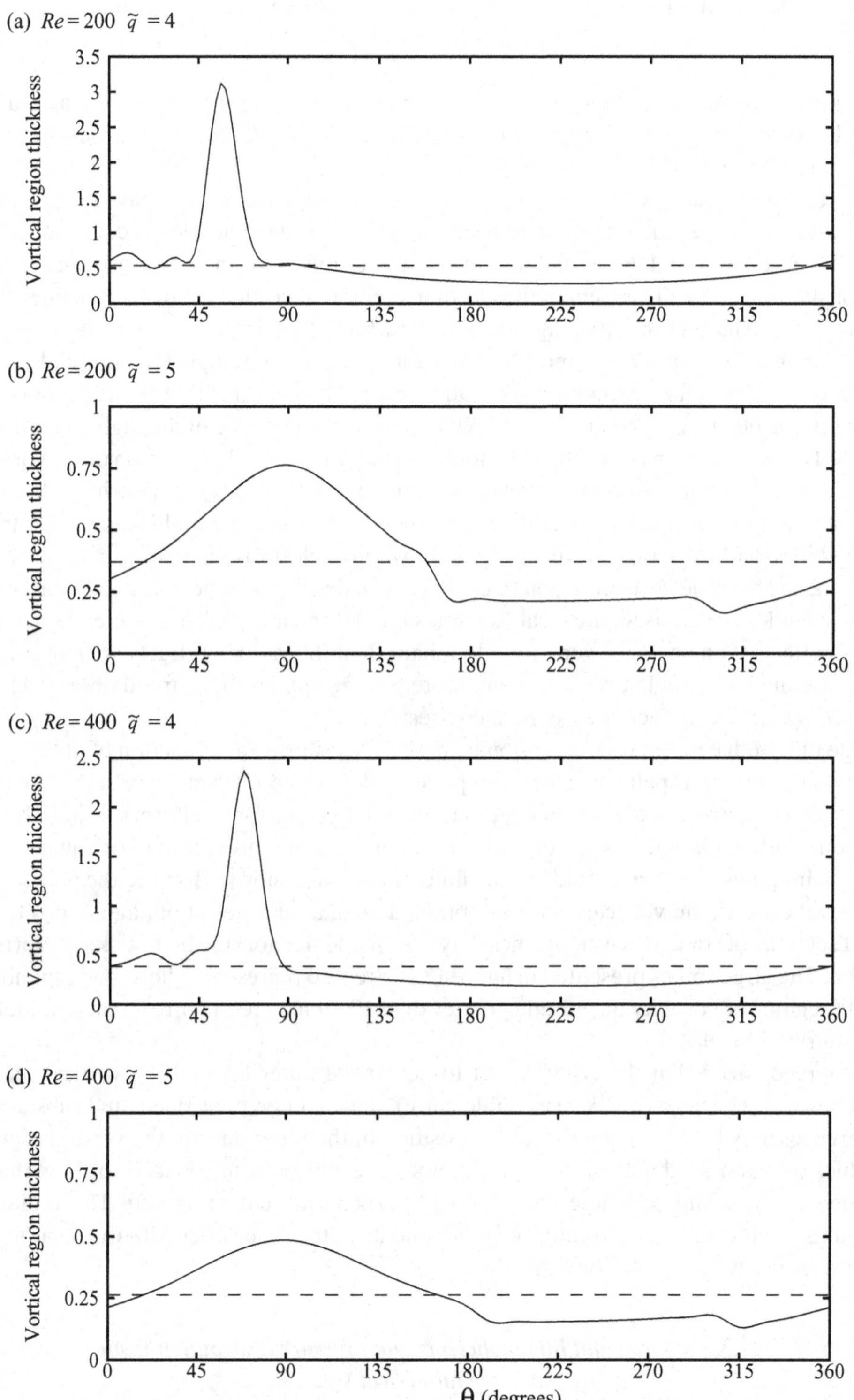

Figure 18.14. Variable vortical region thickness as a function of the angular position θ (solid curve) for various pairs of *Re* and $\tilde{q}$ obtained by application of the 1% criterion. In addition, the uniform vortical region thickness $\delta_{1\%}/a$ (dashed line) computed as the average of the profile is included.

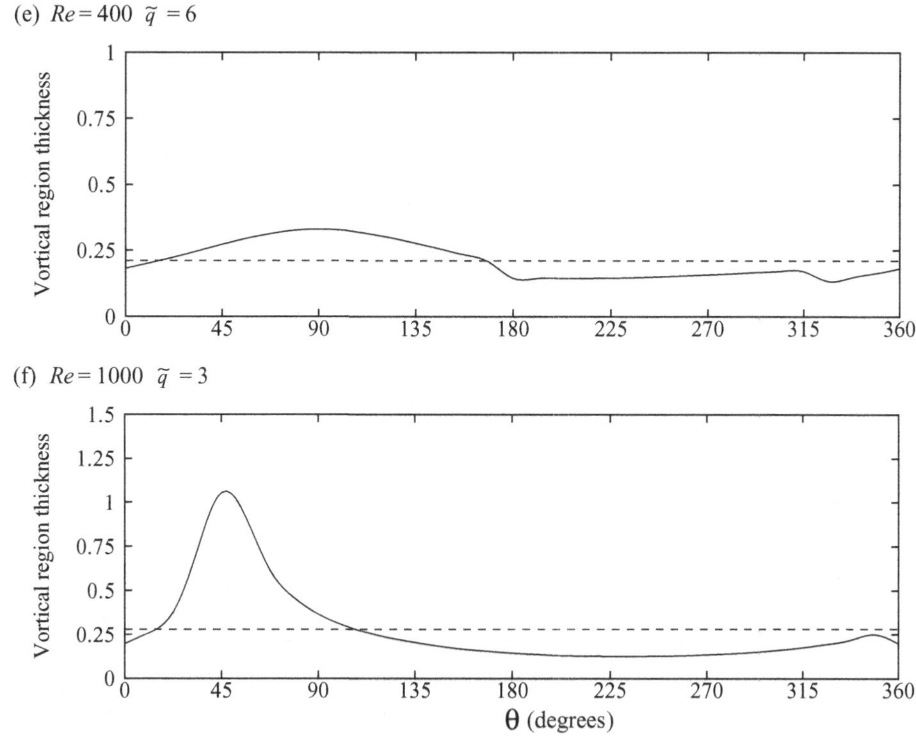

Figure 18.14 (*continued*)

cylinder, these forces have usually been presented in terms of two components mutually perpendicular, a component aligned with the free-stream velocity vector, the drag force, and a component perpendicular to this direction, the lift force.

The numerical results for the total lift and drag coefficients corresponding to the forces that the fluid motion produces on the rotating circular cylinder are presented in table 18.15. The contributions to these values from the pressure and the viscous shear stress are shown as well. The results for $Re = 200$ and $\tilde{q} = 4$ and 5 from table 18.13 are included for

Table 18.14. *Radial position of the outer edge of the vortical region $\tilde{r}_\delta$ and thickness of the vortical region $\delta_{1\%}/a$ based on the 1% of the maximum vorticity magnitude criterion for various pairs of Re and $\tilde{q}$*

Re	$\tilde{q}$	$\tilde{r}_\delta$	$\delta_{1\%}/a$
200	4	0.771	0.541
200	5	0.685	0.369
400	4	0.695	0.390
400	5	0.630	0.260
400	6	0.605	0.210
1000	3	0.638	0.277

Table 18.15. *Numerical results for the lift and drag coefficients, C_L and C_D, corresponding to the forces acting on the cylinder*

The decomposition of these values in their corresponding components from pressure (C_{L_p} and C_{D_p}) and viscous shear stress (C_{L_f} and C_{D_f}) are included.

Re	$\tilde{q}$	C_L	C_{L_p}	C_{L_f}	C_D	C_{D_p}	C_{D_f}
200	4	−17.5820	−16.9612	−0.6208	−0.1240	−0.7278	0.6038
200	5	−27.0287	−26.1826	−0.8460	0.0107	−0.8245	0.8352
400	4	−18.0567	−17.6095	−0.4472	−0.0836	−0.5341	0.4505
400	5	−27.0112	−26.4147	−0.5965	0.0100	−0.5912	0.6012
400	6	−33.7691	−33.0868	−0.6823	0.0136	−0.6677	0.6813
1000	3	−10.6005	−10.4085	−0.1920	−0.0155	−0.2129	0.1974

completeness. The results in table 18.15 indicate that the pressure lift coefficient C_{L_p} represents by far the largest contribution to the total lift C_L, in comparison with the shear stress lift coefficient C_{L_f}. The pressure and shear stress components of the lift force have the same direction of the total lift force, pointing toward the negative direction of the y axis in the current reference frame. By contrast, the pressure drag coefficient C_{D_p} and the shear stress drag coefficient C_{D_f} show a similar order of magnitude; however, these components of the total drag force point to opposite directions, with the pressure drag force pointing toward the upstream boundary (i.e., opposite to the direction of the free-stream

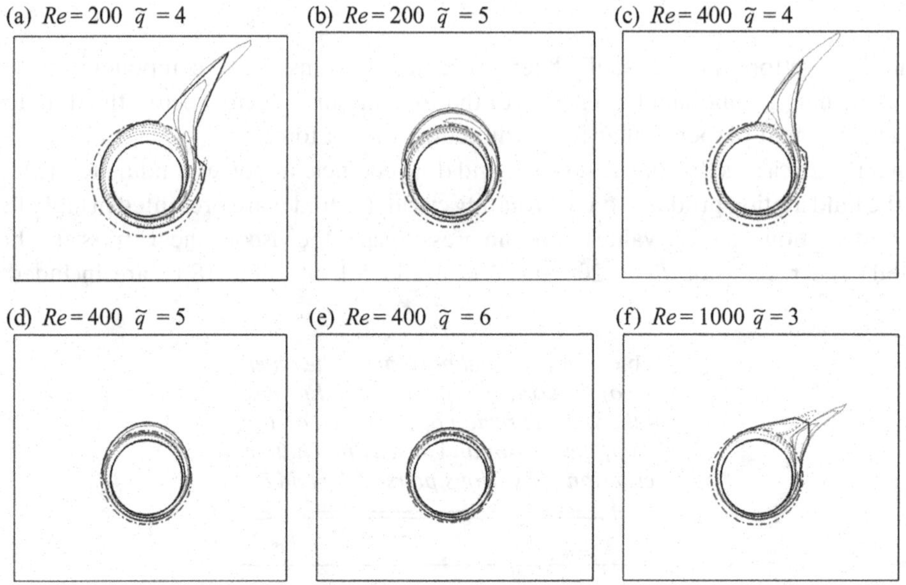

Figure 18.15. Position of the outer edge of the vortical region based on the 1% criterion. The thick-solid curves represent the edge of the vortical region with variable thickness. The thick dash–dotted curves represent the edge of the vortical region with uniform thickness $\delta_{1\%}/a$. The thin-solid curves represent contours of positive vorticity, and the thin-dashed curves represent contours of negative vorticity. The contours show levels of vorticity only with magnitude greater than or equal to 1% of the maximum vorticity magnitude in the fluid domain. The rotation of the cylinder is counterclockwise, and the streaming flow is from left to right.

velocity). The net effect of the pressure and shear stress drag is a relatively small total drag coefficient whose magnitude is, in all the cases considered, within 1% of the magnitude of the total lift coefficient. For a constant value of *Re*, when the peripheral velocity $\tilde{q}$ increases, the magnitude of the total lift coefficient, as well as the pressure and shear stress lift coefficients, increases. The same trend is observed with the magnitude of the pressure and shear stress drag coefficients. When $\tilde{q}$ is kept constant, the increase of *Re* yields an increase of the magnitude of the pressure lift coefficient but a decrease in the magnitude of the shear stress components of the lift and drag and the pressure drag coefficients. It is also of interest to observe and analyze the trend followed by the pressure distribution in the fluid domain, especially in the neighborhood of the rotating cylinder. Figure 18.16 shows the profiles of the pressure coefficient c_p as defined in (18.3.5) as a function of the radial coordinate for various *Re* and $\tilde{q}$. For every pair of *Re* and $\tilde{q}$, the c_p profiles for different angular positions are presented. The pressure coefficient has been computed from the pressure field obtained from the numerical simulations. The pressure coefficient as a result of the exact solution of the equations of motion for purely rotary flow that is due to the spinning cylinder embedded in an infinite fluid domain is presented as a reference level. The corresponding expression is

$$c_p = -\frac{\tilde{q}^2}{4\tilde{r}^2}, \tag{18.3.7}$$

independent of the angular position θ. In all the cases considered, this solution for c_p always lies inside the extreme profiles corresponding to $\theta = 90°$ and $\theta = 270°$. The graphs demonstrate that the pressure coefficient changes strongly near the wall, inside the vortical region, resembling the tendency described by the purely rotary flow solution, while becoming flat and tending slowly to zero as the radial coordinate $\tilde{r}$ approaches the outer boundaries. For a fixed peripheral speed $\tilde{q}$, it is observed that increasing *Re* has little effect on the pressure coefficient profiles. In contrast, for a fixed *Re* increasing $\tilde{q}$ expands the range of values that the pressure coefficient takes for a given radial position. For instance, this trend can be monitored on the surface of the rotating cylinder, $\tilde{r} = 0.5$, and at the outer edge of the vortical region, $\tilde{r} = \tilde{r}_\delta$, in figure 18.16. This tendency may be addressed in the frame of the irrotational flow theory. The expression for the pressure coefficient distribution for potential flow past a circular cylinder with circulation (dimensionless) $\tilde{\Gamma}$ $(= \Gamma/2aU_0)$ is recalled here:

$$c_p = \frac{\cos 2\theta}{2\tilde{r}^2} - \frac{1}{16\tilde{r}^4} - \frac{\tilde{\Gamma}^2}{4\pi^2\tilde{r}^2} + \frac{\tilde{\Gamma}}{\pi}\sin\theta\left(\frac{1}{\tilde{r}} + \frac{1}{4\tilde{r}^3}\right). \tag{18.3.8}$$

It is clear from this expression that, for a fixed radial position, the amplitude of the $(\sin\theta)$ term increases when the circulation increases. Also, this theory predicts the decrease of the mean value of c_p when the circulation rises. The tendencies described by the classical irrotational theory for the pressure coefficient distribution are followed by the numerical solution obtained in this investigation.

The results from the numerical experiments carried out in this investigation as well as from previous publications represent reliable information that can be used to evaluate the prediction capabilities of theoretical approaches. In addition, theoretical models can be used to achieve a better understanding of the numerical results and to extract relevant information from the computations, which is not evident at first sight. For these purposes,

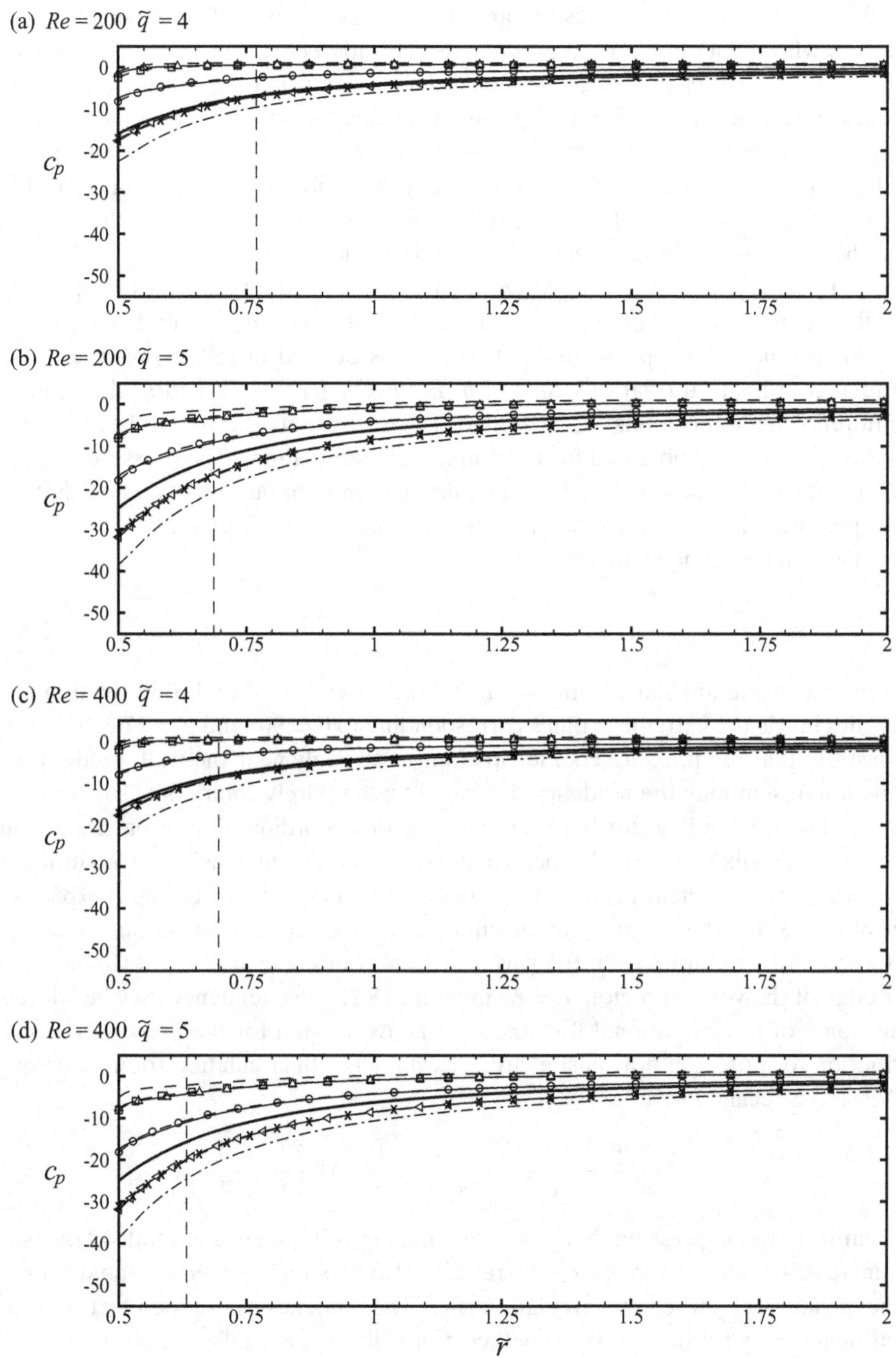

Figure 18.16. Pressure coefficient c_p as a function of the radial position $\tilde{r}$ from the surface of the rotating cylinder for various pairs of Re and $\tilde{q}$ for a fixed angle θ: 0°, thin solid curve; 45°, solid curve with △; 90°, dashed curve; 135°, dashed curve with □; 180°, dashed curve with ○; 225°, solid curve with ◁; 270°, dashed–dotted curve; 315°, dashed curve with ×. The pressure coefficient profile given in (18.3.7) from the exact solution of the equations of motion for a purely rotary flow that is due to the spinning of the cylinder under the absence of the free stream is also presented (thick solid curve). This pressure profile is independent of θ. The average position of the outer edge of the vortical region $\tilde{r}_\delta$ corresponding to the 1% criterion is included (vertical dashed line).

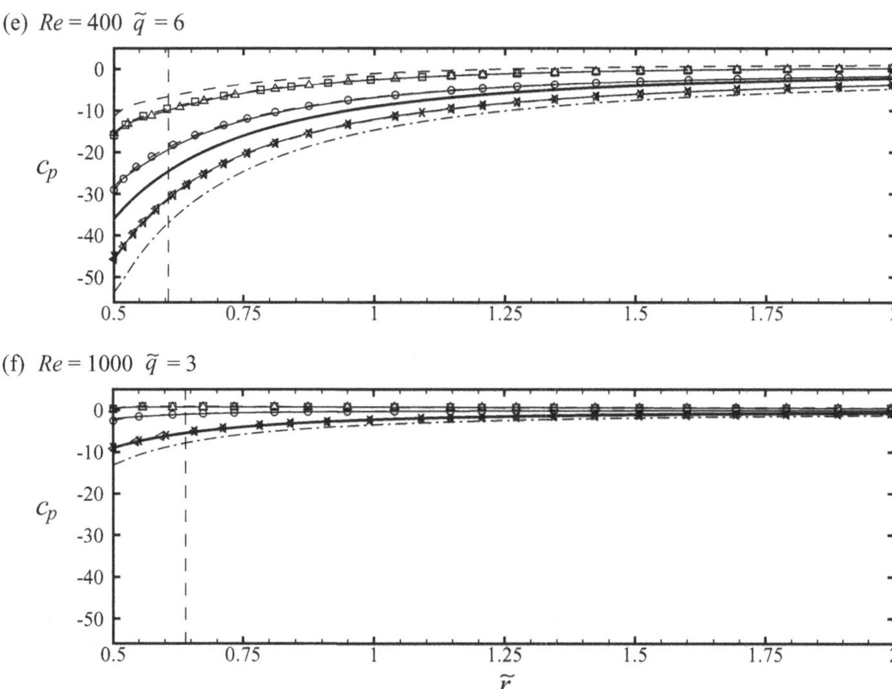

Figure 18.16 *(continued)*

the results from the numerical simulations have been compared with theoretical studies in §§18.1 and 18.2.

18.3.3.4 *Viscous dissipation*

The dissipation represents the work done by the internal viscous stress. The dissipation may be defined by

$$\mathcal{D} \equiv \int_V 2\mu \mathbf{D} : \mathbf{D} \, dV. \tag{18.3.9}$$

In dimensionless form, this expression becomes,

$$\tilde{\mathcal{D}} \equiv \int_{\tilde{V}} \frac{2}{Re} \tilde{\mathbf{D}} : \tilde{\mathbf{D}} \, d\tilde{V}, \tag{18.3.10}$$

where $\tilde{\mathbf{D}}$ is the dimensionless rate-of-strain tensor. The evaluation of (18.3.10) is performed numerically with the velocity field determined from the numerical experiments. The domain is split into two zones: inside the vortical region and outside the vortical region. The dissipation values computed in each region add up to the total dissipation. In §18.1, it is shown that the dissipation for the irrotational flow of a viscous fluid past a circular cylinder with circulation can be computed as

$$\tilde{\mathcal{D}} = \frac{8\pi}{Re} \left(1 + 2\tilde{r}_\delta^2 \tilde{Q}_\delta^2\right), \tag{18.3.11}$$

in dimensionless form.

Table 18.16. *Comparison between the numerical results for the dissipation $\tilde{D}$ determined through numerical integration of (18.3.10) and the predictions from (18.3.11) assuming irrotational flow of a viscous fluid in the entire domain for various Re and $\tilde{q}$*

Computed contributions from inside and outside the vortical region (VR) are included. We determine the radial position of the outer edge of the vortical region $\tilde{r}_\delta$ by matching C_{D_p} given by the VCVPF analysis in §18.1 and the corresponding numerical results.

Re	$\tilde{q}$	$\tilde{D}$ from numerical simulations (inside VR)	$\tilde{D}$ from numerical simulations (outside VR)	$\tilde{D}$ from numerical simulations (Total)	$\tilde{D}$ by (18.3.11)
200	4	1.1867	0.5594	1.7461	0.7213
200	5	1.2701	1.3000	2.5701	1.3001
400	4	0.6659	0.3881	1.0541	0.3718
400	5	0.7608	0.7191	1.4798	0.6547
400	6	0.8289	1.0533	1.8822	0.9997
1000	3	0.2322	0.1128	0.3451	0.0669

Table 18.16 shows the comsputed results for the viscous dissipation obtained with (18.3.10) for the whole computational domain along with the predictions of (18.3.11). In this case, the radial position of the outer edge of the vortical region $\tilde{r}_\delta$ is determined by matching C_{D_p} computed in §18.1 with the corresponding numerical results. The results are presented for various *Re* and $\tilde{q}$. The contributions from inside and outside the vortical region to the dissipation as computed through (18.3.10) are included as well. The theoretical results included in table 18.16 are closer to the viscous dissipation computed numerically in the region outside the vortical region than to the total numerical dissipation computed in the whole fluid domain. Moreover, the numerical viscous dissipation computed in the whole fluid domain is significantly greater than the theoretical prediction in all the cases. This trend may be explained by the fact that we obtain (18.3.11) by considering irrotational flow of a viscous fluid in the whole domain, thus neglecting the thickness of the vortical region. Then the predicted values are anticipated to be lower than the computed values for the total viscous dissipation, which incorporates the contribution from the vortical region. Furthermore, for a fixed *Re*, the dissipation values computed from the numerical simulations and the predicted ones increase when $\tilde{q}$ increases. The trend is reversed for a fixed $\tilde{q}$ while increasing *Re*, yielding attenuation of the viscous effects.

18.3.4 Concluding remarks

We presented results from the numerical simulations of the 2D incompressible Navier–Stokes equations for streaming flow past a rotating circular cylinder. The numerical solution of the governing equation is accomplished by means of a commercial computational fluid dynamics package. The numerical experiments are performed for various pairs of Reynolds number based on the free-stream velocity *Re* and dimensionless peripheral speed $\tilde{q}$, namely, $Re = 200$ with $\tilde{q} = 4$ and 5; $Re = 400$ with $\tilde{q} = 4, 5$, and 6; and $Re = 1000$ with $\tilde{q} = 3$. Based on the literature review and previous computations, these values are selected to avoid separation of the vortical region attached to the rotating cylinder, 3D effects, and transition to turbulence. From the numerical solution, the vorticity field is computed and used to estimate, through an *ad hoc* criterion, the annular region with thickness $\delta_{1\%}/a$ around the cylinder where vorticity is significant.

We have shown that, with the choice of an effective thickness of the vortical layer, the simple modification of Glauert's boundary-layer analysis and VCVPF lead to expressions that exhibit better general agreement with the numerical results than Glauert's original solution. This work provides a novel approach for future studies that attempt to focus on the analysis of boundary layers through computational fluid dynamics.

19

Irrotational flows that satisfy the compressible Navier–Stokes equations

If one assumes that the flow is isentropic, the body force has a potential χ, and the dynamic viscosity is constant, then the compressible Navier–Stokes equations become

$$\frac{\partial \mathbf{u}}{\partial t} + \mathbf{u} \cdot \nabla \mathbf{u} = \nabla \int \mathrm{d}p/\rho(p) + \nu \nabla^2 \mathbf{u} + \frac{1}{3} \nu \nabla \operatorname{div} \mathbf{u} - \nabla \chi \tag{19.0.1}$$

and

$$\frac{\mathrm{d}\rho}{\mathrm{d}t} + \rho \operatorname{div} \mathbf{u} = 0.$$

The kinematic viscosity ν cannot be constant in a compressible fluid with a constant dynamic viscosity. If we ignore the variation of density in the viscous term, but not elsewhere, then $\operatorname{curl} \mathbf{u} = \boldsymbol{\omega} = 0$ is a solution, then $\mathbf{u} = \nabla \phi$ and we find that

$$\frac{\partial \phi}{\partial t} + \frac{1}{2} |\nabla \phi|^2 + \chi + \int \frac{\mathrm{d}p}{\rho} - \frac{4}{3} \nu \nabla^2 \phi = 0, \tag{19.0.2}$$

$$\frac{\mathrm{d}\rho}{\mathrm{d}t} + \rho \nabla^2 \phi = 0. \tag{19.0.3}$$

Following conventional procedures of gas dynamics, we may eliminate explicit dependence on ρ from this system:

$$C^2 \nabla^2 \phi = \mathbf{u} \cdot (\mathbf{u} \cdot \nabla \mathbf{u}) + \frac{\partial}{\partial t} \left(\frac{\partial \phi}{\partial t} + |\mathbf{u}|^2 \right) - \frac{4}{3} \nu \frac{\mathrm{d}}{\mathrm{d}t} \nabla^2 \phi, \tag{19.0.4}$$

where $\mathrm{d}/\mathrm{d}t$ is the substantial derivative and C is the speed of sound. The viscous term is large when $Ul/\nu = Re$ is small, where U and l are the free-stream velocity and a body diameter.

The molecular and kinematic viscosities are not constant in general flows; isentropic flow cannot be achieved because viscous dissipation generates entropy (Sirakov *et al.*, 2005) and entropy generation generates vorticity. For gases, Re is typically large, but if we add one more assumption, even more severe than others, that ν is the eddy viscosity of a turbulent gas flow, then we could expect to see large effects of the turbulent viscosity on the gas flow.

Equation (19.0.4) is exact when $\nu = 0$, it is exact for the acoustic-wave problem subsequently described, and it is exact for the analysis of the effects of viscosity on shock structure in the one-dimensional approximation. For potential flow solutions of the Navier–Stokes equations, it is necessary that $\operatorname{curl} \mathbf{u} = 0$ be a solution (see Joseph and Liao, 1994a) of the vorticity equation. The gradients of density and viscosity that are spoilers for the general vorticity equation do not enter into the equations that perturb the state of rest with uniform pressure p_0 and density ρ_0.

19.1 Acoustics

The stress for a compressible viscous fluid is given by

$$T_{ij} = -\left(p + \frac{2}{3}\mu \operatorname{div}\mathbf{u}\right)\delta_{ij} + \mu\left(\frac{\partial u_i}{\partial x_j} + \frac{\partial u_j}{\partial x_i}\right). \tag{19.1.1}$$

Here, the second coefficient of viscosity is selected so that $T_{ii} = -3p$. (The results to follow apply also to the case in which other choices are made for the second coefficient of viscosity.)

The equations of motion are given by

$$\rho\left(\frac{\partial \mathbf{u}}{\partial t} + \mathbf{u}\cdot\nabla\mathbf{u}\right) = \operatorname{div}\mathbf{T}, \tag{19.1.2}$$

together with

$$\frac{\partial \rho}{\partial t} + \mathbf{u}\cdot\nabla\rho + \rho \operatorname{div}\mathbf{u} = 0. \tag{19.1.3}$$

To study acoustic propagation, these equations are linearized; putting

$$[\mathbf{u},\ p,\ \rho] = [\mathbf{u}',\ p_0 + p',\ \rho_0 + \rho'], \tag{19.1.4}$$

where $\mathbf{u}'$, p', and ρ' are small quantities, we get

$$T_{ij} = -\left(p_0 + p' + \frac{2}{3}\mu_0 \operatorname{div}\mathbf{u}'\right)\delta_{ij} + \mu_0\left(\frac{\partial u_i'}{\partial x_j} + \frac{\partial u_j'}{\partial x_i}\right), \tag{19.1.5}$$

$$\rho_0\frac{\partial \mathbf{u}'}{\partial t} = -\nabla p' + \mu_0\left(\nabla^2\mathbf{u}' + \frac{1}{3}\nabla \operatorname{div}\mathbf{u}'\right), \tag{19.1.6}$$

$$\frac{\partial \rho'}{\partial t} + \rho_0 \operatorname{div}\mathbf{u}' = 0, \tag{19.1.7}$$

where p_0, ρ_0, and μ_0 are constants. For acoustic problems, we assume that a small change in ρ induces small changes in p by fast adiabatic processes; hence

$$p' = C_0^2\rho', \tag{19.1.8}$$

where C_0 is the speed of sound.

Forming now the curl of (19.1.6), we find that

$$\rho_0\frac{\partial \boldsymbol{\zeta}}{\partial t} = \mu_0\nabla^2\boldsymbol{\zeta},\quad \boldsymbol{\zeta} = \operatorname{curl}\mathbf{u}'. \tag{19.1.9}$$

Hence $\boldsymbol{\zeta} = 0$ is a solution of the vorticity equation and we may introduce a potential

$$\mathbf{u}' = \nabla\phi. \tag{19.1.10}$$

Combining next (19.1.10) and (19.1.6), we get

$$\nabla\left(\rho_0\frac{\partial\phi}{\partial t} + p' - \frac{4}{3}\mu_0\nabla^2\phi\right) = 0. \tag{19.1.11}$$

The quantity in the parentheses is equal to an arbitrary function of the time that may be absorbed in ϕ.

A viscosity-dependent Bernoulli equation,

$$\rho_0 \frac{\partial \phi}{\partial t} + p' - \frac{4}{3}\mu_0 \nabla^2 \phi = 0, \tag{19.1.12}$$

is implied by (19.1.11). The stress, (19.1.5), is given in terms of the potential ϕ by

$$T_{ij} = -\left(p_0 - \rho_0 \frac{\partial \phi}{\partial t} + 2\mu_0 \nabla^2 \phi\right)\delta_{ij} + 2\mu_0 \frac{\partial^2 \phi}{\partial x_i \partial x_j}. \tag{19.1.13}$$

To obtain the equation satisfied by the potential ϕ, we eliminate ρ' in (19.1.7) with p', using (19.1.8); then we eliminate $\mathbf{u}' = \nabla\phi$ and p' in terms of ϕ, using (19.1.12) to find

$$\frac{\partial^2 \phi}{\partial t^2} = \left(C_0^2 + \frac{4}{3}\nu_0 \frac{\partial}{\partial t}\right)\nabla^2 \phi, \tag{19.1.14}$$

where the potential ϕ depends on the speed of sound and the kinematic viscosity $\nu_0 = \mu_0/\rho_0$.

A dimensionless form for potential equation (19.1.14),

$$\frac{\partial^2 \phi}{\partial T^2} = \left(1 + \frac{\partial}{\partial T}\right)\nabla^2 \phi, \quad \nabla^2 \phi = \frac{\partial^2 \phi}{\partial X^2} + \frac{\partial^2 \phi}{\partial Y^2} + \frac{\partial^2 \phi}{\partial Z^2}, \tag{19.1.15}$$

arises from a change of variables

$$t = \frac{4\nu_0}{3C_0^2}T, \quad \mathbf{x} = \frac{4}{3}\frac{\nu_0}{C_0}\mathbf{X}. \tag{19.1.16}$$

The classical theory of sound (see Landau and Lifshitz, 1987, chapter VIII) is governed by a wave equation, which may be written in dimensionless form as

$$\frac{\partial^2 \phi}{\partial T^2} = \nabla^2 \phi. \tag{19.1.17}$$

The time derivative on the right-hand side of (19.1.15) leads to a decay of the waves not present in the classical theory. Many, if not all, of the results obtained with (19.1.17) may be redone by use of (19.1.15).

The simplest problem in the theory of sound waves (see Landau and Lifshitz, 1987, 253) is the case of plane monochromatic traveling waves. The one-dimensional version of (19.1.15),

$$\frac{\partial^2 \phi}{\partial T^2} = \left(1 + \frac{\partial}{\partial T}\right)\frac{\partial^2 \phi}{\partial X^2}, \tag{19.1.18}$$

can be solved by separation of variables, $\phi = F(T)G(X)$. We obtain

$$\frac{F''}{F + F'} = \frac{G''}{G} = -k^2. \tag{19.1.19}$$

If $k^2 > 4$, the solution is

$$\phi = \left(Ae^{-\omega_1 T} + Be^{-\omega_2 T}\right)\cos\left(-kX + \alpha\right), \tag{19.1.20}$$

Table 19.1. *Representative dimensional parameters of sound waves*

$\frac{4\nu_0}{3C_0^2}$ and $\frac{4}{3}\frac{\nu_0}{C_0}$ are time and length scales, respectively [equation (19.1.16)]. $k\frac{3C_0}{4\nu_0}$ is the cutoff wavenumber and $w_1\frac{3C_0^2}{4\nu_0} = w_2\frac{3C_0^2}{4\nu_0}$ is the frequency when k is the cutoff wavenumber [equation (19.2.1)]. $k_1\frac{3C_0}{4\nu_0}$ and $k_2\frac{3C_0}{4\nu_0}$ are the wavenumbers when the frequency of the sound wave $w\frac{3C_0^2}{4\nu_0} = 10^9\ \text{sec}^{-1}$ [equation (19.2.2)]. The values are calculated with the properties of the liquids at 15 °C: $\rho_w = 1\ \text{g/cm}^3$, $\mu_w = 0.0114$ g/(cm sec), $C_w = 1.48 \times 10^5$ cm/sec for water; $\rho_w = 1.26\ \text{g/cm}^3$, $\mu_w = 23.3$ g/(cm sec), $C_w = 1.9 \times 10^5$ cm/sec for glycerin.

Liquid	$\frac{4\nu_0}{3C_0^2}$ (sec)	$\frac{4}{3}\frac{\nu_0}{C_0}$ (cm)	$k\frac{3C_0}{4\nu_0}$ (cm^{-1})	$w_1\frac{3C_0^2}{4\nu_0} = w_2\frac{3C_0^2}{4\nu_0}$ (sec^{-1})	$k_1\frac{3C_0}{4\nu_0}$ (cm^{-1})	$k_2\frac{3C_0}{4\nu_0}$ (cm^{-1})
Water	6.94×10^{-13}	1.03×10^{-7}	1.95×10^7	2.88×10^{12}	2.34	6.76×10^3
Glycerin	6.83×10^{-10}	1.30×10^{-4}	1.54×10^4	2.93×10^9	1.41×10^3	4.57×10^3

where A, B, and α are undetermined constants and

$$\begin{bmatrix} \omega_1 \\ \omega_2 \end{bmatrix} = \frac{k^2}{2}\begin{bmatrix} 1 \\ 1 \end{bmatrix} + \frac{1}{2}\begin{bmatrix} \sqrt{k^4 - 4k^2} \\ -\sqrt{k^4 - 4k^2} \end{bmatrix}. \tag{19.1.21}$$

The solution is a standing periodic wave with a decaying amplitude.

If $k^2 < 4$, the solution,

$$\phi = e^{-\frac{k^2}{2}T}\begin{bmatrix} A\cos\left[-kX - \frac{1}{2}\left(4k^2 - k^4\right)^{1/2} T + \alpha\right] \\ +B\cos\left[-kX + \frac{1}{2}\left(4k^2 - k^4\right)^{1/2} T + \alpha\right] \end{bmatrix}, \tag{19.1.22}$$

represents decaying waves propagating to the left and right. Traveling plane-wave solutions that are periodic in T and grow or decay in X are also easily derived by separation of variables.

19.2 Spherically symmetric waves

For spherically symmetric waves,

$$\frac{\partial^2\phi}{\partial T^2} = \left(1 + \frac{\partial}{\partial T}\right)\frac{1}{R^2}\frac{\partial}{\partial R}\left(R^2\frac{\partial\phi}{\partial R}\right), \tag{19.2.1}$$

and, following Landau and Lifshitz (1987, 269), we note that

$$\psi(R, T) = R\phi \tag{19.2.2}$$

satisfies equation (19.1.18) for plane waves with R replacing X. It follows then that solutions (19.1.20), (19.1.22), and (19.2.1) hold for spherically symmetric waves when X is replaced with R and $\phi(X, T)$ with $R\phi(R, T)$.

The properties of these and many other solutions to equation (19.1.15) are for future research. For the present, we indicate in table 19.1 representative values of physical parameters of sound waves which are suppressed in the dimensionless version of our problem. The properties of liquids may be useful in studies of cavitation of liquids that is due to ultrasound.

19.3 Liquid jet in a high-Mach-number airstream

The instability of circular liquid jet immersed in a coflowing high-velocity airstream is studied assuming that the flow of the viscous gas and liquid is irrotational. The basic velocity profiles are uniform and different. The instabilities are driven by KH instability that is due to a velocity difference and neck-down that is due to capillary instability. Capillary instabilities dominate for large Weber numbers. KH instability dominates for small Weber numbers. The wavelength for the most unstable wave decreases strongly with the Mach number and attains a very small minimum when the Mach number is somewhat larger than one. The peak growth rates are attained for axisymmetric disturbances ($n = 0$) when the viscosity of the liquid is not too large. The peak growth rates for the first asymmetric mode ($n = 1$) and the associated wavelength are very close to the $n = 0$ mode; the peak growth rate for $n = 1$ modes exceeds $n = 0$ when the viscosity of the liquid jet is large. The effects of viscosity on the irrotational instabilities are very strong. The analysis predicts that breakup fragments of liquids in high-speed airstreams may be exceedingly small, especially in the transonic range of Mach numbers.

19.3.1 Introduction

The problem of an inviscid liquid jet in an inviscid compressible airstream was studied by Chang and Russel (1965), Nayfeh and Saric (1973), Zhou and Lin (1992), and Li and Kelly (1992). Chawla (1975) studied the stability of a sonic gas jet submerged in a liquid. Chang and Russel (1965) and Nayfeh and Saric (1973) consider temporal instability and found that a singularity in the growth rate occurs as the Mach number tends to unity. Chawla (1975) did not find a singular growth rate, but he restricted his attention to Mach number one ($M = 1$). Li and Kelly (1992) found that the growth rates reach a sharp maximum when the gas velocity is slightly larger than the one, giving $M = 1$ for both the axisymmetric and the first nonaxisymmetric mode of instability. Lin (2003) cites Li and Kelly (1992) for the growth rate near $M = 1$ in the case of temporal stability.

Funada, Joseph, Saitoh, and Yamashita (2006, hereafter FJSY) extended the theory of VPF of a viscous compressible gas given by Joseph (2003b) to the case of perturbations in a compressible gas moving with uniform velocity. We derive a dispersion relation for the perturbations that depends on all the material properties of the incompressible liquid and compressible gas. The effects of shear are neglected, consistent with the assumption that the basic flow can support a discontinuous velocity. We find a sharply peaking growth rate at a slightly supersonic value of the gas Mach number under the conditions that Li and Kelly (1992) find steep changes for both axisymmetric and first asymmetric modes. The analysis of Li and Kelly (1992) differs from the one given here in the way that the isentropic flow is represented. They assume that $\mathrm{d}p/\mathrm{d}\rho = c^2$, as in isentropic flow, but they do not account for the usual isentropic relations that tie the density, pressure, and velocity together as in our equation (19.3.13).

The first application of VPF to the problem of capillary instability was made by Funada and Joseph (2002). The problem of combined KH and capillary instability for an incompressible liquid and gas was investigated by Funada, Joseph, and Yamashita (2004), who treated also the problem of convective and absolute instability in a comprehensive manner.

The effects of compressibility are very important for transonic and supersonic flow, as has already been noted by Lin (2003). These effects include very great increases in growth rates and very sharp decreases in the wavelength for maximum growth. This feature may possibly play a role in the breakup of liquid droplets into fine drops that is observed in shock tube and wind tunnel experiments (Engel, 1958; Joseph *et al.*, 1999; Joseph *et al.*, 2002; Theofanous *et al.*, 2003; Varga, Lasheras, and Hopfinger, 2003).

Chen and Li (1999) did a linear stability analysis for a viscous liquid jet issued into an inviscid moving compressible gas medium. Their analysis differs from ours; they do not assume that motion of the liquid is irrotational; they give results only for the case in which the gas is at rest so that the effects of the basic flow gas velocity is not connected to the basic flow density and pressure as in the case of isotropic flow considered here. They do not compute growth rates for temporal instabilities for supersonic values $M > 1$. Their growth-rate curves do not exhibit the same great increases in transonic and supersonic flow found by other authors and here.

The assumption that the gas is inviscid is not justified for jets of liquids into air, especially when the air velocity is large. What matters here are the ratios of kinematic viscosities [see equation (4.2) in Funada, Joseph, and Yamashita (2004) and figure 4 in Funada and Joseph, 2001] and the kinematic viscosity of high-speed air in isentropic flow can be much greater than the kinematic viscosity of water.

Experimental results on liquid jets in a high-speed gas suitable for comparison with this and other analytical studies are not available. The coaxial jet experiments of Varga, Lasheras, and Hopfinger (2003) discussed in subsection 11.3.10 are suitable, but they do not present data for transonic and supersonic conditions. Dunne and Cassen (1954, 1956) did some experiments on supersonic liquid jets. They injected high-speed jets into air with a spring-loaded injector (1954) and subjected the liquid reservoir to a shock-wave pressure (1956). These jets are transients, and they appear to give rise to RT instabilities on the front face of the jet as in the problem of drop breakup in high-speed air and to KH waves at sides of the jets where the velocity is discontinuous. The data presented by them are not suitable for comparison on the analysis given here.

19.3.2 Basic partial differential equations

For isentropic compressible fluids, the equation of continuity, the viscous stress tensor $\mathbf{T}$, and the equation of motion are expressed in usual notation with the velocity potential ϕ for which $\mathbf{v} = \nabla\phi$ and $\nabla \times \mathbf{v} = 0$, as

$$\frac{\partial \rho}{\partial t} + \nabla \cdot (\rho \mathbf{v}) = 0, \quad \text{hence} \quad \frac{\partial \rho}{\partial t} + (\nabla\phi \cdot \nabla)\,\rho + \rho \nabla^2 \phi = 0, \tag{19.3.1}$$

$$T_{ij} = \mu \left(\frac{\partial v_i}{\partial x_j} + \frac{\partial v_j}{\partial x_i} \right) - \frac{2\mu}{3} (\nabla \cdot \mathbf{v})\, \delta_{ij} = 2\mu \frac{\partial^2 \phi}{\partial x_i \partial x_j} - \frac{2\mu}{3} (\nabla^2 \phi)\, \delta_{ij}, \tag{19.3.2}$$

$$\frac{\partial \mathbf{v}}{\partial t} + (\mathbf{v} \cdot \nabla)\, \mathbf{v} = -\frac{1}{\rho} \nabla p + \frac{1}{\rho} \nabla \cdot \mathbf{T} \;\rightarrow\; \frac{\partial \phi}{\partial t} + \frac{1}{2} |\nabla\phi|^2 + \frac{\gamma}{\gamma - 1} \frac{p}{\rho} - \frac{4}{3} \frac{\mu}{\rho} \nabla^2 \phi = B(t). \tag{19.3.3}$$

The isentropic relation is given by

$$p \rho^{-\gamma} = \text{constant} \equiv A, \quad \text{hence} \quad p = A \rho^{\gamma}, \tag{19.3.4}$$

with the adiabatic exponent γ and the sound velocity c:

$$c^2 = \frac{\mathrm{d}p}{\mathrm{d}\rho} = \gamma \frac{p}{\rho}. \tag{19.3.5}$$

These are used for VPF, which reduces to IPF when the viscosity vanishes.

19.3.3 Cylindrical liquid jet in a compressible gas

A cylindrical liquid jet is surrounded by a compressible gas and is addressed in $0 \le r < a$ (where a is the radius of the cylindrical jet in an undisturbed state) and $-\infty < z < \infty$ in the cylindrical frame (r, θ, z). The equation of continuity, the viscous stress tensor, and Bernoulli's function are given for the compressible gas as

$$\begin{aligned}
&\frac{\partial \rho_a}{\partial t} + (\nabla \phi_a \cdot \nabla)\, \rho_a + \rho_a \nabla^2 \phi_a = 0, \\
&T_{ij}^{(a)} = 2\mu_a \frac{\partial^2 \phi_a}{\partial x_i \partial x_j} - \frac{2\mu_a}{3} \left(\nabla^2 \phi_a\right) \delta_{ij}, \\
&\frac{\partial \phi_a}{\partial t} + \frac{1}{2} |\nabla \phi_a|^2 + \frac{\gamma}{\gamma - 1} \frac{p_a}{\rho_a} - \frac{4}{3} \frac{\mu_a}{\rho_a} \nabla^2 \phi_a = B_a(t), \\
&p_a = A \rho_a^{\gamma} \;\rightarrow\; \frac{\mathrm{d}p_a}{\mathrm{d}\rho_a} = \gamma A \rho_a^{\gamma - 1} = \gamma \frac{p_a}{\rho_a} = c_a^2,
\end{aligned} \tag{19.3.6}$$

and for the liquid as

$$\begin{aligned}
&\rho_\ell = \text{const}, \quad \nabla^2 \phi_\ell = 0, \quad T_{ij}^{(\ell)} = 2\mu_\ell \frac{\partial^2 \phi_\ell}{\partial x_i \partial x_j}, \\
&\frac{\partial \phi_\ell}{\partial t} + \frac{1}{2} |\nabla \phi_\ell|^2 + \frac{p_\ell}{\rho_\ell} = B_\ell(t).
\end{aligned} \tag{19.3.7}$$

Boundary conditions at the interface $r = a + \eta$ [where $\eta = \eta(\theta, z, t)$ is the interface displacement] are the kinematic conditions:

$$\frac{\partial \eta}{\partial t} + (\nabla \phi_a \cdot \nabla)\, \eta = \mathbf{n} \cdot \nabla \phi_a, \quad \frac{\partial \eta}{\partial t} + (\nabla \phi_\ell \cdot \nabla)\, \eta = \mathbf{n} \cdot \nabla \phi_\ell, \tag{19.3.8}$$

with the outer normal vector $\mathbf{n}$,

$$\mathbf{n} = \left(1, \frac{-1}{a + \eta} \frac{\partial \eta}{\partial \theta}, -\frac{\partial \eta}{\partial z}\right) \Bigg/ \sqrt{1 + \left(\frac{1}{a + \eta} \frac{\partial \eta}{\partial \theta}\right)^2 + \left(\frac{\partial \eta}{\partial z}\right)^2}, \tag{19.3.9}$$

and the normal stress balance,

$$p_\ell - p_a + \left[n_i T_{ij}^{(a)} n_j\right] - \left[n_i T_{ij}^{(\ell)} n_j\right] = \sigma \nabla \cdot \mathbf{n}, \tag{19.3.10}$$

where σ is the interfacial tension coefficient.

19.3.4 Basic isentropic relations

A basic state of the gas is with a uniform flow $\bar{\mathbf{v}}_a = \nabla \bar{\phi}_a = (0, 0, U_a)$ in the frame (r, θ, z) and with constant density ρ_{a1} and pressure p_{a1}, and a basic state of the liquid is with a uniform flow $\bar{\mathbf{v}}_\ell = \nabla\bar{\phi}_\ell = (0, 0, U_\ell)$ and with constant density $\rho_{\ell 1}$ and pressure $p_{\ell 1}$. The isentropic relation and Bernoulli's function lead for the gas to

$$p_{a1} = A\rho_{a1}^{\gamma} = p_{a0}\left(\frac{\rho_{a1}}{\rho_{a0}}\right)^{\gamma}, \quad \frac{dp_{a1}}{d\rho_{a1}} = \gamma A\rho_{a1}^{\gamma-1} = \gamma\frac{p_{a1}}{\rho_{a1}} = c_a^2,$$
$$\frac{1}{2}U_a^2 + \frac{\gamma}{\gamma-1}\frac{p_{a1}}{\rho_{a1}} = \frac{1}{2}U_a^2 + \frac{c_a^2}{\gamma-1} = \frac{c_{a0}^2}{\gamma-1} = B_a, \text{ hence } \left[\frac{\gamma-1}{2}M_a^2+1\right]\frac{\gamma p_{a1}}{\rho_{a1}} = \frac{\gamma p_{a0}}{\rho_{a0}}, \tag{19.3.11}$$

where the Mach number M_a is defined as

$$M_a = \frac{U_a}{c_a}, \tag{19.3.12}$$

and ρ_{a0}, p_{a0}, and c_{a0} ($c_{a0}^2 = \gamma p_{a0}/\rho_{a0}$) are defined when $M_a = 0$.

Using (19.3.11), we have

$$\bar{\rho}_a = \frac{\rho_{a1}}{\rho_{a0}} = \left[\frac{\gamma-1}{2}M_a^2+1\right]^{-1/(\gamma-1)}, \quad \bar{p}_a = \frac{p_{a1}}{p_{a0}} = \left(\frac{\rho_{a1}}{\rho_{a0}}\right)^{\gamma},$$
$$\frac{c_a^2}{c_{a0}^2} = \left[\frac{\gamma-1}{2}M_a^2+1\right]^{-1} \text{ or } c_a^2 = c_{a0}^2 - \frac{\gamma-1}{2}U_a^2, \tag{19.3.13}$$

in which the sound velocity c_a is given as a function of U_a. The thermodynamic properties of the ambient gas depend on the Mach number and the reference state when $M_a = 0$. For air of $\rho_{a0} = 1.2\ \text{kg/m}^3$, $p_{a0} = 1\ \text{atm} = 1.013 \times 10^5\ \text{Pa}$, $c_{a0} = 340\ \text{m/sec}$, and $\gamma = 1.4$. When $M_a = 1$, (19.3.13) gives $c_a^2 = 2c_{a0}^2/(\gamma+1)$ for which $c_a = 310.38\ \text{m/sec}$. The third equation in (19.3.13) shows that $c_a = 0\ \text{m/sec}$ when $M_a \to \infty$. Then $U_a^2 = U_{am}^2 = 2c_{a0}^2/(\gamma-1)$, where $U_{am} = 760.26\ \text{m/sec}$ is the maximum air velocity.

Bernoulli's function for the liquid leads to

$$\frac{1}{2}U_\ell^2 + \frac{p_{\ell 1}}{\rho_{\ell 1}} = B_\ell. \tag{19.3.14}$$

The kinematic conditions are satisfied for the unidirectional flows and the interface given by $r = a$. The normal stress balance is given by

$$p_{\ell 1} - p_{a1} = \frac{\sigma}{a}, \tag{19.3.15}$$

where σ/a denotes the capillary pressure.

19.3.5 Linear stability of the cylindrical liquid jet in a compressible gas; dispersion equation

On the basic flows, small disturbances are superimposed as

$$\phi_\ell = U_\ell z + \tilde{\phi}_\ell, \quad \rho_\ell = \rho_{\ell 1} \text{ (no perturbation)}, \quad p_\ell = p_{\ell 1} + \tilde{p}_\ell,$$
$$\phi_a = U_a z + \tilde{\phi}_a, \quad \rho_a = \rho_{a1} + \tilde{\rho}_a, \quad p_a = p_{a1} + \tilde{p}_a. \tag{19.3.16}$$

The isentropic relation gives

$$p_a = A\rho_a^\gamma, \text{ hence } p_{a1} = A\rho_{a1}^\gamma, \ \tilde{p}_a \approx A\rho_{a1}^\gamma \gamma \frac{\tilde{\rho}_a}{\rho_{a1}} = c_a^2 \tilde{\rho}_a, \tag{19.3.17}$$

$$\frac{\gamma}{\gamma - 1}\frac{p_a}{\rho_a} = \frac{c_a^2}{\gamma - 1} + c_a^2 \frac{\tilde{\rho}_a}{\rho_{a1}} = \frac{c_a^2}{\gamma - 1} + \frac{\tilde{p}_a}{\rho_{a1}}. \tag{19.3.18}$$

For the gas, we have the equations for the disturbance:

$$\left(\frac{\partial}{\partial t} + U_a \frac{\partial}{\partial z}\right)\tilde{\rho}_a + \rho_{a1}\nabla^2\tilde{\phi}_a = 0, \ \left(\frac{\partial}{\partial t} + U_a \frac{\partial}{\partial z}\right)\tilde{\phi}_a + c_a^2\frac{\tilde{\rho}_a}{\rho_{a1}} - \frac{4}{3}\frac{\mu_a}{\rho_{a1}}\nabla^2\tilde{\phi}_a = 0; \tag{19.3.19}$$

hence

$$\left(\frac{\partial}{\partial t} + U_a \frac{\partial}{\partial z}\right)^2 \tilde{\phi}_a = \left[c_a^2 + \frac{4}{3}\frac{\mu_a}{\rho_{a1}}\left(\frac{\partial}{\partial t} + U_a \frac{\partial}{\partial z}\right)\right]\nabla^2\tilde{\phi}_a. \tag{19.3.20}$$

For the liquid, we have the equations for the disturbance:

$$\nabla^2\tilde{\phi}_\ell = 0, \ \left(\frac{\partial}{\partial t} + U_\ell \frac{\partial}{\partial z}\right)\tilde{\phi}_\ell + \frac{\tilde{p}_\ell}{\rho_{\ell 1}} = 0. \tag{19.3.21}$$

At the interface $r = a + \tilde{\eta} \approx a$, where $\tilde{\eta} \equiv \tilde{\eta}(\theta, z, t)$ is the interface displacement, the kinematic conditions are given by

$$\frac{\partial \tilde{\eta}}{\partial t} + U_\ell \frac{\partial \tilde{\eta}}{\partial z} = \frac{\partial \tilde{\phi}_\ell}{\partial r}, \ \frac{\partial \tilde{\eta}}{\partial t} + U_a \frac{\partial \tilde{\eta}}{\partial z} = \frac{\partial \tilde{\phi}_a}{\partial r}, \tag{19.3.22}$$

and the normal stress balance is given, on elimination of the pressures by use of the Bernoulli functions, by

$$-\rho_{a1}\left(\frac{\partial}{\partial t} + U_a \frac{\partial}{\partial z}\right)\tilde{\phi}_a + 2\mu_a\left(\nabla^2\tilde{\phi}_a - \frac{\partial^2\tilde{\phi}_a}{\partial r^2}\right) + \rho_{\ell 1}\left(\frac{\partial}{\partial t} + U_\ell \frac{\partial}{\partial z}\right)\tilde{\phi}_\ell + 2\mu_\ell \frac{\partial^2\tilde{\phi}_\ell}{\partial r^2}$$
$$= \sigma\left(\frac{\partial^2\tilde{\eta}}{\partial z^2} + \frac{1}{a^2}\frac{\partial^2\tilde{\eta}}{\partial \theta^2} + \frac{\tilde{\eta}}{a^2}\right). \tag{19.3.23}$$

The solution to the preceding stability problem formulated is expressed by normal modes

$$\tilde{\phi}_\ell = -i\frac{\omega - kU_\ell}{kI_n'(ka)}HI_n(kr)E + \text{c.c.}, \ \tilde{\phi}_a = -i\frac{\omega - kU_a}{\kappa K_n'(\kappa a)}HK_n(\kappa r)E + \text{c.c.},$$
$$\tilde{\eta} = HE + \text{c.c.}, \tag{19.3.24}$$

where $E \equiv \exp(ikz + in\theta - i\omega t)$ with the complex angular frequency $\omega = \omega_R + i\omega_I$ and the real wavenumber k, n denotes the azimuthal mode. $I_n(kr)$ and $K_n(\kappa r)$ are the modified Bessel functions, where the prime denotes the derivative $I_n'(ka) = \mathrm{d}I_n(ka)/\mathrm{d}(ka)$. The Bessel functions satisfy

$$\nabla^2\tilde{\phi}_\ell = \frac{\partial^2\tilde{\phi}_\ell}{\partial r^2} + \frac{1}{r}\frac{\partial\tilde{\phi}_\ell}{\partial r} + \frac{1}{r^2}\frac{\partial^2\tilde{\phi}_\ell}{\partial\theta^2} + \frac{\partial^2\tilde{\phi}_\ell}{\partial z^2} = \frac{\partial^2\tilde{\phi}_\ell}{\partial r^2} + \frac{1}{r}\frac{\partial\tilde{\phi}_\ell}{\partial r} - \frac{n^2}{r^2}\tilde{\phi}_\ell - k^2\tilde{\phi}_\ell = 0,$$
$$\nabla^2\tilde{\phi}_a - \left(\kappa^2 - k^2\right)\tilde{\phi}_a = \frac{\partial^2\tilde{\phi}_a}{\partial r^2} + \frac{1}{r}\frac{\partial\tilde{\phi}_a}{\partial r} - \frac{n^2}{r^2}\tilde{\phi}_a - \kappa^2\tilde{\phi}_a = 0, \tag{19.3.25}$$

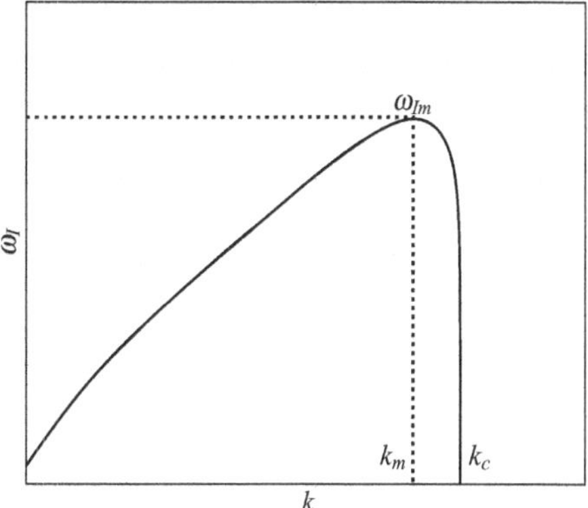

Figure 19.1. The form of a typical graph of the growth rate ω_I versus k. ω_{Im} is the maximum growth, $\lambda_m = 2\pi/k_m$ is the wavelength of the fastest-growing wave, k_c is the cutoff wavenumber, and ω_{Im} and k_m are called peak values.

with

$$\kappa = \sqrt{k^2 - \frac{(\omega - kU_a)^2}{c_a^2 - \dfrac{4i\mu_a}{3\rho_{a1}}(\omega - kU_a)}}, \tag{19.3.26}$$

which arise from (19.3.20) and (19.3.21). Substituting (19.3.24) into (19.3.22) and (19.3.23), we find the dispersion relation:

$$\left[\rho_{a1}(\omega - kU_a)^2 - 2i\mu_a(\kappa^2 - k^2)(\omega - kU_a)\right]\frac{kK_n(\kappa a)}{\kappa K_n'(\kappa a)} - \rho_{\ell 1}(\omega - kU_\ell)^2\frac{I_n(ka)}{I_n'(ka)}$$
$$+ 2i\mu_a k\kappa(\omega - kU_a)\frac{K_n''(\kappa a)}{K_n'(\kappa a)} - 2i\mu_\ell k^2(\omega - kU_\ell)\frac{I_n''(ka)}{I_n'(ka)} + \sigma\left(k^2 - \frac{1-n^2}{a^2}\right)k = 0. \tag{19.3.27}$$

The wavenumber k_m and maximum growth rate ω_{Im} given by $\omega_{Im} = \max \omega_I(k) = \omega_I(k_m)$ define the disturbance that is expected to appear in experiments. A typical dispersion relation is shown in figure 19.1. The cutoff wavenumber is the border of instability $\omega_I(k_c) = 0$.

19.3.6 Stability problem in dimensionless form

The scaling is made as

$$[\text{length, velocity, time}] = [d,\ c_a,\ d/c_a], \tag{19.3.28}$$

with $d = 2a$. The dimensionless variables are

$$(\hat{r}, \theta, \hat{z}) = \left(\frac{r}{d}, \theta, \frac{z}{d}\right), \quad \hat{t} = \frac{c_a}{d}t. \tag{19.3.29}$$

The carets on the independent variables are omitted for brevity. Then we may scale as

$$\frac{\phi_\ell}{c_a d} = \frac{U_\ell}{c_a}\hat{z} + \frac{\tilde{\phi}_\ell}{c_a d} = M_\ell\hat{z} + \hat{\phi}_\ell, \quad \frac{\rho_\ell}{\rho_{\ell 1}} = 1, \quad \frac{p_\ell}{\rho_{\ell 1}c_a^2} = \bar{p}_{\ell 1} + \frac{\tilde{p}_\ell}{\rho_{\ell 1}c_a^2} = \bar{p}_{\ell 1} + \hat{p}_\ell,$$
$$\frac{\phi_a}{c_a d} = \frac{U_a}{c_a}\hat{z} + \frac{\tilde{\phi}_a}{c_a d} = M_a\hat{z} + \hat{\phi}_a, \quad \frac{\rho_a}{\rho_{a1}} = \frac{\rho_{a1}}{\rho_{a1}} + \frac{\tilde{\rho}_a}{\rho_{a1}} = 1 + \hat{\rho}_a, \quad \hat{p}_a = \frac{\tilde{p}_a}{\rho_{\ell 1}c_a^2} = \frac{\tilde{\rho}_a}{\rho_{\ell 1}} = \ell\hat{\rho}_a, \tag{19.3.30}$$

where ¯ denotes the normalized basic flow and ˆ denotes the normalized disturbances, and the parameters are defined as

$$\ell = \frac{\rho_{a1}}{\rho_{\ell 1}}, \quad m = \frac{\mu_a}{\mu_\ell}, \quad \nu = \frac{\mu_a \, \rho_{\ell 1}}{\mu_\ell \, \rho_{a1}} = \frac{m}{\ell}, \quad M_\ell = \frac{U_\ell}{c_a},$$
$$M_a = \frac{U_a}{c_a}, \quad Re = \frac{\rho_{\ell 1} c_a d}{\mu_\ell}, \quad We = \frac{\sigma}{\rho_{\ell 1} d c_a^2}, \tag{19.3.31}$$

where the basic state of the gas is the function of the Mach number, with

$$\frac{\rho_{a1}}{\rho_{a0}} = Q(M_a)^{-1/(\gamma-1)}, \quad \text{where } Q(M_a) \equiv \frac{\gamma-1}{2} M_a^2 + 1,$$
$$\ell = \ell_0 Q(M_a)^{-1/(\gamma-1)}, \quad \ell_0 = \rho_{a0}/\rho_{\ell 1}, \quad \frac{p_{a1}}{p_{a0}} = \left(\frac{\rho_{a1}}{\rho_{a0}}\right)^\gamma, \quad \frac{c_a^2}{c_{a0}^2} = \frac{1}{Q(M_a)},$$
$$Re = Re_0/Q(M_a), \quad Re_0 = \frac{\rho_{\ell 1} c_{a0} d}{\mu_\ell}, \quad We = We_0 \, Q(M_a), \quad We_0 = \frac{\sigma}{\rho_{\ell 1} d c_{a0}^2},$$
$$\frac{1}{2} M_\ell^2 + \bar{p}_{\ell 1} = \text{const.} \tag{19.3.32}$$

For the gas,

$$\left(\frac{\partial}{\partial t} + M_a \frac{\partial}{\partial z}\right) \hat{\rho}_a + \nabla^2 \hat{\phi}_a = 0, \quad \ell \left(\frac{\partial}{\partial t} + M_a \frac{\partial}{\partial z}\right) \hat{\phi}_a + \hat{p}_a - \frac{4m}{3Re} \nabla^2 \hat{\phi}_a = 0. \tag{19.3.33}$$

The combination leads to

$$\left(\frac{\partial}{\partial t} + M_a \frac{\partial}{\partial z}\right)^2 \hat{\phi}_a = \left[1 + \frac{4m}{3\ell Re} \left(\frac{\partial}{\partial t} + M_a \frac{\partial}{\partial z}\right)\right] \nabla^2 \hat{\phi}_a. \tag{19.3.34}$$

For the liquid,

$$\nabla^2 \hat{\phi}_\ell = 0, \quad \left(\frac{\partial}{\partial t} + M_\ell \frac{\partial}{\partial z}\right) \hat{\phi}_\ell + \hat{p}_\ell = 0. \tag{19.3.35}$$

At the interface $r = 1/2 + \hat{\eta} \approx 1/2$, where $\hat{\eta} \equiv \hat{\eta}(\theta, z, t)$ is the interface displacement, the kinematic conditions are given by

$$\frac{\partial \hat{\eta}}{\partial t} + M_\ell \frac{\partial \hat{\eta}}{\partial z} = \frac{\partial \hat{\phi}_\ell}{\partial r}, \quad \frac{\partial \hat{\eta}}{\partial t} + M_a \frac{\partial \hat{\eta}}{\partial z} = \frac{\partial \hat{\phi}_a}{\partial r}, \tag{19.3.36}$$

and the normal stress balance is given by

$$-\ell \left(\frac{\partial}{\partial t} + M_a \frac{\partial}{\partial z}\right) \hat{\phi}_a + \frac{2m}{Re} \left(\nabla^2 \hat{\phi}_a - \frac{\partial^2 \hat{\phi}_a}{\partial r^2}\right) + \left(\frac{\partial}{\partial t} + M_\ell \frac{\partial}{\partial z}\right) \hat{\phi}_\ell + \frac{2}{Re} \frac{\partial^2 \hat{\phi}_\ell}{\partial r^2}$$
$$= We \left(\frac{\partial^2 \hat{\eta}}{\partial z^2} + \frac{\partial^2 \hat{\eta}}{\partial \theta^2} + \hat{\eta}\right). \tag{19.3.37}$$

The solution to the stability problem previously formulated is expressed as

$$\hat{\phi}_\ell = -i \frac{\omega - k M_\ell}{k I_n'(k/2)} \hat{H} I_n(kr) E + \text{c.c.}, \quad \hat{\phi}_a = -i \frac{\omega - k M_a}{\kappa K_n'(\kappa/2)} \hat{H} K_n(\kappa r) E + \text{c.c.},$$
$$\hat{\eta} = \hat{H} E + \text{c.c.}, \tag{19.3.38}$$

where $E \equiv \exp(ikz + in\theta - i\omega t)$ with the complex angular frequency $\omega = \omega_R + i\omega_I$ and the real wavenumber k, $I_n(kr)$, and $K_n(\kappa r)$ are the modified Bessel functions, the prime

Table 19.2. *Properties of air–water*

Diameter of liquid jet, d	0.001 m
Air viscosity μ_a	1.8×10^{-5} N sec/m^2
Air density ρ_{a0}	1.2 kg/m^3
Water density $\rho_{\ell 1}$	1000 kg/m^3
Surface-tension coefficient σ	0.075 N/m
Ratio of the specific heats γ (air)	1.4

denotes the derivative: $I'_n(k/2) = \mathrm{d}I_n(k/2)/\mathrm{d}(k/2)$, and κ is defined as

$$\kappa = \sqrt{k^2 - \frac{\theta^2}{1 - \dfrac{4im}{\ell Re}\theta}}, \tag{19.3.39}$$

where

$$\theta = \omega - kM_a, \quad \theta_\ell = \omega - kM_\ell. \tag{19.3.40}$$

Therefore the dispersion relation is expressed as

$$\left[\ell\theta^2 - \frac{2im}{Re}\left(\kappa^2 - k^2\right)\theta\right]\frac{k}{\kappa}\alpha_{an} + \theta_\ell^2\alpha_n + \frac{2imk\kappa}{Re}\theta\beta_{an} + \frac{2ik^2}{Re}\theta_\ell\beta_n = We\left(k^2 + 4n^2 - 4\right)k, \tag{19.3.41}$$

with

$$\alpha_{\ell n} = \frac{I_n(k/2)}{I'_n(k/2)}, \quad \alpha_{an} = -\frac{K_n(\kappa/2)}{K'_n(\kappa/2)}, \quad \beta_{n\ell} = \frac{I''_n(k/2)}{I'_n(k/2)}, \quad \beta_{an} = -\frac{K''_n(\kappa/2)}{K'_n(\kappa/2)}, \tag{19.3.42}$$

and $\ell = \ell_0 Q(M_a)^{-1/(\gamma-1)}$, $Re = Re_0/Q(M_a)$, $We = We_0 Q(M_a)$ defined under (19.3.32).

It is sometimes convenient to change the frame of the analysis to one moving with the liquid velocity U_ℓ. In this frame the undisturbed liquid jet is at rest and the gas moves with velocity $U_A = U_a - U_\ell$. This is a Gallilean change of frame in which the new coordinates are

$$z' = z + U_\ell t, \tag{19.3.43}$$

$$E = \exp(ikz + in\theta - i\omega_R t + \omega_I t) = \exp(ikz' + in\theta - i\Omega_R t + \omega_I t), \tag{19.3.44}$$

where

$$\Omega_R = \omega_R + U_\ell k \tag{19.3.45}$$

is a new frequency. However, the density, pressure, and sound speed of the gas are determined by the ambient conditions and gas velocity, as in (19.3.13), and these quantities do not change in a Galilean change of frame. For this reason, problems of stability of liquid jets in which U_a and U_ℓ are given, as in the experiments of Varga, Lasheras, and Hopfinger (2003) discussed in subsection 11.3.10, are not simplified by a Galilean change of frame. In the analysis given in subsections 19.3.7–19.3.11 we put $M_\ell = 0$ and $M_a = M$. This is the case of a static liquid cylinder in a moving gas.

In nearly all the computations to follow, ℓ_0, Re_0 and We_0 are evaluated under standard conditions for air–water given in table 19.2. In subsection 19.3.11 we allow We_0 to vary; this can be thought to be the effect of changing surface tension.

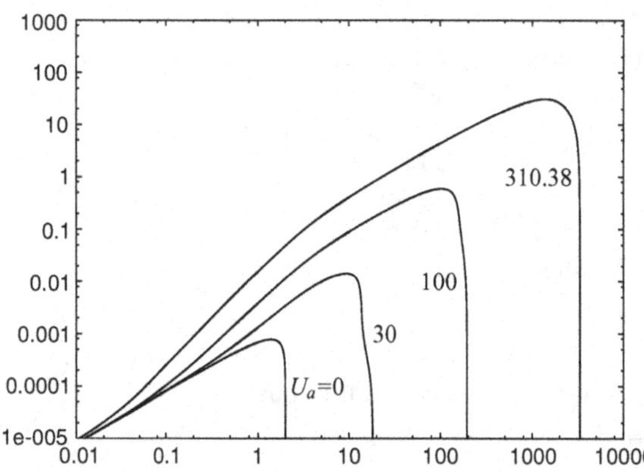

Figure 19.2. The growth rate ω_I versus k in the axisymmetric $n = 0$ mode for IPF, using the material parameters (table 19.2) for stationary water and air with $U_a = 0, 30, 100$, and 310.38 m/sec. The values can be converted into dimensionless form (M, We) by use of table 19.3.

19.3.7 Inviscid potential flow

The problem of an inviscid liquid jet moving in an inviscid compressible gas was considered by Li and Kelly (1992). The dispersion relation for this problem is (19.3.41) with $Re \to \infty$ and $m/Re = 0$:

$$\kappa = \sqrt{k^2 - (\omega - kM)^2}, \quad \ell\,(\omega - kM)^2 \frac{k}{\kappa}\alpha_{an} + \omega^2\alpha_{\ell n} = We\,\left(k^2 + 4n^2 - 4\right) k, \quad (19.3.46)$$

The parameters of this problem are ℓ, n, M, and We.

Pure capillary instability arises in the axisymmetric $n = 0$ mode for large We and in the inviscid case is independent of the gas. The case $We = 0$ is associated with pure KH instability for every n mode, and it cannot occur in a vacuum ($\ell \neq 0$). The variation of ω_I versus k for an inviscid water in air is given in figure 19.2 with the $n = 0$ mode.

The variation of growth rates with M for particular values of k, σ, U_a was given by Li and Kelly (1992); they did not present graphs of peak growth rates $\omega_{Im}(k_m)$ and k_m as functions of M. At very high values of the Mach number,

$$\ell = \ell_0/Q^{2.5} \;\to\; \ell_0/M^5, \quad (19.3.47)$$

and the first term of (19.3.46) may be neglected. Growth-rate curves for IPF under standard conditions and different Mach numbers are shown in figure 19.2.

19.3.8 Growth-rate parameters as functions of M for different viscosities

In this section the data for stability computations are assembled in table 19.3 and figure 19.3. In the table, we list the parameters for air under standard conditions and the liquid density, surface-tension coefficient, and the jet radius used in all the computations. By use of the parameters of table 19.2, the viscosity ratio is evaluated as $m = 1.8 \times 10^{-2}$, and the other basic nondimensional parameters that depend on U_a are shown in table 19.3. Negative values $\omega_I < 0$ in table 19.4 arise for nonaxisymmetric $n = 1$ disturbances when U_a is small. The entries in the column $n = 1$ in the tables are left blank.

Table 19.4 gives the values ω_{Im}, k_m, and k_c of the growth-rate curves like figure 19.1 for $\mu_\ell = 0$, 1 cP, 300 cP, and 8000 cP. In figure 19.4 we blow up the sharply peaking growth-rate curves ω_I versus k for stationary liquid jets of small viscosity in high-speed transonic air.

Table 19.3. *Typical values and nondimensional parameters for various U_a ($U_\ell = 0$)*

U_a (m/sec)	c_a	p_{a1}	ρ_{a1}	ℓ	M	Re	We
0.00	3.400e+02	1.013e+05	1.200e+0	1.200e−03	0.000e+0	3.400e+05	6.488e−07
0.10	3.400e+02	1.013e+05	1.200e+0	1.200e−03	2.941e−04	3.400e+05	6.488e−07
0.20	3.400e+02	1.013e+05	1.200e+0	1.200e−03	5.882e−04	3.400e+05	6.488e−07
0.50	3.400e+02	1.013e+05	1.200e+0	1.200e−03	1.471e−03	3.400e+05	6.488e−07
1.00	3.400e+02	1.013e+05	1.200e+0	1.200e−03	2.941e−03	3.400e+05	6.488e−07
5.00	3.400e+02	1.013e+05	1.200e+0	1.200e−03	1.471e−02	3.400e+05	6.488e−07
10.00	3.400e+02	1.012e+05	1.199e+0	1.199e−03	2.941e−02	3.400e+05	6.489e−07
20.00	3.399e+02	1.011e+05	1.198e+0	1.198e−03	5.884e−02	3.399e+05	6.492e−07
50.00	3.393e+02	9.977e+04	1.187e+0	1.187e−03	1.474e−01	3.393e+05	6.516e−07
70.00	3.386e+02	9.833e+04	1.175e+0	1.175e−03	2.068e−01	3.386e+05	6.543e−07
100.00	3.370e+02	9.530e+04	1.149e+0	1.149e−03	2.967e−01	3.370e+05	6.602e−07
150.00	3.333e+02	8.816e+04	1.087e+0	1.087e−03	4.500e−01	3.333e+05	6.751e−07
200.00	3.280e+02	7.881e+04	1.003e+0	1.003e−03	6.097e−01	3.280e+05	6.970e−07
250.00	3.211e+02	6.787e+04	9.014e−01	9.014e−04	7.786e−01	3.211e+05	7.274e−07
300.00	3.124e+02	5.602e+04	7.860e−01	7.860e−04	9.603e−01	3.124e+05	7.684e−07
310.38	3.104e+02	5.351e+04	7.607e−01	7.607e−04	1.000e+0	3.104e+05	7.786e−07
350.00	3.018e+02	4.401e+04	6.616e−01	6.616e−04	1.160e+0	3.018e+05	8.233e−07
400.00	2.891e+02	3.258e+04	5.337e−01	5.337e−04	1.383e+0	2.891e+05	8.971e−07
450.00	2.740e+02	2.239e+04	4.082e−01	4.082e−04	1.642e+0	2.740e+05	9.987e−07
500.00	2.561e+02	1.395e+04	2.911e−01	2.911e−04	1.952e+0	2.561e+05	1.143e−06
550.00	2.347e+02	7.573e+03	1.882e−01	1.882e−04	2.343e+0	2.347e+05	1.361e−06
600.00	2.088e+02	3.338e+03	1.048e−01	1.048e−04	2.873e+0	2.088e+05	1.720e−06
650.00	1.764e+02	1.023e+03	4.505e−02	4.505e−05	3.686e+0	1.764e+05	2.412e−06
700.00	1.327e+02	1.395e+02	1.085e−02	1.085e−05	5.276e+0	1.327e+05	4.261e−06
750.00	5.568e+01	3.199e−01	1.413e−04	1.413e−07	1.347e+01	5.568e+04	2.419e−05
760.00	8.944e+0	8.832e−07	1.512e−08	1.512e−11	8.497e+01	8.944e+03	9.375e−04
760.20	4.381e+0	5.973e−09	4.262e−10	4.262e−13	1.735e+02	4.381e+03	3.908e−03

The Mach numbers for the seven curves are $[1, 2, 3, 4, 5, 6, 7] = [0.92, 0.96, 0.98, 1.01, 1.05, 1.08, 1.25]$ when $\mu_\ell = 0.15$ cP and $\mu_\ell = 0.175$ cP and for the nine curves $[1, 2, 3, 4, 5, 6, 7, 8, 9] = [0.92, 0.96, 1.00, 1.05, 1.06, 1.12, 1.15, 1.20, 1.25]$ when $\mu_\ell = 0.5$ cP.

19.3.9 Azimuthal periodicity of the most dangerous disturbance

Batchelor and Gill (1962) argued that the conditions at the origin of a cylinder are such as to make the axisymmetric ($n = 0$) mode and the $n = 1$ mode of azimuthal periodicity most dangerous; all the modes except $n = 1$ require that the radial and azimuthal components of the disturbance velocity vanish. The axial component of the disturbance velocity is single valued only when $n = 0$ (see Joseph, 1976, pp. 73, 74). Typical graphs showing the variation of these quantities with the Reynolds number for $M = 0.5$ and $M = 2$ are shown in figures 19.5 and 19.6. Only the axisymmetric ($n = 0$) mode gives rise to instability when $M = 0$. Inspection of figures 19.5 and 19.6 shows that the most dangerous mode is $n = 1$ only for Reynolds numbers smaller than a number near 100; for larger Reynolds numbers the maximum growth rate and the most dangerous wavenumber are nearly the same for $n = 0$ and $n = 1$.

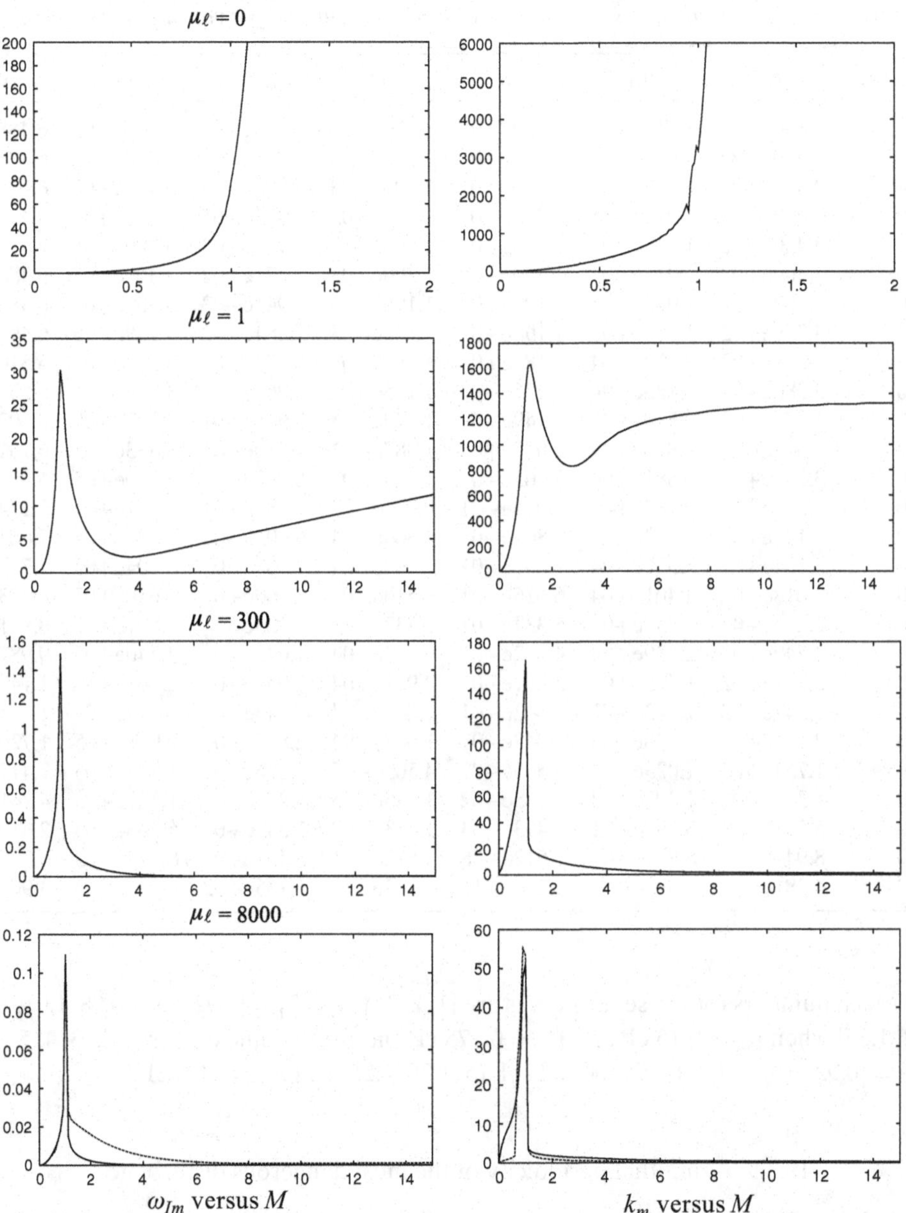

Figure 19.3. Maximum growth rate ω_{Im} and the associated wavenumber k_m as a function of M for $\mu_\ell = 0$, 1 cP, 300 cP, 8000 cP. The solid curve denotes the axisymmetric case ($n = 0$), and the dashed curve for $\mu_\ell = 8000$ cP denotes $n = 1$.

19.3.10 Variation of the growth-rate parameters with the Weber number

Graphs of ω_{Im} and k_m versus We^{-1} are displayed for typical cases in figure 19.7. When We is large, the instability is dominated by capillarity; when We is small, KH instability dominates. This behavior is characteristic also for the liquid jet in an incompressible gas, which was studied by Funada, Joseph and Yamashita (2004); they used We^{-1} rather than We following earlier literature.

Table 19.4. *Maximum growth-rate parameters (figure 19.3) for different values of M: (a) IPF; (b) $\mu_\ell = 1$ cP, $m = 1.8 \times 10^{-2}$; (c) $\mu_\ell = 300$ cP, $m = 6 \times 10^{-5}$; (d) $\mu_\ell = 8000$ cP, $m = 2.25 \times 10^{-6}$*

		$n = 0$			$n = 1$		
	M	ω_{Im}	k_m	k_c	ω_{Im}	k_m	k_c
(a)	0.00	7.821e−04	1.396e+00	2.008e+00	–	–	–
	0.50	3.000e+00	2.989e+02	4.483e+02	3.000e+00	2.989e+02	4.483e+02
	0.75	1.149e+01	7.192e+02	1.081e+03	1.149e+01	7.147e+02	1.081e+03
	1.00	7.853e+01	3.169e+03	4.933e+03	7.853e+01	3.169e+03	4.933e+03
	1.10	2.302e+02	1.540e+04	2.107e+04	2.302e+02	1.540e+04	2.107e+04
	1.50	1.120e+03	2.656e+05	2.755e+05	1.120e+03	2.656e+05	2.755e+05
	2.00	1.576e+03	8.560e+05	8.632e+05	1.576e+03	8.560e+05	8.632e+05
(b)	0.00	7.790e−04	1.387e+00	2.008e+00	–	–	–
	0.50	2.759e+00	2.845e+02	5.869e+02	2.759e+00	2.845e+02	5.869e+02
	0.75	1.007e+01	6.670e+02	1.666e+03	1.007e+01	6.670e+02	1.666e+03
	1.00	3.033e+01	1.432e+03	3.799e+03	3.033e+01	1.432e+03	3.799e+03
	1.10	2.673e+01	1.621e+03	3.871e+03	2.673e+01	1.621e+03	3.871e+03
	1.50	1.161e+01	1.297e+03	2.998e+03	1.161e+01	1.297e+03	2.998e+03
	2.00	6.363e+00	9.703e+02	2.314e+03	6.363e+00	9.703e+02	2.314e+03
(c)	0.00	3.711e−04	9.325e−01	2.008e+00	–	–	–
	0.50	1.457e−01	4.123e+01	4.492e+02	1.453e−01	4.186e+01	4.492e+02
	0.75	3.735e−01	7.651e+01	1.072e+03	3.731e−01	7.723e+01	1.072e+03
	1.00	1.515e+00	1.657e+02	1.531e+03	1.515e+00	1.657e+02	1.531e+03
	1.10	3.803e−01	2.458e+01	1.603e+02	3.897e−01	2.413e+01	1.603e+02
	1.50	1.497e−01	1.495e+01	6.733e+01	1.501e−01	1.477e+01	6.733e+01
	2.00	8.965e−02	1.117e+01	4.492e+01	8.934e−02	1.090e+01	4.483e+01
(d)	0.00	2.656e−05	2.431e−01	2.008e+00	–	–	–
	0.50	6.081e−03	1.153e+01	4.492e+02	5.959e−03	1.666e+01	4.492e+02
	0.75	1.507e−02	2.260e+01	1.072e+03	1.494e−02	2.989e+01	1.072e+03
	1.00	1.096e−01	5.113e+01	1.522e+03	1.095e−01	5.338e+01	1.522e+03
	1.10	1.485e−02	3.511e+00	9.685e+01	2.818e−02	3.268e+00	9.712e+01
	1.50	4.978e−03	2.251e+00	3.223e+01	2.056e−02	1.270e+00	2.989e+01
	2.00	2.301e−03	1.801e+00	2.080e+01	1.604e−02	8.641e−01	1.927e+01

We have shown in subsection 19.3.9 that the most dangerous mode is typically axisymmetric when the Reynolds number is larger than about 100, as is true for the cases considered here. The graphs are all similar; for small values of We^{-1} in which capillarity dominates, the values of $\log \omega_{Im}$ decrease linearly with $\log We^{-1}$, giving rise to a power law $\omega_{Im} = a(We^{-1})^p$, where a and p may be determined from the graphs. The most dangerous wavenumber, $k_m = 1.396$, is a universal value that maximizes ω_I when surface tension dominates. All the growth-rate curves have a minimum value that marks the place where KH instability starts to be important; after this minimum ω_{Im} and k_m increase with We^{-1}. In all case $k_m \to \infty$ as $We^{-1} \to \infty$, but ω_{Im} is lowered as $We^{-1} \to \infty$ when the liquid viscosity is not zero.

19.3.11 Convective/absolute instability

C/A instability is used to determine when the spatial theory of instability makes sense. Practically, this comes down to a determination of the conditions under which a disturbance

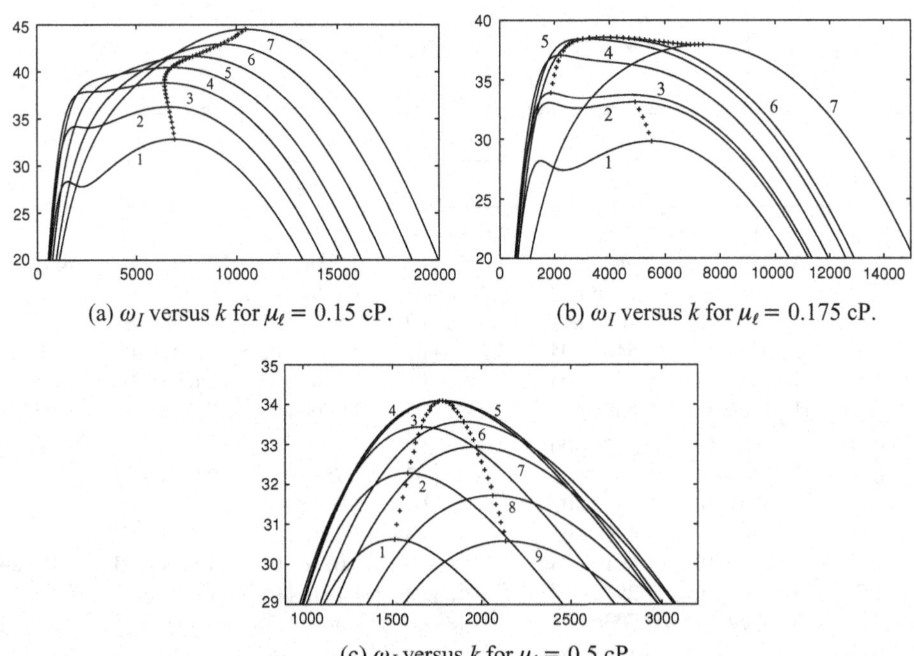

(a) ω_I versus k for $\mu_\ell = 0.15$ cP. (b) ω_I versus k for $\mu_\ell = 0.175$ cP.

(c) ω_I versus k for $\mu_\ell = 0.5$ cP.

Figure 19.4. Growth rate versus wavenumber for stationary liquid jet ($U_\ell = 0$) in transonic air. $U_a = [1, 2, 3, 4, 5, 6, 7] = [290.08, 302.08, 314.08, 326.08, 340.08, 356.08, 370.08]$ m/sec. (a) $\mu_\ell = 0.15$ cP: as U_a increases, the maximum growth rate marked by + increases monotonically without limit; (b) $\mu_\ell = 0.175$ cP: as U_a increases, the maximum growth rate marked by + increases, changes to another peak, attains the maximun near $U_a = 310.38$ m/sec ($M = 1$), and then decreases; (c) $\mu_\ell = 0.5$ cP: as U_a increases, the maximum growth rate marked by + increases, attains the maximun near $U_a = 310.38$ m/sec ($M = 1$), and then decreases.

from a localized source will propagate downstream without corrupting the source. The disturbance may grow as it propagates, but after it passes over a fixed point it leaves the flow undisturbed. This is the case for convectively unstable flows, but absolutely unstable flows propagate both upstream and downstream. Propagation of disturbances from a vibrating ribbon in a boundary-layer or Poiseuille flow are examples. For such propagation these flows must be convectively and not absolutely unstable.

The study of stability of disturbances issuing from a fixed source, leading to C/A theory, is not a complete stability theory; the traditional temporal theory of instability needs also to be considered. The temporal theory determines the conditions under which disturbances at a fixed point will grow or decay. If these conditions are such that all disturbance decay is stable, then disturbances from a fixed point will decay. Disturbances that are convectively or absolutely unstable are also temporally unstable. The propagation of impulses from a source in a convectively unstable flow can be realized, provided the growth rates and amplitudes of temporally unstable flows do not corrupt the flow first. This is why experiments with vibrating ribbons are always done with care to suppress background noise that may amplify in time a fixed point.

Li and Kelly (1992) considered the convective and absolute instability of an inviscid liquid jet in an airstream of an inviscid compressible gas. They motivated their study by experiments on the instability and breakup of liquid fuel jets in cross flow. They note that

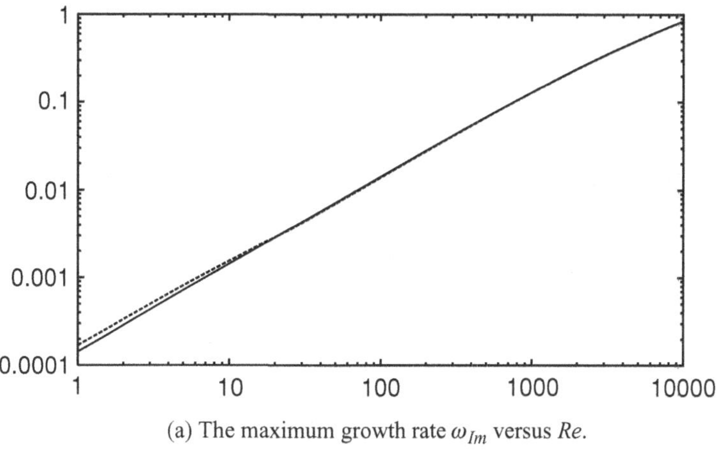

(a) The maximum growth rate ω_{Im} versus *Re*.

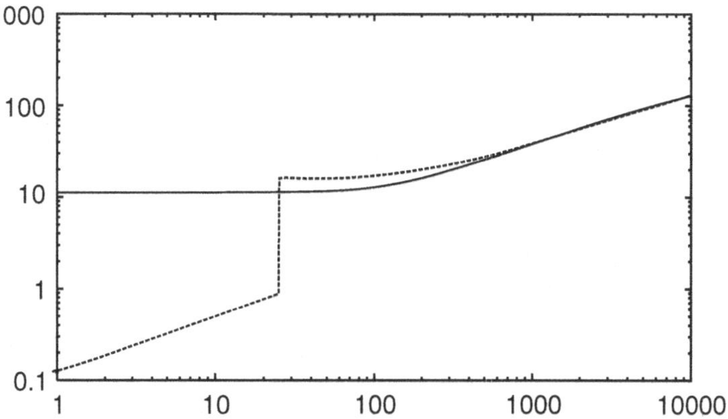

(b) k_m versus *Re;* the growth rate curve ω_I vs. k which is shown in figure 19.5 (c) has two relative maxima; the absolute maxima changes for *Re* between 25 and 26; this is seen as a jump in k in 19.5 (b) .

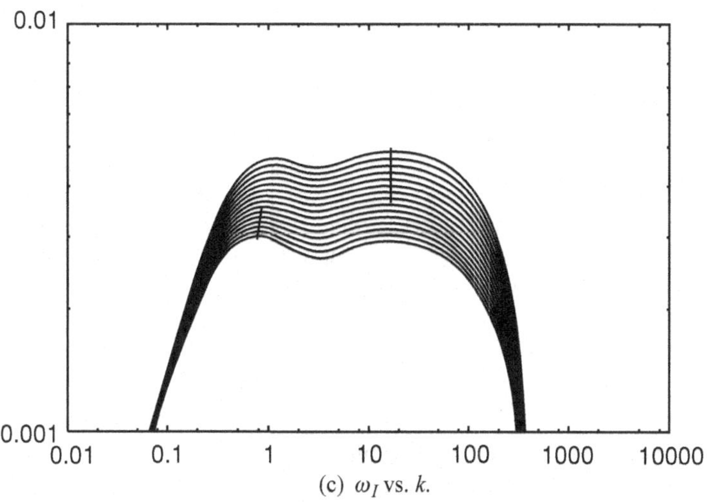

(c) ω_I vs. k.

Figure 19.5. (a) The maximum growth rate ω_{Im} versus *Re* and (b) k_m versus *Re*, for $M = 0.5$; $n = 0$ in compressible gas (solid curve) and $n = 1$ in compressible gas (dashed curve).

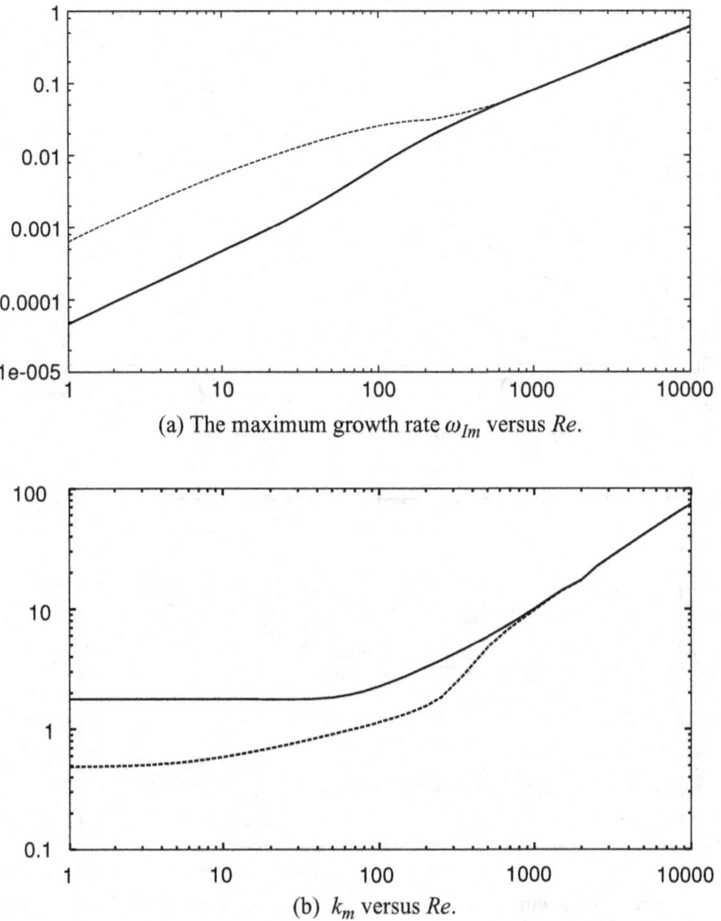

(a) The maximum growth rate ω_{Im} versus Re.

(b) k_m versus Re.

Figure 19.6. (a) The maximum growth rate ω_{Im} versus Re and (b) k_m versus Re, for $M = 2; n = 0$ in compressible gas (solid curve) and $n = 1$ in compressible gas (dashed curve).

> ... the breakup of the jet, however, does not seem to proceed in the gradual manner typical of the capillary instability of a liquid jet issuing into air at rest. Here the jet gradually bends over toward the direction of the free stream so that both tangential and cross-flow components of the gas flow are seen by the jet. When the jet is at an angle of about 30° form the normal to the free stream, the jet breaks into large columns in a manner so sudden that Schetz and co-workers (Sherman and Schetz, 1971; Schetz *et al.*, 1980) have used the phrase "fracture" to describe the phenomenon. At this angle, the component of the gas velocity parallel to the jet's direction is approximately sonic.

Li and Kelly (1992) did not study a jet in cross flow; they studied a liquid in a coflowing airstream. They also considered the convective and absolute instability of a plane jet and found transition to absolute instability in the transonic region. They speculate that fracture coincides with the transition to absolute instability.

The problem considered by Li and Kelly (1992), C/A instability of a plane jet, is rather far from experiments on cross flow of Sherman and Schetz (1971), and it could be considered within the framework of temporal combined RT, KH instability, which should

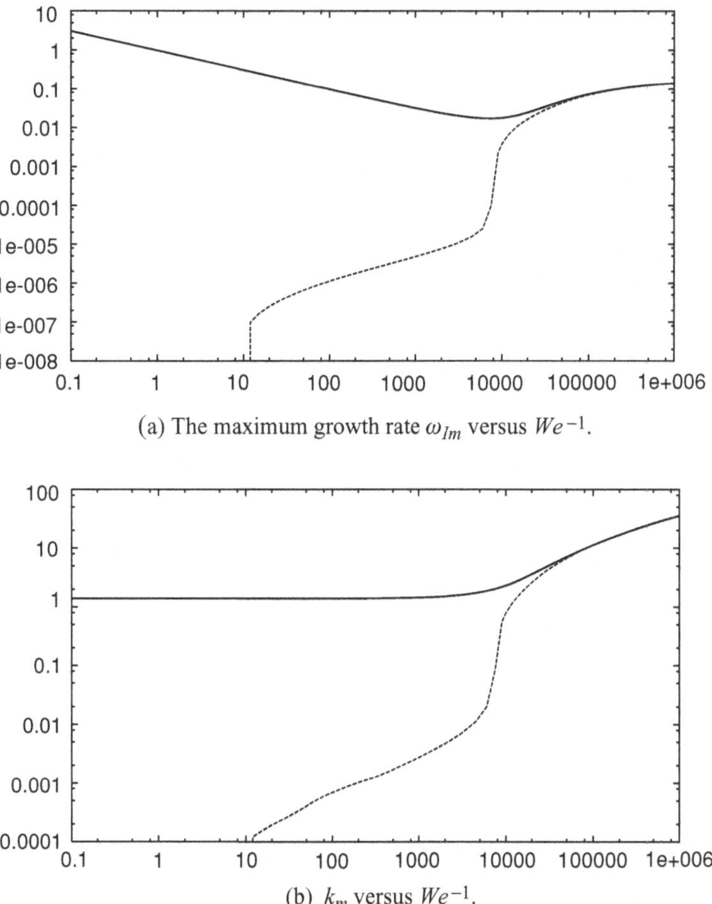

(a) The maximum growth rate ω_{Im} versus We^{-1}.

(b) k_m versus We^{-1}.

Figure 19.7. (a) The maximum growth rate ω_{Im} versus We^{-1}, (b) k_m versus We^{-1}. VPF, $M = 0.5$ for $\mu_\ell = 300$ cP; $n = 0$ in compressible gas (solid curve) and $n = 1$ in compressible gas (dashed curve).

show rather exceptional behavior in transonic flow. This kind of explanation of rapid breakup of a coflowing jet was given by Varga, Lasheras, and Hopfinger (2003), and it is based on a secondary RT istability not associated with absolute instability. An analysis using VPF of KH instability in the coaxial jet studied by Varga *et al.* has been given by FJSY (2006).

19.3.12 Conclusions

We studied the temporal instability of a liquid jet in a high-speed compressible airstream by using VPF. Because the shear stress is ignored in VPF, the analysis is compatible with the discontinuous profile used in all studies of KH instability. This discontinuity is not allowable for real viscous fluids for which shear layers develop. Disturbances with high wavenumbers might see the details of the shear layer and alter the stability results in ways that are currently unknown.

In our analysis, which neglects shear layers, this instability is dominated by capillarity when the Weber number is large and by KH instability when the Weber number is small.

The peak growth rates and the associated wavelengths depend strongly on the Mach number and on the viscosity of the liquid. The growth rates are dramatically larger in the transonic region and the wavelengths of the peak values are much smaller. Viscosity reduces the magnitude of the growth. The growth rate for IPF monotonically increases as the Mach number increases. For $0 < \mu_\ell < 0.168$ cP, the growth rate for VPF monotonically increases as the Mach number increases. For 0.168 cP$< \mu_\ell$, the growth rate for VPF has a peak value when the Mach number is nearly one. For more viscous liquids, the growth rate for VPF has a peak value when the Mach number is nearly one. The peak value decreases as the viscosity μ_ℓ increases. The growth rates are very sharply peaked near $M_a = 1$ when the viscosity is larger than some value near 0.2 cP. The dramatic change in the stability of liquid jets in transonic flow predicted by analysis is possibly related to the dramatic increases in the drag coefficient of spheres and disks in transonic flow observed in experiments (Howarth, 1953, p. 724). It is not known if jet breakup in transonic and supersonic flow is caused directly by KH instability or through a secondary RT instability. In either case, this analysis suggests that the drop fragments would be very small.

20

Irrotational flows of viscoelastic fluids

The coupling of rotational and irrotational flows is more complicated than in Newtonian fluids because in most models of a viscoelastic fluid the stress is not linear. The equation of motion of an incompressible fluid with velocity $\mathbf{u} = \mathbf{v} + \nabla\phi$, where $\operatorname{curl}\mathbf{u} = \operatorname{curl}\mathbf{v}$, can be expressed as

$$\nabla\left(\rho\frac{\partial\phi}{\partial t} + \frac{\rho}{2}|\nabla\phi|^2 + p\right) + \rho\operatorname{div}(\mathbf{v}\otimes\nabla\phi + \nabla\phi\otimes\mathbf{v} + \mathbf{v}\otimes\mathbf{v}) = \operatorname{div}\boldsymbol{\tau}[\mathbf{v} + \nabla\phi], \quad (20.0.1)$$

where $\boldsymbol{\tau}[\mathbf{u}] = \boldsymbol{\tau}[\mathbf{v} + \nabla\phi]$ is modeled by a constitutive equation. The pressure p is an unknown; together with three components of $\mathbf{u}$, it is the fourth unknown for three components of the momentum equation and $\operatorname{div}\mathbf{u} = 0$. In the decomposed equations, ϕ is an additional unknown, and $\nabla^2\phi = 0$ is an additional equation. The pressure p has no thermodynamic significance. If $\operatorname{tr}\boldsymbol{\tau} = 0$, then p is the average stress.

Many models of non-Newtonian fluids have been proposed. No model works for every kind of flow. Intelligent choices for models tune the model to the application.

20.1 Oldroyd B model

The Oldroyd B model is very often used to describe the response of viscoelastic fluids. Like other models, it has features that restrict its applicability (see Tanner, 1985, p. 222), but it appears to combine effects of relaxation and nonlinearity with relative ease of execution better than many other models. The stress $\boldsymbol{\tau}$ in the Oldroyd B model is governed by a rate equation,

$$\lambda\overset{\nabla}{\boldsymbol{\tau}} + \boldsymbol{\tau} = \mu\left(\mathbf{A} + \tilde{\lambda}\overset{\nabla}{\mathbf{A}}\right), \quad (20.1.1)$$

where

$$\overset{\nabla}{\boldsymbol{\tau}} = \frac{\mathrm{d}\boldsymbol{\tau}}{\mathrm{d}t} - \mathbf{L}\boldsymbol{\tau} - \boldsymbol{\tau}\mathbf{L}^T \quad (20.1.2)$$

is an upper convected derivative and

$$\frac{\mathrm{d}\boldsymbol{\tau}}{\mathrm{d}t} = \frac{\partial\boldsymbol{\tau}}{\partial t} + \mathbf{u}\cdot\nabla\boldsymbol{\tau},$$

$$\mathbf{L}[\mathbf{u}] = \nabla\mathbf{u}, \quad L_{ij} = \frac{\partial u_i}{\partial x_j}, \quad \mathbf{A} = \mathbf{L} + \mathbf{L}^T, \quad A_{ij} = L_{ij} + L_{ij}^T, \quad (20.1.3)$$

where λ is a relaxation time and $\tilde{\lambda}$ is a retardation time. The Oldroyd B model reduces to a Newtonian fluid when $\lambda = \tilde{\lambda}$ and to an upper-convected Maxwell (UCM) model when $\tilde{\lambda} = 0$. Usually $0 < \tilde{\lambda} < \lambda$.

The term $\operatorname{div} \boldsymbol{\tau} [\mathbf{v} + \nabla\phi]$ is much more complicated than for a Newtonian fluid; the term $\operatorname{div} \mathbf{A} [\mathbf{u}] = \operatorname{div} \mathbf{A} [\mathbf{v} + \nabla\phi] = \nabla^2 \mathbf{v}$, but the other terms in (20.1.1) and (20.1.2) retain a strong dependence on the potential ϕ.

It can be said that the flow of viscoelastic fluids modeled as Oldroyd B has a strong irrotational component but that effective procedures to extract the irrotational component are not yet known.

20.2 Asymptotic form of the constitutive equations

The retarded motion expansion can be applied if the flow is slow and slowly varying. The antisymmetric part of the velocity gradient tensor $\mathbf{L}$ is defined as

$$\mathbf{W} = \frac{1}{2}(\mathbf{L} - \mathbf{L}^T). \tag{20.2.1}$$

The substantial, upper-convected, and corotational derivatives are defined as

$$\dot{\mathbf{A}} = \frac{\partial \mathbf{A}}{\partial t} + \mathbf{u} \cdot \nabla \mathbf{A}, \tag{20.2.2}$$

$$\overset{\triangledown}{\mathbf{A}} = \dot{\mathbf{A}} - \mathbf{L} \cdot \mathbf{A} - \mathbf{A} \cdot \mathbf{L}^T, \tag{20.2.3}$$

$$\overset{\circ}{\mathbf{A}} = \dot{\mathbf{A}} - \mathbf{W} \cdot \mathbf{A} + \mathbf{A} \cdot \mathbf{W}. \tag{20.2.4}$$

It can be shown that the relation between $\overset{\triangledown}{\mathbf{A}}$ and $\overset{\circ}{\mathbf{A}}$ is

$$\overset{\circ}{\mathbf{A}} = \overset{\triangledown}{\mathbf{A}} + \mathbf{A}^2. \tag{20.2.5}$$

20.2.1 Retarded motion expansion for the UCM model

The UCM model is

$$\lambda \overset{\triangledown}{\boldsymbol{\tau}} + \boldsymbol{\tau} = \mu \mathbf{A}. \tag{20.2.6}$$

Following Larson (1988), we ignore $\lambda \overset{\triangledown}{\boldsymbol{\tau}}$ and obtain at first order

$$\boldsymbol{\tau}^{(1)} = \mu \mathbf{A}. \tag{20.2.7}$$

It follows that

$$\overset{\triangledown}{\boldsymbol{\tau}}^{(1)} = \mu \overset{\triangledown}{\mathbf{A}}. \tag{20.2.8}$$

Then we can obtain the stress tensor to second order in λ by substituting (20.2.8) into (20.2.6):

$$\boldsymbol{\tau}^{(2)} = \mu \mathbf{A} - \lambda\mu \overset{\triangledown}{\mathbf{A}} = \mu \mathbf{A} - \lambda\mu \overset{\circ}{\mathbf{A}} + \lambda\mu \mathbf{A}^2 \tag{20.2.9}$$

$$= \mu \mathbf{A} - \lambda\mu(\dot{\mathbf{A}} - \mathbf{W} \cdot \mathbf{A} + \mathbf{A} \cdot \mathbf{W}) + \lambda\mu \mathbf{A}^2. \tag{20.2.10}$$

This can be compared with the second-order fluid model:

$$\boldsymbol{\tau} = \mu \mathbf{A} + \alpha_1(\dot{\mathbf{A}} + \mathbf{L} \cdot \mathbf{A} + \mathbf{A} \cdot \mathbf{L}^T) + \alpha_2 \mathbf{A}^2. \tag{20.2.11}$$

The same tensors appear with different coefficients.

20.2.2 The expanded UCM model in potential flow

In potential flows, $\nabla\mathbf{u} = \dfrac{\partial^2\phi}{\partial x_i \partial x_j}$. Therefore

$$\mathbf{L} = \mathbf{L}^T = \frac{1}{2}\mathbf{A}, \qquad \mathbf{W} = 0. \tag{20.2.12}$$

The expanded UCM model reduces to

$$\boldsymbol{\tau} = \mu\mathbf{A} - \lambda\mu\dot{\mathbf{A}} + \lambda\mu\mathbf{A}^2. \tag{20.2.13}$$

The momentum equation is

$$\rho\left[\partial\mathbf{u}/\partial t + (\mathbf{u}\cdot\nabla)\mathbf{u}\right] = -\nabla p + \nabla\cdot\boldsymbol{\tau}. \tag{20.2.14}$$

When $\mathbf{u} = \nabla\phi$ (see Joseph, 1992),

$$\operatorname{div}(\mathbf{u}\cdot\nabla\mathbf{A}) = \operatorname{grad}\chi, \quad \operatorname{div}\mathbf{A}^2 = 2\operatorname{grad}\chi, \Rightarrow \nabla\cdot\boldsymbol{\tau} = \lambda\mu\operatorname{grad}\chi, \tag{20.2.15}$$

where

$$\chi = \frac{\partial^2\phi}{\partial x_i\partial x_j}\frac{\partial^2\phi}{\partial x_j\partial x_i} = \frac{1}{4}\operatorname{tr}\mathbf{A}^2, \quad A_{ij} = 2\frac{\partial^2\phi}{\partial x_i\partial x_j}.$$

Combining (20.2.14) and (20.2.15), we find a Bernoulli equation:

$$\rho\frac{\partial\phi}{\partial t} + \frac{\rho}{2}|\nabla\phi|^2 + p - \lambda\mu\chi = C(t). \tag{20.2.16}$$

Now the total stress can be written as

$$\mathbf{T} = -\left(C + \lambda\mu\chi - \rho\frac{\partial\phi}{\partial t} - \frac{\rho}{2}|\nabla\phi|^2\right)\mathbf{1} + \mu\mathbf{A} - \lambda\mu\left(\frac{\partial}{\partial t} + \mathbf{u}\cdot\nabla\right)\mathbf{A} + \lambda\mu\mathbf{A}^2. \tag{20.2.17}$$

20.2.3 Potential flow past a sphere calculated with the expanded UCM model

The potential of a uniform flow past a sphere is given by

$$\phi = -Ur\cos\theta\left(1 + \frac{a^3}{2r^3}\right), \tag{20.2.18}$$

where U is the velocity of the uniform stream and a is the radius of the sphere. The velocities are

$$u_r = -U\left(1 - \frac{a^3}{r^3}\right)\cos\theta, \quad u_\theta = U\left(1 + \frac{a^3}{2r^3}\right)\cos\theta. \tag{20.2.19}$$

The tensor $\mathbf{A}$ can be evaluated as

$$\mathbf{A} = \frac{3a^3U}{r^4}\begin{bmatrix} -2\cos\theta & -\sin\theta & 0 \\ -\sin\theta & \cos\theta & 0 \\ 0 & 0 & \cos\theta \end{bmatrix}. \tag{20.2.20}$$

Then Bernoulli's equation (20.2.16) is used to obtain the pressure at the surface of the sphere:

$$p = p_\infty + \lambda\mu\frac{9U^2}{2a^2}\left(1 + 2\cos^2\theta\right) + \frac{\rho}{2}U^2\left(1 - \frac{9}{4}\sin^2\theta\right), \tag{20.2.21}$$

where p_∞ is the pressure at infinity. The normal stress T_{rr} at the surface of the sphere is calculated from (20.2.17) and expressed in a dimensionless form:

$$T_{rr}^* = \frac{T_{rr} + p_\infty}{\rho U^2/2} = \left(\frac{9}{4}\sin^2\theta - 1\right) - \frac{12}{Re}\cos\theta + \frac{\lambda\mu}{\rho a^2}\left(45 - 72\sin^2\theta\right), \quad (20.2.22)$$

where $Re = \rho U a/\mu$ is the Reynolds number. At the stagnation points of a sphere $[r = a,\ \theta = 0 \text{ or } \pi]$, the normal stresses are, respectively,

$$\frac{T_{rr} + p_\infty}{\rho U^2/2} = \left[-1 + \frac{45\lambda\mu}{\rho a^2}\right] \mp \frac{12}{Re}. \quad (20.2.23)$$

The viscous contribution gives rise to compression $-12/Re$ at the front stagnation point and to tension $12/Re$ at the rear. The stress that is due to inertia and viscoelasticity is the same at $\theta = 0$ and $\theta = \pi$ and is a tension when $45\lambda\mu > \rho a^2$. This condition represents a competition between the inertia and the viscoelasticity. If ρa^2 is not too large, the stress at the stagnation points is a tension, reversing the compression because of inertia.

Coleman and Noll (1960) developed a theory for a general stress functional for incompressible fluids that depends on only the history of the first spatial gradient of the deformation. Fluids of this type are called simple; they are rather too general to be applied. Useful forms of a simple fluid can be obtained in certain special asymptotic limits in which the range of fluid response is severely limited.

One such limit is defined for small-amplitude, but possibly high-frequency (Joseph, 1990, p. 539), motions. In this limit, the present value of stress at a point $\mathbf{x}$,

$$\boldsymbol{\tau}(x,t) = \int_0^\infty G(s)\mathbf{A}\left[\mathbf{u}(x, t - s)\right] ds, \quad (20.2.24)$$

is determined by the history of strain rate $\mathbf{A}$. The relaxation function is assumed to be rapidly decaying, say proportional to $\exp(-\lambda s)$; this weights the present value of the stain more strongly in the recent history than in the past history.

Fluids satisfying (20.2.24) are called linear viscoelastic with fading memory. For these fluids, the force density

$$\operatorname{div}\boldsymbol{\tau} = \int_0^\infty G(s)\nabla^2\mathbf{v}(x, t - s)\, ds \quad (20.2.25)$$

does not depend on the irrotational flow.

20.3 Second-order fluids

Another general class of models, namely the second-order fluid model, introduced by Coleman and Noll (1960), arises from an expansion of the general stress functional for slow and slowly varying motions (Rivlin and Ericksen, 1955; Bird, Armstrong, and Hassager, 1987; Joseph, 1990). It has been used in many studies of viscoelastic behavior with varying degrees of success; the predictions of fluid mechanical response to rapidly varying motions in which fluid memory is important have not been satisfactory, but the predictions for slow steady motions are excellent. We regard the results of analysis by using the second-order fluid model as tentative and subject to ultimate validation by experiment and by comparison with direct numerical simulation by use of other constitutive equations.

If slow motions are defined as those for which $\mathbf{u} = \varepsilon\hat{\mathbf{u}}$ for finite $\hat{\mathbf{u}}$ as ε tends to zero, then slowly varying flows are such that

$$\frac{\partial}{\partial t} = \varepsilon \frac{\partial}{\partial \hat{t}}$$

are nearly steady in the sense that

$$\frac{\mathrm{d}}{\mathrm{d}t}(\cdot) = \varepsilon^2 \left[\frac{\partial}{\partial \hat{t}}(\cdot) + \hat{\mathbf{u}} \cdot \nabla(\cdot) \right].$$

Nearly steady flows are not useful for the description of relaxation effects.

The asymptotic form of the constitutive equation for a viscoelastic fluid in a slow, slowly varying motion, is a Newtonian fluid at order ε and a second-order fluid at order ε^2. Higher-order approximations can be derived; they have unknown coefficients and are rarely used.

At first order

$$\boldsymbol{\tau} = \mu \mathbf{A}[\mathbf{u}]. \tag{20.3.1}$$

At second order

$$\boldsymbol{\tau} = \mu \mathbf{A}[\mathbf{u}] + \alpha_1 \mathbf{B}[\mathbf{u}] + \alpha_2 \mathbf{A}^2[\mathbf{u}], \tag{20.3.2}$$

where

$$\mathbf{B} = \frac{\partial \mathbf{A}}{\partial t} + (\mathbf{u} \cdot \nabla)\mathbf{A} + \mathbf{A}\mathbf{L} + \mathbf{L}^T \mathbf{A}, \tag{20.3.3}$$

$\alpha_1 = -n_1/2$, and $\alpha_2 = n_1 + n_2$, where $[n_1, n_2] = \left[N_1\left(\overset{o}{\gamma}{}^2\right),\ N_2\left(\overset{o}{\gamma}{}^2\right)\right]/\overset{o}{\gamma}{}^2$ as $\overset{o}{\gamma} \to 0$ are constants obtained from the first and second normal stress differences. The first normal stress difference is generally positive; the second normal stress difference N_2 can be negative but much smaller than N_1. For a typical liquid M1 (Hu *et al.*, 1990), $\rho = 0.895$ g/cm^3,

$$[\alpha_1,\ \alpha_2] = [-3,\ 5.34]\ \text{g/cm}.$$

The equations of motion for a second-order incompressible fluid are

$$\rho \left[\frac{\partial \mathbf{u}}{\partial t} + \mathbf{u} \cdot \nabla \mathbf{u} \right] = -\nabla p + \mu \nabla^2 \mathbf{u} + \operatorname{div}\left[\alpha_1 \mathbf{B} + \alpha_2 \mathbf{A}^2\right] \tag{20.3.4}$$

or

$$\operatorname{div} \boldsymbol{\tau}[\mathbf{u}] = \operatorname{div} \boldsymbol{\tau}[\mathbf{v} + \nabla \phi] = \mu \nabla^2 \mathbf{v} + \operatorname{div}[\alpha_1 \mathbf{B} + \alpha_2 \mathbf{A}^2]. \tag{20.3.5}$$

The nonlinear tensors $\mathbf{B}$ and $\mathbf{A}^2$ depend on both $\mathbf{v}$ and ϕ.

The second-order fluid form of viscoelastic models can also be obtained by the method of successive approximations. Consider the Oldroyd B model. The first approximation, $\overset{\nabla}{\boldsymbol{\tau}} = \mu \overset{\nabla}{\mathbf{A}}$, may be inserted into (20.1.1), which becomes

$$\left(\lambda - \tilde{\lambda}\right) \mu \overset{\nabla}{\mathbf{A}}[\mathbf{u}] + \boldsymbol{\tau} = \mu \mathbf{A}. \tag{20.3.6}$$

Equation (20.3.6) may be expressed as a corotational model:

$$\boldsymbol{\tau} = \mu\mathbf{A} - \left(\lambda - \tilde{\lambda}\right)\mu\left[\frac{\mathrm{d}\mathbf{A}}{\mathrm{d}t} - \mathbf{WA} + \mathbf{AW} - \mathbf{A}^2\right] \tag{20.3.7}$$

$$\tau_{ij} = \mu A_{ij} - \left(\lambda - \tilde{\lambda}\right)\mu\left[\frac{\mathrm{d}A_{ij}}{\mathrm{d}t} - W_{il}A_{lj} + A_{il}W_{lj} - A_{il}A_{lj}\right], \tag{20.3.8}$$

where **W** vanishes for potential flow.

20.4 Purely irrotational flows

A sufficient condition for a purely irrotational flow is that every flow given by a potential $\mathbf{u} = \nabla\phi$, $\mathbf{v} = 0$, is compatible with solutions of equations of motion (20.0.1); that is,

$$\mathrm{curl}\left(\nabla\cdot\boldsymbol{\tau}\,[\nabla\phi]\right) = 0, \tag{20.4.1}$$

$$\varepsilon_{lmn}\frac{\partial^2\tau_{lm}}{\partial x_l\partial x_m}\,[\nabla\phi] = 0. \tag{20.4.2}$$

Although this condition is not satisfied for $\mathbf{u} = \nabla\phi$ by most constitutive equations (Joseph and Liao, 1994a, 1994b), it is satisfied for invicid fluids, for viscous fluids with constant properties, and for linear viscoelastic and second-order fluids.

20.5 Purely irrotational flows of a second-order fluid

When $\mathbf{u} = \nabla\phi$ (see Joseph, 1992),

$$\mathrm{div}(\mathbf{u}\cdot\nabla\mathbf{A}) = \mathrm{grad}\,\chi,\quad \mathrm{div}(\mathbf{AL}) = \mathrm{grad}\,\chi,\quad \mathrm{div}\,\mathbf{A}^2 = 2\,\mathrm{grad}\,\chi \tag{20.5.1}$$

$$\Rightarrow \mathrm{div}[\alpha_1\mathbf{B} + \alpha_2\mathbf{A}^2] = \mathrm{grad}\,(3\alpha_1 + 2\alpha_2)\,\chi, \tag{20.5.2}$$

where

$$\chi = \frac{\partial^2\phi}{\partial x_i\partial x_j}\frac{\partial^2}{\partial x_j\partial x_i} = \frac{1}{4}\mathrm{tr}\,\mathbf{A}^2,\quad A_{ij} = 2\frac{\partial^2\phi}{\partial x_i\partial x_j}. \tag{20.5.3}$$

Combining (20.2.14) and (20.5.2), we find a Bernoulli equation:

$$\rho\frac{\partial\phi}{\partial t} + \frac{1}{2}\rho\,|\nabla\phi|^2 + p - \hat{\beta}\chi = C(t), \tag{20.5.4}$$

where $\hat{\beta} = 3\alpha_1 + 2\alpha_2 \geq 0$ is the climbing constant. Returning now to the stress with the pressure (20.5.4), we obtain

$$T = -\left[C + \hat{\beta}\chi - \rho\frac{\partial\phi}{\partial t} - \frac{1}{2}\rho\,|\nabla\phi|^2\right]\mathbf{1} + \left[\mu + \alpha_1\left(\frac{\partial}{\partial t} + \mathbf{u}\cdot\nabla\right)\right]\mathbf{A} + (\alpha_1 + \alpha_2)\,\mathbf{A}^2. \tag{20.5.5}$$

20.6 Reversal of the sign of the normal stress at a point of stagnation

The Cartesian components of stress (20.5.5) are given by

$$T_{ij} = -[C + \hat{\beta}\phi_{,il}\phi_{,il} - \rho\phi_{,t} - \rho|\mathbf{u}|^2/2]\delta_{ij} + 2[\mu + \alpha_1(\partial_t + \mathbf{u}\cdot\nabla)]\phi_{ij} + 4(\alpha_1 + \alpha_2)\phi_{,il}\phi_{,lj}. \tag{20.6.1}$$

In the diagonal coordinates x_1, x_2, x_3 of the frame in which $\phi_{,ij}$ is diagonal,

$$[\phi_{,ij}] = \begin{bmatrix} \lambda_1 & 0 & 0 \\ 0 & \lambda_2 & 0 \\ 0 & 0 & \lambda_3 \end{bmatrix}, \tag{20.6.2}$$

we have

$$\begin{bmatrix} \sigma_{11} & 0 & 0 \\ 0 & \sigma_{22} & 0 \\ 0 & 0 & \sigma_{33} \end{bmatrix} = -[C - \rho\phi_t - \rho|\mathbf{u}|^2/2 + \hat{\beta}(\lambda_1^2 + \lambda_2^2 + \lambda_3^2)] \begin{bmatrix} 1 & 0 & 0 \\ 0 & 1 & 0 \\ 0 & 0 & 1 \end{bmatrix} + 2[\mu + \alpha_1(\partial_t + \nabla\phi\cdot\nabla)] \begin{bmatrix} \lambda_1 & 0 & 0 \\ 0 & \lambda_2 & 0 \\ 0 & 0 & \lambda_3 \end{bmatrix} + 4(\alpha_1 + \alpha_2) \begin{bmatrix} \lambda_1^2 & 0 & 0 \\ 0 & \lambda_2^2 & 0 \\ 0 & 0 & \lambda_3^2 \end{bmatrix}. \tag{20.6.3}$$

The case of flow at the stagnation points of a body in steady flow in an arbitrary direction is of special interest. The steady streaming past a stationary body is equivalent, under a Galilean transformation, to the steady motion of a body in an otherwise quiet fluid. The potential flow of a fluid near a point $(x_1, x_2, x_3) = (0, 0, 0)$ of stagnation is a purely extensional motion with

$$[\lambda_1, \lambda_2, \lambda_3] = \frac{U}{L}\dot{S}[2, -1, -1], \tag{20.6.4}$$

where $\dot{S}$ is the dimensionless rate of stretching in the direction x_1, L is the scale of length, and

$$[u_1, u_2, u_3] = \frac{U}{L}\dot{S}\,[2x_1, -x_2, -x_3]. \tag{20.6.5}$$

In this case,

$$\begin{bmatrix} T_{11} & 0 & 0 \\ 0 & T_{22} & 0 \\ 0 & 0 & T_{33} \end{bmatrix} = \frac{\rho}{2}\left[\frac{U^2}{L^2}\dot{S}^2(4x_1^2 + x_2^2 + x_3^2) - U^2\right] \begin{bmatrix} 1 & 0 & 0 \\ 0 & 1 & 0 \\ 0 & 0 & 1 \end{bmatrix} + \mu\frac{U}{L}\dot{S} \begin{bmatrix} 2 & 0 & 0 \\ 0 & -1 & 0 \\ 0 & 0 & -1 \end{bmatrix} + 2\frac{U^2}{L^2}\dot{S}^2 \begin{bmatrix} -\alpha_1 + 2\alpha_2 & 0 & 0 \\ 0 & -7\alpha_1 - 4\alpha_2 & 0 \\ 0 & 0 & -7\alpha_1 - 4\alpha_2 \end{bmatrix}. \tag{20.6.6}$$

At the stagnation point itself,

$$\sigma_{11} = -\frac{\rho}{2}U^2 + 2\mu\dot{S} + 2(2\alpha_2 - \alpha_1)\frac{U^2}{L^2}\dot{S}^2. \tag{20.6.7}$$

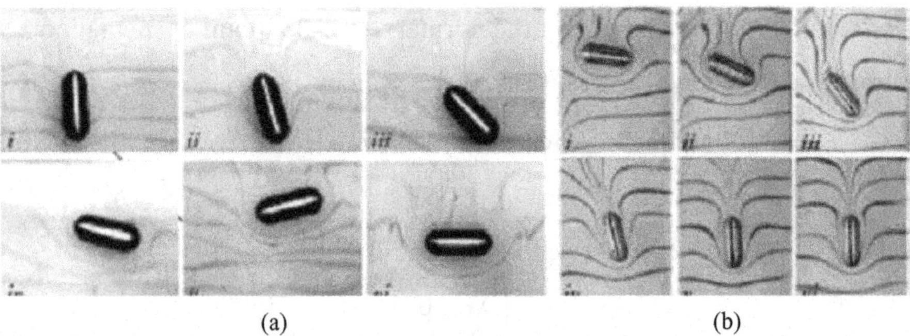

(a) (b)

Figure 20.1. Cylinders falling in (a) Newtonian fluid (glycerin), and (b) viscoelastic fluid (2% aqueous PEO solution). In (a) the cylinder is turned horizontal by inertia; in (b) it is turned vertical by viscoelastic pressures

Because $\alpha_1 < 0$, $2\alpha_2 - \alpha_1 = \frac{5}{2}n_1 + 2n_2 > 0$, the normal stress term $2(2\alpha_2 - \alpha_1)\frac{U^2}{L^2}\dot{S}^2$ in (20.6.7) is positive, independent of the sign of $\dot{S}$, but $2\mu\dot{S}$ is negative at the front side of a falling body and is positive at the rear.

The motion of bodies in viscoelastic fluids can be said to be maximally different than the motion of the same bodies in a Newtonian fluid. As a rule of thumb, contrary thinking is appropriate. This difference is perfectly correlated with a change in the sign of the normal stress, from compression to tension, in the particulate flows subsequently described.

20.7 Fluid forces near stagnation points on solid bodies

20.7.1 Turning couples on long bodies

It is surprising that turning couples on long bodies determine the stable configurations of suspensions of spherical bodies (Liu and Joseph (1993), Liu (1995)). A long body is an ellipsoid or a cylinder; a broad body is a flat plate. When such bodies are dropped in Newtonian fluids, they put their broadside perpendicular to the stream. This is an effect of inertia that is usually explained by turning couples at points of stagnation. The mechanism is the same one that causes an aircraft at a high angle of attack to stall. The settling orientation of long particles is indeterminate in Stokes flow; however, no matter how small the Reynolds number may be, the body will turn its broadside to the stream; inertia will eventually have its way. When the same long bodies fall slowly in a viscoelastic liquid, they do not put their broadside perpendicular to the stream; they do the opposite, aligning the long side parallel to the stream (see figure 20.1, Galdi *et al.* (2002), Wang *et al.* (2004)).

20.7.2 Particle–particle interactions

The flow-induced anisotropy of a sedimenting or fluidized suspension of spheres is determined by the pair interactions between neighboring spheres. These pair interactions are largely determined by normal stresses at points of stagnation on the spheres that are compressive in Newtonian fluids and tensile in viscoelastic fluid; the stagnation points are on side between spheres falling side by side and are pushed apart by compression and pulled together by tension. The principal interactions can be described as drafting, kissing, and tumbling in Newtonian liquids and as drafting, kissing, and chaining in viscoelastic liquids.

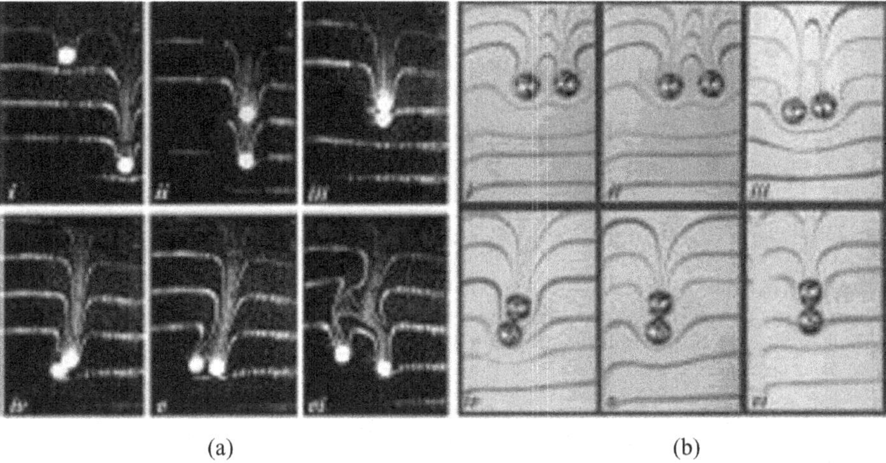

(a) (b)

Figure 20.2. (Joseph 1996, 2000). (a) Spheres in Newtonian fluids. (i), (ii) Spheres settling in glycerin draft, (iii) kiss, and (iv)–(vi) tumble. They tumble because a pair of kissing spheres acts like a long body, which is unstable when its long axis is parallel to the stream. The forces in a Newtonian fluid are dispersive; the tumbling spheres are pushed apart by pressures at stagnation points between the spheres (v)–(vi). (b) Spheres in non-Newtonian fluids. Spheres falling in a 2% aqueous polyethylene oxide solution draft, kiss, and chain. They chain because the forces in a viscoelastic fluid are aggregative. A chain of spheres turns just like the solid cylinder in figure 20.1 (b) (i)–(vi). Reversing time, we see that chaining, kissing, and drafting in (b)(vi)–(i) are like drafting, kissing, and tumbling in (a)(i)–(vi).

The drafting and kissing cases are surely different, despite similar appearances. Kissing spheres align with the stream; they are then momentarily long bodies (see figure 20.2).

The long bodies momentarily formed by kissing spheres are unstable in Newtonian liquids to the same turning couples that turn long bodies broadside-on. Therefore they tumble. This is a local mechanism that implies that, globally, the only stable configuration is the one in which the most probable orientation between any pair of neighboring spheres is across the stream. The consequence of this microstructural property is a flow-induced anisotropy, which leads ubiquitously to lines of spheres across the stream; these are always in evidence in 2D fluidized beds of finite-size spheres. Although they are less stable, planes of spheres in 3D beds can also be found by anyone who cares to look.

If two touching spheres are launched side-by-side in a Newtonian fluid, they will be pushed apart until a stable separation distance between centers across the stream is established; then the spheres fall together without further lateral migration [see figure 20.3(a1)].

On the other hand, if the same two spheres are launched in a viscoelastic fluid from an initial side-by-side configuration in which the two spheres are separated by a smaller than critical gap, as in figure 20.3(a2), the spheres will attract, turn, and chain. One might say that we get dispersion in the Newtonian liquid and aggregation in the viscoelastic liquid.

20.7.3 Sphere–wall interactions

If a sphere is launched near a vertical wall in a Newtonian liquid, it will be forced away from the wall to an equilibrium distance at which lateral migration stops [see figure 20.3(b1)]; in the course of its migration it will acquire a counterclockwise rotation that appears to

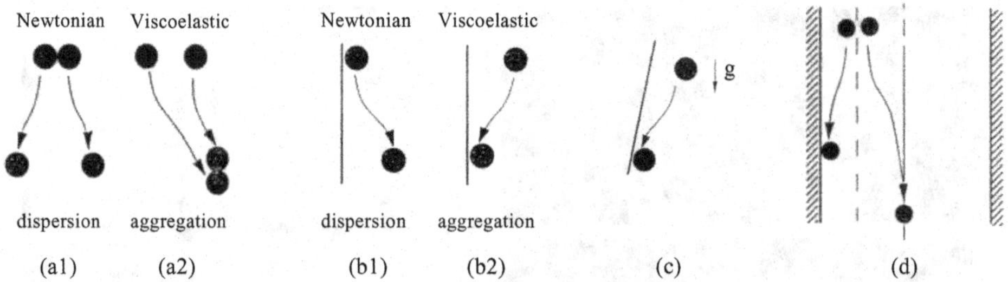

Figure 20.3. (a1), (a2) Side-by-side sphere–sphere interactions; (b1), (b2) sphere–wall interactions; (c) a sphere in a viscoelastic liquid is sucked to a tilted wall; (d) spheres dropped between widely spaced walls. The dotted line is the critical distance d_{cr} for wall–sphere interaction. When $d < d_{cr}$, the sphere goes to the wall. When $d > d_{cr}$, the sphere seeks the center.

stop when the sphere stops migrating. The rotation is anomalous in that clockwise rotation would be induced from shear at the wall. The anomalous rotation seems to be generated by blockage in which high stagnation pressures force the fluid to flow around the outside of the spheres. If the same sphere is launched near a vertical wall in a viscoelastic liquid, it will be sucked all the way to the wall [see figure 20.3(b2)]. It rotates anomalously as it falls. This is very strange because the sphere appears to touch the wall where friction would make it rotate in the other sense. Closer consideration shows that there is a gap between the sphere and the wall. The anomalous rotation is again due to blocking that forces liquid to flow around the outside of the sphere. The pulling action of the wall can be so strong that even if the wall is slightly tilted from the vertical so that the sphere would fall away, it will still be sucked to the wall [see figure 20.3(c)].

If the launching distance between a sphere and a vertical wall is large enough, the wall will not attract a sphere falling in a viscoelastic fluid. This means that there is a critical distance d_{cr} for attraction. Of course, this distance is smaller when the wall is tilted as in figure 20.3(c). In this case, if the sphere is launched at a distance greater than the critical one, it will fall away from the wall.

20.7.4 Flow-induced microstructure

The tendency to aggregate into tight chains of particles all in a row is omnipresent. It is definitely a property of viscoelastic fluids; the same particles disperse rather than chain in Newtonian fluids. This tendency is encountered for particles of all sizes (from microns to centimeters) and in different kinds of motion (subsequently listed). We regard chaining of particles as a form of self-assembly generated by flow stresses in a fluid in motion.

(1) Shear flow (Michele *et al.*, 1977; Petit and Noetinger, 1988)
(2) Extensional flow (Michele *et al.*, 1977)
(3) Sedimentation (Liu and Joseph, 1993, see figure 20.6)
(4) Fluidization (see figure 20.6)

The experiments by Michele *et al.* demonstrated that chains of small spheres may be created and aligned in the direction of the motion in shear flows. A droplet of a suspension

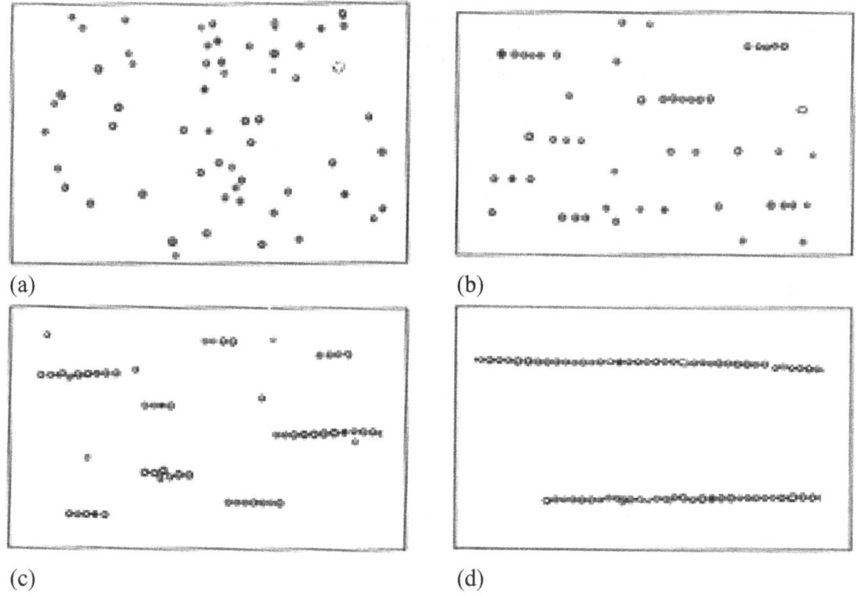

Figure 20.4. A 2% suspension of glass spheres (60–70 μm in size) in a highly viscoelastic polymer solution of 0.5% polyacrylamide in deionized water. (a) After loading, the particles are randomly distributed; (b) after a sideways movement on the top plate of about 3 cm; (c) after the top plate had been moved back and forth several times; (d) after further and faster movement of the top plate. (From Michele *et al.*, 1977.)

of glass spheres was placed between two glass plates that were then pushed together as close as possible. The aggregation of the particles was then accomplished by moving the top plate sideways parallel to the bottom plate to generate an approximate planar shear flow.

The chains of particles shown in figures 20.4 and 20.5 occurred when the distance between the glass was 1 mm. Shear-induced structures in macroscopic dispersions were also reported by Petit and Noetinger (1988). A suspension of glass spheres of diameter 40 μm in silicon oil or in highly viscoelastic polyisobutylene was sandwiched between two parallel glass plates with gaps ranging between 200 and 800 μm and oscillated by

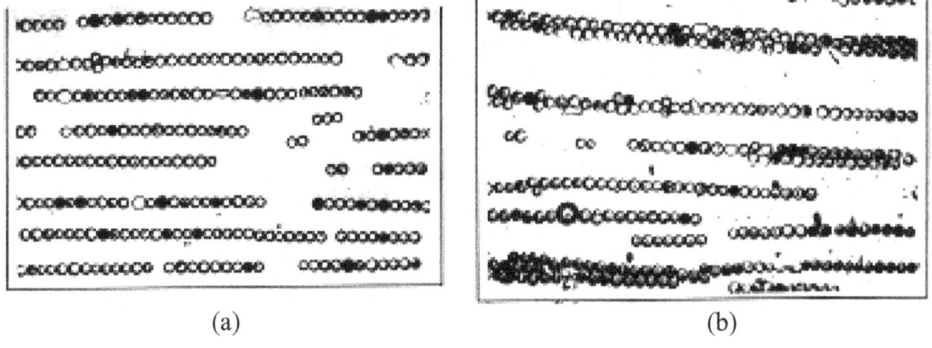

Figure 20.5. 10% suspension of glass spheres (60–70 μm) in a polyisobutylene solution. (a) After movement on the top plate back and forth; (b) after further movement. The lines of spheres in (a) are more or less equally spaced. Further association is observed in (b) where two lines come together. (Michele *et al.*, 1977.)

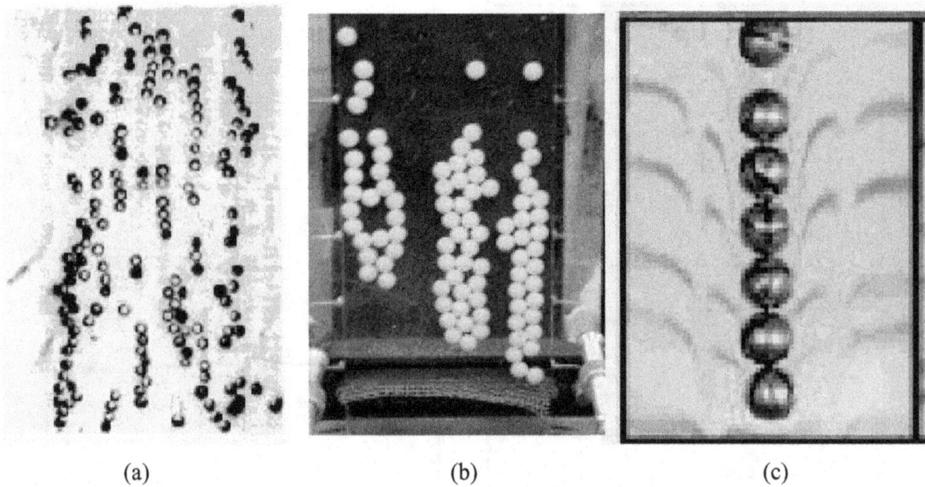

(a) (b) (c)

Figure 20.6. Flow-induced microstructure. Spheres line up in the direction of flow: (a) extensional flow (60–70-μm spheres), (b) fluidization (3-cm spheres), and (c) sedimentation (3-cm spheres) in a 1% aqueous polyethylene oxide (PEO) solution.

shear from the in-plane oscillation of the top plate. The spheres lined up across the velocity direction in the Newtonian silicon solution and with the velocity direction in the viscoelastic solution.

The robust tendency of small particles from microns to centimeters to chain in all kinds of motions must be associated with a powerful and local feature of particle–fluid interactions (figures 20.5 and 20.6). We argue that the chaining of spherical particles is a consequence of the same dynamics that controls the orientation of long bodies moving relative to the stream, across the stream in Newtonian fluids, and along the stream in viscoelasatic fluids. This dynamics is mainly controlled by a reversal of the normal stress at a point of stagnation.

20.8 Potential flow over a sphere for a second-order fluid

We consider the potential of a uniform flow of a second-order fluid past a sphere with radius a. The calculation is similar to that in subsection 20.2.3. Bernoulli's equation (20.5.4) is used to obtain the pressure at the surface of the sphere:

$$p = p_\infty + \left(\frac{27}{2}\alpha_1 + 9\alpha_2\right)\frac{U^2}{a^2}\left(1 + 2\cos^2\theta\right) + \frac{\rho}{2}U^2\left(1 - \frac{9}{4}\sin^2\theta\right), \qquad (20.8.1)$$

where p_∞ is the pressure at infinity. The normal stress T_{rr} at the surface of the sphere is calculated from (20.5.5) and expressed in a dimensionless form:

$$T^*_{rr} = \frac{T_{rr} + p_\infty}{\rho U^2/2} = \left(\frac{9}{4}\sin^2\theta - 1\right) - \frac{12}{Re}\cos\theta + \frac{\alpha_1}{\rho a^2}\left(36\sin^2\theta - 9\right) + 18\frac{\alpha_2}{\rho a^2}\cos^2\theta, \qquad (20.8.2)$$

where $Re = \rho U a/\mu$ is the Reynolds number. The viscous normal stress should be zero at a solid boundary in a Newtonian fluid or an Oldroyd B fluid; the viscous effect on the

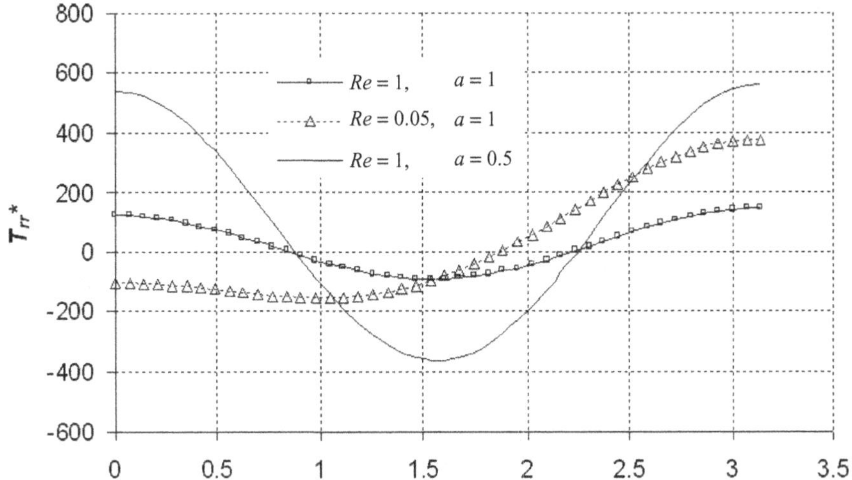

Figure 20.7. The dimensionless normal stress T^*_{rr} as a function of the angle θ. Parameters of the liquid M1 are used in the calculation: $\rho = 0.895$ g/cm^3, $\alpha_1 = -3$, and $\alpha_2 = 5.34$ g/cm. The three curves in the figure correspond to $Re = 1, a = 1$ cm; $Re = 0.05, a = 1$ cm; and $Re = 1, a = 0.5$ cm, respectively.

normal stress is hidden in the pressure. The viscous contribution to the pressure at the surface of a sphere in a Stokes flow of Newtonian fluid is (Panton, 1984, 646)

$$\frac{p - p_\infty}{\rho U^2/2} = \frac{3}{Re}\cos\theta. \tag{20.8.3}$$

In (20.8.2) we have a viscous contribution to the normal stress $(12/Re)\cos\theta$ that is four times the Stokes value $(3/Re)\cos\theta$.

We compute the normal stress (20.8.2) for the liquid M1 with a density $\rho = 0.895$ g/cm^3, $\alpha_1 = -3$, and $\alpha_2 = 5.34$ g/cm (Hu *et al.*, 1990) as an example. We plot the dimensionless normal stress T^*_{rr} as a function of the angle θ in figure 20.7.

At the stagnation points of a sphere $[r = a,\ \theta = 0 \text{ or } \pi]$, the normal stresses are, respectively,

$$\frac{T_{rr} + p_\infty}{\rho U^2/2} = \left[-1 + \frac{9(2\alpha_2 - \alpha_1)}{\rho a^2}\right] \mp \frac{12}{Re}. \tag{20.8.4}$$

The viscous contribution gives rise to compression $-12/Re$ at the front stagnation point and to tension $12/Re$ at the rear. The stress that is due to inertia and viscoelasticity is the same at $\theta = 0$ and $\theta = \pi$ and is a tension when $9(2\alpha_2 - \alpha_1) > \rho a^2$. The quantity $2\alpha_2 - \alpha_1$ is strongly positive; for example, $2\alpha_2 - \alpha_1 = 13.68$ g/cm for the liquid M1. Hence, if a^2 is not too large, the stress at the stagnation points is a tension, reversing the compression that is due to inertia.

Figure 20.7 shows that, when viscous effects are dominant ($Re = 0.05$, $a = 1$ cm), the stress is compression at the leading edge and tension at the trailing edge. When the viscoelastic effects are important ($Re = 1$, $a = 1$ cm and $Re = 1$, $a = 0.5$ cm), the stress is tension at both stagnation points. The normal stress on the sphere with $a = 0.5$ cm is much stronger than that on the sphere with $a = 1$ cm, because the viscoelastic effects are proportional to $1/a^2$. The distribution of normal stresses, especially the tension at the trailing edge, shown in figure 20.7, is compatible with the cusp shape of gas bubbles rising in viscoelastic fluids (see §20.11).

A point of stagnation on a stationary body in potential flow is a unique point at the end of a dividing streamline at which the velocity vanishes. In a viscous fluid all the points on the boundary of a stationary body have a zero velocity but the dividing streamline can be found and it marks the place of zero shear stress near which the velocity is small. The stagnation pressure makes sense even in a viscous fluid where the high pressure of the potential flow outside the boundary layer is transmitted right through the boundary layer to the body. It is a good idea to look for the dividing streamlines where the shear stress vanishes in any analysis of the flow pattern around the body.

20.9 Potential flow over an ellipse

Potential flow over an ellipse is a classical problem in airfoil theory. The solutions are most easily expressed in terms of complex functions of a complex variable (Lamb, 1932; Milne-Thomson, 1968). Hence we use this potential flow solution and obtain the pressure and the normal stress for a second-order fluid as a composition of derivatives of that solution. 2D potential flows around bodies admit the addition of circulation that we have here put to zero.

The complex potential for the flow over an ellipse given by

$$\frac{x^2}{a^2}+\frac{y^2}{b^2}=1 \tag{20.9.1}$$

is (Milne-Thomson, 1968, §6.31)

$$\omega=-\frac{1}{2}U(a+b)\left[\frac{e^{-i\alpha}\left(z+\sqrt{z^2-c^2}\right)}{a+b}+\frac{e^{i\alpha}\left(z-\sqrt{z^2-c^2}\right)}{a-b}\right], \tag{20.9.2}$$

where z is the complex variable, α is the angle of attack, a and b are the semiaxes of the ellipse, and $c^2=a^2-b^2$. We plot the streamlines of the flow with the angle of attack $\alpha=0°$ and $60°$ in figure 20.8.

The velocities are

$$u=\frac{1}{2}\left(\frac{\mathrm{d}\omega}{\mathrm{d}z}+\frac{\mathrm{d}\bar{\omega}}{\mathrm{d}\bar{z}}\right),\quad v=\frac{i}{2}\left(\frac{\mathrm{d}\omega}{\mathrm{d}z}-\frac{\mathrm{d}\bar{\omega}}{\mathrm{d}\bar{z}}\right). \tag{20.9.3}$$

$$\mathbf{L}=\nabla\mathbf{u}=\begin{bmatrix}\partial u/\partial x & \partial v/\partial x\\ \partial u/\partial y & \partial v/\partial y\end{bmatrix}, \tag{20.9.4}$$

$$\mathbf{A}=\mathbf{L}+\mathbf{L}^T=\begin{bmatrix}n & s\\ s & -n\end{bmatrix}, \tag{20.9.5}$$

where $n=\frac{\mathrm{d}^2\omega}{\mathrm{d}z^2}+\frac{\mathrm{d}^2\bar{\omega}}{\mathrm{d}\bar{z}^2}$ and $s=i\left(\frac{\mathrm{d}^2\omega}{\mathrm{d}z^2}-\frac{\mathrm{d}^2\bar{\omega}}{\mathrm{d}\bar{z}^2}\right)$. It follows that

$$\mathbf{A}^2=\left(n^2+s^2\right)\mathbf{1},\ \operatorname{tr}\mathbf{A}^2=2\left(n^2+s^2\right). \tag{20.9.6}$$

Letting U and p_∞ be the velocity and pressure at infinity, respectively, we find the pressure by (20.5.4):

$$p=p_\infty+\frac{\rho}{2}U^2-\frac{\rho}{2}\frac{\mathrm{d}\omega}{\mathrm{d}z}\frac{\mathrm{d}\bar{\omega}}{\mathrm{d}\bar{z}}+(3\alpha_1+2\alpha_2)\frac{1}{2}\left(n^2+s^2\right). \tag{20.9.7}$$

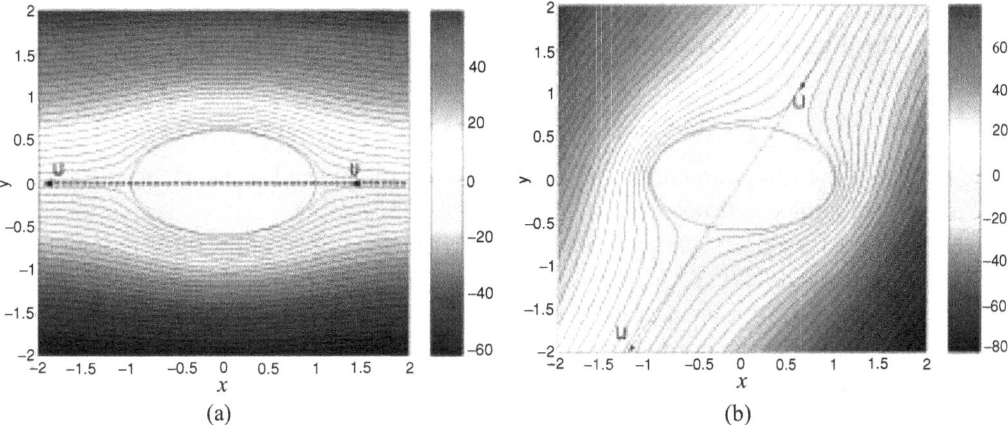

Figure 20.8. The streamlines of the flow over an ellipse. (a) The angle of attack $\alpha = 0°$; (b) $\alpha = 60°$.

The stress can then be calculated with (20.5.5); after some arrangement, we find

$$\mathbf{T} = \left[-p_\infty - \frac{\rho}{2}U^2 + \frac{\rho}{2}\frac{d\omega}{dz}\frac{d\bar{\omega}}{d\bar{z}} - \frac{1}{2}\alpha_1(n^2+s^2) \right]\mathbf{1} + \mu \begin{bmatrix} n & s \\ s & -n \end{bmatrix}$$
$$+ \alpha_1 u \begin{bmatrix} k & q \\ q & -k \end{bmatrix} + \alpha_1 v \begin{bmatrix} q & -k \\ -k & -q \end{bmatrix}, \tag{20.9.8}$$

where $k = \frac{d^3\omega}{dz^3} + \frac{d^3\bar{\omega}}{d\bar{z}^3}$ and $q = i(\frac{d^3\omega}{dz^3} - \frac{d^3\bar{\omega}}{d\bar{z}^3})$. Equation (20.9.8) applies to any 2D flow that can be represented by a complex potential ω. We draw the readers' attention to the fact that α_2 does not appear in the expression for the stress in 2D cases, which has been reported in Joseph (1992).

We are interested in the normal stress on the surface of the ellipse. The unit normal vector on the surface is

$$\mathbf{n} = \frac{\frac{x}{a^2}\mathbf{e}_x + \frac{y}{b^2}\mathbf{e}_y}{(x^2/a^4 + y^2/b^4)^{1/2}}. \tag{20.9.9}$$

The normal stress is calculated from $T_{nn} = \mathbf{n} \cdot \mathbf{T} \cdot \mathbf{n}$.

20.9.1 Normal stress at the surface of the ellipse

We present the results for the normal stress on the ellipse in this section. Besides the angle of attack, there are six relevant parameters, ρ, U, μ, a, b, and α_1, in this problem. Three dimensionless parameters can be constructed:

$$Re = \frac{\rho U a}{\mu}, \quad \frac{-\alpha_1}{\rho a^2}, \quad \frac{a}{b}. \tag{20.9.10}$$

Note that the Deborah number can be defined as $De = \frac{-\alpha_1}{\mu}\frac{U}{a}$ and $\frac{-\alpha_1}{\rho a^2} = De/Re$. We shall see later that the parameter $-\alpha_1/(\rho a^2)$ appears in the expressions for the normal stresses at the stagnation points. Therefore we use the parameter $-\alpha_1/(\rho a^2)$ rather than the Deborah number. The dimensionless normal stress is

$$T^*_{nn} = \frac{T_{nn} + p_\infty}{\rho U^2/2}. \tag{20.9.11}$$

The effects of the three dimensionless parameters on the normal stress at the surface of the ellipse are studied in flows of a zero attack angle. We can obtain explicit expressions for the normal stresses at the stagnation points, from which the effects of the three parameters can be understood readily. Such expressions are not obtained for an arbitrary point on the ellipse surface; instead, we calculate the numerical values of stress and present the plots for the distribution of the normal stress.

At the front stagnation point where $z = a$, we have

$$u = v = 0, \quad n = -2U(a+b)/b^2, \quad s = 0. \tag{20.9.12}$$

Inserting (20.9.12) into (20.9.8) and noting that $T_{nn} = T_{xx}$, we obtain the dimensionless normal stress at the front stagnation point:

$$T^*_{nn}(\theta = 0) = -1 + \frac{-\alpha_1}{\rho a^2} 4\left(1 + \frac{b}{a}\right)^2 \frac{a^4}{b^4} - \frac{4}{Re}\left(1 + \frac{b}{a}\right)\frac{a^2}{b^2}. \tag{20.9.13}$$

Similarly, we can find the dimensionless normal stress at the rear stagnation point:

$$T^*_{nn}(\theta = \pi) = -1 + \frac{-\alpha_1}{\rho a^2} 4\left(1 + \frac{b}{a}\right)^2 \frac{a^4}{b^4} + \frac{4}{Re}\left(1 + \frac{b}{a}\right)\frac{a^2}{b^2}. \tag{20.9.14}$$

The difference between the two stresses is

$$T^*_{nn}(\theta = \pi) - T^*_{nn}(\theta = 0) = \frac{8}{Re}\left(1 + \frac{b}{a}\right)\frac{a^2}{b^2}. \tag{20.9.15}$$

Normal stresses (20.9.12) and (20.9.13) are analogous to the normal stress (20.8.4) at the stagnation points in the sphere case, in the sense that they are all composed of the inertia, viscous, and viscoelastic terms. Here the viscoelastic term $\frac{-\alpha_1}{\rho a^2} 4\left(1 + \frac{b}{a}\right)^2 \frac{a^4}{b^4}$ gives rise to extension at both of the stagnation points; the viscous term $\frac{4}{Re}\left(1 + \frac{b}{a}\right)\frac{a^2}{b^2}$ leads to compression at the front stagnation point and extension at the rear stagnation point.

20.9.2 The effects of the Reynolds number

We calculate the normal stress at the surface of the ellipse in flows where $-\alpha_1/(\rho a^2)$ and a/b are fixed at 3 and 1.67, respectively, and the Reynolds number changes from 0.01 to 100. In figure 20.9, we plot T^*_{nn} at the front and rear stagnation points as functions of the Reynolds number. The stress at the front stagnation point changes from compression to extension as the Reynolds number increases. The critical Reynolds number, at which $T^*_{nn} = 0$ at the front stagnation point, is 0.075, as shown in figure 20.9. Equation (20.9.15) indicates that the difference between the two normal stresses vanishes as the Reynolds number tends to infinity; the asymptotic value of the two stresses is

$$T^*_{nn}[Re \to \infty,\ -\alpha_1/(\rho a^2) = 3,\ a/b = 1.67] = 236.07.$$

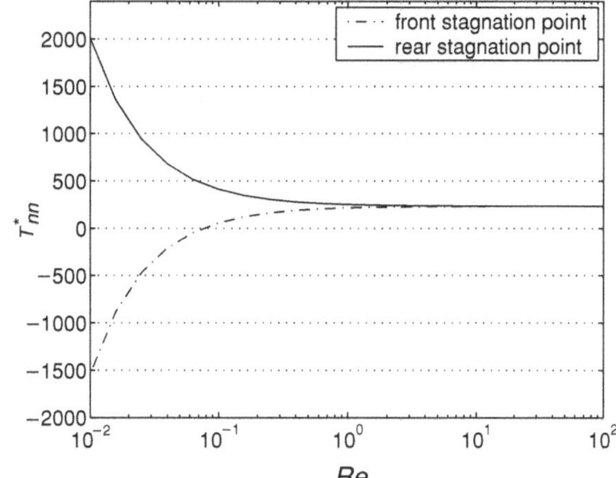

Figure 20.9. The dimensionless normal stresses T^*_{nn} at the front and rear stagnation points as functions of the Reynolds number. The other two parameters are fixed: $-\alpha_1/(\rho a^2) = 3$ and $a/b = 1.67$.

The distribution of the normal stress at the surface is plotted in figure 20.10 for flows with Reynolds numbers 0.05 and 1. We note that the stress is compression at the front stagnation point and extension at the rear stagnation point when $Re = 0.05$; however, the stress is tension at both of the two stagnation points when $Re = 1$.

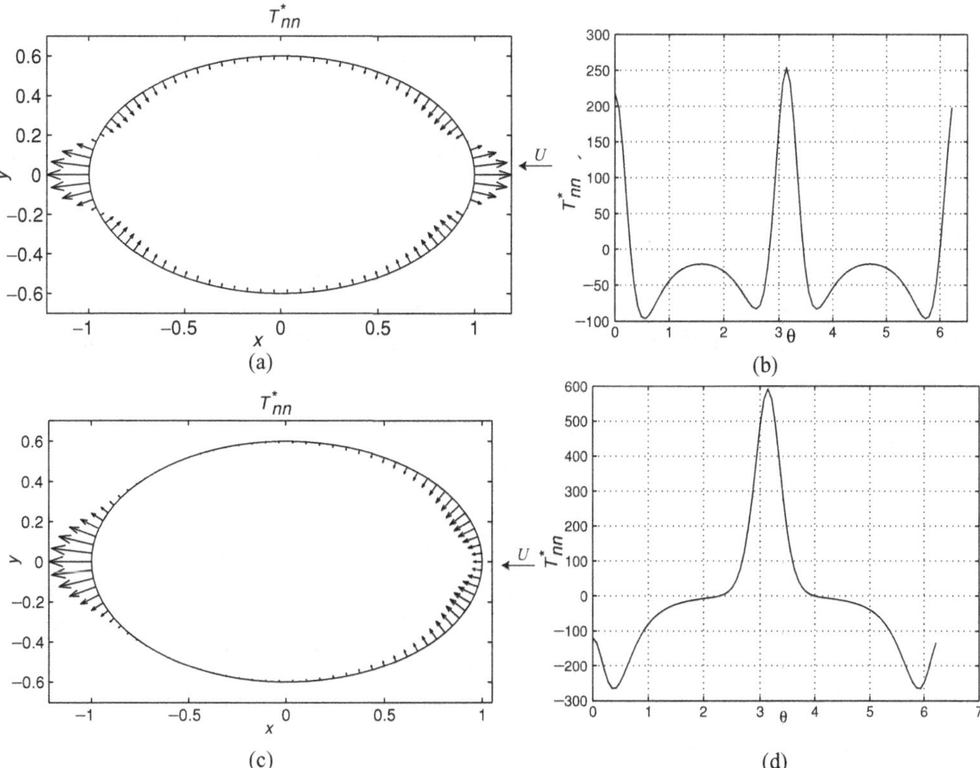

Figure 20.10. The distribution of the dimensionless normal stress T^*_{nn} at the surface of the ellipse in flows with $-\alpha_1/(\rho a^2) = 3$ and $a/b = 1.67$. The Reynolds number is 1.0 in (a) and (b) and 0.05 in (c) and (d). The normal stress is represented by vectors at the surface of the ellipse in (a) and (c) and is plotted against the polar angle θ in (b) and (d).

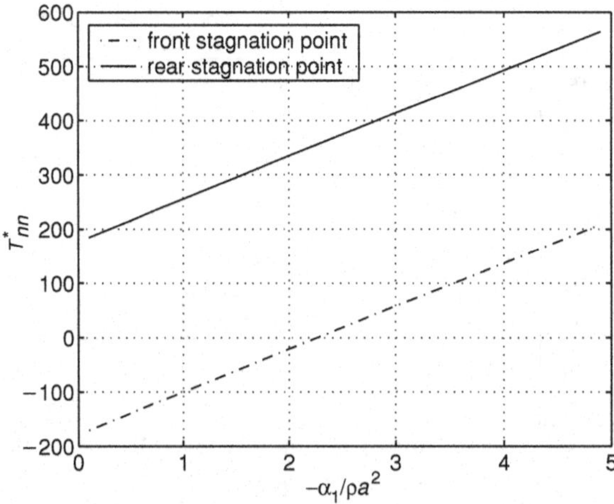

Figure 20.11. The dimensionless normal stresses T^*_{nn} at the front and rear stagnation points as functions of the parameter $-\alpha_1/(\rho a^2)$. The other two parameters are fixed: $Re = 0.1$ and $a/b = 1.67$.

20.9.3 The effects of $-\alpha_1/(\rho a^2)$

The two normal stresses at stagnation points are plotted against the parameter $-\alpha_1/(\rho a^2)$ in figure 20.11; the other two parameters are fixed: $Re = 0.1$ and $a/b = 1.67$. The difference between the two stresses is independent of the parameter $-\alpha_1/(\rho a^2)$, as can be seen from (20.9.15). This difference is 355.6 when $Re = 0.1$ and $a/b = 1.67$. The critical value of $-\alpha_1/(\rho a^2)$, at which $T^*_{nn} = 0$ at the front stagnation point, is 2.26.

20.9.4 The effects of the aspect ratio

We change the aspect ratio from 1.1 to 10 and compute the normal stress on the surface of the ellipse; the other two parameters are fixed: $Re = 0.1$ and $-\alpha_1/(\rho a^2) = 3$. The two stresses at the stagnation points are plotted against the aspect ratio in figure 20.12; The values of the stresses change dramatically with the aspect ratio because of the a^4/b^4 term in (20.9.13) and (20.9.14). The stress at the front stagnation point changes from compression

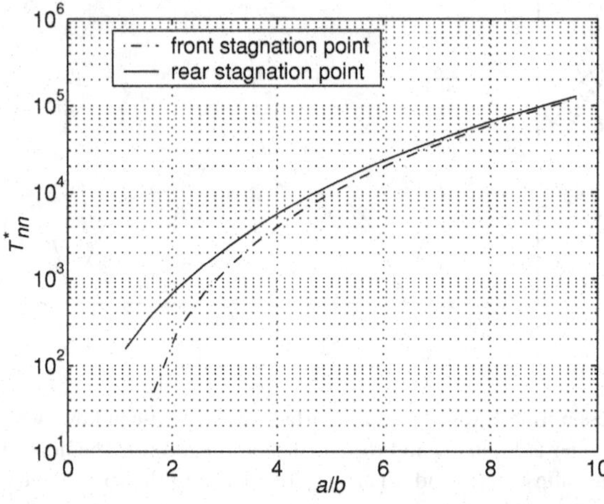

Figure 20.12. The dimensionless normal stresses T^*_{nn} at the front and rear stagnation points as functions of the aspect ratio a/b. The other two parameters are fixed: $Re = 0.1$ and $-\alpha_1/(\rho a^2) = 3$.

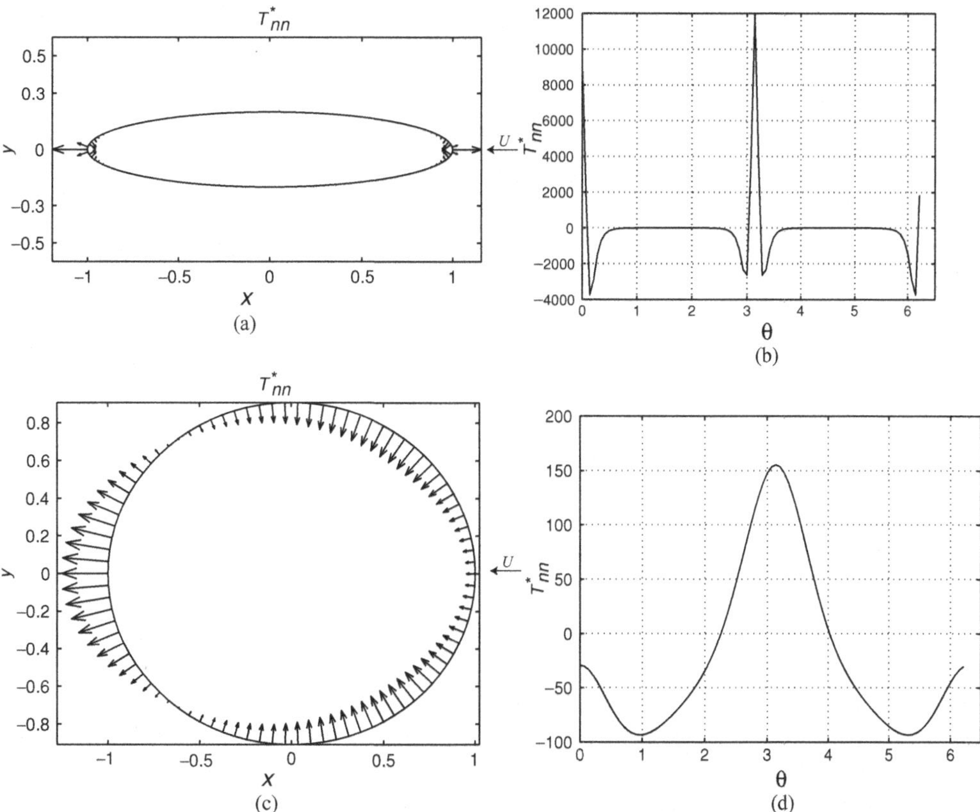

Figure 20.13. The distribution of the dimensionless normal stress T^*_{nn} at the surface of the ellipse in flows with $Re = 0.1$ and $-\alpha_1/(\rho a^2) = 3$. The aspect ratio is 5.0 in (a) and (b) and is 1.1 in (c) and (d). The normal stress is represented by vectors at the surface of the ellipse in (a) and (c) and is plotted against the polar angle θ in (b) and (d).

to extension as a/b increases; when $a/b = 1.40$, $T^*_{nn} = 0$ at the front stagnation point. (The negative values are not shown on the semilog plot in figure 20.12.)

The normal stress distribution at the surface is plotted in figure 20.13 for flows with $a/b = 5.0$ and 1.1. It can be seen that, in the flow with the higher aspect ratio, the ellipse is under very high extensional stresses at both of the stagnation points. The stress at the front stagnation point is compression when $a/b = 1.1$, which implies that the front nose of a gas bubble will be flattened.

20.10 The moment on the ellipse

Long bodies falling in a viscoelastic fluid often turn into the streamwise direction (Liu and Joseph, 1993). We calculate the dimensionless moment by the normal stress on the ellipse:

$$M^* = \frac{M}{\rho U^2 a^2/2} = \frac{\oint \mathbf{x} \wedge (T_{nn}\mathbf{n})\ \mathrm{d}l}{\rho U^2 a^2/2}. \qquad (20.10.1)$$

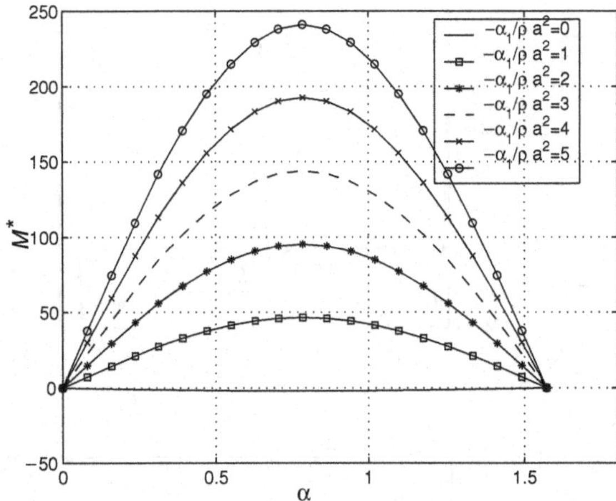

Figure 20.14. The moment on the ellipse by the normal stress as a function of the attack angle α in the range $[0,\ \pi/2]$. The six curves correspond to six values of the parameter $-\alpha_1/\left(\rho a^2\right)$: 0, 1, 2, 3, 4, and 5; the aspect ratio is fixed at $a/b = 1.67$.

In a Newtonian fluid, the moment can be calculated with the theorem of Blasius, and the dimensionless moment is

$$M^* = -2\pi(1 - a^2/b^2)\sin\alpha\cos\alpha, \tag{20.10.2}$$

which does not depend on the Reynolds number. Our calculation shows that the conclusion is also true in a second-order fluid: The dimensionless moment does not depend on the Reynolds number; the parameter $-\alpha_1/(\rho a^2)$ and the aspect ratio a/b are relevant parameters when the moment is concerned.

We plot the dimensionless moment on the ellipse by the normal stress as a function of the attack angle α in the range $[0, \pi/2]$ in figure 20.14. Six values of the parameter $-\alpha_1/(\rho a^2)$, from 0 to 5, are investigated. The curve corresponding to $-\alpha_1/(\rho a^2) = 0$ in figure 20.14 is in agreement with (20.10.2), which is the moment in a Newtonian fluid. The moment on the ellipse is negative in a Newtonian fluid and turns the ellipse broadside-on to the stream. When the parameter $-\alpha_1/(\rho a^2)$ is larger, the moment on the ellipse becomes positive and tends to align the broad side of the ellipse with the streamwise direction. Figure 20.14 also shows that the magnitude of the moment reaches its largest value when $\alpha = \pi/4$, which also occurs in a Newtonian fluid.

We show the effects of the aspect ratio on the moment in figure 20.15. The five curves correspond to five values of the aspect ratio a/b: 1.1, 4, 6, 8, and 10; the parameter $-\alpha_1/\left(\rho a^2\right)$ is fixed at 3. It can be seen that, as the aspect ratio increases, the magnitude of the moment increases to huge values. Hence long slim bodies turn into the streamwise direction quickly in viscoelastic fluids.

20.11 The reversal of the sign of the normal stress at stagnation points

In §§20.8 and 20.9, we showed that the normal stresses at the stagnation points on a sphere or an ellipse in a second-order fluid can be tension, opposite to the high compressive pressures at the stagnation points in Newtonian fluid. This reversal of the sign of the normal stress has significant effects on the behavior of particles and bubbles in Newtonian and viscoelastic fluids.

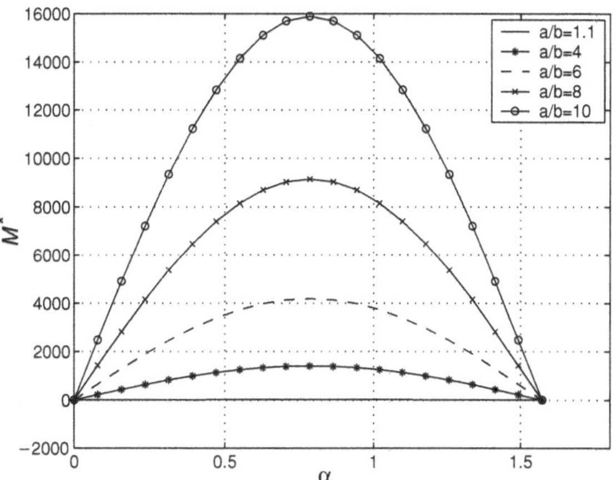

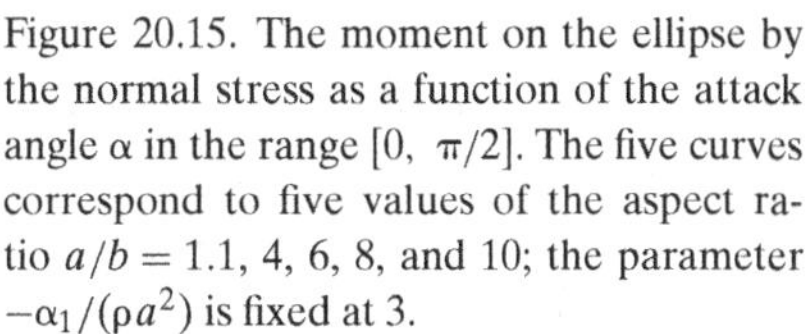

Figure 20.15. The moment on the ellipse by the normal stress as a function of the attack angle α in the range $[0, \ \pi/2]$. The five curves correspond to five values of the aspect ratio $a/b = 1.1$, 4, 6, 8, and 10; the parameter $-\alpha_1/(\rho a^2)$ is fixed at 3.

In Newtonian fluids, long bodies in a uniform flow are turned to the orientation in which their long sides or broadsides are perpendicular to the stream by the high pressures at stagnation points [see figure 20.16(a)].

In viscoelastic fluids, long bodies in a uniform flow often turn into the streamwise direction, as we discussed in §20.10 (see also Liu and Joseph, 1993; Joseph and Feng, 1996; Huang, Hu, and Joseph, 1997). The extensional normal stresses at the stagnation points in viscoelastic fluids contribute to the moment that turns the long bodies into the stream [see figure 20.16(b) and the caption]. Long chains of spherical bodies parallel with the stream are in evidence in sedimentation and fluidization flows of viscoelastic

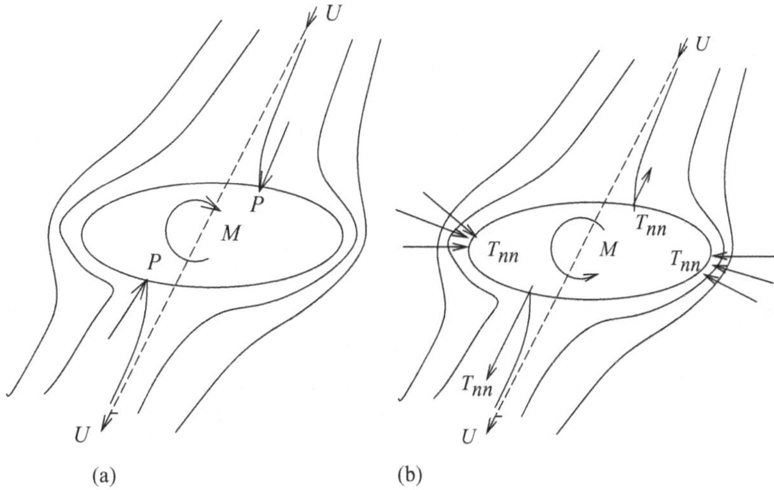

Figure 20.16. The moment on the ellipse in potential flow. (a) In an inviscid fluid, the high pressures at the stagnation points turn the ellipse broadside-on (across the stream); (b) in a second-order fluid, the normal stresses at the two edges where the streamlines are most crowded are compressive and tend to turn the ellipse into the stream. At the two stagnation points, the stresses may change from compression to tension. Here we illustrate the situation in which the stresses are tension at the stagnation points; this pair of stresses gives rise to the moment that tends to turn the ellipse into the stream. Our calculation shows that the resultant moment of the normal stress tends to turn the broadside of the ellipse into the stream when inertia is not dominant.

fluids; such configurations are opposite to those observed in Newtonian fluids. Another unusual phenomenon in viscoelastic fluids is the 2D cusp at the trailing edge of a rising air bubble (Liu *et al.*, 1995). Below a critical capillary number, an air bubble rising in a viscoelastic fluid adopts the shape with a cusp point in one view and a spade edge in the orthogonal view. Figures 20.7, 20.10(c), and 20.13(c) show situations in which the normal stress is compression at the leading edge and tension at the trailing edge; the leading edge is flattened and the trailing edge is extended, tending to the cusped trailing edge observed in experiments. Our calculation on a smooth sphere or an ellipse cannot lead to the exact cusp shape. However, the calculation shows that the normal stress computed on the viscoelastic potential agrees with the experiment qualitatively, much better than the pressure that is the only normal force that can act on the body in IPF.

20.12 Flow past a flat plate

The flow past an ellipse degenerates to the flow past a flat plate when $b = 0$. The complex potential is

$$\omega = \begin{cases} -U(z\cos\alpha - i\sqrt{z^2 - a^2}\sin\alpha) & \text{upstream to the plate} \\ -U(z\cos\alpha + i\sqrt{z^2 - a^2}\sin\alpha) & \text{downstream to the plate} \end{cases}. \tag{20.12.1}$$

The velocities at the upper and lower surfaces of the plate are, respectively,

$$u = -U\left(\cos\alpha - \frac{x\sin\alpha}{\sqrt{a^2 - x^2}}\right), \quad v = 0, \quad \text{upper surface;} \tag{20.12.2}$$

$$u = -U\left(\cos\alpha + \frac{x\sin\alpha}{\sqrt{a^2 - x^2}}\right), \quad v = 0, \quad \text{lower surface.} \tag{20.12.3}$$

The stagnation points at the upper and lower surfaces are $x = a\cos\alpha$ and $x = -a\cos\alpha$, respectively. The dimensionless normal stresses at the two stagnation points are

$$\frac{T_{nn} + p_\infty}{\rho U^2/2}(x = a\cos\alpha) = -1 + \frac{-\alpha_1}{\rho a^2}\frac{4}{\sin^4\alpha} - \frac{4}{Re\sin^2\alpha}, \tag{20.12.4}$$

$$\frac{T_{nn} + p_\infty}{\rho U^2/2}(x = -a\cos\alpha) = -1 + \frac{-\alpha_1}{\rho a^2}\frac{4}{\sin^4\alpha} + \frac{4}{Re\sin^2\alpha}. \tag{20.12.5}$$

These stresses are degenerate cases of (20.9.13) and (20.9.14), which are the stresses at stagnation points in the flow past an ellipse. In (20.12.4) and (20.12.5), the viscoelastic term $\frac{-\alpha_1}{\rho a^2}\frac{4}{\sin^4\alpha}$ gives rise to extension at both of the stagnation points; the viscous term $\frac{4}{Re}\frac{1}{\sin^2\alpha}$ leads to compression at the front stagnation point and extension at the rear stagnation point.

20.13 Flow past a circular cylinder with circulation

The complex potential for the flow past a circular cylinder with circulation is (Milne-Thomson, 1968, §7.12)

$$\omega = -U(z + a^2/z) - i\kappa\log(z/a), \tag{20.13.1}$$

where κ is the strength of the circulation. A dimensionless parameter κ/aU can be introduced.

We calculate the stress by using potential (20.13.1) in a second-order fluid. The stress at the surface of the cylinder, where $z = ae^{i\theta}$, is of interest. We find

$$u = -\sin\theta(2U\sin\theta + \kappa/a), \qquad v = \cos\theta(2U\sin\theta + \kappa/a), \tag{20.13.2}$$

$$n = -4U\cos 3\theta/a + 2\kappa\sin 2\theta/a^2, \quad s = -(4U\sin 3\theta/a + 2\kappa\cos 2\theta/a^2), \tag{20.13.3}$$

$$k = 12U\cos 4\theta/a^2 - 4\kappa\sin 3\theta/a^3, \quad q = 12U\sin 4\theta/a^2 + 4\kappa\cos 3\theta/a^3, \tag{20.13.4}$$

at $z = ae^{i\theta}$ and we obtain the stress tensor by inserting (20.13.2)–(20.13.4) into (20.9.8). The dimensionless normal stress is given by

$$\frac{T_{nn} + p_\infty}{\rho U^2/2} = \left[-1 + 4\sin^2\theta + \left(\frac{\kappa}{aU}\right)^2 + \frac{\kappa}{aU}4\sin\theta\right] - \frac{8}{Re}\cos\theta + \frac{4\alpha_1}{\rho a^2}\left[-4 + 12\sin^2\theta + \left(\frac{\kappa}{aU}\right)^2 + \frac{\kappa}{aU}6\sin\theta\right], \tag{20.13.5}$$

where the first term on the right-hand side is the inertia term, which is the same as the inviscid pressure; the second term is the viscous term, and the third term is the viscoelastic term.

The force and moment on the cylinder can be obtained by direct integration of the normal stress over the surface of the cylinder:

$$F_x = \int_0^{2\pi} T_{nn}\cos\theta\, a d\theta = -4\pi\mu U, \tag{20.13.6}$$

$$F_y = \int_0^{2\pi} T_{nn}\sin\theta\, a d\theta = 2\pi\kappa U(\rho + 6\alpha_1/a^2). \tag{20.13.7}$$

The moment is obviously zero. Equation (20.13.6) shows that the cylinder experiences a drag that is due to the viscosity; the drag would be zero if the shear stress were included in the integration. Equation (20.13.7) shows that the lift has a contribution from the viscoelastic effect in addition to the inviscid lift $2\pi\kappa\rho U$. Because α_1 is negative, the viscoelastic lift is opposite to the inviscid lift. When $-\alpha_1/(\rho a^2) = 1/6$, the total lift force is zero.

20.14 Potential flow of a second-order fluid over a triaxial ellipsoid

Viana *et al.* (2005) solved the problem of potential flow of a second-order fluid around a 3D ellipsoid by using the general expressions in Lamb (1932) and computed the flow and stress fields. The flow fields are determined by the harmonic potential but the stress fields depend on viscosity and the parameters of the second-order fluid. The stress fields on the surface of a triaxial ellipsoid depend strongly on the ratios of principal axes and are such as to suggest the formation of gas bubble with a round flat nose and a 2D cusped trailing edge. A thin flat trailing edge gives rise to a large stress that makes the thin trailing edge thinner.

The main goal of the calculations of Viana *et al.* (2005) for the second-order fluid model was to identify mechanisms that lead to "two-dimensional cusps" (see figure 20.17) at the trailing edge of a gas bubble rising in an unbounded liquid where axisymmetric solutions might be expected. They calculated the effects of viscosity, second-order viscoelasticity,

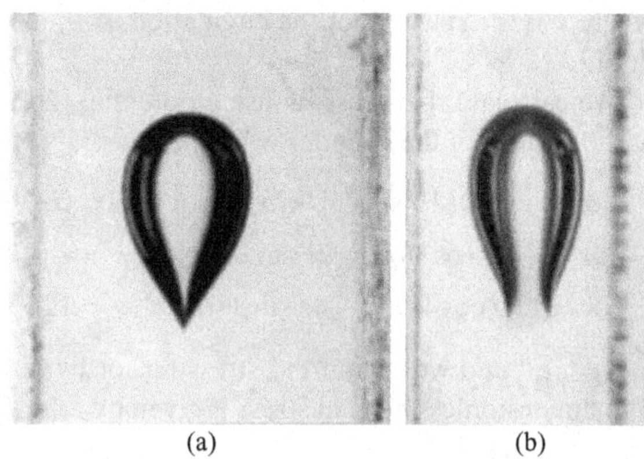

Figure 20.17. Two orthogonal views showing the cusped (a) and broad (b) shapes of the trailing edge of an air bubble (2 cm^3), rising in a viscoelastic liquid (S1). The two photographs are from Liu, Liao, and Joseph (1995).

and inertia. The effects of viscoelasticity is opposite to the effects of inertia; under modest and realizable assumptions about the values of the second-order fluid parameters, the normal stresses at points of stagnation change from compression to tension. The effect of inertia and elasticity is essentially symmetric in that they depend on squares of velocity and velocity gradients but the effects of viscosity are asymmetric.

For the rising gas bubbles, the effects of the second-order and viscous terms on the normal stress are such as to extend and flatten the trailing edge. These calculations suggest that "two-dimensional cusping" can be viewed as an instability in which a thin flat trailing edge gives rise to a large stress that makes the thin trailing edge even thinner.

20.15 Motion of a sphere normal to a wall in a second-order fluid

The motion of a sphere normal to a wall was studied by Ardekani, Rangel, and Joseph (2007). They calculated the normal stress distributions on the surface of the sphere and showed how these distributions were influenced by viscoelastic effects. For small separation distances, a tensile normal stress exists at the trailing edge of the particle as it moves away from the wall in a Newtonian fluid. A larger tensile stress exists at the trailing edge of a particle moving away from a wall in a second order fluid. When the particle moves toward the wall, the stress is compressive in a Newtonian fluid and tensile in a second order fluid. The contributions of the normal stress are toward the wall whether the particle moves toward or away from the wall. Ardekani *et al.* (2007) obtained results for Stokes flow and viscous potential flow which are described below.

The interaction between a sphere and a vertical wall has been extensively studied in experiments and numerical simulations (Goldman, Cox, and Brenner (1967); Joseph, Liu, and Poletto (1994); Becker, McKinley, and Stone (1996); Singh and Joseph (2000)). These studies show that a sedimenting particle migrates to a wall when the fluid is viscoelastic and the particle is close to the wall. On the other hand, when a particle in a Newtonian fluid sediments near a wall it will be pushed away from the wall. Studies targeting particle-particle interactions not directly related to particle-wall interactions are discussed in the Ardekani *et al.* paper. Here, we focus on forces on a sphere moving in a second order fluid as it moves toward or away from a wall. These forces are evaluated when the velocity is computed on Stokes flow and when it is computed on potential flow. In both cases the forces are attractive in sharp contrast to the repulsive force which is generated in Newtonian fluid as it moves toward a wall.

20.15.1 Low Reynolds numbers

The governing equations for a second-order fluid are given by (20.3.1) to (20.3.5). In this study, a spherical particle moving normal to a wall with constant velocity is considered. For low Reynolds number flows, in two dimensions or when $\alpha_1 + \alpha_2 = 0$, the velocity field for a second-order fluid is the same as the one predicted by the Stokes flow while pressure is modified as

$$p = p_N + \frac{\alpha_1}{\mu_f}\frac{Dp_N}{Dt} + \frac{\beta}{4} tr\mathbf{A}^2, \tag{20.15.1}$$

where $\beta = 3\alpha_1 + 2\alpha_2$ is the climbing constant and p_N is the Stokes pressure. In this case, the non-linearities in the constitutive equation affect only the distribution of normal stress (Tanner (1985)). $\alpha_1 + \alpha_2$ is positive for the fluids known to us and for simplification this constraint is applied to the fluid in this section. However, different methods are utilized in the following sections and the constraint on normal stress coefficients is removed.

The boundary conditions on the surface of the spheres are more easily expressed in terms of bispherical coordinates. Cylindrical coordinates can be transformed to bispherical coordinates as

$$r = \tilde{c}\frac{\sin\eta}{\cosh\xi - \cos\eta}, \qquad z = \tilde{c}\frac{\sinh\xi}{\cosh\xi - \cos\eta}, \tag{20.15.2}$$

where the surface of the sphere is at $\xi = \alpha$, $\cosh\alpha = h/a$, $\tilde{c} = a\sinh\alpha$, and h is the separation distance between the particle and the wall and a is the particle radius. Let $\mu = \cos\eta$, then the velocity gradient $\nabla\mathbf{u}$ in bispherical coordinates can be written as

$$\nabla\mathbf{u}|_{\text{on particle}} = \frac{\cosh\xi - \mu}{\tilde{c}} \times \begin{pmatrix} \frac{\partial u_\xi}{\partial\xi} - u_\eta\frac{\sin\eta}{\cosh\xi - \mu} & \frac{\partial u_\xi}{\partial\eta} + u_\eta\frac{\sinh\xi}{\cosh\xi-\mu} & 0 \\ \frac{\partial u_\eta}{\partial\xi} + u_\xi\frac{\sin\eta}{\cosh\xi-\mu} & \frac{\partial u_\eta}{\partial\eta} - u_\xi\frac{\sinh\xi}{\cosh\xi-\mu} & 0 \\ 0 & 0 & \frac{-u_\xi\sin\eta\sinh\xi + u_\eta(\mu\cosh\xi - 1)}{\sin\eta(\cosh\xi-\mu)} \end{pmatrix} \tag{20.15.3}$$

The axisymmetric motion of a sphere towards a wall in Stokes flow has been studied by Brenner (1961) and Maude (1961). Here we briefly summarize the results.

The stream function can be written as

$$\psi = (\cosh\xi - \mu)^{-\frac{3}{2}}\sum_{n=1}^{\infty} UX_n(P_{n-1}(\mu) - P_{n+1}(\mu)), \tag{20.15.4}$$

where U is the velocity of the particle and $P_n(\mu)$ is the Legendre polynomial of degree n and

$$X_n = \hat{A}_n\cosh\left(n-\frac{1}{2}\right)\xi + \hat{B}_n\sinh\left(n-\frac{1}{2}\right)\xi + \hat{C}_n\cosh\left(n+\frac{3}{2}\right)\xi + \hat{D}_n\sinh\left(n+\frac{3}{2}\right)\xi. \tag{20.15.5}$$

The coefficients $\hat{A}_n - \hat{D}_n$ are described by Brenner (1961). The pressure P_N can be expressed as an infinite summation of spherical harmonics as follows (Pasol *et al.* (2005))

$$p_N = \frac{\mu_f}{\tilde{c}^3}(\cosh\xi - \mu)^{\frac{1}{2}}\sum_{n=0}^{\infty}\left[A_n\cosh\left(n+\frac{1}{2}\right)\xi + B_n\sinh\left(n+\frac{1}{2}\right)\xi\right]P_n(\mu). \tag{20.15.6}$$

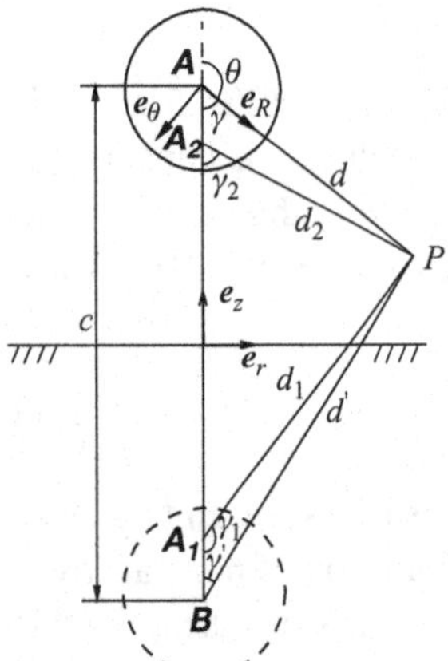

Figure 20.18. A spherical particle moving perpendicular to a wall.

The coefficients A_n and B_n are defined by Pasol *et al.* (2005). Calculating u_ξ, u_η, and P_N and using equations (20.3.2) and (20.15.3) the stress tensor in bispherical coordinates ($\mathbf{T}_b$) can be calculated. Using the rotation matrix from cylindrical to bispherical coordinates we have

$$R_1 = \begin{pmatrix} \frac{\cosh\xi-\mu}{\bar{c}}\frac{\partial r}{\partial \xi} & \frac{\cosh\xi-\mu}{\bar{c}}\frac{\partial z}{\partial \xi} & 0 \\ \frac{\cosh\xi-\mu}{\bar{c}}\frac{\partial r}{\partial \eta} & \frac{\cosh\xi-\mu}{\bar{c}}\frac{\partial z}{\partial \eta} & 0 \\ 0 & 0 & 1 \end{pmatrix}, \quad \mathbf{T}_{cyl} = R_1^T \mathbf{T}_b R_1. \tag{20.15.7}$$

To calculate the stress tensor in spherical coordinates centered at the sphere center, we have

$$R_2 = \begin{pmatrix} \sin\theta & \cos\theta & 0 \\ \cos\theta & -\sin\theta & 0 \\ 0 & 0 & 1 \end{pmatrix}, \quad \mathbf{T}_{sph} = R_2^T \mathbf{T}_{cyl} R_2, \tag{20.15.8}$$

where θ is the polar angle as shown in figure 20.18. Figure 20.19(a) shows the dimensionless normal stress as a function of θ for different separation distances when the particle is moving away from the wall. The stress is non-dimensionalized by $1/2\rho U^2$ and $c^* = c/(2a) = h/a$. Results for low Reynolds and Deborah numbers are shown ($Re = 0.05$ and $De = |\alpha_1 U|/(\mu_f a) = 0.168$). It can be seen that for small separation distances, a tensile normal stress occurs at the trailing edge when the fluid is Newtonian, while a larger tensile stress is observed for a second-order fluid. In figure 20.19(b), the particle is moving towards the wall. The stress is compressive at the leading edge for a Newtonian fluid whereas a large tensile stress is observed for a second-order fluid. This tensile stress arises from the *modified pressure*, $(\alpha_1/\mu_f)(\partial p_N/\partial t)$. If one calculates $T_n + p$ at this point, the result is zero due to the fact that the shear rate is zero at this point. This is in agreement with the results by Joseph and Feng (1996). For a large separation distance, the last term in equation (20.15.1), which is related to the shear rate, and second term of the right hand side of equation (20.15.1), which generates an extensional normal stress, are of the same order and relatively small compared to the overall force applied to the particle. For a particle nearly touching the wall, $(\alpha_1/\mu_f)/(Dp_N/Dt)$ is much larger than $\frac{\beta}{4} tr\mathbf{A}^2$ and this results in a large deviation from the Newtonian case. The shear stress in a

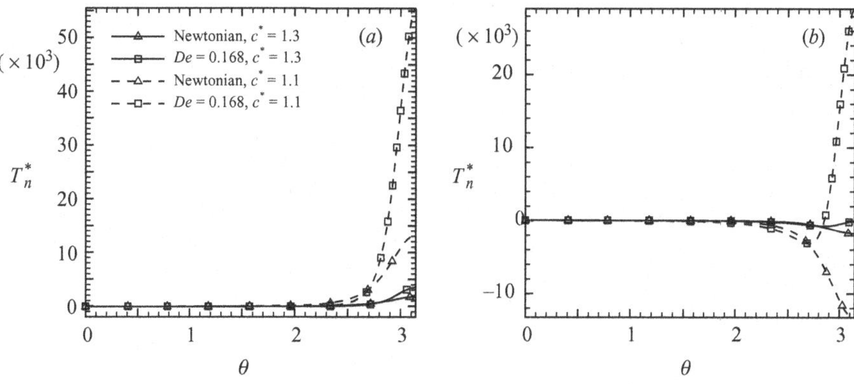

Figure 20.19. A spherical particle moving normal to a wall at $Re = 0.05$, $De = 0.168$: (a) away from the wall; (b) towards the wall.

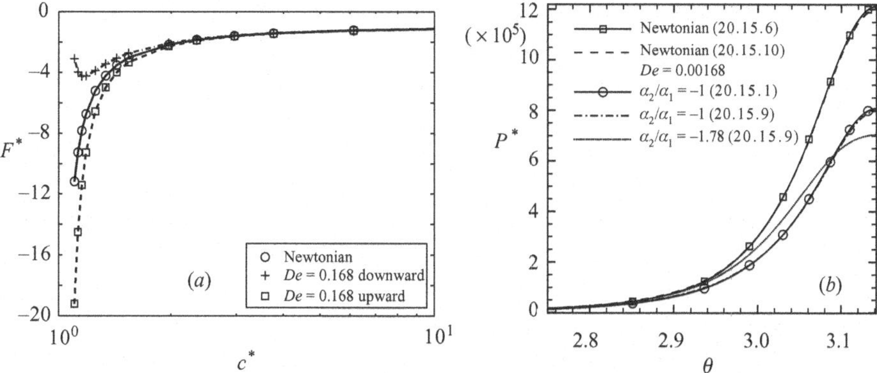

Figure 20.20. (a) Force $F^* = F/(6\pi\mu_f aU)$ acting on a spherical particle moving away from or towards the wall at $Re = 0.05$. (b) Dimensionless pressure for a sphere moving towards the wall at $Re = 0.05$ and $c^* = 1.01$.

second-order fluid is similar to the one in the Newtonian fluid when the separation distance is large. However, as the separation distance decreases, the shear stress differs from that of a Newtonian fluid and this difference arises from the term $\alpha_1(\partial A/\partial t)$. The total force acting on the particle moving away from or towards the wall is plotted in figure 20.20(a). It can be observed that the contribution of the second-order fluid to the overall force applied to the particle is towards the wall independently of the direction of motion of the particle.

When the particle is very close to the wall, $\epsilon = h/a - 1$ is small and there is no constraint on α_1 and α_2. In this case we may construct a perturbation analysis for De small. The dimensionless stretched cylindrical coordinates are defined as $Z = z/a\epsilon$ and $R = r/a\sqrt{\epsilon}$. It can be shown that the dimensionless velocity and pressure are scaled as:

$$u_r^* = \sqrt{\epsilon}\left[U_{1r}^* + \frac{De}{\epsilon}U_{2r}^* + h.o.t.\right], \quad u_z^* = U_{1z}^* + \frac{De}{\epsilon}U_{2z}^* + h.o.t.,$$

$$p^* = \epsilon^{-2}\left[P_1^* + \frac{De}{\epsilon}P_2^* + h.o.t.\right] \tag{20.15.9}$$

and De/ϵ is a small quantity compared to one. Superscript $*$ refers to dimensionless variables. The surface of the sphere is described by $Z = H = 1 + \frac{1}{2}R^2 + O(\epsilon)$. The terms

U_{1r}^*, U_{1z}^*, and P_1^* are determined by Jeffrey and Corless (1988) as:

$$U_{1r}^* = \frac{3R(Z^2 - HZ)}{H^3}, \quad U_{1z}^* = -\frac{2Z^3 - 3HZ^2}{H^3} + \frac{3R^2(Z^3 - HZ^2)}{H^4}, \quad P_1^* = -\frac{3}{H^2}. \tag{20.15.10}$$

The solution at the first order of *De* can be determined using the following equations:

$$\partial P_2^*/\partial Z = -\partial\left(\mathbf{B}_{1zz}^* + \frac{\alpha_2}{\alpha_1}\mathbf{A}_{1zz}^{*^2}\right)\Big/\partial Z$$

$$\partial P_2^*/\partial R = -\frac{1}{R}\left[\partial R\left(\mathbf{B}_{1rr}^* + \frac{\alpha_2}{\alpha_1}\mathbf{A}_{1rr}^{*^2}\right)\Big/\partial R\right] - \partial\left(\mathbf{B}_{1rz}^* + \frac{\alpha_2}{\alpha_1}\mathbf{A}_{1rz}^{*^2}\right)\Big/\partial Z + \partial^2 U_{2r}^*/\partial Z^2$$

$$\partial(RU_{2r}^*)/\partial R + R\partial U_{2z}^*/\partial Z = 0 \tag{20.15.11}$$

where $\mathbf{B}_1^*$ is the dimensionless tensor $\mathbf{B}$ (defined in (20.3.3)) for the velocity field U_1^*. The boundary conditions to be applied are: $U_2^*(Z=0) = U_2^*(Z=H) = 0$. Since $\mathbf{u}^*$ is a function of R, Z and ϵ, the term $\partial \mathbf{A}_1^*/\partial t^*|_{r,z}$ can be calculated as follows:

$$\left.\frac{\partial \mathbf{A}_1^*}{\partial t^*}\right|_{r,z} = -\left.\frac{R}{2\epsilon}\frac{\partial \mathbf{A}_1^*}{\partial R}\right|_{Z,\epsilon} - \left.\frac{Z}{\epsilon}\frac{\partial \mathbf{A}_1^*}{\partial Z}\right|_{R,\epsilon} + \left.\frac{\partial \mathbf{A}_1^*}{\partial \epsilon}\right|_{R,Z}. \tag{20.15.12}$$

Solving equations (20.15.11), U_{2r}^*, U_{2z}^*, and P_2^* can be determined. The dimensionless pressure is plotted in figure 20.20(b). As can be seen, the results from the previous section and this section are in agreement when $\alpha_1 + \alpha_2 = 0$, as expected. The difference between the pressure in second-order and Newtonian fluids at the stagnation point is more pronounced as $|\alpha_2/\alpha_1|$ is increased. The pressure at the stagnation point can be written as:

$$p|_{Z=H,R=0} = -3\frac{\mu_f U}{a}\left[\frac{1}{\epsilon^2} + \frac{De}{10\epsilon^3}\left(14 - 6\frac{\alpha_2}{\alpha_1}\right)\right]. \tag{20.15.13}$$

The total force acting on the particle can be written as:

$$F^* = -\frac{1}{\epsilon}\left[1 + \frac{De}{10\epsilon^3}\left(2 - 3\frac{\alpha_2}{\alpha_1}\right)\right]. \tag{20.15.14}$$

Equation (20.15.14) is in agreement with the results shown in figure 20.20(a).

20.15.2 Viscoelastic Potential Flow

Irrotational normal stresses produced by potential flow of a second-order fluid give rise to a motion of solid bodies which agrees with experimental observations as explained in the introduction. The shear stress and tangential velocity on the boundaries are in general discontinuous in viscous and viscoelastic irrotational flows. Potential flow of a viscous or viscoelastic liquid is incompatible with the no-slip condition at the boundary of the liquid and solid. However, to consider particle interaction in viscoelastic flow, we could look at viscoelastic potential flow locally. The literature shows that the sedimenting particles chain robustly in all flows, sedimentation, fluidization, shear flows, oscillating shear flows, and elongational flows. This chaining occurs for particles in sizes ranging from microns to centimeters (see subsection 20.7.4). Therefore the cause must be local and we believe the local mechanism is due to the change in the normal stress which we compute in the second-order fluid. Locally, near the stagnation point, the flow is slow and it could be argued that for this reason the local behavior is second order. Takagi *et al.* (2003) similarly use the idea of a local Stokes flow at the boundary of a moving particle. In addition, at the stagnation point, the no-slip condition is satisfied exactly while the slip velocity is small in the vicinity of the stagnation point. This is a valid argument to look at the normal stresses in the neighborhood of the stagnation point in a second-order fluid using viscoelastic potential flow.

It has been shown that for potential flow where $\mathbf{u} = \nabla\phi$ (Joseph (1992))

$$\nabla\cdot(\alpha_1\mathbf{B} + \alpha_2\mathbf{A}^2) = (3\alpha_1 + 2\alpha_2)\nabla\chi, \quad \chi = \frac{\partial^2\phi}{\partial x_i \partial x_j}\frac{\partial^2\phi}{\partial x_i \partial x_j} = \frac{1}{4}tr\mathbf{A}^2. \tag{20.15.15}$$

Thus, divergence of the stress is irrotational. Using equation (20.15.15), the pressure can be calculated using the Bernoulli equation. Thus, the stress tensor for viscoelastic potential flow can be written as

$$\mathbf{T} = \left[\rho\frac{\partial\phi}{\partial t} + \frac{1}{2}\rho|\nabla\phi|^2 - \beta\chi - C(t)\right]\mathbf{I} + \left[\mu_f + \alpha_1\left(\frac{\partial}{\partial t} + \mathbf{u}\cdot\nabla\right)\right]\mathbf{A} + (\alpha_1 + \alpha_2)\mathbf{A}^2. \tag{20.15.16}$$

For a spherical particle moving perpendicularly to a wall as shown in Figure 20.18, the potential-flow solution can be obtained using the image of a doublet source in a sphere and is given as the following series (Lamb (1932))

$$\begin{aligned}\phi = {} & U\left(\frac{\mu_0\cos\gamma}{d^2} + \frac{\mu_1\cos\gamma_1}{d_1^2} + \frac{\mu_2\cos\gamma_2}{d_2^2} + \cdots\right) \\ & + U\left(\frac{\mu_0\cos\gamma'}{d'^2} + \frac{\mu_1\cos\gamma_1'}{d_1'^2} + \frac{\mu_2\cos\gamma_2'}{d_2'^2} + \cdots\right),\end{aligned} \tag{20.15.17}$$

where $\mu_0 = \frac{1}{2}a^3$, a is the particle radius, U is the particle velocity, A is the center of a sphere moving towards the wall, B is the center of the imaginary sphere on the other side of the wall, $d = AP$, $d' = BP$, $d_1 = A_1P$, $d_1' = B_1P$, etc., are the distances between the doublets and a fixed point P. $AA_1 = f_1$, $AA_2 = f_2$, etc. can be defined using

$$\begin{aligned} & f_1 = c - \frac{a^2}{c}, && f_2 = \frac{a^2}{f_1}, && \frac{\mu_1}{\mu_0} = -\frac{a^3}{c^3}, && \frac{\mu_2}{\mu_1} = -\frac{a^3}{f_1^3}, \\ & f_3 = c - \frac{a^2}{c - f_2}, && f_4 = \frac{a^2}{f_3}, && \frac{\mu_3}{\mu_2} = -\frac{a^3}{(c - f_2)^3}, && \frac{\mu_4}{\mu_3} = -\frac{a^3}{f_3^3}, \cdots \end{aligned} \tag{20.15.18}$$

where c is twice the separation distance between the sphere and the wall. Using equations (20.15.17) and (20.15.18), we have

$$\phi = U\varphi(r, z, c) \Rightarrow \frac{\partial\phi}{\partial t} = a_c\varphi + 2U^2\frac{\partial\varphi}{\partial c}, \tag{20.15.19}$$

where a_c is the sphere acceleration. Also,

$$\mathbf{A} = U\tilde{\mathbf{A}} = 2U\begin{pmatrix} \frac{\partial^2\varphi}{\partial r^2} & \frac{\partial^2\varphi}{\partial r\partial z} & 0 \\ \frac{\partial^2\varphi}{\partial r\partial z} & \frac{\partial^2\varphi}{\partial z^2} & 0 \\ 0 & 0 & \frac{1}{r}\frac{\partial\varphi}{\partial r} \end{pmatrix}. \tag{20.15.20}$$

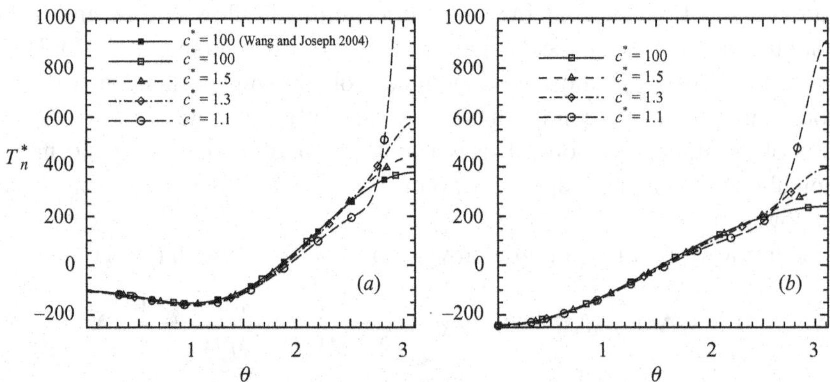

Figure 20.21. A spherical particle moving away from the wall at $Re = 0.05$, $De = 0.168$, and $\alpha_2/\alpha_1 = -1.78$ (a) Second-order fluid, (b) Newtonian fluid.

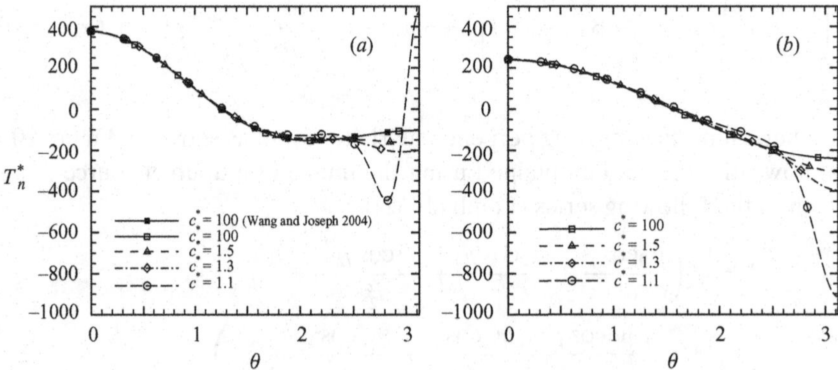

Figure 20.22. A spherical particle moving towards the wall at $Re = 0.05$, $De = 0.168$, and $\alpha_2/\alpha_1 = -1.78$ (a) Second-order fluid, (b) Newtonian fluid.

Thus, the stress tensor can be written as

$$\mathbf{T} + C\mathbf{I} = \mu_f \tilde{\mathbf{A}} U + (\rho\varphi\mathbf{I} + \alpha_1\tilde{\mathbf{A}})a_c + \left(\left[2\rho\frac{\partial\varphi}{\partial c} + \frac{1}{2}\rho\left\{ \left(\frac{\partial\varphi}{\partial r}\right)^2 + \left(\frac{\partial\varphi}{\partial z}\right)^2 \right\} \right.\right.$$
$$\left. - \beta\left\{ \left(\frac{\partial^2\varphi}{\partial r^2}\right)^2 + \left(\frac{1}{r}\frac{\partial\varphi}{\partial r}\right)^2 + \left(\frac{\partial^2\varphi}{\partial z^2}\right)^2 + 2\left(\frac{\partial^2\varphi}{\partial r\partial z}\right)^2 \right\} \right]\mathbf{I}$$
$$\left. + 2\alpha_1\frac{\partial\tilde{\mathbf{A}}}{\partial c} + \alpha_1\tilde{\mathbf{u}}\cdot\nabla\tilde{\mathbf{A}} + (\alpha_1+\alpha_2)\tilde{\mathbf{A}}^2 \right) U^2. \qquad (20.15.21)$$

The normal stress T_n and the shear stress T_t are

$$T_n = T_{rr}\sin^2\theta + T_{zz}\cos^2\theta + T_{rz}\sin 2\theta,$$
$$T_t = \frac{T_{rr} - T_{zz}}{2}\sin 2\theta + T_{rz}\cos 2\theta. \qquad (20.15.22)$$

Using equations (20.15.17), (20.15.21), and (20.15.22), the normal stress is computed at the surface of a sphere moving with constant velocity U perpendicularly to the wall.

Figure 20.21 shows the dimensionless normal stress as a function of θ for different separation distances when the particle is moving away from the wall. Results for $Re = 0.05$ and $De = 0.168$ are shown, which agree with the published results by Wang and Joseph (2004) when $c \to \infty$. It can be seen that for small separation distances, a tensile normal stress occurs at the trailing edge when the fluid is Newtonian, while a larger tensile stress is observed for a second-order fluid. In figure 20.22, the particle is moving towards the wall. The stress is compressive at the leading edge for a Newtonian fluid whereas a large tensile stress is observed for a second-order fluid. This behavior can be explained by examination of equation (20.15.21). The first term in this equation is the same for Newtonian and second-order fluids while for a non-accelerating particle, the second term is zero. The third term, which strongly depends on viscoelasticity, is proportional to U^2 and is independent of the direction of motion. Thus, a tensile stress is observed on the sphere surface at $\theta = \pi$ in both cases when the particle is moving away or towards the wall. The Stokes and potential flows give different but complimentary results when the motion is steady, due to the shear rate and extensional normal stresses, respectively. For unsteady flows, the Stokes flow evaluation of the stresses also gives rise to tension at a point of stagnation.

20.15.3 Conclusions

The second-order fluid is a valid asymptotic representation of a viscoelastic fluid for slow and slowly varying motions. There are two different cases in which the stresses in a second-order fluid may be evaluated a-priori on solutions of the Navier-Stokes equations without further calculations of the effects of viscoelasticity on the velocity field. The first case is Stokes flow (see Tanner (1985) for details). The second case is potential flow which is slow and slowly varying near stagnation points but not elsewhere. Results following from the potential flow analysis are in broad agreement with all the known facts about the response of particles in viscoelastic fluids observed in experiments suggesting that the main dynamics underway are controlled by the reversal of the sign of the normal stress at a point of stagnation. The forces predicted by Stokes equations for a sphere moving perpendicularly toward or away from a wall in a second order fluid with $\alpha_1 + \alpha_2 = 0$ are independent of direction and are always attractive toward the wall. The modified pressure associated with the time dependence of the Newtonian pressure in (20.15.1) is responsible for the sign reversal of the normal stress in Stokes flow. A perturbation analysis for small De when the sphere is close to the wall leads again to a tensile stress at both stagnation points when the sphere moves toward or away from the wall even when $\alpha_1 + \alpha_2 \neq 0$.

21

Purely irrotational theories of stability of viscoelastic fluids

As in the case of viscous fluids, very good approximations to exact results for viscoelastic fluids can be obtained from purely irrotational studies of stability. Here we consider RT instability (§21.1) and capillary instability (§21.2) of an Oldroyd B fluid. Viscoelastic effects enter into the irrotational analysis of RT instability through the normal stress at the free surface. For capillary instability, the short waves are stabilized by surface tension, and an irrotational viscoelastic pressure must be added to achieve excellent agreements with the exact solution. The extra pressure gives the same result as the dissipation method as is true in viscous fluids where VPF works for short waves and VCVPF and DM give the same results for capillary instability.

21.1 Rayleigh–Taylor instability of viscoelastic drops at high Weber numbers

Movies of the breakup of viscous and viscoelastic drops in the high-speed airstream behind a shock wave in a shock tube have been reported by Joseph, Belanger, and Beavers (1999; hereafter JBB). They performed a RT stability analysis for the initial breakup of a drop of Newtonian liquid and found that the most unstable RT wave fits nearly perfectly with waves measured on enhanced images of drops from the movies, but the effects of viscosity cannot be neglected. Snapshots from these movies are displayed in figures 21.1 to 21.4 and 21.9; data for the experiments are shown in table 21.1. Here we construct a RT stability analysis for an Oldroyd-B fluid by using measured data for acceleration, density, viscosity, and relaxation time λ_1. The most unstable wave is a sensitive function of the retardation time λ_2 that fits experiments when $\lambda_2/\lambda_1 = O(10^{-3})$. The growth rates for the most unstable wave are much larger than for the comparable viscous drop, which agrees with the surprising fact that the breakup times for viscoelastic drops are shorter. We construct an approximate analysis of RT instability based on viscoelastic potential flow that gives rise to nearly the same dispersion relation as the unapproximated analysis.

21.1.1 Introduction

Aitken and Wilson (1993) studied the problem of the stability to small disturbances of an incompressible elastic fluid above a free surface. They derived dispersion relations for an Oldroyd B fluid in the case in which the fluid is bounded below by a rigid surface. When the retardation time and inertia are neglected the analysis predicts an unbounded growth rate at a certain Weissenberg number. The addition of inertia or retardation smooths

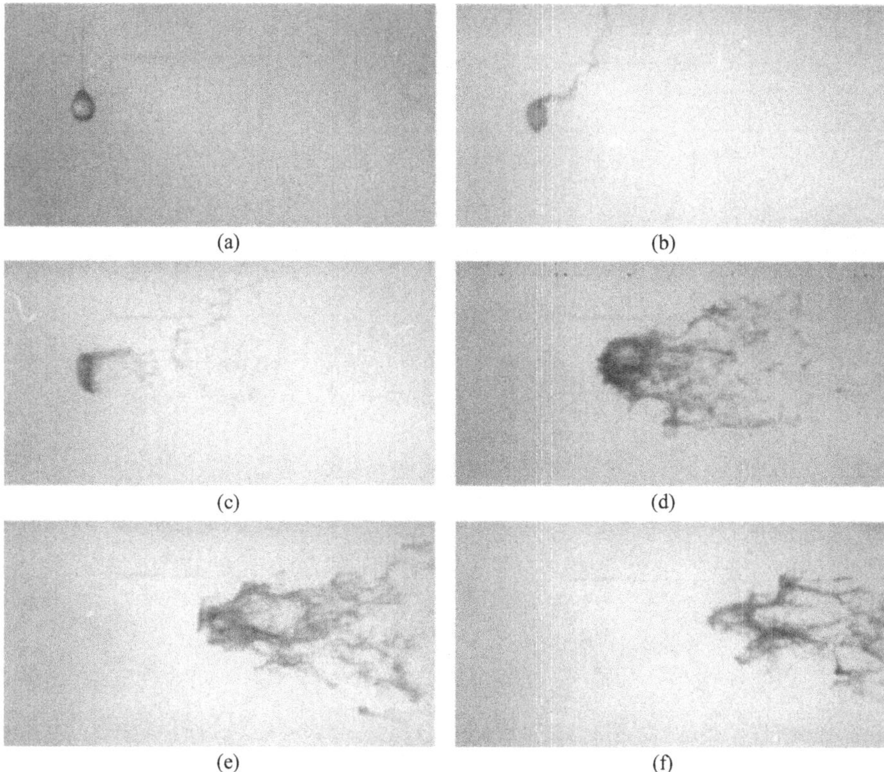

Figure 21.1. Stages in the breakup of a drop of 2% aqueous solution of polyox (WSR 301; diameter = 2.9 mm) in the flow behind a Mach 2 shock wave. Air velocity = 432 m $\sec^{-1}$; dynamic pressure = 165.5 kPa; Weber number = 15,200. Time (μs): (a) 0, (b) 55, (c) 95, (d) 290, (e) 370, (f) 435.

this singularity. The work presented here differs from that of Aitken and Wilson in the following ways: In our work the two fluids are unbounded; we construct both an exact analysis and an approximate analysis based on potential flow; we aim to apply the analysis of RT instability of viscoelastic drops by using measured data; we compute and present dispersion relations emphasizing the role of the most dangerous wave associated with the maximum growth rate, thereby emphasizing the role of the huge acceleration in the drop breakup problem that is due to RT instability (see figure 21.12); and we use the maximum growth rate to define a breakup time.

21.1.2 Experiments

21.1.2.1 *Displacement-time graphs and accelerations*

Displacement vs. time graphs for the Mach 3 experiments are shown in figure 21.5. The Mach 2 graphs are of similar form. The distance refers to the slowest moving drop fragment (the windward stagnation point); other parts of the fragmenting drop accelerate from rest even more rapidly. The graphs are nearly perfect parabolas for about the first 200 μsec of the motion, which allows the initial acceleration to be obtained by fitting a curve of the form $x - x_0 = \alpha(t - t_0)^2$. Values of the parameters α, t_0, x_0, and the initial acceleration

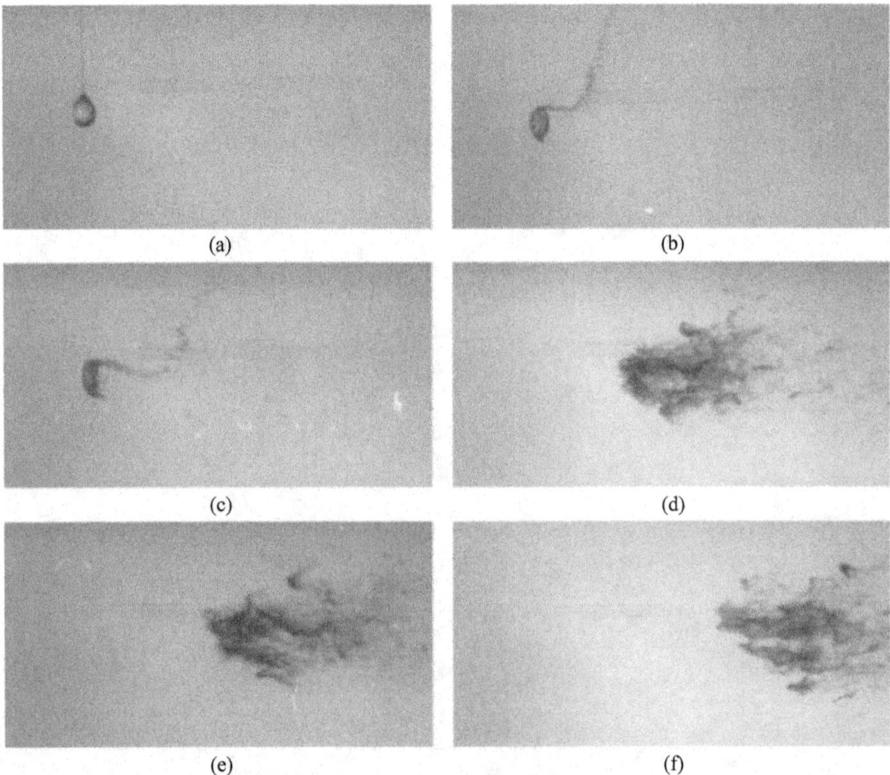

Figure 21.2. Stages in the breakup of a drop of 2% aqueous solution of polyox (WSR 301; diameter = 2.9 mm) in the flow behind a Mach 3 shock wave. Air velocity = 755 m sec^{-1}; dynamic pressure = 587.2 kPa; Weber number = 54,100. Time (μs): (a) 0, (b) 30, (c) 45, (d) 170, (e) 195, (f) 235.

are listed in table 21.2. It is noteworthy that in these graphs the acceleration is constant, independent of time for small times, and about 10^4–10^5 times the acceleration, that is due to gravity, depending on the shock-wave Mach number. In general there is a moderate decrease in acceleration with time over the course of the several hundred microseconds that it takes to totally fragment the drop.

The initial accelerations are an increasing function of the shock Mach number; the dynamic pressure that accelerates the drop increases with the free-stream velocity. At a fixed free-stream dynamic pressure there appears to be a tendency for the acceleration to decrease with drop size. If we take the drag on a spherical drop to be proportional to the drop diameter squared and the mass to the diameter cubed, then the acceleration is proportional to D^{-1} and decreases with increasing D.

21.1.3 Theory

The fluid mechanics of RT instability in an Oldroyd B fluid is controlled by acceleration, as is true for viscous fluids discussed in chapter 9. The theoretical background discussed there works well for the problem here. Ortiz *et al.* (2004) developed a correlation for the acceleration of a liquid drop suddenly exposed to a high-speed airstream. The correlation

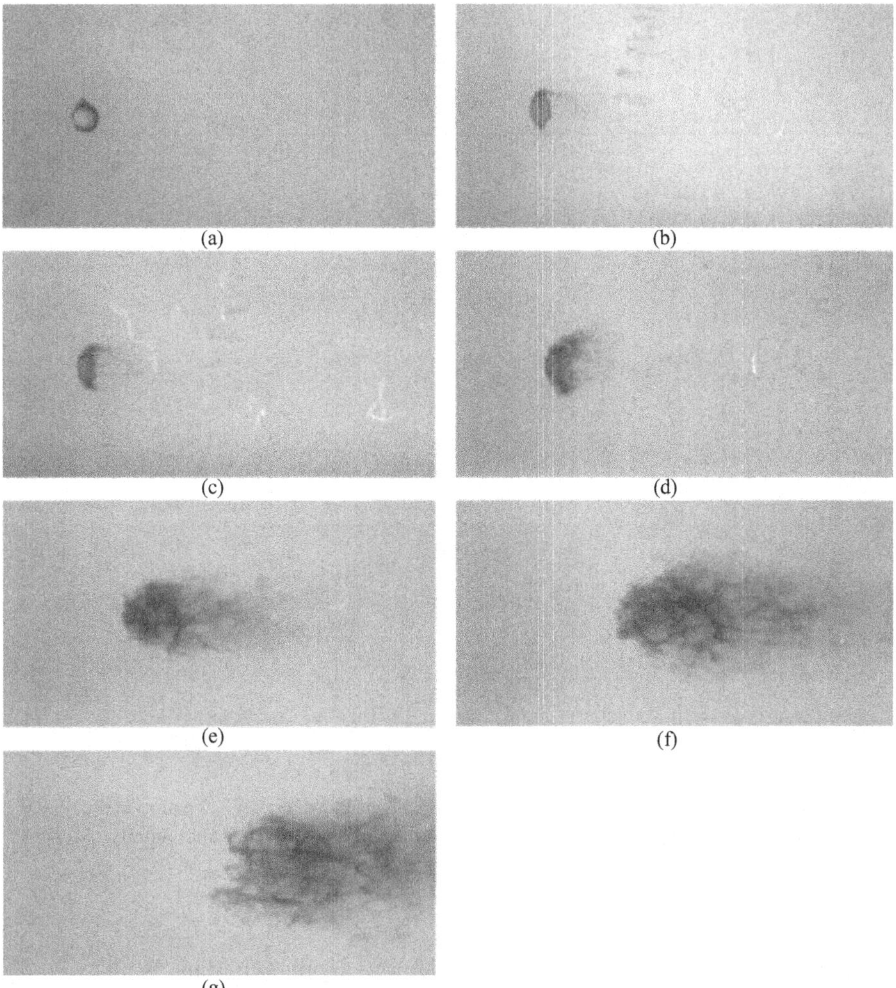

Figure 21.3. Stages in the breakup of a drop of 2% aqueous solution of polyacrylamide (Cyanamer N-300LMW; diameter = 3.2 mm) in the flow behind a Mach 3 shock wave. Air velocity = 771 m sec^{-1}; dynamic pressure = 578.1 kPa; Weber number = 82, 200. Time (μs): (a) 0, (b) 45, (c) 60, (d) 90, (e) 145, (f) 185, (g) 225.

depends weakly on the viscosity through the Ohnesorge number, and though it works also very well for viscoelastic drops, no viscoelastic parameter enters.

21.1.3.1 *Stability analysis*

The undisturbed interface between two fluids is located at $z = 0$, with a system of Cartesian coordinates $\mathbf{x} = (x, y, z) = (x_1, x_2, x_3)$ moving with acceleration $\mathbf{a}$:

$$\mathbf{a} = \mathbf{g} - \dot{\mathbf{V}} = \left(0, \ -g, \ -\dot{V}\right) = (0, \ -g, \ -a) \,. \tag{21.1.1}$$

For the conditions of the experiments described in this chapter the drop moves in a horizontal plane, and we may neglect g as at least 4 orders of magnitude smaller than $\dot{V}$. The undisturbed rest state is given by the pressure $\bar{p}^{(2)}$ in the heavy non-Newtonian fluid

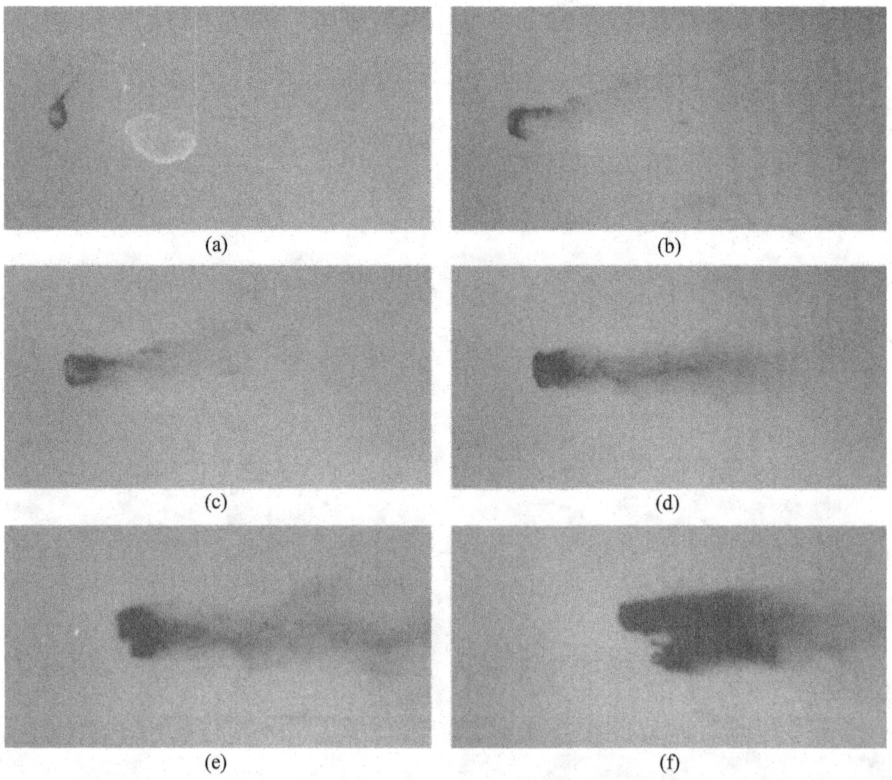

Figure 21.4. Stripping breakup of a drop of 1 kg m^{-1} sec^{-1} silicone oil (diameter = 2.6 mm) in the flow behind a Mach 3 shock wave. Air velocity = 767 m sec^{-1}; dynamic pressure = 681.0 kPa Weber number =168,600. Time (μs): (a) 15, (b) 40, (c) 50, (d) 80, (e) 115, (f) 150.

(the Oldroyd B fluid) in $z > 0$ and $\bar{p}^{(1)}$ in the light Newtonian fluid in $z < 0$:

$$\bar{p}^{(2)} = p_0 - \rho_2 a z, \quad \bar{p}^{(1)} = p_0 - \rho_1 a z, \tag{21.1.2}$$

where p_0 is the pressure at the interface, ρ_2 denotes the density of the heavy fluid, and ρ_1 is the density of the light fluid. Small disturbances are superimposed on the undisturbed state to give rise to the RT instability, for which the equations in the heavy fluid (in $0 < z$) are given by

$$\rho_2 \frac{\partial \mathbf{u}^{(2)}}{\partial t} = -\nabla p^{(2)} + \nabla \cdot \boldsymbol{\tau}^{(2)}, \tag{21.1.3a}$$

$$\nabla \cdot \mathbf{u}^{(2)} = 0, \tag{21.1.3b}$$

$$\tau_{ij}^{(2)} + \lambda_1 \frac{\partial \tau_{ij}^{(2)}}{\partial t} = 2\mu_2 \left[e_{ij}^{(2)} + \lambda_2 \frac{\partial e_{ij}^{(2)}}{\partial t} \right], \tag{21.1.3c}$$

$$e_{ij}^{(2)} = \frac{1}{2} \left[\frac{\partial u_i^{(2)}}{\partial x_j} + \frac{\partial u_j^{(2)}}{\partial x_i} \right], \tag{21.1.3d}$$

Table 21.1. *Experimental parameters: (a) liquid properties, (b) free-stream conditions*

(a) Liquid	Diameter (mm)	Viscosity (kg m^{-1} sec^{-1})	Surface tension (N m^{-1})	Density (kg m^{-3})	Relaxation time (sec)	*Oh*
Newtonian						
SO 1000	2.6	1	0.021	969		4.3
Viscoelastic						
2% PO	2.9	35	0.063	990	0.21	82.3
2% PAA	3.2	0.96	0.045	990	0.039	2.5

(b) Liquid	Velocity (m sec^{-1})	Density (kg m^{-3})	Pressure (kPa)	Dynamic pressure (kPa)	T_2 temp (K)	Weber no. ($\times 10^3$)	Reynolds no. ($\times 10^3$)	Shock *M*
Newtonian								
SO 1000	438.8	1.876	269.2	180.6	502	44.7	80.6	2.03
SO 1000	767.4	2.312	523.7	681.0	792	168.6	129.1	3.02
Viscoelastic								
2% PO	431.7	1.776	252.3	165.5	497	15.2	84.2	2.01
2% PO	754.8	2.061	458.7	587.2	778	54.1	127.6	2.98
2% PAA	770.6	1.947	442.9	578.1	795	82.2	134.0	3.03

Relaxation time for polyox (PO) and polyacrylamide (PAA) are computed from measured values taken on the wave-speed meter, PO in the tables given by Joseph (1990) and PAA in Liu (1995).

Table 21.2. *Curve-fitting parameters and initial accelerations for the liquid drops specified in table 21.1*

Parameter	Silicone oil		2% Aqueous PO		2% Aqueous PAA
Viscosity (kg m^{-1} sec^{-1})	1	1	35	35	0.96
Shock Mach no.	2	3	2	3	3
α (m sec^{-2})	1.463×10^5	5.561×10^5	0.687×10^5	3.240×10^5	2.461×10^5
x_0 (m)	-28.5×10^{-5}	7.45×10^{-5}	-17.7×10^{-5}	-0.046×10^{-5}	-6.16×10^{-5}
t_0(sec)	-3.43×10^{-5}	0.21×10^{-5}	-5.07×10^{-5}	-0.12×10^{-5}	-1.49×10^{-5}
Initial acceleration (m sec^{-2})	2.92×10^5	11.12×10^5	1.37×10^5	6.48×10^5	4.92×10^5
Max. accel. ($c = 0$) (m sec^{-2})	1.07×10^5	4.05×10^5	0.86×10^5	3.07×10^5	2.74×10^5
Mean accel./max. accel.	2.7	2.7	1.6	2.1	1.8

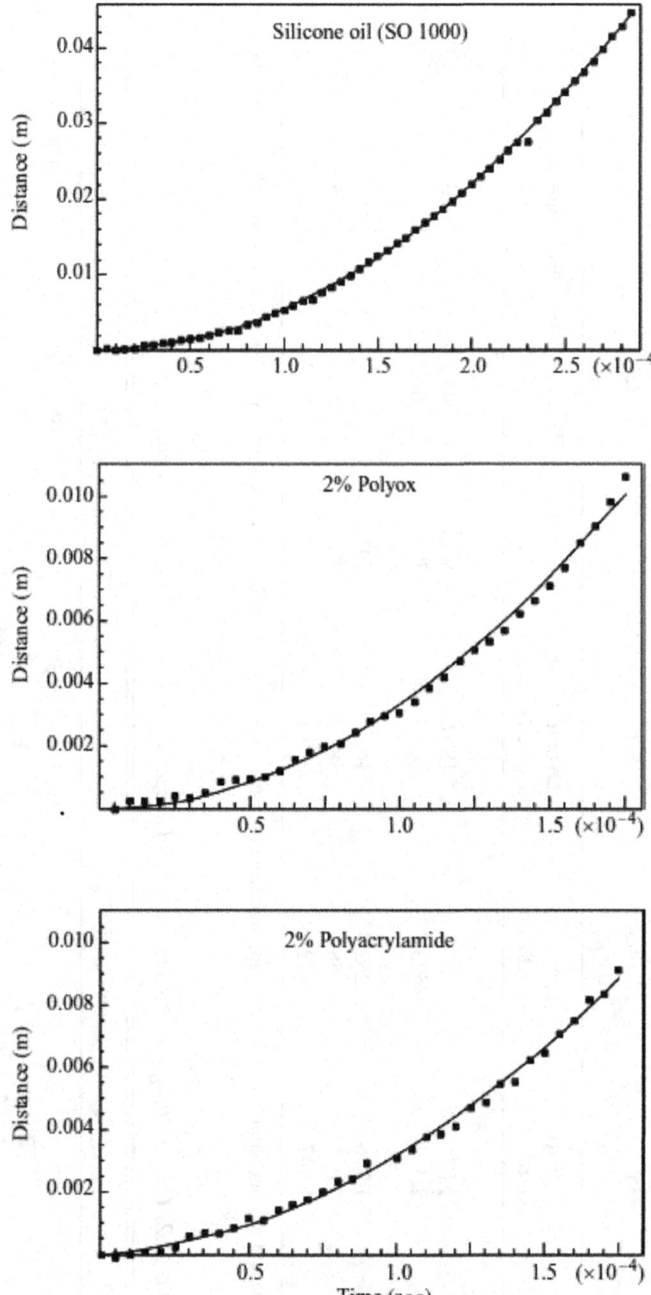

Figure 21.5. Distance traveled vs. time. $x - x_0 = \alpha(t - t_0)^2$, where x_0 and t_0 are the extrapolated starting values from the curve-fitting technique. The starting values x_0 and t_0 are uncertain within several pixels and several frames (5 μs per frame).

where $\mathbf{u}^{(2)} = \left[u^{(2)}, v^{(2)}, w^{(2)}\right] = \left[u_1^{(2)}, u_2^{(2)}, u_3^{(2)}\right]$ is the velocity disturbance; the viscous stress tensor $\tau_{ij}^{(2)}$ of the Oldroyd B fluid is expressed as constitutive equation (21.1.3c) with the strain tensor $e_{ij}^{(2)}$ and the viscosity μ_2; λ_1 is the relaxation time and λ_2 the retardation time; the conventional tensor notation is used here. Then, equations for disturbances in the light fluid (in $z < 0$) are given by

$$\rho_1 \frac{\partial \mathbf{u}^{(1)}}{\partial t} = -\nabla p^{(1)} + \nabla \cdot \tau^{(1)}, \tag{21.1.4a}$$

$$\nabla \cdot \mathbf{u}^{(1)} = 0, \tag{21.1.4b}$$

$$\tau_{ij}^{(1)} = 2\mu_1 e_{ij}^{(1)}, \tag{21.1.4c}$$

$$e_{ij}^{(1)} = \frac{1}{2}\left[\frac{\partial u_i^{(1)}}{\partial x_j} + \frac{\partial u_j^{(1)}}{\partial x_i}\right], \tag{21.1.4d}$$

where the viscous stress tensor $\tau_{ij}^{(1)}$ of the Newtonian fluid is expressed as (21.1.4c) with the strain tensor $e_{ij}^{(1)}$ and the viscosity μ_1.

Boundary conditions at the interface with its displacement h (at $z = h \approx 0$) are given by the continuity of velocity, the kinetic condition, and the continuity of the stress:

$$\mathbf{u}^{(1)} = \mathbf{u}^{(2)}, \tag{21.1.5a}$$

$$\frac{\partial h}{\partial t} = w^{(1)} = w^{(2)}, \tag{21.1.5b}$$

$$\tau_{13}^{(1)} = \tau_{13}^{(2)}, \tag{21.1.5c}$$

$$\tau_{23}^{(1)} = \tau_{23}^{(2)}, \tag{21.1.5d}$$

$$-p^{(2)} + \tau_{33}^{(2)} + \rho_2 a h - \left[-p^{(1)} + \tau_{33}^{(1)} + \rho_1 a h\right] = -\gamma \Delta h, \tag{21.1.5e}$$

where γ is the surface tension and Δ is the horizontal Laplacian:

$$\Delta = \frac{\partial^2}{\partial x^2} + \frac{\partial^2}{\partial y^2}. \tag{21.1.6}$$

Further, the boundary conditions require that the disturbances vanish, respectively, as $z \to \pm\infty$.

The solution to the system of the disturbances may take the following form:

$$\left[\mathbf{u}^{(2)}, p^{(2)}, h, \mathbf{u}^{(1)}, p^{(1)}\right] = \left[\hat{\mathbf{u}}^{(2)}(z), \hat{p}^{(2)}(z), \hat{h}, \hat{\mathbf{u}}^{(1)}(z), \hat{p}^{(1)}(z)\right] \exp\left(nt + ik_x x + ik_y y\right) + \text{c.c.}, \tag{21.1.7}$$

where n denotes the complex growth rate, $(k_x, k_y, 0)$ is the wavenumber vector of magnitude $k = \sqrt{k_x^2 + k_y^2}$, and c.c. stands for the complex conjugate of the preceding expression. Using (21.1.7), we can now write constitutive equation (21.1.3c) as

$$\tau_{ij}^{(2)} = 2\hat{\alpha} e_{ij}^{(2)}, \tag{21.1.8a}$$

with $\hat{\alpha}$ defined by

$$\hat{\alpha} = \mu_2 \frac{1 + \lambda_2 n}{1 + \lambda_1 n}. \tag{21.1.8b}$$

Taking this into account and taking rotation of (21.1.3a) and (21.1.4a), using $\nabla \times \nabla \times \mathbf{u} = -\nabla^2\mathbf{u}$ for incompressible fluid, we get the following equations:

$$\left(\nabla^2 - \frac{n\rho_1}{\mu_1}\right)\nabla^2 w^{(1)} = 0 \quad \text{in } z < 0, \quad \left(\nabla^2 - \frac{n\rho_2}{\hat{\alpha}}\right)\nabla^2 w^{(2)} = 0 \quad \text{in } z > 0, \tag{21.1.9}$$

for which the boundary conditions at the disturbed interface are written in terms of $w^{(1)}$ and $w^{(2)}$ as

$$\frac{\partial w^{(1)}}{\partial z} = \frac{\partial w^{(2)}}{\partial z}, \tag{21.1.10a}$$

$$\frac{\partial h}{\partial t} = w^{(1)} = w^{(2)}, \tag{21.1.10b}$$

$$\mu_1\left(\Delta - \frac{\partial^2}{\partial z^2}\right)w^{(1)} = \hat{\alpha}\left(\Delta - \frac{\partial^2}{\partial z^2}\right)w^{(2)}, \tag{21.1.10c}$$

$$\begin{aligned} -\left[\rho_2\frac{\partial^2 w^{(2)}}{\partial t\partial z} - \hat{\alpha}\nabla^2\frac{\partial w^{(2)}}{\partial z}\right] + 2\hat{\alpha}\Delta\frac{\partial w^{(2)}}{\partial z} + \left[\rho_1\frac{\partial^2 w^{(1)}}{\partial t\partial z} - \mu_1\nabla^2\frac{\partial w^{(1)}}{\partial z}\right] \\ - 2\mu_1\Delta\frac{\partial w^{(1)}}{\partial z} + (\rho_2 - \rho_1)\,a\Delta h + \gamma\Delta^2 h = 0, \end{aligned} \tag{21.1.10d}$$

and the conditions away from the interface are

$$w^{(1)} \to 0 \quad \text{as } z \to -\infty, \quad w^{(2)} \to 0 \quad \text{as } z \to \infty. \tag{21.1.10e}$$

To satisfy (21.1.10e), the solutions to equations (21.1.9) are expressed as

$$w^{(1)} = A^{(1)}\exp(kz) + B^{(1)}\exp(q_1 z), \quad w^{(2)} = A^{(2)}\exp(-kz) + B^{(2)}\exp(-q_2 z), \tag{21.1.11}$$

with q_1 and q_2 defined by

$$q_1 = \sqrt{k^2 + \frac{n\rho_1}{\mu_1}}, \quad q_2 = \sqrt{k^2 + \frac{n\rho_2}{\hat{\alpha}}}. \tag{21.1.12}$$

After substituting (21.1.11) into boundary conditions (21.1.10a)–(21.1.10d), we obtain an inhomogeneous system of linear equations for $A^{(1)}$, $B^{(1)}$, $A^{(2)}$, and $B^{(2)}$ that is solvable if and only if the determinant of the coefficient matrix vanishes. After a straightforward but tedious analysis we have the dispersion relation:

$$\begin{aligned} -\left[1 + \frac{1}{n^2}\left((\alpha_1 - \alpha_2)\,ak + \frac{\gamma k^3}{\rho_1 + \rho_2}\right)\right](\alpha_2 q_1 + \alpha_1 q_2 - k) - 4k\alpha_1\alpha_2 + 4\frac{k^2}{n}\frac{\mu_1 - \hat{\alpha}}{\rho_1 + \rho_2} \\ \times\left[\alpha_2 q_1 - \alpha_1 q_2 + (\alpha_1 - \alpha_2)\,k\right] + 4\frac{k^3}{n^2}\left(\frac{\mu_1 - \hat{\alpha}}{\rho_1 + \rho_2}\right)^2 (q_1 - k)(q_2 - k) = 0, \end{aligned} \tag{21.1.13}$$

where

$$\alpha_1 = \frac{\rho_1}{\rho_1 + \rho_2}, \quad \alpha_2 = \frac{\rho_2}{\rho_1 + \rho_2}. \tag{21.1.14}$$

Then the experiment shows $\rho_2 \gg \rho_1$, for which $\alpha_2 \to 1$ and $\alpha_1 \to 0$. Moreover, $\mu_1 \ll \hat{\alpha}$ in the experiment, so that (21.1.13) reduces to

$$-\left[1+\frac{1}{n^2}\left(-ak+\frac{\gamma k^3}{\rho_2}\right)\right]-4\frac{k^2}{n}\frac{\hat{\alpha}}{\rho_2}+4\frac{k^3}{n^2}\left(\frac{\hat{\alpha}}{\rho_2}\right)^2(q_2-k)=0. \tag{21.1.15}$$

Equation (21.1.15) approximates (21.1.13) with only a small error; it is appropriate for RT instability in a vacuum.

The solution of (21.1.13) gives rise to a dispersion relation of the type shown in figure 9.2. The border of stability is given by a critical wavenumber with stability only when

$$k > k_c = \sqrt{\frac{\rho \dot{V}}{\gamma}} \tag{21.1.16}$$

independent of viscosity, relaxation, or retardation time. Dispersion relations for our experiments are presented in figures 21.6–21.8 of §21.1.3.3.

21.1.3.2 *Viscoelastic potential flow analysis of stability*

RT instability at an air–liquid or vacuum–liquid surface is one of the many cases for which accurate results may be obtained by use of potential flow. For VPF the viscosity enters only in the normal component of the viscous stress. The dispersion relations for viscous flow and VPF derived in JBB (see chapter 9), though different, give values for the wavenumber and the growth rate of the most dangerous wave that are in good agreement. Viscous potential theory yields values for the wavenumber that are about 2% higher and values for the growth rate that are about 8.8% higher than the corresponding values from fully viscous theory (JBB, table 3). This shows that the main physical effect of viscosity is on the normal stress balance.

The results given in JBB carry over to viscoelastic potential flows, as we now show. We now require for each fluid that the potential ϕ give the velocity disturbance ($\mathbf{u} = \nabla\phi$) and satisfy the Laplace equation,

$$\nabla^2\phi = 0. \tag{21.1.17}$$

The pressure disturbance is given by Bernoulli's equation

$$\rho\frac{\partial\phi}{\partial t} + p + \rho a z = -\frac{\rho}{2}|\nabla\phi|^2 \approx 0, \tag{21.1.18}$$

for the same undisturbed state that is given in §21.1.3.1. Then the boundary conditions are given by (21.1.5a) and (21.1.5c) at the disturbed interface and (21.1.10e) away from the interface. The normal stress balance (21.1.10d) is now written, by use of (21.1.18), as

$$\rho_2\frac{\partial\phi^{(2)}}{\partial t} + \tau_{33}^{(2)} + \rho_2 a h - \left[\rho_1\frac{\partial\phi^{(1)}}{\partial t} + \tau_{33}^{(1)} + \rho_1 a h\right] = -\gamma\Delta h, \tag{21.1.19}$$

where

$$\frac{\tau_{33}}{2\mu} = e_{33} = \frac{\partial w}{\partial z} = \frac{\partial^2\phi}{\partial z^2} = k^2\phi. \tag{21.1.20}$$

Thus the solutions to (21.1.17) that vanish respectively as $z \to \pm\infty$ may be expressed as

$$w^{(1)} = A^{(1)} \exp(kz) \text{ in } z < 0, \quad w^{(2)} = A^{(2)} \exp(-kz) \quad \text{in } z > 0. \qquad (21.1.21)$$

Substitution of these into the boundary conditions by use of (21.1.10b) leads to the dispersion relation:

$$1 = \frac{\alpha_2 - \alpha_1}{n^2} ka - \frac{k^3 \gamma}{n^2 (\rho_2 + \rho_1)} - \frac{2k^2}{n} \frac{\hat{\alpha} + \mu_1}{\rho_2 + \rho_1}. \qquad (21.1.22)$$

Without much loss of generality, we may put $\alpha_1 = 0$, $\alpha_2 = 1$ and $\hat{\alpha} \gg \mu_1$, so that the dispersion relation becomes

$$1 = \frac{ka}{n^2} - \frac{k^3\gamma}{n^2\rho_2} - \frac{2k^2}{n}\frac{\hat{\alpha}}{\rho_2}, \qquad (21.1.23)$$

which can then be written as a cubic equation for the growth rate n.

It is interesting to note here that (21.1.23) for viscoelastic potential flow gives the same growth rate as the dispersion relation (21.1.15) for fully viscous flow if q_2 in (21.1.15) is approximated as

$$q_2 - k = \sqrt{k^2 + \frac{n\rho_2}{\hat{\alpha}}} - k \approx \frac{n\rho_2}{2k\hat{\alpha}}, \qquad (21.1.24)$$

i.e., under the condition that

$$\frac{n\rho_2}{2k\hat{\alpha}} \ll 1. \qquad (21.1.25)$$

Thus, under this condition, the theory of viscoelastic potential flow may provide a good approximation of the fully viscous theory.

21.1.3.3 *Comparison of the exact and the potential flow analysis*

Based on the data for the experimental conditions cited in tables 21.1 and 21.2, dispersion relation (21.1.15) is used to calculate the stability conditions, and the results are shown in figure 21.6: 2% polyox (PO) ($M_s = 2$); figure 21.7: 2% PO ($M_s = 3$); and figure 21.8: 2% polyacrylamide (PAA) ($M_s = 3$). In each of the figures several plots of dispersion relation (21.1.15) are shown for a fixed (known) value of the relaxation time and various assumed values of the retardation time λ_2. The growth rates are computed at increments in the wavenumber of 200 m^{-1} from $k = 0$ to the critical value. Dispersion relation (21.1.23) from viscoelastic potential theory gives rise to graphs that are nearly identical to those in figures 21.6–21.8.

For comparison of (21.1.23) and (21.1.15), values of the wavenumber k, wavelength l, and growth rate n of the most dangerous wave are shown in table 21.3, 2% PO ($M_s = 2$); table 21.4, 2% PO ($M_s = 3$); and table 21.5, 2% PAA ($M_s = 3$). These results show that the set of values of the growth rate and the wavenumber given by the viscoelastic potential analysis and the corresponding set of values obtained from the exact stability analysis are at the same level of good agreement as in the Newtonian case. The wavenumber predicted from viscoelastic potential analysis is greater than the corresponding value from fully viscoelastic theory by between 0 and 5.4% (with two exceptions): the growth rates from

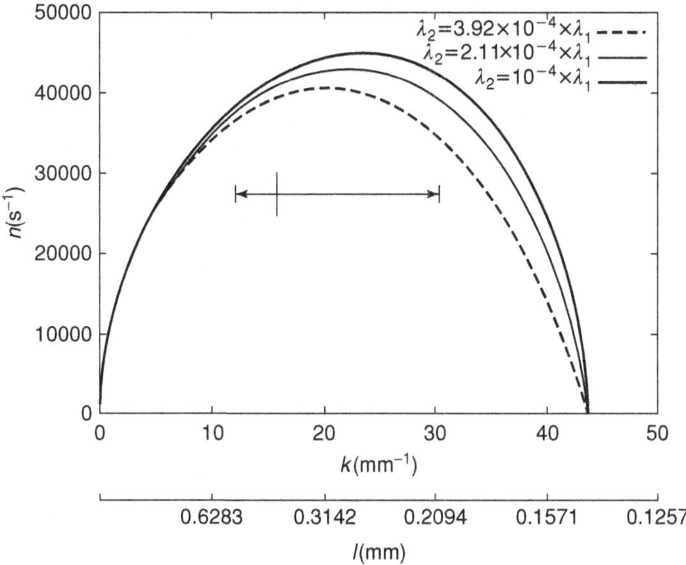

Figure 21.6. The growth rate n vs. the wavenumber k from (21.1.15) for 2% PO ($M_s = 2$); $\lambda_1 = 0.21$ sec. The average wavelength and scatter from a very early time in the experiment are indicated.

viscoelastic potential analysis are between 8.5% and 9.0% higher than predicted by fully viscoelastic theory, except at the smallest values of λ_2.

21.1.4 Comparison of theory and experiment

We now compare the RT stability theory with experiments on drop breakup for the three viscoelastic cases discussed in §21.1.2 (see table 21.6). For comparison, we repeat results

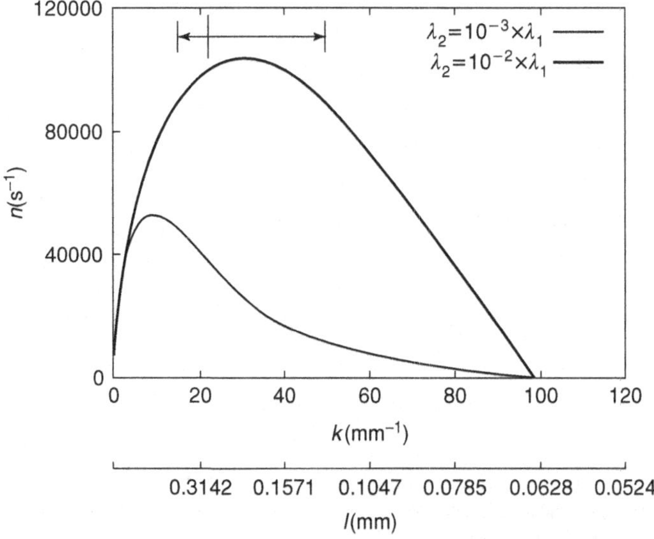

Figure 21.7. The growth rate n vs. the wavenumber k from (21.1.15) for 2% PO ($M_s = 3$); $\lambda_1 = 0.21$ sec. The average wavelength and scatter from a very early time in the experiment are indicated.

Table 21.3. *2% PO ($M_s = 2$)*

Values of the wavenumber k, wavelength l, and growth rate n of the most dangerous wave for the experimental conditions given in tables 21.1 and 21.2; the retardation time λ_2 is changed against the relaxation time λ_1. The values of k and n predicted by viscoelastic potential theory are higher than the corresponding fully viscoelastic predictions. The differences are indicated as a percentage of the fully viscoelastic values

	Exact			Viscoelastic potential			Percent difference	
λ_2/ (sec^{-1})	k (m^{-1})	l (mm)	n (sec^{-1})	k (m^{-1})	l (mm)	n (sec^{-1})	k	n
$\lambda_1/5$	600	10.472	6331.7	800	7.8539	6870.9	33.3	8.5
$\lambda_1/8$	1000	6.2832	7425.1	1000	6.2832	8077.7	0	8.8
$\lambda_1/10$	1000	6.2832	7991.5	1200	5.2359	8684.8	20.0	8.2
$\lambda_1/20$	1800	3.4907	10061.5	1800	3.4907	10945.9	0	8.8
$\lambda_1/100$	4800	1.3090	17000.0	5000	1.2566	18489.8	4.2	8.8
$\lambda_1/1000$	15 000	0.4189	32238.7	15 800	0.3977	34849.6	5.3	8.1
$\lambda_1/10000$	22 400	0.2805	43036.2	23 600	0.2662	45074.9	5.4	4.7
0	24 200	0.2596	45697.3	25 000	0.2513	47119.7	3.3	3.1

from JBB for a 1.0 kg m^{-1} sec^{-1} silicone oil whose viscosity nearly matches the 0.96 kg m^{-1} sec^{-1} PAA. Figure 21.9, taken from JBB, shows the waves on drops of this Newtonian liquid at very early times in the motion at shock Mach numbers of 2 and 3.

The waves on both the polyox and polyacrylamide drops were smaller and more difficult to identify than the waves on the Newtonian liquids shown, for example, in Engel (1958, Fig. 9), Hwang *et al.* (1996, Fig. 8), and JBB. For example, the measured average wavelengths for the 1.0 kg m^{-1} sec^{-1} silicone oil are about 2.0 and 1.25 mm for shock Mach

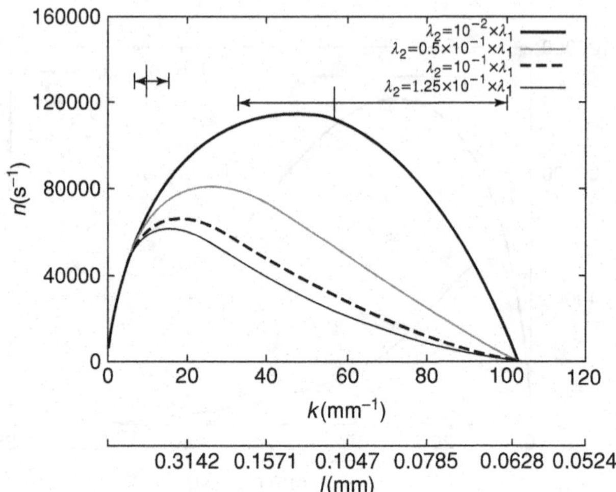

Figure 21.8. The growth rate n vs. the wavenumber k from (21.1.15) for 2% PAA ($M_s = 3$); $\lambda_1 = 0.034$ sec. The average wavelength and scatter from a very early time in the experiment are indicated. Also shown by dotted lines are the average wavelength and scatter for the set of waves of small wavelength that appear to be superimposed on the long-wavelength waves.

Table 21.4. *As table 21.3 but for 2% PO ($M_s = 3$)*

	Exact			Viscoelastic potential			Percent difference	
λ_2 (sec^{-1})	k (m^{-1})	l (mm)	n (sec^{-1})	k (m^{-1})	l (mm)	n (sec^{-1})	k	n
$\lambda_1/5$	1200	5.2359	17925.3	1200	5.2359	19496.0	0	8.8
$\lambda_1/8$	1600	3.9269	20968.4	1600	3.9269	22801.0	0	8.7
$\lambda_1/10$	1800	3.4907	22584.4	1800	3.4907	24549.0	0	8.7
$\lambda_1/20$	3000	2.0944	28424.1	3000	2.0944	30915.1	0	8.8
$\lambda_1/100$	8200	0.7662	48320.5	8400	0.7480	52541.6	2.4	8.7
$\lambda_1/1000$	29 200	0.2152	96037.0	30 600	0.2053	103 960	4.8	8.2
$\lambda_1/10000$	49 400	0.1272	138 925	51 600	0.1218	145 138	4.5	4.4
0	55 600	0.1130	152 570	56 600	0.1110	155 111	1.8	1.7

numbers of 2 and 3, respectively, and the corresponding values for the 2% PO solution are 0.39 and 0.20 mm. In an attempt to identify the waves more clearly on the computer screen, Adobe Photoshop™ was used to exaggerate the contrast. We then measured the lengths of the waves by first locating the troughs across the front of the drop on the computer screen and then measuring the distance between troughs in pixels that we finally converted to millimeters using a predetermined scaling factor for each frame. The enhanced contrast images are shown in figure 21.10 for the 2% aqueous PO and Figure 21.11 for the 2% PAA. The tick marks identify the wave troughs. Like the Newtonian liquids in JBB, the troughs are easier to identify on the computer screen than in the printed figure. The length of the waves increases with time because the waves are ultimately forced apart by high pressures in the wave troughs; from this it follows that the length of unstable waves should be measured at the earliest times for which all the waves can be identified.

The early appearance and short life of distinctly identifiable RT waves is illustrated in figure 9.9(a), which shows contrast-enhanced images from a repeat movie of the breakup of a drop (2.9-mm diameter) of a 2.0% aqueous solution of PO at a shock Mach number of 2.9. The four images in figure 9.9 (a) show the drop at 5-μsec intervals starting at 30 μsec after the passage of the shock wave. As before, the images are clearer and the waves are much easier to identify on the computer screen than in the printed version, where they appear pixelated. The waves have wavelengths of five pixels, which translates to 0.2 mm

Table 21.5. *As table 21.3 but for 2% PAA ($M_s = 3$)*

	Exact			Viscoelastic potential			Percent difference	
λ_2/ (sec^{-1})	k (m^{-1})	l (mm)	n (sec^{-1})	k (m^{-1})	l (mm)	n (sec^{-1})	k	n
$\lambda_1/5$	11 200	0.5610	49081.4	11 400	0.5512	53350.0	1.8	8.7
$\lambda_1/8$	14 800	0.4245	57009.4	15 200	0.4134	61948.0	2.7	8.7
$\lambda_1/10$	17 000	0.3696	61112.1	17 400	0.3611	66389.7	2.4	8.6
$\lambda_1/20$	24 400	0.2575	75051.4	25 400	0.2474	81393.6	4.1	8.5
$\lambda_1/100$	43 200	0.1454	108 441	45 400	0.1384	115 536	5.1	6.5
$\lambda_1/1000$	56 400	0.1114	133 671	57 800	0.1087	136 333	2.5	2.0
$\lambda_1/10000$	59 000	0.1065	138 403	59 400	0.1058	139 290	0.7	0.6
0	59 400	0.1058	139 007	59 800	0.1051	139 633	0.7	0.5

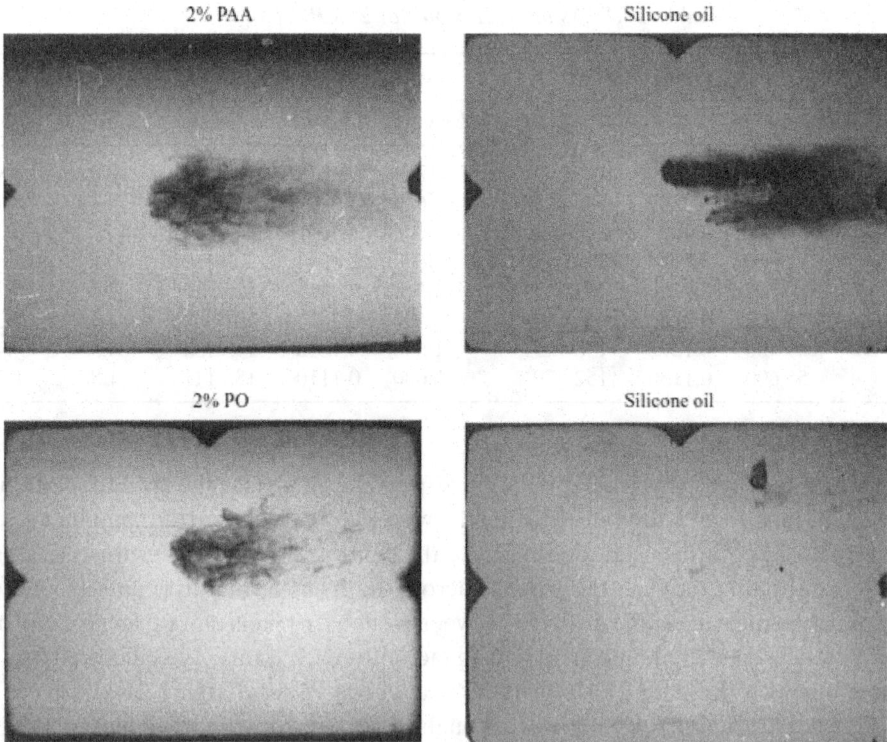

Figure 21.9. Droplet configurations for 2% PAA, 2% PO, and two different silicone oils 170 μsec after passing of the shock over the drop under the same conditions ($M_s = 3$). The top pair of photographs compares PAA with a silicone oil of approximately the same viscosity (1 kg m^{-1} sec^{-1}). The bottom pair compares 2% PO (viscosity = 35 kg m^{-1} sec^{-1}) with a silicone oil that has a viscosity of about one-third that of the PO (10 kg m^{-1} sec^{-1}).

on the scaling used for this movie, in frames (i), (ii), and (iii) but in (iv) the waves are becoming less distinct and only a few five-pixel wavelengths could be found. For times greater than that of frame (iv) the front face of the drop becomes very irregular as the drop sheds liquid and begins to break up.

The time interval in which the waves can be identified appears to correspond to the interval in which the original almost-spherical drop is undergoing severe deformation as

Table 21.6. *Comparison of measured breakup times (defined as the time at which liquid first starts to "blow" off the perimeter of the drop) with predicted times $\hat{t}_b$ calculated from equation $\hat{t}_b = \ln M/n$ with $M = 10$ and using values of n from the fully viscoelastic analysis*

Liquid	Shock Mach number	Approx. n (sec^{-1})	Time for $A = 10A_0$ (μsec)	Experimental blow-off time (μsec)
SO 1000	3	48,769	47	40
2% PO	3	90,000	25	30
2% PO	2	38,000	60	50
2% PAA: Short	3	110,000	20	35
Long		75,000	30	

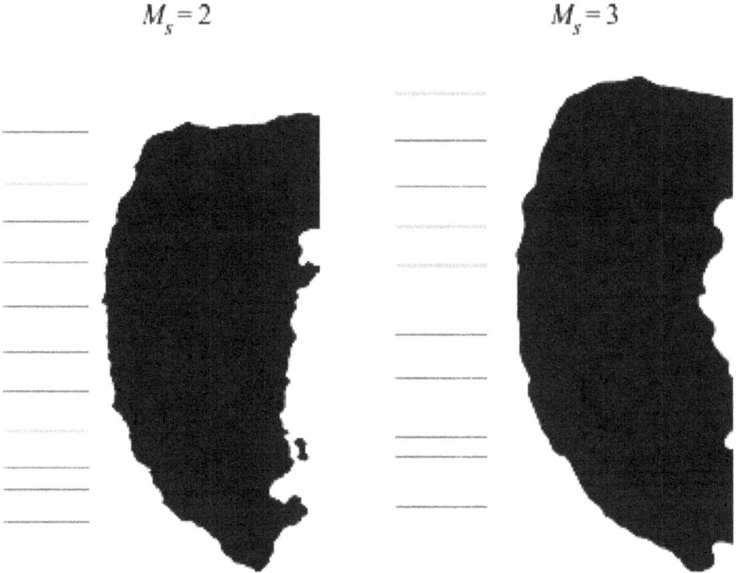

Figure 21.10. RT waves in 2% PO.

the front and back faces are being flattened and the cross-sectional area to the flow is increasing. This deformation is shown in figure 9.9(b), which presents the movie images corresponding to the contrast-enhanced images of figure 9.9(a). When the drop of PO is injected into the test section of the shock tube it leaves a thin, trailing thread of liquid connecting it to the injection needle. The disintegration of the thread is visible in the

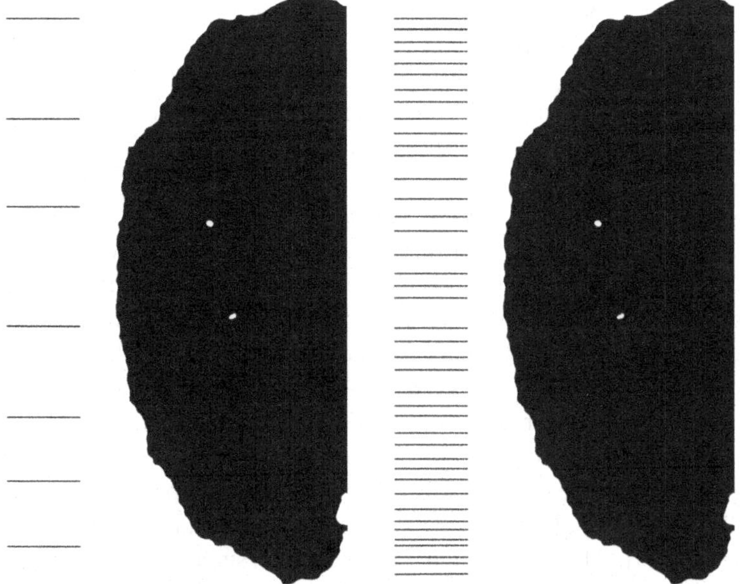

Figure 21.11. RT waves in 2% PAA.

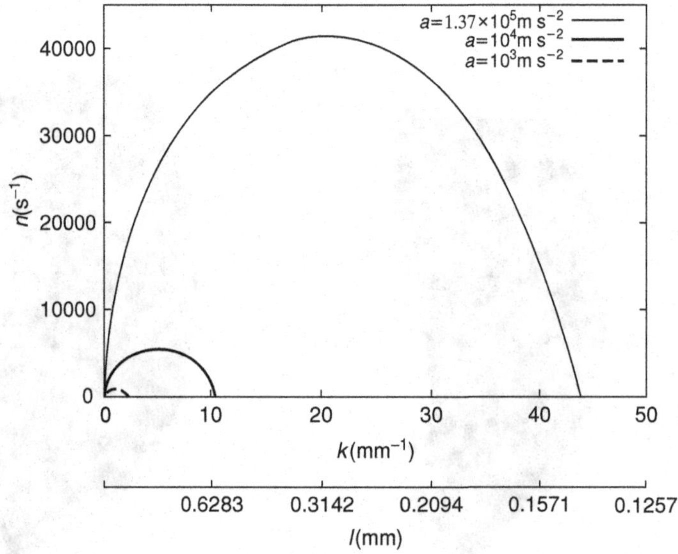

Figure 21.12. The effect of acceleration on the dispersion relation for 2% PO; $\lambda_1 = 0.21$ sec, $\lambda_2 = 3.3 \times 10^{-4}\lambda_1$.

frames of figure 9.9(b). The dark area that moves downstream from the top of the drop is the liquid that formed the small web at the top of the drop where the thread was attached. Figure 9.9(b) also indicates that liquid starts to be torn from the equator of the drop about 30 μsec after exposure to the high-speed flow.

Returning to figure 21.11, there is some uncertainty in the measurements of the wavelengths from the 2% PAA picture because there appeared to be two sets of waves, a distinct set with an average wavelength of 0.70 mm with a second set of smaller waves superimposed on the larger waves. The wavelengths of the smaller waves were very irregular, with values between approximately 0.05 and 0.24 mm. Smaller, but less distinct, waves could also be identified over parts of the front face of the PO drops.

In figures 21.6–21.8 we graph dispersion relations corresponding to measured data given in tables 21.1 and 21.2. The retardation time λ_2 is a fitting parameter. The dispersion graphs are sensitive to values of λ_2 as is shown in figures 21.6–21.8, where for each figure values for λ_2 have been chosen to yield curves such that the wavelengths of maximum growth are close to the interval of instability defined from the experiments, that is also included on the figures. From these we may estimate a λ_2 that centers the wavelength of maximum growth in the interval of instability. The estimated values of λ_2 needed to achieve agreement are uniformly small, ranging from $\lambda_2 \approx \lambda_1/5000$ for 2% aqueous PO at $M_s = 2$, to $\lambda_2 \approx \lambda_1/100$ for 2% aqueous PAA at $M_s = 3$. Boltzmann described the viscosity of a fluid as an effect of relaxed elastic modes, and it is given as the area under the shear relaxation modulus. Joseph (1990, chapter 18) interpreted the retardation time as representing the effect of the most rapidly relaxing modes; it depends on the time of observation as well as the material. The small value of the retardation time that matches theory and experiment reported here is as might be expected in such an explosive and short-time (10–50-μsec) event as produces RT waves on drops suddenly exposed to a high-speed airstream.

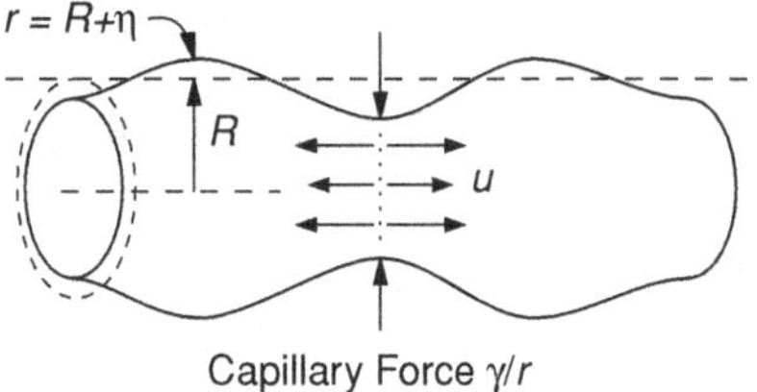

Figure 21.13. Capillary instability. The force γ/r drives fluid away from the throat, leading to collapse.

In the previous paragraph we argued that λ_2 is not fixed but depends at least on some conditions of external excitation and that RT disturbances are so fast that the response of the drops is highly elastic (small values of λ_2).

21.2 Purely irrotational theories of the effects of viscosity and viscoelasticity on capillary instability of a liquid cylinder

Capillary instability of a liquid cylinder can arise when either the interior or exterior fluid is a gas of negligible density and viscosity. The shear stress must vanish at the gas–liquid interface, but it does not vanish in irrotational flows. Joseph and Wang (2004) derived an additional viscous correction to the irrotational pressure. They argued that this pressure arises in a boundary layer induced by the unphysical discontinuity of the shear stress. Wang, Joseph, and Funada (2005a) showed that the dispersion relation for capillary instability in the Newtonian case is almost indistinguishable from the exact solution when the additional pressure contribution is included in the irrotational theory. Here we extend the formulation for the additional pressure to potential flows of viscoelastic fluids in flows governed by linearized equations and apply this additional pressure to capillary instability of viscoelastic liquid filaments of Jeffreys type. The shear stress at the gas–liquid interface cannot be made to vanish in an irrotational theory, but the explicit effect of this uncompensated shear stress can be removed from the global equation for the evolution of the energy of disturbances. This line of thought allows us to present the additional pressure theory without appeal to boundary layers. The validity of this purely irrotational theory can be judged by comparison with the exact solutions. Here we show that our purely irrotational theory is in remarkably good agreement with the exact solution in linear analysis of the capillary instability of a viscoelastic liquid cylinder.

21.2.1 Introduction

Capillary instability of a liquid cylinder of mean radius R leading to capillary collapse can be described as a neck-down that is due to surface tension γ in which fluid is ejected from the throat of the neck, leading to a smaller neck and greater neckdown capillary force, as seen in the diagram in figure 21.13.

Capillary instability of Newtonian fluids was studied by Rayleigh (1879) following earlier work by Plateau (1873), who showed that a long cylinder of liquid is unstable to disturbances with wavelengths greater than $2\pi R$. The analysis of Rayleigh is based on potential flow of an inviscid liquid. Tomotika (1935) studied the capillary instability and gave an exact normal-mode solution of the linearized Navier–Stokes equations.

The linear analysis of capillary instability of viscoelastic fluids has been done by Middleman (1965), Goldin *et al.* (1969), and Goren and Gottlieb (1982). They showed that the growth rates are larger for the viscoelastic fluids than for the equivalent Newtonian fluids.

Funada and Joseph (2002, 2003) presented potential flow analyses of capillary instability of viscous and viscoelastic fluids. In their studies, the flow is assumed to be irrotational but the viscous and viscoelastic effects are retained (viscous or viscoelastic potential flow, VPF). The viscous and viscoelastic stresses enter into the analyses through the normal stress balance at the interface. Funada and Joseph compared their results based on potential flow with the unapproximated normal-mode results (Tomotika, 1935). They showed that the results with viscous and viscoelastic effects retained are in better agreement with the unapproximated results than those assuming inviscid fluids.

The capillary instability can be viewed as a free-surface problem when either the interior or the exterior fluid is a gas of negligible density and viscosity. One difficulty in the potential flow analyses of free-surface problems is that the nonzero irrotational shear stress violates the zero-shear-stress condition at the free surface. Joseph and Wang (2004) derived an additional viscous correction for the irrotational pressure, which arises in the boundary layer induced by the unphysical discontinuity of the shear stress. Wang, Joseph, and Funada (2005a) applied this additional pressure contribution to the potential flow analysis of capillary instability of Newtonian fluids. They showed that the results computed with the additional pressure contribution are almost indistinguishable from the exact results. Here we extend the formulation for the additional pressure correction to potential flows of viscoelastic fluids in flows governed by linearized equations (viscoelastic correction of viscoelastic potential flow, VCVPF), and apply this additional pressure correction to capillary instability of viscoelastic liquid filaments of Jeffreys type. The results are in remarkably good agreement with those obtained from the unapproximated normal-mode analysis for viscoelastic fluids.

The linear stability analysis given here and elsewhere indicates that the liquid jets are less stable with increasing elasticity, which contradicts the observation in experiments. A possible explanation of this contradiction is related to the linear stability analysis of a stressed filament at rest (Entov, 1978). One difficulty is that a stressed filament at rest is not a permanent solution.

21.2.2 Linear stability equations and the exact solution

In an undisturbed rest state, the long cylinder of a viscoelastic liquid is surrounded by a gas of negligible density and viscosity. We use cylindrical coordinates (r, θ, z) and consider small axisymmetric disturbances. The linearized governing equations of the interior liquid are

$$\nabla \cdot \mathbf{u} = 0, \tag{21.2.1}$$

$$\rho \frac{\partial \mathbf{u}}{\partial t} = -\nabla p + \nabla \cdot \boldsymbol{\tau}, \tag{21.2.2}$$

where $\mathbf{u} = u\mathbf{e}_r + w\mathbf{e}_z$ is the velocity, ρ is the density, p is the pressure, and $\boldsymbol{\tau}$ is the extra stress. The extra stress may be modeled by Jeffreys model

$$\boldsymbol{\tau} + \lambda_1 \frac{\partial \boldsymbol{\tau}}{\partial t} = 2\mu \left(\mathbf{D} + \lambda_2 \frac{\partial \mathbf{D}}{\partial t} \right), \tag{21.2.3}$$

where $\mathbf{D}$ is the rate-of-strain tensor, μ is the viscosity, and λ_1 and λ_2 are the relaxation and retardation times, respectively. Suppose that we have normal-mode solutions with the growth rate σ:

$$\tau = \exp(\sigma t)\tilde{\tau}, \quad \mathbf{D} = \exp(\sigma t)\tilde{\mathbf{D}}; \tag{21.2.4}$$

then (21.2.3) leads to

$$\tilde{\tau} = \frac{1+\lambda_2\sigma}{1+\lambda_1\sigma}2\mu\tilde{\mathbf{D}} \quad \Rightarrow \quad \tau = \frac{1+\lambda_2\sigma}{1+\lambda_1\sigma}2\mu\mathbf{D}. \tag{21.2.5}$$

Momentum equation (21.2.2) becomes

$$\rho\frac{\partial \mathbf{u}}{\partial t} = -\nabla p + \nabla\cdot\left(\frac{1+\lambda_2\sigma}{1+\lambda_1\sigma}2\mu\mathbf{D}\right) = -\nabla p + \frac{1+\lambda_2\sigma}{1+\lambda_1\sigma}\mu\nabla^2\mathbf{u}. \tag{21.2.6}$$

The shear and normal stress boundary conditions are

$$\frac{1+\lambda_2\sigma}{1+\lambda_1\sigma}\mu\left(\frac{\partial u}{\partial z} + \frac{\partial w}{\partial r}\right) = 0\,, \tag{21.2.7}$$

$$-p + \frac{1+\lambda_2\sigma}{1+\lambda_1\sigma}2\mu\frac{\partial u}{\partial r} = \gamma\left(\frac{\partial^2\eta}{\partial z^2} + \frac{\eta}{R^2}\right), \tag{21.2.8}$$

where η is the varicose displacement. Governing equations (21.2.1) and (21.2.6) and boundary conditions (21.2.8) and (21.2.7) are the same as those for a Newtonian fluid except that $\frac{1+\lambda_2\sigma}{1+\lambda_1\sigma}\mu$ replaces μ.

The following scales are used to construct dimensionless governing equations: the cylinder diameter D for length, $U = \sqrt{\gamma/(\rho D)}$ for velocity, $T = D/U$ for time, and $p_0 = \rho U^2$ for pressure. The dimensionless momentum equation is (we use the same symbols for dimensionless variables)

$$\frac{\partial \mathbf{u}}{\partial t} = -\nabla p + \frac{\hat{\mu}}{\sqrt{J}}\nabla^2\mathbf{u}, \tag{21.2.9}$$

where

$$\hat{\mu} = \frac{1+\hat{\lambda}_2\sigma}{1+\hat{\lambda}_1\sigma}, \tag{21.2.10}$$

with

$$\hat{\lambda}_1 = \lambda_1\frac{U}{D} = \lambda_1\sqrt{\frac{\gamma}{\rho D^3}} \quad \text{and} \quad \hat{\lambda}_2 = \lambda_2\frac{U}{D} = \lambda_2\sqrt{\frac{\gamma}{\rho D^3}}, \tag{21.2.11}$$

and

$$J = \rho\gamma D/\mu^2 \tag{21.2.12}$$

is the Reynolds number and $J^{-1/2}$ is the Ohnesorge number. The dimensionless boundary conditions at the cylinder surface $R = 0.5$ are

$$-p + 2\frac{\hat{\mu}}{\sqrt{J}}\frac{\partial u}{\partial r} = \frac{\partial^2\eta}{\partial z^2} + \frac{\eta}{R^2}, \tag{21.2.13}$$

$$\frac{\hat{\mu}}{\sqrt{J}}\left(\frac{\partial u}{\partial z} + \frac{\partial w}{\partial r}\right) = 0. \tag{21.2.14}$$

A solution of (21.2.9) that satisfies both boundary conditions (21.2.13) and (21.2.14) takes the following form:

$$\psi = [A_1 r I_1(kr) + A_2 r I_1(k_v r)]\exp(\sigma t + ikz), \quad u = \frac{1}{r}\frac{\partial\psi}{\partial z}, \quad w = -\frac{1}{r}\frac{\partial\psi}{\partial r}, \tag{21.2.15}$$

$$\eta = H\exp(\sigma t + ikz), \tag{21.2.16}$$

where k is the wavenumber and I_1 denotes the first-kind modified Bessel function of the first order. Substitution of (21.2.15) and (21.2.16) into (21.2.13) and (21.2.14) leads to the solvability condition, which is given as the dispersion relation of σ:

$$\begin{vmatrix} 2k^2 I_1(kR) & (k^2 + k_v^2)\, I_1(k_v R) \\ F_1 & F_2 \end{vmatrix} = 0, \tag{21.2.17}$$

where

$$F_1 = \sigma I_0(kR) + 2\frac{\hat{\mu}k^2}{\sqrt{J}}\left[\frac{\mathrm{d}I_1(kR)}{\mathrm{d}(kR)}\right] - \left(\frac{1}{R^2} - k^2\right)\frac{k}{\sigma} I_1(kR), \tag{21.2.18}$$

$$F_2 = 2\frac{\hat{\mu}kk_v}{\sqrt{J}}\left[\frac{\mathrm{d}I_1(k_v R)}{\mathrm{d}(k_v R)}\right] - \left(\frac{1}{R^2} - k^2\right)\frac{k}{\sigma} I_1(k_v R), \tag{21.2.19}$$

with $k_v = \sqrt{k^2 + \frac{\sqrt{J}}{\hat{\mu}}\sigma}$. This solution satisfies the governing equations and all the boundary conditions and is an exact solution.

21.2.3 Viscoelastic potential flow

It is easy to show that momentum equation (21.2.9) admits potential flow solutions. Taking the curl of equation (21.2.9) and using $\mathbf{u} = \nabla\phi$, we obtain

$$\nabla\wedge\frac{\partial\nabla\phi}{\partial t} = \nabla\wedge(-\nabla p) + \frac{\hat{\mu}}{\sqrt{J}}\nabla\wedge\nabla^2\nabla\phi. \tag{21.2.20}$$

Both sides of (21.2.20) are zero; therefore potential flow solutions are compatible in this problem. The pressure integral can also be easily obtained from (21.2.9),

$$\nabla\left(\frac{\partial\phi}{\partial t}\right) = -\nabla p_p + \frac{\hat{\mu}}{\sqrt{J}}\nabla\nabla^2\phi \quad\Rightarrow\quad p_p = -\frac{\partial\phi}{\partial t}, \tag{21.2.21}$$

where p_p denotes the pressure from the potential flow solution, and it is equal to the pressure from the inviscid potential flow.

The potential flow solution is given by

$$\phi = Ai I_0(kr)\exp(\sigma t + ikz), \quad u = \frac{\partial\phi}{\partial r}, \quad w = \frac{\partial\phi}{\partial z}, \tag{21.2.22}$$

$$\eta = H\exp(\sigma t + ikz). \tag{21.2.23}$$

Substitution of the potential flow solution into normal stress balance (21.2.13) leads to the dispersion relation:

$$\frac{I_0(kR)}{I_1(kR)}\sigma^2(1+\hat{\lambda}_1\sigma) + (1+\hat{\lambda}_2\sigma)\sigma\frac{2k^2}{\sqrt{J}}\left[\frac{I_0(kR)}{I_1(kR)} - \frac{1}{kR}\right] - k\left(\frac{1}{R^2} - k^2\right)(1+\hat{\lambda}_1\sigma) = 0, \tag{21.2.24}$$

which is a cubic equation of σ and has explicit solutions.

When $J \to \infty$, equation (21.2.24) reduces to

$$\frac{I_0(kR)}{I_1(kR)}\sigma^2 = k\left(\frac{1}{R^2} - k^2\right), \tag{21.2.25}$$

which is the dispersion relation for the IPF solution. The IPF solution does not allow viscous or viscoelastic effects.

21.2.4 Dissipation and the formulation for the additional pressure contribution

Joseph and Wang (2004) derived a viscous pressure contribution in addition to the irrotational pressure for the potential flow solutions of Newtonian fluids by considering the dissipation of energy. Here we extend the analysis to a viscoelastic fluid of Jeffreys type in flows governed by linearized equations. We start from the momentum equation:

$$\rho\frac{d\mathbf{u}}{dt} = \nabla\cdot\mathbf{T} \quad \Rightarrow \quad \mathbf{u}\cdot\rho\frac{d\mathbf{u}}{dt} = (\nabla\cdot\mathbf{T})\cdot\mathbf{u}, \tag{21.2.26}$$

where $\mathbf{T}$ is the total stress. It then follows that

$$\begin{aligned}\rho\frac{d}{dt}\left(\frac{1}{2}\mathbf{u}\cdot\mathbf{u}\right) &= \nabla\cdot(\mathbf{T}\cdot\mathbf{u}) - \nabla\mathbf{u}:\mathbf{T}\\ &= \nabla\cdot(\mathbf{T}\cdot\mathbf{u}) - (\mathbf{D}+\Omega):(-p\mathbf{1}+2\hat{\mu}\mu\mathbf{D})\\ &= \nabla\cdot(\mathbf{T}\cdot\mathbf{u}) - \mathbf{D}:(-p\mathbf{1}+2\hat{\mu}\mu\mathbf{D})\\ &= \nabla\cdot(\mathbf{T}\cdot\mathbf{u}) - 2\hat{\mu}\mu\mathbf{D}:\mathbf{D}.\end{aligned}$$

It then follows that

$$\frac{d}{dt}\int_V\left(\frac{\rho}{2}\mathbf{u}\cdot\mathbf{u}\right)dV = \int_\Omega \mathbf{n}\cdot(\mathbf{T}\cdot\mathbf{u})\,d\Omega - 2\hat{\mu}\mu\int_V \mathbf{D}:\mathbf{D}dV, \tag{21.2.27}$$

where V is the volume occupied by the viscoelastic fluid, Ω is the boundary of V, and $\mathbf{n}$ is the outward normal of V on Ω. We have shown that the potential flow is a solution of the momentum equation in this problem. Thus we can insert the velocity and the stress tensor evaluated on the potential flow into (21.2.27) to obtain

$$\frac{d}{dt}\int_V\left(\frac{\rho}{2}\mathbf{u}\cdot\mathbf{u}\right)dV = \int_\Omega [(-p_p+\tau_{rr})u + \tau_{rz}w]\,d\Omega - 2\hat{\mu}\mu\int_V \mathbf{D}:\mathbf{D}\,dV. \tag{21.2.28}$$

At the free surface, the potential flow leads to a nonzero irrotational shear stress and does not satisfy the zero-shear-stress condition. We introduce a pressure contribution p_c in addition to the irrotational pressure p_p; p_c cancels out the power because of the unphysical irrotational shear stress in the energy equation, and (21.2.27) becomes

$$\frac{d}{dt}\int_V\left(\frac{\rho}{2}\mathbf{u}\cdot\mathbf{u}\right)dV = \int_\Omega [(-p_p-p_c+\tau_{rr})u]\,d\Omega - 2\hat{\mu}\mu\int_V \mathbf{D}:\mathbf{D}\,dV. \tag{21.2.29}$$

Comparing (21.2.28) and (21.2.29), we obtain

$$\int_\Omega \tau_{rz}w\,d\Omega = \int_\Omega(-p_c)u\,d\Omega, \tag{21.2.30}$$

which is the same as the formulation for the additional pressure contribution as in the potential flow of a viscous Newtonian fluid (Joseph and Wang, 2004). However, the calculation of τ_{rz} in viscoelastic fluids is different than in Newtonian fluids. The additional pressure contribution p_c depends strongly on viscoelastic parameters and is determined solely by the irrotational flow.

21.2.5 The additional pressure contribution for capillary instability

Now we consider the additional pressure contribution for the potential flow analysis of capillary instability. Joseph and Wang (2004) showed that, in linearized problems, the governing equation for the additional pressure contribution is

$$\nabla^2 p_c = 0. \tag{21.2.31}$$

It is easy to show that (21.2.31) holds for the viscoelastic fluid under consideration here. Solving (21.2.31), we obtain

$$-p_c = \sum_{j=0}^{\infty} C_j i I_0(jr)\exp(\sigma t + ijz), \tag{21.2.32}$$

where C_j are constants. With the additional pressure contribution, the normal stress balance becomes

$$-p_p - p_c + 2\hat{\mu}\frac{1}{\sqrt{J}}\frac{\partial u}{\partial r} = \frac{\partial^2 \eta}{\partial z^2} + \frac{\eta}{R^2}, \tag{21.2.33}$$

which gives rise to

$$\begin{aligned}&\left\{A\sigma I_0(kR) + C_k I_0(kR) + \frac{2\hat{\mu}k^2}{\sqrt{J}}A\left[I_0(kR) - \frac{I_1(kR)}{kR}\right]\right\}\exp(\sigma t + ikz)\\ &+\sum_{j\neq k} C_j I_0(jR)\exp(\sigma t + ijz) = A\frac{k}{\sigma}I_1(kR)\left(\frac{1}{R^2} - k^2\right)\exp(\sigma t + ikz).\end{aligned} \tag{21.2.34}$$

By orthogonality of Fourier series, $C_j = 0$ if $j \neq k$. The coefficient C_k can be determined with (21.2.30). The left-hand side of (21.2.30) is

$$\int_{\Omega} \tau_{rz} w^* \mathrm{d}\Omega = \frac{\hat{\mu}}{\sqrt{J}} 4\pi l RAA^* k^3 I_0(kR) I_1(kR)\exp(\sigma + \sigma^*)t, \tag{21.2.35}$$

where l is the length of one wave period and * denotes conjugate variables. On the other hand,

$$\int_{\Omega} (-p_c) u^* \mathrm{d}\Omega = 2\pi l RC_k A^* k I_0(kR) I_1(kR)\exp(\sigma + \sigma^*)t. \tag{21.2.36}$$

It follows that $C_k = 2\frac{\hat{\mu}}{\sqrt{J}}Ak^2$ and

$$-p_c = iAk^2\frac{2\hat{\mu}}{\sqrt{J}}I_0(kr)\exp(\sigma t + ikz). \tag{21.2.37}$$

Inserting C_k into (21.2.34), we obtain

$$\sigma I_0(kR) + \frac{2\hat{\mu}k^2}{\sqrt{J}}I_0(kR) + \frac{2\hat{\mu}k^2}{\sqrt{J}}\left[I_0(kR) - \frac{I_1(kR)}{kR}\right] = \frac{k}{\sigma}I_1(kR)\left(\frac{1}{R^2} - k^2\right),$$

which can be written as

$$\frac{I_0(kR)}{I_1(kR)}\sigma^2\left(1+\hat{\lambda}_1\sigma\right)+\left(1+\hat{\lambda}_2\sigma\right)\sigma\frac{2k^2}{\sqrt{J}}\left[\frac{2I_0(kR)}{I_1(kR)}-\frac{1}{kR}\right]-k\left(\frac{1}{R^2}-k^2\right)\left(1+\hat{\lambda}_1\sigma\right)=0. \tag{21.2.38}$$

Equation (21.2.38) is the dispersion relation from the VCVPF.

If pressure correction (21.2.37) is inserted back into governing equation (21.2.9), we obtain

$$\frac{\partial \mathbf{u}_c}{\partial t}=-\nabla p_c+\frac{\hat{\mu}}{\sqrt{J}}\nabla^2\mathbf{u}_c, \tag{21.2.39}$$

where $\mathbf{u}_c$ is the velocity correction induced by the pressure correction p_c. We can find a potential flow solution $\mathbf{u}_c = \nabla\phi_c$, such that $\nabla^2\mathbf{u}_c = \nabla\nabla^2\phi_c = 0$ and

$$\nabla\frac{\partial}{\partial t}\phi_c=-\nabla p_c. \tag{21.2.40}$$

It can be readily shown that

$$\phi_c=\frac{i}{\sigma}Ak^2\frac{2\hat{\mu}}{\sqrt{J}}I_0(kr)\exp(\sigma t+ikz). \tag{21.2.41}$$

Thus the pressure correction p_c, which is proportional to $J^{-1/2}$, induces a velocity correction proportional to $J^{-1/2}$. This velocity correction gives rise to an uncompensated shear stress proportional to J^{-1}, which may induce a new pressure correction now proportional to J^{-1}. In this way we may generate, successively, irrotational solutions proportional to increasing powers of $J^{-1/2}$. We believe that only the first pressure correction proportional to $J^{-1/2}$ is of physical significance; the higher-order corrections are not considered in normal stress balance (21.2.33).

21.2.6 Comparison of the growth rate

We compare dispersion relation (21.2.38) from VCVPF with (21.2.24) from VPF, (21.2.25) from IPF, and (21.2.17) from the ES. Equations (21.2.17), (21.2.24), (21.2.25), and (21.2.38) are solved by numerical methods for the growth rate σ, and the values of σ are compared.

First we examine two practical cases: 2% PAA in air and 2% polyethylene oxide (PEO) in air (following Funada and Joseph, 2003). We choose the diameter of the fluid cylinder to be 1 cm. The σ vs. k plots for 2% PAA and 2% PEO are shown in figures 21.14 and 21.15, respectively. These figures show that the results from VCVPF are almost indistinguishable from the ES, whereas IPF and VPF overestimate σ significantly.

Capillary instability is controlled by three dimensionless numbers: J, $\hat{\lambda}_1$, and $\hat{\lambda}_2$. We vary these parameters and present the computed growth rate in figures 21.16–21.19. The Reynolds number J ranges from 10^{-4} to 10^4, $\hat{\lambda}_1$ ranges from 0.1 to 1000, and $\hat{\lambda}_2$ ranges from 0 to 100. In all the cases, the growth rates from VCVPF are in excellent agreement with the ES, indicating that our additional pressure contribution is valid for a wide range of controlling parameters.

Figures 21.16 and 21.17 show that the growth rates increase with $\hat{\lambda}_1$ when J and $\hat{\lambda}_2$ are fixed. Comparing figures 21.17 and 21.18, we can see that the effect of $\hat{\lambda}_2$ is opposite to

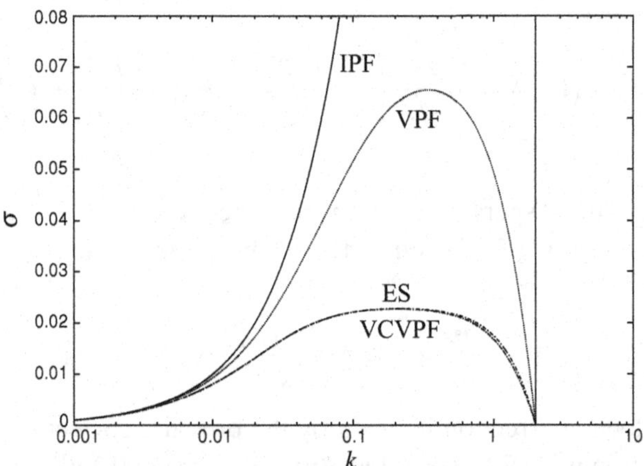

Figure 21.14. The growth rate σ vs. k from IPF, VPF, VCVPF, and the exact solution (ES). The growth rates for the ES and VCVPF are almost the same. The fluid is 2% PAA, $\rho = 0.99$ g cm^{-1}, $\mu = 96$ P, $\gamma = 45.0$ dyn cm^{-1}, $\lambda_1 = 0.039$ sec, $\lambda_2 = 0$ sec, $J = 4.834 \times 10^{-3}$, $\hat{\lambda}_1 = 0.263$.

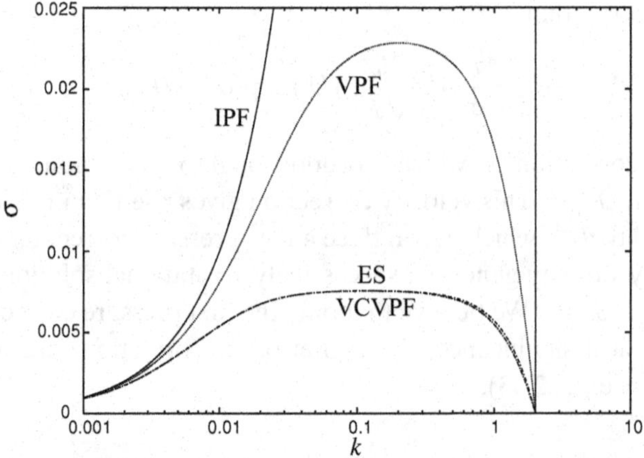

Figure 21.15. The growth rate σ vs. k from IPF, VPF, VCVPF, and the exact solution (ES). The results of the ES and VCVPF are almost the same. The fluid is 2% PEO, $\rho = 0.99$ g cm^{-1}, $\mu = 350$ P, $\gamma = 63.0$ dyn cm^{-1}, $\lambda_1 = 0.21$ sec, $\lambda_2 = 0$ sec, $J = 5.091 \times 10^{-4}$, $\hat{\lambda}_1 = 1.676$.

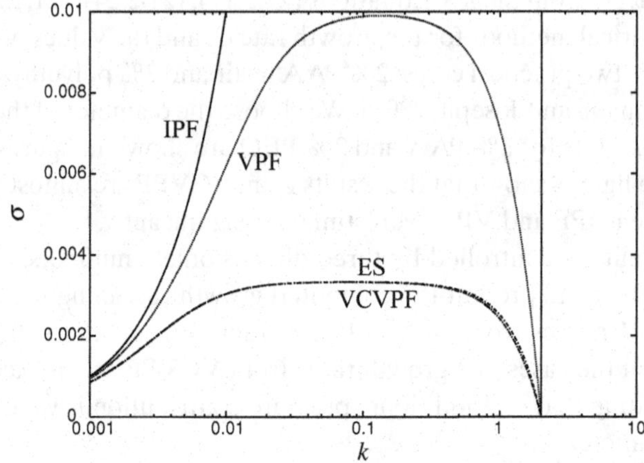

Figure 21.16. The growth rate σ vs. k from IPF, VPF, VCVPF, and the exact solution (ES). $J = 10^{-4}$, $\hat{\lambda}_1 = 0.1$, $\hat{\lambda}_2 = 0$.

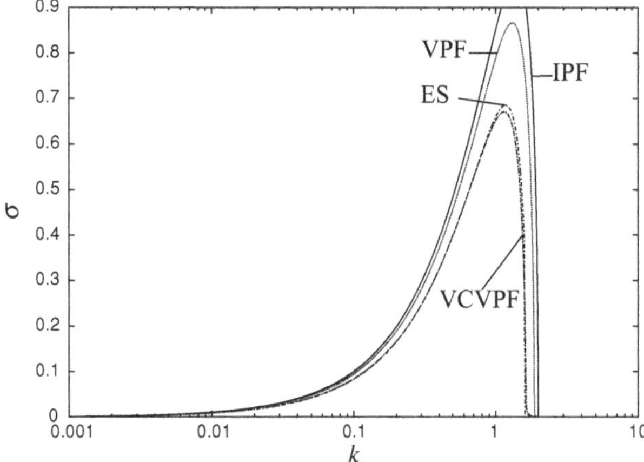

Figure 21.17. The growth rate σ vs. k from IPF, VPF, VCVPF, and the exact solution (ES). $J = 10^{-4}, \hat{\lambda}_1 = 1000$, $\hat{\lambda}_2 = 0$.

that of $\hat{\lambda}_1$; the growth rates decreases with $\hat{\lambda}_2$. When $\hat{\lambda}_1 = \hat{\lambda}_2$, the fluid becomes Newtonian. When the Reynolds number is as high as 10^4 (figure 21.19), IPF and VPF slightly overestimate the maximum growth rate whereas the VCVPF results are almost the same as the ES.

In table 21.7 we present the maximum growth rate σ_m and the associated wavenumber k_m computed from VPF, VCVPF, and the ES. The value of σ_m given by VPF is several times larger than the exact result when J is small. VCVPF gives excellent approximation to the values of σ_m and k_m in all the cases.

21.2.7 Comparison of the stream functions

Next we compare the stream functions from VPF, VCVPF, and the ES at the same wavenumber. The wavenumber chosen for the comparison is k_m at which the maximum

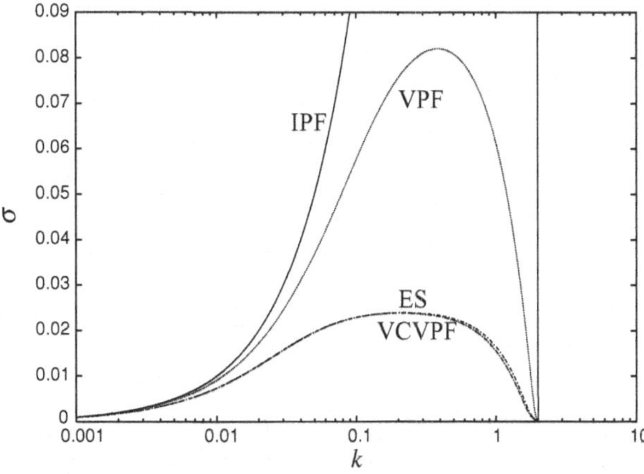

Figure 21.18. The growth rate σ vs. k from IPF, VPF, VCVPF, and the exact solution (ES). $J = 10^{-4}, \hat{\lambda}_1 = 1000$, $\hat{\lambda}_2 = 100$.

Table 21.7. *Maximum growth rate σ_m and the associated wavenumber k_m for VPF, VCVPF, and the ES in figures 21.14–21.19*

For the IPF solution, $k_m = 1.394$ and $\sigma_m = 0.9711$ in all six cases.

	VPF		VCVPF		ES	
Figure	k_m	σ_m	k_m	σ_m	k_m	σ_m
21.14	3.439×10^{-1}	6.557×10^{-2}	2.052×10^{-1}	2.274×10^{-2}	2.135×10^{-1}	2.278×10^{-2}
21.15	2.025×10^{-1}	2.283×10^{-2}	1.183×10^{-1}	7.554×10^{-3}	1.229×10^{-1}	7.559×10^{-3}
21.16	1.331×10^{-1}	9.899×10^{-3}	7.831×10^{-2}	3.322×10^{-3}	8.154×10^{-2}	3.323×10^{-3}
21.17	1.309	8.665×10^{-1}	1.144	6.703×10^{-1}	1.170	6.850×10^{-1}
21.18	3.848×10^{-1}	8.200×10^{-2}	2.101×10^{-1}	2.384×10^{-2}	2.186×10^{-1}	2.390×10^{-2}
21.19	1.386	9.618×10^{-1}	1.374	9.447×10^{-1}	1.375	9.458×10^{-1}

growth rate σ_m occurs in the ES. The relation between the constants A_1 and A_2 in exact stream function (21.2.15) and A in potential flow solution (21.2.22) must be established before we can compare the stream functions. Here we obtain this relation by assuming that the magnitude of the disturbance H is the same in the ES and in the potential flow solution.

We use a superscript E for quantities appearing in the ES, and (21.2.15) and (21.2.16) are rewritten as

$$\psi^E = \left[A_1^E r I_1(kr) + A_2^E r I_1(k_v r)\right] \exp(\sigma^E t + ikz), \tag{21.2.42}$$

$$\eta^E = H^E \exp(\sigma^E t + ikz). \tag{21.2.43}$$

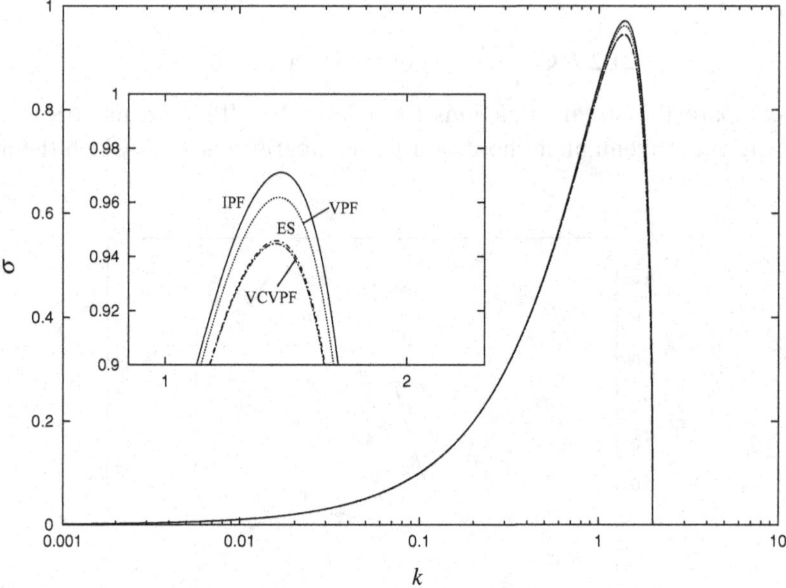

Figure 21.19. The growth rate σ vs. k from IPF, VPF, VCVPF, and the exact solution (ES). $J = 10^4$, $\hat{\lambda}_1 = 0.1$, $\hat{\lambda}_2 = 0$. When the Reynolds number J is large, viscoelastic effects are relatively small, and the four curves are close; but differences among them can be seen near the peak growth rate. The inset is the amplified plot for the region near the peak growth rate. VCVPF is the best approximation to the ES.

The relation between A_1^E and A_2^E is determined by the zero-shear-stress condition at $r \approx R$:

$$A_2^E = \frac{-2k^2 I_1(kR)}{(k^2+k_v^2)I_1(k_v R)} A_1^E. \tag{21.2.44}$$

Therefore we can write the stream function as

$$\psi^E = A_1^E r \left[I_1(kr) - \frac{2k^2 I_1(kR)}{(k^2+k_v^2)I_1(k_v R)} I_1(k_v r) \right] \exp(\sigma^E t + ikz). \tag{21.2.45}$$

The amplitude of the disturbance H^E is related to A_1^E through the kinematic condition:

$$H^E = \frac{ik}{\sigma^E}\left(1 - \frac{2k^2}{k^2+k_v^2}\right) I_1(kR) A_1^E. \tag{21.2.46}$$

Now we consider the potential flow solution that is indicated by the a superscript P. The stream function and the disturbance are given by

$$\psi^P = A^P r I_1(kr) \exp(\sigma^P t + ikz), \tag{21.2.47}$$

$$\eta^P = H^P \exp(\sigma^P t + ikz), \tag{21.2.48}$$

respectively. The amplitude of the disturbance H^P is related to A^P through the kinematic condition:

$$H^P = \frac{ik}{\sigma^P} A^P I_1(kR). \tag{21.2.49}$$

We assume that the amplitude of the disturbance is the same in the ES and the potential flow solution. Thus $H^E = H^P$, and it follows that

$$A^P = A_1^E \frac{\sigma^P}{\sigma^E}\left(1 - \frac{2k^2}{k^2+k_v^2}\right). \tag{21.2.50}$$

Then the stream function of the potential flow can be written as

$$\psi^P = A_1^E \frac{\sigma^P}{\sigma^E}\left(1 - \frac{2k^2}{k^2+k_v^2}\right) r I_1(kr) \exp(\sigma^P t + ikz). \tag{21.2.51}$$

Now we can compare (21.2.45) and (21.2.51). The stream function is decomposed into two parts, the exponential function depending on t and z and the rest depending on r. Because we are comparing the stream functions at the same wavenumber k_m, the comparison of the exponential function is equivalent to the comparison of the growth rate. In table 21.8, we list the values of the growth rate σ computed from VPF, VCVPF, and the ES. In all the cases, the growth rate from VPF is larger than the exact result, whereas the growth rate from VCVPF is very close to the exact result. The rest of the stream function depends on r, and we define

$$\mathrm{SF}(r) = \frac{\sigma^{\mathrm{VPF}}}{\sigma^E}\left(1 - \frac{2k^2}{k^2+k_v^2}\right) r I_1(kr) \quad \text{for} \quad \text{VPF}; \tag{21.2.52}$$

$$\mathrm{SF}(r) = \frac{\sigma^{\mathrm{VCVPF}}}{\sigma^E}\left(1 - \frac{2k^2}{k^2+k_v^2}\right) r I_1(kr) \quad \text{for} \quad \text{VCVPF}; \tag{21.2.53}$$

$$\mathrm{SF}(r) = r\left[I_1(kr) - \frac{2k^2 I_1(kR)}{(k^2+k_v^2)I_1(k_v R)} I_1(k_v r) \right] \quad \text{for} \quad \text{the ES}. \tag{21.2.54}$$

Table 21.8. *The growth rate σ computed from VPF, VCVPF, and the ES at the same wavenumber k_m*

In the ES, k_m is the wavenumber for the maximum growth rate.

J	$\hat{\lambda}_1$	$\hat{\lambda}_2$	k_m	σ^{VPF}	σ^{VCVPF}	σ^{E}
4.834×10^{-3}	0.263	0	0.2135	0.06345	0.02274	0.02278
5.091×10^{-4}	1.676	0	0.1229	0.02252	0.007554	0.007559
10^{-4}	0.1	0	0.08154	0.009843	0.003322	0.003323
10^{-4}	1000	0	1.170	0.8495	0.6696	0.6850
10^{-4}	1000	100	0.2186	0.07718	0.02384	0.02390
10^{4}	0.1	0	1.375	0.9617	0.9447	0.9458

Three examples for the comparison of the function SF(r) are shown in figures 21.20, 21.21, and 21.22. The curves for SF(r) are very close to straight lines, indicating power functions. This can also be seen from (21.2.52), (21.2.53), and (21.2.54). The expansion of the modified Bessel function gives

$$I_1(kr) = \frac{kr}{2} + \frac{k^3r^3}{16} + \frac{k^5r^5}{384} + O(r^7). \tag{21.2.55}$$

Higher-order terms of r may be neglected because $0 \leq r \leq 0.5$ inside the cylinder. If we keep only the first term in the expansion, stream functions (21.2.52) and (21.2.53) become, respectively,

$$\text{SF}(r) = \frac{\sigma^{VPF}}{\sigma^{E}}\left(1 - \frac{2k^2}{k^2 + k_v^2}\right)\frac{kr^2}{2} + O(r^4) \quad \text{for} \quad \text{VPF}; \tag{21.2.56}$$

$$\text{SF}(r) = \frac{\sigma^{VCVPF}}{\sigma^{E}}\left(1 - \frac{2k^2}{k^2 + k_v^2}\right)\frac{kr^2}{2} + O(r^4) \quad \text{for} \quad \text{VCVPF}. \tag{21.2.57}$$

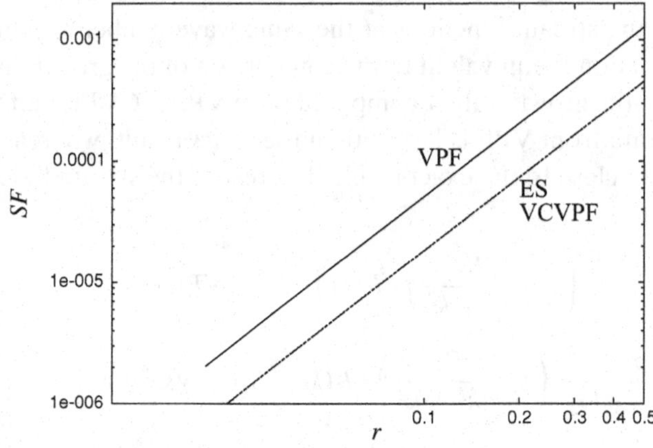

Figure 21.20. The part of the stream function depending on r defined in (21.2.52), (21.2.53), and (21.2.54) for VPF, VCVPF, and the exact solution (ES), respectively. The fluid is 2% PAA, $J = 4.834 \times 10^{-3}$, $\hat{\lambda}_1 = 0.263$, $\hat{\lambda}_2 = 0$. The wavenumber for the maximum growth rate $k_m = 0.2135$ is chosen for the comparison.

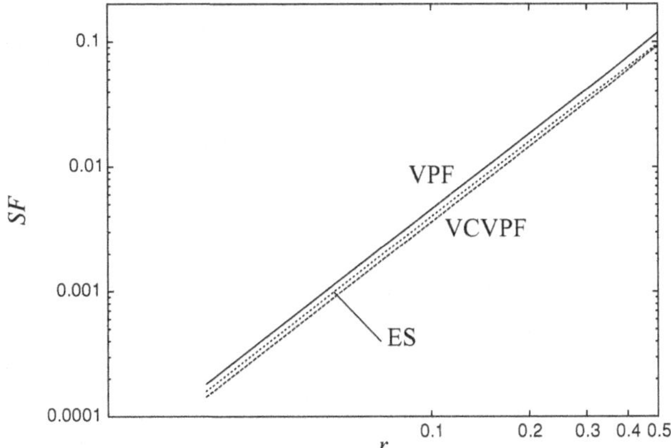

Figure 21.21. The part of the stream function depending on r defined in (21.2.52), (21.2.53), and (21.2.54) for VPF, VCVPF, and the exact solution (ES), respectively. In this case, $J = 10^{-4}$, $\hat{\lambda}_1 = 1000$, $\hat{\lambda}_2 = 0$. The wavenumber for the maximum growth rate $k_m = 1.170$ is chosen for the comparison.

For (21.2.54), we also expand $I_1(kR)$ and $I_1(k_\upsilon R)$ and keep only the leading term, which gives rise to

$$\begin{aligned} \mathrm{SF}(r) &= \frac{kr^2}{2} - \frac{2k^2 \frac{kR}{2}}{(k^2+k_\upsilon^2)\frac{k_\upsilon R}{2}} \frac{k_\upsilon r^2}{2} + O(r^4) \\ &= \left(1 - \frac{2k^2}{k^2+k_\upsilon^2}\right)\frac{kr^2}{2} + O(r^4) \quad \text{for} \quad \text{the ES.} \end{aligned} \tag{21.2.58}$$

Equations (21.2.56), (21.2.57), and (21.2.58) show that the functions $\mathrm{SF}(r)$ are approximately quadratic functions for small r, and this is confirmed in figures 21.20, 21.21, and 21.22. The comparison of the leading terms of $\mathrm{SF}(r)$ depends directly on the growth rate σ^{VPF}, σ^{VCVPF}, and σ^E. Because $\sigma^{\mathrm{VPF}} > \sigma^E$, the curves for $\mathrm{SF}(r)$ of VPF are higher than those for the exact solution. On the other hand, σ^{VCVPF} is very close to σ^E, and the curves for VCVPF and the ES almost overlap. Combining the comparison of the growth rate in

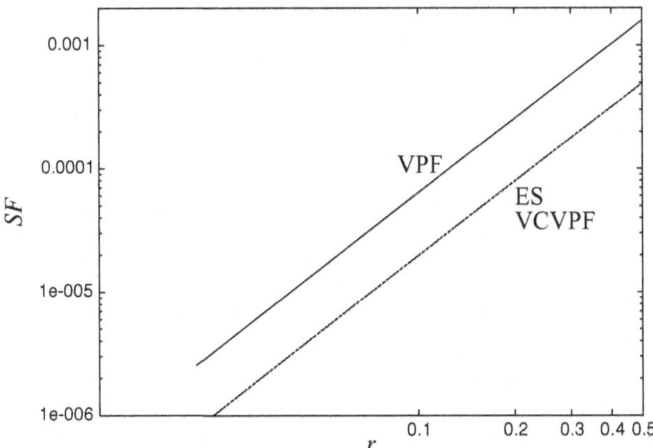

Figure 21.22. The part of the stream function depending on r defined in (21.2.52), (21.2.53), and (21.2.54) for VPF, VCVPF, and the exact solution (ES), respectively. In this case, $J = 10^{-4}$, $\hat{\lambda}_1 = 1000$, $\hat{\lambda}_2 = 100$. The wavenumber for the maximum growth rate $k_m = 0.2186$ is chosen for the comparison.

table 21.8 and the comparison of the function SF(r) in figures 21.20–21.22, we show that the stream function given by VCVPF is in remarkably good agreement with the ES. This result indicates that the vorticity plays a small role in the exact solution and our VCVPF solution, which is based solely on potential flow, can give an excellent approximation to the flow.

21.2.8 Discussion

Chang, Demekhin, and Kalaidin (1999) made a long-wave study of the stretching dynamics of bead-string filaments for FENE and Oldroyd B fluids. They made a long-wave study of linear stability, and their results can be compared with ours. To this end, we first convert the parameters used by Chang *et al.* (1999) to the parameters used by us. In the notation of Chang *et al.* (1999), Ca is the capillary number, We is the Weissenberg number, and S is the retardation number. We linearize the stress equation of Chang *et al.* and reduce it to a form comparable with our Jeffreys model (21.2.3); then the relation between We and S used by Chang *et al.* and $\hat{\lambda}_1$ and $\hat{\lambda}_2$ used by us is revealed. After taking the different length and time scales into account, we can express the parameters in Chang *et al.* in terms of our parameters:

$$Ca = 2/J, \quad We = 4\hat{\lambda}_1/\sqrt{J}, \quad S = \hat{\lambda}_2/\hat{\lambda}_1. \tag{21.2.59}$$

Then the dispersion relation given by the linear stability analysis of Chang *et al.* [their Equation (16)] can be written as

$$\hat{\lambda}_1\sigma^3 + \left(1 + 3k^2\frac{\hat{\lambda}_2}{\sqrt{J}}\right)\sigma^2 + \left[\frac{3k^2}{\sqrt{J}} - \frac{k^2\hat{\lambda}_1}{4}(4-k^2)\right]\sigma - \frac{k^2}{4}(4-k^2) = 0. \tag{21.2.60}$$

Now we consider dispersion relation (21.2.38) from the VCVPF method. The dimensionless radius $R = 1/2$ and the Bessel functions can be expanded for small k:

$$\frac{I_0(kR)}{I_1(kR)} = \frac{4}{k} + \frac{k}{8} - \frac{k^3}{768} + O(k^5), \quad \frac{2I_0(kR)}{I_1(kR)} - \frac{1}{kR} = \frac{6}{k} + \frac{k}{4} - \frac{k^3}{384} + O(k^5). \tag{21.2.61}$$

Inserting (21.2.61) into (21.2.38), we can obtain

$$\begin{aligned}&\left(1+\frac{k^2}{32}\right)\hat{\lambda}_1\sigma^3 + \left[1 + \frac{k^2}{32} + 3k^2\frac{\hat{\lambda}_2}{\sqrt{J}}\left(1+\frac{k^2}{24}\right)\right]\sigma^2 \\ &\quad + \left[\frac{3k^2}{\sqrt{J}}\left(1+\frac{k^2}{24}\right) - \frac{k^2\hat{\lambda}_1}{4}(4-k^2)\right]\sigma - \frac{k^2}{4}(4-k^2) + O(k^4) = 0.\end{aligned} \tag{21.2.62}$$

The expansion of the Bessel functions can also be applied to the ES, and the result is compared with (21.2.60) and (21.2.62). After some arrangement, dispersion relation (21.2.17) of the ES can be written as

$$\begin{aligned}&\frac{4k^3k_v}{\sqrt{J}}\hat{\mu}\left[\frac{I_0(k_vR)}{I_1(k_vR)} - \frac{1}{k_vR}\right] - \frac{2k^2(k^2+k_v^2)}{\sqrt{J}}\hat{\mu}\left[\frac{I_0(kR)}{I_1(kR)} - \frac{1}{kR}\right] \\ &\quad - \frac{2k^3}{\sigma}\left(\frac{1}{R^2} - k^2\right) + (k^2+k_v^2)\frac{k}{\sigma}\left(\frac{1}{R^2} - k^2\right) - (k^2+k_v^2)\sigma\frac{I_0(kR)}{I_1(kR)} = 0.\end{aligned} \tag{21.2.63}$$

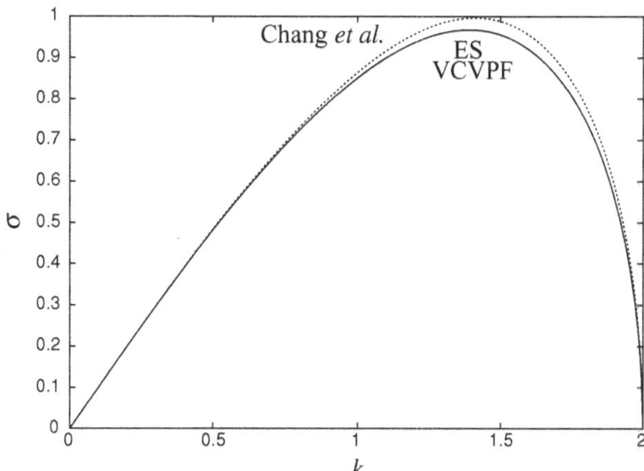

Figure 21.23. The growth rate σ as a function of k computed with (21.2.60) given by Chang *et al.*, (21.2.38) given by our VCVPF method, and (21.2.17) given by the exact solution (ES). The fluid is water with $\rho = 1000$ kg/m^3, $\mu = 0.001$ kg/ (m sec), $\gamma = 0.0728$ N/m. The diameter of the liquid cylinder is assumed to be 0.01 m, and the Reynolds number is $J = 7.28 \times 10^5$.

After expanding the Bessel functions as power series of k, we obtain

$$\hat{\lambda}_1\left(1+\frac{k^2}{32}\right)\sigma^3+\left(1+\frac{k^2}{32}+3k^2\frac{\hat{\lambda}_2}{\sqrt{J}}\right)\sigma^2+\left[\frac{3k^2}{\sqrt{J}}-\frac{k^2\hat{\lambda}_1}{4}(4-k^2)\right] \times\sigma-\frac{k^2}{4}(4-k^2)+O(k^4)=0. \tag{21.2.64}$$

Dispersion relation (21.2.64) given by the ES is different from both (21.2.60) given by Chang *et al.* and (21.2.62) given by our VCVPF method; the first-order differences are $O(k^2)$ in both cases. The differences between (21.2.64) and (21.2.60) are two $k^2/32$ terms in the coefficients of σ^3 and σ^2; the differences between (21.2.64) and (21.2.62) are two $k^2/24$ terms in the coefficients of σ^2 and σ.

We can obtain the limit of a Newtonian fluid by letting $\hat{\lambda}_1 = \hat{\lambda}_2 = 0$. Then dispersion relations (21.2.60), (21.2.62), and (21.2.64) reduce to, respectively,

$$\sigma^2+\frac{3k^2}{\sqrt{J}}\sigma-\frac{k^2}{4}(4-k^2)=0 \quad \text{for} \quad \text{Chang } et\ al.; \tag{21.2.65}$$

$$\left(1+\frac{k^2}{32}\right)\sigma^2+\frac{3k^2}{\sqrt{J}}\left(1+\frac{k^2}{24}\right)\sigma-\frac{k^2}{4}(4-k^2)+O(k^4)=0 \quad \text{for} \quad \text{VCVPF}; \tag{21.2.66}$$

$$\left(1+\frac{k^2}{32}\right)\sigma^2+\frac{3k^2}{\sqrt{J}}\sigma-\frac{k^2}{4}(4-k^2)+O(k^4)=0 \quad \text{for} \quad \text{the ES}. \tag{21.2.67}$$

The first-order differences among dispersion relations (21.2.65), (21.2.66), and (21.2.67) are $O(k^2)$. The difference between (21.2.65) and (21.2.67) is a $k^2/32$ term in the coefficient of σ^2; the difference between (21.2.66) and (21.2.67) is a $k^2/24$ term in the coefficient of σ.

In figures 21.23–21.27, we plot the growth rate σ as a function of k computed using (21.2.60) given by Chang *et al.* (21.2.38), given by our VCVPF method, and (21.2.17) given

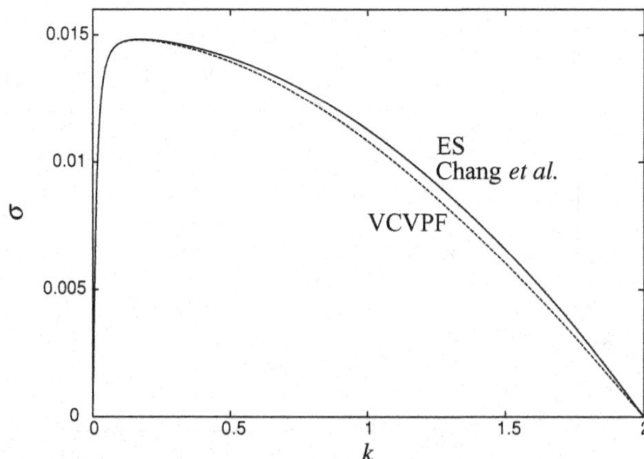

Figure 21.24. The growth rate σ as a function of k computed with (21.2.60) given by Chang *et al.*, (21.2.38) given by our VCVPF method, and (21.2.17) given by the exact solution (ES). The fluid is a Newtonian fluid SO10000 oil with $\rho = 969$ kg/m^3, $\mu = 10$ kg/(m sec), $\gamma = 0.021$ N/m. The Reynolds number is $J = 2.04 \times 10^{-3}$.

by the ES. Both Newtonian fluids and viscoelastic fluids are compared. We achieve the limit of Newtonian fluids by setting $\hat{\lambda}_1$ and $\hat{\lambda}_2$ to be zero in (21.2.60), (21.2.38), and (21.2.17). There is almost no difference among the three curves when k is close to zero, and small differences can be seen when k is close to 2. The dispersion relation of Chang *et al.* is in better agreement with the ES when the Reynolds number J is small (figures 21.24–21.27), whereas our VCVPF is in better agreement with the ES when J is large (figure 21.23).

In this section, linear stability analysis of the capillary instability of a viscoelastic thread is carried out under the assumption that the flow is irrotational. The nonzero irrotational shear stress at the surface of the liquid thread does not agree with the zero-shear-stress condition. We derive a pressure contribution in addition to the irrotational pressure. This

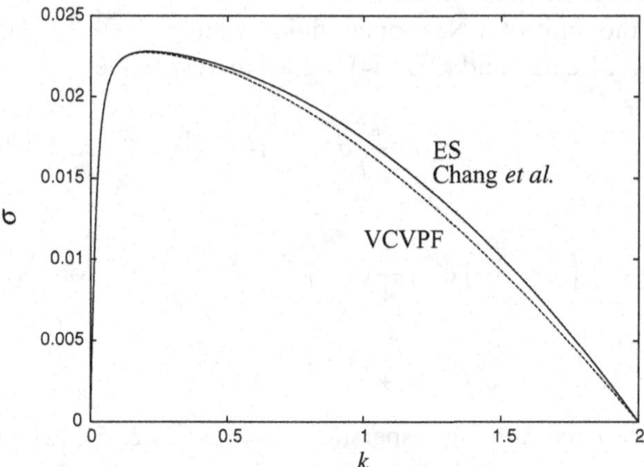

Figure 21.25. The growth rate σ as a function of k computed with (21.2.60) given by Chang *et al.*, (21.2.38) given by our VCVPF method, and (21.2.17) given by the exact solution (ES). The fluid is 2% PAA with $J = 4.834 \times 10^{-3}$, $\hat{\lambda}_1 = 0.263, \hat{\lambda}_2 = 0$.

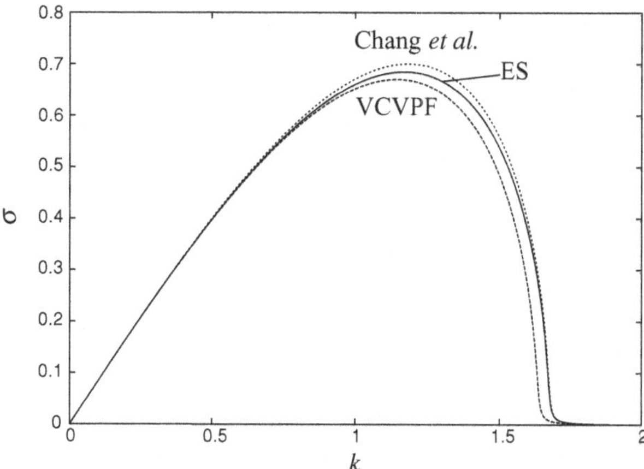

Figure 21.26. The growth rate σ as a function of k computed with (21.2.60) given by Chang *et al.*, (21.2.38) given by our VCVPF method, and (21.2.17) given by the exact solution (ES). The fluid is a viscoelastic fluid with $J = 10^{-4}, \hat{\lambda}_1 = 1000, \hat{\lambda}_2 = 0$.

additional pressure contribution depends on the viscoelastic parameters and cancels out the power because of the uncompensated irrotational shear stress in the energy equation. We include the additional pressure contribution, the irrotational pressure, and the extra stress we evaluated by using the irrotational flow in the normal stress balance at the surface; then a dispersion relation is obtained. We call this approach the viscoelastic correction of the viscoelastic potential flow (VCVPF). A comparison of the growth rate and the stream function shows that the VCVPF solution is an excellent approximation to the ES. The dispersion relation given by VCVPF is also compared with that obtained by Chang *et al.* (1999) by use of a long-wave approximation. The differences between the two dispersion relations are negligible when the wavenumber k is small, and both dispersion relations are in remarkably good agreement with the ES.

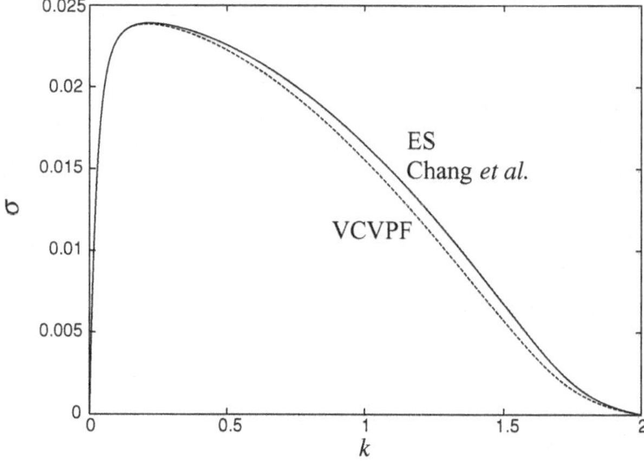

Figure 21.27. The growth rate σ as a function of k computed with (21.2.60) given by Chang *et al.*, (21.2.38) given by our VCVPF method, and (21.2.17) given by the exact solution (ES). The fluid is a viscoelastic fluid with $J = 10^{-4}, \hat{\lambda}_1 = 1000, \hat{\lambda}_2 = 100$.

21.3 Steady motion of a deforming gas bubble in a viscous potential flow

Miksis, Vanden-Broeck, and Keller (1982; hereafter MVK) computed the shape of an axisymmetric rising bubble, or a falling drop, in an incompressible fluid assuming that the flow in the liquid is irrotational but viscous. The boundary condition for the normal stress including surface tension is satisfied but, as in other problems of VPF, the tangential stress is neglected. The shape function is obtained from the gravitational potential evaluated on the free surface; two shape functions are computed, one on the top and one on the bottom of the bubble. The shape is single valued on each function. The potential function is obtained from the values of the potential on the free surface, by use of a Green's function approach following ideas introduced by Longuet-Higgins and Cokelet (1976), Vanden-Broeck and Keller (1980), and MVK (1982). The system of differential and integral equations is solved within a framework in which the bubble is stationary and the velocity at infinity is U, which is calculated by a drag balance in two ways. The first calculation is like that of Moore (1959) in which the drag comes from the normal irrotational viscous stress, leading to $32/Re$. This direct method should not be used because of the additional contribution that is due to the irrotational viscous pressure.

This pressure is not easy to calculate in general, but the correct drag leading to $48/Re$ can be obtained, and was obtained by MVK in a second calculation by use of the DM.

The solution of the system of governing equations was obtained as a power series in the Weber number and Re^{-1} and is therefore restricted to low Weber numbers (large surface tension) and high Reynolds numbers (small viscosity).

22

Numerical methods for irrotational flows of viscous fluid

Problems of potential flow in irregular domains bounded by rigid solids and satisfying perhaps conditions at infinity require numerical methods. Computers and software are now so powerful that it can be easier to compute a solution than to find the exact one in a reference book. There are many techniques that may be used to solve Laplace's equation with prescribed boundary conditions. These techniques are readily available even in "search" on the web.

The numerical simulation of the deformation of interfaces between two immiscible fluids or in gas–liquid flows is currently an active topic of research and many options are available for researchers. Level-set methods associated with the names of S. Osher, R. Fedkiw, and J. Sethian, volume-of-fluid methods associated with the name of S. Zaleski, and front-tracking methods associated with the name of G. Trygvasson, are high among the most popular methods. Readers can find references in the comprehensive reviews by Yeung (1982), Tsai and Yue (1996), and Scardovelli and Zaleski (1999), or in "search" on Google.

22.1 Perturbation methods

The problem of numerical simulation of the shape of free surfaces in potential flows of inviscid fluids has been considered by various authors. Perturbation methods for nonlinear irrotational waves on an inviscid fluid were introduced by Stokes (1847). He expanded the solution in powers of the amplitude. Many authors have worked with these series, and proofs of convergence and nonconvergence have been considered (Schwartz, 1974). A mapping method for the perturbation series was presented by Joseph (1973). Yoo (1973) computed many terms. The mapping method was devised to justify an apparent problem of the Stokes method in which boundary conditions at $z = \eta$ are enforced on the unperturbed surface at $z = 0$.

The problem of computer-aided studies and analytical continuation of the perturbation series has been considered by Schwartz (1974). The main contribution of this paper is summation-protocol-based Padé approximations, which in many cases improve convergence.

Stokes waves cannot be permanently maintained in the presence of viscosity. This fact places certain limits on the utility of the perturbation methods.

22.2 Boundary-integral methods for inviscid potential flow

A major objective in the solution of interfacial flow problems is a highly accurate description of the interface. Therefore boundary-integral techniques are a fitting choice for the analysis, as they seek solutions of integral equations involving information only on the interface. This feature reduces the dimensionality of the problem by one. Thus, when an approximate solution is sought, a fine mesh can be afforded on the interface, especially in regions with high curvature, without having to discretize the neighboring domain. This attribute is particularly important for an unbounded domain, in which case appropriate boundary conditions at infinity can be satisfied automatically by the governing integral equations. A survey of the literature on free-surface or interfacial flows indicates that boundary-integral methods have been applied mostly to problems in two dimensions or three dimensions with axial symmetry because domain discretization simply takes place over a curve in the plane for these cases.

The application of the boundary-integral method relies on the existence of a "fundamental solution" for the PDE on hand. In particular, this solution is known for Laplace's equation. Therefore, boundary-integral methods become a useful tool for incompressible potential flow problems because the velocity potential is harmonic. Boundary-integral formulations have also been developed for Stokes flow (Pozrikidis 1992).

In general, boundary-integral methods can be grouped into two major categories, namely, the indirect and the direct formulations (Banerjee and Butterfield, 1981; Brebbia, Telles, and Wrobel, 1984). In the indirect formulation an integral equation is written in terms of the density distribution over the boundary of a unit singular solution of the PDE of interest. Numerical techniques are then applied to compute this density profile that may have no explicit physical connotation. Once the density function over the boundary has been determined, the physical variables of the problem inside the domain can be obtained by integration. On the other hand, the direct formulation poses integral equations on the boundaries in terms of the physical parameters of interest. By enforcing the boundary conditions, one can solve for the unknown field on the boundary first and then at particular locations on the interior. For instance, the normal derivative of the potential at the boundary can be computed in the Dirichlet problem or the potential over the boundary in the Neumann problem. Also, mixed boundary conditions can be easily handled.

For most of the problems of interest in science and engineering, finding a solution of the boundary-integral equations is possible only in an approximate manner, by use of numerical techniques. A widely used approach is the boundary-element method (BEM). In general terms, this method sets marker points or nodes on the boundary. A number of segments or "elements" connecting the nodes are used to approximate this boundary. In two dimensions, these elements may be straight segments or, if greater accuracy is desired, of higher order, including circular, parabolic or cubic representations. In three dimensions, triangular or quadrilateral elements may be chosen. Regarding the fields functions taken to the boundary (e.g., the potential and its normal derivative), they are approximated with a truncated polynomial over each element. For instance, the simplest choice is to hold the function constant on the element. To improve accuracy, the linear expansion or higher-order approximations may be used. The coefficients of the expansion correspond to the values of the function at particular locations on the element, which may be those of the nodes. Next, the integrals on the boundary-integral equation are split into integrals evaluated over each element, and the local expansions for the field functions are substituted. Then the discretized equation is satisfied at a set of collocation points on the boundary. This process gives rise to a set of algebraic equations that may be solved for the vector of unknowns. Detailed descriptions on the implementation of BEM can be found in monographs on the subject, such as those by Brebbia *et al.* (1984), Pozrikidis (1992), and Wrobel (2002), among others. A fairly rigorous mathematical treatment of boundary-integral equations and their numerical solution is given by Jaswon and Symm (1977).

For problems involving the deformation of a free surface or interface, the application of BEM to solve Laplace's equation as previously described is carried out at a given time with known geometry and boundary conditions. To advance the position of the boundary and the conditions on it to the next time level, the boundary-integral solution algorithm has to be coupled with a time-marching scheme. This scheme is obtained from the time integration of the kinematic and dynamic boundary conditions, together with the transport of momentum for points on the interface written in Lagrangian form. Starting from the initial conditions, the coupling of boundary-integral and time-marching algorithms is repeated until the final time is reached.

Examples of the applications of boundary-integral methods to problems involving free-surface flows are abundant. For instance, the inviscid analysis of steady motion of free surfaces is carried out by Byatt-Smith and Longuet-Higgins (1976) for a steep solitary wave, by Miksis, Vanden-Broeck, and Keller (1981) to study the deformation of an axisymmetric bubble in a uniform flow by use of a direct formulation, and by Meiron and Saffman (1983) for interfacial gravity waves by application of the indirect formulation.

The accurate description of the unsteady motion of free surfaces or fluid–fluid interfaces bounding regions of inviscid irrotational flow can be performed with boundary-integral techniques. An approach that arises from the indirect double-layer potential formulation is the generalized vortex method. This method was presented and developed by Baker, Meiron, and Orszag (1980, 1982, 1984) in a series of publications. They obtained an integral equation for the Lagrangian time derivative of the dipole density distribution over the boundary. Solving for this time derivative allows updating the dipole density that is used to march the interface forward in time. Baker *et al.* (1980) used the vortex method to simulate the RT instability in its classical form, and Verdon *et al.* (1982) considered the acceleration of a thin fluid layer. Baker *et al.* (1982) applied the method to the breaking of surface waves and interacting triads of surface and interfacial waves. Lundgren and Mansour (1988, 1991) studied the oscillations of an inviscid axisymmetric drop in a dynamically inactive fluid (e.g., vacuum) and the motion of toroidal gas bubbles in an inviscid liquid, respectively, with a modified version of the generalized vortex method. Mansour and Lundgren (1990) also applied this approach to model satellite formation in capillary jet breakup. The vortex method based on the indirect approach was used to study the dynamic breakup of an inviscid liquid bridge by Chen and Steen (1997) and capillary pinchoff of an inviscid drop surrounded by an ambient inviscid fluid by Leppinen and Lister (2003).

The direct formulation of the boundary-integral method for inviscid fluids was used by Keller and Miksis (1983) to model a breaking sheet of liquid and the flow near the intersection of a solid boundary with the free surface of a liquid. Several papers by Oguz and Prosperetti (1989, 1990, 1993) on the effect of surface tension in the contact of liquid surfaces, bubble entrainment by the impact of drops on liquid surfaces, and dynamics of bubble growth with detachment from a needle, respectively, use the direct formulation of the boundary-integral method. This formulation is also applied by Machane and Canot (1997) for various 2D and axisymmetric free-surface problems and by Day, Hinch, and Lister (1998) and Rodriguez-Rodriguez, Gordillo, and Martinez-Bazan (2006) to problems of capillary pinchoff.

An interesting application of boundary-integral methods is in the description of the motion of nonlinear waves. Longuet-Higgins and Cokelet (1976) used the direct formulation to study the deformation of steep, periodic, solitary waves in deep liquid. A condensed review of BEM analysis of nonlinear waves was given by Ligget and Liu (1984). The phenomena of generation, propagation, shoaling, breakup, and absorption of 2D nonlinear waves have been studied by Grilli *et al.* (1989, 1994, 1996, 1997) through the direct formulation of BEM. The numerical modeling of a 3D nonlinear wave over complex bottom topography was carried out by Grilli, Guyenne, and Dias (2001). Their model is able to predict wave overturning. Boundary conditions of the absorbing or reflective type can be indicated on lateral boundaries.

22.3 Boundary-integral methods for viscous potential flow

BEM has been extended to accommodate the effects of viscosity in a purely irrotational flow by Georgescu, Achard, and Canot (2002) to study a gas bubble bursting at a free surface and by Canot *et al.* (2003) in their numerical simulation of the buoyancy-driven bouncing of a 2D bubble at a horizontal wall, using the direct formulation. Another kind of VPF analysis of the deformation of a rising 3D bubble was given by Miksis, Vanden-Broeck, and Keller (1982). They converted their problem into a system of integrodifferential equations that they solved under the conditions of small Weber numbers and large Reynolds numbers.

Lundgren and Mansour (1988) also included the effect of a small viscosity by decomposing the velocity field into the sum of an irrotational and a rotational velocity, in which the former is expressed as the gradient of a potential and the latter is written as the curl of a vector potential. Substitution of this decomposition into the incompressible Navier–Stokes equations and applying order-of-magnitude arguments under the assumption of a thin vortical layer at the free surface of the drop yields a new set of differential equations for the potentials. These equations carry weak viscous effects and are coupled with the boundary-integral formulation for potential flow based on the vortex method.

Weak viscous effects emanating from boundary layers in which vorticity does not vanish are associated with the quasi-potential methodology discussed in §14.1.7.

The BEM calculations of Georgescu *et al.* (2002) and Canot *et al.* (2003) neglect vorticity but the viscous effects in the purely irrotational flow are not restricted to small viscosity.

APPENDIX A

Equations of motion and strain rates for rotational and irrotational flow in Cartesian, cylindrical, and spherical coordinates

A1.1 Cartesian coordinates

In Cartesian coordinates $(\xi_1, \xi_2, \xi_3) = (x, y, z)$ with unit vectors $\mathbf{e}_x$, $\mathbf{e}_y$, and $\mathbf{e}_z$, equations for $\mathbf{u} = (u_1, u_2, u_3) = (u_x, u_y, u_z)$ are given as follows,

$$\mathbf{n} \cdot \nabla \mathbf{u} = \mathbf{e}_x (\mathbf{n} \cdot \nabla) u_x + \mathbf{e}_y (\mathbf{n} \cdot \nabla) u_y + \mathbf{e}_z (\mathbf{n} \cdot \nabla) u_z, \tag{A1.1.1}$$

$$\nabla \cdot \mathbf{u} = \frac{\partial u_x}{\partial x} + \frac{\partial u_y}{\partial y} + \frac{\partial u_z}{\partial z}, \tag{A1.1.2}$$

$$\begin{aligned} \nabla \times \mathbf{u} &= \begin{vmatrix} \mathbf{e}_x & \mathbf{e}_y & \mathbf{e}_z \\ \dfrac{\partial}{\partial x} & \dfrac{\partial}{\partial y} & \dfrac{\partial}{\partial z} \\ u_x & u_y & u_z \end{vmatrix} \\ &= \left(\frac{\partial u_z}{\partial y} - \frac{\partial u_y}{\partial z} \right) \mathbf{e}_x + \left(\frac{\partial u_x}{\partial z} - \frac{\partial u_z}{\partial x} \right) \mathbf{e}_y + \left(\frac{\partial u_y}{\partial x} - \frac{\partial u_x}{\partial y} \right) \mathbf{e}_z, \end{aligned} \tag{A1.1.3}$$

$$\nabla^2 \mathbf{u} = \mathbf{e}_x \nabla^2 u_x + \mathbf{e}_y \nabla^2 u_y + \mathbf{e}_z \nabla^2 u_z = \mathbf{e}_x \Delta u_x + \mathbf{e}_y \Delta u_y + \mathbf{e}_z \Delta u_z, \tag{A1.1.4}$$

with

$$\nabla^2 = \Delta = \frac{\partial^2}{\partial x^2} + \frac{\partial^2}{\partial y^2} + \frac{\partial^2}{\partial z^2}, \tag{A1.1.5}$$

$$T_{ij} = -p\delta_{ij} + S_{ij}, \quad S_{ij} = 2\mu e_{ij}, \tag{A1.1.6}$$

$$L_{ij} = \frac{\partial u_i}{\partial x_j}, \quad L_{ij} = \frac{1}{2}\left(\frac{\partial u_i}{\partial x_j} + \frac{\partial u_j}{\partial x_i} \right) + \frac{1}{2}\left(\frac{\partial u_i}{\partial x_j} - \frac{\partial u_j}{\partial x_i} \right) = e_{ij} + \Omega_{ij}, \tag{A1.1.7}$$

$$e_{ij} = \frac{1}{2}(L_{ij} + L_{ji}), \quad e_{ij} = e_{ji}, \quad \Omega_{ij} = \frac{1}{2}(L_{ij} - L_{ji}), \quad \Omega_{ij} = -\Omega_{ji}, \tag{A1.1.8}$$

$$\begin{aligned} &e_{xx} = \frac{\partial u_x}{\partial x}, \quad e_{yy} = \frac{\partial u_y}{\partial y}, \quad e_{zz} = \frac{\partial u_z}{\partial z}, \\ &e_{xy} = \frac{1}{2}\left(\frac{\partial u_x}{\partial y} + \frac{\partial u_y}{\partial x} \right), \quad e_{yz} = \frac{1}{2}\left(\frac{\partial u_y}{\partial z} + \frac{\partial u_z}{\partial y} \right), \quad e_{zx} = \frac{1}{2}\left(\frac{\partial u_z}{\partial x} + \frac{\partial u_x}{\partial z} \right), \end{aligned} \tag{A1.1.9}$$

$$\frac{\partial u_x}{\partial t} + \mathbf{u} \cdot \nabla u_x = -\frac{1}{\rho}\frac{\partial p}{\partial x} + \nu \Delta u_x, \tag{A1.1.10}$$

$$\frac{\partial u_y}{\partial t} + \mathbf{u} \cdot \nabla u_y = -\frac{1}{\rho}\frac{\partial p}{\partial y} + \nu \Delta u_y, \tag{A1.1.11}$$

$$\frac{\partial u_z}{\partial t} + \mathbf{u} \cdot \nabla u_z = -\frac{1}{\rho}\frac{\partial p}{\partial z} + \nu \Delta u_z. \tag{A1.1.12}$$

For $\mathbf{u} = \nabla\phi$, equations for ϕ are given as follows:

$$\nabla\phi = \mathbf{e}_x \frac{\partial \phi}{\partial x} + \mathbf{e}_y \frac{\partial \phi}{\partial y} + \mathbf{e}_z \frac{\partial \phi}{\partial z}, \tag{A1.1.13}$$

$$\nabla^2\phi = \Delta\phi = \frac{\partial^2 \phi}{\partial x^2} + \frac{\partial^2 \phi}{\partial y^2} + \frac{\partial^2 \phi}{\partial z^2}, \tag{A1.1.14}$$

$$e_{ij} = \frac{1}{2}(L_{ij} + L_{ji}) = \frac{\partial^2 \phi}{\partial x_i \partial x_j}, \quad e_{ij} = e_{ji}, \tag{A1.1.15}$$

$$e_{xx} = \frac{\partial^2 \phi}{\partial x^2}, \quad e_{yy} = \frac{\partial^2 \phi}{\partial y^2}, \quad e_{zz} = \frac{\partial^2 \phi}{\partial z^2},$$

$$e_{xy} = \frac{\partial^2 \phi}{\partial x \partial y}, \quad e_{yz} = \frac{\partial^2 \phi}{\partial y \partial z}, \quad e_{zx} = \frac{\partial^2 \phi}{\partial z \partial x}, \tag{A1.1.16}$$

$$\frac{\partial \phi}{\partial t} + \frac{1}{2}|\nabla\phi|^2 + \frac{p}{\rho} = \frac{\partial \phi}{\partial t} + \frac{1}{2}\left[\left(\frac{\partial \phi}{\partial x}\right)^2 + \left(\frac{\partial \phi}{\partial y}\right)^2 + \left(\frac{\partial \phi}{\partial z}\right)^2\right] + \frac{p}{\rho} = f(t). \tag{A1.1.17}$$

A1.1.1 2D flow

$$\mathbf{v} = \nabla \times (\psi(x, y)\mathbf{e}_z), \tag{A1.1.18}$$

$$v_x = \frac{\partial \psi}{\partial y}, \quad v_y = -\frac{\partial \psi}{\partial x}, \quad v_z = 0, \tag{A1.1.19}$$

$$\nabla \times \mathbf{v} = -(\nabla^2\psi)\mathbf{e}_z, \tag{A1.1.20}$$

$$\mathrm{curl}^3\mathbf{v} = \nabla \times \nabla \times \nabla \times \mathbf{v} = [\nabla^2(\nabla^2\psi)]\mathbf{e}_z. \tag{A1.1.21}$$

A1.2 Cylindrical coordinates

In cylindrical coordinates $(\xi_1, \xi_2, \xi_3) = (r, \theta, z)$ with unit vectors $\mathbf{e}_r$, $\mathbf{e}_\theta$, $\mathbf{e}_z$, equations for $\mathbf{u} = (u_1, u_2, u_3) = (u_r, u_\theta, u_z)$ are given as follows:

$$\frac{\partial \mathbf{e}_r}{\partial \theta} = \mathbf{e}_\theta, \quad \frac{\partial \mathbf{e}_\theta}{\partial \theta} = -\mathbf{e}_r, \quad \frac{\partial \mathbf{e}_z}{\partial \theta} = 0, \tag{A1.2.1}$$

$$\mathbf{n} \cdot \nabla\mathbf{u} = \mathbf{e}_r\left[(\mathbf{n} \cdot \nabla)\, u_r - \frac{n_\theta}{r}u_\theta\right] + \mathbf{e}_\theta\left[(\mathbf{n} \cdot \nabla)\, u_\theta + \frac{n_\theta}{r}u_r\right] + \mathbf{e}_\varphi\, (\mathbf{n} \cdot \nabla)\, u_\varphi, \tag{A1.2.2}$$

$$\nabla \cdot \mathbf{u} = \frac{1}{r}\frac{\partial}{\partial r}(ru_r) + \frac{1}{r}\frac{\partial u_\theta}{\partial \theta} + \frac{\partial u_z}{\partial z}, \tag{A1.2.3}$$

$$\nabla \times \mathbf{u} = \mathbf{e}_r\left[\frac{1}{r}\frac{\partial u_z}{\partial \theta} - \frac{\partial u_\theta}{\partial z}\right] + \mathbf{e}_\theta\left[\frac{\partial u_r}{\partial z} - \frac{\partial u_z}{\partial r}\right] + \mathbf{e}_z\left[\frac{1}{r}\frac{\partial (ru_\theta)}{\partial r} - \frac{1}{r}\frac{\partial u_r}{\partial \theta}\right], \tag{A1.2.4}$$

$$\nabla^2 \mathbf{u} = \mathbf{e}_r\left[\Delta u_r - \frac{u_r}{r^2} - \frac{2}{r^2}\frac{\partial u_\theta}{\partial \theta}\right] + \mathbf{e}_\theta\left[\Delta u_\theta + \frac{2}{r^2}\frac{\partial u_r}{\partial \theta} - \frac{u_\theta}{r^2}\right] + \mathbf{e}_z \Delta u_z, \tag{A1.2.5}$$

with

$$\nabla^2 = \Delta = \frac{\partial^2}{\partial r^2} + \frac{1}{r}\frac{\partial}{\partial r} + \frac{1}{r^2}\frac{\partial^2}{\partial \theta^2} + \frac{\partial^2}{\partial z^2}, \tag{A1.2.6}$$

$$T_{ij} = -p\delta_{ij} + 2\mu e_{ij}, \tag{A1.2.7}$$

$$e_{rr} = \frac{\partial u_r}{\partial r}, \quad e_{\theta\theta} = \frac{1}{r}\frac{\partial u_\theta}{\partial \theta} + \frac{u_r}{r}, \quad e_{zz} = \frac{\partial u_z}{\partial z}, \tag{A1.2.8}$$

$$e_{r\theta} = \frac{r}{2}\frac{\partial}{\partial r}\left(\frac{u_\theta}{r}\right) + \frac{1}{2r}\frac{\partial u_r}{\partial \theta}, \quad e_{\theta z} = \frac{1}{2r}\frac{\partial u_z}{\partial \theta} + \frac{1}{2}\frac{\partial u_\theta}{\partial z}, \quad e_{zr} = \frac{1}{2}\frac{\partial u_r}{\partial z} + \frac{1}{2}\frac{\partial u_z}{\partial r}, \tag{A1.2.9}$$

$$\frac{\partial u_r}{\partial t} + \mathbf{u}\cdot\nabla u_r - \frac{u_\theta^2}{r} = -\frac{1}{\rho}\frac{\partial p}{\partial r} + \nu\left[\Delta u_r - \frac{u_r}{r^2} - \frac{2}{r^2}\frac{\partial u_\theta}{\partial \theta}\right], \tag{A1.2.10}$$

$$\frac{\partial u_\theta}{\partial t} + \mathbf{u}\cdot\nabla u_\theta + \frac{u_r u_\theta}{r} = -\frac{1}{\rho r}\frac{\partial p}{\partial \theta} + \nu\left[\Delta u_\theta + \frac{2}{r^2}\frac{\partial u_r}{\partial \theta} - \frac{u_\theta}{r^2}\right], \tag{A1.2.11}$$

$$\frac{\partial u_z}{\partial t} + \mathbf{u}\cdot\nabla u_z = -\frac{1}{\rho}\frac{\partial p}{\partial z} + \nu\Delta u_z. \tag{A1.2.12}$$

For $\mathbf{u} = \nabla\phi$, equations for ϕ are given as follows:

$$\nabla\phi = \mathbf{e}_r\frac{\partial \phi}{\partial r} + \frac{\mathbf{e}_\theta}{r}\frac{\partial \phi}{\partial \theta} + \mathbf{e}_z\frac{\partial \phi}{\partial z}, \tag{A1.2.13}$$

$$\nabla^2\phi = \Delta\phi = \frac{1}{r}\frac{\partial}{\partial r}\left(r\frac{\partial \phi}{\partial r}\right) + \frac{1}{r^2}\frac{\partial^2 \phi}{\partial \theta^2} + \frac{\partial^2 \phi}{\partial z^2}, \tag{A1.2.14}$$

$$e_{rr} = \frac{\partial^2 \phi}{\partial r^2}, \quad e_{\theta\theta} = \frac{1}{r^2}\frac{\partial^2 \phi}{\partial \theta^2} + \frac{1}{r}\frac{\partial \phi}{\partial r}, \quad e_{zz} = \frac{\partial^2 \phi}{\partial z^2}, \tag{A1.2.15}$$

$$e_{r\theta} = \frac{1}{r}\frac{\partial^2 \phi}{\partial r \partial \theta} - \frac{1}{r^2}\frac{\partial \phi}{\partial \theta}, \quad e_{\theta z} = \frac{1}{r}\frac{\partial^2 \phi}{\partial z \partial \theta}, \quad e_{zr} = \frac{\partial^2 \phi}{\partial z \partial r}, \tag{A1.2.16}$$

$$\frac{\partial \phi}{\partial t} + \frac{1}{2}|\nabla\phi|^2 + \frac{p}{\rho} = \frac{\partial \phi}{\partial t} + \frac{1}{2}\left[\left(\frac{\partial \phi}{\partial r}\right)^2 + \left(\frac{1}{r}\frac{\partial \phi}{\partial \theta}\right)^2 + \left(\frac{\partial \phi}{\partial z}\right)^2\right] + \frac{p}{\rho} = f(t). \tag{A1.2.17}$$

A1.2.1 Axisymmetric flow (cylindrical)

$$\mathbf{v} = \nabla \times \left(\frac{\psi(r, z)}{r}\mathbf{e}_\theta\right), \tag{A1.2.18}$$

$$v_r = -\frac{1}{r}\frac{\partial \psi}{\partial z}, \quad v_\theta = 0, \quad v_z = \frac{1}{r}\frac{\partial \psi}{\partial r}, \tag{A1.2.19}$$

$$\nabla \times \mathbf{v} = -\left(\frac{1}{r}E^2\psi\right)\mathbf{e}_\theta, \tag{A1.2.20}$$

$$\mathrm{curl}^3\mathbf{v} = \nabla \times \nabla \times \nabla \times \mathbf{v} = \left[\frac{1}{r}E^2(E^2\psi)\right]\mathbf{e}_\theta, \tag{A1.2.21}$$

where

$$E^2\psi = \frac{\partial^2\psi}{\partial r^2} - \frac{1}{r}\frac{\partial\psi}{\partial r} + \frac{\partial^2\psi}{\partial z^2}. \tag{A1.2.22}$$

The equation $\nabla^2\theta = 0$ arises from the condition that the irrotational flow is solenoidal. The equation $E^2\psi = 0$ arises from the condition that the solenoidal flow is irrotational. The equation for the stream function for axisymmetric flow in cylindrical coordinates is

$$\frac{\partial E^2\psi}{\partial t} - \frac{1}{r}\frac{\partial(\psi, E^2\psi)}{\partial(r,z)} + \frac{2E^2\psi}{r^2}\frac{\partial(\psi, r)}{\partial(r,z)} = \upsilon E^4\psi. \tag{A1.2.23}$$

The solution of this equation is $\psi = \psi_p + \psi_v$.

A1.3 Polar-spherical coordinates

In polar-spherical coordinates $(\xi_1, \xi_2, \xi_3) = (r, \theta, \varphi)$ with unit vectors $\mathbf{e}_r, \mathbf{e}_\theta, \mathbf{e}_\varphi$, equations for $\mathbf{u} = (u_1, u_2, u_3) = (u_r, u_\theta, u_\varphi)$ are given as follows:

$$\frac{\partial \mathbf{e}_r}{\partial r} = 0, \quad \frac{\partial \mathbf{e}_r}{\partial \theta} = \mathbf{e}_\theta, \quad \frac{\partial \mathbf{e}_r}{\partial \varphi} = \mathbf{e}_\varphi \sin\theta, \tag{A1.3.1}$$

$$\frac{\partial \mathbf{e}_\theta}{\partial r} = 0, \quad \frac{\partial \mathbf{e}_\theta}{\partial \theta} = -\mathbf{e}_r, \quad \frac{\partial \mathbf{e}_\theta}{\partial \varphi} = \mathbf{e}_\varphi \cos\theta, \tag{A1.3.2}$$

$$\frac{\partial \mathbf{e}_\varphi}{\partial r} = 0, \quad \frac{\partial \mathbf{e}_\varphi}{\partial \theta} = 0, \quad \frac{\partial \mathbf{e}_\varphi}{\partial \varphi} = -\mathbf{e}_r \sin\theta - \mathbf{e}_\theta\cos\theta, \tag{A1.3.3}$$

$$\begin{aligned}\mathbf{n}\cdot\nabla\mathbf{u} = {} & \mathbf{e}_r\left[(\mathbf{n}\cdot\nabla)\,u_r - \frac{n_\theta}{r}u_\theta - \frac{n_\varphi}{r}u_\varphi\right] + \mathbf{e}_\theta\left[(\mathbf{n}\cdot\nabla)\,u_\theta + \frac{n_\theta}{r}u_r - \frac{n_\varphi}{r}u_\varphi\cot\theta\right] \\ & + \mathbf{e}_\varphi\left[(\mathbf{n}\cdot\nabla)\,u_\varphi + \frac{n_\varphi}{r}u_r + \frac{n_\varphi}{r}u_\theta\cot\theta\right],\end{aligned} \tag{A1.3.4}$$

$$\nabla\cdot\mathbf{u} = \frac{1}{r^2}\frac{\partial}{\partial r}\left(r^2u_r\right) + \frac{1}{r\sin\theta}\frac{\partial}{\partial\theta}(u_\theta\sin\theta) + \frac{1}{r\sin\theta}\frac{\partial u_\varphi}{\partial\varphi}, \tag{A1.3.5}$$

$$\begin{aligned}\nabla\times\mathbf{u} = {} & \mathbf{e}_r\frac{1}{r\sin\theta}\left[\frac{\partial}{\partial\theta}(\sin\theta u_\varphi) - \frac{\partial u_\theta}{\partial\varphi}\right] - \frac{1}{r\sin\theta}\mathbf{e}_\theta\left[\sin\theta\frac{\partial(ru_\varphi)}{\partial r} - \frac{\partial u_r}{\partial\varphi}\right] \\ & + \frac{1}{r}\mathbf{e}_\varphi\left[\frac{\partial(ru_\theta)}{\partial r} - \frac{\partial u_r}{\partial\theta}\right],\end{aligned} \tag{A1.3.6}$$

$$\begin{aligned}\nabla^2\mathbf{u} = {} & \mathbf{e}_r\left[\Delta u_r - 2\frac{u_r}{r^2} - \frac{2}{r^2}\frac{\partial u_\theta}{\partial\theta} - 2\frac{u_\theta}{r^2}\cot\theta - \frac{2}{r^2\sin\theta}\frac{\partial u_\varphi}{\partial\varphi}\right] \\ & + \mathbf{e}_\theta\left[\Delta u_\theta + \frac{2}{r^2}\frac{\partial u_r}{\partial\theta} - \frac{u_\theta}{r^2\sin^2\theta} - \frac{2\cos\theta}{r^2\sin^2\theta}\frac{\partial u_\varphi}{\partial\varphi}\right] \\ & + \mathbf{e}_\varphi\left[\Delta u_\varphi + \frac{2}{r^2\sin\theta}\frac{\partial u_r}{\partial\varphi} + \frac{2\cos\theta}{r^2\sin^2\theta}\frac{\partial u_\theta}{\partial\varphi} - \frac{u_\varphi}{r^2\sin^2\theta}\right],\end{aligned} \tag{A1.3.7}$$

with

$$\nabla^2 = \Delta = \frac{\partial^2}{\partial r^2} + \frac{2}{r}\frac{\partial}{\partial r} + \frac{1}{r^2}\frac{\partial^2}{\partial \theta^2} + \frac{1}{r^2}\cot\theta\frac{\partial}{\partial \theta} + \frac{1}{r^2 \sin^2\theta}\frac{\partial^2}{\partial \varphi^2}, \tag{A1.3.8}$$

$$T_{ij} = -p\delta_{ij} + 2\mu e_{ij}, \tag{A1.3.9}$$

$$e_{rr} = \frac{\partial u_r}{\partial r}, \quad e_{\theta\theta} = \frac{1}{r}\frac{\partial u_\theta}{\partial \theta} + \frac{u_r}{r}, \quad e_{\varphi\varphi} = \frac{1}{r\sin\theta}\frac{\partial u_\varphi}{\partial \varphi} + \frac{u_r}{r} + \frac{u_\theta}{r}\cot\theta, \tag{A1.3.10}$$

$$e_{r\theta} = \frac{r}{2}\frac{\partial}{\partial r}\left(\frac{u_\theta}{r}\right) + \frac{1}{2r}\frac{\partial u_r}{\partial \theta}, \quad e_{\theta\varphi} = \frac{\sin\theta}{2r}\frac{\partial}{\partial \theta}\left(\frac{u_\varphi}{\sin\theta}\right) + \frac{1}{2r\sin\theta}\frac{\partial u_\theta}{\partial \varphi},$$

$$e_{\varphi r} = \frac{1}{2r\sin\theta}\frac{\partial u_r}{\partial \varphi} + \frac{r}{2}\frac{\partial}{\partial r}\left(\frac{u_\varphi}{r}\right),$$

$$\frac{\partial u_r}{\partial t} + \mathbf{u}\cdot\nabla u_r - \frac{u_\theta^2}{r} - \frac{u_\varphi^2}{r}$$
$$= -\frac{1}{\rho}\frac{\partial p}{\partial r} + \nu\left[\Delta u_r - 2\frac{u_r}{r^2} - \frac{2}{r^2}\frac{\partial u_\theta}{\partial \theta} - 2\frac{u_\theta}{r^2}\cot\theta - \frac{2}{r^2\sin\theta}\frac{\partial u_\varphi}{\partial \varphi}\right], \tag{A1.3.11}$$

$$\frac{\partial u_\theta}{\partial t} + \mathbf{u}\cdot\nabla u_\theta + \frac{u_r u_\theta}{r} - \frac{u_\varphi^2}{r}\cot\theta$$
$$= -\frac{1}{\rho r}\frac{\partial p}{\partial \theta} + \nu\left[\Delta u_\theta + \frac{2}{r^2}\frac{\partial u_r}{\partial \theta} - \frac{u_\theta}{r^2\sin^2\theta} - \frac{2\cos\theta}{r^2\sin^2\theta}\frac{\partial u_\varphi}{\partial \varphi}\right], \tag{A1.3.12}$$

$$\frac{\partial u_\varphi}{\partial t} + \mathbf{u}\cdot\nabla u_\varphi + \frac{u_\varphi u_r}{r} + \frac{u_\theta u_\varphi}{r}\cot\theta$$
$$= -\frac{1}{\rho r\sin\theta}\frac{\partial p}{\partial \varphi} + \nu\left[\Delta u_\varphi + \frac{2}{r^2\sin\theta}\frac{\partial u_r}{\partial \varphi} + \frac{2\cos\theta}{r^2\sin^2\theta}\frac{\partial u_\theta}{\partial \varphi} - \frac{u_\varphi}{r^2\sin^2\theta}\right]. \tag{A1.3.13}$$

For $\mathbf{u} = \nabla\phi$, equations for ϕ are given as follows:

$$\nabla\phi = \mathbf{e}_r\frac{\partial\phi}{\partial r} + \frac{\mathbf{e}_\theta}{r}\frac{\partial\phi}{\partial\theta} + \frac{\mathbf{e}_\varphi}{r\sin\theta}\frac{\partial\phi}{\partial\varphi}, \tag{A1.3.14}$$

$$\nabla^2\phi = \Delta\phi = \frac{1}{r^2}\frac{\partial}{\partial r}\left(r^2\frac{\partial\phi}{\partial r}\right) + \frac{1}{r^2\sin\theta}\frac{\partial}{\partial\theta}\left(\sin\theta\frac{\partial\phi}{\partial\theta}\right) + \frac{1}{r^2\sin^2\theta}\frac{\partial^2\phi}{\partial\varphi^2}, \tag{A1.3.15}$$

$$e_{rr} = \frac{\partial^2\phi}{\partial r^2}, \quad e_{\theta\theta} = \frac{1}{r^2}\frac{\partial^2\phi}{\partial\theta^2} + \frac{1}{r}\frac{\partial\phi}{\partial r},$$

$$e_{\varphi\varphi} = \frac{1}{r^2\sin^2\theta}\frac{\partial^2\phi}{\partial\varphi^2} + \frac{1}{r}\frac{\partial\phi}{\partial r} + \frac{1}{r^2}\frac{\partial\phi}{\partial\theta}\cot\theta, \tag{A1.3.16}$$

$$e_{r\theta} = \frac{1}{r}\frac{\partial^2\phi}{\partial r\partial\theta} - \frac{1}{r^2}\frac{\partial\phi}{\partial\theta}, \quad e_{\theta\varphi} = \frac{1}{r^2\sin\theta}\left(\frac{\partial^2\phi}{\partial\theta\partial\varphi} - \cot\theta\frac{\partial\phi}{\partial\varphi}\right),$$

$$e_{\varphi r} = \frac{1}{r\sin\theta}\left(\frac{\partial^2\phi}{\partial r\partial\varphi} - \frac{1}{r}\frac{\partial\phi}{\partial\varphi}\right),$$

$$\frac{\partial\phi}{\partial t} + \frac{1}{2}|\nabla\phi|^2 + \frac{p}{\rho} = \frac{\partial\phi}{\partial t} + \frac{1}{2}\left[\left(\frac{\partial\phi}{\partial r}\right)^2 + \left(\frac{1}{r}\frac{\partial\phi}{\partial\theta}\right)^2 + \left(\frac{1}{r\sin\theta}\frac{\partial\phi}{\partial\varphi}\right)^2\right] + \frac{p}{\rho} = f(t). \tag{A1.3.17}$$

A1.3.1 Axisymmetric flow (spherical)

$$\mathbf{v} = \nabla \times \left(\frac{\psi(r,\theta)}{r \sin\theta} \mathbf{e}_\varphi \right), \tag{A1.3.18}$$

$$v_r = \frac{1}{r^2 \sin\theta} \frac{\partial \psi}{\partial \theta}, \qquad v_\theta = -\frac{1}{r \sin\theta} \frac{\partial \psi}{\partial r}, \qquad v_\varphi = 0, \tag{A1.3.19}$$

$$\nabla \times \mathbf{v} = -\left(\frac{1}{r \sin\theta} E^2 \psi \right) \mathbf{e}_\varphi, \tag{A1.3.20}$$

$$\text{curl}^3 \mathbf{v} = \nabla \times \nabla \times \nabla \times \mathbf{v} = \left[\frac{1}{r \sin\theta} E^2 (E^2 \psi) \right] \mathbf{e}_\varphi, \tag{A1.3.21}$$

where

$$E^2 \psi = \frac{\partial^2 \psi}{\partial r^2} + \frac{\sin\theta}{r^2} \frac{\partial}{\partial \theta} \left(\frac{1}{\sin\theta} \frac{\partial \psi}{\partial \theta} \right). \tag{A1.3.22}$$

The equation $\nabla^2 \phi = 0$ arises from the condition that the irrotational flow is solenoidal. The equation $E^2 \psi = 0$ arises from the condition that the solenoidal flow is irrotational.

After eliminating the pressure from the equation for axially symmetric flow in spherical coordinates, we get

$$\frac{\partial E^2 \psi}{\partial t} - \frac{1}{r^2 \sin\theta} \frac{\partial(\psi, E^2 \psi)}{\partial(r,\theta)} + \frac{2E^2 \psi}{r^3 \sin^2\theta} \frac{\partial(\psi, r \sin\theta)}{\partial(r,\theta)} = \nu E^4 \psi. \tag{A1.3.23}$$

The solution of this equation is $\psi = \psi_p + \psi_v$.

APPENDIX B

List of frequently used symbols and concepts

$K = \frac{p_u - p_d}{p_d - p_v}$, cavitation number
$Re = \frac{\rho \ell U}{\mu}$, Reynolds number
$f(z) = \phi + i\psi$, complex velocity potential
k, wavenumber
k_m, wavenumber for maximum growth for "most dangerous wave"
$\mathbf{L}$, $L_{ij} \equiv \frac{\partial u_i}{\partial x_j}$
$\mathbf{A}$, $A_{ij} \equiv L_{ij} + L_{ji} = \frac{\partial u_i}{\partial x_j} + \frac{\partial u_j}{\partial x_i}$
p, pressure
(r, θ, z), cylindrical coordinates
t, time
We, Weber number
(r, θ, φ), spherical-polar coordinates
(x, y, z), Cartesian coordinates
$z = x + iy$, complex variable
$\mathbf{u} = (u_1, u_2, u_3)$, velocity vector
$\mathbf{v} = (v_1, v_2, v_3)$, the rotational part of velocity vector
$\mathbf{D}$, D_{ij}, rate-of-strain tensor, $D_{ij} = \frac{1}{2}(L_{ij} + L_{ji})$
D, drag force
V, volume
∂V, surface element of S
$\mathbf{g}$, acceleration due to gravity
$\mathbf{S}$, S_{ij}, extra stress tensor
M, moment
$\mathbf{T}$, T_{ij}, stress tensor
$\mathbf{1}$, unit diagonal matrix
δ_{ij}, Kronecker's delta
γ, coefficient of surface/interface tension
Γ, circulation
$\lambda = 2\pi/k$, wavelength
ϕ, irrotational velocity potential
Φ, irrotational viscous dissipation $\Phi = 2\mu \int \mathbf{D} : \mathbf{D} dV$
α_1, α_2, extra stress tensor
$\hat{\beta}$, climbing constant $\hat{\beta} = 3\alpha_1 + 2\alpha_2$

λ_1, λ_2, extra stress tensor
ψ, stream function
ρ, density of fluid
μ, viscosity of fluid
ν, kinematic viscosity $\nu = \mu/\rho$
$\boldsymbol{\omega}$, vorticity, $\boldsymbol{\omega} = \text{curl}\,\mathbf{u}$
Ω, gravitational potential
σ_{ij}, dynamic stress tensor
$\boldsymbol{\Omega}$, $\Omega_{ij} = L_{ij} - L_{ji}$
σ $e^{\sigma t}$ complex growth; $\sigma = \sigma_R + i\sigma_I$; σ_R, growth rate; $\sigma_I = \omega$, oscillation of frequency
$\sigma(k)$, dispersion relation; $\sigma_R(k)$ growth-rate curve; $\sigma_R(k) = 0$, neutral curve or neutral stability curve, $\sigma_{Rm} = \max_{k \in \mathbb{R}} \sigma_R(k) = \sigma_R(k_m)$
$\boldsymbol{\tau}$, τ_{ij}, viscous or viscoelastic part of the stress tensor
$\tau_{ij} = 2\mu D_{ij}$, Newtonian fluid; $\boldsymbol{\tau} = 2\mu \nabla \otimes \nabla\phi$; $\tau_{ij} = 2\mu \partial^2\phi/\partial x_i \partial x_j$, irrotational viscous stress

References

The numbers in the square brackets at the end of each reference locate the page where that reference is cited.

Ackeret J. (1952). Über exakte Lösungen der Stokes-Navier-Gleichungen inkompressibler Flüssigkeiten bei veränderten Grenzbedingungen. *Z. Angew. Math. Phys.* **3**, 259–271. [127, 144, 145, 153, 316, 322]

Aitken L. S., Wilson S. D. R. (1993). Rayleigh–Taylor instability in elastic liquids. *J. Non-Newtonian Fluid Mech.* **49**, 13–22. [426, 427]

Andritsos N., Hanratty T. J. (1987). Interfacial instabilities for horizontal gas–liquid flows in pipelines. *Int. J. Multiphase Flow* **13**, 583–603. [119–121]

Andritsos N., Williams L., Hanratty T. J. (1989). Effect of liquid viscosity on the stratified-slug transition in horizontal pipe flow. *Int. J. Multiphase Flow* **15**, 877–892. [118–120]

Apfel R. E. (1970). The role of impurities in cavitation-threshold determination. *J. Acoust. Soc. Am.* **48**, 1179–1186. [277]

Archer L. A., Ternet D., Larson R. G. (1997). 'Fracture' phenomena in shearing flow of viscous liquids. *Rheol. Acta* **36**, 579–584. [277]

Ardekani A. M., Rangel R. H., Joseph D. D. (2007). Motion of a sphere normal to a wall in a second-order fluid. *J. Fluid Mech.* **587**, 163–172. [418]

Ashgritz N., Mashayek F. (1995). Temporal analysis of capillary jet breakup. *J. Fluid Mech.* **291**, 163–190. [304]

Badr H. M., Coutanceau M., Dennis S. C. R., Menard C. 1990 Unsteady flow past a rotating circular cylinder at Reynolds numbers 10^3 and 10^4. *J. Fluid Mech.* **220**, 459–484. [353]

Bair S., Winer W. O. (1990). The high shear stress rheology of liquid lubricants at pressure of 2 to 200 MPa. *J. Tribol.* **114**, 246–253. [273, 277]

Bair S., Winer W. O. (1992). The high pressure high shear stress rheology of liquid lubricants. *J. Tribol.* **114**, 1–13. [277]

Baker G. R., Meiron D. I., Orszag S. A. (1980). Vortex simulations of the Rayleigh-Taylor instability. *Phys. Fluids* **23**, 1485–1490. [463]

Baker G. R., Meiron D. I., Orszag S. A. (1982). Generalized vortex methods for free-surface flow problems. *J. Fluid Mech.* **123**, 477–501. [463]

Baker G. R., Meiron D. I., Orszag S. A. (1984). Boundary integral methods for axisymmetric and three-dimensional Reyleigh–Taylor instability problems. *Physica D* **12**, 19–31. [463]

Banerjee P. K., Butterfield R. (1981). *Boundary Element Methods in Engineering Science.* McGraw-Hill. [462]

Barnea D. (1991). On the effect of viscosity on stability of stratified gas–liquid flow – application to flow pattern transition at various pipe inclinations. *Chem. Eng. Sci.* **46**, 2123–2131. [120]

Barnea D., Taitel Y. (1993). Kelvin–Helmholtz stability criteria for stratified flow: Viscous versus non-viscous (inviscid) approaches. *Int. J. Multiphase Flow* **19**, 639–649. [119, 120]

Barr G. (1926). The air bubble viscometer. *Philos. Mag.* Ser. **7**, 395. [56]

Basaran O. (1992) Nonlinear oscillations of liquid drops. *J. Fluid Mech.* **241**, 169–198. [172]

Basset A. B. (1888) *A Treatise on Hydrodynamics with Numerous Examples*, Vol. 2, reprinted by Dover in 1961. [194]

Batchelor G. K. (1967). *An Introduction to Fluid Dynamics*. Cambridge University Press. [1–3, 19, 31, 32, 39, 42, 43, 53, 54, 127, 128]

Batchelor G. K. (1987). The stability of a large gas bubble rising through liquid. *J. Fluid Mech.* **184**, 399–422. [48]

Batchelor G. K., Gill A. E. (1962). Analysis of the stability of axisymmetric jets. *J. Fluid Mech.* **14**, 529–551. [267, 269, 270, 387]

Becker L. E., McKinley G. H., Stone H. A. (1996). Sedimentation of a sphere near a plane wall: Weak non-Newtonian and inertial effects. *J. Non-Newtonian Fluid Mech.* **63**, 201–233. [418]

Benjamin T. B., Ursell F. (1954). The stability of the plane free surface of a liquid in vertical periodic motion. *Proc. R. Soc. London Ser. A* **225**, 505–515. [197]

Bergwerk W. (1959). Flow patterns in diesel nozzle spray holes. *Proc. Inst. Mech. Eng.* **173**, 655–660. [278, 280]

Bers A. (1975). Linear waves and instabilities. In *Physique des Plasmas*, C. DeWitt and J. Peyraud, eds. Gordon and Breach, pp. 117–213. [254, 259, 267]

Bhaga T., Weber M. (1981). Bubbles in viscous liquids: Shapes, wakes and velocities. *J. Fluid Mech.* **105**, 61–85. [42, 46, 48, 50, 53]

Bi Q. C., Zhao T. S. (2001). Taylor bubbles in miniaturized circular and noncircular channels. *Int. J. Multiphase Flow* **27**, 561–570. [56]

Billet M. L. (1985). Cavitation nuclei measurements – A review. ASME Cavitation and Multiphase Flow Forum–1985 **23**, 31–38. [277]

Binnie A. M. (1953). The stability of the surface of a cavitation bubble. *Proc. Cambridge Philos. Soc.* **49**, 151–155. [40]

Bird R., Armstrong R., Hassager O. (1987). *Dynamics of Polymeric Liquids*. Wiley. [398]

Birkhoff G. (1954). Note on Taylor instability. Q. *Appl. Math.* **12**, 306–309. [40, 41]

Boulton-Stone J. M., Robinson P. B., Blake J. R. (1995). A note on the axisymmetric interaction of pairs of rising, deforming gas bubbles. *Int. J. Multiphase Flow* **21**, 1237–1241. [48]

Bowman J. J., Senior T. B. A., Uslenghi P. L. E. (eds.) (1987) *Electromagnetic and Acoustic Scattering by Simple Shapes*. Hemisphere. [178]

Brebbia C. A., Telles J. C. F., Wrobel L. C. (1984). *Boundary Element Techniques: Theory and Applications in Engineering*. Springer-Verlag. [462, 463]

Brennen C. E. (1995). *Cavitation and Bubble Dynamics*. Oxford University Press. [280]

Brenner H. (1961). The slow motion of a sphere through a viscous fluid towards a plane surface. *Chem. Eng. Sci.* **16**, 242–251. [419]

Brenner M. P., Eggers J., Joseph K., Nagel S. R., Shi X. D. (1997). Breakdown of scaling in droplet fission at high Reynolds number. *Phys. Fluids* **9**, 1573–1590. [304]

Bretherton F. P. (1961). The motion of long bubbles in tubes. *J. Fluid Mech.* **10**, 166–188. [56]

Briggs L. J. (1950). Limiting negative pressure of water. *J. Appl. Phys.* **21**, 721–722. [276]

Briggs R. J. (1964). Electron-stream interaction with plasmas. MIT Press, Research Monograph No. 29. [254, 267]

Brown R. A. S. (1965). The mechanics of large gas bubbles in tubes. *Can. J. Chem. Eng.* **2**, 217–223. [54, 55, 67]

Byatt-Smith J. G. B., Longuet-Higgins M. S. (1976). On the speed and profile of steep solitary waves. *Proc. R. Soc. London Ser. A* **350**, 175–189. [463]

Canot E., Davoust L., Hammoumi M. E., Lachkar D. (2003). Numerical simulation of the buoyancy-driven bouncing of a 2-D bubble at a horizontal wall. *Theor. Comput. Fluid Dyn.* **17**, 51–72. [152, 464]

Cerda E. A., Tirapegui E. L. (1998). Faraday's instability in viscous fluid. *J. Fluid Mech.* **368**, 195–228. [205]

Chandrasekhar S. (1961). *Hydrodynamic and Hydromagnetic Stability*. Oxford University Press. [70, 73, 74, 76, 171, 172, 182, 216]

Chang H. C., Demekhin E. A., Kalaidin E. (1999). Interated stretching of viscoelastic jets. *Phys. Fluids* **11**, 1717–1737. [307, 456–459]

Chang I. D., Russel P. E. (1965). Stabiity of a liquid layer adjacent to a high-speed gas stream. *Phys. Fluids* **8**, 1018–1026. [378]

Charru F., Hinch E. J. (2000). "Phase diagram" of interfacial instabilities in a two-layer Couette flow and mechanism of the long-wave instability. *J. Fluid Mech.* **414**, 195–223. [105]

Chaves H., Knapp M., Kubitzek A., Obermeier F., Schneider T. (1995). Experimental study of cavitation in the nozzle hole of diesel injectors using transparent nozzles. Tech. Rep. 950290, Warrendale, PA: Soc. Automotive Engs. Tech. paper. [278, 285, 287]

Chawla T. C. (1975). The Kelvin–Helmholtz instability of the gas–liquid interface of a sonic gas jet submerged in a liquid. *J. Fluid Mech.* **67**, 513–537. [378]

Chen A. U., Notz P. K., Basaran O. A. (2002). Computational and experimental analysis of pinch-off and scaling. *Phys. Rev. Lett.* **88**, 1745011–1745014. [306, 307]

Chen J. L. S. (1974). Growth of the boundary layer on a spherical gas bubble. *Trans. ASME. J. Appl. Mech.* **41**, 873–878. [157]

Chen T., Li X. (1999). Liquid jet atomization in a compressible gas stream. *J. Propul. Power* **15**, 369–376. [379]

Chen Y., Israelachvili J. (1991). New mechanism of cavitation damage. Science **252**, 1157–1160. [277]

Chen Y. J., Steen P. H. (1997). Dynamics of inviscid capillary breakup: Collapse and pinchoff of a film bridge. *J. Fluid Mech.* **341**, 245–267. [463]

Chew Y. T, Cheng M., Luo S. C. (1995) A numerical study of flow past a rotating circular cylinder using a hybrid vortex scheme. *J. Fluid Mech.* **299**, 35–71. [353, 358, 362]

Chorin A. J. (1973). Numerical study of slightly viscous flow. *J. Fluid Mech.* **57**, 785–796. [33]

Chorin A. J. (1978). Vortex sheet approximation of boundary layers. *J. Comput. Phys.* **27**, 428–442. [33]

Chou M. H. (2000) Numerical study of vortex shedding from a rotating cylinder immersed in a uniform flow field, *Int. J. Numer. Meth. Fluids* **32**, 545–567. [354]

Christiansen R. M., Hixson A. N. (1957). Breakup of a liquid jet in a denser liquid. *Ind. Eng. Chem.* **49**, 1017. [219]

Ciliberto S., Gollub J. P. (1985). Chaotic mode competition in parametrically forced surface waves. *J. Fluid Mech.* **158**, 381–398. [197, 213]

Coleman B., Noll W. (1960) An approximation theorem for functionals, with applications in continuum mechanics. *Arch. Ration. Mech. Anal.* **6**, 355–370. [398]

Crowley C. J., Wallis G. B., Barry J. J. (1992). Validation of a one-dimensional wave model for the stratified-to-slug flow regime transition, with consequences for wave growth and slug frequency. *Int. J. Multiphase Flow* **18**, 249–271. [120, 121]

Crum L. A. (1982). Nucleation and stabilization of microbubbles in liquids. *Appl. Sci. Res.* **38**, 101–115. [277]

Currie I. G. (1974). *Fundamental Mechanics of Fluids*, 1st ed. McGraw-Hill. [279]

Dabiri S., Sirignano W. A., Joseph D. D. (2007) Two-dimensional viscous aperture flow: Navier–Stokes and viscous potential flow solutions. [287] http://www.aem.umn.edu/people/faculty/joseph/ViscousPotentialFlow/

Darwin, G. H. 1878 On the bodily tides of viscous and semi-elastic spheroids. *Philos. Trans.* **cixx** 1. [171]

Davies R. M., Taylor G. I. (1950). The mechanics of large bubbles rising thorough liquids in tubes. *Proc. R. Soc. London Ser. A* **200**, 375–390. [42–47, 49, 50, 52, 53, 55]

Day R. F., Hinch E. J., Lister J. R. (1998). Self-similar capillary pinchoff of an inviscid fluid. *Phys. Rev. Lett.* **80**, 704–707. [464]

de Chizelle Y. K., Ceccio S. L., Brennen C. E. (1995). Observation and scaling of travelling bubble cavitation. *J. Fluid Mech.* **293**, 99–126. [290]

Dryden H., Murnaghan F., Bateman H. (1932). *Hydrodynamics*. Dover. (A complete unabridged reprinting of National Research Council Bulletin 84, 1956.) [127]

Dias F., Kharif C. (1999). Nonlinear gravity and capillary-gravity waves. *Annu. Rev. Fluid Mech.* **31**, 301–346. [198]

Dizés S. L. (1997). Global modes in falling capillary jets. *Eur. J. Mech. B/Fluids* **16**, 761–778. [239]

Douady S., Fauve S. (1988). Pattern selection in Faraday instability. *Europhys. Lett.* **6**, 221–226. [213]

Drazin P. G., Reid W. H. (1981). *Hydrodynamic Stability*. Cambridge University Press. [100, 101, 123, 253]

Dumitrescue D. T. (1943). Strömung and einer Luftblase in senkrechten Rohr. *Z. Angew. Math. Mech.* **23**, 139–149. [52, 55]

Dunne B., Cassen B. (1954). Some phenomena associated with supersonic liquid jets. *J. Appl. Phys.* **25**, 569–572. [379]

Dunne B., Cassen B. (1956). Velocity discontinuity instability of a liquid jet. *J. Appl. Phys.* **27**, 577–582. [379]

Edwards W. S., Fauve S. (1993). Parametrically excited quasicrystalline surface waves. *Phys. Rev. E* **47**, R788–R791. [211, 212]

Eggers J. (1993). Universal pinching of 3D axisymmetric free-surface flow. *Phys. Rev. E* **71**, 3458–3460. [305–308]

Eggers J. (1997). Nonlinear dynamics and breakup of free-surface flows. *Rev. Mod. Phys.* **69**, 865–930. [216, 304, 307]

Engel O. G. (1958). Fragmentation of waterdrops in the zone behind an air shock. *J. Res. Natl. Bur. Stand.* **60**, 245–280. [72, 84–86, 379, 438]

Entov V. M. (1978). On the stability of capillary jets of elastoviscous fluids. *J. Eng. Phys. Thermophys*. **34**, 243. [444]

Faeth G. M. (1996). Spray combustion phenomena. In the *Twenty-Sixth Symposium on Combustion*. The Combustion Institute, pp. 1593–1611. [86]

Faraday M. (1831). On the forms and states assumed by fluids in contact with vibrating elastic surfaces. *Philos. Trans. R. Soc.* **121**, 319–340. [197]

Feng Z. C., Sethna P. R. (1989). Symmetry breaking bifurcations in resonant surface waves. *J. Fluid Mech.* **199**, 495–518. [197, 198, 213]

Fisher J. C. (1948). The fracture of liquids. *J. Appl. Phys.* **19**, 1062–1067. [276]

Fluent Inc. (2003). *Fluent 6. 1. User's Guide*. Fluent Inc., Lebanon, NH. [355–357]

Foteinopoulou K., Mavrantzas V. G., Tsamopoulos J. (2004). Numerical simulation of bubble growth in Newtonian and viscoelastic filaments undergoing stretching. *J. Non-Newtonian Fluid Mech.* 122, 177–200. [274]

Francis J. R. D. (1951). The aerodynamic drag of a free water surface. *Proc. R. Soc. London Ser. A* 206, 387–406. [119]

Funada T., Joseph D. D. (2001). Viscous potential flow analysis of Kelvin–Helmholtz instability in a channel. *J. Fluid Mech.* **445**, 263–283. [101, 103, 122, 240, 246, 247, 266, 379]

Funada T., Joseph D. D. (2002). Viscous potential flow analysis of capillary instability. *Int. J. Multiphase Flow* **28**, 1459–1478. [174, 216, 218, 235, 239, 240, 250, 262, 298, 378, 444]

Funada T., Joseph D. D. (2003). Viscoelastic potential flow analysis of capillary instability. *J. Non-Newtonian Fluid Mech.* **111**, 87–105. [444, 449]

Funada T., Joseph D. D., Maehara T., Yamashita S. (2004). Ellipsoidal model of the rise of a Taylor bubble in a round tube. *Int. J. Multiphase Flow* **31**, 473–491. [46]

Funada T., Joseph D. D., Saitoh M., Yamashita S. (2006). Liquid jet into a high Mach number air stream. *Int. J. Multiphase Flow* **32**, 20–50. [378]

Funada T., Joseph D. D., Yamashita S. (2004). Stability of a liquid jet into incompressible gases and liquids. *Int. J. Multiphase Flow* **30**, 1279–1310. [254, 378, 379, 388]

Funada T., Padrino J. C., Joseph D. D. (2006). Viscous potential flow of the Kelvin–Helmholtz instability of a cylindrical jet of one fluid into the same fluid. In preparation. <http://www.aem.umn.edu/people/faculty/joseph/ViscousPotentialFlow/> [149 (footnote)]

Funada T., Saitoh M., Wang J., Joseph D. D. (2005). Stability of a liquid jet into incompressible gases and liquids: Part 2. Effects of the irrotational viscous pressure. *Int. J. Multiphase Flow* **31**, 1134–1154. [131]

Funada T., Wang J., Joseph D. D. (2006). Viscous potential flow analysis of stress induced cavitation in an aperture flow. *Atomization Sprays* **16** (7),763–776. [273, 284, 287, 290]

Funada T., Wang J., Joseph D. D., Tashiro N. (2005). Irrotational Faraday waves on a viscous fluid. <http://www.aem.umn.edu/people/faculty/joseph/ViscousPotentialFlow/> [198]

Galdi P. G., Pokomy M., Vaidya A., Joseph D. D., Feng J. (2002). Orientation of symmetric bodies falling in a second-order liquid at non-zero Reynolds number. *Math. Models Methods Appl. Sci.* **12**, 1653–1690. [402]

Georgescu S. C., Achard J. L., Canot E. (2002). Jet drops ejection in bursting gas bubbles processes. *Eur. J. Mech. B/Fluids* **21**, 265–280. [152, 464]

Gibson A. H. (1913). Long air bubbles in a vertical tube. *Philos. Mag. Ser.* **6**, 952. [56]

Glauert M. B. (1957). The flow past a rapidly rotating circular cylinder. *Proc. R. Soc. London Ser. A* **242**, 108–115. [316, 330, 332, 335, 351–353]

Goldin M., Yerushalmi J., Pfeffer R., Shinnar R. (1969). Breakup of a laminar capillary jet of a viscoelastic fluid. *J. Fluid Mech.* **38**, 689–711. [443]

Goldman A. J., Cox R. G., Brenner H. (1967). Slow viscous motion of a sphere parallel to a plane wall. I. Motion through quiescent fluid. *Chem. Eng. Sci.* **22**, 637–651. [418]

Goren S. L., Gottlieb M. (1982). Surface-tension-driven breakup of viscoelstic liquid threads. *J. Fluid Mech.* **120**, 245–266. [443]

Grace J. R., Harrison D. (1967). The influence of bubble shape on the rise velocities of large bubbles. *Chem. Eng. Sci.* **22**, 1337–1347. [56]

Grace J. R., Wairegi T., Brophy J. (1978). Breakup of drops and bubbles in stagnant media. *Can. J. Chem. Eng.* **65**, 3–8. [48]

Grilli S. T., Guyenne P., Dias F. (2001) A fully non-linear model for three-dimensional overturning waves over an arbitrary bottom. *Int. J. Numer. Meth. Fluids* **35**, 829–867. [464]

Grilli S. T., Horrillo J. (1997) Numerical generation and absorption of fully nonlinear periodic waves. *J. Eng. Mech.* October, 1060–1069. [464]

Grilli S. T., Skourup J., Svendsen I. A. (1989) An efficient boundary element method for nonlinear water waves. *Eng. Anal. Boundary Elements* **6**, 97–107. [464]

Grilli S. T., Subramanya R. (1994) Quasi-singular integrals in the modeling of nonlinear water waves in shallow water. *Eng. Anal. Boundary Elements* **13**, 181–191. [464]

Grilli S. T., Subramanya R. (1996) Numerical modeling of wave breaking induced by fixed or moving boundaries. *Comput. Mech.* **17**, 374–391. [464]

Gu X. M., Sethna P. R. (1987). Resonant surface waves and chaotic phenomena. *J. Fluid Mech.* **183**, 543–565. [197, 198]

Hamel G. (1917). Spiralförmige Bewgungen zäher Flüssigkeiten. *Jahresber. Deutsch. Math. Vereinigung* **25**, 34–60. [19, 31, 32]

Hanson A. R., Domich E. G. (1956). The effect of viscosity on the breakup of droplets by air blasts – a shock tube study. Research Rep. No. 130, Dept. of Aerospace Engineering, University of Minnesota, Minneapolis, MN. [85]

Hanson A. R., Domich E. G., Adams H. S. (1963). Shock tube investigation of the breakup of drops by air blasts. *Phys. Fluids* **6**, 1070–1080. [85]

Harper J. F. (1972). The motion of bubbles and drops through liquids. *Adv. Appl. Mech.* **12**, 59–129. [46, 48]

Harper J. F., Moore D. W. (1968). The motion of a spherical liquid drop at high Reynolds number. *J. Fluid Mech.* **32**, 367–391. [137, 139, 162]

Harrison W. J. (1908). The influence of viscosity on the stability of superposed fluids. *Proc. London Math. Soc.* **6**, 396–405. [70]

Harvey E. N., Barnes D. K., McElroy W. D., Whiteley A. H., Pease D. C., Cooper K. W. (1944a). Bubble formation in animals, I. Physical factors. *J. Cell. Comp. Physiol.* **44**, 1–22. [276]

Harvey E. N., McElroy W. D., Whiteley A. H. (1947). On cavity formation in water. *J. Appl. Phys.* **18**, 162–172. [276, 277]

Harvey E. N., Whiteley A. H., McElroy W. D., Pease D. C., Barnes D. K. (1944b). Bubble formation in animals, II. Gas nuclei and their distribution in blood and tissues. *J. Cell. Comp. Physiol.* **44**, 23–34. [276]

Hattori S. (1935). On motion of cylindrical bubble in tube and its application to measurement of surface tension of liquid. *Rep. Aeronaut. Res. Inst. Tokyo Imp. Univ.* **115**. [56]

Hesla T., Huang A., Joseph D. D. (1993). A note on the net force and moment on a drop due to surfaces forces. *J. Colloid Interface Sci.* **158**, 255–257. [47]

Hicks W. M. (1884). On the steady motion and small vibrations of a hollow vortex. *Philos. Trans. R. Soc. London* **175**, 161–195. [143]

Hiemenz K. (1911). Die Grenzschicht an einem in den gleichförmigen Flüssigkeitsstrom eingetauchten geraden Kreiszylinder. Göttingen Dissertation; and *Dingler's Polytech. J.* **326**, 321. [19, 29–31]

Higuera M., Knobloch E. (2006). Nearly inviscid Faraday waves in slightly rectangular containers. *Prog. Theor. Phys. Suppl.* **161**, 53–67. [197]

Higuera M., Knobloch E., Vega J. M. (2005). Dynamics of nearly inviscid Faraday waves in almost circular containers. *Physica D* **201**, 83–120. [197]

Hirahara H., Kawahashi M. (1992). Experimental investigation of viscous effects upon a breakup of droplets in high-speed air flow. *Exp. Fluids* **13**, 423–428. [85]

Homann F. (1936). Der Einfluss grosser Zähigkeit bei der Strömmung um den Zylinder und um die Kugel. *Z. Angew. Math. Mech.* **16**, 153–164. [31]

Howarth L. (1953). Modern developments in fluid dynamics: High speed flow. Clarendon. [394]

Hsiang L. P., Faeth G. M. (1992). Near-limit drop deformation and secondary breakup. *Int. J. Multiphase Flow* **18**, 635–652. [84, 86]

Hu H. H., Riccius O., Chen K. P., Arney M., Joseph D. D. (1990). Climbing constant, second-order correction of Trouton's viscosity, wave speed and delayed die swell for M1. *J. Non-Newtonian Fluid Mech.* **35**, 287–307. [399, 407]

Huang P. Y., Hu H. H., Joseph D. D. (1997). Direct simulation of the sedimentation of elliptic particles in Oldroyd-B fluids. *J. Fluid Mech.* **362**, 297–325. [415]

Huerre P. (2000). Open shear flow instabilities. In *Perspectives in Fluid Dynamics,* G. K. Batchelor, H. K. Moffatt, M. G. Worster, eds. Cambridge University Press, Chapter 4, pp. 159–229. [253]

Huerre P., Monkewitz P. A. (1985). Absolute and convective instabilities in free shear layers. *J. Fluid Mech.* **159**, 151–168. [254, 257]

Hwang S. S., Liu Z., Reitz R. D. (1996). Breakup mechanisms and drag coefficients of high speed vaporizing liquid drops. *Atomization Sprays* **6**, 353–376. [86, 438]

Ingham D. B. 1983 Steady flow past a rotating cylinder. *Comput. Fluids* **11**, 351–366. [353]

Ingham D. B., Tang T. 1990 A numerical investigation into the steady flow past a rotating circular cylinder at low and intermediate Reynolds numbers. *J. Comput. Phys.* **87**, 91–107. [353]

Jaswon M. A., Symm G. T. (1977). *Integral Equation Methods in Potential Theory and Elastostatics*. Academic. [463]

Jeffrey D. J., Corless R. M. (1988). Forces and stresslets for the axisymmetric motion of nearly touching unequal spheres. *PCH PhysicoChem. Hydrodyn.* **10**, 461–470. [422]

Jeffreys, H., (1928) Equations of viscous fluid motion and the circulation theorem. *Proc. Cambridge Philos. Soc.* **24**, 477–479. [2]

Joseph D. D. (1973). Domain perturbations: The higher order theory of infinitesimal water waves. *Arch. Ration. Mech. Anal.* **51**(4), 295–303. [461]

Joseph D. D. (1976). *Stability of Fluid Motions, I and II*. Vols. **27** and **28** of Springer Tracts in Natural Philosophy. [387]

Joseph D. D. (1990). *Fluid Dynamics of Viscoelastic Liquids*. Vol. **84** of Springer-Verlag Series in Applied Mathematical Sciences. [244, 398, 431, 442]

Joseph D. D. (1992). Bernoulli equation and the competition of elastic and inertial pressures in the potential flow of a second-order fluid. J. Non-Newtonian Fluid Mech. **42**, 385–389. [397, 400, 409, 423]

Joseph D. D. (1995). Cavitation in a flowing liqud. *Phys. Rev. E* **51**, 1649–1650. [273, 283, 298, 419]

Joseph D. D. (1996). Flow induced microstructure in Newtonian and viscoelastic fluids. *In Proceedings of the Fifth World Congress of Chemical Engineering, Particle Technology Track*. American Institute of Chemical Engineers, Vol. 6, pp. 3–16. [403]

Joseph D. D. (1998). Cavitation and the state of stress in a flowing liquid. *J. Fluid Mech.* **366**, 367–378. [273, 274, 283, 287, 298]

Joseph D. D. (2000). Interrogations of direct numerical simulation of solid-liquid flow. Available at http://www.efluids.com. [403]

Joseph D. D. (2003a). Rise velocity of a spherical cap bubble. *J. Fluid Mech.* **488**, 213–223. [42, 52, 58, 59, 63]

Joseph D. D. (2003b). Viscous potential flow. *J. Fluid Mech.* **479**, 191–197. [378]

Joseph D. D. (2006a). (Helmholtz decomposition:) An addendum to the paper "Potential flow of viscous fluids: Historical notes." *Int. J. Multiphase Flow* **32**, 285–310. [8, 17, 19]

Joseph D. D. (2006b). Potential flow of viscous fluids: Historical notes. *Int. J. Multiphase Flow* **32**, 285–310. [8, 19]

Joseph D. D. (2006c). Helmholtz decomposition coupling rotational to irrotational flow of a viscous fluid. In *Proceedings of the National Academy of Sciences PNAS*, published September 18, 2006, 10.1073/pnas.0605792103 (Applied Mathematics), pp. 14272–14277. http://www.pnas.org/papbyrecent.shtml. [17]

Joseph D. D., Beavers G. S., Funada T. (2002). Rayleigh–Taylor instability of viscoelastic drops at high Weber numbers. *J. Fluid Mech.* **453**, 109–132. [81, 379]

Joseph D. D., Belanger J., Beavers G. S. (1999). Breakup of a liquid drop suddenly exposed to a high-speed airstream. *Int. J. Multiphase Flow* **25**, 1263 –1303. [72, 74, 207, 379, 426]

Joseph D. D., Feng J. (1996). A note on the forces that move particles in a second-order fluid. *J. Non-Newtonian Fluid Mech.* **64**, 299–302. [415, 420]

Joseph D. D., Huang A., Candler G. V. (1996). Vaporization of a liquid drop suddenly exposed to a high-speed airstream. *J. Fluid Mech.* **318**, 223–239. [85]

Joseph D. D., Liao T. Y. (1994a). Potential flow of viscous and viscoelastic fluids. *J. Fluid Mech.* **265**, 1–23. [135, 374, 400]

Joseph D. D., Liao T. Y. (1994b). Viscous and viscoelastic potential flow. In *Trends and Perspectives in Applied Mathematics, Applied Mathematical Sciences*, L. Sirovich, ed. Springer-Verlag, Chapter 5, pp. 109–154. Also in Army HPCRC preprint 93-010, 100, 1–54. [135, 400]

Joseph D. D., Liao T. Y., Saut J. C. (1992). Kelvin–Helmholtz mechanism for side branching in the displacement of light with heavy fluid under gravity. *Eur. J. Mech. B/Fluids* **11**, 253–264. [124]

Joseph D. D., Liu Y. J., M. Poletto J. F. (1994). Aggregation and dispersion of spheres falling in viscoelastic liquids. *J. Non-Newtonian Fluid Mech.* **54**, 45–86. [418]

Joseph D. D., Renardy Y. Y. (1991). *Fundamentals of two-Fluid Dynamics*. Springer-Verlag. [13, 14, 105]

Joseph D. D., Saut J. C. (1990). Short-wave instabilities and ill-posed initial-value problems. *Theor. Comput. Fluid Dyn.* **1**, 191–227. [70, 244, 271]

Joseph D. D., Wang J. (2004). The dissipation approximation and viscous potential flow. *J. Fluid Mech.* **505**, 365–377. [128, 156, 159, 161, 162, 164, 183, 219, 220, 298, 443, 444, 447, 448]

Kang I. S., Leal L. G. (1988a). The drag coefficient for a spherical bubble in a uniform streaming flow. *Phys. Fluids* **31**, 233–237. [127, 135, 136, 314]

Kang I. S., Leal L. G. (1988b). Small-amplitude perturbations of shape for a nearly spherical bubble in an inviscid straining flow (steady shapes and oscillatory motion). *J. Fluid Mech.* **187**, 231–266. [127, 130]

Kang S., Choi H., Lee S. (1999). Laminar flow past a rotating circular cylinder. *Phys. Fluids* **11**, 3312–3321. [350, 351, 354, 356–358]

Keller J. B., Miksis M. J. (1983). Surface tension driven flows. *SIAM J. Appl. Math.* **43**, 268–277. [307, 463]

Keller J. B., Rubinow S. I., Tu Y. O. (1973). Spatial instability of a jet. *Phys. Fluids* **16**, 2052–2055. [239, 251, 253, 261]

Kellogg O. D. (1929). *Foundations of Potential Theory*. Springer-Verlag. [1]

Kelvin Lord (1890) Oscillations of a liquid sphere. In *Mathematical and Physical Papers*. Clay and Sons, London, Vol. **3**, pp. 384 386. [171]

Kitscha J., Kocamustafaogullari G. (1989). Breakup criteria for fluid particles. *Int. J. Multiphase Flow* **15**, 573–588. [86]

Knapp R. T. (1958). Cavitation and nuclei. *Trans. ASME* **80**, 1315–1324. [277]

Knapp R. T., Daily J. W., Hammit F. G. (1970). *Cavitation.* McGraw-Hill. [275, 276]

Kojo T., Ueno I. (2006). Bubble behavior under ultrasonic vibration – nonlinear oscillation and multiple bubble interaction. *Space Utiliz. Res.* **22**, 67–69. [191, 192]

Kordyban E. S., Ranov T. (1970). Mechanism of slug formation in horizontal two-phase flow. *Trans. ASME J. Basic Eng.* **92**, 857–864. [114, 117–121, 123]

Kottke P. A., Bair S., Winer W. O. (2003). The measurement of viscosity of liquids under tension. *J. Tribol.* **125**, 260–266. [276]

Kottke P. A., Bair S., Winer W. O. (2005). Cavitation in creeping shear flows. *AIChE J.* **51**, 2150–2170. [273, 274, 277]

Krzeczkowski S. A. (1980). Measurement of liquid droplet disintegration mechanisms. *Int. J. Multiphase Flow* **6**, 227–239. [86]

Kuhl T., Ruths M., Chen Y., Israelachvili J. (1994). Direct visualization of cavitation and damage in ultrathin liquid films. *J. Heart Valve Disease* **3**, 117–127. [275, 277]

Kumar K., Tuckerman L. S. (1994). Parametric instability of the interface between two fluids. *J. Fluid Mech.* **279**, 49–68. [198]

Kumar S. (2000). Mechanisms for the Faraday instability in viscous liquids. *Phys. Rev. E* **62**, 1416–419. [206, 208]

Lagerstrom A. P., Cole D. J. (1955). Examples illustrating expansion procedures for the Navier–Stokes equations. *J. Rat. Mech.* **4**, 817–882. [347]

Lagrange J. (1781). Memoire sur la theorie du mouvement des fluides. *Nour. Mem. Acad. Berlin*, 151–198, Oeuvres **4**, 695–748. [2, 9]

Lamb H. (1881). On the oscillations of a viscous spheroid. *Proc. London Math. Soc.* **13**, 51–66. [171]

Lamb H. (1932). *Hydrodynamics*, 6th ed. Cambridge University Press (Reprinted by Dover 1945). [3, 9, 10, 23, 25, 32, 36, 126–128, 132, 137, 143, 156, 159, 161, 170, 171, 175, 176, 179, 180, 183–185, 193–195, 408, 417, 423]

Landau L. D., Lifshitz E. M. (1987). *Fluid Mechanics*, 2nd ed. Butterworth-Heinemann. [3, 31, 198, 199, 376, 377]

Lane W. R. (1951). Shatter of drops in stream of air. *Ind. Eng. Chem.* **43**, 1312–1317. [87]

Laoonual Y., Yule A. J., Walmsley S. J. (2001). Internal fluid flow and spray visualization for a large scale valve covered orifice (VCO) injector nozzle. ILASS-Europe, Zurich, September, 2001. [278]

Larson R. G. (1988). *Constitutive Equations for Polymer Melts and Solutions*. Butterworths Series in Chemical Engineering. [396]

Leib S. J., Goldstein M. E. (1986a). Convective and absolute instability of a viscous liquid jet. *Phys. Fluids* **29**, 952–954. [239, 261, 264, 265]

Leib S. J., Goldstein M. E. (1986b). The generation of capillary instability on a liquid jet. *J. Fluid Mech.* **168**, 479–500. [239, 253, 264, 265]

Leppinen, D. M., Lister, J. R. 2003 Capillary pinch-off in inviscid fluids. *Phys. Fluids* **15**, 568–578. [463]

Levich V. G. (1949). The motion of bubbles at high Reynolds numbers. *Zh. Eksp. Teor. Fiz.* 19, 18. see also in *Physiochemical Hydrodynamics* (English transl. by Scripta Technica), Prentice-Hall, 1962. [24, 43, 46, 127, 142, 145, 156, 312]

Lewis D. J. 1950 The instability of liquid surfaces when accelerated in a direction perpendicular to their planes: Part II. *Proc. R. Soc. London Ser. A* **202**, 81–96. [72]

Li H. S., Kelly R. E. (1992). The instability of a liquid jet in a compressible airstream. *Phys. Fluids A* **4**, 2162–2168. [250, 378, 386, 390, 392]

Ligget J. A., Liu P. L.-F. (1984) *Application of Boundary Element Methods in Fluid Mechanics*. Vol. 1 of Topics in Boundary Element Research (ed. C. A. Brebbia.) Springer-Verlag, Chapter 4. [464]

Lighthill M. J. (1963). Introduction. Boundary layer theory. In *Laminar Boundary Layers*, L. Rosenhead, ed. Oxford University Press, Chapter II, 56–58. [32, 33, 333]

Lighthill J., 1978. *Waves in Fluids*. Cambridge University Press. [10]

Lin P. Y., Hanratty T. J. (1986). Prediction of the initiation of slugs with linear stability theory. *Int. J. Multiphase Flow* **12**, 79–98. [118, 120, 121]

Lin S. P. (2003). *Breakup of Liquid Sheets and Jets*. Cambridge University Press. [240, 266, 378, 379]

Lin S. P., Lian Z. W. (1989). Absolute instability in a gas. *Phys. Fluids A* **1**, 490–493. [239, 264, 265]

Liu J. Y., Joseph D. D. (1993). Sedimentation of particles in polymer solutions. *J. Fluid Mech.* **255**, 565–595. [402, 404, 413, 415]

Liu J. Y., Liao Y. T., Joseph D. D. (1995). A two-dimensional cusp at the trailing edge of an air bubble rising in a viscoelastic liquid. *J. Fluid Mech.* **304**, 321–342. [416, 418]

Liu Y. J. (1995). Particle motions in non-Newtonian fluids. Ph.D. dissertation, University of Minnesota. [402, 431]

Liu Z., Brennen C. E. (1998). Cavitation nuclei population and event rates. *J. Fluids Eng.* **120**, 728–737. [290]

Liu Z., Reitz R. D. (1997). An analysis of the distortion and breakup mechanisms of high-speed liquid drops. *Int. J. Multiphase Flow* **23**, 631–650. [86]

Loiseleux T., Chomaz J. M., Huerre P. (1998). The effect of swirl on jets and wakes: Linear instability of the Rankine vortex with axial flow. *Phys. Fluids* **10**, 1120–1134. [269]

Longuet-Higgins M. S. (1992). Theory of weakly damped Stokes waves: A new formulation and its physical interpretation. *J. Fluid Mech.* **235**, 319 – 324. [170]

Longuet-Higgins M. S. (1997). Viscous dissipation in steep capillary–gravity waves. *J. Fluid Mech.* **344**, 271–289. [11, 170, 196]

Longuet-Higgins M. S., Cokelet E. D. (1976). The deformation of steep surface waves on water. I. A numerical method of computation. *Proc. R. Soc. London Ser. A* **350**, 1–26. [157, 460, 464]

Lundgren T. S., Joseph D. D. (2007). Symmetric model of capillary collapse and rupture. FEDSM2007-37262, 5th joint ASME/JSME Fluids Engineering Conf. July 30-August 2, 2007. San Diego, CA. [297, 308] <http://www.aem.umn.edu/people/faculty/joseph/ViscousPotentialFlow/>

Lundgren T. S., Mansour N. N. (1988). Oscillations of drops in zero gravity with weak viscous effects. *J. Fluid Mech.* **194**, 479–510. [172, 463, 464]

Lundgren T. S., Mansour N. N. (1991). Vortex ring bubbles. *J. Fluid Mech.* **224**, 177–196. [139, 141, 152, 153, 155, 463]

Machane R., Canot E. (1997). High-order schemes in boundary element methods for transient non-linear free surface problems. *Int. J. Numer. Methods Fluids* **24**, 1049–1072. [464]

Magnaudet J., Eames I. (2000). The motion of high-Reynolds-number bubbles in inhomogeneous flows. *Annu. Rev. Fluid Mech.* **32**, 659–708. [156, 157]

Mansour N. M., Lundgren T. S. (1990). Satellite formation in capillary jet breakup, *Phys. Fluids* A **2**, 1141–1144. [463]

Maslen H. S. (1963). Second-order effects in laminar boundary layers. *AIAA J.* **1**, 33–40. [348]

Mata C., Pereyra E., Trallero J. L., Joseph D. D. (2002). Stability of stratified gas-liquid flows. *Int. J. Multiphase Flow* **28**, 1249–1268. [100]

Matsumoto Y., Kunugi T., Serizawa A. (1999). Thirteenth Symposium, Japan Society of CFD. Available at http://www.nucleng.kyoto-u-ac.jp/Groups/F-group/gallery/pdf/C03-2.pdf. [139, 141]

Matta J. E., Tytus R. P. (1982). Viscoelastic breakup in a high velocity airstream. *J. Appl. Polymer Sci.* **27**, 397–405. [87]

Matta J. E., Tytus R. P., Harris J. (1983). Aerodynamic atomization of polymeric solutions. *Chem. Eng. Commun.* **19**, 191–204. [87]

Maude A. D. (1961). End effects in a falling-sphere viscometer. *Br. J. Appl. Phys.* **12**, 293–295. [419]

McKinley G. H., Tripathi A. (2000). How to extract the Newtonian viscosity from capillary breakup measurements in a filament rheometer. *J. Rheol.* **44**, 653–671. [307]

Meiron D. I., Saffman P. G. (1983). Overhanging interfacial gravity waves of large amplitude. *J. Fluid Mech.* **129**, 213–218. [463]

Michele J., Pazold R., Donis R. (1977). Alignment and aggregation effects in suspensions of spheres in non-Newtonian media. *Rheol. Acta.* **16**, 317–321. [404, 405]

Middleman S. (1965). Stability of a viscoelastic jet. *Chem. Eng. Sci.* **20**, 1937. [443]

Miksis M., Vanden-Broeck J. M., Keller J. B. (1981). Axisymmetric bubble or drop in a uniform flow. *J. Fluid Mech.* **108**, 89–100. [157, 463]

Miksis M., Vanden-Broeck J. M., Keller J. B. (1982). Rising bubbles. *J. Fluid Mech.* **123**, 31–41. [43, 50, 152, 157, 460, 464]

Miles J. W. (1967). Surface-wave damping in closed basins. *Proc. R. Soc. London Ser. A* **297**, 459–475. [197, 198]

Miles J. W. (1984). Nonlinear Faraday resonance. *J. Fluid Mech.* **146**, 285–302. [198]

Miles J. W., Henderson D. M. (1990). Parametrically forced surface waves. *Annu. Rev. Fluid Mech.* **22**, 143–165. [198, 213]

Miller C. A., Scriven L. E. (1968). The oscillations of a droplet immersed in another fluid. *J. Fluid Mech.* **32**, 417–435. [172, 181–183, 185, 186]

Milne-Thomson L. M. (1968). *Theoretical Hydrodynamics*, 5th ed. Macmillan. [3, 25, 58, 90, 92, 93, 97, 148, 408, 416]

Mishima K., Ishii M. (1980). Theoretical prediction of onset of horizontal slug flow. *Trans. ASME J. Fluids Eng.* **102**, 441–445. [120]

Mittal S., Kumar B. (2003). Flow past a rotating cylinder. *J. Fluid Mech.* **476**, 303–334. [350, 351, 353, 354, 356–358, 360, 361, 362]

Moore D. W. (1957). The flow past a rapidly rotating circular cylinder in a uniform stream. *J. Fluid Mech.* **2**, 541–550. [331, 332, 351]

Moore D. W. (1959). The rise of a gas bubble in a viscous liquid. *J. Fluid Mech.* **6**, 113–130. [43, 45, 46, 127, 157, 314, 460]

Moore D. W. (1963). The boundary layer on a spherical gas bubble. *J. Fluid Mech.* **16**, 161–176. [43, 46, 127, 130, 140, 156, 157, 162]

Moore D. W. (1965). The velocity of rise of distorted gas bubbles in a liquid of small viscosity. *J. Fluid Mech.* **23**, 749–766. [136, 137]

Morrison F. A. (1970). Electrophoresis of a particle of arbitrary shape. *J. Colloid. Interface Sci.* **34**, 210–214. [32]

Nair M. T., Sengupta T. K., Chauhan U. S. (1998). Flow past rotating cylinders at high Reynolds numbers using higher order upwind scheme. *Comput. Fluids* **27**, 47–70. [354]

Nayfeh A. H., Saric W. S. (1973). Nonlinear stability of a liquid film adjacent to a supersonic stream. *J. Fluid Mech.* **58**, 39–51. [378]

Navier C. L. M. H. (1822). On the laws of motion of fluids taking into consideration the adhesion of the molecules. *Ann. Chim. Phys.* **19**, 234–245. [1, 9]

Ockendon J. R., Ockendon H. (1973). Resonant surface waves. *J. Fluid Mech.* **59**, 397–413. [197]

Oguz H., Prosperetti A. (1989). Surface-tension effects in the contact of liquid surfaces. *J. Fluid Mech.* **203**, 149–171. [463]

Oguz H., Prosperetti A. (1990). Bubble entrainment by the impact of drops on liquid surfaces. *J. Fluid Mech.* **219**, 143–179. [463]

Oguz H., Prosperetti A. (1993). Dynamics of bubble growth and detachment from a needle. *J. Fluid Mech.* **257**, 111–145. [463]

Ortiz C., Joseph D. D., Beavers G. S. (2004). Acceleration of a liquid drop suddenly exposed to a high speed air stream. *Int. J. Multiphase Flow* **30**, 217–224. [428]

Padrino J. C., Funada, T., Joseph, D. D. (2007a). Purely irrotational theories for the viscous effects on the oscillations of drops and bubbles. *Int. J. Multiphase Flow,* Accepted. [172, 191] http://www.aem.umn.edu/people/faculty/joseph/ViscousPotentialFlow/

Padrino J. C., Joseph D. D. (2006). Numerical study of the steady state uniform flow past a rotating cylinder. *J. Fluid Mech.* **557**, 191–223. [311, 320, 328, 330, 350, 351]

Padrino J. C., Joseph D. D. (2007). Correction of Lamb's dissipation calculation for the effects of viscosity on capillary-gravity waves. *Phys. Fluids* **19**, 082105. [10, 19, 127, 171, 195]

Padrino J. C., Joseph D. D., Funada T., Wang J., Sirignano W. A. (2007b). Stress induced cavitation for the streaming motion of a viscous liquid past a sphere. *J. Fluid Mech.* **578**, 381–411. [272, 287, 289]

Pan T. W., Joseph D. D., Glowinski R. (2001). Modelling Rayleigh–Taylor instability of a sedimenting suspension of several thousand circular particles in a direct numerical simulation. *J. Fluid Mech.* **434**, 23–37. [88]

Panton L. R. (1984). *Incompressible Flow*. Wiley-Interscience. [348, 363, 407]

Papageorgiou D. T. (1995). On the breakup of viscous liquid threads. *Phys. Fluids* **7**, 1529–1544. [306, 307]

Park W. C., Klausner J. F., Mei R. (1995). Unsteady forces on spherical bubbles. *Exp. Fluids* **19**, 167–172. [156]

Pasol L., Chaoui M., Yahiaoui S., Feuillebois F. (2005). Analytical solutions for a spherical particle near a wall in axisymmetrical polynomial creeping flows. *Phys. Fluids* **17**, 073602. [419, 420]

Patzek T. W., Benner R. E., Basaran O. A., Scriven L. E. (1991). Nonlinear oscillations of inviscid free drops. *J. Comput. Phys.* **97**, 489–515. [172]

Pedley T. J. (1967). The stability of rotating flows with a cylindrical free surface. *J. Fluid Mech.* **30**, 127–147. [148–150]

Pedley T. J. (1968). The toroidal bubble. *J. Fluid Mech.* **32**, 97–112. [139–146]

Peregrine D. H., Shoker G., Symon A. (1990). The bifurcation of liquid bridges. *J. Fluid Mech.* **212**, 25–39. [303]

Pereira A., McGrath G., Joseph D. D. (2001). Flow and stress induced cavitation in a journal bearing with axial throughput. *J. Tribol.* **123**, 742–754. [273]

Perlin M., Schultz W. W. (2000). Capillary effects on surface waves. *Annu. Rev. Fluid Mech.* **32**, 241–274. [198]

Petit L., Noetinger B. (1988). Shear-induced structure in macroscopic dispersions. *Rheol. Acta.* **27**, 437–441. [404, 405]

Pilch M., Erdman C. (1987). Use of break-up time data and velocity history data to predict the maximum size of stable fragments for acceleration-induced break-up of a liquid drop. *Int. J. Multiphase Flow* **13**, 741–757. [84]

Plateau J. A. F. (1873). *Statique expérimentale et Théorique des Liquides*. Gauthier-Villars. [216, 443]

Plesset M. (1949). The dynamics of cavitation bubbles. *ASME J. Appl. Mech.* **16**, 228–231. [39, 40]

Plesset M. (1969). Tensile strength of liquids. Tech. Rep., Office of Naval Res. Rep. [276]

Plesset M. S. (1954). On the stability of fluid flows with spherical symmetry. *J. Appl. Phys.* **25**(1), 96–98. [40]

Plesset M., and Whipple G. (1974). Viscous effects in Rayleigh–Taylor instability. *Phys. Fluids* **17**, 1–7. [73]

Poisson S. D. (1829). Mémoire sur les equations générales de l'équilibre et du mouvement des corps solides élastiques et des fluides, *J. École Poly.* **13**, cahier 20, 1–174. [8]

Ponstein J. (1959). Instability of rotating cylindrical jets. *Appl. Sci. Res. A* 8, 425–456. [148, 150]

Poritsky M. (1951). The collapse on growth of a spherical bubble on cavity in a viscous fluid. In *Proceedings of the First U.S. National Congress of Applied Mechanics*, held at the Illinois Institute of Technology. ASME, pp. 813–821. [39, 40]

Pouliquen O., Chomaz J. M., Huerre P. (1994). Propagating Holmboe waves at the interface between two immiscible fluids. *J. Fluid Mech.* **226**, 277–302. [121]

Pozrikidis C. (1992). *Boundary Integral and Singularity Methods for Linearized Viscous Flow*. Cambridge University Press. [462, 463]

Prandtl L., Tietjens G. O. (1934). *Applied Hydro- and Aero-Mechanics*. McGraw-Hill. [330]

Pretsch J. (1938). Zur theoretischen Berechnung des Profilwiderstandes. Jahrb. Deutsch. Luftfahrtforschung **I**, 61. Engl. transl., NACA TM 1009. [333]

Prosperetti A. (1976). Viscous effects on small-amplitude surface waves. *Phys. Fluids* **19**, 195–203. [162, 185, 194]

Prosperetti A. (1977) Viscous effects on perturbed spherical flows. *Q. Appl. Math.* **35**, 339–352. [172, 180]

Prosperetti A. (1980a) Normal-mode analysis for the oscillations of a viscous liquid drop in an immiscible liquid. *J. Mec.* **19**, 149–182. [172, 181, 185, 187, 193]

Prosperetti A. (1980b) Free oscillations of drops and bubbles: The initial value problem. *J. Fluid Mech.* **19**, 1, 149–182. [172, 185, 187]

Ranger A. A., Nicholls J. A. (1969). Aerodynamic shattering of liquid drops. *AIAA J.* **7**, 285–290. [85, 86]

Rayleigh L. (1878). On the instability of jets. *Proc. London Math. Soc.* **10**, 4–13. [239]

Rayleigh L. (1879). On the capillary phenomena of jets. *Proc. R. Soc. London Ser. A* **29**, 71. [216, 443]

Rayleigh L. (1883a). On maintained vibrations. *Philos. Mag.* **15**, 229–235. Reprinted, 1900, in *Scientific Papers*, Vol. 2. Cambridge University Press, pp.188–193. [197]

Rayleigh L. (1883b). On the crispations of fluid resting upon a vibrating support. *Philos. Mag.* **16**, 50–58. Reprinted, 1900, in *Scientific Papers*, Vol. 2. Cambridge University Press, pp. 212–219. [197]

Rayleigh L. (1890). On the theory of surface forces. *Philos. Mag. Sci.* **5**, 30 (185), 285–298. [70–72]

Rayleigh L. (1892). On the instability of a cylinder of viscous liquid under capillary force. *Philos. Mag.* **34**, 145. [148, 216]

Rayleigh L. (1917). On the pressure developed in a liquid during the collapse of a spherical cavity. *Philos. Mag.* **34**, 94–98. [39]

Rayleigh L. (1896) *The Theory of Sound*, 2nd ed. MacMillan. Reprinted by Dover, 1945. Vol. 2, p. 371. [171]

Reid E. G. (1924) *Tech. Notes Natl. Adv. Comm. Aero.*, Washington, no. 209. [319]

Reid W. H. (1960) The oscillations of a viscous liquid drop. *Q. Appl. Math.* **18**, 86–89. [171, 181, 182]

Reinecke W. G., McKay W. L. (1969). Experiments on water drop breakup behind Mach 3 to 12 shocks. Tech. Rep., AVTD-0172–69-RR, AVCO Corp. [85]

Reinecke W. G., Waldman G. D. (1970). A study of drop breakup behind strong shocks with applications to flight. Tech. Rep., SAMSO-TR-70–142, Avco Systems Division. [85, 86]

Reinecke W. G., Waldman G. D. (1975). Shock layer shattering of cloud drops in reentry flight. AIAA Paper 75–152. [85]

Rivlin R. S., Ericksen J. L. (1955). Stress deformation relations for isotropic materials. *J. Rat. Mech. Anal.* **4**, 323–425. [398]

Rodriguez-Rodriguez J., Gordillo J. M., Martinez-Bazan C. (2006). Breakup time and morphology of drops and bubbles in a high-Reynolds-number flow. *J. Fluid Mech.* **548**, 69–86. [464]

Romberg G. (1967). Über die Dissipation in dichtebeständiger Strömung. *Z. Angew. Math. Mech.* **47**, 155–162. [310]

Rothert A., Richter R., Rehberg I. (2001). Transition from symmetric to asymmetric scaling function before drop pinch-off. *Phys. Rev. Lett.* **87**, 084501–1–4. [306]

Ruvinsky K. D., Feldstein F. I., Freidman G. I. (1991). Numerical simulations of the quasi-stationary stage of ripple excitation by steep capillary-gravity waves. *J. Fluid Mech.* **230**, 339–353. [169]

Ruvinsky K. D., Freidman G. I. (1985a). Improvement of the first Stokes method for the investigation of finite-amplitude potential gravity-capillary waves. In *Ninth All Union Symposium on Diffraction and Propagation Waves*, Vol. 2, pp. 22–25. Theses of Reports. [169]

Ruvinsky K. D., Freidman G. I. (1985b). Ripple generation on gravity-capillary wave crests and its influence on wave propagation. Tech. Rep., Inst. Appl. Acad. Sci. USSR Gorky. Preprint 132. [169]

Ruvinsky K. D., Freidman G. I. (1987). The fine structure of strong gravity-capillary waves. In *Nonlinear Waves: Structures and Bifurcations*, A. V. Gaponov-Grekhov and M. I. Rabinovich, eds. Moscow: Nauka, 304–326. pp. [169]

Ryskin G., Leal L. G. (1984). Numerical solution of free-boundary problems in fluid mechanics. Part 1: The finite-difference technique. *J. Fluid Mech.* **148**, 19–35. [43, 49]

Saint-Venant A.-J.-C.-B. De. (1843) Note to be added to the memoir on the dynamics of fluids. *Compt. Rend.* **17**: 1240–1243. [8]

Saint-Venant A.-J.-C.-B., De. (1869). Probleme des mouvements que peuvent prendre les divers points d'une masse liquide, ou solide ductile, continue dans un vase a parois verticals, pendant son ecoulement par un orifice horizontal interieur. *C. R. Acad. Sci. Paris* **68**, 221–237. [2]

Sangani A. S. (1991). A pairwise interaction theory for determining the linear acoustic properties of dilute bubbly liquids. *J. Fluid Mech.* **232**, 221–284. [157]

Sangani A. S., Didwania A. K. (1993). Dynamic simulations of flows of bubbly liquids at large Reynolds numbers. *J. Fluid Mech.* **250**, 307–337. [157]

Scardovelli R., Zaleski S. (1999). Direct numerical simulation of free-surface and interfacial flow. *Annu. Rev. Fluid Mech.* **31**, 567–603. [461]

Schetz J. A., Kush E. A., Joshi P. B. (1980). Wave phenomena in liquid jet breakup in a supersonic crossflow. *AIAA J.* **18**, 774–778. [392]

Schlichting H. (1960). *Boundary Layer theory*, 4th ed. Translated by J. Kestin. McGraw-Hill. [333–335]

Schmid P., Henningson D. S. (2001). *Stability and Transition in Shear Flows*. Springer-Verlag. [254]

Schwartz L. W. (1974). Computer extension and analytic continuation of Stokes' expansion for gravity waves. *J. Fluid Mech.* **62**, 553–578. [461]

Sherman A., Schetz J. A. (1971). Breakup of liquid sheets and jets in a supersonic gas stream. *AIAA J.* **9**, 666–673. [392]

Shi X. D., Brenner M. P., Nagel S. R. (1994). A cascade of structure in a drop falling from a faucet. *Science* **265**, 219–222. [306]

Simonelli F., Gollub J. P. (1989). Surface wave mode interactions: Effects of symmetry and degeneracy. *J. Fluid Mech.* **199**, 471–494. [197, 213]

Simpkins P. G., Bales E. L. (1972). Water-drop response to sudden accelerations. *J. Fluid Mech.* **55**, 629–639. [85]

Singh P., Joseph D. D. (2000). Sedimentation of a sphere near a vertical wall in an Oldroyd-B fluid. *J. Non-Newtonian Fluid Mech.* **94**, 179–203. [418]

Singhal A. K., Athavale M. M., Li H., Jiang Y. (2002). Mathematical basis and validation of the full cavitation model. *J. Fluids Eng.* **124**, 617–624. [275, 277]

Sirakov B. T., Greitzer E. M., Tan C. S. (2005). A note on irrotational viscous flow. *Phys. Fluids* **17**, 108102. [374]

Soteriou C., Andrews R., Smith M. (1995). Direct injection diesel sprays and the effect of cavitation and hydraulic flip on atomization. Tech. Rep. 950080, SAE Technical Paper. [278, 280]

Spiegelberg S. H., Gaudet S., McKinley G. H. (1994). In extensional deformation of non-Newtonian materials: Liquid bridge studies. In *Second Microgravity Fluid Physics Conference*, Vol. CP-3276 (U.S. Govt. Printing Office 1994-553-454), p. 311. [304,306]

Spivak B., Vanden-Broeck J. M., Miloh T. (2002). Free-surface wave damping due to viscosity and surfactants. *Eur. J. Mech. B/Fluids* **21**, 207–224. [169]

Squire H. B., Young A. D. (1838). The calculation of the profile drag of airfoils. ARC R & M. [333]

Stokes G. G. (1845). On the theories of the internal friction of fluids in motion. *Trans. Cambridge Philos. Soc.* **8**, 287–319; *Math. Phys. Papers* **1**, 75–129. [1, 8]

Stokes G. G. (1847). On the theory of oscillatory waves. *Trans. Camb. Philos. Soc.* **8**, 441–455. [461]

Stokes G. G. (1851). On the effect of the internal friction of fluids on the motion of pendulums. *Trans. Cambridge. Philos. Soc.* **IX**, 8–106 (read on Dec. 9, 1850); *Math. Phys. Papers* **3**, 1–141. [1, 9, 12, 20, 126, 193]

Stokes G. G. (1880). Considerations relative to the greatest height of oscillatory irrotational waves which can be propagated without change of form. *Math. Phys. Papers* **I**, 225–228. [12]

Stone H. A. (1994). Dynamics of drop deformation and breakup in viscous fluids. *Annu. Rev. Fluid Mech.* **26**, 65–102. [86]

Strasberg M. (1959). Onset of ultrasonic cavitation in tap water. *J. Acoust. Soc. Am.* **31**, 163–176. [276, 277]

Strauss W. A. (1992). *Partial Differential Equations, an Introduction*. Wiley. [174]

Tait P. G. (1890). Note on ripples in a viscous liquid. *Proc. R. Soc. Edinburgh* **xvii**, **110**, 322–337. Also in *Scientific Papers*, Cambridge, 1898–1900, ii, p. 313. [167]

Taitel Y., Dukler A. E. (1976). A model for predicting flow regime transitions in horizontal and near horizontal gas-liquid flow. *AIChE J.* **22**, 47–55. [120–123]

Takagi S., Oguz H. N., Zhang Z., Prosperetti A. (2003). A new method for particle simulation – part ii: Two-dimensional Navier-Stokes flow around cylinders. *J. Comput. Phys.* **187**, 371–390. [423]

Tang T., Ingham, D. B. (1991). On steady flow past a rotating circular cylinder at Reynolds numbers 60 and 100. *Comput. Fluids* **19**, 217–230. [353, 357]

Tani I. (1977). History of boundary-layer theory. *Annu. Rev. Fluid Mech.* **9**, 87–111. [348]

Tanner R. I. (1985). *Engineering Rheology*, Oxford University Press. [395, 419, 425]

Taylor G. I. (1923). On the decay of vortices in a viscous fluid. *Philos. Mag.* **XLVI**, 671–674. [27]

Taylor G. I. (1950). The instability of liquid surfaces when accelerated in a direction perpendicular to their planes I. *Proc. R. Soc. London Ser. A* **201**, 192–196. Also in *The Scientific Papers of G.I. Taylor*, 3rd ed. (ed. G.K.

Batchelor), Cambridge University Press (1993). [40, 70, 71]

Taylor T. D., Acrivos A. (1964). On the deformation and drag of a falling viscous drop at low Reynolds number. *J. Fluid Mech.* **18**, 466–476. [43, 48]

Theofanous T. G., Li G. J., Dinh T. N. (2003). Aerobreakup in rarefied supersonic gas flows. *J. Fluid Eng.* **126** (4), 516. [379]

Tokumaru P. T. & Dimotakis, P. E. (1993). The lift of a cylinder executing rotary motions in a uniform flow. *J. Fluid Mech.* **255**, 1–10. [353]

Tomotika S. (1935). On the instability of a cylindrical thread of a viscous liquid surrounded by another viscous fluid. *Proc. R. Soc. London Ser. A* **150**, 322–337. [169, 215, 216, 218, 234, 443, 444]

Truesdell C. (1954). *The Kinematics of Vorticity*. Indiana University Press. [1, 9]

Tsai W., Yue D. K. P. (1996). Computation of non-linear free-surface flows. *Annu. Rev. Fluid. Mech.* **28**, 249–278. [461]

Tsamopoulus J. A., Brown R. A. (1983). Nonlinear oscillations of inviscid drops and bubbles. *J. Fluid Mech.* **127**, 519–537. [172]

Turner T. S. (1957). Buoyant vortex rings. *Proc. R. Soc. London Ser. A* **239**, 61–75. [140, 141, 146]

Valentine R. S., Sather N. F., & Heideger W. J. (1965). The motion of drops in viscous media. *Chem. Eng. Sci.* **20**, 719–728. [172, 180]

Van Dyke M. (1962a). Higher approximations in boundary-layer theory. Part 1. General analysis. *J. Fluid Mech.* **14**, 161–177. [348, 350]

Van Dyke, M. (1962b). Higher approximations in boundary-layer theory. Part 2. Application to leading edges. *J. Fluid Mech.* **14**, 481–495. [349, 350]

Van Dyke, M. (1964). Higher approximations in boundary-layer theory. Part 3. Parabola in uniform stream. *J. Fluid Mech.* **19**, 145–159. [349]

Van Dyke M. (1969). Higher-order boundary-layer theory. *Annu. Rev. Fluid Mech.* **1**, 265–292. [348, 349]

Vanden-Broeck, J. M., Keller, J. B. (1980). A new family of capillary instability. *J. Fluid Mech.* **98**, 161–169. [157, 460]

Varga C. M., Lasheras J. C., Hopfinger E. J. (2003). Initial breakup of a small-diameter liquid jet by a high-speed gas stream. *J. Fluid Mech.* **497**, 405–434. [123–125, 379, 385, 393]

Vaynblat D., Lister J. R., Witelski T. P. (2001). Symmetry and self-similarity in rupture and pinchoff: A geometric bifurcation. *Eur. J. Appl. Math.* **12**, 209–232. [307]

Vega J. M., Knobloch E., Martel C. (2001). Nearly inviscid Faraday waves in annular containers of moderately large aspect ratio. *Physica D* **154**, 313–336. [197]

Verdon C. P., McCory R. L., Morse R. L., Baker G. R., Meiron D. I., Orszag S. A. (1982). Nonlinear effects of multifrequency hydrodynamic instabilities on ablatively accelerated thin shells. *Phys. Fluids* **25**, 1653–1674. [463]

Viana F., Funada T., Joseph D. D., Tashiro N., Sonoda Y. (2005). Potential flow of a second-order fluid over a tri-axial ellipsoid. *J. Appl. Math.* **4**, 341–364. [417]

Viana F., Pardo R., Yánez R., Trallero J., Joseph D. (2003). Universal correlation for the rise velocity of Taylor bubbles in round pipes. *J. Fluid Mech.* **494**, 379–398. [46, 47, 51, 53, 56, 57, 62, 63, 66, 68]

Waldman G. D., Reinecke W. G., Glenn D. (1972). Raindrop breakup in the shock layer of a high-speed vehicle. *AIAA J.* **10**, 1200–1204. [85]

Wallis G. B. (1969). *One Dimensional Two-Phase Flow*. McGraw-Hill. [46, 120]

Wallis G. B., Dobson J. E. (1973). The onset of slugging in horizontal stratified air-water flow. *Int. J. Multiphase Flow* **1**, 173–193. [114, 117–121–123]

Walters J. K., Davidson J. F. (1963). The initial motion of a gas bubble formed in an inviscid liquid. II. *J. Fluid Mech.* **17**, 321–336. [140, 144, 145, 148]

Wang J., Bai R., Lewandowski C., Galdi P. G., Joseph D. D. (2004). Sedimentation of cylindrical particles in a viscoelastic liquid: Shape-tilting. *China Particuology* **2**(1), 13–18. [402]

Wang J., Joseph D. D. (2004). Potential flow of a second order fluid over a sphere or an ellipse. *J. Fluid Mech.* **511**, 201–215. [425]

Wang J., Joseph D. D. (2006a). Purely irrotational theories of the effect of the viscosity on the decay of free gravity waves. *J. Fluid Mech.* **559**, 461–472. [10, 25, 128, 131, 192]

Wang J., Joseph D. D. (2006b). Pressure corrections for the effects of viscosity on the irrotational flow outside Prandtl's boundary layer. *J. Fluid. Mech.* **557**, 145–165. [19, 153, 310]

Wang J., Joseph D. D. (2006c). Boundary layer analysis for effects of viscosity of the irrotational flow on the flow induced by a rapidly rotating cylinder in a uniform stream. *J. Fluid Mech.* **557**, 167–190. [19, 310]

Wang J., Joseph D. D., Funada T. (2005a). Pressure corrections for potential flow analysis of capillary instability of viscous fluids. *J. Fluid Mech.* **522**, 383–394. [169, 216, 218, 235, 298, 443]

Wang J., Joseph D. D., Funada T. (2005b). Purely irrotational theories of the effects of viscosity and viscoelastacity on capillary instability of a liquid cylinder. *J. Non-Newtonian Fluid Mech.* **129**, 106–116. [131]

Wang J., Joseph D. D., Funada T. (2005c). Viscous contributions to the pressure for potential flow analysis of capillary instability of two viscous fluids. *Phys. Fluids* **17**, 052105a. [176, 219, 444]

Weber C. (1931). Zum Zerfall eines Flüssigkeitsstrahles. *Z. Angew. Math. Mech.* **11**, 136. [216]

Wegener P. P., Parlange J. Y. (1973). Spherical-cap bubbles. *Annu. Rev. Fluid Mech.* **5**, 79–100. [46]

Weinbaum S., Kolansky M. S., Gluckman M. J., Pfeffer R. (1976). An approximate theory for incompressible viscous flow past two-dimensional bluff bodies in the intermediate Reynolds number regime $O(1) < Re < O(10^2)$. *J. Fluid Mech.* **77**, 129–152. [350]

White E. T., Beardmore R. H. (1962). The velocity of rise of single cylindrical air bubbles through liquids contained in vertical tubes. *Chem. Eng. Sci.* **17**, 351–361. [56]

Wierzba A. (1990). Deformation and breakup of liquid drops in a gas stream at nearly critical Weber numbers. *Exp. Fluids* **9**, 59–64. [86]

Wierzba A., Takayama K. (1988). Experimental investigation of the aerodynamic breakup of liquid drops. *AIAA J.* **26**, 1329–1335. [85]

Wilcox J. D., June R. K., Braun H. A., Kelly R. C. (1961). The retardation of drop breakup in high-velocity airstreams by polymeric modifiers. *J. Appl. Polymer Sci.* **13**, 1–6. [87]

Winer W. O., Bair S. (1987). The influence of ambient pressure on the apparent shear thinning of liquid lubricants-an overlooked phenomena. In *Proceedings of the Institution of Mechanical Engineers-Tribology 50 Years*, pp. 395–398. [273]

Wood W. W. 1957 Boundary layers whose streamlines are closed. *J. Fluid Mech.* **2**, 77–87. [331,332]

Wrobel L. C. (2002). *The Boundary Element Method.* Vol. 1, Applications in Thermo-Fluids and Acoustics. Wiley. [463]

Wu H. L., Pots B. F. M., Hollenberg J. F., Meerhoff R. (1987). Flow pattern transitions in two-phase gas/condensate flow at high pressure in an 8-inch horizontal pipe. In *Third International Conference on Multiphase Flow*, Vol. A2, pp. 13–21. [120]

Yang H. Q. (1992). Asymmetric instability of a liquid jet. *Phys. Fluids A* **4**, 681–689. [250]

Yang S., Leal G. L. (1991). A note on memory-integral contributions to the force on an accelerating spherical drop at low Reynolds number. *Phys. Fluids A* **3**, 1822–1823. [156]

Yariv E., Brenner H. (2003). Near-contact electrophoretic motion of a sphere parallel to a planar wall. *J. Fluid Mech.* **484**, 85–111. [32]

Yariv E., Brenner H. (2004). The electrophoretic mobility of a closely fitting sphere in a cylindrical pore. *SIAM Appl. Math.* **64**, 423–441. [32]

Yeung R. W. (1982). Numerical methods in free-surface flows. *Annu. Rev. Fluid Mech.* **14**, 395–442. [461]

Yoo J. (1973). Water waves on a channel of finite depth. Master's thesis, Aerospace Engineering and Mechanics, University of Minnesota. [461]

Yoshida T., Takayama K. (1990). Interaction of liquid droplets with planar shock waves. *Trans. ASME J. Fluids Eng.* **112**, 481–486. [85]

Young F. R. (1989). *Cavitation.* McGraw-Hill. [275]

Zhou Z. W., Lin S. P. (1992). Absolute and convective instability of a compressible jet. *Phys. Fluids A* **4**, 277–282. [378]

Zierep J. (1984). Viscous potential flow. *Arch. Mech.* **36**, 127–133 [135]

Index

Printed in the United States
by Baker & Taylor Publisher Services